Springer Series in Computational Mathematics

32

Springer
Berlin
Heidelberg
New York
Hong Kong
London
Milan
Paris
Tokyo

Michel Rappaz
Michel Bellet
Michel Deville

Numerical Modeling in Materials Science and Engineering

Translated by Ray Snyder

Springer

Michel Rappaz

Laboratory of
Physical Metallurgy
Swiss Federal Institute of
Technology
1015 Lausanne, Switzerland

e-mail: Michel. Rappaz@epfl.ch

Michel Deville

Laboratory of
Computational Engineering
Swiss Federal Institute of
Technology
1015 Lausanne, Switzerland

e-mail: Michel.Deville@epfl.ch

Michel Bellet

Ecole des Mines de Paris
Center for Materials Processing
(CEMEF)
06904 Sophia Antipolis, France

e-mail: Michel.Bellet@ensmp.fr

Translator

Ray Snyder
142 rue du Faubourg St. Denis
75010 Paris, France

e-mail:
snyder@alumni.caltech.edu

Originally published in French under the title
"Modélisation numérique en science et génie des matériaux (TM vol. 10)"

Cataloging-in-Publication Data applied for
Bibliographic information published by Die Deutsche Bibliothek
Die Deutsche Bibliothek lists this publication in the Deutsche
Nationalbibliografie; detailed bibliographic data is available in the
Internet at <http://dnb.ddb.de>.

Mathematics Subject Classification (2000): 65M06, 65M60, 65N06, 65N30

ISSN 0179-3632
ISBN 3-540-42676-0 Springer-Verlag Berlin Heidelberg New York

Springer-Verlag Berlin Heidelberg New York
a part of Springer Science + Business Media

http://www.springer.de

Cover design: *design&production*, Heidelberg
Typesetting: EDV-Beratung Frank Herweg, Leutershausen
Printed on acid-free paper SPIN: 11017325 46/3111-54321

Ce qui se conçoit bien
s'énonce clairement
et les équations pour le dire
arrivent aisément

The authors

PREFACE

This book is devoted to the numerical simulation and modeling in Materials Science and Engineering.

The technological revolution of the second half of the last century has been without doubt based on the impressive development of computers (hardware and software). Apart from changes in communications systems or information processing, this technology brought profound and irreversible modifications in instrumentation (data acquisition, device control and feedback loops) and in scientific computing. In parallel with the computer evolution, more and more accurate and refined numerical methods have been invented and employed by engineers, scientists and mathematicians in order to solve systems of complex equations which are very often non linear and coupled. The finite difference (FDM) and finite element (FEM) methods which are already available as very efficient commercial softwares are currently used in many industrial applications for the development and optimisation of materials processing and transformation. Among the many examples of modeling applications in materials science and processing are the production of materials from raw minerals, the elaboration of finished or semi-finished products in the solid state (forging, extrusion, stamping, etc.) or from the liquid state (casting, injection moulding, single crystal growth, etc), assembling techniques such as welding or brazing.

Nevertheless, the last two decades have seen an increasing diversification of numerical methods and their application in the field of materials science, in particular the Monte Carlo methods, cellular automata, random walkers, atomistic methods related to molecular dynamics, boundary element methods, homogeneization techniques based on average conservation laws, etc.

The aim of this monograph is to acquaint the materials science student or the engineer with the numerical methods which are state-of-the-art in this subject, with some insight into more specific methods. The bulk of the effort was therefore devoted to FDM and FEM methods. After a reminder of the conservation laws and constitutive relations in the first chapter, these two methods are presented in chapters 2 and 3, respectively. Chapter 4 considers more practical aspects related to FDM and FEM, namely, mesh generation, solution of linear and non linear systems, storage of the matrices. In order to avoid the rapid obsolescence of the book, even before its publication, any aspect treating computer hardware implementation has been deliberately omitted!

The next three chapters which constitute the heart of the book are dedicated to the most important topics in materials science: phase transformations, deformation of solids and fluid flows. For each of these chapters, the specific constitutive equations are first introduced followed by a description of specific numerical aspects and finally by a collection of illustrative examples.

The last two chapters of the book are concerned with more specific numerical methods that become more and more popular in materials science. Chapter 8 aims at inverse methods, i.e. the use of numerical methods for the determination of properties or boundary conditions based on experimental results. Finally, chapter 9 introduces the stochastic Monte Carlo methods, random walkers and cellular automata.

Due to the various aspects proposed in this book, it should be emphasized that the text gives a rather succinct view of the numerical simulation methods. The reader who wishes to go deeper in one of the particular items is referred to the specialized literature that is given at the end of each chapter. Even though the three co-authors share the same first name, they have educational background and expertise in different disciplines. They hope that this diversity has contributed to enrich and widen the reader's viewpoint without harming too much the unity of the manuscript.

Acknowledgments

The authors would like to thank those who contributed significantly to this book through their valuable remarks or through figures or illustrative examples. The list includes: Jean-Luc Desbiolles, Raissa D'Souza, Charles-André Gandin, Sylvain Henry, Alain Jacot, Tuomas Jalanti, Nadim Kharraz, Gilmar Mompean, Jean-Daniel Reymond and Rohit Trivedi. Several contributions were generously provided by the companies Calcom S.A., Lausanne and Transvalor, Sophia Antipolis, with which two authors have scientific and friendly links.

The authors also thank Dr. Ray Snyder who translated the French original text into English. Through his perfect command of both languages and his own expertise in engineering, he proved the Italian saying "traduttore, tradittore" wrong and brought many improvements to this English version.

March 2002,
Michel Rappaz,
Michel Bellet,
Michel Deville.

TABLE OF CONTENTS

CHAPTER 1

CONTINUOUS MEDIA

1.1 OBJECTIVES

- Introduce the equations of conservation of mass, momentum, energy and solute.
- Introduce the principal equations for materials behavior, which will be developed in depth in chapters 5 through 7.
- Define the boundary conditions and initial conditions.

The goal of this chapter is to review briefly, but as clearly as possible, the principle equations of conservation and behavior for the complex phenomena occurring in the science and technology of materials. The equations of electromagnetism will not be covered here; we will discuss conservation of mass, of momentum, of energy and of solute. Next, we will cover the most important constitutive equations for material behavior that connect, for example: stress to strain or to velocity in deformation or flow problems; or enthalpy and heat flux to temperature in heat transfer calculations; or, finally, the flux of solute to concentration in chemical diffusion. The derivation of these equations will be done in the perspective of subsequent applications that may involve two different scales. At the macroscopic level, the conservation and constitutive equations allow modeling, or even optimization, of an industrial process or simulation of the behavior of a sample undergoing mechanical testing. At the microscopic scale, the equations can be used to describe the formation and evolution of the microstructure (dendrites, spherulites, lamellae, or eutectic fibers, etc.), the interaction between a fiber and the matrix in composite materials, or the deformation of a crystal lattice around a dislocation. The distinction between these two scales is somewhat arbitrary and reveals the current limitations of simulation applied to materials. Even with the most powerful computers available, it is not possible to model a mechanics problem in continuous media taking into account all scales simultaneously, for example between the microstructure (micron) and the process or component scales (meter) (fig. 1.1)! In any case, we will neglect the atomic scale, and consider the medium to be continuous, as most problems examined in this text deal with structures on a scale many orders of magnitude larger than typical inter-atomic distances. Finally, we will derive the boundary conditions controlling the exchange of mass, momentum or energy between the domain under consideration and the external world.

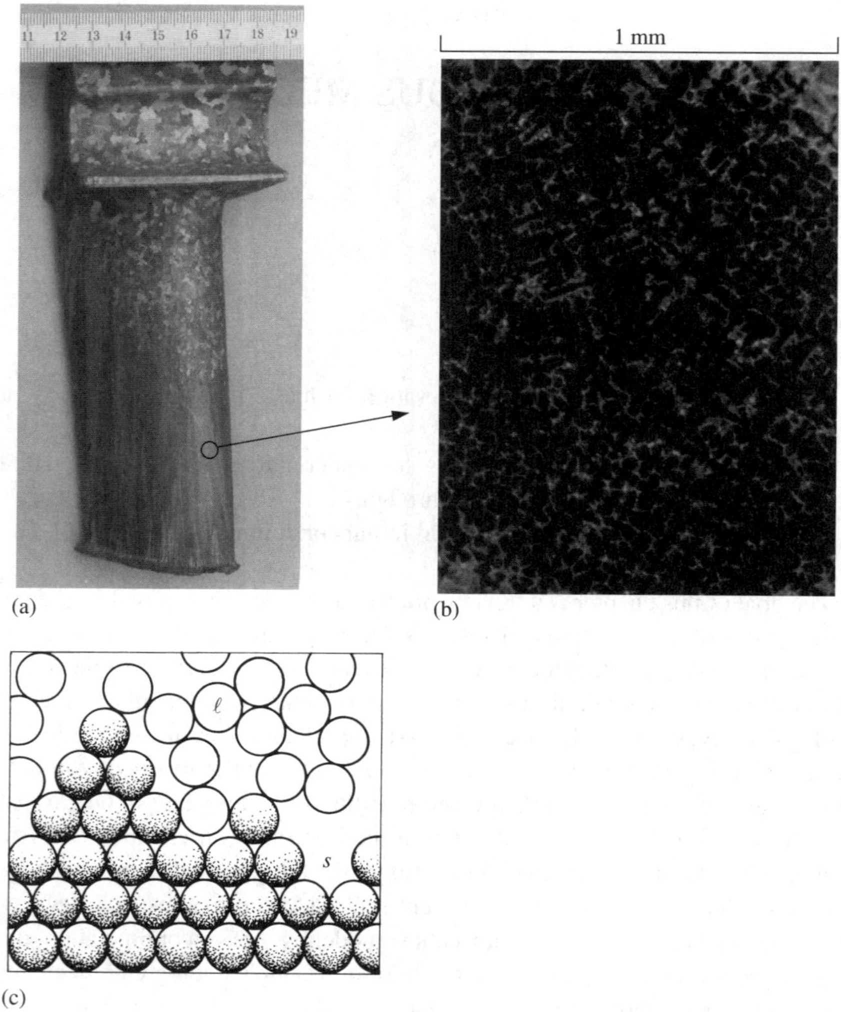

Figure 1.1 The different scales appearing in materials science. A turbine blade (a), solidified in a ceramic mold (investment casting), measures a few dozen centimeters. It is composed of grains which are clearly visible after chemical etching (scale: a few millimeters), which themselves are made up of dendrites which are spaced at a few dozen to hundreds of microns (b). In (c) is shown a schematic at atomic scale of the transformation from a liquid to a solid for a metal alloy during the formation of such a blade by precision casting.

1.2 CONSERVATION AND CONTINUITY EQUATIONS

The semi-continuous casting of aluminum ingots, upstream of the production of sheets by rolling, is shown in figure 1.2. It is a typical process used in industry which shows well, first, the complexity of the phenomena occurring, and second, the different scales that can be distinguished. At the macroscopic scale of the process (fig. 1.2(a)), liquid metal is injected through a nozzle (forced convection), cools on contact with the mold (heat exchange) and solidifies (phase transformation). The temperature gradients induce convection movements in the liquid (natural convection) and stress in the solid (deformation). The convection, in turn, modifies the temperature distribution in the liquid, as the deformation of the solid changes the thermal contact with the mold. One sees then that the problems of mass, heat and momentum exchange are completely coupled. For alloys, one must consider also the redistribution of the different constituents during solidification.

It is apparent, even at this scale, that the problem is far from simple. And yet, the macroscopic scale directly influences the microscopic, namely the formation of microstructures during solidification. As shown in figure 1.2(b), grains of aluminum form in the liquid, grow and can be entrained by convection. Strong convection can

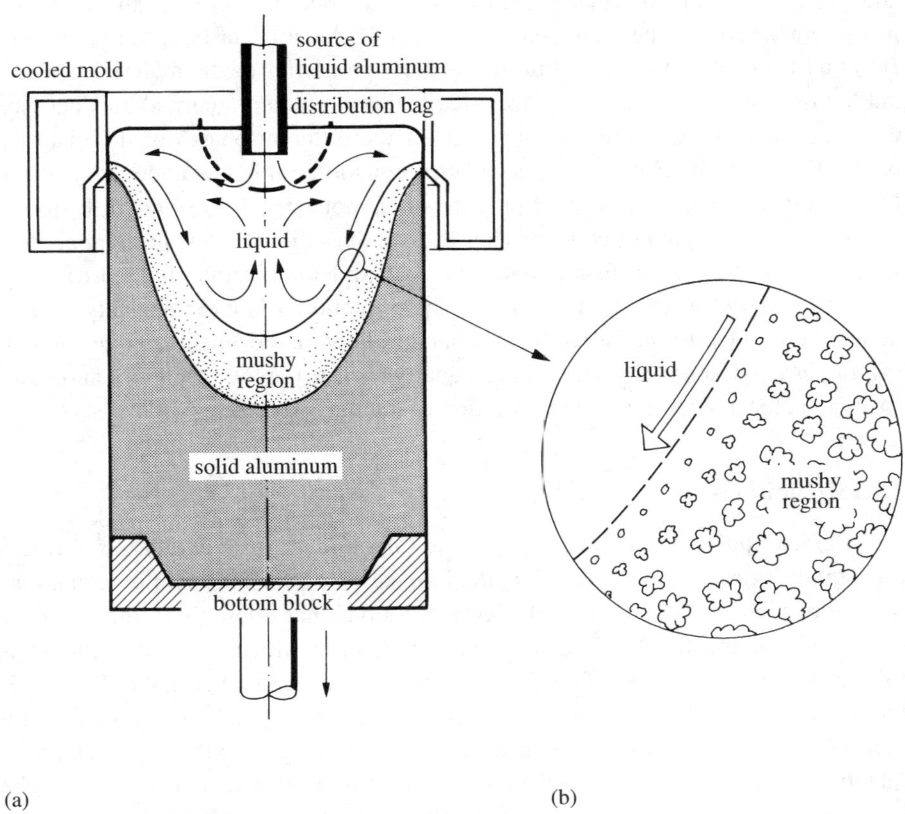

Figure 1.2 In the continuous casting of aluminum (a), liquid metal is injected from a nozzle through a distribution bag which filters out inclusions and oxide skin debris. The metal cools on contact with a mold, which itself is cooled by circulating water and water spraying. Solid metal is extracted continuously by a jack attached to a bottom block. Convection in the liquid (indicated by arrows) can transport growing grains of aluminum in the 'mushy' region (b)

also break up the dendrites that compose a grain, giving birth to new grains. This coupling between macroscopic and microscopic scales works in both directions, as the formation of crystals liberates heat, rejects solute, and slows the convection in the liquid and thus affects the processes at the largest scale.

It should be noticed that the environment considered in figure 1.2 is not homogeneous as it is composed of at least two phases, namely solid and liquid. The conservation equations for multiphase environments are presented later, in chapter 5. Even though very complex, the phenomena occurring at the macroscopic scale of the continuous casting process (fig. 1.2(a)) or those that control the formation of microstructures at the microscopic scale can be described with four conservation equations (mass, momentum, energy and solute)! These equations need to be matched with the appropriate boundary and initial conditions and the constitutive equations controlling the behavior of the material in question in its different phases. By behavior we mean the relation between cause and effect such as between stress and strain for elastic deformation, stress and strain rate in fluid mechanics, or temperature-gradient and heat-flux, etc. The type of behavior (elastic, plastic, viscous, etc.) and also the values of thermo-mechanical properties of the material (specific mass, viscosity, elastic modulus, strain-hardening coefficient, thermal conductivity, etc.) enter into the constitutive equations for the different phases of the material. Notice that the formalism of the conservation equations allows an approach to problems as diverse as the injection of polymers in a mold (fig. 1.3(a)), the diffusion of water during the drying of concrete (fig. 1.3(b)), the strain of a test piece under tension (fig. 1.3(c)) or the diffusion that occurs in a sintered ceramic (fig. 1.3(d)).

The goal of this chapter is to lay out a presentation, if not exhaustively, at least in overview, of the formalism of the mechanics of continuous media, while attempting to show its generality, describing equally well a process such as continuous casting, a mechanical test, or the evolution of the microstructure in a material.

1.2.1 Definitions

Before applying Lavoisier's principle "*nothing is lost, nothing is created, everything is transformed*", which is the foundation of the conservation equations, it is convenient to define some elementary notions that allow the principle to be expressed mathematically. Consider a domain, Ω, over which a conservation law is to be computed (fig. 1.4(a)). This domain could be the aluminum ingot of figure 1.2, the injection mold for the polymer part (fig. 1.3(a)), the sample under tension (fig 1.3(c)), or a portion of a material in which we want to describe the evolution of the quantity of water (fig. 1.3(b)) or the concentration of a component (fig. 1.3(d)). The boundary, $\partial\Omega$, separates the domain from the external world that we do not want to include in the model, but which we are obliged to consider when we formulate the boundary conditions (section 1.4). For the sample undergoing tensile testing, the interaction with the external world is essentially reduced to the jaws of the tensile testing machine, and possibly also the temperature and composition of the sur-

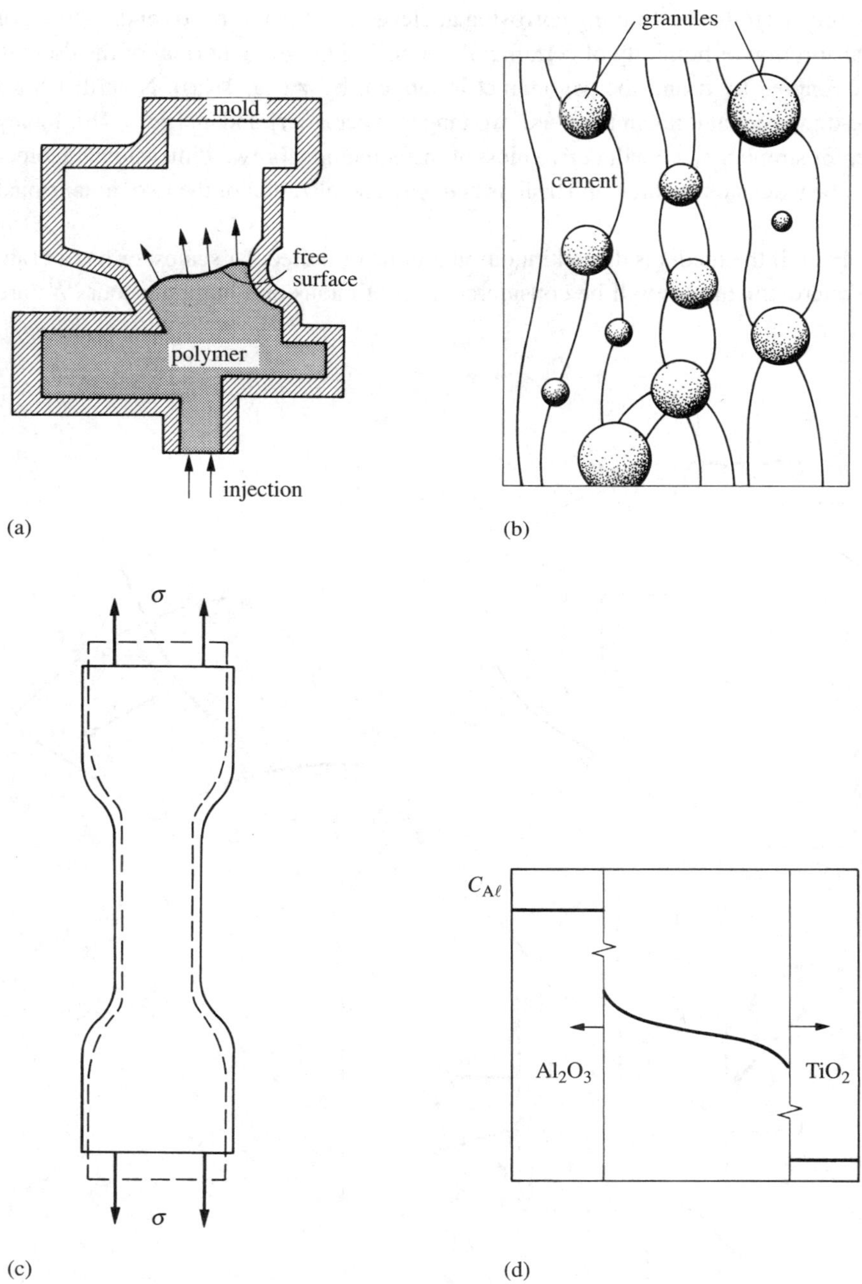

Figure 1.3 A few areas of materials science where modeling plays an important role: polymer injection (a), water diffusion in concrete (b), deformation of a test specimen under tension (c) and diffusion at an interface between Al_2O_3 and TiO_2 (d).

rounding air if one is studying corrosion at elevated temperatures or under stress, or the absorption of humidity of certain polymers. The outgoing normal of the domain Ω is denoted by $\boldsymbol{n}$ and the trigonometric tangent by $\boldsymbol{\tau}$ (fig. 1.4a)). Note that for a three-dimensional domain there are two tangent vectors, $\boldsymbol{\tau}_1$ and $\boldsymbol{\tau}_2$ (fig. 1.4b). In any case, for simplicity, we will derive most of the equations for two dimensional geometries, the extension to three dimensions being a generalization of the two dimensional case.

Even if the media is discontinuous at atomic or molecular scales, or has a granular microstructure, it will be considered as continuous and homogeneous. We are

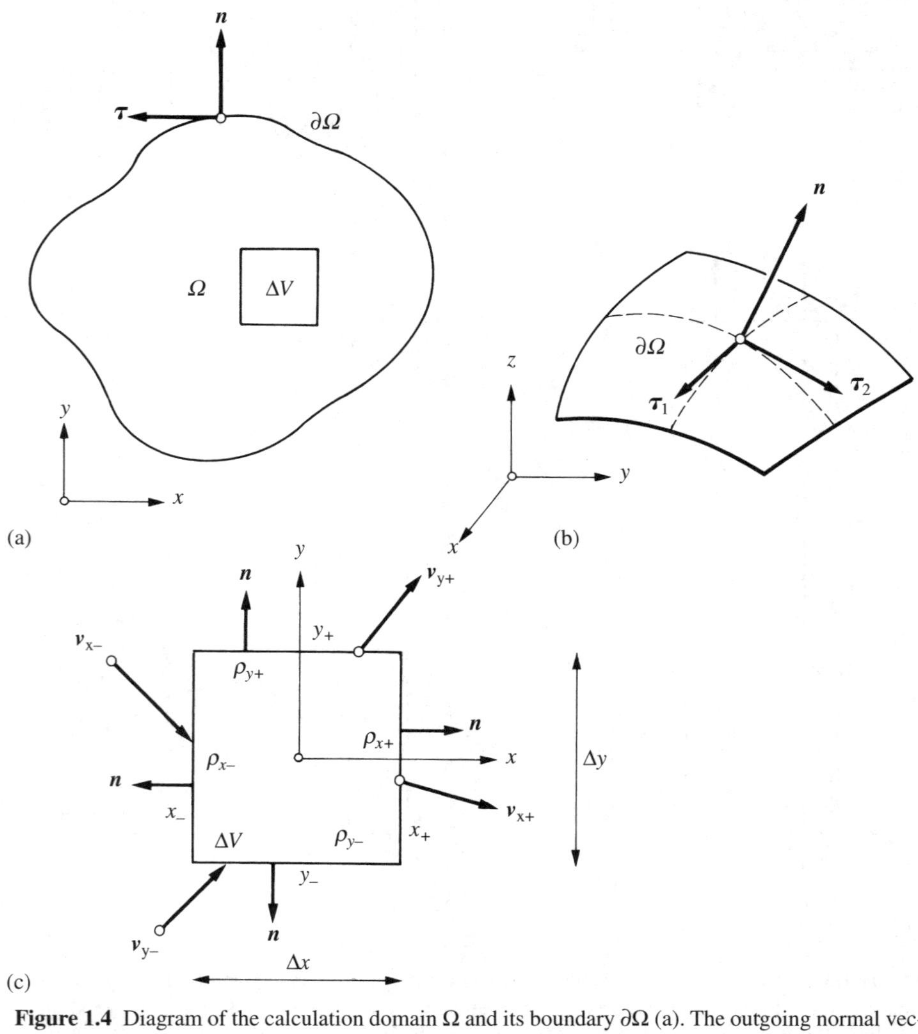

Figure 1.4 Diagram of the calculation domain Ω and its boundary ∂Ω (a). The outgoing normal vector $\boldsymbol{n}$ and the tangent vector $\boldsymbol{\tau}$ are also shown. In three dimensions (b), there are two tangent vectors. In (c), the volume element ΔV is shown with the velocities in the media at each face for the derivation of the conservation equations.

thus placing ourselves at a scale such that a volume element, though considered mathematically as infinitesimally small, contains a statistically significant number of atoms or grains. For such an element it is then possible to define average macroscopic values such as the density, $\rho(\boldsymbol{x}, t)$, the temperature, $T(\boldsymbol{x}, t)$, the enthalpy per unit mass, $H(\boldsymbol{x}, t)$, the volumetric concentration of a component i, $c_i(\boldsymbol{x}, t)$, the velocity, $\boldsymbol{v}(\boldsymbol{x}, t)$, or the stress, $\boldsymbol{\sigma}(\boldsymbol{x}, t)$. We adopt the convention of using normal text symbols for scalars, such as time, t, and bold text symbols for vectors, such as the vector $\boldsymbol{x}$ spatial coordinates, or for tensors, such as the stress tensor $\boldsymbol{\sigma}$.

We now have to choose between two possible approaches to describe conservation (of mass, energy, momentum, etc.). In the first, called the *Lagrangian,* analogous to traditional mechanics, we follow the material element through its movements, as with a raft following the current in a river and having sensors to measure the values that we seek (for example, a thermometer if one wanted to know the water temperature along the line of flow). In the second approach, *Eulerian*, the reference frame is fixed and we watch what happens at a point as a function of time (following the previous example, the thermometer would now be attached to the shore). These two representations are of course equivalent. The Eulerian frame is generally used in fluid mechanics while most solid deformation models use the Lagrangian frame. In the following we will derive the conservation equations in a reference frame attached to the laboratory (Eulerian frame).

Thus the volume element ΔV, shown in figure 1.4(a), is fixed and unchanging in the chosen reference frame. This volume, defined geometrically and not attached to the material, is shown in detail in figure 1.4(c). To compute the conservation sums it is simpler if the faces of the volume are parallel to the coordinate system, but, as we will see, the equations obtained will be applicable for any arbitrary selected volume. We indicate by x_- and x_+ the vertical faces to the left and to the right respectively, and by y_- , y_+ the horizontal faces below and above. The size of these faces are Δy and Δx. The outward normal, $\boldsymbol{n}$, of each face of the domain ΔV is thus oriented in the direction indicated by the name for the face (for example, the outgoing normal of the face x_- is in the negative direction of the x-axis).

As long as the domain ΔV is small, we can make the assumption that, on each face, the values considered are constant (a more exact derivation can be made with integrals over each surface; the same result being obtained as the element size goes to zero). Then, as shown in figure 1.4(c), the density of the material on the face x_+ is indicated by ρ_{x+} or the velocity on the face y_- by $\boldsymbol{v}_{y-}$ (in a more exact derivation these would be $\rho(x_+, y, t)$ and $\boldsymbol{v}(x, y_-, t)$, still a function of y and t on the face x_+ and still a function of x and t on the face y_-). Time dependence is implicitly assumed for all the values.

1.2.2 Equation of conservation of mass

With the notation introduced, we can determine the sum for the conservation of mass in the volume element in figure 1.4(c). This sum includes only the two contributions expressing that the mass variation inside the element must be due to transfer of mass across the faces by the velocity field $\boldsymbol{v}$. There is no diffusion, nor production of mass. Then the mass variation inside the element is given by:

$$\frac{\mathrm{d}m}{\mathrm{d}t} = \frac{\mathrm{d}}{\mathrm{d}t}\int_{\Delta V} \rho \,\mathrm{d}V = \int_{\Delta V} \frac{\partial \rho}{\partial t}\,\mathrm{d}V \tag{1.1}$$

The transfer of the time derivative to the inside of the integral causes no problems as the volume ΔV is *fixed*. The time derivative inside then becomes a partial derivative, $\partial/\partial t$, since the variable is a function of position, $\boldsymbol{x}$, and of time, t. The mass flux of material transported, in units of kg m^{-2} s^{-1}, is defined as the product of the density and velocity. The quantity of material (in kg s^{-1}) going out through each face of ΔV is then given by the normal component of $\rho\,\boldsymbol{v}$, that is $\rho\, v_\mathrm{n} = \rho\,\boldsymbol{v}\cdot\boldsymbol{n}$, multiplied by the corresponding surface area (the symbol "$\cdot$" indicates the scalar product; a list of operators and their symbols is given in chapter 10). The total quantity of material leaving the volume ΔV is then given by an integral over the surface ∂V (or a sum over the four faces if one takes the values to be constant):

$$\int_{\partial V} \rho\boldsymbol{v}\cdot\boldsymbol{n}\,\mathrm{d}S \simeq [(\rho v_\mathrm{x})_{\mathrm{x}+} - (\rho v_\mathrm{x})_{\mathrm{x}-}]\Delta y + [(\rho v_\mathrm{y})_{\mathrm{y}+} - (\rho v_\mathrm{y})_{\mathrm{y}-}]\Delta x \tag{1.2}$$

Expressing the right hand side in terms of the size of the volume, $\Delta V = \Delta x\, \Delta y$, the derivatives of the terms in brackets are seen to appear as the size of the element goes to zero:

$$\left[\frac{(\rho v_\mathrm{x})_{\mathrm{x}+} - (\rho v_\mathrm{x})_{\mathrm{x}-}}{\Delta x} + \frac{(\rho v_\mathrm{y})_{\mathrm{y}+} - (\rho v_\mathrm{y})_{\mathrm{y}-}}{\Delta y}\right]\Delta V \simeq \left[\frac{\partial(\rho v_\mathrm{x})}{\partial x} + \frac{(\partial \rho v_\mathrm{y})}{\partial y}\right]\Delta V \tag{1.3}$$

This operation is equivalent to using the divergence theorem (Gauss' theorem) shown in the appendix at the end of this book. In a form more mathematically rigorous, the integral (1.2) can be manipulated as:

$$\int_{\partial V} \rho\boldsymbol{v}\cdot\boldsymbol{n}\,\mathrm{d}S = \int_{\Delta V} \mathrm{div}(\rho\boldsymbol{v})\,\mathrm{d}V = \int_{\Delta V}\left[\frac{\partial(\rho v_\mathrm{x})}{\partial x} + \frac{\partial(\rho v_\mathrm{y})}{\partial y}\right]\mathrm{d}V \tag{1.4}$$

The mass flux expressed is positive if the net flux is outward (that is $\boldsymbol{v}$ has the same direction as the outgoing normal $\boldsymbol{n}$). Since there is neither loss nor creation of material, the local conservation of mass can be expressed taking the contributions (1.1) and (1.4) and letting the size of the element ΔV go to zero:

$$\frac{\partial \rho}{\partial t} + \mathrm{div}(\rho\boldsymbol{v}) = 0 \tag{1.5}$$

One sees that the variations of density in time are linked to the divergence of the flux of material, $\rho \boldsymbol{v}$. This equation, valid at every point in the domain Ω under consideration, can thus be integrated over any volume element, and in particular, over the whole domain Ω shown in figure 1.4. We have then, still using the divergence theorem:

$$\int_{\Omega} \left[\frac{\partial \rho}{\partial t} + \mathrm{div}(\rho \boldsymbol{v})\right] \mathrm{d}V = \frac{\mathrm{d}}{\mathrm{d}t}\int_{\Omega} \rho \, \mathrm{d}V + \int_{\partial\Omega} \rho \boldsymbol{v} \cdot \boldsymbol{n} \, \mathrm{d}S = 0 \tag{1.6}$$

In other words, the variation of mass inside of a fixed volume can only be due to the flux of material entering or leaving the domain.

In this derivation we made a fundamental assumption, namely that the material making up the domain Ω was *homogeneous*. If another phase appeared, for example bubbles, or yet another phase having a different density, the conservation sum would of course have to be modified (chapter 5).

Note finally that two interesting situations can appear. The first is the *stationary* (*steady state*) case, that is, in the frame of reference, nothing appears to change over time. In this case the partial derivative with respect to time becomes zero, and the conservation equation becomes:

$$\mathrm{div}(\rho \boldsymbol{v}) = 0 \tag{1.7}$$

The second situation is that of a material for which the density is constant. In this case the material is said to be ***incompressible***[1]. Equation (1.5) then becomes:

$$\mathrm{div}\, \boldsymbol{v} = 0 \tag{1.8}$$

1.2.3 Conservation of solute

A completely analogous sum can be computed for the concentration of a chemical solute element or of a constituent in an alloy i, $c_i(\boldsymbol{x}, t)$, in solution in the continuous domain (figure 1.4). We assume in the following that the concentration c_i is expressed per unit volume, for example in mole m^{-3}. As for the mass sum derived above, a temporal variation of the quantity in solution in the volume element as well as a transport term containing the velocity field, $c_i \boldsymbol{v}$ [mole m^{-2} s^{-1}], will appear in the sum. Expressed again for the volume element ΔV of figure 1.4, the two contributions are in this case:

$$\int_{\Delta V} \frac{\partial c_i}{\partial t} \mathrm{d}V + \int_{\partial V} c_i \boldsymbol{v} \cdot \boldsymbol{n} \, \mathrm{d}S = \int_{\Delta V} \left[\frac{\partial c_i}{\partial t} + \mathrm{div}(c_i \boldsymbol{v})\right] \mathrm{d}V \tag{1.9}$$

1 Even though the adjective "incompressible" applies normally to a material for which the density is independent of the pressure, it is used loosely here to mean that the specific mass is constant.

Two new contributions however, need to be taken into account in the sum. The first is a diffusive term associated with concentration gradients (Fick's law); and the second is a source term (or sink) for chemical reactions that can make appear (or disappear) atoms or molecules of the component considered coming from (or transformed into) other chemical components. The diffusive term, associated with a flux $\boldsymbol{j}_i$, yields a contribution analogous to the transport term $c_i\boldsymbol{v}$. The component flux by diffusion across the surfaces of the volume element ΔV is then of the form:

$$\int_{\partial V} \boldsymbol{j}_i \cdot \boldsymbol{n}\, dS = \int_{\Delta V} \operatorname{div} \boldsymbol{j}_i\, dV \simeq \operatorname{div} \boldsymbol{j}_i \Delta V \tag{1.10}$$

The approximation of the volume integral is valid when the volume ΔV is sufficiently small to take the average. The source term, $\dot{Q}_i$, corresponds to the number of moles that appear or disappear locally per unit time and unit volume. The total quantity of solute in the volume ΔV, expressed in moles per second, is then simply:

$$\int_{\Delta V} \dot{Q}_i\, dV \simeq \dot{Q}_i \Delta V \tag{1.11}$$

Regrouping all the contributions (1.9) to (1.11), and taking the limit as the volume element goes to zero, we obtain the local equation for the conservation of solute:

$$\frac{\partial c_i}{\partial t} + \operatorname{div}(c_i \boldsymbol{v}) + \operatorname{div} \boldsymbol{j}_i = \dot{Q}_i \tag{1.12}$$

Placing the source term $\dot{Q}_i$ on the right hand side of the equation highlights the fact that if there is no solute flux, neither by transport nor by diffusion, across the boundaries of the volume element, the variation of the concentration inside the element, given by the first term of (1.12), must equal the source term (that is positive if $\dot{Q}_i$ is a source, and negative if $\dot{Q}_i$ is a sink).

At this point no assumptions have been made to obtain equation (1.12). Note that there will be as many equations as of the form (1.12) as there are species in the material in question. Unlike the mass conservation equation where only the velocity and density fields appear, the solute conservation equation contains two new fields: one a vector field, the solute diffusion flux, $\boldsymbol{j}_i(\boldsymbol{x}, t)$, the other scalar, $\dot{Q}_i(\boldsymbol{x}, t)$. We see that to solve such a problem, it will be necessary to have equations relating the diffusive flux and the concentration (for example, Fick's first law) and others for chemical reactions. The latter will involve not only the concentration of the other solute species, but also, in general, the temperature. These equations, necessary complements of the conservation equations, are the ***constitutive equations*** of the material which will be discussed in section 1.3.

Finally, it should be noted that if one uses mass specific concentration, w_i, rather than volume specific concentration, c_i, the conservation equation (1.12) takes the form:

$$\frac{\partial(\rho w_i)}{\partial t} + \operatorname{div}(\rho w_i \boldsymbol{v}) + \operatorname{div}\boldsymbol{j}_i = \dot{Q}_i \tag{1.13}$$

Using the rules for partial derivation from vector calculus, the first two terms of (1.13) can also be written:

$$\frac{\partial(\rho w_i)}{\partial t} + \operatorname{div}(\rho w_i \boldsymbol{v}) = w_i\left[\frac{\partial\rho}{\partial t} + \operatorname{div}(\rho\boldsymbol{v})\right] + \rho\left[\frac{\partial w_i}{\partial t} + \boldsymbol{v}\cdot\mathbf{grad}\, w_i\right] \tag{1.14}$$

From the conservation of mass (1.5), the term in the first brackets is zero, and the conservation of solute equation becomes:

$$\rho\left[\frac{\partial w_i}{\partial t} + \boldsymbol{v}\cdot\mathbf{grad}\, w_i\right] + \operatorname{div}\boldsymbol{j}_i = \dot{Q}_i \tag{1.15}$$

1.2.4 Equation of conservation of momentum

The velocity field, $\boldsymbol{v}(\boldsymbol{x}, t)$, appears in the equations of conservation of mass and solute above. It can be determined from the conservation of momentum. By analogy with the momentum defined for an element of matter (product of the mass and the velocity), we introduce here a momentum density, $\rho\boldsymbol{v}$. Consider the x component of this momentum density, ρv_x. The variations of this value are due to volume forces acting on the body and the contact forces. These forces are shown in figure 1.5 for a small element of volume ΔV. On each face of the element, the forces exerted by neighboring elements are decomposed into normal and tangential components. The first is a tensile stress applied to the surface of the element, the second a shear stress. If $\boldsymbol{T}_{x+}$ represents the contact force per unit surface on the face x_+, then the two stress components are given by $\sigma_{x,x+}$ and $\sigma_{y,x+}$. Generalizing, in the coordinate system (x_1, x_2, x_3) the notation σ_{ij} refers to the stress acting in direction x_i applied on a surface perpendicular to the direction x_j. The volume force, $\boldsymbol{F}$, is the product of the force density, $\boldsymbol{F}_v$, and the volume ΔV. In the usual case where this force arises from gravity, it is given by $\rho\, \boldsymbol{g}\, \Delta V$ ($\boldsymbol{F}_v = \rho\boldsymbol{g}$), where $\boldsymbol{g}$ is the gravitational acceleration vector (fig. 1.5). The volume force can also arise from electromagnetic forces acting on the interior of the medium. Taking the sum of the contributions to the force on the surface and on the volume, we obtain for the component x:

$$F_{v,x}\Delta V + [\sigma_{x,x+} - \sigma_{x,x-}]\Delta y + [\sigma_{x,y+} - \sigma_{x,y-}]\Delta x \simeq \left[F_{v,x} + \frac{\partial\sigma_{xx}}{\partial x} + \frac{\partial\sigma_{xy}}{\partial y}\right]\Delta V \tag{1.16}$$

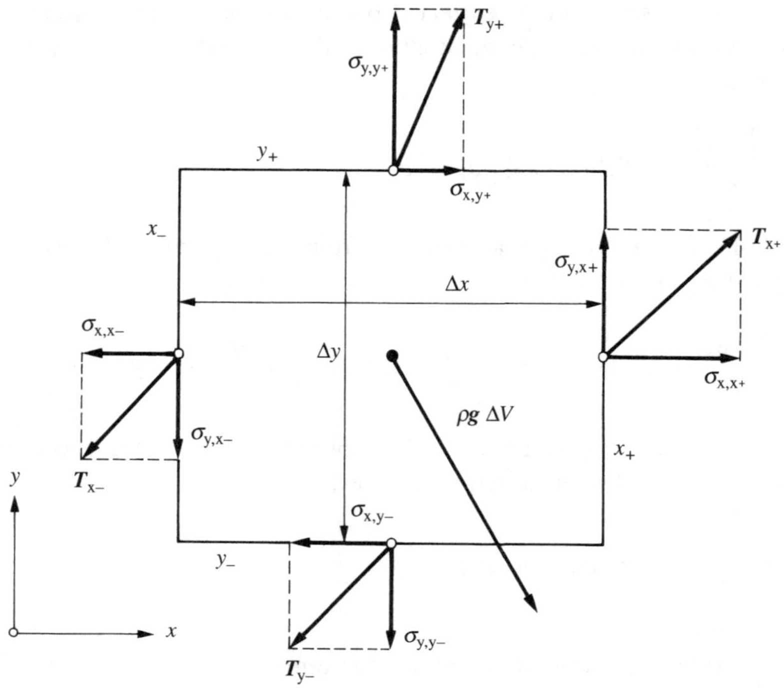

Figure 1.5 The surface forces, ***T***, and the gravity force, $\rho \boldsymbol{g}$, acting on a small volume element ΔV.

We have assumed again that the dimensions Δx and Δy of the volume element are very small, which allows us to approximate the differences by the derivatives. Before setting up an equation of conservation of momentum for the volume element ΔV, that is an expression of the variation of the momentum as being equal to the sum of the external forces, we must take into account the fact that, in the Eulerian coordinates chosen here, we do not follow the material's trajectory. In a reference frame fixed to the stationary observer, the flux of momentum entering or leaving the volume element must be considered. In other words, the velocity field transports not just mass or solute, but also momentum itself. By analogy with the previous definitions, the momentum flux in the x direction, ρv_x, entering or leaving the domain is given by:

$$\int_{\partial V} \rho v_x (\boldsymbol{v} \cdot \boldsymbol{n})\ \mathrm{d}S = \int_{\Delta V} \mathrm{div}(\rho v_x \boldsymbol{v})\ \mathrm{d}V \tag{1.17}$$

As before, the local temporal variation of the momentum in the direction x must be caused by the volume and surface forces (1.16) or by the flux of momentum entering or leaving (1.17). Again taking the limit as ΔV goes to zero, we obtain for the x component of the conservation of momentum:

$$\frac{\partial}{\partial t}(\rho v_x) + \mathrm{div}(\rho v_x \boldsymbol{v}) = F_{v,x} + \frac{\partial \sigma_{xx}}{\partial x} + \frac{\partial \sigma_{xy}}{\partial y} = F_{v,x} + \mathrm{div}(\sigma_{x.}) \tag{1.18}$$

The notation $\mathrm{div}(\sigma_{x.})$ has been introduced to express that the divergence, as commonly employed, is applied to the components of the stress tensor in direction x. This equation is not different in form from that previously obtained for the solute: the volume force $\boldsymbol{F}_v$ takes the place of the source or sink of momentum, and the term involving the stress tensor has the same form as the chemical diffusion term encountered before. We will see when presenting the equations of fluid behavior, that this term corresponds in fact to the diffusion of momentum.

The leftmost term in (1.18) can be transformed as was done before for the solute equation. The equation of conservation of momentum in the direction x then becomes:

$$\rho\left[\frac{\partial v_x}{\partial t} + \boldsymbol{v} \cdot \mathbf{grad}\, v_x\right] = F_{v,x} + \frac{\partial \sigma_{xx}}{\partial x} + \frac{\partial \sigma_{xy}}{\partial y} = F_{v,x} + \mathrm{div}(\sigma_{x.}) \tag{1.19}$$

All three directional components of the momentum (1.19) can be expressed together as follows[2]:

$$\rho\left[\frac{\partial \boldsymbol{v}}{\partial t} + (\mathbf{Grad}\, \boldsymbol{v}) \cdot \boldsymbol{v}\right] = \boldsymbol{F}_v + \mathbf{div}\, \boldsymbol{\sigma} \tag{1.20}$$

with the following notations for the generalized operators **div** and **Grad** acting on the stress tensor, $\boldsymbol{\sigma}$, and the velocity vector, $\boldsymbol{v}$, respectively (a list of operators appears in chapter 10):

$$\mathbf{div}\, \boldsymbol{\sigma} = \begin{bmatrix} \frac{\partial \sigma_{xx}}{\partial x} + \frac{\partial \sigma_{xy}}{\partial y} + \frac{\partial \sigma_{xz}}{\partial z} \\ \frac{\partial \sigma_{yx}}{\partial x} + \frac{\partial \sigma_{yy}}{\partial y} + \frac{\partial \sigma_{yz}}{\partial z} \\ \frac{\partial \sigma_{zx}}{\partial x} + \frac{\partial \sigma_{zy}}{\partial y} + \frac{\partial \sigma_{zz}}{\partial z} \end{bmatrix} \qquad \mathbf{Grad}\, \boldsymbol{v} = \begin{bmatrix} \frac{\partial v_x}{\partial x} & \frac{\partial v_x}{\partial y} & \frac{\partial v_x}{\partial z} \\ \frac{\partial v_y}{\partial x} & \frac{\partial v_y}{\partial y} & \frac{\partial v_y}{\partial z} \\ \frac{\partial v_z}{\partial x} & \frac{\partial v_z}{\partial y} & \frac{\partial v_z}{\partial z} \end{bmatrix} \tag{1.21a}$$

2 In some books, the operation $(\mathbf{Grad}\ \boldsymbol{v}) \cdot \boldsymbol{v}$ is also written $(\boldsymbol{v} \cdot \nabla)\, \boldsymbol{v}$, where ∇ is the "nabla" vector operator defined as $(\partial_x, \partial_y, \partial_z)$. In parallel, the term $\mathbf{div}\, \boldsymbol{\sigma}$ is written $\nabla \cdot \boldsymbol{\sigma}$.

In the coordinate system (x_1, x_2, x_3), these definitions can be expressed with Einstein's notation where repeated indices indicate a sum:

$$[\mathbf{div}\,\boldsymbol{\sigma}]_i = \frac{\partial \sigma_{ij}}{\partial x_j} \qquad [\mathbf{Grad}\,\boldsymbol{v}]_{ij} = \frac{\partial v_i}{\partial x_j} \tag{1.21b}$$

Note that **div** $\boldsymbol{\sigma}$ is a vector while **Grad** $\boldsymbol{v}$ is a tensor. The latter is often decomposed into two new tensors, the ***symmetric velocity gradient tensor***, $\boldsymbol{d}$, and the ***anti-symmetric rotation rate tensor***, $\boldsymbol{\omega}$, defined as:

$$\boldsymbol{d} = \tfrac{1}{2}[(\mathbf{Grad}\,\boldsymbol{v}) + (\mathbf{Grad}\,\boldsymbol{v})^T] \qquad \boldsymbol{\omega} = \tfrac{1}{2}[(\mathbf{Grad}\,\boldsymbol{v}) - (\mathbf{Grad}\,\boldsymbol{v})^T] \tag{1.22a}$$

or expressed in terms of components:

$$d_{ij} = \frac{1}{2}\left[\frac{\partial v_i}{\partial x_j} + \frac{\partial v_j}{\partial x_i}\right] \qquad \omega_{ij} = \frac{1}{2}\left[\frac{\partial v_i}{\partial x_j} - \frac{\partial v_j}{\partial x_i}\right] \tag{1.22b}$$

The velocity gradient, **Grad** $\boldsymbol{v}$**,** is then the sum of the two tensors:

$$[\mathbf{Grad}\,\boldsymbol{v}]_{ij} = \frac{\partial v_i}{\partial x_j} = d_{ij} + \omega_{ij} \tag{1.23}$$

Note that at this stage the equation of conservation of momentum (1.20) is capable in principle of describing any material, be it solid, liquid or gaseous. We have made no assumptions about the nature of the stress tensor, yet this tensor is symmetric. This is a consequence of the conservation of angular momentum that we will not develop here (exercise 1.5.1). The tensor has then six independent components while we have only three equations corresponding to the three components of velocity, which themselves are unknown. It is thus necessary to relate the stress tensor to the strain or strain rates by constitutive equations. In this context we can distinguish two classes of problems: *quasi-static* problems in which inertia is insignificant, and *dynamic* problems in which inertia can not be neglected. In the first case, the left hand side of (1.20) can be assumed to be zero or neglected relative to the forces on the right hand side. Then the equation takes the form:

$$\boldsymbol{F}_v + \mathbf{div}\,\boldsymbol{\sigma} = \mathbf{0} \tag{1.24}$$

Also neglecting gravity, which is often a legitimate assumption, the equation is further reduced to:

$$\mathbf{div}\,\boldsymbol{\sigma} = \mathbf{0} \tag{1.25}$$

In the second case, the dynamic behavior of the material will most often be viscous, which means that the stress tensor will only depend on the velocity field,

$\boldsymbol{v}(\boldsymbol{x}, t)$, and not on the strain. Nonetheless in some complex problems, such as polymer injection or high speed forming, it may be necessary to consider all the terms of (1.20) and at the same time express the stress tensor in terms of both the strain and the strain rate (chap. 7).

It should be noted that the stress field allows the calculation of the surface contact force density, $\boldsymbol{T}$, or the stress vector acting on a surface inside the domain Ω or a portion of the boundary $\partial\Omega$ (fig. 1.6). The surface normal being $\boldsymbol{n}$, it can be shown that:

$$\boldsymbol{T} = \boldsymbol{\sigma} \cdot \boldsymbol{n} \qquad \text{or} \qquad T_{\mathrm{i}} = \sigma_{\mathrm{ij}} n_{\mathrm{j}} \tag{1.26}$$

Consider the small tetrahedron defined by the surface S and by the three coordinate planes x, y and z. Mechanical equilibrium implies that the contact forces acting on all the faces of the element sum to zero. Considering the x component, the force acting on the surface S is T_{x}. On the surface perpendicular to the x axis, the normal stress per unit area σ_{xx} acts on the projection of the surface S, $S\,n_{\mathrm{x}}$. Similarly, for the surfaces perpendicular to the y and z axes, shear forces per unit area σ_{xy} and σ_{xz} act on the surfaces $(S\,n_{\mathrm{y}})$ and $(S\,n_{\mathrm{z}})$ respectively. The equilibrium is thus:

$$T_{\mathrm{x}} S = \sigma_{\mathrm{xx}} S n_{\mathrm{x}} + \sigma_{\mathrm{xy}} S n_{\mathrm{y}} + \sigma_{\mathrm{xz}} S n_{\mathrm{z}} \tag{1.27}$$

Simplifying by removing the common S, and considering also the y and z components, one arrives at (1.26).

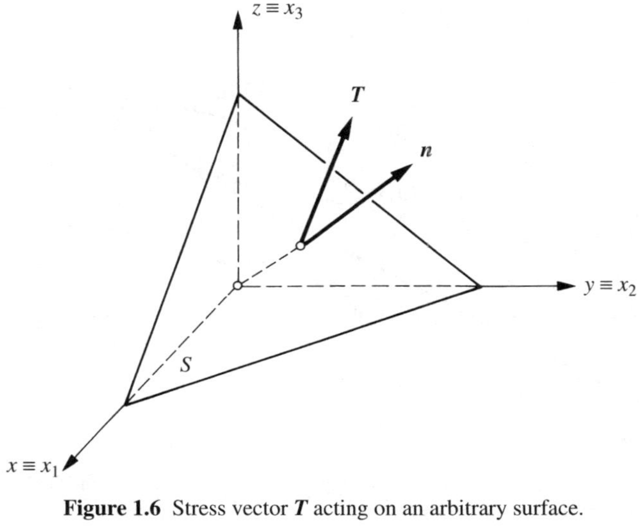

Figure 1.6 Stress vector $\boldsymbol{T}$ acting on an arbitrary surface.

1.2.5 Displacement, strain, strain rate

Figure 1.7a illustrates a forging process where a bar is deformed to obtain the finished part on the right. Over a period of time, displacements of the material occur due to the applied stress and volume forces. In Lagrangian coordinates, the movement of the material is described by the set of trajectories of all the points in the material: $\boldsymbol{x}(t) = \phi(\boldsymbol{x}_0, t)$, where $\boldsymbol{x}_0$ is the initial position of a point at time $t = 0$ (fig. 1.7b). (In more rigorous notation, the position at time t of a point initially at $\boldsymbol{x}_0$ would be $\boldsymbol{x}(\boldsymbol{x}_0, t)$). The displacement $\boldsymbol{u}(\boldsymbol{x}_0, t)$ of a point $\boldsymbol{x}_0$ at time t is naturally defined as the difference between its position at time t and its initial position:

$$\boldsymbol{u}(\boldsymbol{x}_0, t) = \boldsymbol{x}(t) - \boldsymbol{x}_0 = \phi(\boldsymbol{x}_0, t) - \boldsymbol{x}_0 \tag{1.28}$$

The displacement field $\boldsymbol{u}$ includes the translation, rotation and deformation of the material. The strain can be isolated by considering the displacement of a point neighboring $\boldsymbol{x}_0$ (fig. 1.7b). Differentiating (1.28) at a given time t, the new position of a point initially at $(\boldsymbol{x}_0 + \Delta\boldsymbol{x}_0)$ is given by:

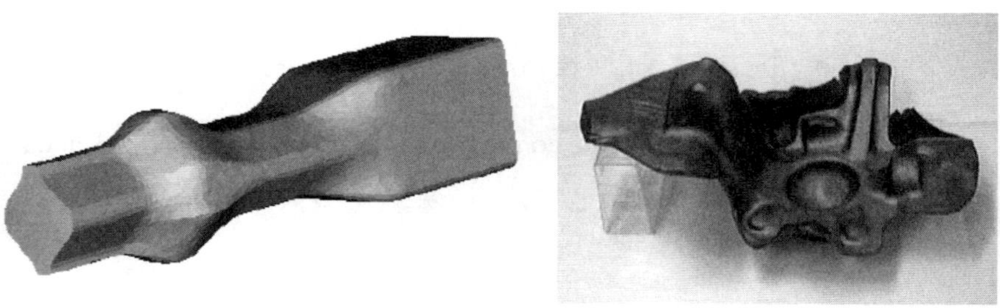

(a) initial form form after forging

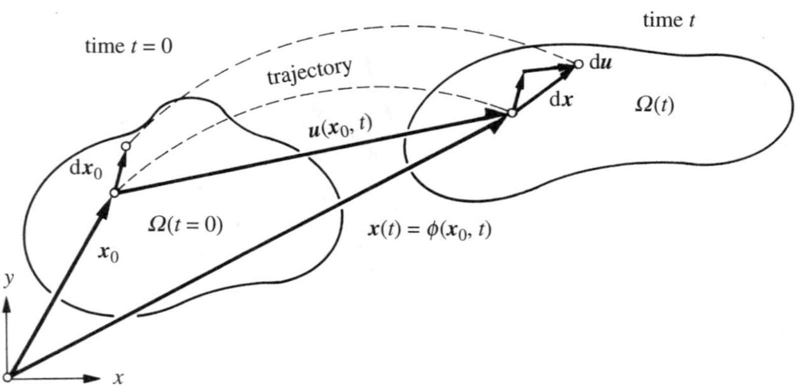

(b)

Figure 1.7 Forging of a solid to obtain a complex form (a) and two-dimensional representation of the displacements (b) for such a process.

$$\boldsymbol{x} + \mathrm{d}\boldsymbol{x} = (\boldsymbol{x}_0 + \mathrm{d}\boldsymbol{x}_0) + (\boldsymbol{u} + \mathrm{d}\boldsymbol{u}) = (\boldsymbol{x}_0 + \boldsymbol{u}) + \mathrm{d}\boldsymbol{x}_0 + \frac{\partial \boldsymbol{u}}{\partial \boldsymbol{x}_0}\mathrm{d}\boldsymbol{x}_0 \tag{1.29}$$

From which we obtain:

$$\mathrm{d}\boldsymbol{x} = \mathrm{d}\boldsymbol{x}_0 + \frac{\partial \boldsymbol{u}}{\partial \boldsymbol{x}_0}\mathrm{d}\boldsymbol{x}_0 \tag{1.30}$$

To simplify the expression, it is necessary at this stage to switch to the coordinate notation (x_1, x_2, x_3) in place of (x, y, z). Equation (1.30) then takes the form:

$$\mathrm{d}x_\mathrm{i} = \mathrm{d}x_{0\mathrm{i}} + \frac{\partial u_\mathrm{i}}{\partial x_{0\mathrm{j}}}\mathrm{d}x_{0\mathrm{j}} = \left[\delta_\mathrm{ij} + \frac{\partial u_\mathrm{i}}{\partial x_{0\mathrm{j}}}\right]\mathrm{d}x_{0\mathrm{j}} = F_\mathrm{ij}\mathrm{d}x_{0\mathrm{j}} \tag{1.31}$$

where δ_ij is the Kronecker delta function ($\delta_\mathrm{ij} = 1$ if $i = j$; $\delta_\mathrm{ij} = 0$ if $i \neq j$). This equation defines the ***gradient tensor*** of the transformation $\boldsymbol{F} = \boldsymbol{I} + \mathbf{Grad}\ \boldsymbol{u}$. Identical relations can be deduced for another point in the neighborhood of $\boldsymbol{x}_\mathrm{o}$, which ends up at $\boldsymbol{x} + \mathrm{d}\boldsymbol{x}'$ after displacement. Using (1.30), the scalar product of the two incremental vectors, $\mathrm{d}\boldsymbol{x}_\mathrm{o}$ and $\mathrm{d}\boldsymbol{x}$, is:

$$\mathrm{d}\boldsymbol{x} \cdot \mathrm{d}\boldsymbol{x}' = \left[\mathrm{d}\boldsymbol{x}_0 + \frac{\partial \boldsymbol{u}}{\partial \boldsymbol{x}_0}\mathrm{d}\boldsymbol{x}_0\right] \cdot \left[\mathrm{d}\boldsymbol{x}_0 + \frac{\partial \boldsymbol{u}}{\partial \boldsymbol{x}_0}\mathrm{d}\boldsymbol{x}_0'\right] \tag{1.32}$$

or in terms of components:

$$\mathrm{d}x_\mathrm{i}\mathrm{d}x_\mathrm{i}' = [F_\mathrm{ij}\mathrm{d}x_{0\mathrm{j}}][F_\mathrm{ik}\mathrm{d}x_{0\mathrm{k}}'] = C_\mathrm{jk}\mathrm{d}x_{0\mathrm{j}}\mathrm{d}x_{0\mathrm{k}}' \tag{1.33}$$

with:

$$\begin{aligned} C_\mathrm{jk} &= F_\mathrm{ij}F_\mathrm{ik} = \left[\delta_\mathrm{ij} + \frac{\partial u_\mathrm{i}}{\partial x_{0\mathrm{j}}}\right]\left[\delta_\mathrm{ik} + \frac{\partial u_\mathrm{i}}{\partial x_{0\mathrm{k}}}\right] \\ &= \delta_\mathrm{jk} + \frac{\partial u_\mathrm{k}}{\partial x_{0\mathrm{j}}} + \frac{\partial u_\mathrm{j}}{\partial x_{0\mathrm{k}}} + \frac{\partial u_\mathrm{i}}{\partial x_{0\mathrm{j}}}\frac{\partial u_\mathrm{i}}{\partial x_{0\mathrm{k}}} \end{aligned} \tag{1.34}$$

or as tensors: $\boldsymbol{C} = \boldsymbol{F}^\mathrm{T}\boldsymbol{F} = \boldsymbol{I} + \mathbf{Grad}\ \boldsymbol{u} + (\mathbf{Grad}\ \boldsymbol{u})^\mathrm{T} + (\mathbf{Grad}\ \boldsymbol{u})^\mathrm{T} \cdot (\mathbf{Grad}\ \boldsymbol{u})$. The result is a symmetric second order tensor, a 3×3 matrix called the ***right Cauchy-Green strain tensor***, or Cauchy's dilatation tensor. Note that the last term in (1.34), which is non linear in the displacement, includes a sum on the index *i*. The ***Green-Lagrange strain tensor***, $\boldsymbol{\varepsilon}$, is defined as:

$$\boldsymbol{\varepsilon} = \tfrac{1}{2}[\boldsymbol{C} - \boldsymbol{I}] \quad \text{or} \quad \varepsilon_\mathrm{jk} = \tfrac{1}{2}[C_\mathrm{jk} - \delta_\mathrm{jk}] = \tfrac{1}{2}\left[\frac{\partial u_\mathrm{k}}{\partial x_{0\mathrm{j}}} + \frac{\partial u_\mathrm{j}}{\partial x_{0\mathrm{k}}} + \frac{\partial u_\mathrm{i}}{\partial x_{0\mathrm{j}}}\frac{\partial u_\mathrm{i}}{\partial x_{0\mathrm{k}}}\right] \tag{1.35}$$

If the material deformations are small, the last term, being of second order, can be neglected. The result is the linearized expression of the strain:

$$\varepsilon_{jk} = \frac{1}{2}\left[\frac{\partial u_k}{\partial x_{0j}} + \frac{\partial u_j}{\partial x_{0k}}\right] \tag{1.36}$$

Returning to the (*x*, *y*, *z*) coordinates, the symmetric strain tensor is written:

$$\begin{aligned} &\varepsilon_{xx} = \frac{\partial u_x}{\partial x} \qquad \varepsilon_{yy} = \frac{\partial u_y}{\partial y} \qquad \varepsilon_{zz} = \frac{\partial u_z}{\partial z} \\ &\varepsilon_{xy} = \frac{1}{2}\left[\frac{\partial u_x}{\partial y} + \frac{\partial u_y}{\partial x}\right] \quad \varepsilon_{yz} = \frac{1}{2}\left[\frac{\partial u_y}{\partial z} + \frac{\partial u_z}{\partial y}\right] \quad \varepsilon_{zx} = \frac{1}{2}\left[\frac{\partial u_z}{\partial x} + \frac{\partial u_x}{\partial z}\right] \end{aligned} \tag{1.37}$$

The different components of the strain tensor are shown in figure 1.8 for the deformation of an infinitesimal parallelepiped. The symmetric velocity gradient tensor $\boldsymbol{d}$ introduced in (1.22) appears as the material or total derivative of the linearized strain tensor $\boldsymbol{\varepsilon}$ defined in (1.36). This is the reason why it is often called the ***strain rate tensor***. However, this is only true for the approximation of small deformation; in general, $\boldsymbol{d}$ is not the total derivative of the Green-Lagrange tensor (1.35). Adding to the confusion, the symbol $\dot{\boldsymbol{\varepsilon}}$ is often used in place of $\boldsymbol{d}$, even for large deformations:

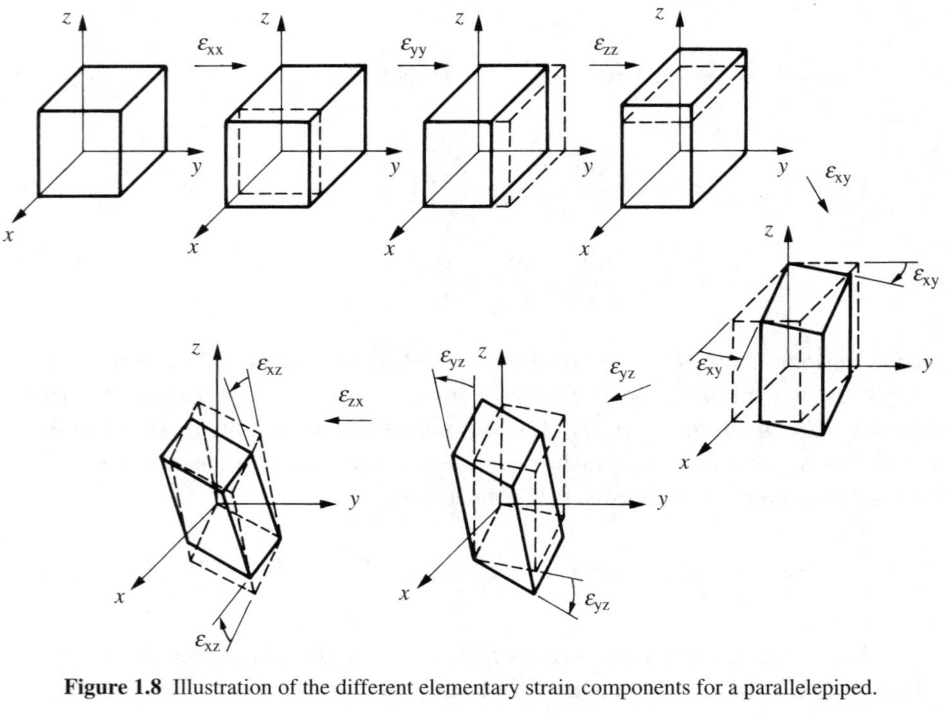

Figure 1.8 Illustration of the different elementary strain components for a parallelepiped.

$$\begin{aligned}
\dot{\varepsilon}_{xx} &= d_{xx} = \frac{\partial \dot{u}_x}{\partial x} \qquad \dot{\varepsilon}_{yy} = d_{yy} = \frac{\partial \dot{u}_y}{\partial y} \qquad \dot{\varepsilon}_{zz} = d_{zz} = \frac{\partial \dot{u}_z}{\partial z} \\
\dot{\varepsilon}_{xy} &= d_{xy} = \frac{1}{2}\left[\frac{\partial \dot{u}_x}{\partial y} + \frac{\partial \dot{u}_y}{\partial x}\right] \qquad \dot{\varepsilon}_{yz} = d_{yz} = \frac{1}{2}\left[\frac{\partial \dot{u}_y}{\partial z} + \frac{\partial \dot{u}_z}{\partial y}\right] \\
\dot{\varepsilon}_{zx} &= d_{zx} = \frac{1}{2}\left[\frac{\partial \dot{u}_z}{\partial x} + \frac{\partial \dot{u}_x}{\partial z}\right]
\end{aligned} \tag{1.38}$$

where $\dot{\boldsymbol{u}}$ designates the ***velocity vector*** associated with the displacement vector $\boldsymbol{u}$. This velocity is simply that of a point in Lagrangian coordinates:

$$\dot{\boldsymbol{u}} = \frac{\mathrm{d}\boldsymbol{u}(\boldsymbol{x}_0, t)}{\mathrm{d}t} = \frac{\partial \boldsymbol{u}(\boldsymbol{x}_0, t)}{\partial t} = \boldsymbol{v}(\boldsymbol{x}_0, t) \tag{1.39}$$

With the introduction of a velocity field, the ***total*** (or material) ***derivative*** of a scalar field $f(\boldsymbol{x}, t)$ (for example, the density) consists of the temporal variations of the variable under considerations, following it as it moves. Although this poses no difficulties in Lagrangian coordinates, in Eulerian coordinates the displacement $\boldsymbol{x}(t) = \boldsymbol{x}_o + \boldsymbol{u}(\boldsymbol{x}_o, t)$ (1.28) must be taken into account. Thus:

$$\frac{\mathrm{d}f}{\mathrm{d}t} = \frac{\partial f}{\partial t} + \frac{\partial f}{\partial \boldsymbol{x}}\frac{\partial \boldsymbol{x}}{\partial t} = \frac{\partial f}{\partial t} + \frac{\partial f}{\partial \boldsymbol{x}}\frac{\partial \boldsymbol{u}}{\partial t} = \frac{\partial f}{\partial t} + \frac{\partial f}{\partial \boldsymbol{x}}\boldsymbol{v} = \frac{\partial f}{\partial t} + \boldsymbol{v} \cdot \mathbf{grad}\, f \tag{1.40}$$

It is seen that this expression occurs naturally in the conservation equations derived in Eulerian coordinates (both mass and solute). It can be extended to vectors such as velocity, or even tensors. For the velocity field, it is not difficult to show that the ***acceleration vector*** is given by:

$$\boldsymbol{a} = \frac{\mathrm{d}\boldsymbol{v}}{\mathrm{d}t} = \frac{\partial \boldsymbol{v}}{\partial t} + \frac{\partial \boldsymbol{v}}{\partial \boldsymbol{x}}\frac{\partial \boldsymbol{x}}{\partial t} = \frac{\partial \boldsymbol{v}}{\partial t} + (\mathbf{Grad}\,\boldsymbol{v}) \cdot \boldsymbol{v} \tag{1.41}$$

and that the equation of conservation of momentum (1.20) can also be written:

$$\rho\left[\frac{\partial \boldsymbol{v}}{\partial t} + (\mathbf{Grad}\,\boldsymbol{v}) \cdot \boldsymbol{v}\right] = \rho \boldsymbol{a} = \boldsymbol{F}_{\mathrm{v}} + \mathbf{div}\,\boldsymbol{\sigma} \tag{1.42}$$

Equation (1.42) is called the ***fundamental equation of dynamics,*** none other than the generalization of Newton's law written locally for a deformable material.

Finally, following a volume element in Lagrangian coordinates through its trajectory, the boundaries will evolve in time. The total derivative of a variable integrated over such a changing volume has two contributions: the first corresponds to the time derivative of the variable integrated as if the volume did not change its form; and the second corresponds to the surface integral of the normal flux of the variable in question (there will be a flux, as the domain surface is moving). Thus:

$$\frac{d}{dt}\int_{\Omega(t)} f(\boldsymbol{x},t)\,dV = \int_{\Omega(t)} \frac{\partial f(\boldsymbol{x},t)}{\partial t}\,dV + \int_{\partial\Omega(t)} f(\boldsymbol{x},t)\boldsymbol{v}\cdot\boldsymbol{n}\,dS \tag{1.43}$$

This equation can be easily proven for a one dimensional problem (exercise 1.5.2). Using the divergence theorem, the expression becomes:

$$\frac{d}{dt}\int_{\Omega(t)} f(\boldsymbol{x},t)\,dV = \int_{\Omega(t)} \left\{\frac{\partial f(\boldsymbol{x},t)}{\partial t} + \mathrm{div}[f(\boldsymbol{x},t)\boldsymbol{v}]\right\}dV \tag{1.44}$$

1.2.6 Virtual power

It is well known for non deforming solid mechanics that the forces applied to a body are converted to kinetic energy (translation and rotation of the body). In the case of a deformable solid, the deformation energy must be taken into account, as will be shown.

We start with (1.42) and multiply it by an arbitrary or *virtual* velocity field, $\boldsymbol{v}^*(\boldsymbol{x}, t)$, and integrate over the domain Ω. We obtain:

$$\int_{\Omega} \rho\boldsymbol{a}\cdot\boldsymbol{v}^*\,dV - \int_{\Omega} \boldsymbol{F}_{\mathrm{v}}\cdot\boldsymbol{v}^*\,dV = \int_{\Omega} \mathbf{div}\boldsymbol{\sigma}\cdot\boldsymbol{v}^*\,dV \tag{1.45}$$

The generalized divergence given by (1.21) can be used to develop the term in the integral on the right hand side as follows (in the coordinates (x_1, x_2, x_3) with summation over repeated indices):

$$\mathbf{div}\boldsymbol{\sigma}\cdot\boldsymbol{v}^* = \frac{\partial\sigma_{ij}}{\partial x_j}v_i^* = \frac{\partial(\sigma_{ij}v_i^*)}{\partial x_j} - \sigma_{ij}\frac{\partial v_i^*}{\partial x_j} \tag{1.46}$$

Because the stress tensor is symmetric, this expression can also be written:

$$\mathbf{div}\,\boldsymbol{\sigma}\cdot\boldsymbol{v}^* = \frac{\partial(\sigma_{ij}v_i^*)}{\partial x_j} - \sigma_{ij}\frac{1}{2}\left[\frac{\partial v_i^*}{\partial x_j} + \frac{\partial v_j^*}{\partial x_i}\right] = \frac{\partial(\sigma_{ji}v_i^*)}{\partial x_j} - \sigma_{ij}\dot{\varepsilon}_{ij}^*$$
$$= \mathrm{div}\,(\boldsymbol{\sigma}\cdot\boldsymbol{v}^*) - \boldsymbol{\sigma}:\dot{\boldsymbol{\varepsilon}}^* \tag{1.47}$$

We have introduced the strain rate, $\dot{\varepsilon}_{ij}^*$, as well as the contracted product, with the notation ':', of this tensor with the stress tensor (chap. 10). With these definitions and the divergence theorem, equation (1.45) becomes:

$$\int_{\Omega} \rho\boldsymbol{a}\cdot\boldsymbol{v}^*\,dV - \int_{\Omega} \boldsymbol{F}_{\mathrm{v}}\cdot\boldsymbol{v}^*\,dV = \int_{\partial\Omega} (\boldsymbol{\sigma}\cdot\boldsymbol{n})\cdot\boldsymbol{v}^*\,dS - \int_{\Omega} \boldsymbol{\sigma}:\dot{\boldsymbol{\varepsilon}}^*\,dV \tag{1.48}$$

At the surface $\partial\Omega$ of the domain, the product $\boldsymbol{\sigma} \cdot \boldsymbol{n}$ is none other than the stress vector $\boldsymbol{T}$, the local contact force density (fig. 1.6). One then has the following balance of ***virtual power***:

$$\int_{\Omega} \boldsymbol{F}_{\mathrm{v}} \cdot \boldsymbol{v}^* \,\mathrm{d}V + \int_{\partial\Omega} \boldsymbol{T} \cdot \boldsymbol{v}^* \,\mathrm{d}S = \int_{\Omega} \rho \boldsymbol{a} \cdot \boldsymbol{v}^* \,\mathrm{d}V + \int_{\Omega} \boldsymbol{\sigma} : \dot{\boldsymbol{\varepsilon}}^* \,\mathrm{d}V \tag{1.49}$$

In other words, the virtual power due to the applied exterior forces on the volume and its surface is transformed into a virtual kinetic power and a virtual strain power. It must be noted that this relation is valid regardless of the virtual velocity chosen. In the particular case where the field is identically the real velocity field ($\boldsymbol{v}^* = \boldsymbol{v}$), the kinetic term becomes equal to the derivative of the kinetic energy (exercise 1.5.3):

$$\int_{\Omega} \rho \boldsymbol{a} \cdot \boldsymbol{v} \,\mathrm{d}V = \int_{\Omega} \rho \frac{\mathrm{d}\boldsymbol{v}}{\mathrm{d}t} \cdot \boldsymbol{v} \,\mathrm{d}V = \int_{\Omega} \rho \frac{\mathrm{d}}{\mathrm{d}t}\left[\frac{v^2}{2}\right] \mathrm{d}V = \frac{\mathrm{d}}{\mathrm{d}t} \int_{\Omega} \tfrac{1}{2} \rho v^2 \,\mathrm{d}V \tag{1.50}$$

This kinetic energy term will be seen again for the conservation of energy in the following section.

1.2.7 Equation of conservation of energy

In many situations encountered in materials science, a heat conservation calculation is performed rather than one for energy; but heat is not a state variable. This amounts to neglecting viscous dissipation in fluid dynamics, strain energy in solid mechanics or even potential energy. It is true that these energies are generally weak in comparison to the latent heats of transformation or the specific heat. For example, the potential energy variation due to gravity between the top and bottom of a 5 meter tall aluminum ingot in continuous casting (fig. 1.2a) is equivalent, in terms of specific heat, to a temperature difference of 0.05 °C! Nonetheless, in some situations the conversion of work into heat can not be neglected, as, for example, for a supersonic gas jet out of a nozzle in plasma deposition, or more simply, for large deformations (forging).

In general, the first law of thermodynamics tells us that the variation of total energy of a domain under consideration is due to the mechanical power of the external forces, P_{mech}, and the caloric power applied, P_{cal}. The total energy is composed of the kinetic energy, E_{k}, and the internal energy, E_{i}. Putting all the terms together:

$$\frac{\mathrm{d}E_{\mathrm{k}}}{\mathrm{d}t} + \frac{\mathrm{d}E_{\mathrm{i}}}{\mathrm{d}t} = P_{\mathrm{mech}} + P_{\mathrm{cal}} \tag{1.51}$$

where the time derivatives are total derivatives. The mechanical power is applied to the system by volume forces and contact forces on the surface $\partial\Omega$ of the domain:

$$P_{\text{mech}} = \int_{\Omega} \boldsymbol{F}_{\text{v}} \cdot \boldsymbol{v} \, \mathrm{d}V + \int_{\partial\Omega} \boldsymbol{T} \cdot \boldsymbol{v} \, \mathrm{d}S \tag{1.52}$$

The caloric power into a system is also composed of two parts. The volume contribution, $\dot{Q}_{\mathrm{T}}$, can be due to chemical reactions, radioactive decay or the Joule effect. At the surface, a heat flux, $-\boldsymbol{j}_{\mathrm{T}} \cdot \boldsymbol{n}$, can enter the domain by radiation, convection, or conduction (in the same way as solute flux appears in section 1.2.3). The minus sign is present because a flux in (i.e., $\boldsymbol{j}_{\mathrm{T}}$ in the opposite sense of the outward normal $\boldsymbol{n}$ of the domain) increases the internal energy. The caloric power thus takes the form:

$$P_{\text{cal}} = \int_{\Omega} \dot{Q}_{\mathrm{T}} \, \mathrm{d}V - \int_{\partial\Omega} \boldsymbol{j}_{\mathrm{T}} \cdot \boldsymbol{n} \, \mathrm{d}S = \int_{\Omega} (\dot{Q}_{\mathrm{T}} - \operatorname{div} \boldsymbol{j}_{\mathrm{T}}) \, \mathrm{d}V \tag{1.53}$$

Taking the kinetic power from (1.50):

$$\frac{\mathrm{d}E_{\mathrm{k}}}{\mathrm{d}t} = \frac{\mathrm{d}}{\mathrm{d}t} \int_{\Omega} \tfrac{1}{2} \rho v^2 \, \mathrm{d}V = \int_{\Omega} \rho \frac{\mathrm{d}}{\mathrm{d}t}\left[\frac{\boldsymbol{v}^2}{2}\right] \mathrm{d}V \tag{1.54}$$

the internal energy balance becomes:

$$\frac{\mathrm{d}E_{\mathrm{i}}}{\mathrm{d}t} = -\int_{\Omega} \rho \frac{\mathrm{d}}{\mathrm{d}t}\left[\frac{\boldsymbol{v}^2}{2}\right] \mathrm{d}V + \int_{\Omega} \boldsymbol{F}_{\text{v}} \cdot \boldsymbol{v} \, \mathrm{d}V + \int_{\partial\Omega} \boldsymbol{T} \cdot \boldsymbol{v} \, \mathrm{d}S + \int_{\Omega} (\dot{Q}_{\mathrm{T}} - \operatorname{div} \boldsymbol{j}_{\mathrm{T}}) \, \mathrm{d}V \tag{1.55}$$

The sum of the first three terms on the right hand side of the equation is the deformation power, $\boldsymbol{\sigma}\colon \dot{\boldsymbol{\varepsilon}}$, integrated over the volume of the domain ((1.49) with $\boldsymbol{v}^* = \boldsymbol{v}$). The internal energy of the volume under consideration being:

$$E_{\mathrm{i}} = \int_{\Omega} \rho e_{\mathrm{i}} \, \mathrm{d}V \tag{1.56}$$

where e_{i} is the mass specific internal energy, the total derivative of this equation with respect to time is (1.44):

$$\frac{\mathrm{d}E_{\mathrm{i}}}{\mathrm{d}t} = \int_{\Omega} \left\{ \frac{\partial(\rho e_{\mathrm{i}})}{\partial t} + \operatorname{div}(\rho e_{\mathrm{i}} \boldsymbol{v}) \right\} \mathrm{d}V \tag{1.57}$$

Taking all the contributions, the local internal energy balance becomes:

$$\frac{\partial(\rho e_\mathrm{i})}{\partial t} + \mathrm{div}\,(\rho e_\mathrm{i}\boldsymbol{v}) = \boldsymbol{\sigma}\colon\dot{\boldsymbol{\varepsilon}} + \dot{Q}_\mathrm{T} - \mathrm{div}\,\boldsymbol{j}_\mathrm{T} \qquad (1.58)$$

or, again taking into account the conservation of mass:

$$\rho\frac{\partial e_\mathrm{i}}{\partial t} + \rho\boldsymbol{v}\cdot\mathbf{grad}\,e_\mathrm{i} = \rho\frac{\mathrm{d}e_\mathrm{i}}{\mathrm{d}t} = \boldsymbol{\sigma}\colon\dot{\boldsymbol{\varepsilon}} + \dot{Q}_\mathrm{T} - \mathrm{div}\,\boldsymbol{j}_\mathrm{T} \qquad (1.59)$$

Using the ***enthalpy*** per unit mass, H, given by:

$$H = e_\mathrm{i} + \frac{p}{\rho} \qquad (1.60)$$

where p is the pressure, (1.58) can also be written:

$$\frac{\partial(\rho H)}{\partial t} + \mathrm{div}\,(\rho H\boldsymbol{v}) = \left[\frac{\partial p}{\partial t} + \mathrm{div}\,(p\boldsymbol{v})\right] + \boldsymbol{\sigma}\colon\dot{\boldsymbol{\varepsilon}} + \dot{Q}_\mathrm{T} - \mathrm{div}\,\boldsymbol{j}_\mathrm{T} \qquad (1.61)$$

The variation of enthalpy can be caused by four different kinds of contributions on the right hand side of (1.61). The first two correspond to the total derivatives of pressure and thus to enthalpy variation due to local compression and decompression. For example, following a volume element containing gas that does not exchange heat with the exterior (local adiabatic conditions), it is well known that compression increases the free enthalpy and thus heats the gas. The term $\boldsymbol{\sigma}\colon\dot{\boldsymbol{\varepsilon}}$ is the increase in the enthalpy due to strain. As for the last two terms, they have the same signification as for the conservation of solute (1.12): $\dot{Q}_\mathrm{T}$ is a volumetric ***enthalpy production*** term (for example by the Joule effect or chemical reaction) and $\boldsymbol{j}_\mathrm{T}$ corresponds to ***heat diffusion***.

When mechanical contributions (compression, strain) to the energy balance can be ignored, we find the heat balance in analogous form as the equation of conservation of solute:

$$\frac{\partial(\rho H)}{\partial t} + \mathrm{div}(\rho H\boldsymbol{v}) + \mathrm{div}\boldsymbol{j}_\mathrm{T} = \dot{Q}_\mathrm{T} \qquad (1.62)$$

or in non conservative form:

$$\rho\frac{\partial H}{\partial t} + \rho\boldsymbol{v}\cdot\mathbf{grad}\,H + \mathrm{div}\boldsymbol{j}_\mathrm{T} = \dot{Q}_\mathrm{T} \qquad (1.63)$$

1.2.8 Unified form of the conservation equations

The conservation equations derived above are similar in that they all contain a temporal variation term, an advective transport term, and some have a diffusion term and a source term. They can be reduced to a single general equation:

$$\frac{\partial}{\partial t}(\rho\xi) + \mathrm{div}(\rho\xi\boldsymbol{v}) + \mathrm{div}\,\boldsymbol{j}_\xi = \dot{Q}_\xi \tag{1.64}$$

the variables, $\boldsymbol{j}_\xi$ and $\dot{Q}_\xi$, are given in table 1.1 for the conservation of mass ($\xi = 1$), of solute ($\xi = w_i$), of heat, ($\xi = H$) or the momentum along the x axis, ($\xi = v_x$). The non conservative form of these equations can also be used:

$$\rho\frac{\partial\xi}{\partial t} + \rho\boldsymbol{v}\cdot\mathbf{grad}\,\xi + \mathrm{div}\,\boldsymbol{j}_\xi = \dot{Q}_\xi \tag{1.65}$$

This form employs the total derivative of the variable ξ, and can equivalently be written:

$$\rho\frac{\mathrm{d}\xi}{\mathrm{d}t} + \mathrm{div}\,\boldsymbol{j}_\xi = \dot{Q}_\xi \tag{1.66}$$

Table 1.1 Terms appearing in the conservation equations

Conservation Equation	ξ		j_ξ		$\dot{Q}_\xi$	
mass	1	[–]	0	[–]	0	[–]
solute	w_i	[mole kg^{-1}]	j_i	[mole m^{-2} s^{-1}]	$\dot{Q}_i$	[mole m^{-3} s^{-1}]
energy (heat)	H	[J kg^{-1}]	j_T	[W m^{-2}]	$\dot{Q}_T$	[W m^{-3}]
x momentum	v_x	[m s^{-1}]	$-\sigma_{x.}$	[N m^{-2}]	$F_{v,x}$	[N m^{-3}]

The conservation equations can be integrated over a fixed domain, Ω, or a variable domain, $\Omega(t)$, attached to the material (Lagrangian reference frame). In the first case, we start from (1.64) to write:

$$\int_\Omega \frac{\partial(\rho\xi)}{\partial t}\,\mathrm{d}V + \int_\Omega \mathrm{div}\,(\rho\xi\boldsymbol{v})\,\mathrm{d}V + \int_\Omega \mathrm{div}\,\boldsymbol{j}_\xi\,\mathrm{d}V = \int_\Omega \dot{Q}_\xi\,\mathrm{d}V \tag{1.67}$$

The volume being fixed, the derivative with respect to time can be taken outside the first integral, while the next two are transformed by the divergence theorem. One then has:

$$\frac{\mathrm{d}}{\mathrm{d}t}\int_\Omega \rho\xi\,\mathrm{d}V = -\int_{\partial\Omega} \rho\xi\boldsymbol{v}\cdot\boldsymbol{n}\,\mathrm{d}S - \int_{\partial\Omega} \boldsymbol{j}_\xi\cdot\boldsymbol{n}\,\mathrm{d}S + \int_\Omega \dot{Q}_\xi\,\mathrm{d}V \tag{1.68}$$

When the integration domain, $\Omega(t)$, evolves with time in the Lagrangian representation, the starting point is (1.43) applied to the field $f(\boldsymbol{x}, t) = \rho\xi(\boldsymbol{x}, t)$, from which:

$$\begin{aligned}\frac{\mathrm{d}}{\mathrm{d}t}\int_{\Omega(t)} \rho\xi \,\mathrm{d}V &= \int_{\Omega(t)} \frac{\partial(\rho\xi)}{\partial t}\,\mathrm{d}V + \int_{\Omega(t)} \mathrm{div}(\rho\xi\boldsymbol{v})\,\mathrm{d}V \\ &= -\int_{\Omega(t)} \mathrm{div}\,\boldsymbol{j}_\xi \,\mathrm{d}V + \int_{\Omega(t)} \dot{Q}_\xi \,\mathrm{d}V\end{aligned} \tag{1.69}$$

or, again using the divergence theorem:

$$\frac{\mathrm{d}}{\mathrm{d}t}\int_{\Omega(t)} \rho\xi \,\mathrm{d}V = -\int_{\partial\Omega(t)} \boldsymbol{j}_\xi \cdot \boldsymbol{n}\,\mathrm{d}S + \int_{\Omega(t)} \dot{Q}_\xi \,\mathrm{d}V \tag{1.70}$$

In the first case (1.68), the temporal variation of $(\rho\xi)$ inside a fixed volume is given by the diffusive and advective flux integrated on the surface as well as the internal production term. If the domain $\Omega(t)$ follows the material (1.70), the advective term at the boundary naturally disappears.

1.3 CONSTITUTIVE EQUATIONS

As we indicated in the previous section, other equations are needed before we can "close" the problem. For example, it is necessary to link the solute diffusive flux, $\boldsymbol{j}_i$, to the concentration, c_i, or to link the enthalpy, H, to the temperature, T. These missing equations, called the ***constitutive equations*** of the material, are in fact the most difficult part of a problem; but, at the same time, they are the interesting part, that which gives their diversity. Since the conservation equations are universal, something else is needed to distinguish a liquid from a solid, a polymer from a metal, etc. This is the point where the knowledge of a materials science engineer intervenes. The field being vast, we restrict ourselves in this section to a presentation of the simplest and most often used constitutive equations; chapters 5 to 7 give more details for the three major classes of problems in materials science; namely: phase changes, solid deformations and fluid mechanics. Nonetheless, we will try to show in this section some of the coupling between the different fields that appear in the constitutive equations.

1.3.1 Constitutive equations for mass

In the simplest conservation equation, that for mass, two fields appear already: the density, $\rho(\boldsymbol{x}, t)$, and the velocity, $\boldsymbol{v}(\boldsymbol{x}, t)$. Typically, in a gas the density is a function of the temperature and pressure. Whereas the pressure in a gas is, as will be shown later, a function of the velocity field through the conservation of momentum, for homogeneous condensed matter we can make the hypothesis that the pressure has little effect on the density. Then the material equation for the density takes the form:

$$\rho(\boldsymbol{x}, t) = \rho_0\left[1 - \beta_T(T(\boldsymbol{x}, t) - T_0) - \sum_i \beta_i(c_i(\boldsymbol{x}, t) - c_{0,i})\right] \quad (1.71)$$

where β_T and β_i are the ***thermal*** and ***solute*** volumetric ***dilatation coefficients*** respectively. ρ_o is a reference density at temperature T_o and concentration $c_{o,i}$. One sees already in this simple case that the density field is a function of the temperature and concentration fields. This coupling can result in natural convection, either thermal or solutal, in the case of a liquid, or in stress in the case of a solid. If the material is not homogeneous (there are more than one phase, or pores), one can average the conservation equations (chap. 5). In many applications the incompressible hypothesis (ρ = constant) is made to simplify the problem, in which case (1.8) is applicable. Nevertheless, the density known, the mass conservation equation is insufficient to calculate the velocity field because it is a *scalar* equation (with first-order partial derivatives), while $\boldsymbol{v}(\boldsymbol{x}, t)$ is a *vector* field. We will see in section 1.3.5 that the conservation of momentum must be considered to treat a fluid dynamics problem.

1.3.2 Constitutive equations for solute

Assuming that the velocity field is known in the domain Ω (for example $\boldsymbol{v} = 0$), (1.12) does not allow the determination of the concentration field, $c_i(\boldsymbol{x}, t)$, if the other two variables appearing in the solute equation, the solute flux, $\boldsymbol{j}_i$, and the volumetric source term, $\dot{Q}_i$, are not specified. The flux can generally be expressed in terms of the gradient of the concentration by ***Fick's first law***:

$$\boldsymbol{j}_i(\boldsymbol{x}, t) = -D_i\,\mathbf{grad}\,c_i(\boldsymbol{x}, t) \quad (1.72)$$

where D_i is the ***diffusion coefficient***. Nevertheless, it must be noted that this is a constitutive equation under the assumption that the material is an ideal, dilute solution. Figure 1.9 shows the case of a concentration profile in one dimension and the resulting solute flux.

In the case where there is neither transport by advection nor a source term ($\boldsymbol{v} = 0$ and $\dot{Q}_i = 0$), the constitutive equation (1.72) and the solute conservation equation (1.12) give ***Fick's second law***:

$$\frac{\partial c_i}{\partial t} - \mathrm{div}\,(D_i\,\mathbf{grad}\,c_i) = 0 \quad (1.73)$$

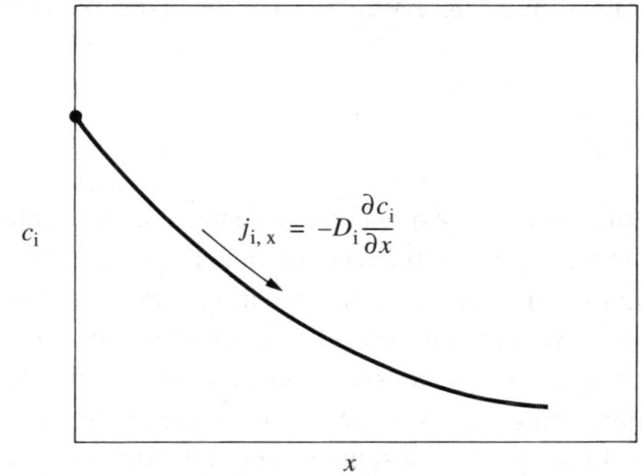

Figure 1.9 Diagram of diffusion according to Fick's first law relating the chemical flux to the concentration gradient.

With this equation, the concentration field, $c_i(\boldsymbol{x}, t)$, can be found in the domain Ω subject to two conditions: first, that the initial conditions (at time $t = 0$) and those at the boundary $\partial\Omega$ are given (sect. 1.4); second, the diffusion coefficient must be known. The latter could be a function itself of the concentration, making the problem non linear, or a function of temperature, introducing a link to the temperature field.

Taking into consideration the transport term for an incompressible fluid, Fick's second law takes the form:

$$\frac{\partial c_i}{\partial t} + \boldsymbol{v} \cdot \mathbf{grad}\, c_i - \mathrm{div}\,(D_i\,\mathbf{grad}\, c_i) = \frac{\mathrm{d}c_i}{\mathrm{d}t} - \mathrm{div}(D_i\,\mathbf{grad}\, c_i) = 0 \qquad (1.74)$$

Equation (1.73) is an example of a *diffusion* problem, whereas with a non zero material velocity, (1.74) becomes an ***advection-diffusion problem***. Numerical solution of (1.73) or (1.74) by the finite difference or finite element methods will be taken as examples in chapters 2 and 3 to illustrate the specific numerical issues that arise in these problems.

Finally, the volumetric source term, $\dot{Q}_i$, depends on the type and number of chemical reactions that occur in the domain, and thus on the solute species present. In the simple case of a weak concentration of a single chemical species in a solvent where only a single, first-order reaction is possible, for example, a substance dissolving in the solute, the source term becomes:

$$\dot{Q}_i = -a_r c_i \qquad (1.75)$$

where a_r is the ***reaction rate***. Note that the negative sign in front of the rate indicates that the substance is disappearing. If the concentration is uniform in the domain Ω, inserting (1.75) in the solute conservation equation gives ***Wilhelmy's law*** for a first-order reaction:

$$\frac{\mathrm{d}c_{\mathrm{i}}}{\mathrm{d}t} = -a_{\mathrm{r}}c_{\mathrm{i}} \tag{1.76}$$

which, after integration, gives an exponentially decreasing concentration c_{i} with a time constant equal to the inverse of the reaction rate. In general, the source term for a chemical species in the solute conservation equation is expressed as a sum over all the possible reactions, reversible and irreversible, in which the species in question takes part, thus coupling the solute conservation equations for the different species.

Finally, reactions acting on a surface, for example in the neighborhood of a corroding electrode, do not produce a source term in the diffusion equation; rather, they appear in the boundary conditions (sect. 1.4).

1.3.3 Constitutive equations for energy

Besides the temperature and velocity fields, the heat conservation equation (1.62) contains three other variables that must be described with constitutive equations. They are: the heat flux, $\boldsymbol{j}_{\mathrm{T}}$, the volume specific enthalpy, ρH, and the source term, $\dot{Q}_{\mathrm{T}}$. The diffusive heat flux, $\boldsymbol{j}_{\mathrm{T}}$, is given by an equation similar to Fick's first law (***Fourier's law***):

$$\boldsymbol{j}_{\mathrm{T}}(\boldsymbol{x}, t) = -\kappa\,\mathbf{grad}\,T(\boldsymbol{x}, t) \tag{1.77}$$

where κ is the ***thermal conductivity*** of the material. Note that (1.62) takes into account the heat transfer by advection with the term $\mathrm{div}(\rho H\,\boldsymbol{v})$, but not the exchanges by radiation if the material is transparent or semi transparent. In the latter case, if the material is optically thick, that is, when its thickness is much greater than the mean free path of radiation, an equivalent ***radiative conductivity*** κ_{R} can be introduced:

$$\kappa_{\mathrm{R}} = \frac{16}{3}\frac{\sigma_{\mathrm{B}} n^2 T^3}{\alpha_{\mathrm{r}}} \tag{1.78}$$

where σ_{B} is the Stefan-Boltzmann radiation constant, n is the index of refraction of the material and α_{r} its absorption coefficient. This approximation (***Rosseland's approximation***) can be applied to glass in melting furnaces (chap. 7).

Considering the pressure constant, which is a good approximation for condensed matter, the specific enthalpy of a phase, H, is given by:

$$H(\boldsymbol{x}, t) = \int_0^{T(\boldsymbol{x},t)} c_{\mathrm{p}}(T')\mathrm{d}T' \tag{1.79}$$

where $c_p(T)$ is the ***specific heat*** at constant pressure of the material. The specific heat can be a function of the temperature, but not explicitly on position or time. Since only the upper limit of integration depends on position or time, the integral is easily differentiated and the differential form of the heat conservation equation (1.63) becomes:

$$\rho c_p \frac{\partial T}{\partial t} + \rho c_p \boldsymbol{v} \cdot \mathbf{grad}\, T - \mathrm{div}(\kappa \mathbf{grad}\, T) = \rho c_p \frac{\mathrm{d}T}{\mathrm{d}t} - \mathrm{div}(\kappa \mathbf{grad}\, T) = \dot{Q}_T \tag{1.80}$$

which has essentially the same form as (1.74) does for the solute.

The source term for heat, $\dot{Q}_T$, can have various origins. For example it can be associated with a volumetric emission of heat due to an electrical current in the material (Joule effect). In that case we have:

$$\dot{Q}_T = \boldsymbol{E} \cdot \boldsymbol{j}_E = \rho_E \boldsymbol{j}_E^2 \tag{1.81}$$

where $\boldsymbol{E}$ is the electrical field, $\boldsymbol{j}_E$ the electrical current density, and ρ_E the electrical resistivity. In the case of chemical reactions, the heat source term becomes a sum over all the *chemical species*:

$$\dot{Q}_T = \sum_i \frac{\mathrm{d}c_i}{\mathrm{d}t} H_i^0 \tag{1.82}$$

where H_i^0 is the ***molar enthalpy of formation*** of component (i) (units: J mole^{-1} if c_i is expressed in mole m^{-3}). Here too we see appear a coupling between the heat equation and *all* the equations for the various solutes that exist in a process.

The source term introduced by a phase transformation will be treated in detail in chapter 5. Strictly speaking, the ***latent heat*** per unit mass, $L^{\alpha/\beta}$, associated with a phase transformation $\alpha \rightarrow \beta$, is not a volumetric source term as it is produced at the moving *interface* α/β. Nonetheless, in many materials problems the interface between two phases is finely dispersed with a characteristic scale determined by the microstructure (fig. 1.1). We will see in chapter 5 that in such situations the latent heat given off in a transformation can be put in a form equivalent to a volumetric source term:

$$\dot{Q}_T = (\rho L)^{\alpha/\beta} \frac{\mathrm{d}f^\alpha}{\mathrm{d}t} \tag{1.83}$$

where f^α is the ***volume fraction*** of the phase α. For a phase transformation $\alpha \rightarrow \beta$ occurring at a well defined temperature $T^{\alpha/\beta}$ (for example, solidification of a pure substance), $f^\alpha = 1$ for $T < T^{\alpha/\beta}$ and $f^\alpha = 0$ for $T > T^{\alpha/\beta}$ (fig. 1.10). The source term defined in (1.83) comes from the definition of the average enthalpy weighted over

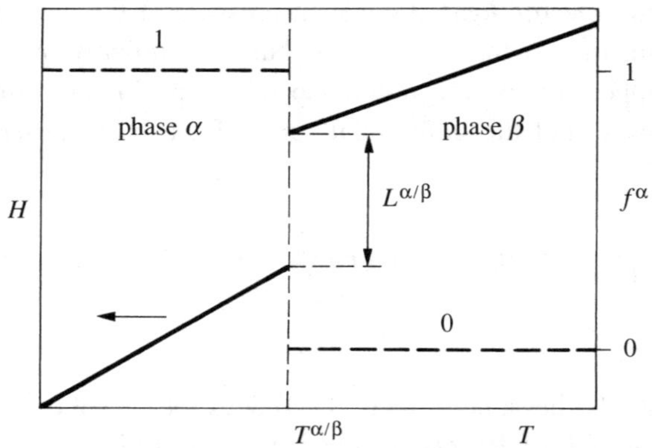

Figure 1.10 Enthalpy of a material undergoing a phase transformation $\alpha \rightarrow \beta$ at a discrete temperature, $T^{\alpha/\beta}$. Also shown in the figure is the volumetric fraction of the phase, α, which jumps from one to zero at this temperature.

its phases:

$$H(\boldsymbol{x}, t) = \int_0^{T(\boldsymbol{x}, t)} c_p(T')\, dT' + L^{\alpha/\beta}(1 - f^\alpha) \tag{1.84}$$

and is shown in figure 1.10.

1.3.4 Constitutive equations for materials: quasi-static case

As we have said in the previous section, many problems of solid deformation can be solved making the static approximation (1.24). The volume forces, $\boldsymbol{F}_v$, are most often related to gravity, for which:

$$\boldsymbol{F}_v = \rho \boldsymbol{g} \tag{1.85}$$

In the simplest case, that of linear elasticity, we try to relate the stress tensor $\boldsymbol{\sigma}$ to the strain $\boldsymbol{\varepsilon}$ (x, y, z). Being symmetric, ($\sigma_{ij} = \sigma_{ji}$ and $\varepsilon_{ij} = \varepsilon_{ji}$), these two tensors only have six independent components. In the coordinates (x, y, z), ***Hooke's law*** relating the six components takes the form:

$$\begin{bmatrix}\sigma_{xx}\\ \sigma_{yy}\\ \sigma_{zz}\\ \sigma_{xy}\\ \sigma_{yz}\\ \sigma_{zx}\end{bmatrix} = \begin{bmatrix} D^{e\ell}_{xxxx} & D^{e\ell}_{xxyy} & D^{e\ell}_{xxzz} & D^{e\ell}_{xxxy} & D^{e\ell}_{xxyz} & D^{e\ell}_{xxzx}\\ D^{e\ell}_{yyxx} & D^{e\ell}_{yyyy} & D^{e\ell}_{yyzz} & D^{e\ell}_{yyxy} & D^{e\ell}_{yyyz} & D^{e\ell}_{yyzx}\\ D^{e\ell}_{zzxx} & D^{e\ell}_{zzyy} & D^{e\ell}_{zzzz} & D^{e\ell}_{zzxy} & D^{e\ell}_{zzyz} & D^{e\ell}_{zzzx}\\ D^{e\ell}_{xyxx} & D^{e\ell}_{xyyy} & D^{e\ell}_{xyzz} & D^{e\ell}_{xyxy} & D^{e\ell}_{xyyz} & D^{e\ell}_{xyzx}\\ D^{e\ell}_{yzxx} & D^{e\ell}_{yzyy} & D^{e\ell}_{yzzz} & D^{e\ell}_{yzxy} & D^{e\ell}_{yzyz} & D^{e\ell}_{yzzx}\\ D^{e\ell}_{zxxx} & D^{e\ell}_{zxyy} & D^{e\ell}_{zxzz} & D^{e\ell}_{zxxy} & D^{e\ell}_{zxyz} & D^{e\ell}_{zxzx}\end{bmatrix} \begin{bmatrix}\varepsilon^{e\ell}_{xx}\\ \varepsilon^{e\ell}_{yy}\\ \varepsilon^{e\ell}_{zz}\\ \varepsilon^{e\ell}_{xy}\\ \varepsilon^{e\ell}_{yz}\\ \varepsilon^{e\ell}_{zx}\end{bmatrix} \tag{1.86}$$

where the index "$e\ell$" has been added to specify that we are dealing with elastic deformation components. With the coordinate notation (x_1, x_2, x_3), and the convention that sums are taken when an index is repeated, Hooke's law condenses down to:

$$\sigma_{ij} = D^{e\ell}_{ijkl}\varepsilon^{e\ell}_{kl} \tag{1.87}$$

The matrix $[\boldsymbol{D}^{e\ell}]$ is the ***elasticity matrix***. In the simple case of an isotropic material, this matrix takes the particular form:

$$[\boldsymbol{D}^{e\ell}] = \begin{bmatrix} c_{11} & c_{12} & c_{12} & 0 & 0 & 0\\ c_{12} & c_{11} & c_{12} & 0 & 0 & 0\\ c_{12} & c_{12} & c_{11} & 0 & 0 & 0\\ 0 & 0 & 0 & c_{44} & 0 & 0\\ 0 & 0 & 0 & 0 & c_{44} & 0\\ 0 & 0 & 0 & 0 & 0 & c_{44}\end{bmatrix}$$

$$= \frac{E}{(1+\nu_P)(1-2\nu_P)}\begin{bmatrix} (1-\nu_P) & \nu_P & \nu_P & 0 & 0 & 0\\ \nu_P & (1-\nu_P) & \nu_P & 0 & 0 & 0\\ \nu_P & \nu_P & (1-\nu_P) & 0 & 0 & 0\\ 0 & 0 & 0 & (1-2\nu_P) & 0 & 0\\ 0 & 0 & 0 & 0 & (1-2\nu_P) & 0\\ 0 & 0 & 0 & 0 & 0 & (1-2\nu_P)\end{bmatrix} \tag{1.88}$$

where E is the ***elastic modulus*** and ν_P is ***Poisson's coefficient***.

Inserting (1.87) and (1.36) into the quasi-static equation (1.24) makes the second derivatives of the displacement (u_x, u_y, u_z) or (u_1, u_2, u_3) appear. Being a vector equation, it should be possible to calculate the three components of the displacement field, $\boldsymbol{u}(\boldsymbol{x})$, given the boundary conditions at $\partial\Omega$. In the coordinates (x_1, x_2, x_3), the equation in the x_i direction becomes the following:

$$\rho g_i + \frac{\partial \sigma_{ij}}{\partial x_j} = \rho g_i + \frac{\partial}{\partial x_j}[D^{e\ell}_{ijkl}\varepsilon_{kl}] = \rho g_i + \frac{D^{e\ell}_{ijkl}}{2}\left[\frac{\partial^2 u_k}{\partial x_j \partial x_l} + \frac{\partial^2 u_l}{\partial x_j \partial x_k}\right] = 0 \tag{1.89}$$

In the coordinates (x, y, z), the characteristic equations of a linear elastic and isotropic material become:

$$\rho g_x + \frac{\partial}{\partial x}\left[c_{11}\frac{\partial u_x}{\partial x} + c_{12}\frac{\partial u_y}{\partial y} + c_{12}\frac{\partial u_z}{\partial z}\right] + \frac{\partial}{\partial y}\left[\frac{c_{44}}{2}\left(\frac{\partial u_x}{\partial y} + \frac{\partial u_y}{\partial x}\right)\right] + \frac{\partial}{\partial z}\left[\frac{c_{44}}{2}\left(\frac{\partial u_x}{\partial z} + \frac{\partial u_z}{\partial x}\right)\right] = 0 \tag{1.90}$$

$$\rho g_y + \frac{\partial}{\partial y}\left[c_{12}\frac{\partial u_x}{\partial x} + c_{11}\frac{\partial u_y}{\partial y} + c_{12}\frac{\partial u_z}{\partial z}\right] + \frac{\partial}{\partial x}\left[\frac{c_{44}}{2}\left(\frac{\partial u_x}{\partial y} + \frac{\partial u_y}{\partial x}\right)\right] + \frac{\partial}{\partial z}\left[\frac{c_{44}}{2}\left(\frac{\partial u_y}{\partial z} + \frac{\partial u_z}{\partial y}\right)\right] = 0 \tag{1.91}$$

$$\rho g_z + \frac{\partial}{\partial z}\left[c_{12}\frac{\partial u_x}{\partial x} + c_{12}\frac{\partial u_y}{\partial y} + c_{11}\frac{\partial u_z}{\partial z}\right] + \frac{\partial}{\partial x}\left[\frac{c_{44}}{2}\left(\frac{\partial u_x}{\partial z} + \frac{\partial u_z}{\partial x}\right)\right] + \frac{\partial}{\partial y}\left[\frac{c_{44}}{2}\left(\frac{\partial u_y}{\partial z} + \frac{\partial u_z}{\partial y}\right)\right] = 0 \tag{1.92}$$

The problem encountered in materials science is that most of the time the material does not have linear elastic behavior; rather, it can undergo ***plastic strain***, $\boldsymbol{\varepsilon}^{pl}$. In addition to this component, others of a thermal nature, $\boldsymbol{\varepsilon}^{th}$, or associated with phase transformations, $\boldsymbol{\varepsilon}^{tr}$ (volume changes), can contribute to the local ***total strain***, $\boldsymbol{\varepsilon}$, given by (1.36). In general, one has:

$$\boldsymbol{\varepsilon} = \boldsymbol{\varepsilon}^{e\ell} + \boldsymbol{\varepsilon}^{pl} + \boldsymbol{\varepsilon}^{th} + \boldsymbol{\varepsilon}^{tr} \tag{1.93}$$

A constitutive equation is necessary for each of these components. They will be given in chapter 6 for a plastic material. As an example, the strain $\boldsymbol{\varepsilon}^{th}$ due to thermal dilation in the material is expressed as:

$$\varepsilon^{th}_{xx} = \varepsilon^{th}_{yy} = \varepsilon^{th}_{zz} = \frac{\beta_T}{3}(T(\boldsymbol{x}) - T_0) \qquad \varepsilon^{th}_{xy} = \varepsilon^{th}_{yz} = \varepsilon^{th}_{zx} = 0 \tag{1.94}$$

where β_T is the thermal dilatation coefficient and T_o a reference temperature at which the strain field of the material is known. In a *thermo-elastic* problem one sees the coupling appear between the temperature field, $T(\boldsymbol{x}, t)$, and the deformation field, $\boldsymbol{\varepsilon}(\boldsymbol{x}, t) = \boldsymbol{\varepsilon}^{e\ell}(\boldsymbol{x}, t) + \boldsymbol{\varepsilon}^{th}(\boldsymbol{x}, t)$.

Assuming that the plastic strain occurs without change in the volume (chap. 6), the relative volume variation, $\Delta V/V$, is given by the trace of the other strain tensor components:

$$\frac{\Delta V}{V} = [\varepsilon_{xx}^{e\ell} + \varepsilon_{yy}^{e\ell} + \varepsilon_{zz}^{e\ell}] + [\varepsilon_{xx}^{tr} + \varepsilon_{yy}^{tr} + \varepsilon_{zz}^{tr}] + [\varepsilon_{xx}^{th} + \varepsilon_{yy}^{th} + \varepsilon_{zz}^{th}]$$

$$= \mathrm{Tr}(\boldsymbol{\varepsilon}^{e\ell} + \boldsymbol{\varepsilon}^{tr} + \boldsymbol{\varepsilon}^{th}) \quad (1.95)$$

where Tr is the trace operator of a tensor (chap. 10).

1.3.5 Constitutive equations for materials: dynamic case

In dynamic problems (fluid flow, high speed deformation of solids, etc.), the inertial terms in the momentum equation (1.20) can not be neglected. If the volume force term is the same as that given for the quasi-static case (1.85), the stress tensor $\boldsymbol{\sigma}$ is generally related to the strain and the rate of strain. In the following we assume that the fluid is "ideally" viscous or ***Newtonian***, that is, the stress tensor is a linear function of the rate of strain and is not dependent on the displacements, $\boldsymbol{u}$. The velocity field will be given by the components (v_x, v_y, v_z) or (v_1, v_2, v_3) according to the coordinate system adopted.

Figure 1.11 shows the behavior of a Newtonian fluid when it is between two solid plane surfaces. The upper surface moves at a velocity v_x^s with respect to the

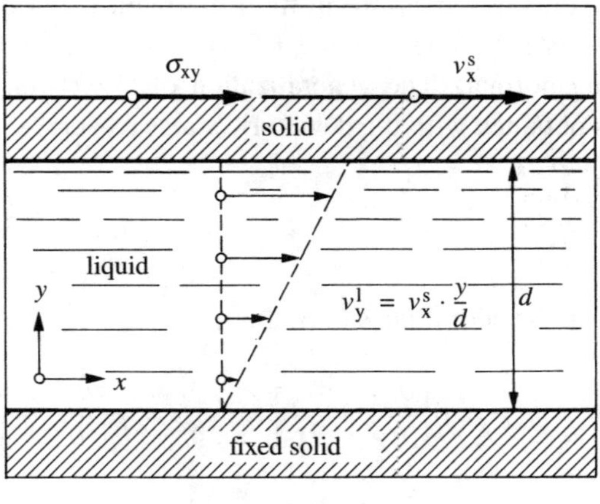

Figure 1.11 Velocity profile obtained in a fluid when the upper surface is displaced at a fixed horizontal velocity.

lower; a velocity profile results in the fluid layer of thickness d:

$$v_x = v_x^s y/d \tag{1.96}$$

It should be noted that the non-slip assumption (sect. 1.4) is implicit: the liquid in direct contact with a surface has the same velocity as that surface. The tangential stress (or shear), σ_{xy}, that is applied on the upper surface is proportional to the velocity gradient:

$$\sigma_{xy} = \mu \frac{\partial v_x}{\partial y} \tag{1.97}$$

The constant appearing in the equation is the ***dynamic viscosity*** of the fluid μ (units: Pa s or Poiseuille). Note that the ***kinematic viscosity*** ν, defined by the ratio (μ/ρ), has the same units as a chemical diffusion coefficient or a thermal diffusivity ($m^2 s^{-1}$).

In general, the stress tensor is expressed as a function of the strain rates defined in (1.38). Earlier it was convenient to decompose this tensor in two parts: the ***pressure***, p, and the ***deviatoric stress tensor***, **s** with components s_{ij}. Thus:

$$\sigma_{ij} = s_{ij} - p\delta_{ij} \tag{1.98}$$

where δ_{ij} is the Kronecker delta function. The pressure in an incompressible fluid is defined as the mean of the trace of the tensor $\boldsymbol{\sigma}$:

$$p = -\tfrac{1}{3}[\sigma_{xx} + \sigma_{yy} + \sigma_{zz}] = -\tfrac{1}{3}[\sigma_{11} + \sigma_{22} + \sigma_{33}] = -\tfrac{1}{3}\,\mathrm{Tr}\,\boldsymbol{\sigma} \tag{1.99}$$

Its value is invariant under any coordinate transformation. The minus sign reflects the fact that the hydrostatic pressure in fluid mechanics ($p > 0$) corresponds to a negative normal stress according to figure 1.5 ($\sigma_{11} = \sigma_{22} = \sigma_{33} < 0$ as they are opposite to the outgoing normal). The generalization of (1.97) for an incompressible Newtonian fluid is obtained with the deviatoric stress tensor:

$$s_{ij} = \mu\left[\frac{\partial v_i}{\partial x_j} + \frac{\partial v_j}{\partial x_i}\right] \tag{1.100}$$

or more explicitly, in the coordinates (x, y, z):

$$\begin{aligned} s_{xx} &= 2\mu\frac{\partial v_x}{\partial x} & s_{xy} &= \mu\left[\frac{\partial v_x}{\partial y} + \frac{\partial v_y}{\partial x}\right] & s_{xz} &= \mu\left[\frac{\partial v_x}{\partial z} + \frac{\partial v_z}{\partial x}\right] \\ s_{yy} &= 2\mu\frac{\partial v_y}{\partial y} & s_{yz} &= \mu\left[\frac{\partial v_y}{\partial z} + \frac{\partial v_z}{\partial y}\right] & s_{zz} &= 2\mu\frac{\partial v_z}{\partial z} \end{aligned} \tag{1.101}$$

Using (1.98) and (1.100) in the conservation of momentum equation (1.18), one can show (exercise 1.5.4) that for the x component:

$$\begin{aligned}\frac{\partial}{\partial t}(\rho v_{\mathrm{x}}) + \operatorname{div}(\rho v_{\mathrm{x}} \boldsymbol{v}) &= F_{\mathrm{v,x}} - \frac{\partial p}{\partial x} + \mu\left[\frac{\partial^2 v_{\mathrm{x}}}{\partial x^2} + \frac{\partial^2 v_{\mathrm{x}}}{\partial y^2} + \frac{\partial^2 v_{\mathrm{x}}}{\partial z^2}\right] \\ &= F_{\mathrm{v,x}} - \frac{\partial p}{\partial x} + \mu \Delta v_{\mathrm{x}} \end{aligned} \quad (1.102)$$

which is none other than the x component of the ***Navier-Stokes equation***. We introduce here the Laplacian, Δ, of the velocity component v_{x}. Generalizing the Laplacian for the vector $\boldsymbol{v}$ as applied to each of the components, and taking the non conservative form of the momentum equation (1.20), we obtain the following vector equation:

$$\rho \boldsymbol{a} = \rho\left[\frac{\partial \boldsymbol{v}}{\partial t} + (\mathbf{Grad}\,\boldsymbol{v}) \cdot \boldsymbol{v}\right] = \boldsymbol{F}_{\mathrm{v}} - \mathbf{grad}\, p + \mu \Delta \boldsymbol{v} \quad (1.103)$$

In a fluid dynamics problem we then see appear a vector field, the velocity $\boldsymbol{v}(\boldsymbol{x}, t)$, and a scalar field, the pressure in the liquid, $p(\boldsymbol{x}, t)$. We have a scalar equation, the mass equation (1.8), and a vector equation (1.103) (for an incompressible Newtonian fluid). There are thus as many equations as unknowns, and the problem should be solvable as long as initial and boundary conditions are specified.

1.4 BOUNDARY AND INITIAL CONDITIONS

1.4.1 Generalities

Consider again the general conservation equation (1.68) integrated over the domain Ω of figure 1.4(a), which is assumed invariant in time:

$$\frac{\mathrm{d}}{\mathrm{d}t}\int_{\Omega} \rho \xi \,\mathrm{d}V + \int_{\partial\Omega} \rho \xi \boldsymbol{v} \cdot \boldsymbol{n}\,\mathrm{d}S = -\int_{\partial\Omega} \boldsymbol{j}_{\xi} \cdot \boldsymbol{n}\,\mathrm{d}S + \int_{\Omega} \dot{Q}_{\xi}\,\mathrm{d}V \quad (1.104)$$

In a *non stationary* problem, the first term of the equation requires the specification of the values of the field $\xi(\boldsymbol{x}, t = 0)$ at every point in the domain Ω at time $t = 0$. This is the ***initial condition***. These are generally simple to express, even though they do not always correspond to reality! Most often, it is assumed at the start of a calculation that the field: concentration, temperature, velocity, etc., is *uniform*, i.e. independent of position.

In the same equation, there are two integrals on the boundary $\partial\Omega$ of the domain: the first involves the normal velocity component, $v_{\mathrm{n}} = \boldsymbol{v} \cdot \boldsymbol{n}$, and corresponds to the transport of the quantity ξ across the surface, whereas the second is related to the flux entering or leaving the domain at the surface by diffusion, $j_{\xi,n} = \boldsymbol{j}_\xi \cdot \boldsymbol{n}$. It can be seen that it will be necessary to specify these terms at the surface to determine what happens in the interior. These are ***boundary conditions***.

In the case of a *closed* system, or when a Lagrangian frame is employed (1.70), the normal component of velocity is zero everywhere at the boundary of the domain, and the boundary conditions for the value under consideration, ξ, only involve the diffusive term. The boundary conditions for diffusive problems are divided into two types:

I. A ***natural condition*** expresses directly the value of $j_{\xi,n}$ on the boundary. Three cases of this type are generally considered:

I.1 The ***homogeneous Neumann condition*** is a null flux:

$$j_{\xi,\,n} = \boldsymbol{j}_\xi \cdot \boldsymbol{n} = 0 \tag{1.105}$$

I.2 The ***Neumann condition*** corresponds to a non zero flux given on the boundary, q_ξ:

$$-j_{\xi,\,n} = -\boldsymbol{j}_\xi \cdot \boldsymbol{n} = q_\xi \tag{1.106}$$

By convention, the flux q_ξ is negative when the associated value, ξ, is going out of the domain, that is $j_{\xi,n} = \boldsymbol{j}_\xi \cdot \boldsymbol{n} > 0$, thus the minus sign in the equation above.

I.3 The ***Cauchy*** or ***mixed condition*** consists of a linear relation between the flux and the value of the variable itself:

$$-j_{\xi,\,n} + a\xi = b \tag{1.107}$$

where a and b are given constants.

It must be noted that for the conditions of types (1.106) and (1.107), the imposed flux, q_ξ, or the constants a and b can be functions of position and time. In some cases, these parameters can also depend on the variable ξ itself, which makes the boundary condition non linear.

II. An ***essential condition*** or a ***Dirichlet condition*** consists of specifying the variable directly rather than the diffusive flux, $j_{\xi,n}$:

$$\xi = \bar{\xi} \tag{1.108}$$

The bar over the variable on the right indicates that it is an imposed value. This value can be a function of position and time.

The equation of conservation of mass (1.5) plays a particular role in that there is no diffusion term, only a transport term. For problems with transport but without diffusion, two cases need to be distinguished:

1) When the velocity $\boldsymbol{v}$ points into the domain ($v_n = \boldsymbol{v} \cdot \boldsymbol{n} < 0$), the values of the variables should be specified on the boundary by a condition of type (1.108).
2) If, on the other hand, $v_n = \boldsymbol{v} \cdot \boldsymbol{n} \geq 0$ (outgoing velocity, or velocity parallel to the boundary), no boundary condition should be specified.

In other words, in a transport problem without diffusion, it is necessary to specify what goes into the domain in order to determine what occurs within the domain, including what goes out. For problems with both advection and diffusion, we will see that there can be conflict in the boundary conditions if an essential boundary condition is applied on the boundary $\partial\Omega$ where the velocity is outgoing.

Having covered these generalities, we will briefly discuss the conditions that are encountered in practice for the equations of solute, heat, deformation and fluid dynamics.

1.4.2 Solute transfer

The boundary conditions introduced above are shown in figure 1.12 for the solute conservation equation. An essential condition means specifying the value of concentration of solute i on a part $\partial\Omega$ of the domain boundary (fig. 1.12(a)) :

$$c_i = \bar{c}_i \tag{1.109}$$

This condition occurs, for example, in the analysis of the deposition of a thin graphite layer on a steel surface (cementation process).

The natural homogeneous Neumann condition (fig. 1.12(b)) corresponds, when $v_n = 0$, to the assumption of a *closed* system as the diffusive flux is null. Taking the expression (1.72) of Fick's first law for the diffusion flux as a function of the concentration gradient, this condition is equivalent to:

$$-j_{i,n} = -\boldsymbol{j}_i \cdot \boldsymbol{n} = D_i \frac{\partial c_i}{\partial n} = D_i \,\mathbf{grad}\, c_i \cdot \boldsymbol{n} = 0 \tag{1.110}$$

where $\partial/\partial n$ is the derivative in the direction of the outgoing normal ($= \boldsymbol{n} \cdot \mathbf{grad}$). A natural non homogeneous Neumann condition (fig. 1.12(c)) corresponds to a fixed, non zero, solute flux at the surface. This would be the case, for example, of a reaction occurring at the surface of a plate in a corrosion cell. Again according to Fick's law, this condition is of the type:

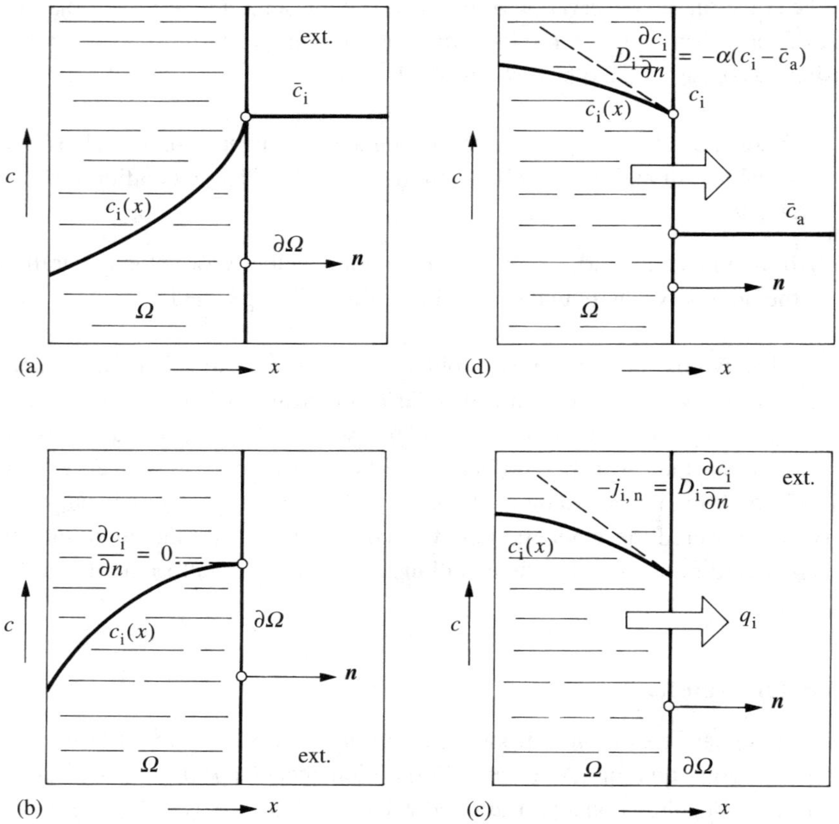

Figure 1.12 Boundary conditions encountered for the solute diffusion equation: (a) imposed concentration (Dirichlet condition); (b) zero flux (homogeneous Neumann condition); (c) imposed solute flux (Neumann condition); (d) mixed (Cauchy) condition for solute transfer.

$$D_i \frac{\partial c_i}{\partial n} = D_i \,\mathbf{grad}\, c_i \cdot \boldsymbol{n} = q_i \tag{1.111}$$

A mixed, Cauchy, boundary condition can be expressed in the form:

$$D_i \frac{\partial c_i}{\partial n} = -\alpha(c_i - \bar{c}_a) \tag{1.112}$$

where α is a ***chemical transfer coefficient*** between the surface of the domain (concentration c_i) and the ambient exterior environment (concentration $\bar{c}_a$).

1.4.3 Thermal transfer

A surface temperature imposed by an ideal heat source (or cold sink) corresponds to an essential condition (fig. 1.13a) :

$$T = \bar{T}_a \tag{1.113}$$

where $\bar{T}_a$ is the imposed exterior (or ambient) temperature. This type of condition, although never exactly obtained in nature, corresponds, for example, to the situation when a surface is sprayed with an intense water jet at a fixed temperature.

An imposed flux, possibly zero at an *adiabatic* boundary, corresponds to a natural Neumann condition (fig. 1.13b), or a homogeneous Neumann condition in the adiabatic case:

$$-j_{T,n} = \kappa\frac{\partial T}{\partial n} = q_T \tag{1.114a}$$

$$-j_{T,n} = \kappa\frac{\partial T}{\partial n} = 0 \tag{1.114b}$$

Finally, a natural mixed (or Cauchy) boundary condition is frequently encountered in thermal problems when there is heat transfer with the external surroundings (fig. 1.13c). Assuming that the heat flux is proportional to the temperature difference between the surface of the domain and that of the surroundings, one is led to write the condition in the form:

$$-j_{T,n} = \kappa\frac{\partial T}{\partial n} = -h(T - \bar{T}_a) \qquad \text{(with } h > 0) \tag{1.115}$$

where h is a ***heat transfer coefficient*** which encompasses exchange by conduction, convection and radiation (chap. 5). In most cases, the heat exchange is not linear, (1.115) is then used with a transfer coefficient which is a function of the temperature. Such is the case, notably, for radiation where the heat flux is of the form:

$$-j_{T,n} = \kappa\frac{\partial T}{\partial n} = -\sigma_B\varepsilon_r(T^4 - \bar{T}_a^4) \tag{1.116}$$

σ_B is the Stefan-Boltzmann constant ($\sigma_B = 5{,}67 \cdot 10^{-8}$ W m^{-2} K^{-4}) and ε_r the surface emissivity ($0 < \varepsilon_r < 1$) if the material is not an ideal black body. The mixed boundary condition expressed by this equation is clearly non linear, but can be rearranged in the form:

$$\kappa\frac{\partial T}{\partial n} = -\sigma_B\varepsilon_r[(T^2 + \bar{T}_a^2)(T + \bar{T}_a)](T - \bar{T}_a) = -h(T, \bar{T}_a)(T - \bar{T}_a) \tag{1.117}$$

so that a condition analog to (1.115) appears, but with a third power function $h(T, \bar{T}_a)$.

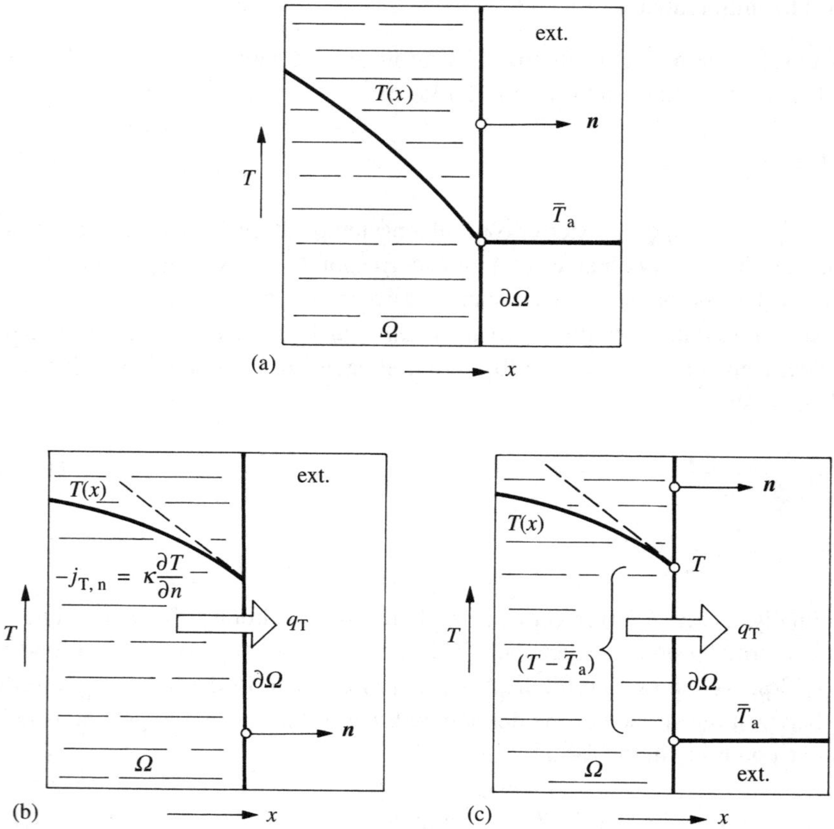

Figure 1.13 Boundary conditions encountered for the heat diffusion equation: (a) imposed temperature (Dirichlet condition); (b) imposed heat flux (Neumann condition) and (c) imposed heat transfer coefficient (Cauchy condition).

1.4.4 Quasi-static

When the static equation (1.24) is integrated over the whole volume, a surface integral appears through the divergence theorem:

$$\int_{\partial\Omega} \boldsymbol{\sigma} \cdot \boldsymbol{n} \, \mathrm{d}S + \int_{\Omega} \boldsymbol{F}_\mathrm{v} \, \mathrm{d}V = \int_{\partial\Omega} \boldsymbol{T} \, \mathrm{d}S + \int_{\Omega} \boldsymbol{F}_\mathrm{v} \, \mathrm{d}V = 0 \tag{1.118}$$

To write this equation, we have used the fact that locally on a surface $\partial\Omega$, the product of the tensor $\boldsymbol{\sigma}$ with the surface normal $\boldsymbol{n}$ of an element is equal to the surface force, $\boldsymbol{T}$. We arrive then at the usual equation that says that the sum of the forces applied to the surface of a solid body (the *contact forces*) and of the volume forces is zero at equilibrium. The imposition of a force at the surface of a domain appears *naturally* in (1.118) while an *essential* condition consists of fixing the displacement of the surface.

The different boundary conditions above can now be applied to *each* of the components *x*, *y* or *z* of this surface force. In this way, in figure 1.14a, the two vertical surfaces $\partial\Omega_2$ and $\partial\Omega_4$ of the rectangular domain Ω are assumed to be free of all contact forces. In this case we have a homogeneous Neumann condition for the two components:

$$T_x = [\boldsymbol{\sigma}\cdot\boldsymbol{n}]_x = 0 \quad \text{and} \quad T_y = [\boldsymbol{\sigma}\cdot\boldsymbol{n}]_y = 0 \qquad (\text{or } T_i = \sigma_{ij}n_j) \qquad (1.119)$$

The upper surface $\partial\Omega_3$ on which a traction force, $\bar{T}_y$, is applied, is free to move in the horizontal direction. We have then on this surface:

$$T_x = [\boldsymbol{\sigma}\cdot\boldsymbol{n}]_x = 0 \quad \text{and} \quad T_y = [\boldsymbol{\sigma}\cdot\boldsymbol{n}]_y = \bar{T}_y \qquad (1.120)$$

Finally, for the lower surface, $\partial\Omega_1$, the points of the domain Ω are assumed to be rigidly fixed such that they can not move. This assumption of zero displacement corresponds to an essential boundary condition:

$$u_x = u_y = 0 \qquad (1.121)$$

Note that a surface can have an essential condition in one direction and a natural condition in another: this would be the case, for example, of the surface $\partial\Omega_1$ in figure 1.14 if it was pulled down by an imposed vertical force. Mixed boundary conditions can also be encountered in static problems: for a surface attached by a spring

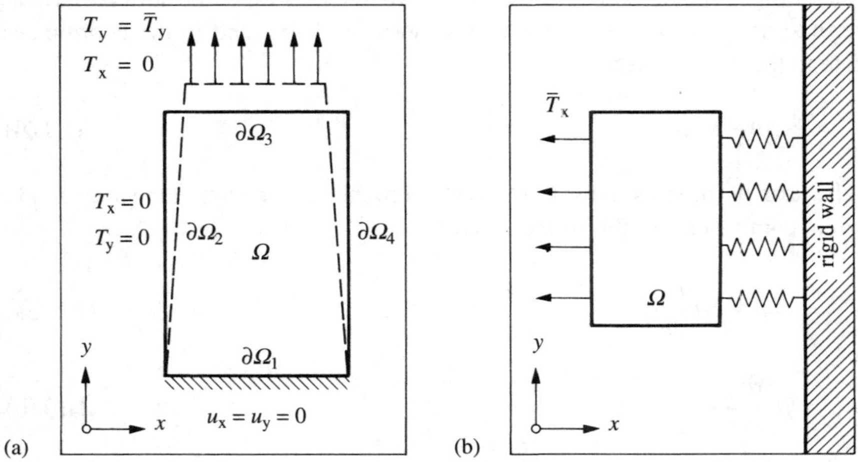

Figure 1.14 Natural and essential conditions imposed for mechanical deformation of a rectangle (a). In (b), the surface attached to a rigid wall by springs has Cauchy boundary conditions.

to a rigid wall (fig. 1.14b), the component T_x is proportional to the displacement of the boundary in the x direction and corresponds to the following mixed condition:

$$T_x = [\boldsymbol{\sigma} \cdot \boldsymbol{n}]_x = -K_r(u_x - u_x^0) \tag{1.122}$$

K_r is the spring constant (normalized by the surface) and u_x^0 is a reference displacement at which the spring exerts no force.

It should finally be noted that a surface force $\overline{T}_y$ on the horizontal surface $\partial\Omega_2$ (fig. 1.14a), corresponds to normal tension while on a vertical surface the same condition would be a shear or friction force. For surfaces that are not parallel to the coordinate axes, the applied forces are generally specified in terms of the normal and tangential stress components by means of a change of axes.

1.4.5 Fluid dynamics

We finish this chapter with the boundary conditions valid for fluid dynamics problems. The same as for static problems, the boundary conditions can be written for each of the components x, y and z, or for the normal and tangential components. An essential condition consists of the imposition of a velocity at the boundary. For example, for a fluid in contact with a fixed, solid wall, a Dirichlet condition can be applied to each component of velocity at the boundary: the so called ***no slip condition***. In two dimensions (fig. 1.15a):

$$v_n = \boldsymbol{v} \cdot \boldsymbol{n} = 0 \quad \text{and} \quad v_\tau = \boldsymbol{v} \cdot \boldsymbol{\tau} = 0 \tag{1.123}$$

A surface which is completely free to move corresponds to a homogeneous Neumann condition; there is no applied force. As in static problems, the surface force, $\boldsymbol{T}$, is expressed in fluid dynamics as the stress tensor times the surface normal. Taking into account the pressure which appears in (1.98), and in the coordinates attached to the surface $(\boldsymbol{n},\boldsymbol{\tau})$:

$$\boldsymbol{T} = -p\boldsymbol{n} + \boldsymbol{s} \cdot \boldsymbol{n} \tag{1.124}$$

The form of the stress tensor in (1.100) allows the expression of the normal and tangential components of the surface force:

$$T_n = -p + 2\mu\frac{\partial v_n}{\partial n} \tag{1.125a}$$

$$T_\tau = \mu\left[\frac{\partial v_n}{\partial \tau} + \frac{\partial v_\tau}{\partial n}\right] \tag{1.125b}$$

where $\partial/\partial n$ and $\partial/\partial\tau$ are the derivatives with respect to the normal and tangential coordinates, respectively, of the components v_n and v_τ of the velocity. A homoge-

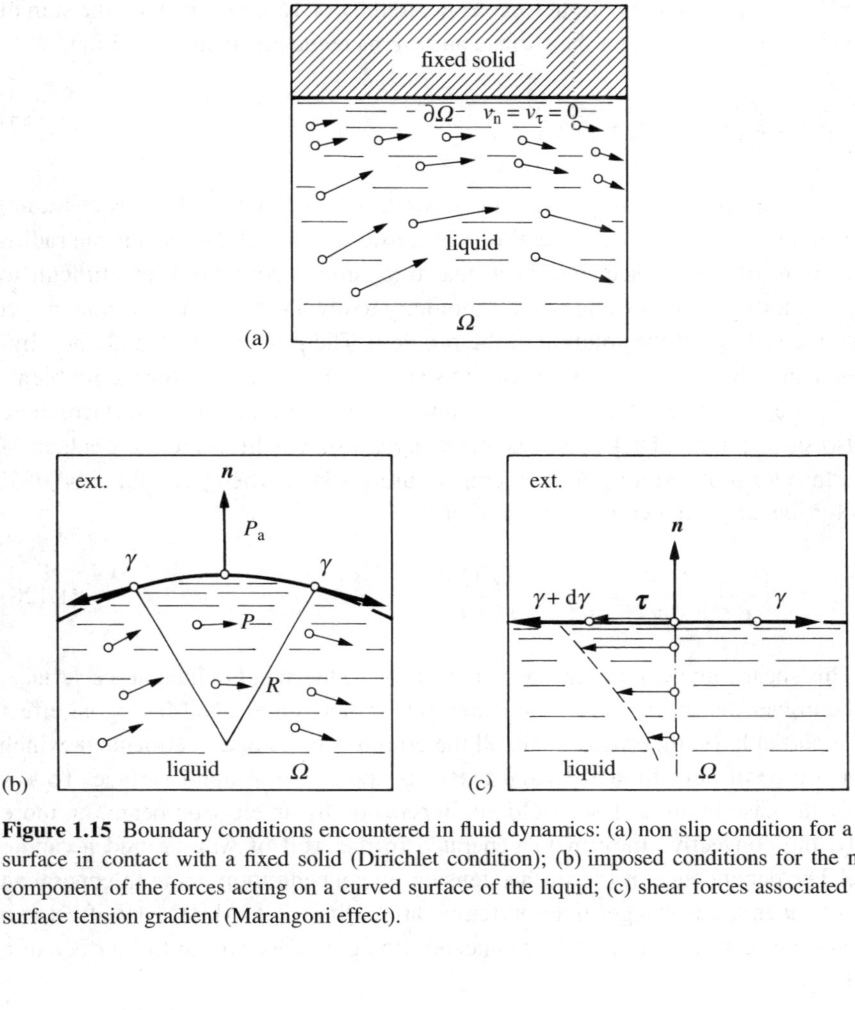

Figure 1.15 Boundary conditions encountered in fluid dynamics: (a) non slip condition for a liquid surface in contact with a fixed solid (Dirichlet condition); (b) imposed conditions for the normal component of the forces acting on a curved surface of the liquid; (c) shear forces associated with a surface tension gradient (Marangoni effect).

neous Neumann condition is then equivalent to saying, on this surface:

$$-p + 2\mu\frac{\partial v_n}{\partial n} = 0 \tag{1.126a}$$

$$\mu\left[\frac{\partial v_n}{\partial \tau} + \frac{\partial v_\tau}{\partial n}\right] = 0 \tag{1.126b}$$

Remark that a liquid surface is rarely free as it is subject to capillary forces and to the exterior pressure, p_a. The latter, although applied by another fluid or a gas, is assumed known and homogeneous. For the case where the liquid surface has a *cur-*

vature Ξ, (fig. 1.15b), the normal force expressed by (1.125a) is given by the sum of the exterior forces acting on the surface and we have the Neumann condition:

$$-p+2\mu\frac{\partial v_{\mathrm{n}}}{\partial n} = -p_{\mathrm{a}}+\gamma\Xi \tag{1.127}$$

where γ is the surface tension of the liquid. In the case of figure 1.15b,the curvature Ξ is negative, given by $-R^{-1}$, where R is the absolute value of the local mean radius of curvature of the surface. It is seen that free surface conditions are difficult to apply, as they imply knowledge of the geometry to obtain the surface normal, $\boldsymbol{n}$, yet the geometry is itself the solution of the problem. The problem is severely non linear and can only be treated with algorithms specifically conceived for the problem.

For the case where the surface tension is not uniform along the surface, there are also shear forces (fig. 1.15c). The ***Marangoni effect*** is linked to the gradient of the surface tension resulting from a temperature gradient when γ is a function of T. Then for the tangential component (1.125b):

$$T_{\tau} = \mu\left[\frac{\partial v_{\mathrm{n}}}{\partial \tau}+\frac{\partial v_{\tau}}{\partial n}\right] = \frac{\partial \gamma}{\partial \tau} = \frac{\partial \gamma}{\partial T}\frac{\partial T}{\partial \tau} \tag{1.128}$$

This shear force will be stronger the greater the thermal gradient at the surface, and the higher the sensitivity of the material to temperature. The Marangoni effect plays a particularly important role in all material processes and treatments in which a heat source of very limited spatial extent is applied to a liquid surface. This is notably the case in arc or laser welding, in refusion by an electron beam, or, more simply, the convective movement generated in the well of wax around a candle flame! The dependence of the surface tension on concentration being in general an order of magnitude stronger than on temperature, the *solute* Marangoni effect can play an even more important role if concentration gradients exist at the surface of a liquid.

1.5 EXERCISES

1.5.1. The gravity forces $\rho \boldsymbol{g}\Delta V$ associated with the element ΔV of figure 1.5 give moments that are proportional to the fourth power of the characteristic dimension, δ, of ΔV. The contact forces represented by the tensor $\boldsymbol{\sigma}$ being proportional to the surfaces of ΔV, the associated moments are of order δ^3, lower than those associated with the mass. If the equilibrium of the moments about an axis is considered, and the limit as $\delta \to 0$ is taken, only the moments due to the surface forces need to be considered as those associated with the volume forces converge more rapidly. Show then that mechanical equilibrium implies $\sigma_{ij} = \sigma_{ji}$, for all indices i and j.

1.5.2. For a function $f(x, t)$ in one spatial dimension, show graphically and mathematically that the total derivative of the integral over a variable domain $[x_1(t), x_2(t)]$ is given by (Leibniz' formula):

$$\frac{\mathrm{d}}{\mathrm{d}t}\int_{x_1(t)}^{x_2(t)} f(x,t)\,\mathrm{d}x = \int_{x_1(t)}^{x_2(t)} \frac{\partial f(x,t)}{\partial t}\,\mathrm{d}x + f(x_2,t)\frac{\mathrm{d}x_2}{\mathrm{d}t} - f(x_1,t)\frac{\mathrm{d}x_1}{\mathrm{d}t}$$

Show it for a finite time increment, Δt, which can then be taken to zero.

1.5.3. Using (1.44), show that for a material volume Ω attached to the material (thus time dependent), we have:

$$\int_{\Omega} \rho\frac{\mathrm{d}}{\mathrm{d}t}\left[\frac{v^2}{2}\right]\mathrm{d}V = \frac{1}{2}\frac{\mathrm{d}}{\mathrm{d}t}\int_{\Omega} \rho v^2\,\mathrm{d}V$$

1.5.4. Using (1.98) express the stress tensor, $\boldsymbol{\sigma}$, in terms of pressure and a deviatoric stress tensor, $\boldsymbol{s}$ (1.100), and show that the equation of conservation of momentum (1.19) in the x direction reduces to (1.102).

1.6 BIBLIOGRAPHY

P. GERMAIN, *Cours de mécanique des milieux continus, tome 1, Théorie générale*, Masson, Paris, 1973.

P. GERMAIN, *Mécanique,* vol. 1 and 2, Cours de l'Ecole Polytechnique, Edition Marketing, Ellipses, Paris, 1986.

W.M. LAI, D. RUBIN, E. KREMPL, *Introduction to Continuum Mechanics*, 3rd edition, Butterworth-Heinemann, Oxford, 1996.

J. MANDEL, *Mécanique des milieux continus*, vol. 1 and 2, Gauthier-Villars, Paris, 1966.

J. SALENÇON, *Handbook of Continuum Mechanics*, Springer, Heidelberg, 2001.

R.L. PANTON, *Incompressible flow*, J. Wiley, New York, 1984.

J.E. MARSDEN, T.J.R. HUGHES, *Mathematical Foundations of Elasticity*, Dover, New York, 1994.

H. ZIEGLER, *An introduction to thermomechanics*, North Holland, Amsterdam, 1977.

CHAPTER 2

THE FINITE DIFFERENCE METHOD

2.1 OBJECTIVES

- Illustrate the concept of the finite difference method for the simple one dimensional case of solute diffusion, with and without advective transport, for both stationary and non stationary cases.
- Generalize the concept of the finite difference method for two and three dimensional geometries.
- Show how the boundary and initial conditions are applied.
- Establish a link between the finite difference method and the similar finite volume and direct finite difference methods.
- Give an idea of the extension of the method to more complicated cases (non-Cartesian coordinate systems, alternating direction implicit method).

As was seen in the preceding chapter, the simulation of a process where microscopic scale phenomena take place is generally treated with continuum mechanics. The first step of every simulation is then to pose the problem physically, chemically or thermodynamically with the mathematical formalism of the conservation equations, the appropriate state equations and the necessary boundary and initial conditions. Next the problem of solving these *partial differential equations* on the domain under consideration appears. Most of the time, the problem is non-linear and an analytical solution does not exist; one then calls upon a numerical method for solution. Among the many methods developed so far, the *"**Finite Difference Method**"*, or FDM, is certainly the simplest from the point of view of mathematical formalism and implementation in a programming language. That probably explains its success among materials science engineers, typically not inclined towards heavy mathematics! Nonetheless, if the simplicity of the FDM is an advantage in comparison with more elaborate methods when it is necessary to deal with complex and coupled phenomena, its lack of flexibility for use with complicated geometries has impeded its implementation in industry. In fact, the majority of large commercial codes used in practice, notably for solving deformation problems, are based on the finite element method which is presented in the following chapter.

2.2 ONE DIMENSIONAL CASE

To illustrate the finite difference method, we consider solute diffusion with convective transport in this section. This type of problem is encountered, for example, during crystal growth of a planar interface (fig. 2.1). One process actually used in single crystal fabrication is to move a cell, initially containing a liquid, at constant speed through a temperature gradient (fig. 2.1a). The gradient is created with hot and cold regions separated by an insulator. After an initial stage during which the growth starts, the growth front stabilizes and propagates at a speed v with respect to the cell. If the liquid initially contains a component at concentration $\bar{c}_0$, the solid formed in a stationary process has the same concentration. In any case, the concentration on the liquid side of the interface is $\bar{c}_0/k$, where k is imposed by the thermodynamic conditions from the phase diagram (fig. 2.1b). The coefficient k is called the ***partition coefficient*** ($k < 1$ for the example in figure 2.1). Because of this difference, there is diffusion in the liquid. In a coordinate system that moves with the interface, the material enters on the right and exits to the left of figure 2.1. It has thus a velocity $-v$ along the x axis. The equation of conservation of solute (1.74) becomes in this case:

$$D\frac{d^2c}{dx^2} + v\frac{dc}{dx} = 0 \tag{2.1}$$

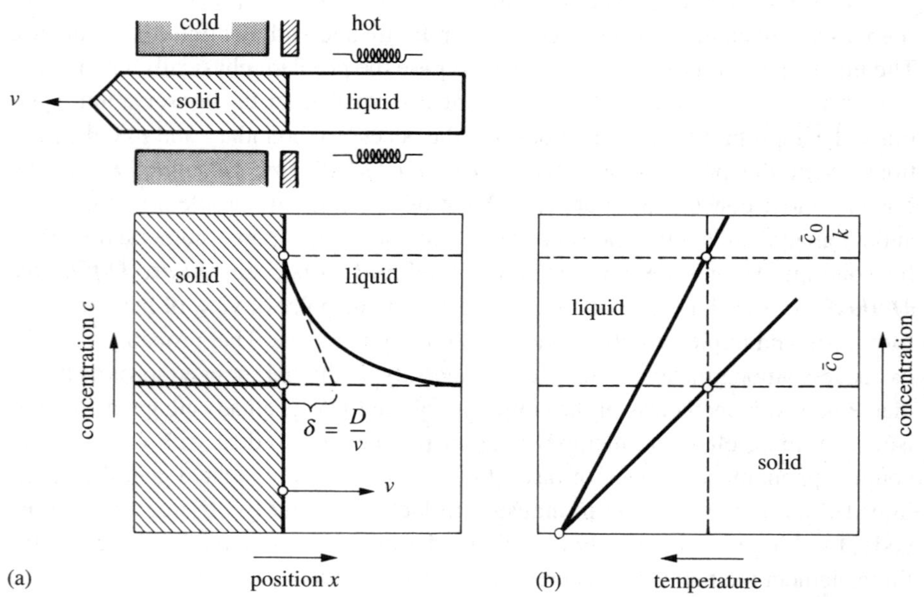

Figure 2.1 Bridgman growth of a single crystal in a temperature gradient with an imposed velocity and the solute profile in the liquid for a stationary process (a). Figure (b) shows the corresponding phase diagram.

where D is the diffusion coefficient, which we assume is uniform. Note that the index "i" which was used in the preceding chapter to distinguish the components of solute is omitted in this section. Note also that the partial derivatives have been replaced with ordinary derivatives as the solution $c(x)$ is a function of only one variable. The solution will have the form:

$$c(x) = A + B \exp\left[\frac{-vx}{D}\right] \tag{2.2}$$

To determine the integration constants A and B, the boundary conditions must be considered:

$$c(x=0) = \frac{\bar{c}_0}{k} \qquad \text{at the interface} \tag{2.3}$$

$$c(x=\infty) = \bar{c}_0 \qquad \text{far from the interface} \tag{2.4}$$

The solution then becomes:

$$c(x) = \frac{\bar{c}_0}{k}\left\{k + (1-k)\exp\left[\frac{-vx}{D}\right]\right\} \tag{2.5}$$

Note that the liquid domain in which (2.1) is solved is semi-infinite. The solution introduces the notion of a ***boundary layer*** at the solid-liquid interface: in a thickness δ given by the ratio (D/v), the solute concentration varies considerably. This layer can be geometrically constructed by taking the intersection of the slope of the concentration profile at $x = 0$ with the horizontal $c = \bar{c}_0$ (fig. 2.1).

2.2.1 Derivatives and differences

Now take equation (2.1) as representative of an advection-diffusion problem, but for a finite interval $[0, L]$ (fig. 2.2a). In the following, we take the material to be flowing in the direction of the x axis, whereas in the preceding example the interface moved in that direction. The stationary equation then becomes:

$$-D\frac{d^2c}{dx^2} + v\frac{dc}{dx} = 0 \tag{2.6}$$

Applying the Dirichlet boundary conditions:

$$c(x=0) = \bar{c}_0 \quad \text{and} \quad c(x=L) = \bar{c}_L \tag{2.7}$$

we have the solution:

$$\frac{c(x) - \bar{c}_0}{\bar{c}_L - \bar{c}_0} = \frac{\exp[vx/D] - 1}{\exp[vL/D] - 1} \tag{2.8}$$

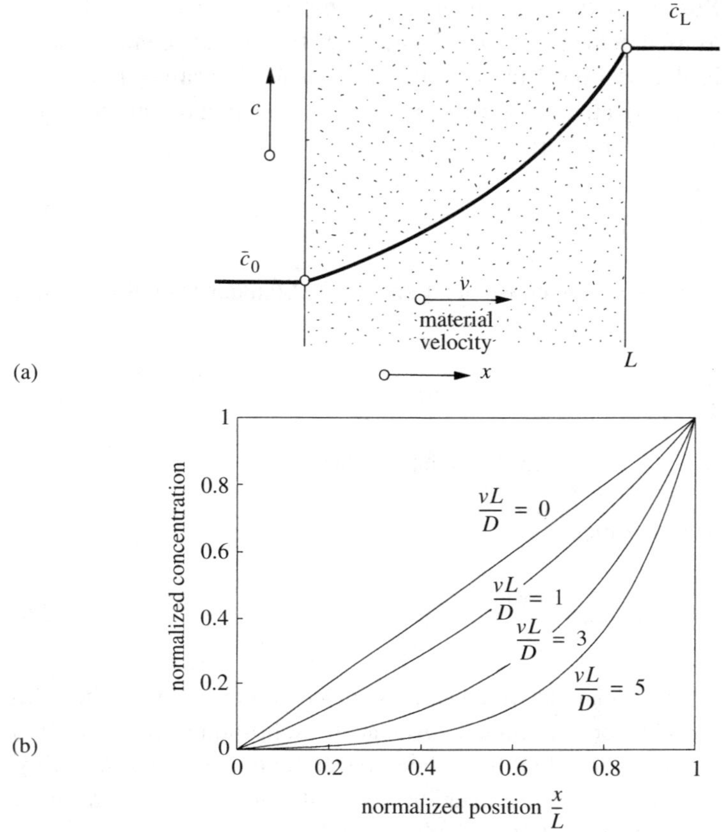

Figure 2.2 Normalized concentration profiles as a function of normalized distance, x/L, for different values of the Péclet number Pe_L.

In addition to the boundary layer developing downstream in the domain, that is at $x = L$, the solution reveals a non dimensional number, the ***Péclet number*** Pe_L, associated with the liquid thickness L:

$$Pe_L = \frac{vL}{D} = \frac{L}{\delta} \tag{2.9}$$

This number is the ratio between a geometric dimension, here the thickness L of the region, and the thickness of the diffusion boundary layer δ previously defined. Figure 2.2b shows the profiles of solute concentration (2.8) for different values of Pe_L as a function of the non dimensional spatial variable x/L.

Suppose now that we do not know the analytic solution (2.8) and that we want to solve the stationary equation (2.6) by the finite difference method (FDM). As the name suggests, the FDM consists of approximating the derivatives that appear in an equation by differences between the values of the function taken at fixed points, called nodes, of a grid. Consider, for simplicity, a set of points on a regular grid where the spacing is given by $h = L/(N + 1)$, where $(N + 1)$ is the number of intervals dividing the length L (fig. 2.3 for which $(N + 1) = 11$ intervals). As the concentrations at the two extremes of the domain, at $x = 0$ and $x = L$, are known from the Dirichlet boundary conditions (2.7), we do not include these points in the numbering. We are then trying to calculate the concentrations, $c(x_k)$, at the N internal points of the domain having the coordinates: $x_k = k\,h$ $(k = 1, N)$. The function $c(x)$ can be expanded in a Taylor series about the points x_k of the grid. We have then:

$$c(x) = c(x_k) + \left[\frac{dc}{dx}\right]_{x_k}(x - x_k) + \left[\frac{d^2c}{dx^2}\right]_{x_k}\frac{(x - x_k)^2}{2!} + \left[\frac{d^3c}{dx^3}\right]_{x_k}\frac{(x - x_k)^3}{3!} + \ldots \tag{2.10}$$

Adopting the notation:

$$c_k = c(x_k) \qquad c_{k\pm1} = c(x_{k\pm1}) \quad \text{values at the nodes } k,\ k \pm 1 \tag{2.11a}$$

$$c_k' = \left(\frac{dc}{dx}\right)_{x_k} \quad \text{first derivative in } k \tag{2.11b}$$

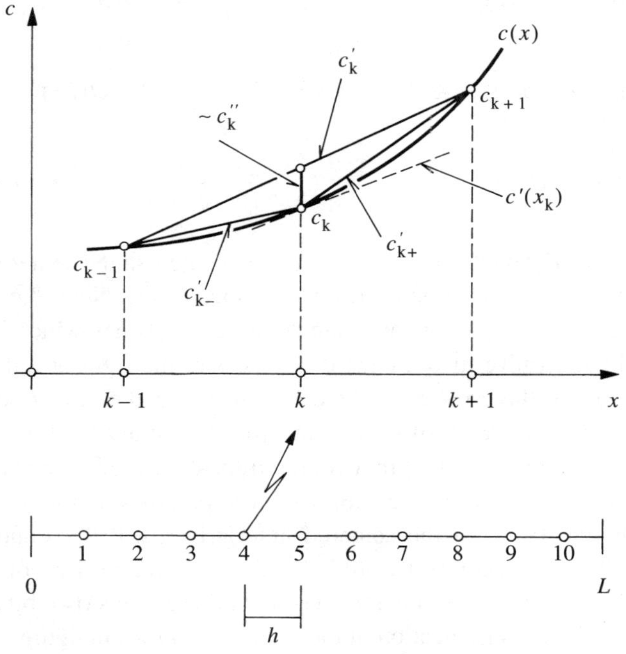

Figure 2.3 Graphical representation of the approximations of the first and second derivatives by differences.

$$c''_k = \left(\frac{d^2c}{dx^2}\right)_{x_k} \qquad \text{second derivative in } k \tag{2.11c}$$

$$c'''_k = \left(\frac{d^3c}{dx^3}\right)_{x_k} \qquad \text{third derivative in } k \tag{2.11d}$$

the values of the concentration at the two points neighboring the node k are given by:

$$c_{k+1} = c_k + c'_k h + c''_k \frac{h^2}{2} + c'''_k \frac{h^3}{6} + O(h^4) \tag{2.12a}$$

$$c_{k-1} = c_k - c'_k h + c''_k \frac{h^2}{2} - c'''_k \frac{h^3}{6} + O(h^4) \tag{2.12b}$$

Linearly combining these two equations, we easily obtain approximations for the first and second derivatives appearing in (2.6) for all points k of the grid. We define the following:

the right first derivative: $$c'_{k+} = \frac{c_{k+1} - c_k}{h} + O(h) \tag{2.13a}$$

the left first derivative: $$c'_{k-} = \frac{c_k - c_{k-1}}{h} + O(h) \tag{2.13b}$$

the centered first derivative: $$c'_k = \frac{c_{k+1} - c_{k-1}}{2h} + O(h^2) \tag{2.13c}$$

the second derivative: $$c''_k = \frac{c_{k+1} - 2c_k + c_{k-1}}{h^2} + O(h^2) \tag{2.13d}$$

(The indices "+" and "–" have been used to distinguish the centered first derivative from the right and left first derivatives, respectively). Note that the first two equations (2.13a) and (2.13b) are accurate "to order h" ($O(h)$), which is to say, that in replacing the derivative in x_k by the differences to the right or left of the point under consideration, the error committed by truncating the Taylor series goes to zero as h when the number N of grid points goes to infinity (and h goes to zero). Thus when the spacing is divided by two, the truncation errors are also divided by the same factor. In other words, we gain one bit of precision in the mantissa on the computer! The equations for the centered first derivative (2.13c) and the second derivative (2.13d) are accurate to order h^2: the division by two of the spacing reduces by a factor of four the truncation errors of these terms (two bits in the computer). The geometrical signification of these terms is shown in figure 2.3. Note that similar equations can be found for higher order derivatives if one considers higher order terms in the expansion of the function (2.10) and "more distant" neighbors of the point under consideration (for example, the terms x_{k+2} and x_{k-2} in addition to the immediate neighbors already used (exercise 2.6.1).

2.2.2 Stationary diffusion

Now take the approximation of the second derivative obtained in (2.13d) to solve the diffusion problem in the interval [0, L] for the case where v is zero (no first derivative) and with the Dirichlet boundary conditions at $x = 0$ and $x = L$. Multiplying through by h^2, and grouping on the right hand side of the equations the known terms $\bar{c}_0$ and $\bar{c}_L$, we have a set of N linear algebraic equations relating the N unknown concentrations c_k. For the particular case $N = 5$, shown in figure 2.4, we have:

$$\left.\begin{array}{llllll} +2c_1 & -c_2 & & & & = \bar{c}_0 \\ -c_1 & +2c_2 & -c_3 & & & = 0 \\ & -c_2 & +2c_3 & -c_4 & & = 0 \\ & & -c_3 & +2c_4 & -c_5 & = 0 \\ & & & -c_4 & +2c_5 & = \bar{c}_L \end{array}\right\} \qquad (2.14)$$

To solve such a system, we have simply to apply Gaussian elimination (chap. 4): we find c_1 as a function of c_2 from the first equation, which we then insert into the second; then we find c_2 as a function of c_3 from the second equation, which we then insert into the third, and so on until the last equation which directly yields

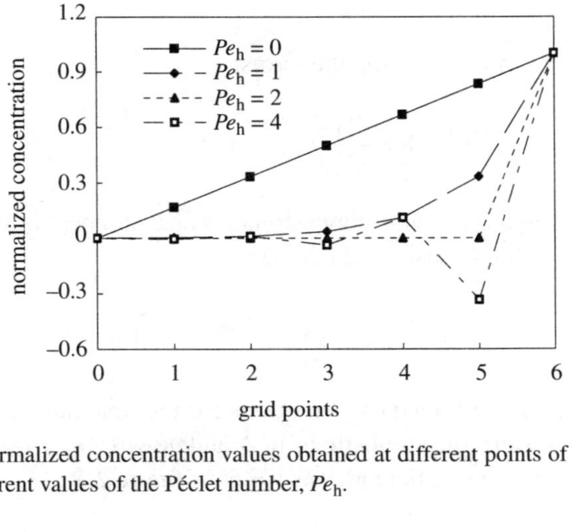

Figure 2.4 Normalized concentration values obtained at different points of a grid composed of five points for different values of the Péclet number, Pe_h.

c_5. Subsequently ascending back up through the equations, we determine the other concentrations. This method of elimination, which can be applied to any matrix system, is especially efficient when the matrix is *tridiagonal*, that is, when the only non zero terms are those on the upper, principal and lower diagonals. Such is the case for the system of equations (2.14), and more generally for the systems encountered in numerical simulation of one dimensional problems. In these cases we use the ***Tri-Diagonal Matrix Algorithm*** (TDMA, chap. 4).

It is simple to show that the values of concentration at the five nodes resulting from the system of equations (2.14) are given by:

$$c_1 = \frac{5\bar{c}_0 + \bar{c}_L}{6} \qquad c_2 = \frac{4\bar{c}_0 + 2\bar{c}_L}{6} \qquad c_3 = \frac{3\bar{c}_0 + 3\bar{c}_L}{6}$$

$$c_4 = \frac{2\bar{c}_0 + 4\bar{c}_L}{6} \qquad c_5 = \frac{\bar{c}_0 + 5\bar{c}_L}{6} \tag{2.15}$$

This is none other than linear interpolation between the imposed values $\bar{c}_0$ and $\bar{c}_L$. This numerical solution of the diffusion problem corresponds to the exact solution of a linear concentration profile (the solution of (2.6) when v equals zero (fig. 2.4)).

2.2.3 Diffusion with stationary transport

Now take again the diffusion equation with transport and express the first and second derivatives of concentration by (2.13c) and (2.13d). The first derivatives are approximated by the differences centered on the nodes of the grid. For a point on the grid that is not an immediate neighbor of the boundary, we obtain:

$$-D\frac{c_{k+1} - 2c_k + c_{k-1}}{h^2} + v\frac{c_{k+1} - c_{k-1}}{2h} = 0 \tag{2.16}$$

or, multiplying by h^2/D and regrouping the terms:

$$\left(-1 - \frac{Pe_h}{2}\right)c_{k-1} + 2c_k + \left(-1 + \frac{Pe_h}{2}\right)c_{k+1} = 0 \tag{2.17}$$

This equation reveals again the non dimensional Péclet number; in this case it is associated with the spacing h between the nodes:

$$Pe_h = \frac{h}{\delta} = \frac{hv}{D} \tag{2.18}$$

Taking into account the boundary conditions for the first and last grid points, one obtains again a system of N equations in N unknowns $\{c_1, c_2, \ldots, c_{N-1}, c_N\}$. Expressed as a matrix for the particular case $N = 5$ of figure 2.4:

$$\begin{bmatrix} +2 & -1+Pe_h/2 & & & \\ -1-Pe_h/2 & +2 & -1+Pe_h/2 & & \\ & -1-Pe_h/2 & +2 & -1+Pe_h/2 & \\ & & -1-Pe_h/2 & +2 & -1+Pe_h/2 \\ & & & -1-Pe_h/2 & +2 \end{bmatrix} \cdot \begin{bmatrix} c_1 \\ c_2 \\ c_3 \\ c_4 \\ c_5 \end{bmatrix}$$

$$= \begin{bmatrix} \bar{c}_0(1+Pe_h/2) \\ 0 \\ 0 \\ 0 \\ \bar{c}_L(1-Pe_h/2) \end{bmatrix} \tag{2.19}$$

The known terms have again been regrouped on the right hand side. Applying the same elimination procedure as before, we find the solution:

$$c_1 = \bar{c}_0\left[1+\frac{Pe_h}{2}\right]\frac{A}{D} \qquad + \bar{c}_L\left[1-\frac{Pe_h}{2}\right]^5\frac{1}{D} \tag{2.20a}$$

$$c_2 = \bar{c}_0\left[1+\frac{Pe_h}{2}\right]^2\frac{B}{D} \qquad + \bar{c}_L\left[1-\frac{Pe_h}{2}\right]^4\frac{2}{D} \tag{2.20b}$$

$$c_3 = \bar{c}_0\left[1+\frac{Pe_h}{2}\right]^3\frac{C}{D} \qquad + \bar{c}_L\left[1-\frac{Pe_h}{2}\right]^3\frac{C}{D} \tag{2.20c}$$

$$c_4 = \bar{c}_0\left[1+\frac{Pe_h}{2}\right]^4\frac{2}{D} \qquad + \bar{c}_L\left[1-\frac{Pe_h}{2}\right]^2\frac{B}{D} \tag{2.20d}$$

$$c_5 = \bar{c}_0\left[1+\frac{Pe_h}{2}\right]^5\frac{1}{D} \qquad + \bar{c}_L\left[1-\frac{Pe_h}{2}\right]\frac{A}{D} \tag{2.20e}$$

with $A = \left[5+\frac{5Pe_h^2}{2}+\frac{Pe_h^4}{16}\right]$ $\qquad B = (4+Pe_h^2)$

$C = \left[3+\frac{Pe_h^2}{4}\right]$ $\qquad D = 2\left[3+\frac{Pe_h^2}{4}\right]\left[1+\frac{3Pe_h^2}{4}\right]$

Figure 2.4 shows graphically the numerical solution obtained for five equidistant grid points and increasing Péclet numbers. The normalized concentration $(c_k - \bar{c}_0)/(\bar{c}_L - \bar{c}_0)$ is shown on the vertical axis to permit comparison with the analytic solution (2.8) (fig. 2.2). It is seen that the errors become more pronounced as the Péclet number increases and as the analytic solution differs increasingly from a linear relation. For $Pe_h > 2$, the coefficients multiplying $\bar{c}_L$ in the equations (2.20) become negative, and the numerical solution no longer makes physical sense as the concentrations obtained at the nodes, notably the third and fifth, are outside the interval $[\bar{c}_0, \bar{c}_L]$. This numerical problem is due to the grid chosen: it is not fine enough to correctly represent the thin boundary layer δ downstream in the flow (at $x = L$). In order to maintain the correct slope imposed by the convective transport term at this boundary, it is necessary to have the concentration at the neighboring node, c_5, less than $\bar{c}_0$ for the chosen spacing, h. In other words, h is too large with respect to δ. This overshoot propagates as wiggles of decreasing intensity moving upstream away from the interface. There are two ways to eliminate this problem. The most direct way is to *refine* the grid, at least close to the interface in question, so that there are many points which fall within the boundary layer, that is with $h < 2D/v$ or $Pe_h < 2$. Although this remedy is the only one that permits the correct representation of the variation of the concentration in the boundary layer, it is often preferable to employ an ***upwind scheme*** that allows the use of a coarser grid. The principles of this method are as follows.

Since the boundary condition imposed at $x = L$ perturbs the numerical solution for high values of velocity, a rudimentary procedure is to simply use the left first derivative (upstream in the solute flow) rather than the centered derivative. Applying this technique to (2.6), we obtain the system of equations (still for 5 nodes):

$$\begin{bmatrix} 2+Pe_h & -1 & & & \\ -(1+Pe_h) & 2+Pe_h & -1 & & \\ & -(1+Pe_h) & 2+Pe_h & -1 & \\ & & -(1+Pe_h) & 2+Pe_h & -1 \\ & & & -(1+Pe_h) & 2+Pe_h \end{bmatrix} \cdot \begin{bmatrix} c_1 \\ c_2 \\ c_3 \\ c_4 \\ c_5 \end{bmatrix}$$

$$= \begin{bmatrix} \bar{c}_0(1+Pe_h) \\ 0 \\ 0 \\ 0 \\ \bar{c}_L \end{bmatrix} \tag{2.21}$$

We see that the Péclet number no longer appears as multiplying the concentration $\bar{c}_L$ imposed downstream on the domain. That resolves some of the problems as seen in figure 2.5 where the numerical solutions employing the upwind scheme and with central differences are shown for comparison at Péclet number $Pe_h = 4$. Nonetheless, the solution also diverges from the desired solution; in decentering the first derivative in the equation for diffusion with transport, an error of order h has been introduced in place of one of order h^2. The difference between a centered derivative (c_k') and a left derivative (c_{k-}') causes a term proportional to the second derivative and the spacing to appear as seen in the following:

$$c_{k-}' = \frac{c_k - c_{k-1}}{h} = \frac{c_{k+1} - c_{k-1}}{2h} - \frac{c_{k+1} - 2c_k + c_{k-1}}{2h} = c_k' - \frac{h}{2}c_k'' \tag{2.22}$$

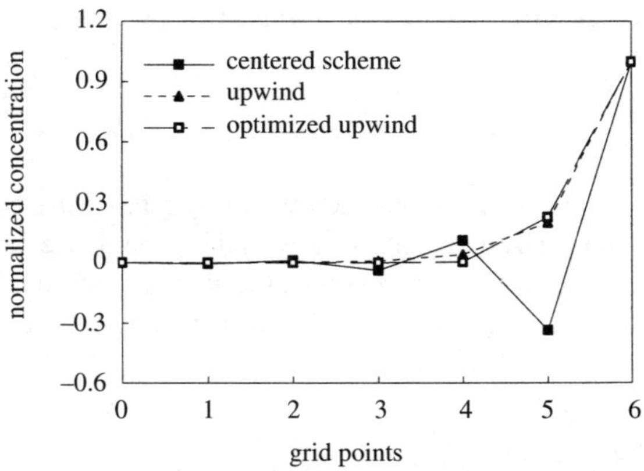

Figure 2.5 Normalized concentration values obtained with different upwinding schemes for a Péclet number, $Pe_h = 4$.

Inserting this relation in the discretized equation for diffusion with transport (2.6), it is seen that, with respect to the centered scheme, the upwind scheme introduces a supplementary term:

$$-Dc''_k + vc'_{k-} = -\left[D + \frac{vh}{2}\right]c''_k + vc'_k \tag{2.23}$$

Employing the left (upstream) derivative in the advection-diffusion equation clears the solution of the oscillations that were seen near the downstream boundary, but at the price of introducing ***numerical diffusion***. In other words, the same result could have been obtained keeping the centered derivative but replacing the diffusion coefficient of the material D by a fictitious value D^* given by:

$$D^* = D + \frac{vh}{2} \tag{2.24}$$

A more clever scheme would be to pose the problem as follows: the boundary layer, especially its slope, cannot in any case be correctly represented with the coarse grid chosen; we would like to find a numerical solution for which the *values* of the concentration at the nodes are exact. To achieve that we introduce a first derivative skewed upwind by a factor θ, $c'_{k\theta}$, and expressed as a linear combination of the right and left derivatives:

$$c'_{k\theta} = (1-\theta)c'_{k+} + \theta c'_{k-} \tag{2.25}$$

The particular values $\theta = 0$, 0.5 and 1 correspond to the right, centered and left derivatives, respectively. It is not difficult to show that the factor θ that yields the analytic solution of (2.8) at the nodes of the grid is given by (exercise 2.6.2):

$$2(\theta - \tfrac{1}{2}) = \coth\left[\frac{Pe_h}{2}\right] - \frac{2}{Pe_h} \tag{2.26}$$

The values of the numerical solution obtained with this scheme are also shown in figure 2.5. It must be emphasized that although with this scheme we have found the values of the analytic solution at the grid points, the slope in the boundary layer is still distorted by numerical diffusion. The effective diffusion coefficient is in this case:

$$D^* = D + vh[\theta - \tfrac{1}{2}] = D + vh\left[\tfrac{1}{2}\coth\left[\frac{Pe_h}{2}\right] - \frac{1}{Pe_h}\right] \tag{2.27}$$

The numerical problems encountered with advection-diffusion, seen here in a simple one dimensional example with the FDM, are also to be found in multi-dimensional problems and with the finite element method. They play a particularly important role not only in all advection problems (chap. 7), but also in phase transformations problems when the speed at which the interface advances is large with respect to the chemical or thermal diffusion (when the corresponding Péclet number is high). However, it must be noted that for multi-dimensional problems, the equivalent of equation (2.26) is no longer as accurate.

2.2.4 Non stationary diffusion

Non stationary phenomena being particularly important in materials science, consider now the solute diffusion equation, without transport (1.73) but non stationary on the domain $[0, L]$:

$$-D\frac{\partial^2 c}{\partial x^2} + \frac{\partial c}{\partial t} = 0 \tag{2.28}$$

An example of this type of problem is encountered in the carburizing or nitriding treatment of steel. The thermal analog of this equation (eq. (1.80)) is the cooling of a large plate through its lateral surfaces.

The ordinary differential equation seen before for stationary diffusion has become a *partial differential equation* as the concentration, $c(x, t)$, is now a function of x and of time t. This type of equation is *parabolic* (second order derivative in space, and first-order in time), for which the initial conditions and the boundary conditions on the domain are needed. For this example, we choose:

Initial condition: $$c(x, 0) = \bar{c}_0 \tag{2.29a}$$

Boundary conditions: $$c(0, t) = c(L, t) = \bar{c}_L \tag{2.29b}$$

The traditional method for analytically solving this linear equation is separation of variables, with superpostion of the solutions and determination of the integration constants with the conditions (2.29) (exercise 2.6.3). This solution is given by a Fourier series:

$$c(x,t) = \bar{c}_L + \frac{4(\bar{c}_0 - \bar{c}_L)}{\pi} \sum_{n=0}^{\infty} \frac{1}{(2n+1)} \sin\left[\frac{2n+1}{L}\pi x\right] \exp\left[-\frac{(2n+1)^2}{L^2}\pi^2 D t\right] \tag{2.30}$$

The normalized concentration is shown in figure 2.6 for various values of nondimensional time $Fo_L = Dt/L^2$. Fo_L is a nondimensional number associated with the plate thickness L, called the ***Fourier number***. It can also be viewed as the ratio of the ***diffusion distance***, $\delta \sim \sqrt{Dt}$, and the thickness L.

Suppose again that one does not know this analytic solution and that one wants to employ the FDM to find an approximate solution at N points of a grid and for M given points in time. We take the time points to be spaced regularly at Δt. As indicated in figure 2.7, where space-time domain $\Omega \times t$ is shown, there are then NM values of the concentration to be determined. The values along the lower edge of the domain are given by the initial conditions whereas those along the two vertical edges of space-time are set by the boundary conditions. In addition to the subscript "k", used to distinguish the nodes in space, the superscript "j" is used for the time steps. As indicated by the thick horizontal segments in figure 2.7, the numerical expression of the second derivative with respect to x (2.13d) at a point (k, j) of space-time involves the values of concentration of that node and of the two neigh-

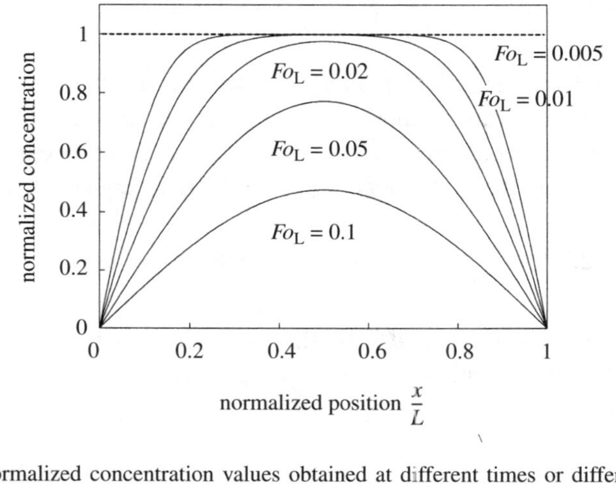

Figure 2.6 Normalized concentration values obtained at different times or different Fourier numbers.

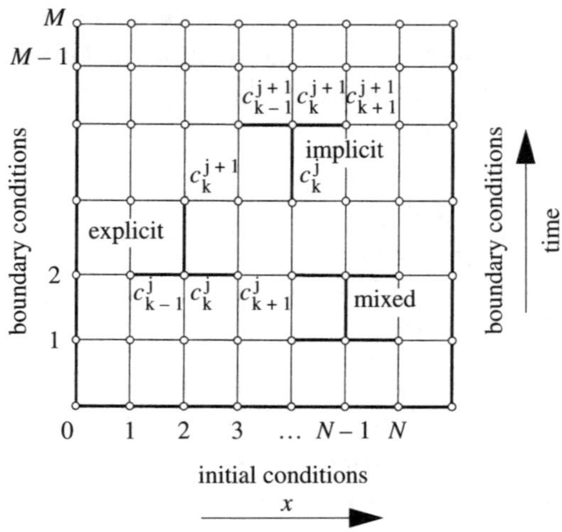

Figure 2.7 : Graphical representation in space-time of the explicit, implicit and mixed schemes.

boring nodes $(k-1, j)$ and $(k+1, j)$. The time derivative appearing in the non stationary diffusion equation (2.28) can be approximated at the node (k, j) by differences, as for the spatial derivative. Two situations can exist:

- In the first case, the known concentrations at time j $(c_{k-1}^{j}, c_{k}^{j}, c_{k+1}^{j})$ are used to determine the net solute flux, and thus the variation of concentration Δc_k^j, at the center node k. This is referred to as a ***forward difference*** or an ***explicit scheme***. The characteristic equation for such a scheme at a node which is not a direct neighbor of the boundary can be written:

$$-c_{k-1}^{j} + 2c_{k}^{j} - c_{k+1}^{j} + \frac{1}{Fo_h}[c_k^{j+1} - c_k^j] = 0 \qquad j = 0, M-1 \tag{2.31}$$

- The second scheme, called ***implicit***, uses ***backward differences***: considering the concentrations to be known at time j, the solute flux is calculated with the *unknown* concentrations at time step $(j+1)$ to determine the variation of concentration at the center node. The characteristic equation is:

$$-c_{k-1}^{j+1} + 2c_{k}^{j+1} - c_{k+1}^{j+1} + \frac{1}{Fo_h}[c_k^{j+1} - c_k^j] = 0 \quad j = 0, M-1 \tag{2.32}$$

The two equations were divided by the term (D/h^2), which reintroduces the nondimensional Fourier number, Fo_h, this time associated with grid spacing h and the time step Δt:

$$Fo_h = \frac{D\Delta t}{h^2} \tag{2.33}$$

The explicit time discretization corresponds to the inverted "*T*" in the sketch in figure 2.7. Taking a step forward in time from the initial conditions ($j = 0$), the temporal variation of the solute at each node is simply deduced from the known concentrations at the current time step. This method is very simple to implement as there is no system of equations to be solved. Rearranging (2.31) and also considering the nodes at the edges of the domain, the concentrations at time $(j+1)$ are given as a function of the known concentrations c_k^j by the following equations:

$$\left.\begin{aligned} c_k^{j+1} &= Fo_h c_{k-1}^j + (1-2Fo_h)c_k^j + Fo_h c_{k+1}^j \qquad k = 2, N-1 \\ c_1^{j+1} &= Fo_h \bar{c}_L + (1-2Fo_h)c_1^j + Fo_h c_2^j \\ c_N^{j+1} &= Fo_h c_{N-1}^j + (1-2Fo_h)c_N^j + Fo_h \bar{c}_L \end{aligned}\right\} \tag{2.34}$$

The implicit scheme corresponds to the upright "*T*" in figure 2.7. As the solute flux is calculated with the unknown concentrations at the end of the time step, we are led to solve a tridiagonal system of *N* equations and *N* unknowns at each time step. We have in this case:

$$\begin{bmatrix} 1+2Fo_h & -Fo_h & & & & \\ -Fo_h & 1+2Fo_h & -Fo_h & & & \\ & -Fo_h & 1+2Fo_h & -Fo_h & & \\ & & \dots & \dots & \dots & \\ & & & -Fo_h & 1+2Fo_h & -Fo_h \\ & & & & -Fo_h & 1+2Fo_h \end{bmatrix} \cdot \begin{bmatrix} c_1^{j+1} \\ c_2^{j+1} \\ \dots \\ \dots \\ c_{N-1}^{j+1} \\ c_N^{j+1} \end{bmatrix}$$

$$= \begin{bmatrix} Fo_h \bar{c}_L \\ 0 \\ 0 \\ \dots \\ \dots \\ 0 \\ Fo_h \bar{c}_L \end{bmatrix} + \begin{bmatrix} c_1^j \\ c_2^j \\ \dots \\ \dots \\ c_{N-1}^j \\ c_N^j \end{bmatrix} \tag{2.35}$$

It must be emphasized that the implicit and explicit schemes both have the same temporal precision: it is $O(\Delta t)$; as for the left or right spatial derivatives, the time derivative is either forward or backward relative to the current time. As will be seen in the next section, the implicit scheme, more difficult to implement than the explicit scheme, offers nonetheless the advantage of being *unconditionally stable*, that is, stable regardless of the time step chosen, which is not true of the explicit scheme, which is *conditionally stable*.

The concept of the stability of a scheme is important. We can also connect it to the concepts of consistency and of convergence with Lax's theorem. The latter is as follows: given a well posed non stationary diffusion problem, with initial and boundary conditions. Then, with a consistent finite difference scheme, that is, one for which the truncation error goes to zero as Δt and h go to zero, we can show that the stability of the scheme is a necessary and sufficient condition for it to converge. This notion implies that the numerical solution approximates the exact solution closer and closer as the spacing and time steps are refined.

It is of course possible to use *mixed* temporal schemes: they weight the forward and backward time derivatives with a factor θ. Multiplying (2.31) by $(1-\theta)$ and adding it to (2.32) multiplied by θ, one obtains:

$$(1-\theta)[-c_{k-1}^{j}+2c_{k}^{j}-c_{k+1}^{j}]+\theta[-c_{k-1}^{j+1}+2c_{k}^{j+1}-c_{k+1}^{j+1}] + \frac{1}{Fo_h}[c_k^{j+1}-c_k^j] = 0 \tag{2.36}$$

This type of scheme involves the concentrations at the node and its two neighbors at time steps $(j+1)$ and (j) and corresponds graphically to the sideways "H" in figure 2.7. The solution of the set of equations (2.36) as k varies from 1 to N requires matrix calculations (matrix inversion by Gaussian elimination or by an iterative method) as for the implicit scheme. The special case $\theta = 0.5$, called the ***Crank-Nicolson*** scheme, or the ***trapezoidal rule***, corresponds effectively to a *centered* temporal derivative between the time steps (j) et $(j+1)$. As for the centered spatial derivative (2.13c), this scheme has a higher temporal precision than the explicit or implicit schemes, being $O(\Delta t^2)$.

Note that more precise methods of temporal integration such as Runge-Kutta could have also been used to solve the system of differential equations obtained after formulation of the second derivatives by the FDM. However, these methods do not offer real advantages in view of the error already committed by the approximation of the spatial derivatives.

2.2.5 Stability of the explicit and implicit schemes

The explicit scheme is ***conditionally stable***, that is, the spacing between grid points and the time step must satisfy a condition to avoid obtaining oscillations in the concentration profiles that will rapidly grow as time advances (instabilities). A simple way to show that the explicit method is conditionally stable is to watch what happens at the first time step ($j = 0$) at the first node ($k = 1$) (fig. 2.8). Replacing the values of c_k^0 in (2.34) by the initial condition $\bar{c}_0$, the variation of concentration at this node, $\Delta c_1 = c_1^1 - \bar{c}_0$, is given by:

$$\Delta c_1 = c_1^1 - \bar{c}_0 = Fo_h[\bar{c}_L - \bar{c}_0] \tag{2.37}$$

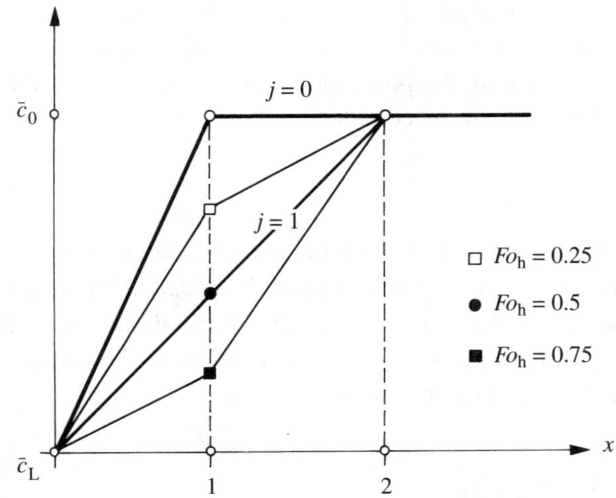

Figure 2.8 Concentration profiles at the first time step and first node of the grid obtained with an explicit scheme for different values of the Fourier number, Fo_h.

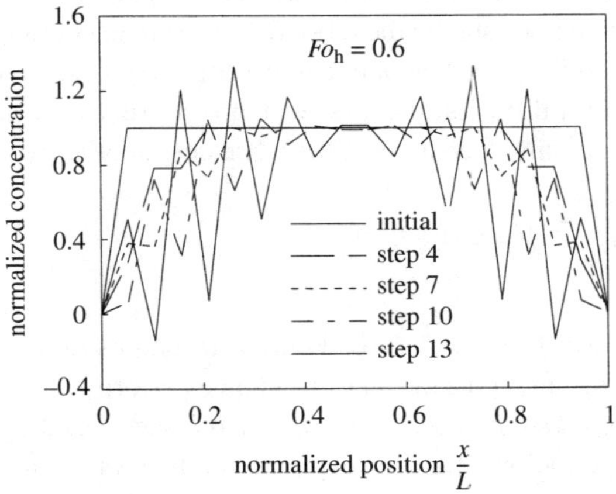

Figure 2.9 Normalized concentration values obtained with an explicit scheme and for Fo_h = 0.6 (unstable conditions).

Figure 2.8 shows the initial profile of concentration ($j = 0$) and three profiles obtained at time step $j = 1$ for increasing values of the Fourier number, Fo_h. For $Fo_h = {}^1/_4$, the profile is such that derivative c_1'' at point 1 is still negative at time step $j = 1$. However, when $Fo_h = {}^3/_4$, the concavity of the profile is inverted. This would be contrary to the second law of thermodynamics as a positive value of c_1'' would have the effect of increasing the concentration at this point at the next time step. In such a situation, the concentration profile becomes more and more oscillatory with time, as shown in figure 2.9, and the numerical values rapidly exceed the maximum or minimum allowed by the computer. These oscillations are clearly *instabilities*.

The limiting case at which the explicit scheme is still stable is given by $Fo_h = {}^1/_2$: in this case, the concentration profile at $j = 1$ is flat ($c_1'' = 0$), which gives zero variation of the concentration at this point for the next time step (fig. 2.8). The explicit scheme is thus stable only if the condition:

$$Fo_h = \frac{D\Delta t}{h^2} \leq \frac{1}{2} \tag{2.38}$$

is satisfied. This stability condition for the explicit scheme, called *Fourier's criterion* or *condition* or also *von Neumann's condition*, can be demonstrated more rigorously with the help of *Fourier's method.* This method consists of first writing the concentrations c_k^j at time j and at node k, in terms of the Fourier components, C_α^j. For a regular grid, one obtains the decomposition:

$$c_k^j = \int_0^\pi C_\alpha^j \exp(\underline{i}k\alpha)\, d\alpha \tag{2.39}$$

where $\underline{i}$ is the square root of (–1). At each time step Δt, each component, C_α^j, of the Fourier decomposition (2.39) is transformed into a new component, C_α^{j+1}. One can show that for the scheme to be stable, the ratio, R_α, of the new Fourier component to the old must have a modulus less than or equal to 1 for every value of α. This can be easily understood as in a diffusion process, which is a process of dissipation, the Fourier components can only decrease as time advances. In other words:

$$|R_\alpha| = \left|\frac{C_\alpha^{j+1}}{C_\alpha^j}\right| \leq 1 \qquad \text{for } \alpha \in [0, \pi] \tag{2.40}$$

As the functions $\exp(\underline{i}k\alpha)$ used in the Fourier decomposition constitute a linearly independent set of functions, the analysis can be performed independently for each component $C_\alpha \exp(\underline{i}k\alpha)$. It is then sufficient to replace c_k^j and $c_{k\pm1}^j$ in the previously encountered equations for the explicit, implicit or mixed schemes by $C_\alpha^j \exp(\underline{i}k\alpha)$ and $C_\alpha^j \exp(\underline{i}(k \pm 1)\alpha)$, respectively. Taking first the explicit scheme (2.31), one has:

$$\begin{aligned} C_\alpha^{j+1}\exp(\underline{i}k\alpha) = {} & Fo_h C_\alpha^j \exp(\underline{i}(k+1)\alpha) + (1 - 2Fo_h) C_\alpha^j \exp(\underline{i}k\alpha) \\ & + Fo_h C_\alpha^j \exp(\underline{i}(k-1)\alpha) \end{aligned} \tag{2.41}$$

Rearranging the term on the right and dividing through by $\exp(\underline{i}k\alpha)$, one obtains the relation for R_α:

$$|R_\alpha| = |1 - 2Fo_h(1 - \cos\alpha)| \tag{2.42}$$

The most "unfavorable" case for the stability of the scheme is when $\alpha = \pi$ (the values of c_k^j corresponding to this Fourier component alternate between $+C_\alpha^j$ and

$-C_\alpha^j$ between two neighboring nodes of the grid). In order that R_π be less than or equal to 1 in the relation above, it is necessary that the condition (2.38) is satisfied, that is, that $Fo_h \leq 1/2$.

Using the same method of analysis, one can easily show that the value of R_α for the implicit scheme (2.32) is given by:

$$R_\alpha = \frac{1}{1 + 2Fo_h(1 - \cos\alpha)} \tag{2.43}$$

The modulus of R_α is always less than or equal to 1, regardless of the value of α; therefore, the implicit scheme is *unconditionally stable.* The Fourier number can take any value, which means that the time step can be chosen independently of the grid spacing h. One must not confuse stability and precision: if the implicit scheme remains stable regardless of the time step, its precision, linked to the temporal truncation, is identical to that of the explicit scheme, varying as $O(\Delta t)$.

Using again the same method, one can show that the mixed scheme (2.36), is stable if the following condition is satisfied:

$$|R_\alpha| = \left|\frac{1 - 2Fo_h(1 - \cos\alpha)(1 - \theta)}{1 + 2Fo_h(1 - \cos\alpha)\theta}\right| \leq 1 \qquad \text{for } \alpha \in [0, \pi] \tag{2.44}$$

Analyzing the variation of R_α as a function of α, one finds that the condition (2.44) is satisfied if:

$$Fo_h(1 - 2\theta) \leq \tfrac{1}{2} \tag{2.45}$$

Thus, if θ is zero, one finds again the stability condition (2.38) of the explicit scheme. If $\theta \geq 1/2$, the scheme is unconditionally stable, while if $\theta < 1/2$ the stability of the scheme depends on a condition. Mixed schemes with $\theta < 1/2$ are thus not very interesting: they require solving a system of linear equations (inversion of a matrix) without offering a free choice for the time step. The Crank-Nicolson scheme with $\theta = 1/2$ is interesting because it is unconditionally stable while offering a temporal resolution of order Δt^2. Remark in this case that if Δt becomes too large ($Fo_h \to \infty$), then $R_\alpha \to -1$. The Crank-Nicolson scheme tends thus to oscillate although it remains stable.

To conclude this section on the stability of temporal schemes, it must be noted that the conditions given above constitute essential knowledge before embarking on any modeling. If one wants to model, for example, chemical diffusion phenomena in the solid phase, where D is of the order of 1 $\mu m^2/s$, at the scale of the microstructure ($h \cong 1$ μm) with an explicit scheme, the time step must be less than a second, which is probably acceptable, even for modeling a thermal treatment that lasts a day. But if, in the same problem, one also takes thermal processes into account at the same scale, a thermal diffusivity of 1 mm^2/s reduces the time step to 1 μs, which is certainly not desirable!

Finally, the stability conditions will be modified if one considers:

i) An irregular grid. In this case it is the size of the smallest spacing that, unfortunately, appears in the stability criterion.

ii) A two or three dimensional problem. In this case one can show that for equal grid spacing in each spatial direction, the stability criteria of the explicit scheme are given by:

$$Fo_h \leq \frac{1}{4} \text{ or } Fo_h \leq \frac{1}{6}, \text{ respectively} \tag{2.46}$$

iii) The boundary conditions (section 2.2.6).

2.2.6 Formulation for natural boundary conditions

The boundary conditions encountered up until now for stationary or non stationary problems were Dirichlet (essential) boundary conditions. The concentration being imposed on the boundary, it was not necessary to have a variable at the node on the boundary. As was shown, such a condition introduces a term on the right hand side for the nodes immediately neighboring the boundary. We will treat next the two situations where a Neumann and a Cauchy natural boundary conditions are imposed at the boundary $x = 0$ (sect. 1.4.2). The concentration at this boundary no longer being imposed, it is necessary to define the first node of the grid as coincident with this boundary: the concentration c_1 becomes now an unknown of the problem (fig. 2.10). The finite difference formulation remaining identical for all the other nodes of the grid, only the formulation at this point is considered in the following.

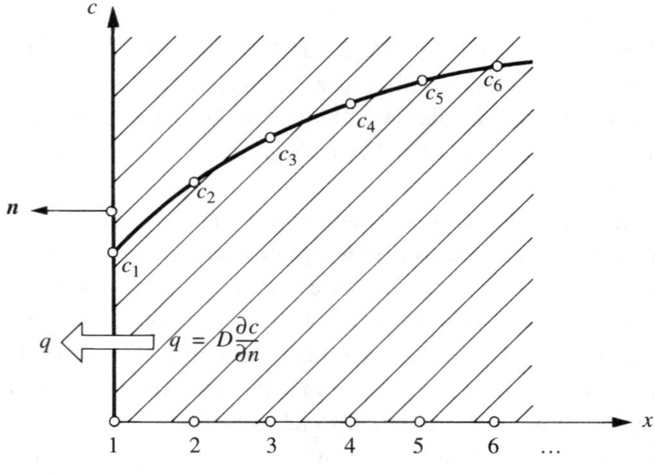

Figure 2.10 Neumann natural condition at $x = 0$ and the concentration profile.

Neumann condition

Returning to the Taylor expansion in (2.10), the first derivative at $x = 0$ is determined by the boundary condition (1.111) :

$$c_1' = \frac{\partial c}{\partial x} = -\frac{\partial c}{\partial n} = \frac{-q}{D} \tag{2.47}$$

(the outgoing normal derivative is opposite to the direction x; and the index "i" of the outgoing solute flux has been omitted for simplicity). Therefore, the second derivative at this point, which appears in the diffusion term, is approximated by:

$$c_1'' = \frac{(c_2 - c_1)/h + q/D}{h/2} + O(h) = \frac{c_{1+}' + q/D}{h/2} + O(h) \tag{2.48}$$

One sees that the second derivative is expressed as the difference between the right derivative of the point 1, c_{1+}', and the derivative at the point itself written in terms of the boundary condition. This difference is divided by half the grid spacing, the derivative c_{1+}' being "centered" in the middle of the interval separating the first two points. (The first derivative associated with the advective transport in (2.6) can be approximated by c_{1+}' or by $-q/D$). One should, note however, that the insertion of the second derivative (2.48) in the Taylor series expansion (2.10) does not eliminate the third derivative; the approximation at such points is thus of order h. The temporal variation term is identical to that derived for a node inside the volume.

When inserted in the conservation equation as formulated for the first grid point, (2.48) introduces the concentrations c_1 and c_2 as well as a term on the right hand side linked to the solute flux. This term is of course zero for the case of a homogeneous Neumann condition (1.110). Then the non stationary equation at point 1, without convective transport and with an implicit scheme, takes the form:

$$(1 + 2Fo_h)c_1^{j+1} - 2Fo_h c_2^{j+1} = c_1^j + \frac{2q\Delta t}{h} \tag{2.49}$$

This equation should be compared with that deduced above for a Dirichlet boundary condition. Note that the explicit stability criterion given in (2.38) is not modified by the introduction of this boundary condition.

Cauchy condition

Such a natural condition is written (chap. 1):

$$D\frac{\partial c}{\partial n} = -D\left(\frac{\partial c}{\partial x}\right)_{x = x_1} = -Dc_1' = -\alpha(c_1 - \bar{c}_a) \tag{2.50}$$

where c_1 is the concentration at the domain surface, $\bar{c}_a$ the concentration of the external medium and α is a solute transfer coefficient at the boundary. Again, inserting this in the Taylor expansion (2.10) at point 1, the second derivative becomes:

$$c_1'' = \frac{(c_2 - c_1)/h - \alpha/D(c_1 - \bar{c}_a)}{h/2} + O(h) = \frac{c_{1+}' - \alpha/D(c_1 - \bar{c}_a)}{h/2} + O(h) \tag{2.51}$$

The non stationary conservation equation in the implicit scheme takes the following form at point 1:

$$[1 + 2Fo_h(1 + Bi_h)]c_1^{j+1} - 2Fo_h c_2^{j+1} = c_1^j + 2Fo_h Bi_h \bar{c}_a \tag{2.52}$$

A second non dimensional number appears in this equation associated with the grid spacing h, the ***Biot number***, Bi_h, defined as:

$$Bi_h = \frac{\alpha h}{D} \tag{2.53}$$

For the explicit scheme, it is easy to show that the Fourier stability criterion associated with such a boundary condition is more constraining as it is given by the condition (exercise 2.6.4):

$$Fo_h \leq \frac{1}{2}\frac{1}{1 + Bi_h} \tag{2.54}$$

2.3 TWO DIMENSIONAL PROBLEMS

2.3.1 Problem definition

We now propose to generalize what was seen in the previous section for the one dimensional case with regular spacing, to a two dimensional geometry with an irregular grid. For this we consider the non stationary chemical diffusion problem in the rectangular domain, Ω, shown in figure 2.11, with a uniform transport velocity in the x direction and a constant diffusion coefficient. The equation governing the phenomenon is then given by (1.74):

$$\frac{\partial c}{\partial t} - D\,\mathrm{div}(\mathbf{grad}\,c) + \mathbf{v} \cdot \mathbf{grad}\,c = \frac{\partial c}{\partial t} - D\left[\frac{\partial^2 c}{\partial x^2} + \frac{\partial^2 c}{\partial y^2}\right] + v\frac{\partial c}{\partial x} = 0 \tag{2.55}$$

The concentration, $c(x, y, t)$, is now a function of two spatial variables, x and y, and of time t. The boundary conditions on the four boundaries are of different nature to illustrate how they intervene in the calculations. In this way, we have:

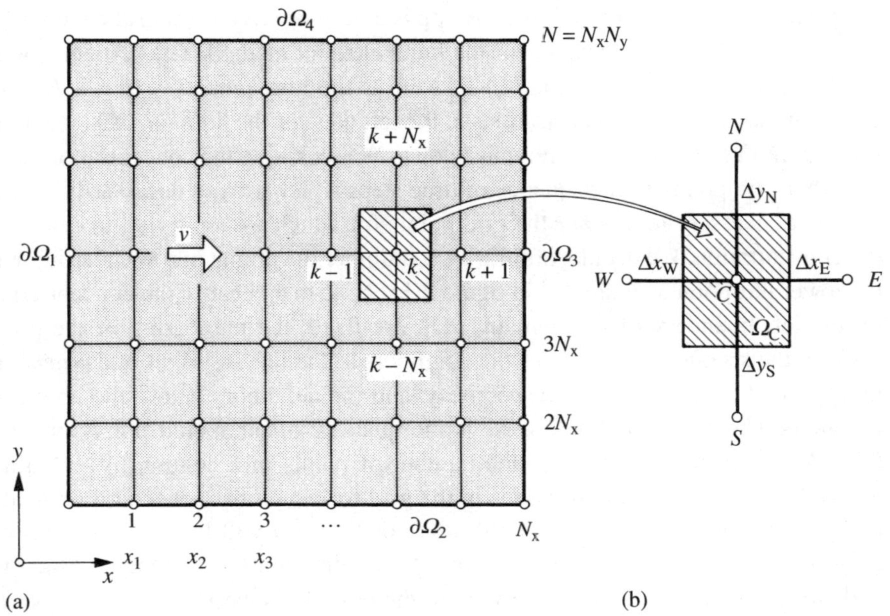

Figure 2.11 Domain and grid considered for the non stationary calculation of the concentration field (a) and detail of a particular grid element around the point C (b).

$$\text{On } \partial\Omega_1: \quad c(x,t) = \bar{c} \qquad \text{(Dirichlet condition)} \tag{2.56a}$$

$$\text{On } \partial\Omega_2: \quad D\frac{\partial c}{\partial n} = -D\frac{\partial c}{\partial y} = q \qquad \text{(Neumann condition, imposed flux)} \tag{2.56b}$$

$$\text{On } \partial\Omega_3: \quad D\frac{\partial c}{\partial n} = D\frac{\partial c}{\partial x} = 0 \qquad \text{(homogeneous Neumann condition)} \tag{2.56c}$$

$$\text{On } \partial\Omega_4: \quad D\frac{\partial c}{\partial n} = D\frac{\partial c}{\partial y} = -\alpha(c - \bar{c}_a) \qquad \text{(mixed Cauchy condition)} \tag{2.56d}$$

The initial condition is given by a uniform concentration everywhere in the domain Ω:

$$c(x,y,0) = \bar{c}_0 \tag{2.57}$$

2.3.2 Griding, local and global numbering of the nodes

To apply the FDM, the rectangular domain of dimensions L_x by L_y is first subdivided into rectangular elements, not necessarily regular, parallel to the coordinate axes. This is then a *structured grid*, as opposed to more free form grids that will be defined in the following chapter for the finite element method. The vertical column positions of the grid, x_v ($v = 1, N_v$) as well as the horizontals, y_h ($h = 1, N_h$), are stored in memory. The concentrations at the left edge of the domain, $\partial\Omega_1$, including the extremities, are imposed and thus known. We seek now the concentrations at all the other grid points for a sequence of time steps which are not necessarily regular. It is first necessary to *number* the grid points for which we are trying to determine the concentration. We decide, arbitrarily, to number the points in rows, starting with the lower left point as indicated in figure 2.11. Note that because the concentrations on the left boundary of the domain, $\partial\Omega_1$, are fixed, the numbering of the points starts at the second column. Thus the neighbors to the left and right of a point k are numbered $(k - 1)$ and $(k + 1)$ respectively, and the neighbors below and above are numbered $(k - N_x)$ and $(k + N_x)$. N_x is the quantity of numbered points in a row ($N_x = N_v - 1$ in this case), N_y is the number of points in a column ($N_y = N_h$) and $N = N_x N_y$ the total number of nodes in the grid for which the concentration field is to be determined. In order to simplify the notation, we will label the four nearest neighbors of a specific node k according to the directions of a compass; then the node to the right is labeled "*E*" (East), to the left, "*W*" (West), above "*N*" (North) and below "*S*" (South); the node itself will have the index "*C*" (Center) (fig. 2.11). Note that the grid points on the boundaries $\partial\Omega_2$, $\partial\Omega_3$, $\partial\Omega_4$ do not have neighbors to, respectively, the South, the East or the North. Those neighboring the boundary $\partial\Omega_1$ have a neighbor to the West with a known concentration from the Dirichlet boundary condition (2.56a). If the neighbor exists, the distance separating it from the center node is Δx_E, Δx_W, Δy_N or Δy_S. These distances are calculated directly from the coordinates x_v and y_h of the grid.

As for the one dimensional problem in the preceding section, we must now formulate the partial derivatives (2.55) at any point in the grid, including boundary points, in terms of finite differences of the concentrations at the nodes, c_k^θ. (The index θ is used to denote the time at which the concentration is specified). We will do this first for an internal node, showing the equivalence that exists between the FDM and the Finite Volume Method (FVM). Then we will show the formulation for the nodes at the boundaries before assembling all the parts of the "puzzle".

2.3.3 FDM formulation for an internal node

By analogy with the treatment of the regular one dimensional grid case, the value of $c(x, y, t)$ at the neighboring points of a node C can be expressed using a Taylor series expansion of the function. We have then, omitting the time index θ to simplify the expression:

$$c_E = c_C + \left[\frac{\partial c}{\partial x}\right]_C \Delta x_E + \left[\frac{\partial^2 c}{\partial x^2}\right]_C \frac{\Delta x_E^2}{2} + O(\Delta x_E^3) \tag{2.58a}$$

$$c_{\mathrm{W}} = c_{\mathrm{C}} - \left[\frac{\partial c}{\partial x}\right]_{\mathrm{C}} \Delta x_{\mathrm{W}} + \left[\frac{\partial^2 c}{\partial x^2}\right]_{\mathrm{C}} \frac{\Delta x_{\mathrm{W}}^2}{2} + O(\Delta x_{\mathrm{W}}^3) \tag{2.58b}$$

$$c_{\mathrm{N}} = c_{\mathrm{C}} + \left[\frac{\partial c}{\partial y}\right]_{\mathrm{C}} \Delta y_{\mathrm{N}} + \left[\frac{\partial^2 c}{\partial y^2}\right]_{\mathrm{C}} \frac{\Delta y_{\mathrm{N}}^2}{2} + O(\Delta y_{\mathrm{N}}^3) \tag{2.58c}$$

$$c_{\mathrm{S}} = c_{\mathrm{C}} - \left[\frac{\partial c}{\partial y}\right]_{\mathrm{C}} \Delta y_{\mathrm{S}} + \left[\frac{\partial^2 c}{\partial y^2}\right]_{\mathrm{C}} \frac{\Delta y_{\mathrm{S}}^2}{2} + O(\Delta y_{\mathrm{S}}^3) \tag{2.58d}$$

From the first two equations, we can extract an approximation for the centered first derivative and the second derivative of $c(x, y, t)$ with respect to x:

$$\left[\frac{\partial c}{\partial x}\right]_{\mathrm{C}} \simeq \frac{(c_{\mathrm{E}} - c_{\mathrm{C}})\Delta x_{\mathrm{W}}/\Delta x_{\mathrm{E}} + (c_{\mathrm{C}} - c_{\mathrm{W}})\Delta x_{\mathrm{E}}/\Delta x_{\mathrm{W}}}{2\Delta x} \tag{2.59a}$$

$$\left[\frac{\partial^2 c}{\partial x^2}\right]_{\mathrm{C}} \simeq \frac{(c_{\mathrm{E}} - c_{\mathrm{C}})/\Delta x_{\mathrm{E}} - (c_{\mathrm{C}} - c_{\mathrm{W}})/\Delta x_{\mathrm{W}}}{\Delta x} \tag{2.59b}$$

where Δx is the arithmetic mean of the distances to the neighbors to the East and West, that is: $\Delta x = (\Delta x_{\mathrm{E}} + \Delta x_{\mathrm{W}})/2$. These two relations are accurate to second order as long as the grid is regular locally ($\Delta x_{\mathrm{W}} = \Delta x_{\mathrm{E}}$). If the grid is not too irregular, so that $\Delta x_{\mathrm{E}} \approx \Delta x_{\mathrm{W}}$, the first derivative can be approximated by the equation:

$$\left[\frac{\partial c}{\partial x}\right]_{\mathrm{C}} \simeq \frac{c_{\mathrm{E}} - c_{\mathrm{W}}}{2\Delta x} \tag{2.60}$$

In the same way, the second derivative with respect to y appearing in (2.55) can be approximated by:

$$\left[\frac{\partial^2 c}{\partial y^2}\right]_{\mathrm{C}} \simeq \frac{(c_{\mathrm{N}} - c_{\mathrm{C}})/\Delta y_{\mathrm{N}} - (c_{\mathrm{C}} - c_{\mathrm{S}})/\Delta y_{\mathrm{S}}}{\Delta y} \tag{2.61}$$

where Δy is again the arithmetic mean: $\Delta y = (\Delta y_{\mathrm{N}} + \Delta y_{\mathrm{S}})/2$. Only the temporal variation of the concentration appearing in (2.55) remains to be formulated in terms of differences. Reintroducing the time step index, j, this term is, as for the one dimensional case:

$$\left[\frac{\partial c}{\partial t}\right]_{\mathrm{C}} \simeq \frac{c_{\mathrm{C}}^{j+1} - c_{\mathrm{C}}^{j}}{\Delta t} \tag{2.62}$$

Regrouping all the contributions, we have for the volume node under consideration:

$$\frac{c_{\mathrm{C}}^{j+1} - c_{\mathrm{C}}^{j}}{\Delta t} - D\left[\frac{(c_{\mathrm{E}} - c_{\mathrm{C}})/\Delta x_{\mathrm{E}} - (c_{\mathrm{C}} - c_{\mathrm{W}})/\Delta x_{\mathrm{W}}}{\Delta x} + \frac{(c_{\mathrm{N}} - c_{\mathrm{C}})/\Delta y_{\mathrm{N}} - (c_{\mathrm{C}} - c_{\mathrm{S}})/\Delta y_{\mathrm{S}}}{\Delta y}\right]^{\theta} + v\left[\frac{c_{\mathrm{E}} - c_{\mathrm{W}}}{2\Delta x}\right]^{\theta} = 0 \tag{2.63}$$

As for the one dimensional case, the square bracketed terms can be calculated at time (j), at time (j + 1), or at a linear combination of the two, for which reason they have the index "θ" (mixed scheme). If "θ" = "j" (explicit scheme), the bracketed terms are calculated with the concentration values known at time (j), which leaves c_C^{j+1} as the only unknown in (2.63). For "θ" = "j + 1" (implicit scheme), the terms are expressed with unknown concentrations at time (j + 1), and only the concentration c_C^j is known.

For a regular grid in x and y ($\Delta x_E = \Delta x_W = \Delta y_N = \Delta y_S = h$), the above equation can be reduced to:

$$\frac{1}{Fo_h}(c_C^{j+1} - c_C^j) - (c_E + c_W + c_N + c_S - 4c_C)^\theta + \frac{Pe_h}{2}(c_E - c_W)^\theta = 0 \qquad (2.64)$$

where the Péclet number, $Pe_h = (vh)/D$, and the Fourier number, $Fo_h = (D\Delta t)/h^2$, have again been introduced, thus making the coefficients multiplying the concentration values dimensionless. For the explicit scheme, the stability conditions $Fo_h \leq 1/4$ and $Pe_h \leq 2$ are imposed simultaneously. They lead to the stability criteria $D\Delta t/h^2 \leq 1/4$ and $v\Delta t/h \leq 1/2$. This last relation is called the ***CFL condition***, after the initials of the mathematicians that first derived it: Courant, Friedrichs and Lewy. It expresses that the numerical information should not move a distance greater than half a cell for a single time step.

2.3.4 Equivalence of FDM and FVM for a volume node

One often sees a distinction made in the literature between the finite difference method (FDM) and the ***finite volume method*** (FVM). As the following will show, there is not really such a great distinction, the FVM is only the formulation of an integral over the grid element volume of the conservation equation under consideration. The FVM guarantees automatically the property of conservation of the fundamental equations at the discrete level, which is not necessarily the case for the FDM. The major difference between these two methods is in the definition of the points where values of the fields (here, the concentration) are evaluated with respect to the domain grid (fig. 2.12). For the FDM, the intersection points of the grid are the nodes where the field values are taken (fig. 2.12.a). The same grid in the FVM defines the element volumes over which the conservation sum is evaluated; the field value is effectively defined at the center of gravity of each of the small volumes. The first method results in half elements at the domain edges (or quarter elements at the corners), whereas the grid defined in figure 2.12b only has complete volumes. The latter method has, nonetheless, the inconvenience of displacing by a half element the concentration that appears in a Dirichlet or Cauchy boundary condition. For more details the reader is referred to the classic work on the subject by Patankar mentioned in the bibliography. Whatever volume is chosen, the integral formulation has the advantage, compared to the previously presented method, that the boundary

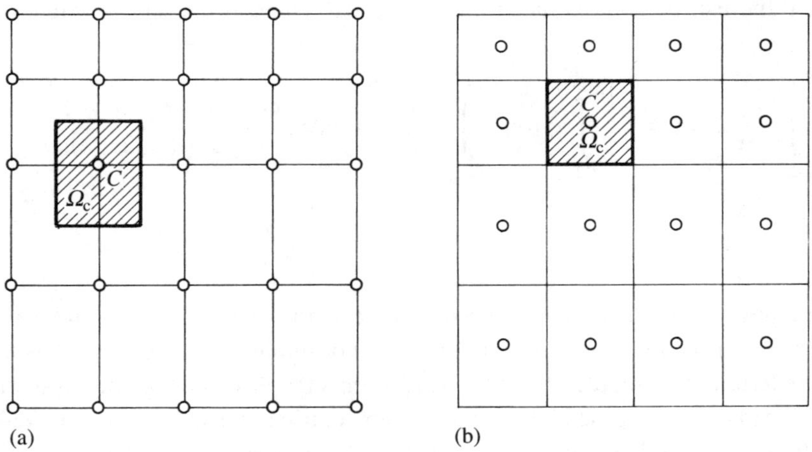

Figure 2.12 Grid and nodes defined for the finite difference method (a) and the finite volume method (b).

conditions can be introduced *naturally*. It guarantees automatically the conservation of the quantities considered on the discrete volumes. In addition, the method also allows the use of ***unstructured grids***, that is on grids that do not necessarily follow the coordinate axes (chap. 4). Note that such grids are used in the direct finite difference method discussed at the end of the chapter. In the following, we consider the situation shown in figure 2.12a; the grid remains the same as that defined for the finite difference method (fig. 2.11).

To evaluate the integral we define the elementary volume of the grid, Ω_C, around the point C as that delimited by the perpendicular bisectors of the four segments connecting the point to its nearest neighbors. The integration of the diffusion equation with transport (2.55) over this elementary volume, using the divergence theorem yields:

$$\frac{\mathrm{d}}{\mathrm{d}t}\int_{\Omega_C} c\ \mathrm{d}x\mathrm{d}y - \int_{\partial\Omega_C} D\,\mathbf{grad}\,c\cdot\boldsymbol{n}\ \mathrm{d}s + \int_{\Omega_C} v\frac{\partial c}{\partial x}\ \mathrm{d}x\mathrm{d}y = 0 \tag{2.65}$$

In fact, we have again an integral of the conservation of solute for the volume Ω_C considered, which is rather comforting! The first term appearing in this conservation balance is approximated by:

$$\frac{\mathrm{d}}{\mathrm{d}t}\int_{\Omega_C} c\ \mathrm{d}x\mathrm{d}y \simeq \frac{c_C^{j+1} - c_C^{j}}{\Delta t}\Delta x\Delta y \tag{2.66}$$

where Δx and Δy are the dimensions of Ω_C previously defined by the arithmetic means of the distances to the first neighbors. The last term can be integrated directly over x and can be approximated on the vertical edges to the left and right of Ω_C by:

$$\int_{\Omega_C} v\frac{\partial c}{\partial x}\,\mathrm{d}x\mathrm{d}y = v\left[\int_{\partial\Omega_E} c\,\mathrm{d}y - \int_{\partial\Omega_W} c\,\mathrm{d}y\right]^{\theta} \simeq v\Delta y\left[\frac{c_E + c_C}{2} - \frac{c_W + c_C}{2}\right]^{\theta}$$

$$= v\Delta y\left[\frac{c_E - c_W}{2}\right]^{\theta} \tag{2.67}$$

(The index "θ" has been reintroduced to indicate that the evaluation can be at time (j) or $(j+1)$ according to whether the explicit or implicit scheme is employed). As for the term in the middle of (2.65), it represents the flux in or out of Ω_C by diffusion. It can be decomposed into a sum of four contributions at the Eastern, Western, Northern and Southern boundaries of the domain, paying attention to the sign appearing in the normal derivatives (for example, the outgoing normal at $\partial\Omega_E$ is toward $+x$, that at $\partial\Omega_W$ toward $-x$). We have then:

$$\int_{\partial\Omega_C} D\,\mathbf{grad}\,c\cdot\boldsymbol{n}\,\mathrm{d}s = \int_{\partial\Omega_E} D\frac{\partial c}{\partial x}\,\mathrm{d}y - \int_{\partial\Omega_W} D\frac{\partial c}{\partial x}\,\mathrm{d}y + \int_{\partial\Omega_N} D\frac{\partial c}{\partial y}\,\mathrm{d}x - \int_{\partial\Omega_S} D\frac{\partial c}{\partial y}\,\mathrm{d}x \tag{2.68}$$

Replacing the first derivatives by the differences between the appropriate nodes:

$$\int_{\partial\Omega_C} D\,\mathbf{grad}\,c\cdot\boldsymbol{n}\,\mathrm{d}s \simeq D\left[\frac{c_E - c_C}{\Delta x_E} - \frac{c_C - c_W}{\Delta x_W}\right]^{\theta}\Delta y + D\left[\frac{c_N - c_C}{\Delta y_N} - \frac{c_C - c_S}{\Delta y_S}\right]^{\theta}\Delta x \tag{2.69}$$

Regrouping the contributions (2.66), (2.67) and (2.69), the integral balance of solute (2.65) is expressed for the volume element Ω_C:

$$\frac{c_C^{j+1} - c_C^{j}}{\Delta t}\Delta x\Delta y - D\left\{\left[\frac{c_E - c_C}{\Delta x_E} - \frac{c_C - c_W}{\Delta x_W}\right]\Delta y + \left[\frac{c_N - c_C}{\Delta y_N} - \frac{c_C - c_S}{\Delta y_S}\right]\Delta x\right\}^{\theta}$$

$$+ v\left[\frac{c_E - c_W}{2}\right]^{\theta}\Delta y = 0 \tag{2.70}$$

It is seen that this equation, when divided by the volume of the grid element $\Delta V_C = \Delta x\Delta y = (\Delta x_E + \Delta x_W)(\Delta y_N + \Delta y_S)/4$, is identical to equation (2.63) obtained by the finite difference method. The FVM is thus equivalent to the FDM, at least in the discrete formulation. Small differences can appear for irregular grids, where the first and second derivatives obtained by the FDM are not of 2nd order in the grid spacing. Otherwise, there is equivalence to this order of approximation; the advantage of FVM being the natural introduction of the boundary conditions, as will be seen. Note finally that dividing (2.70) by the diffusion coefficient and by the square of the characteristic distance of the grid, we obtain a non dimensional equation analogous to (2.64). Although the non dimensional form is preferable from a computational point of view (the coefficients are all of order unity if $Fo_h \approx 1$ and $Pe_h \approx 1$), we will retain the form (2.70) in the following for clarity of recognition of the terms appearing in the solute conservation equation. The equation can be represented in a more condensed notation as:

$$M_{CC}\left[\frac{c_C^{j+1} - c_C^j}{\Delta t}\right] + \sum_K [D_{CK} + V_{CK}] c_K^\theta = 0 \tag{2.71}$$

where the sum over K indicates the four neighbors and the node itself (K = "S", "W", "C", "E", "N"). The elements of the ***mass matrix*** M_{CC}, of the *diffusion* or ***stiffness matrix*** D_{CK} and of the ***transport matrix*** V_{CK} can be deduced directly from (2.70):

$$M_{CC} = \Delta x \Delta y \qquad D_{CE} = -D\frac{\Delta y}{\Delta x_E} \qquad D_{CW} = -D\frac{\Delta y}{\Delta x_W} \qquad D_{CN} = -D\frac{\Delta x}{\Delta y_N}$$

$$D_{CS} = -D\frac{\Delta x}{\Delta y_S} \qquad D_{CC} = -\{D_{CW} + D_{CE} + D_{CN} + D_{CS}\}$$

$$V_{CE} = -V_{CW} = \frac{v\Delta y}{2} \qquad V_{CN} = V_{CS} = V_{CC} = 0 \tag{2.72}$$

As we will see further on, the coefficients D_{CK} or V_{CK} correspond to the matrix elements when all the grid points are considered with the global numbering. Note that the mass matrix is diagonal (i.e., $M_{CK} = M_{CC}\ \delta_{CK}$, where δ_{CK} is the Kroneker delta function for the local node numbering). Having introduced these notations, we can now examine in detail the contributions to the nodes at the boundaries of the domain Ω.

2.3.5 FVM formulation at the boundaries

$\partial\Omega_1$ – Dirichlet condition

As the grid is constructed, there is no node on this boundary since the concentration is imposed (fig. 2.11). However, equation (2.65) for nodes that are direct neighbors of the boundary (here, those for which the neighbor to the West is on the boundary (fig. 2.13a)) will introduce two known terms: $D_{CW}\bar{c}$ and $V_{CW}\bar{c}$, where $\bar{c}$ is the concentration imposed at $\partial\Omega_1$. As for the one dimensional problem, these two terms will be moved to the right hand side of the equation for solving, which gives:

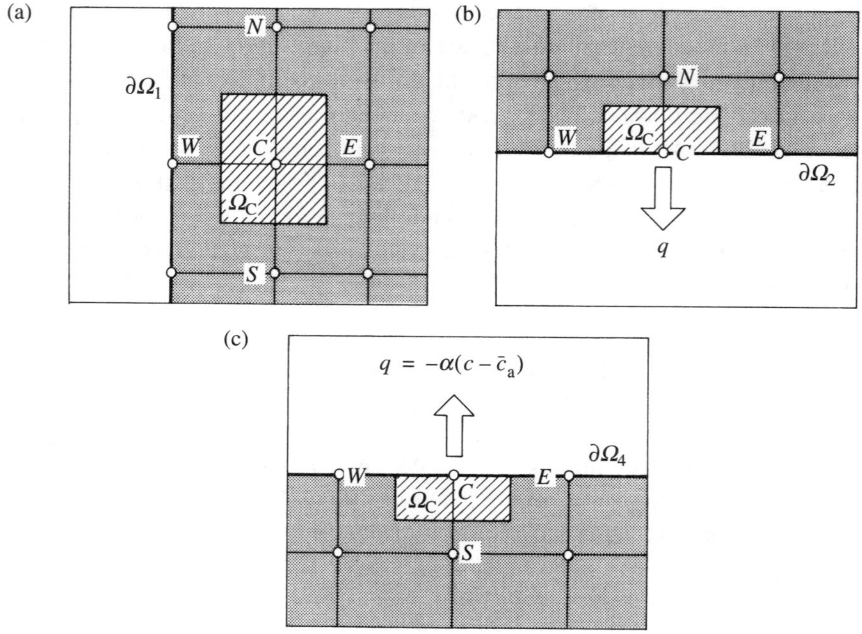

Figure 2.13 Dirichlet boundary conditions on the western boundary (a), Neumann conditions on the southern (b) and Cauchy conditions on the northern (c).

$$M_{CC}\left[\frac{c_C^{j+1} - c_C^j}{\Delta t}\right] + \sum_{K \neq W} [D_{CK} + V_{CK}]c_K^\theta = -[D_{CW} + V_{CW}]\bar{c} = b_C \qquad (2.73)$$

where $b_C = -[D_{CW} + V_{CW}]\bar{c}$, is the right hand side term. The sum is over the neighboring nodes of point C, including itself, but excluding the point to the West for which the concentration is known. The terms M_{CC}, D_{CK} and V_{CK} are always given by (2.72).

It must be noted that the computational procedure that must be implemented at the boundaries where a Dirichlet condition is imposed is more complicated than it seems here. In fact, the distance Δx_W to the direct neighbors of the boundary $\partial\Omega_1$ must be known, thus points on the boundary must be included in the grid, which was explicitly not done here. These nodes must then be removed during the solution as the value of the function is known. For this reason, some programs (finite element or finite difference) keep the nodes where an essential condition is imposed and then employ a ***penalization method*** for solving (chap. 4).

$\partial\Omega_2$ and $\partial\Omega_3$ – Neumann condition (imposed flux)

The boundary $\partial\Omega_3$ (null flux) is a special case of the boundary $\partial\Omega_2$ (imposed flux), thus these two boundaries are treated simultaneously. The situation at node C at the boundary $\partial\Omega_2$ is shown in figure 2.13b: there is no neighbor to the South and the boundary $\partial\Omega_S$ passes through the point. Setting $\Delta y_S = 0$, or, if one prefers, $S \equiv C$, a half grid element volume is defined for the node under consideration. Equations (2.72) previously deduced for a volume node can then be used for the half element volume except for the diffusive flux term at the southern boundary of the grid. There, a division by zero will appear. This term can however be replaced by the boundary condition (2.56b) in equation (2.68) giving for the contribution:

$$\int_{\partial\Omega_S} D\frac{\partial c}{\partial n}\,\mathrm{d}x = -\int_{\partial\Omega_S} D\frac{\partial c}{\partial y}\,\mathrm{d}x = q\Delta x \tag{2.74}$$

It is then sufficient to replace D_{CS} by 0 for all the nodes on this boundary in (2.71) and to move the term (2.74) to the right hand side of the equation. Note that the zeroing of the term D_{CS} has, of course, an effect on the expression for D_{CC} and that the zeroing of Δy_S automatically creates an associated half element at this point.

At the boundary $\partial\Omega_3$, the same procedure can be applied for the diffusive term: in this case, it is the eastern neighbor that is missing ($\Delta x_E = 0$), thus D_{CE} is zero. The flux vanishing at this boundary (homogeneous Neumann condition (2.56c)), the right hand side term is also zero.

As for the transport term, there is no modification to be made at the boundary $\partial\Omega_2$ as the velocity is parallel ($\boldsymbol{v} \cdot \boldsymbol{n} = 0$). Then (2.72) can still be used as Δy_S has been set to zero (half element). At the boundary $\partial\Omega_3$, the velocity is outgoing and could create oscillations if a Dirichlet condition were imposed and the Péclet number of the grid for x were greater than 2 (cf. one dimensional case). The problem can be avoided by the imposition of a flux (here null), but the integral (2.67) is modified in the following way:

$$\int_{\Omega_C} v\frac{\partial c}{\mathrm{d}x}\,\mathrm{d}x\mathrm{d}y = v\left[\int_{\partial\Omega_E} c\,\mathrm{d}y - \int_{\partial\Omega_W} c\,\mathrm{d}y\right] \simeq v\Delta y\left[c_C - \frac{c_W + c_C}{2}\right] = v\Delta y\left[\frac{c_C - c_W}{2}\right] \tag{2.75}$$

We see the concentration c_C appear in the place of the concentration c_E, which in turn modifies the terms V_{CK} written in (2.72). Division by the volume of the half element shows that, in fact, for this boundary to the East of the domain Ω, we have taken a left first derivative at the point and not a centered derivative as in (2.67).

$\partial\Omega_4$ – *mixed condition*

The same reasoning applies for the boundary $\partial\Omega_4$ (fig. 2.13c). There is no northern neighbor ($\Delta y_N = 0$) and the diffusive term on this boundary can be replaced by the boundary condition (2.56d). We then have:

$$\int_{\partial\Omega_N} D\frac{\partial c}{\partial n}\,dx = \int_{\partial\Omega_N} D\frac{\partial c}{\partial y}\,dx = -\alpha(c_C - \bar{c}_a)\Delta x \qquad (2.76)$$

which causes two terms to appear in the global equation for the node. The term $D_{CN}c_N$ is replaced by the known term $-(\alpha\Delta x)\bar{c}_a$ which is moved to the right hand side of the equation, while the term D_{CN} appearing on the diagonal D_{CC} of (2.72) is replaced by $-(\alpha\Delta x)$.

Summary

The temporal variation term in the solute conservation equation is not modified for the nodes on (or neighboring) a boundary of the domain, as long as the distances corresponding to the missing neighbors are set to zero. For a node neighboring a boundary with a Dirichlet condition, the grid elements are complete. For nodes on a boundary where a natural boundary condition is imposed, there can be a half element, or even a quarter element at a corner of the grid.

The diffusive part of the solute conservation is modified as follows by the boundary conditions:

- The essential boundary conditions introduce a term on the right hand side for the neighboring nodes (equation (2.73)).
- The Neumann conditions zero the matrix diffusion term corresponding to the missing neighbor and introduce a term on the right hand side (equation (2.74)).
- The mixed conditions zero the diffusion term corresponding to the missing neighbor by introducing a term on the right hand side and a modified term on the diagonal (equation (2.76)).

The transport term is also affected by the boundary conditions:

- A Dirichlet condition on the boundary where $\boldsymbol{v} \cdot \boldsymbol{n} \neq 0$ introduces a term on the right hand side given by (2.73). As already seen for the one dimensional case, such a condition can cause oscillations near the boundaries where the velocity is outgoing if the corresponding Péclet number for the grid is greater than 2.
- For a boundary where a natural condition is imposed and where $\boldsymbol{v} \cdot \boldsymbol{n} \neq 0$, one is obliged to replace the concentration gradient by the first derivative skewed towards the interior of the domain (to the right, left, above or below according to whether the boundary is to the West, East, South or North of the domain, respectively).

In general, the conservation equation formulated at a point anywhere in the domain can be written:

$$M_{CC}\left[\frac{c_C^{j+1} - c_C^j}{\Delta t}\right] + \sum_K [D_{CK} + V_{CK}]c_K^\theta = b_C \quad K = C, S, E, N, W$$
$$C = 1, 2, \ldots, N \tag{2.77}$$

We can now assemble these contributions to solve the problem.

2.3.6 Assembling the contributions

Equation (2.77) corresponds to solute conservation expressed for any point in the grid $C = 1, N$. The five concentrations defined for the four neighbors and the central node appear. If the scheme is explicit ($\theta =$ "j"), these concentrations are known at time j, and the variation of concentration at the central node $\Delta c_C = c_C^{j+1} - c_C^j$ can be found for each equation. In contrast, for an implicit scheme ($\theta =$ "$j + 1$") or a mixed scheme, the five concentrations c_K^θ appearing in (2.77) are unknown and the solution can be found only with the set of N equations written at each point of the grid. In matrix form, these equations can be written:

$$[\boldsymbol{M}]\left[\frac{\boldsymbol{c}^{j+1} - \boldsymbol{c}^j}{\Delta t}\right] + \{[\boldsymbol{D}] + [\boldsymbol{V}]\}\boldsymbol{c}^\theta = \boldsymbol{b} \tag{2.78}$$

for which the global matrices [M], [D] and [V] have been constructed from the local expressions derived above using the correspondence between the local numbering of the nodes (indices "S", "W", "C", "E", "N") and the global numbering (fig. 2.11). We have the following table for the correspondence of the local and global numbering of the nodes:

$$C \to k \quad S \to k - N_x \quad W \to k - 1 \quad E \to k + 1 \quad N \to k + N_x \tag{2.79}$$

For the matrices, this correspondence becomes:

$$\left.\begin{array}{ll} M_{CC} \to M_{kk} & D_{CC} \to D_{kk} \\ D_{CN} \to D_{k,k+Nx} & D_{CS} \to D_{k,k-Nx} \\ D_{CE} \to D_{k,k+1} & D_{CW} \to D_{k,k-1} \\ V_{CE} \to V_{k,k+1} & V_{CW} \to V_{k,k-1} \end{array}\right\} \tag{2.80}$$

In tabular form we have:

$$M_{ik} = M_{ii}\delta_{ik} = \begin{bmatrix} M_{11} & 0 & 0 & \dots & 0 & 0 \\ 0 & M_{22} & 0 & \dots & 0 & 0 \\ 0 & 0 & M_{33} & \dots & 0 & 0 \\ \dots & \dots & \dots & \dots & \dots & \dots \\ 0 & 0 & 0 & \dots & M_{N-1,N-1} & 0 \\ 0 & 0 & 0 & \dots & 0 & M_{NN} \end{bmatrix} \tag{2.81a}$$

$$D_{ik} = \begin{bmatrix} D_{11} & D_{12} & 0 & \dots & D_{1,Nx+1} & 0 & \dots & 0 \\ D_{21} & D_{22} & D_{23} & \dots & 0 & D_{2,Nx+2} & \dots & 0 \\ 0 & D_{32} & D_{33} & \dots & 0 & 0 & \dots & 0 \\ \dots & \dots & \dots & \dots & \dots & \dots & \dots & \dots \\ D_{Nx+1,1} & 0 & 0 & \dots & D_{Nx+1,Nx+1} & D_{Nx+1,Nx+2} & \dots & 0 \\ 0 & D_{Nx+2,2} & 0 & \dots & D_{Nx+2,Nx+1} & D_{Nx+2,Nx+2} & \dots & 0 \\ \dots & \dots & \dots & \dots & \dots & \dots & \dots & \dots \\ 0 & 0 & 0 & \dots & 0 & 0 & \dots & D_{NN} \end{bmatrix} \tag{2.81b}$$

$$V_{ik} = \begin{bmatrix} V_{11} & V_{12} & 0 & \dots & 0 & 0 \\ V_{21} & V_{22} & V_{23} & \dots & 0 & 0 \\ 0 & V_{32} & V_{33} & \dots & 0 & 0 \\ \dots & \dots & \dots & \dots & \dots & \dots \\ 0 & 0 & 0 & \dots & V_{N-1,N-1} & V_{N-1,N} \\ 0 & 0 & 0 & \dots & V_{N,N-1} & V_{NN} \end{bmatrix} \tag{2.81c}$$

$$\boldsymbol{c} = \begin{pmatrix} c_1 \\ c_2 \\ \dots \\ c_{N-1} \\ c_N \end{pmatrix} \qquad \boldsymbol{b} = \begin{pmatrix} b_1 \\ b_2 \\ \dots \\ b_{N-1} \\ b_N \end{pmatrix} \tag{2.81d}$$

The mass matrix $[\boldsymbol{M}]$, given by the volume of the elements, is thus *diagonal* although the two other matrices, $[\boldsymbol{D}]$ and $[\boldsymbol{V}]$, generally have five non zero diagonals in two dimensions (seven diagonals in three dimensions). In the current case with a horizontal transport velocity, only three diagonals of $[\boldsymbol{V}]$ are non zero: the lower diagonal, corresponding to the coefficients multiplying the concentrations to the West, the central diagonal where the term is non zero for all the boundary points $\partial\Omega_3$ (eqn. (2.75)) and, finally, the upper diagonal that multiplies the concentrations to the East of the point under consideration. The two other non zero diagonals of the matrix $[\boldsymbol{D}]$ correspond to the terms D_{CS} and D_{CN} and are shifted N_x with respect to the principle diagonal for the grid topology shown in figure 2.11. It should be noted that for more complicated forms (fig. 2.14), there are in general always five terms for a given row of these matrices, but no longer necessarily aligned on diagonals.

We see then that these matrices ($N \times N$) have at most $5N$ non zero coefficients. This property, called ***sparseness,*** of the matrices allows them to be compressed for storage in computer memory. The practical storage aspects of matrices and the solving of a system of equations (2.78) will be dealt with in chapter 4. As for the one dimensional case, all the known terms must first be moved to the right hand side and the terms multiplying the unknown concentrations must be regrouped on the left hand side of the equations. For the implicit scheme, one is led to solve the following system of equations:

$$\left\{[\boldsymbol{D}] + [\boldsymbol{V}] + \frac{[\boldsymbol{M}]}{\Delta t}\right\}\boldsymbol{c}^{j+1} = [\boldsymbol{A}]\boldsymbol{c}^{j+1} = \boldsymbol{b} + \frac{[\boldsymbol{M}]}{\Delta t}\boldsymbol{c}^{j} = \boldsymbol{b}' \tag{2.82}$$

or:

$$\boldsymbol{c}^{j+1} = [\boldsymbol{A}]^{-1}\boldsymbol{b}' \tag{2.83}$$

where $[\boldsymbol{A}]^{-1}$ is the inverse of the matrix $[\boldsymbol{A}] = [\boldsymbol{D}] + [\boldsymbol{V}] + [\boldsymbol{M}]/\Delta t$. Note that this matrix has a band and its ***bandwidth*** is determined by the matrix $[\boldsymbol{D}]$: the bandwidth, bw, of a matrix is defined such that $A_{ij} = 0$ if $|i - j| \geq bw$ (chap. 4). In this case, the non zero coefficients of $[\boldsymbol{A}]$ the furthest from the principle diagonal are on the diagonal $(N_x + 1)$, thus $bw = N_x + 1$.

2.4 SOME OTHER ASPECTS OF FDM

2.4.1 Grids

As shown in figure 2.11, the grids used in the FDM must follow a coordinate system (here, Cartesian coordinates) to approximate a geometric form. This is a handicap of the method as a good approximation of a complicated geometry requires many nodes. Curved or skewed boundaries must be approximated by stair

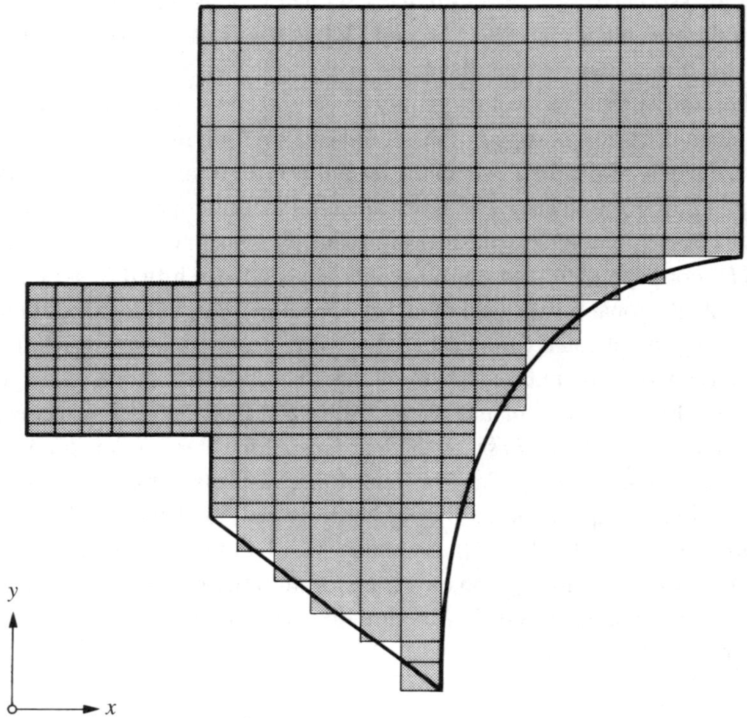

Figure 2.14 FDM grid and domain for a complex form.

steps (fig. 2.14) which can pose a problem for the application of the boundary conditions.

Thus, if a flux is imposed on a boundary at 45° to the grid lines, the stair steps created by the grid at the boundary multiply it's effective length (thus the flux integral) by a factor $\sqrt{2}$ if a correcting factor is not introduced! Similarly, Dirichlet conditions can only be applied at a distance accurate to $\sqrt{2}\,h$, which can cause oscillations in the concentration field near this boundary. Worse, a non slip condition imposed on such a stair-stepped boundary introduces ***numerical viscosity*** for a fluid flow problem: the continued changes of direction of the velocity near the steps has the global effect of making the fluid appear more viscous. The association of reference numbers with the nodes on the boundaries, rather than on the edges, poses a problem at the corners between two boundaries at which natural conditions are imposed. For example, node number N in figure 2.11 at the upper right corner of the grid will have an outgoing flux on the northern edge if it is considered as part of $\partial\Omega_4$. However, if it belongs to $\partial\Omega_3$ on which a homogeneous Neumann condition is applied, there is no flux out of the volume element attached to this node! Finally, it must be noted that the refining of a grid in a zone where there is a strong gradient,

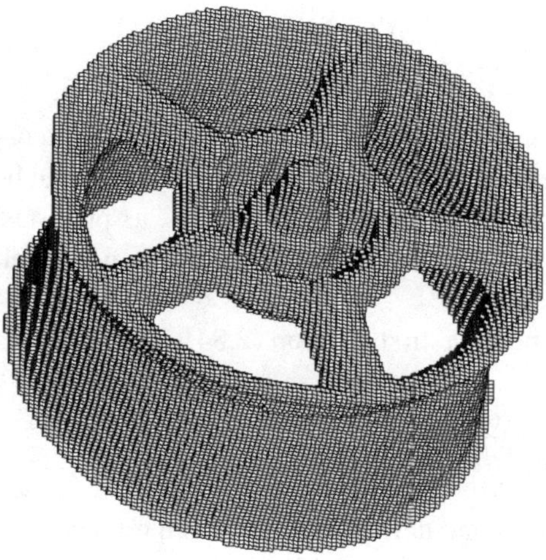

Figure 2.15 FDM grid for a three dimensional part.

that is, where the concentration varies rapidly, is propagated throughout the volume as a structured grid follows the coordinate system. These inconveniences are largely avoided with the finite element method presented in the next chapter.

Nonetheless, grids for the FDM have the advantage of being easy to construct. One starts by defining a rectilinear grid, not necessarily regular, for a rectangle (a rectangular parallelepiped in three dimensions) that completely covers the domain. As before, the coordinates of the rows and columns are known. Subsequently scanning all the points of the grid, only those nodes that fall within the domain are retained and numbered, and the numbers of their neighbors are stored, along with their medium if there are multiple media in the domain. These operations can be performed with a CAD (***computer aided design***) file describing the geometry of the domain in terms of surfaces separating the various domains or phases. For example, figure 2.15 shows a three dimensional grid for a finite difference calculation on a complex form.

2.4.2 Other coordinate systems

For a number of problems encountered in materials science, there exists a symmetry that permits the reduction of the dimensionality of the problem. For example, as a first approximation, the calculation of diffusion of carbon across an austenite layer around a graphite nodule in ductile cast iron can be approximated by a concentration field $c(r, t)$, where r is the radius from the nodule center (fig. 2.16). By spherical symmetry, the problem is independent of the angles θ and ϕ. The non stationary diffusion equation without transport (from 1.12) can be written for this coordinate system (chap. 10):

$$-D\Delta c + \frac{\partial c}{\partial t} = -D\frac{1}{r^2}\frac{\partial}{\partial r}\left[r^2\frac{\partial c}{\partial r}\right] + \frac{\partial c}{\partial t} = 0 \tag{2.84}$$

To discretize such an equation (we will only do so for a node inside the volume and not deal with the boundary conditions), two methods can be employed. The first is to simply replace the derivatives by differences, as previously; the second, more oriented by the physics, is to integrate the equation on a spherical shell around the node under consideration.

For the first method, we first develop (2.84):

$$-D\left[\frac{\partial^2 c}{\partial r^2} + \frac{2}{r}\frac{\partial c}{\partial r}\right] + \frac{\partial c}{\partial t} = 0 \tag{2.85}$$

in which the first and second derivatives have been replaced by the differences as in the Cartesian case. Using equation (2.13) for the position of the nodes of the regular grid in figure 2.16, we obtain for a volume node:

$$-D\left[\frac{c_{k+1}^{\theta} - 2c_k^{\theta} + c_{k-1}^{\theta}}{\Delta r^2} + \frac{2}{r_k}\frac{c_{k+1}^{\theta} - c_{k-1}^{\theta}}{2\Delta r}\right] + \frac{c_k^{j+1} - c_k^{j}}{\Delta t} = 0 \tag{2.86}$$

where Δr is the node spacing (presumed uniform) and r_k the position of the kth node. (The index "θ" is equal to either "j" or "$j + 1$" depending again on whether the scheme is explicit or implicit).

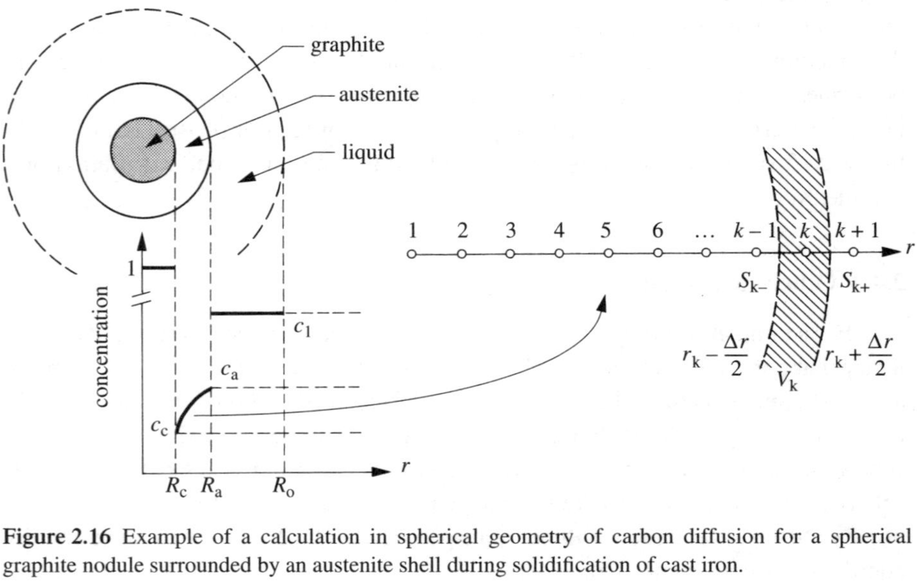

Figure 2.16 Example of a calculation in spherical geometry of carbon diffusion for a spherical graphite nodule surrounded by an austenite shell during solidification of cast iron.

With the second method, (2.84) is integrated on a spherical shell for which the inner and outer radii are $(r_k - \Delta r/2)$ and $(r_k + \Delta r/2)$, respectively. Given that the volume of a shell of thickness dr is $4\pi r^2 dr$, we have:

$$\int_{r_k - (\Delta r/2)}^{r_k + (\Delta r/2)} \left\{-\frac{D}{r^2}\frac{\partial}{\partial r}\left[r^2\frac{\partial c}{\partial r}\right] + \frac{\partial c}{\partial t}\right\} 4\pi r^2 dr = 0 \tag{2.87}$$

The integral of the diffusive term is trivial as the r^2 terms can be simplified, making the solute flux at the boundaries of the shell appear. We have thus:

$$-D\frac{c_{k+1}^{\theta} - c_k^{\theta}}{\Delta r}S_{k+} + D\frac{c_k^{\theta} - c_{k-1}^{\theta}}{\Delta r}S_{k-} + \frac{c_k^{j+1} - c_k^{j}}{\Delta t}V_k = 0 \tag{2.88}$$

To simplify the expression, the internal and external surfaces, S_{k-} and S_{k+}, of the shell and its volume, V_k, have been introduced. These are given by:

$$S_{k+} = 4\pi\left[r_k + \frac{\Delta r}{2}\right]^2 \qquad S_{k-} = 4\pi\left[r_k - \frac{\Delta r}{2}\right]^2$$

$$V_k = \frac{4\pi}{3}\left\{\left[r_k + \frac{\Delta r}{2}\right]^3 - \left[r_k - \frac{\Delta r}{2}\right]^3\right\} \tag{2.89}$$

One can show that to first-order, that is, when the approximations $S_{k\pm} \simeq 4\pi(r_k^2 \pm r_k\Delta r)$ and $V_k \simeq 4\pi r_k^2\Delta r$ are used, the two expressions (2.86) and (2.88) are equivalent (exercise 2.6.5).

Another case often encountered in modeling is a domain with an axis of rotational symmetry: such a problem is called ***axisymmetric*** (fig. 2.17). Note that for the problem to be axisymmetric, the axisymmetry of the geometry is insufficient! It is also necessary that the boundary conditions, the initial conditions and the material properties have rotational symmetry. In such a situation, the problem becomes independent of the angle θ and the fields are functions only of the radius r and height z. The equation of non stationary diffusion (1.73) takes the following form:

$$-D\left\{\frac{1}{r}\frac{\partial}{\partial r}\left[r\frac{\partial c}{\partial r}\right] + \frac{\partial^2 c}{\partial z^2}\right\} + \frac{\partial c}{\partial t} = 0 \tag{2.90}$$

Here also, one can adopt either of the two methods mentioned above, direct replacement of the derivatives by differences or integration on the characteristic volume element of the grid: a torus with rectangular section (fig. 2.17). Employing $2\pi r dr dz$ for the characteristic volume of an axisymmetric element, the integration of (2.90) yields (exercise 2.6.6):

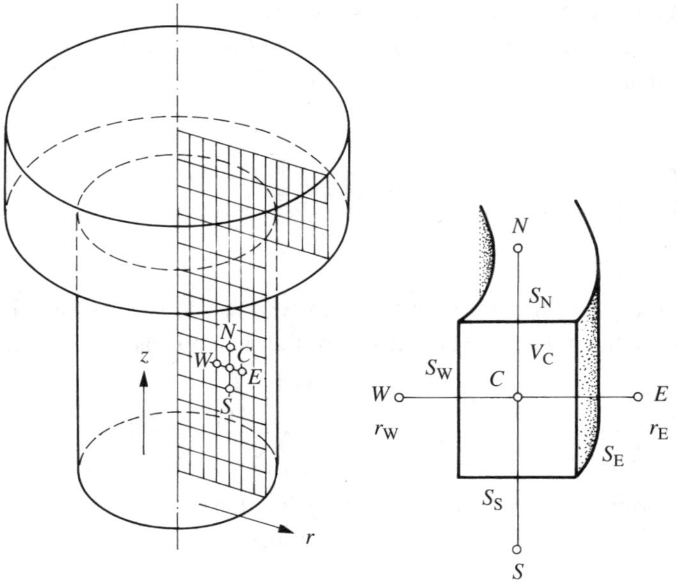

Figure 2.17 Axisymmetric geometry.

$$-D\left[\frac{c_E^\theta - c_C^\theta}{\Delta r_E} S_E + \frac{c_W^\theta - c_C^\theta}{\Delta r_W} S_W + \frac{c_N^\theta - c_C^\theta}{\Delta z_N} S_N + \frac{c_S^\theta - c_C^\theta}{\Delta z_S} S_S\right] + \frac{c_C^{j+1} - c_C^j}{\Delta t} V_C = 0 \tag{2.91}$$

where the surfaces and volume of the element C are defined as follows (fig. 2.17):

$$S_E = 2\pi\left[r_C + \frac{\Delta r_E}{2}\right]\frac{\Delta z_S + \Delta z_N}{2} \qquad S_W = 2\pi\left[r_C - \frac{\Delta r_W}{2}\right]\frac{\Delta z_S + \Delta z_N}{2}$$

$$S_N = S_S = \pi\left\{\left[r_C + \frac{\Delta r_E}{2}\right]^2 - \left[r_C - \frac{\Delta r_W}{2}\right]^2\right\} \qquad V_C = S_N\frac{\Delta z_S + \Delta z_N}{2} \tag{2.92}$$

2.4.3 Direct finite difference method

A method encountered many times in this chapter is to integrate the conservation equations over the volume of a grid element before replacing the derivatives that still appear in the formulation with differences (finite volume method). Going further with this approach, Ohnaka developed the "direct finite difference method" (DFDM), which is equivalent to the finite volume method but with an unstructured grid. Without going into the details, this method uses – as does the finite element

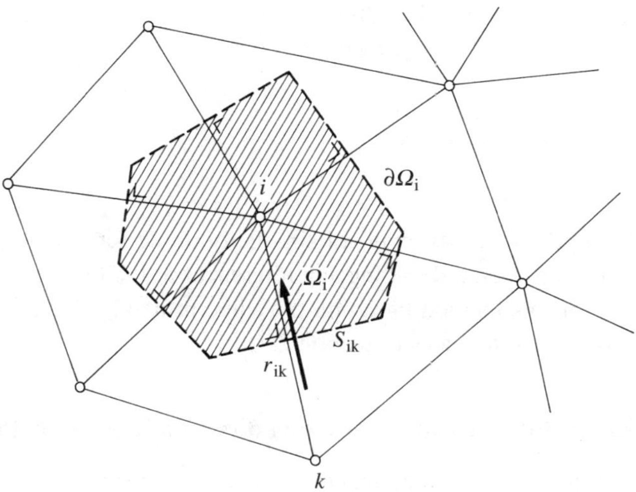

Figure 2.18 Example of a grid used for the direct finite difference method.

method – grids that are much more flexible than the structured grids used in the FDM, as they are not required to follow the coordinate system. First, a triangular mesh is defined as shown in figure 2.18. After having numbered all the nodes, the nearest neighbors are found for each. A node i having $N(i)$ neighbors, $N(i)$ being variable, one can again change from a global numbering to a local scheme $k = 1, N(i)$. Let r_{ik} be the distances to the neighbors. The set of perpendicular bisectors of the segments $(i - k)$ between the nodes defines the element Ω_i associated with node i, whose volume is marked V_i and surfaces S_{ik}. Note that the triangles in the mesh should not have angles greater than 90° to avoid having negative contributions for the surfaces S_{ik}.

One then integrates the conservation equation on the element Ω_i. Using the divergence theorem, the diffusion equation under consideration (1.73) becomes:

$$\int_{\Omega_i} \operatorname{div}(D\,\mathbf{grad}\,c)\,\mathrm{d}V = \int_{\partial\Omega_i} D\,\mathbf{grad}\,c\cdot\boldsymbol{n}\,\mathrm{d}S = \int_{\partial\Omega_i} D\frac{\partial c}{\partial n}\,\mathrm{d}S = \int_{\Omega_i}\frac{\partial c}{\partial t}\,\mathrm{d}V \tag{2.93}$$

where $\mathbf{grad}\,c \cdot \boldsymbol{n} = \partial c/\partial n$ is the chemical gradient normal to the surface of domain $\partial\Omega_i$. This surface integral becomes a sum over the facets S_{ik} where the normal gradient on each is approximated by:

$$\left[\frac{\partial c}{\partial n}\right]_{ik} \simeq \frac{c_k^{\theta} - c_i^{\theta}}{r_{ik}} \tag{2.94}$$

where the index "θ" indicates, as always, the time at which the variables are evaluated ("θ" = "j" or "$j+1$" for the explicit and implicit scheme, respectively). The equation expressed with differences then becomes:

$$\sum_{k=1}^{N(i)} DS_{ik}\frac{c_k^{\theta}-c_i^{\theta}}{r_{ik}} = \frac{c_i^{j+1}-c_i^{j}}{\Delta t}V_i \qquad (2.95)$$

The method offers the advantage of simplicity in the formulation but requires the same tools for unstructured griding and references as the finite element method. Other variations of this method based on the same principles have also been proposed by Ohnaka (see references in section 2.7).

2.4.4 Temporal solution by the alternating direction implicit method

It was seen in section 2.3 that the non stationary diffusion equation with transport led to pentadiagonal matrices for two dimensional problems, and to heptadiagonal matrices for three dimensional problems. An implicit (or mixed) scheme for time requires the solution of a set of equations. As will be seen in chapter 4, the solution can be obtained by a direct method (Gaussian elimination) or an iterative method. In either case, the operation is rather costly in computation time. The other alternative is to use an explicit scheme, thus without inversion of other matrices than the diagonal mass matrix, but, in this case, the time step is constrained by a stability criterion (Fourier criterion).

The ***alternating direction implicit method*** (ADI) involves, as its name indicates, the solution of (2.78) by considering all the field variables along the direction x as known (explicit), with the values along y being implicit, and alternating the roles of these two directions at each time step. The advantage of the method is that the unknown contributions are multiplied by tridiagonal matrices for which Gaussian elimination is trivial and rapid (TDMA algorithm, chap. 4).

To illustrate the method, we retain a double indexing (i, k) of the nodes according to the directions x and y and consider the diffusion equation without transport, the index j being always reserved for the time step. With this indexing, the node (i, k) is surrounded by the nodes $(i \pm 1, k)$ to the right and left (E and W nodes in the previous development) and the nodes $(i, k \pm 1)$ above and below (N and S previously). The equation takes the form:

$$\begin{aligned} M_{i,k}^{i,k}\frac{c_{i,k}^{j+1}-c_{i,k}^{j}}{\Delta t} &+ D_{i+1,k}^{i,k}[c_{i+1,k}^{\theta}-c_{i,k}^{\theta}] + D_{i-1,k}^{i,k}[c_{i-1,k}^{\theta}-c_{i,k}^{\theta}] \\ &+ D_{i,k+1}^{i,k}[c_{i,k+1}^{\theta}-c_{i,k}^{\theta}] + D_{i,k-1}^{i,k}[c_{i,k-1}^{\theta}-c_{i,k}^{\theta}] = 0 \end{aligned} \qquad (2.96)$$

The matrix elements having four indices have been written $D_{i,k+1}^{i,k}$ in order to compact the expression: such an element connects the central node (i, k) and its neighbor $(i, k+1)$. It is thus identical to the term D_{CN} seen in chapter 2.3. The ADI method consists then of separating the time step Δt into two half steps, naming $c^*_{i,k}$ the values at the half step. During the first half step, the value of the index "θ" is set to "j" for the x direction (known values) and to "*" for the other direction. One then has:

$$\begin{aligned} M^{i,k}_{i,k}\frac{c^*_{i,k}-c^j_{i,k}}{\Delta t/2} &+ D^{i,k}_{i+1,k}[c^j_{i+1,k}-c^j_{i,k}] + D^{i,k}_{i-1,k}[c^j_{i-1,k}-c^j_{i,k}] \\ &+ D^{i,k}_{i,k+1}[c^*_{i,k+1}-c^*_{i,k}] + D^{i,k}_{i,k-1}[c^*_{i,k-1}-c^*_{i,k}] = 0 \end{aligned} \tag{2.97}$$

Moving to the left hand side all the known terms, that is, those that include the concentrations at time j, results in a tridiagonal system to solve for the unknown concentrations, $c^*_{i,k}$. Once the latter have been found, the roles of x and y are reversed for the next half step giving:

$$\begin{aligned} M^{i,k}_{i,k}\frac{c^{j+1}_{i,k}-c^*_{i,k}}{\Delta t/2} &+ D^{i,k}_{i+1,k}[c^{j+1}_{i+1,k}-c^{j+1}_{i,k}] + D^{i,k}_{i-1,k}[c^{j+1}_{i-1,k}-c^{j+1}_{i,k}] \\ &+ D^{i,k}_{i,k+1}[c^*_{i,k+1}-c^*_{i,k}] + D^{i,k}_{i,k-1}[c^*_{i,k-1}-c^*_{i,k}] = 0 \end{aligned} \tag{2.98}$$

which again leads to a tridiagonal matrix to solve for the concentrations $c^{j+1}_{i,k}$.

Using a method analogous to that presented for one dimensional problems, it can be shown that this method is *unconditionally stable*. It can also be extended to three dimensions.

2.5 EXAMPLE

Figure 2.19 shows the result of a thermal calculation obtained with the FDM for the melting of a material with a laser. In such a process, a high power laser beam (typically more than a kW with a CO_2 laser) is focused on the material surface (a spot size of the order mm^2). For the case treated here, the laser beam is characterized by an axisymmetric energy distribution, for which the profile, $I(r)$, is shown above the surface. Because of the symmetry of the problem about a vertical plane, the problem can be reduced to a half plane containing half the scanning laser beam. The boundary condition imposed on the symmetry plane is adiabatic. At the upper surface of the material, on which the beam is incident, a mixed boundary condition is imposed:

$$\kappa(T)\frac{\partial T}{\partial n} = \alpha_r(T)I(r) - h(T)(T-\bar{T}_a) \tag{2.99}$$

where $\alpha_r(T)$ is the absorption coefficient of the material, $\kappa(T)$ its thermal conductivity (assumed to be a function of temperature T), $h(T)$ a heat transfer coefficient with the exterior temperature, $\bar{T}_a$.

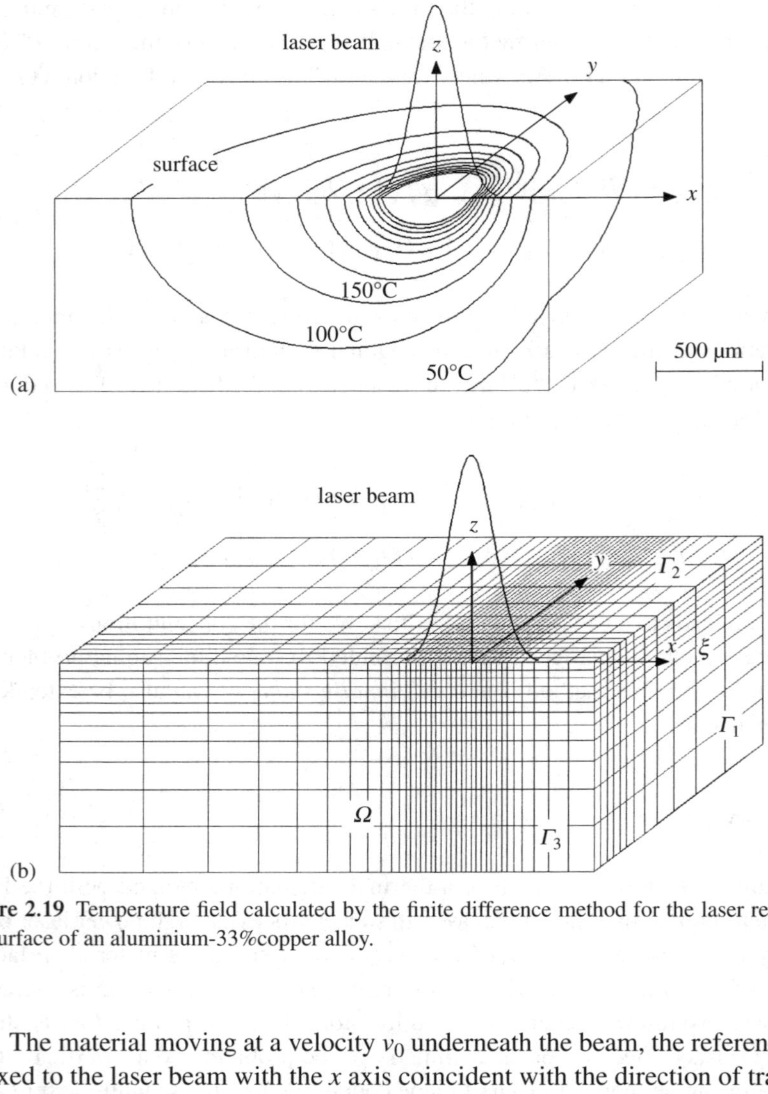

Figure 2.19 Temperature field calculated by the finite difference method for the laser remelting of the surface of an aluminium-33%copper alloy.

The material moving at a velocity v_0 underneath the beam, the reference frame is fixed to the laser beam with the x axis coincident with the direction of translation of the material. Thus the heat equation is not solved in the whole of the material, rather in a parallelepiped of volume V arbitrarily cut out of the material. Dirichlet boundary conditions are applied on the lateral and lower surfaces of this volume: the temperature is fixed at a value given by an analytic solution valid for a point heat source. These conditions are not realistic unless the faces are relatively far from the laser spot heat source.

Under these conditions, the stationary thermal equation including phase transformations is given by (1.71):

$$\mathrm{div}[\kappa(T)\,\mathbf{grad}\,T(\boldsymbol{r},t)] + v_0\frac{\partial(\rho H(\boldsymbol{r},t))}{\partial x} = 0 \tag{2.100}$$

where (ρH) is the enthalpy per unit volume. Aside from the nonlinearities introduced by $\kappa(T)$ and the latent heat liberated at the moment of the phase transformation (chap. 5), which are treated in chapter 4, (2.100) was formulated with the FDM, in the same way as seen with solute diffusion. Given the high velocity that can be used for the movement of a part under the laser beam (up to 1 m/s), an upwind difference scheme has been applied to the transport terms. Also, the grid has been refined in the neighborhood of the laser spot as the temperature gradients are high in that region. As mentioned earlier, the FDM has the disadvantage of propagating this refinement out to regions where it is not necessary.

The three dimensional temperature distribution plotted in figure 2.19 shows clearly the molten zone under the beam spot (the first isotherm marked under the peak of the energy distribution). The strong thermal gradients ahead of the beam appear as a bunching of the isotherms. Other than their value for visualizing the form of the liquid pool and the two and three dimensional isotherms, such calculations are indispensable for thermomechanical calculations of residual stresses, which are often significant in such processes. They are also much utilized for the estimation of the parameters of the microstructure during the solidification downstream of the beam.

2.6 EXERCISES

2.6.1. Show that the centered third and fourth derivatives of a variable at point k of a regular grid can be expressed as:

$$c_{\mathrm{k}}''' = \frac{c_{\mathrm{k}+2} - 2c_{\mathrm{k}+1} + 2c_{\mathrm{k}-1} - c_{\mathrm{k}-2}}{2h^3} + O(h^2)$$

$$c_{\mathrm{k}}^{\mathrm{iv}} = \frac{c_{\mathrm{k}+2} - 4c_{\mathrm{k}+1} + 6c_{\mathrm{k}} - 4c_{\mathrm{k}-1} + c_{\mathrm{k}-2}}{h^4} + O(h^2)$$

2.6.2. Show that with the upwind skew factor (2.26):

$$\theta = \tfrac{1}{2} + \tfrac{1}{2}\coth\left[\frac{Pe_{\mathrm{h}}}{2}\right] - \frac{1}{Pe_{\mathrm{h}}}$$

the analytic solution (2.8) is obtained at the grid points x_{k}.

2.6.3. Using the method of separation of variables, deduce (2.30). To do that, first write the concentration field $c(x, t)$ in the form of a product $(\Xi(x)\ T(t))$ and insert that relation in the diffusion equation. Taking into account the boundary conditions, obtain the particular solutions: $c(x, t) - \bar{c}_{\mathrm{L}} = A \sin(\lambda x) \exp(-D\lambda^2 t)$, where the value of λ is discretized: $\lambda = (n\,\pi)/L$. Superpose the solutions obtained, and apply the initial condition $c(x, 0) = \bar{c}_0$ to obtain the desired relation.

2.6.4. Consider a one dimensional problem with a boundary condition at $x = 0$ given by the Cauchy condition (2.50). Redo the analysis as for a Dirichlet boundary condition (fig. 2.8), to obtain the stability condition of the explicit scheme (2.54) with such a condition.

2.6.5. Using the approximate relations for the surfaces $S_{k\pm}$ and volume V_k of a grid in spherical coordinates, show that (2.86) and (2.88) are equivalent. How do they differ in the neighborhood of $r = 0$?

2.6.6. Derive (2.91) from (2.90) for an axisymmetric geometry by integrating the diffusion equation on a mesh. Compare the relation so obtained with that which would be deduced by replacing the derivatives by differences directly in (2.90).

2.7 BIBLIOGRAPHY

B. CARNAHAN, H. A. LUTHER and J. O. WILKES, *Applied Numerical Methods,* J. Wiley & Sons, New York, 1969.

D. R. CROFT and D. G. LILLEY, *Heat Transfer Calculations using Finite Difference Equations,* Applied Sc. Publ., London, 1977.

C. A. J. FLETCHER, *Computational Techniques for Fluid Dynamics,* Springer Verlag, Springer Series in Computational Physics, Vols 1 and 2, 1988.

I. OHNAKA and T. FUKUSAKO, *Finite Element Method and a Matrix Method in Transient Heat-Conduction Problems*, in VIth Int. Heat Transfer Conference, Toronto, 1978, Vol. 3, p. 251.

S. V. PATANKAR, *Numerical Heat Transfer and Fluid Flow,* Hemisphere Publ., New York, 1980.

CHAPTER 3

THE FINITE ELEMENT METHOD

3.1 OBJECTIVES

- Show the basic principles of the discretization of space using finite elements.
- Establish the means to obtain the integral forms of the conservation equations and to discretize them.
- Develop some aspects of the treatment of non stationary problems; non linear problems are dealt with in chapter 4.

As we have seen in chapter 1, the physical analysis of a phenomenon leads to modeling with local partial differential equations. When it is possible to employ global energy balances, these equations can be integrated over the domain in consideration, making volume contributions and boundary conditions appear.

In chapter 2, we saw that the finite difference method (FDM) was well adapted to the solution of local equations. Integrated over small volume elements, these equations can also be solved by the finite volume method (FVM). As for the finite element method (FEM), it is a powerful method to solve the equations in integral form. Two possibilities exist for application of the FEM. In the first case, there exists an integral form of the physical problem to solve: this can be a functional that results from a variational principle, the minimum of which corresponds to the desired solution, as in elasticity or visco-plasticity, or more generally, an integral equation to solve. In the second case, an integral formulation must be obtained from an initial system of partial differential equations by a weak formulation, also called the ***weighted residual method.***

With respect to the FDM or the FVM, the FEM is distinguished by its more general character, which makes it more capable of dealing with complex geometries, allows it to use non structured grids and allows more natural imposition of the boundary conditions.

Although the formalism of the FEM can appear heavier than that introduced in the previous chapter for the FDM, the principle is simple; the objective of this chapter is to show the fundamentals of the method. In the first section we introduce the basic tools for the formulation of a finite element problem: discretization of the domain, function interpolation, integration and derivation. This formalism will then be applied to a stationary scalar diffusion problem, and then to a vector problem of mechanical equilibrium. The implementation will be discussed next, excepting the general aspects in common with the FDM which are treated in chapter 4. Finally, the method will be extended to non stationary problems.

Note that it is not our intention, in this introduction to the method, to enter into the mathematical formalism, even though it is indispensable to obtain the convergence properties of the numerical solution and the associated error estimates. The interested reader is referred to the bibliography.

3.2 GENERAL PRINCIPLES: GEOMETRIC DISCRETIZATION AND INTEGRATION

3.2.1 The discretization of space into finite elements

The principle of the FEM consists first of decomposing, or tiling, the spatial domain under consideration into a set of ***elements*** of arbitrary shape and size. This decomposition is commonly called a ***grid*** or a ***mesh***. The only restriction is that elements cannot overlap nor leave any zone of the domain uncovered. A schematic example of a two dimensional mesh is shown in figure 3.1 while figure 3.2 shows a three dimensional mesh used to solve a real problem.

As can be seen in figure 3.1, the boundary of the domain is approximated by a polygonal contour. Nonetheless, contrary to the FDM, for which the grid follows a coordinate system, the definition of a mesh for the FEM is much more free. That allows the form of the domain to be approximated with very few elements. As an aside, note that elements with non linear edges (for example, second order curves) can be used.

For each element, a certain number of points must be defined which can be situated on the edges of the element or inside. These points, called the ***nodes***, will be used to construct the approximations of the functions under consideration over the whole domain by *interpolation* (next section). As in the previous chapter, we will be

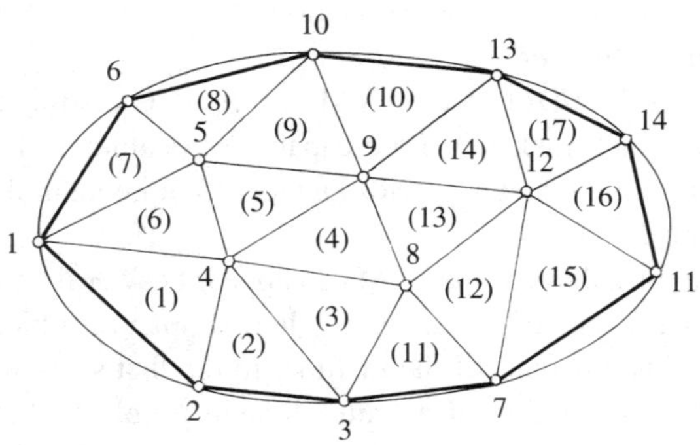

Figure 3.1 Subdivision of a geometric domain into finite elements.

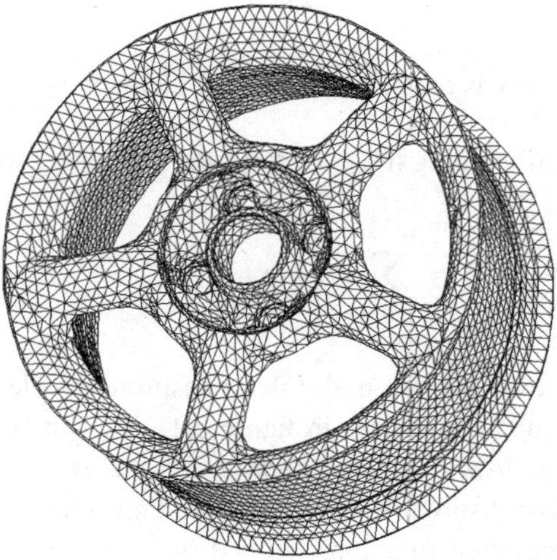

Figure 3.2 Finite element mesh of an automobile wheel rim.

led to determine the value of each unknown function or field in the problem at these points. The number of unknowns at each node is the number of nodal ***degrees of freedom***.

In the case of figure 3.1, we have assumed for simplicity that the nodes are the vertices of the triangle elements. There are then 17 elements and 14 nodes. In the following, the node index will be noted with "*n*" or "*i*", "*j*", "*k*", whereas the elements will be preferentially noted with "*e*".

3.2.2 The approximation of functions by interpolation in the finite elements

The approximation of a function $u(\boldsymbol{x})$ ($\boldsymbol{x}$ is a vector of spatial coordinates) on a geometric domain meshed with finite elements is obtained as a linear combination of ***interpolation functions***, $\psi_{\mathrm{n}}(\boldsymbol{x})$, associated with the mesh, thus known a priori. If $u_{\mathrm{h}}(\mathbf{x})$ is the approximation of the function $u(\boldsymbol{x})$ under consideration, it can be expressed in the form of a sum over the Nn nodes of the domain:

$$u_{\mathrm{h}}(\boldsymbol{x}) = u^{\mathrm{n}}\psi_{\mathrm{n}}(\boldsymbol{x}) \tag{3.1}$$

in which the convention is employed whereby repeated indices are summed (as throughout this chapter, except where noted). The functions $\psi_{\mathrm{n}}(\boldsymbol{x})$, used to approximate the different unknown functions in the problem, can be of diverse forms, with different degrees of continuity and differentiability. However, in the standard finite element method that we present here, the interpolation functions are defined locally at the level of each element. Thus if the node n belongs to element e, and if we use $\psi_{\mathrm{n}}^{\mathrm{e}}$ to denote the restriction of ψ_{n} within the element, we have for every coordinate vector $\boldsymbol{x}$,

$$\psi_{\mathrm{n}}^{\mathrm{e}}(\boldsymbol{x}) = 0 \qquad \text{if } \boldsymbol{x} \text{ is outside (strictly) the element } e, \tag{3.2}$$

and thus for $\boldsymbol{x}$ inside (including the boundary) the element e, we have:

$$u_{\mathrm{h}}(\boldsymbol{x}) = \sum_{\mathrm{n}=1}^{\mathrm{Nn}} u^{\mathrm{n}} \psi_{\mathrm{n}}(\boldsymbol{x}) = \sum_{\mathrm{n} \in \mathrm{e}} u^{\mathrm{n}} \psi_{\mathrm{n}}^{\mathrm{e}}(\boldsymbol{x}) \tag{3.3}$$

The last sum is over the *nne* nodes that constitute the element e (for example *nne* = 3 for the case of the triangles in figure 3.1). The interpolation used for the approximation is thus *locally defined at the level of each finite element.* This elementary decomposition distinguishes the standard finite element method from other methods using interpolation functions defined over the whole of the domain; for example, spectral methods.

Also, in the standard finite element method, the coefficients u^{n} are the values of the function u_{h} at the nodes of the mesh. In consequence, the interpolation functions must respect the following two conditions, in addition to (3.2):

- If n and p are two nodes of the same element e, $\boldsymbol{x}^{\mathrm{p}}$ being the position vector of the latter node,

$$\psi_{\mathrm{n}}^{\mathrm{e}}(\boldsymbol{x}^{\mathrm{p}}) = \delta_{\mathrm{np}} \tag{3.4}$$

  where δ_{np} is the Kronecker delta function.
- A second condition comes from the necessity to exactly represent constant functions:

$$\sum_{\mathrm{n} \in \mathrm{e}} \psi_{\mathrm{n}}^{\mathrm{e}}(\boldsymbol{x}) = 1 \qquad \text{for all } \boldsymbol{x} \text{ inside the element } e \text{ (border included).} \tag{3.5}$$

3.2.3 One dimensional finite elements

We have the conditions for the partition of the domain into finite elements and the fundamental properties of the interpolation functions. To clarify these concepts, we will study in detail a one dimensional example of finite elements. This will permit the introduction of the fundamental concept of *reference elements*, or ***parent elements***. In one dimension, a finite element is necessarily reduced to a straight line segment. An example is shown in figure 3.3, where the domain $[0, L]$ is partitioned into *Ne* elements.

Linear elements

The simplest finite elements have two nodes, at the extremities of each element, and the interpolation functions are linear on these elements. Referring to

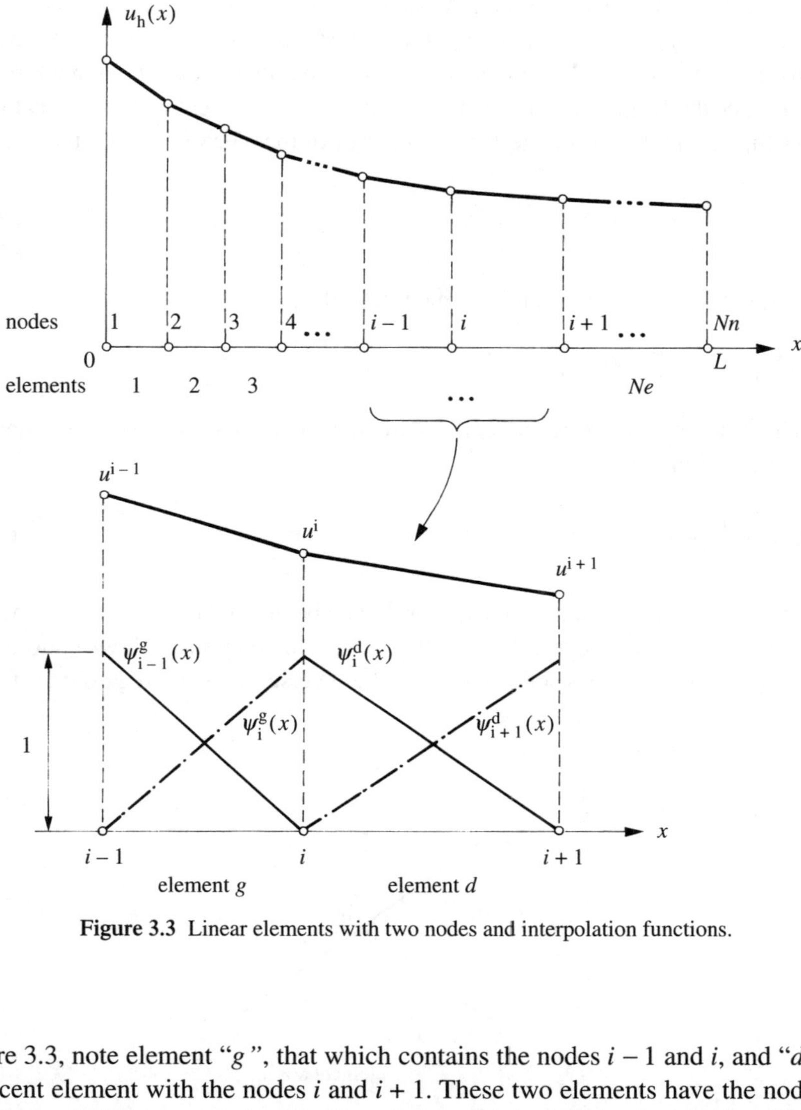

Figure 3.3 Linear elements with two nodes and interpolation functions.

figure 3.3, note element "g", that which contains the nodes $i-1$ and i, and "d" the adjacent element with the nodes i and $i+1$. These two elements have the node i in common, and their respective lengths are Δx_g and Δx_d. Taking into account the linearity of the interpolation functions, and the conditions (3.2, 3.4 and 3.5) that they must satisfy, their expression is easily found. With the notation that x^{i-1}, x^i, x^{i+1} are the coordinates at the nodes i - 1, i and $i+1$:

$$\psi^g_{i-1}(x) = \frac{x^i - x}{\Delta x_g} \text{ and } \psi^g_i(x) = \frac{x - x^{i-1}}{\Delta x_g} \text{ in the element } g: [x^{i-1}, x^i] \quad (3.6)$$

$$\psi^d_i(x) = \frac{x^{i+1} - x}{\Delta x_d} \text{ and } \psi^d_{i+1}(x) = \frac{x - x^i}{\Delta x_d} \text{ in the element } d: [x^i, x^{i+1}] \quad (3.7)$$

The global interpolation function, ψ_i, associated with node i and defined on the entire grid, is obtained by assembling ψ_i^g and ψ_i^d, as illustrated in figure 3.3. It is zero for $x \geq x^{i+1}$ or $x \leq x^{i-1}$, that is, in every element that does not contain the node i. Because of the form it thus takes, the function ψ_i is often called a 'hat function'.

Defining a local coordinate ξ in the interior of the element g, with the relation

$$\xi = \frac{x - x^{i-1}}{\Delta x_g} \tag{3.8}$$

the interpolation functions on this element take the form

$$\psi_{i-1}^g = 1 - \xi \quad \text{et} \quad \psi_i^g = \xi \tag{3.9}$$

This can of course be generalized to the element d and to every element $[x^i, x^{i+1}]$, by defining

$$\xi = \frac{x - x^i}{x^{i+1} - x^i} \tag{3.10}$$

Thus, for every element e, a mapping E_e can be defined between the element in physical space and the segment [0, 1] (fig. 3.4). This mapping allows us to define the interpolation functions universally for the diverse elements regardless of their

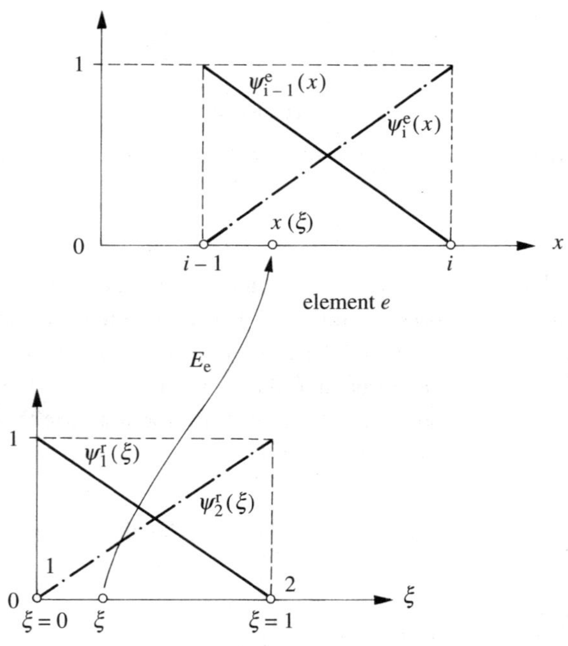

Figure 3.4 The notion of the reference element.

coordinates. The interpolation functions depend only on the nature of the element and the number of points that they contain. This is the fundamental notion of the ***reference element***.

From this reference element composed of nodes 1, at coordinate $\xi = 0$, and 2, at coordinate $\xi = 1$, the interpolation functions in physical space are easily reconstructed; denoting ν_k^e the global numbering of node k (local) in the element e ($k = 1, 2$), we have for the preceding example:

$$\nu_1^g = i-1 \qquad \text{and} \quad \nu_2^g = i \qquad \text{in the element } g \tag{3.11}$$

$$\nu_1^d = i \qquad \text{and} \quad \nu_2^d = i+1 \qquad \text{in the element } d \tag{3.12}$$

Denoting ψ_1^r and ψ_2^r the interpolation functions associated with nodes 1 and 2 of the reference element, and ξ being given by (3.10), we have:

$$\psi_{i-1}^g(x) = \psi_1^r(\xi(x)) = 1-\xi(x) \text{ and } \psi_i^g(x) = \psi_2^r(\xi(x)) = \xi(x) \tag{3.13}$$
(element g)

$$\psi_i^d(x) = \psi_1^r(\xi(x)) = 1-\xi(x) \text{ and } \psi_{i+1}^d(x) = \psi_2^r(\xi(x)) = \xi(x) \tag{3.14}$$
(element d)

The function $u(x)$ is approximated in element g by the following linear combination:

$$u_h(x) = u^{i-1}\psi_{i-1}^g(x) + u^i\psi_i^g(x) = u^{i-1} + \frac{x-x^{i-1}}{\Delta x_g}(u^i - u^{i-1}) \tag{3.15}$$

for $x \in [x^{i-1}, x^i]$.

Similarly, in element d, we have:

$$u_h(x) = u^i\psi_i^d(x) + u^{i+1}\psi_{i+1}^d(x) = u^i + \frac{x-x^i}{\Delta x_d}(u^{i+1} - u^i) \tag{3.16}$$

for $x \in [x^i, x^{i+1}]$.

Using (3.11), the mapping relation between the local (element level) node numbering and the global numbering, we can write for each element e:

$$u_h(x) = u^n\psi_n^e(x) = u^{\nu_k^e}\psi_k^r(\xi(x)) \tag{3.17}$$

the two sums are over *nne* nodes, the number of nodes of the reference finite element, which is 2 here (n global numbering, k local numbering).

The derivative of u_h in the interior of the element e is given in this representation by the following:

$$\frac{du_h}{dx} = u^n \frac{d\psi_n^e}{dx} = u^{v_k^e} \frac{d\psi_k^r}{d\xi} \frac{d\xi}{dx} \tag{3.18}$$

where $d\xi/dx$ comes from (3.10). It can be seen that the derivatives of u_h are not necessarily continuous at the nodes. Thus to express the derivative at a node, a smoothing procedure will be required for the derivatives calculated in each of the elements to which it belongs.

Note here that because every linear function is exactly represented on the elements, it is true in particular for the mappings $\xi(x)$ and $x(\xi)$. Thus, (3.8) and (3.10) can be written as linear combinations of the interpolation functions:

$$x = x^n \psi_n^e(x) = x^{v_k^e} \psi_k^r(\xi(x)) \tag{3.19}$$

In general, the elements for which the *form functions* (those that describe the geometry of the element according to (3.19)) and the interpolation functions (those that are used to approximate a function according to (3.1)) coincide are called ***isoparametric elements***.

Quadratic elements

In the case of a one dimensional element with three nodes, two at the extremities and one in the center, the interpolation functions are of second order which permits the exact representation of quadratic functions. The reference element is classically defined on the segment [–1, +1] with the central node situated at $\xi = 0$, as shown in figure 3.5. The quadratic interpolation functions are defined on this reference element by:

$$\psi_1^r = -\frac{\xi}{2}(1-\xi) \qquad \psi_2^r = 1-\xi^2 \qquad \psi_3^r = \frac{\xi}{2}(1+\xi) \tag{3.20}$$

with $\xi = 2(x - x^i)/(x^{i+1} - x^{i-1})$ on the element $[x^{i-1}, x^{i+1}]$.

It can be verified that these quadratic functions satisfy the conditions (3.4) and (3.5). For isoparametric elements, the mapping between x and ξ is of second order and is defined, for x in $[x^{i-1}, x^{i+1}]$, by the interpolation functions:

$$\begin{aligned} x(\xi) &= x^{i-1}\psi_{i-1}^e(x) + x^i\psi_i^e(x) + x^{i+1}\psi_{i+1}^e(x) \\ &= x^{i-1}\psi_1^r(\xi) + x^i\psi_2^r(\xi) + x^{i+1}\psi_3^r(\xi) \end{aligned} \tag{3.21}$$

Similarly as for the linear elements, the derivative of a function $u_h(x)$ is obtained from the relation:

$$\frac{du_h}{dx} = u^n \frac{d\psi_n^e}{dx} \tag{3.22}$$

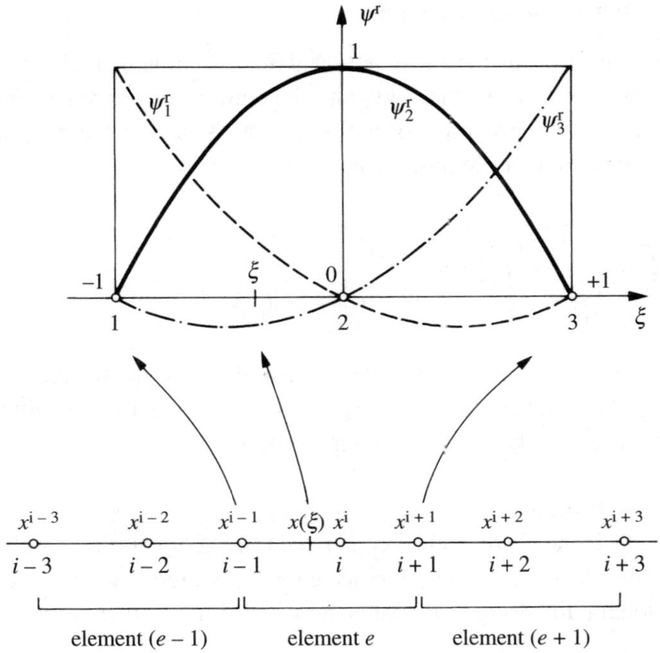

Figure 3.5 Quadratic elements.

in which the derivatives of the interpolation functions are calculated as follows (the node n corresponding in the reference element to the node k; $n = \nu^{\mathrm{e}}_{\mathrm{k}}$):

$$\frac{\mathrm{d}\psi^{\mathrm{e}}_n}{\mathrm{d}x} = \frac{\mathrm{d}\psi^{\mathrm{r}}_{\mathrm{k}}}{\mathrm{d}x} = \frac{\mathrm{d}\psi^{\mathrm{r}}_{\mathrm{k}}}{\mathrm{d}\xi}\frac{\mathrm{d}\xi(x)}{\mathrm{d}x} = \frac{\mathrm{d}\psi^{\mathrm{r}}_{\mathrm{k}}}{\mathrm{d}\xi}\frac{1}{x^{\nu^{\mathrm{e}}_{\mathrm{m}}}\,\mathrm{d}\psi^{\mathrm{r}}_{\mathrm{m}}/\mathrm{d}\xi} \tag{3.23}$$

In the example in figure 3.5, we have:

$$\begin{aligned}\frac{\mathrm{d}x}{\mathrm{d}\xi} &= x^{\mathrm{i}-1}\frac{\mathrm{d}\psi^{\mathrm{r}}_1}{\mathrm{d}\xi} + x^{\mathrm{i}}\frac{\mathrm{d}\psi^{\mathrm{r}}_2}{\mathrm{d}\xi} + x^{\mathrm{i}+1}\frac{\mathrm{d}\psi^{\mathrm{r}}_3}{\mathrm{d}\xi} \\ &= x^{\mathrm{i}-1}(-\tfrac{1}{2} + \xi) + x^{\mathrm{i}}(-2\xi) + x^{\mathrm{i}+1}(\tfrac{1}{2} + \xi) \qquad (3.24) \\ &= \tfrac{1}{2}(x^{\mathrm{i}+1} - x^{\mathrm{i}-1}) - \xi((x^{\mathrm{i}} - x^{\mathrm{i}-1}) - (x^{\mathrm{i}+1} - x^{\mathrm{i}}))\end{aligned}$$

For a regular grid, the right hand side reduces to $\Delta x/2$. In general, the formula (3.23) with a change of variables $x(\xi)$, permits the calculation of all the derivatives of the interpolation functions at all points of the real elements.

3.2.4 Two dimensional finite elements

As a prolongation of the introduction with one dimensional elements, a two dimensional example will further illustrate the concept of reference elements. Figure 3.6 shows the geometric transformation E_e that associates the reference element with any real element (for the triangular case):

$$E_e = \boldsymbol{\xi} = \begin{bmatrix} \xi_1 \\ \xi_2 \end{bmatrix} \rightarrow \boldsymbol{x}(\boldsymbol{\xi}) = \begin{bmatrix} x_1(\boldsymbol{\xi}) \\ x_2(\boldsymbol{\xi}) \end{bmatrix} \tag{3.25}$$

This time $\boldsymbol{x}$ and $\boldsymbol{\xi}$ are vectors with two components, for which we will sometimes employ the notation (x, y) and (ξ, η). Conforming to that which was said before, E_e must satisfy the three following properties:

- Property 1: bidirectional one-to-one mapping.
- Property 2: the correspondence between the geometric node of the real element and the node of the reference element, which is written, denoting by n the global numbering of the kth local node of the element e ($n = \nu_k^e$), by:

$$\boldsymbol{x}(\boldsymbol{\xi}^k) = \boldsymbol{x}^{\nu_k^e} = \boldsymbol{x}^n \tag{3.26}$$

- Property 3: local definition of the boundaries of the real and reference elements; that is, by interpolation with only the nodes actually on the boundary in order to have a uniform representation of the boundary for an element and its neighbor.

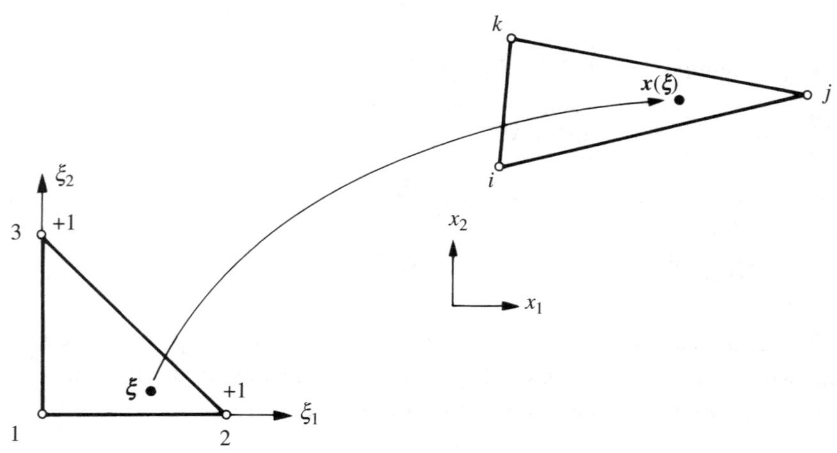

Figure 3.6 Reference element in two dimensions (linear triangle).

Given the fundamental relation

$$x(\xi) = \psi_n^e(x)x^n = \psi_k^r(\xi)x^{v_k^e} \tag{3.27}$$

the second property implies an essential property of nodal interpolation functions:

$$\psi_k^r(\xi^p) = \delta_{kp} \tag{3.28}$$

In other words, the interpolation function associated with node k takes the value 1 at the node, and the value 0 at all other nodes of the reference element. Projected onto the real grid, we see that the interpolation function of a given node takes the value 1 at this node and 0 at all the other nodes of the grid (fig. 3.7).

The approximation of a function u is thus realized in each element e by:

$$u_h(x) = u^n \psi_n^e(x) = u^{v_k^e} \psi_k^r(\xi(x)) \tag{3.29}$$

In the following we consider, except where specified, isoparametric elements. We then have also:

$$x = \psi_k^r(\xi(x))x^{v_k^e} \tag{3.30}$$

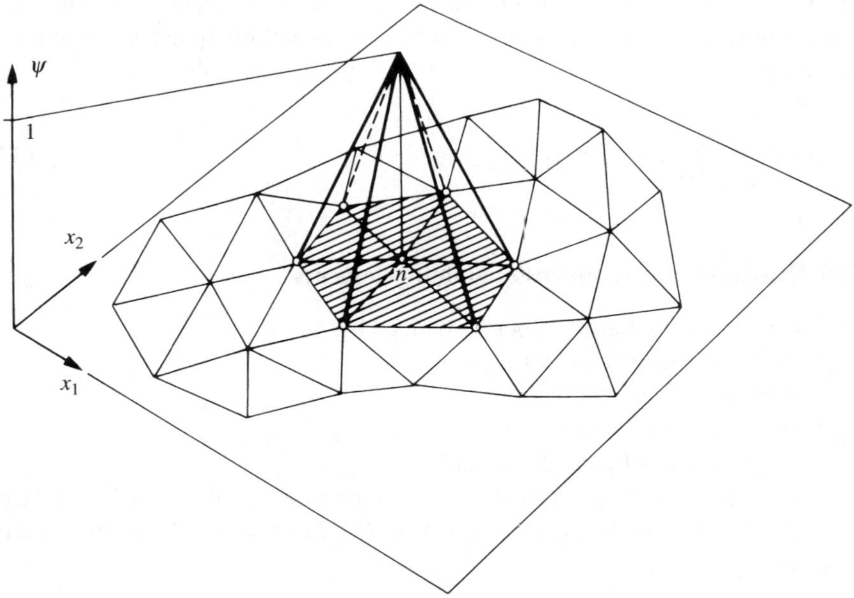

Figure 3.7 Interpolation function associated with the node n (global numbering). In this case, for linear triangular elements, the function decreases linearly from the value 1 at node n to the value 0 at all the nodes neighboring n. It is then 0 over all the elements that do not contain the node n.

Thus the interpolation functions ψ_n only depend on the reference element; and its utility becomes apparent. The approximation u_h is expressed by its nodal values and the interpolation functions of the reference element corresponding to the real element.

Note that it is not always evident how to calculate the inverse transformation $\xi(\boldsymbol{x})$ to use (3.30); in particular, if the interpolation functions are of higher degree than second order. Nonetheless, as we will see later, the calculation of $u(\boldsymbol{x})$, and thus of $\psi_k^r(\xi(\boldsymbol{x}))$, is only necessary at certain points in the elements: those for which the coordinates ξ are known in advance.

REMARK: The interpolation functions need not be identical for the different functions that one wants to approximate in a single problem. Such is the case, for example, of formulations in so called primitive variables, for which one solves in velocity and pressure fields $(\boldsymbol{v}, p)$. These formulations are very often used in incompressible fluid mechanics (chap. 7) and increasingly more often in solid mechanics (chap. 6).

The nodes and interpolation functions used for the approximation of the velocity components and the pressure are generally different, and one writes for example:

$$(v_h)_i(\boldsymbol{x}) = \sum_{k=1}^{nnv} \psi_k^{rv}(\xi(\boldsymbol{x}))v_i^{v_k^e} \tag{3.31}$$

for each component i of the velocity field, nnv being the number of nodes and ψ^{rv} the functions used for interpolation of the velocity in the element. For the scalar pressure field, one has a different number of nodes and interpolation functions for the element:

$$p_h(\boldsymbol{x}) = \sum_{k=1}^{nnp} \psi_k^{rp}(\xi(\boldsymbol{x}))p^{v_k^e} \tag{3.32}$$

3.2.5 Examples of two dimensional finite elements

Each type of reference element is defined by:

- its form (triangular, quadrangular, ...),
- the coordinates of its nodes,
- its interpolation functions,
- the definition of its nodal variables,
- the type of continuity satisfied by the approximation of a function at the element boundary (continuous function C^0, function with a continuous first derivative C^1, ...).

We will show a few of the most classic isoparametric finite elements. The reader is referred to the bibliography for other elements, notably those with higher continuity.

To simplify the figures, in this paragraph we indicate by (ξ, η) the coordinates in the reference element, rather than (ξ_1, ξ_2), and by (x, y) the spatial coordinates, rather than (x_1, x_2).

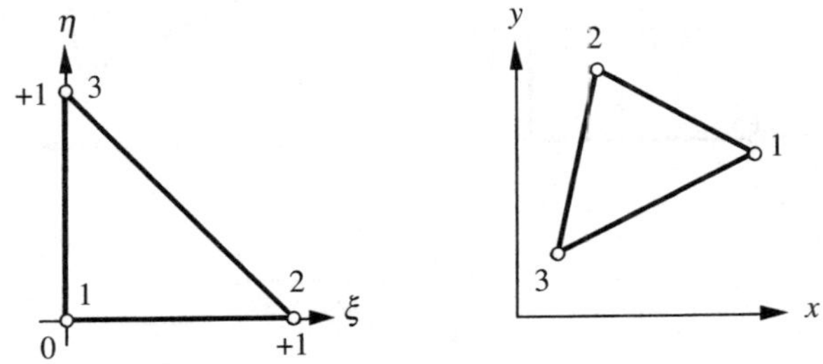

Node k	ψ_k
1	$1-\xi-\eta$
2	ξ
3	η

Figure 3.8 Linear triangular element (element $P1$ according to the taxonomy proposed by Ciarlet): a triangle with three nodes.

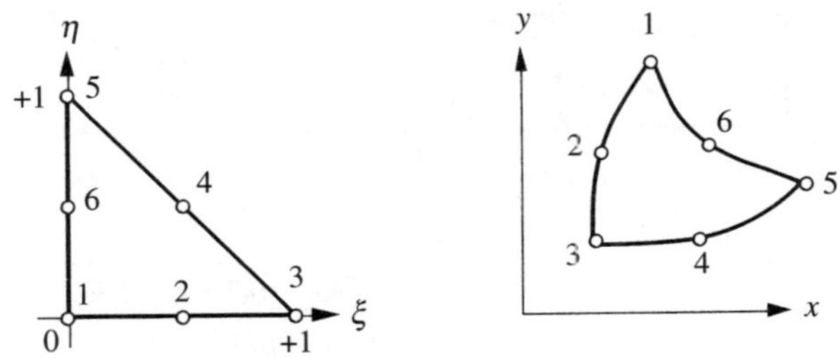

Node k	ψ_k
1	$\lambda\,(2\lambda-1)$
2	$4\xi\lambda$
3	$\xi\,(2\xi-1)$
4	$4\xi\eta$
5	$\eta\,(2\eta-1)$
6	$4\eta\lambda$

$(\lambda = 1-\xi-\eta)$

Figure 3.9 Quadratic triangular element ($P2$): a triangle with six nodes.

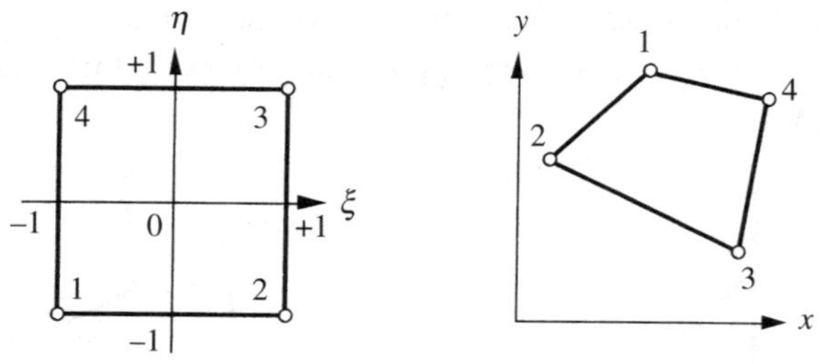

Node k	ψ_k
1	$(1-\xi)(1-\eta)/4$
2	$(1+\xi)(1-\eta)/4$
3	$(1+\xi)(1+\eta)/4$
4	$(1-\xi)(1+\eta)/4$

Figure 3.10 Linear quadrangular element ($Q1$): a quadrangle with four nodes.

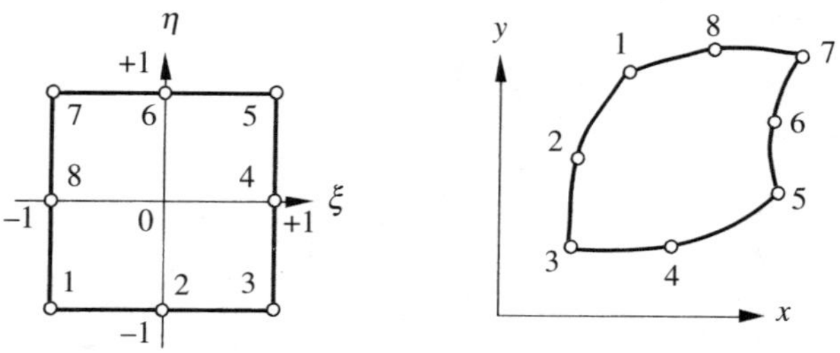

Node k	ψ_k
1	$-(1-\xi)(1-\eta)(1+\xi+\eta)/4$
2	$(1-\xi^2)(1-\eta)/2$
3	$-(1+\xi)(1-\eta)(1-\xi+\eta)/4$
4	$(1+\xi)(1-\eta^2)/2$
5	$-(1+\xi)(1+\eta)(1-\xi-\eta)/4$
6	$(1-\xi^2)(1+\eta)/2$
7	$-(1-\xi)(1+\eta)(1+\xi-\eta)/4$
8	$(1-\xi)(1-\eta^2)/2$

Figure 3.11 Quadratic quadrangular element ($Q2$ incomplete): a quadrangle with eight nodes.

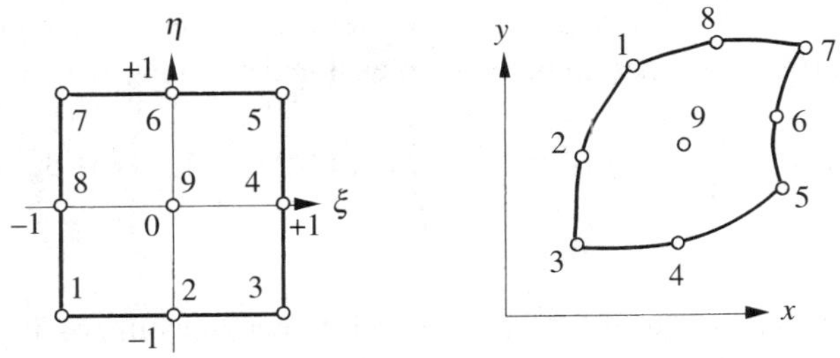

Node k	ψ_k
1	$\xi\eta(1-\xi)(1-\eta)/4$
2	$-\eta(1+\xi)(1-\xi)(1-\eta)/2$
3	$-\xi\eta(1+\xi)(1-\eta)/4$
4	$\xi(1+\xi)(1+\eta)(1-\eta)/2$
5	$\xi\eta(1+\xi)(1+\eta)/4$
6	$\eta(1+\xi)(1-\xi)(1+\eta)/2$
7	$-\xi\eta(1-\xi)(1+\eta)/4$
8	$-\xi(1-\xi)(1+\eta)(1-\eta)/2$
9	$(1+\xi)(1-\xi)(1+\eta)(1-\eta)$

Figure 3.12 Complete quadratic quadrangular element ($Q2$): a quadrangle with nine nodes.

3.2.6 Integration of a function

The integration of a function u on a domain Ω is first decomposed into a sum of integrals over each of the elements, each occupying a domain Ω^e, the union of which is Ω.

$$I = \int_{\Omega} u(\boldsymbol{x})\,\mathrm{d}V \approx \sum_{e=1}^{Ne}\left(\int_{\Omega^e} u_h(\boldsymbol{x})\,\mathrm{d}V\right) \tag{3.33}$$

In each element e, the change of variables $\boldsymbol{x}(\boldsymbol{\xi})$ allows the integration on the reference element, denoted Ω^r, using the determinant of the ***Jacobian matrix*** $[\boldsymbol{J}^e]$ for the change of variables. If d is the dimension of the problem (d is 2 in two dimensions, 3 in 3 dimensions), the matrix $[\boldsymbol{J}^e]$, of size $d \times d$, has the following components:

$$J_{ij}^e = \frac{\partial x_i}{\partial \xi_j} = \frac{\partial \psi_k^r}{\partial \xi_j}\, x_i^{v_k^e} \tag{3.34}$$

The sum k is over the *nne* nodes of the element. We have then for each element of the domain (here assumed to be two dimensional):

$$\int_{\Omega^e} u_h(\boldsymbol{x})\,\mathrm{d}V = \int_{\Omega^e} u_h(\boldsymbol{x})\mathrm{d}x_1\,\mathrm{d}x_2 = \int_{\Omega^r} u_h(\boldsymbol{x}(\boldsymbol{\xi}))\,\det[\boldsymbol{J}^e(\boldsymbol{\xi})]\,\mathrm{d}\xi_1\,\mathrm{d}\xi_2 \tag{3.35}$$

The exact calculation of the integral rapidly becomes impossible for complex forms or nonlinear elements (even if the symbolic mathematical programs can be helpful in some cases); one must turn to numerical integration techniques. The last integral can be calculated by diverse methods; the most common is ***Gauss' method***, which consists of evaluating the function u_h at judiciously chosen points in the reference element, according to the formula:

$$\int_{\Omega^r} u_h(\boldsymbol{x}(\boldsymbol{\xi}))\,\det[\boldsymbol{J}^e(\boldsymbol{\xi})]\,\mathrm{d}\xi_1\,\mathrm{d}\xi_2 \simeq \sum_{p=1}^{npie} \underbrace{u_h(\boldsymbol{x}(\boldsymbol{\xi}^p))}_{\psi_k^r(\boldsymbol{\xi}^p)u^{v_k^e}}\,\det[\boldsymbol{J}^e(\boldsymbol{\xi}^p)]w^p \tag{3.36}$$

where *npie* is the number of ***integration points*** (or ***Gauss points***), with coordinate vectors $\boldsymbol{\xi}^p$ and associated weights w^p. The details of the choice of the number of points and their associated weights, which are functions of the degree of precision desired in the integral, and the formulas most often used for two dimensional finite elements are given in chapter 10. Now it is sufficient to know that, in one dimension, Gauss' method allows the exact integration of a polynomial of degree $2r-1$ with r points. For example, with two points on the segment $[-1, +1]$, having abscissas $-1/\sqrt{3}$ and $+1/\sqrt{3}$ and equal weights (= 1), polynomials of order 3 can be exactly integrated (chap. 10).

3.2.7 Derivatives

In the interior of element e, the derivative of the function u_h with respect to one of the coordinates x_j is written, using (3.29):

$$\frac{\partial u_h}{\partial x_j} = \frac{\partial \psi_n^e(\boldsymbol{x})}{\partial x_j}u^n \tag{3.37}$$

or again using the transformation $\boldsymbol{x}(\boldsymbol{\xi})$,

$$\frac{\partial u_h}{\partial x_j} = \frac{\partial \psi_k^r(\boldsymbol{\xi})}{\partial x_j}u^{v_k^e} = \frac{\partial \psi_k^r}{\partial \xi_i}\frac{\partial \xi_i}{\partial x_j}u^{v_k^e} \tag{3.38}$$

the sum over i being from 1 to d (spatial components). Denoting by $[\boldsymbol{J}^e]^{-1}$ the inverse matrix of $[\boldsymbol{J}^e]$ (which does exist; the transformation E_e is bi-directional), we have finally:

$$\frac{\partial u_h}{\partial x_j} = \frac{\partial \psi_k^r}{\partial \xi_i}(J^e)^{-1}_{ij} u^{v_k^e} \tag{3.39}$$

in which we have somewhat abusively denoted $(J^e)^{-1}_{ij}$ the component *ij* of the matrix $[J^e]^{-1}$, to simplify the expression. In practice, for each element *e*, we calculate the matrix $[J^e]$ at each integration point using (3.34), and take the inverse if necessary. The values of the interpolation functions and their derivatives at the integration points of the reference element are calculated once for the whole computation at the beginning.

3.2.8 Example of integration and derivation on a linear triangle

To illustrate the preceding, take as an example the calculation of an integral of the derivative of a scalar function *u* (the subscript "*h*" is omitted here), over the linear triangular element *e* of figure 3.8. The integral to calculate is the following:

$$I = \int_{\Omega^e} \frac{\partial u}{\partial x_1}\, \mathrm{d}x_1\, \mathrm{d}x_2 \tag{3.40}$$

The linear interpolation functions are expressed in the following manner on the reference element Ω^r (fig. 3.8):

$$\psi_1^r(\boldsymbol{\xi}) = 1 - \xi_1 - \xi_2 \qquad \psi_2^r(\boldsymbol{\xi}) = \xi_1 \qquad \psi_3^r(\boldsymbol{\xi}) = \xi_2 \tag{3.41}$$

The transformation E_e is thus:

$$\begin{aligned} E_e : \boldsymbol{\xi} = \begin{bmatrix} \xi_1 \\ \xi_2 \end{bmatrix} \rightarrow \boldsymbol{x}(\boldsymbol{\xi}) &= \psi_1^r(\boldsymbol{\xi})\boldsymbol{x}^1 + \psi_2^r(\boldsymbol{\xi})\boldsymbol{x}^2 + \psi_3^r(\boldsymbol{\xi})\boldsymbol{x}^3 \\ &= (1 - \xi_1 - \xi_2)\boldsymbol{x}^1 + \xi_1 \boldsymbol{x}^2 + \xi_2 \boldsymbol{x}^3 \end{aligned} \tag{3.42}$$

and, written as components:

$$\left.\begin{aligned} \begin{bmatrix} x_1 \\ x_2 \end{bmatrix} &= \psi_1^r(\boldsymbol{\xi}) \begin{bmatrix} x_1^1 \\ x_2^1 \end{bmatrix} + \psi_2^r(\boldsymbol{\xi}) \begin{bmatrix} x_1^2 \\ x_2^2 \end{bmatrix} + \psi_3^r(\boldsymbol{\xi}) \begin{bmatrix} x_1^3 \\ x_2^3 \end{bmatrix} \\ &= (1 - \xi_1 - \xi_2) \begin{bmatrix} x_1^1 \\ x_2^1 \end{bmatrix} + \xi_1 \begin{bmatrix} x_1^2 \\ x_2^2 \end{bmatrix} + \xi_2 \begin{bmatrix} x_1^3 \\ x_2^3 \end{bmatrix} \\ &= \begin{bmatrix} x_1^1 \\ x_2^1 \end{bmatrix} + \begin{bmatrix} x_1^2 - x_1^1 & x_1^3 - x_1^1 \\ x_2^2 - x_2^1 & x_2^3 - x_2^1 \end{bmatrix} \begin{bmatrix} \xi_1 \\ \xi_2 \end{bmatrix} \end{aligned}\right\} \tag{3.43}$$

The Jacobian matrix of the transformation E_e is given by:

$$[\boldsymbol{J}^e] = \left[\frac{\partial \boldsymbol{x}}{\partial \boldsymbol{\xi}}\right] = \begin{bmatrix} \partial x_1/\partial \xi_1 & \partial x_1/\partial \xi_2 \\ \partial x_2/\partial \xi_1 & \partial x_2/\partial \xi_2 \end{bmatrix} = \begin{bmatrix} x_1^2 - x_1^1 & x_1^3 - x_1^1 \\ x_2^2 - x_2^1 & x_2^3 - x_2^1 \end{bmatrix} \tag{3.44}$$

and its determinant by:

$$\det[\boldsymbol{J}^e] = (x_1^2 - x_1^1)(x_2^3 - x_2^1) - (x_2^2 - x_2^1)(x_1^3 - x_1^1) \tag{3.45}$$

It is seen that the Jacobian matrix is constant on the element, and that its determinant is, except for a factor of 2, the area of the triangle. This is of course a particularity of linear triangular elements. In general, the matrix $[\boldsymbol{J}^e]$ varies over the element and it should be rigorously noted by $[\boldsymbol{J}^e(\boldsymbol{\xi})]$. In the present case we denote it with $[\boldsymbol{J}^e]$, and then the inverse is $[\boldsymbol{J}^e]^{-1}$:

$$[\boldsymbol{J}^e]^{-1} = \begin{bmatrix} \dfrac{\partial \xi_1}{\partial x_1} & \dfrac{\partial \xi_1}{\partial x_2} \\ \dfrac{\partial \xi_2}{\partial x_1} & \dfrac{\partial \xi_2}{\partial x_2} \end{bmatrix} \tag{3.46}$$

which contains in the first column the necessary derivatives for the expression of the derivative of u that interests us. We have:

$$\frac{\partial u}{\partial x_1} = \frac{\partial u}{\partial \xi_1}\frac{\partial \xi_1}{\partial x_1} + \frac{\partial u}{\partial \xi_2}\frac{\partial \xi_2}{\partial x_1} = (J^e)^{-1}_{11}\frac{\partial u}{\partial \xi_1} + (J^e)^{-1}_{21}\frac{\partial u}{\partial \xi_2} \tag{3.47}$$

Moving on to the calculation of the integral I over the reference element, by the change of variables E_e:

$$I = \int_{\Omega^e} \frac{\partial u}{\partial x_1}\, \mathrm{d}x_1\, \mathrm{d}x_2 = \int_{\Omega^r} \left((J^e)^{-1}_{11}\frac{\partial u}{\partial \xi_1} + (J^e)^{-1}_{21}\frac{\partial u}{\partial \xi_2} \right) \det[\boldsymbol{J}^e]\, \mathrm{d}\xi_1\, \mathrm{d}\xi_2 \tag{3.48}$$

in which the expression of the derivatives of u are the following (assuming for simplicity here that the numbering coincides between Ω^e and Ω^r):

$$\frac{\partial u}{\partial \xi_1} = \sum_{n=1}^{3} \frac{\partial \psi_n^r}{\partial \xi_1} u^n \qquad \frac{\partial u}{\partial \xi_2} = \sum_{n=1}^{3} \frac{\partial \psi_n^r}{\partial \xi_2} u^n \tag{3.49}$$

Numerical integration by Gauss' method gives:

$$I = \sum_{p=1}^{npie} w^p \det[J^e]\left[\sum_{n=1}^{3} u^n\left((J^e)^{-1}_{11}\frac{\partial\psi^r_n(\xi^p)}{\partial\xi_1} + (J^e)^{-1}_{21}\frac{\partial\psi^r_n(\xi^p)}{\partial\xi_2}\right)\right] \tag{3.50}$$

It should be noted here that the function u interpolated over the finite element is linear, and thus that the derivatives are constant. As a result, the integration only requires a single point in the triangle (chap. 10). The reader can verify it with a quick calculation whose result is (exercise 3.7.1):

$$I = \tfrac{1}{2}[(x_2^3 - x_2^1)(u^2 - u^1) + (x_2^1 - x_2^2)(u^3 - u^1)] \tag{3.51}$$

In conclusion, such combinations of calculations are very frequently encountered in finite element programs, and require the calculations of the following quantities in each element for each integration point ξ^p:

- The matrix $[J^e(\xi^p)]$, its determinant and its inverse $[J^e]^{-1}$. These values depend on the real coordinates; in the general case they must be calculated for each real element e. If the coordinates of the real element change during the problem (as in a study of solid deformation, for example), it will be necessary to recalculate them every time they change.
- The functions ψ_n and their derivatives in the reference space. Nonetheless, these values only depend on the reference element; they need only be calculated once at the beginning of the computation.

3.3 OBTAINING AND DISCRETIZING THE INTEGRAL FORM FOR A SCALAR PROBLEM: A CHEMICAL DIFFUSION EXAMPLE

Having presented the general principles of the geometric discretization into finite elements and the integration and derivation operations, we advance to the practical application for the solution of partial differential equations. To illustrate the method, we again take the example of stationary chemical diffusion with transport as seen already in chapter 2.

The unknown field is the scalar solute concentration, $c(\boldsymbol{x})$, in the domain Ω. The governing equation for the phenomenon in the absence of a solute source term, and in a stationary incompressible flow (equation (1.74), with $\partial c/\partial t = 0$) is:

$$\boldsymbol{v}\cdot\mathbf{grad}\,c - \mathrm{div}(D\,\mathbf{grad}\,c) = 0 \tag{3.52}$$

where $\boldsymbol{v}$ is the transport velocity vector, assumed uniform, and D the diffusion coefficient. In this example, we only consider two types of boundary conditions (fig. 3.13):

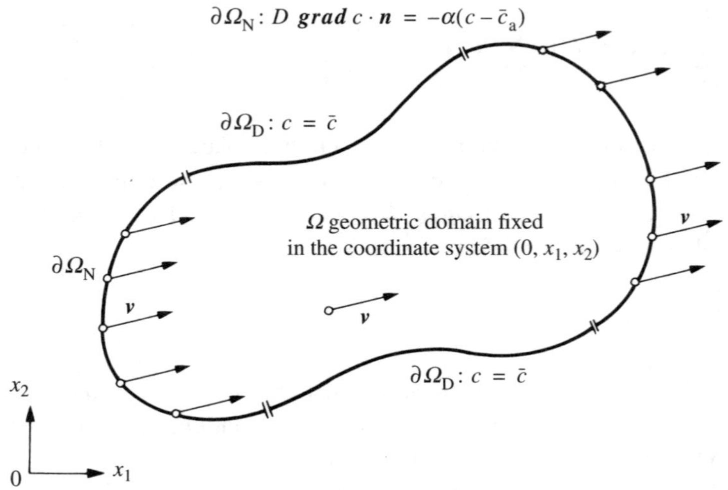

Figure 3.13 Boundary conditions for a chemical diffusion problem on a geometric domain Ω, in two dimensions and stationary in time. The velocity field $\boldsymbol{v}$ is assumed given and uniform.

- A Cauchy, or mixed, condition:

$$D\,\mathbf{grad}\,c \cdot \boldsymbol{n} = -\alpha(c - \bar{c}_a) \qquad \text{on } \partial\Omega_N \tag{3.53}$$

where α is a chemical transfer coefficient with the ambient environment, where the concentration is $\bar{c}_a$.

- A Dirichlet condition (imposed concentration):

$$c = \bar{c} \qquad \text{on } \partial\Omega_D \tag{3.54}$$

3.3.1 Obtaining the integral form: weighted residuals method

We call a ***residual*** the value, scalar in this case, defined by:

$$R(c) = \boldsymbol{v} \cdot \mathbf{grad}\,c - \mathrm{div}(D\,\mathbf{grad}\,c) \tag{3.55}$$

The solution of the problem is clearly that which zeros the residual while simultaneously satisfying the boundary conditions at $\partial\Omega$. The basic idea of the solution method is to search for scalar functions c which zero the following integral form:

$$\Phi(c) = \int_\Omega \varphi R(c)\,\mathrm{d}V = \int_\Omega \varphi[\boldsymbol{v} \cdot \mathbf{grad}\,c - \mathrm{div}(D\,\mathbf{grad}\,c)]\,\mathrm{d}V = 0 \tag{3.56}$$

for every ***weighting function*** (or ***test function***) φ belonging to a set of functions E_φ, while c satisfies the boundary conditions at $\partial\Omega$.

The equivalence between $R(c) = 0$ on Ω and (3.56) is in fact only true if the set E_φ has infinite dimension and is composed of independent functions (for example, the set of Dirac delta functions $\delta(\boldsymbol{x})$ for every point $\boldsymbol{x}$ in Ω). If that is not the case, that is if E_φ is finite, (as is the case for the finite element method), the solution c which satisfies (3.56) is only an *approximate* solution to the problem.

In order to integrate (3.56), we can also write it in the form:

$$\int_\Omega \varphi \boldsymbol{v} \cdot \mathbf{grad}\, c \; \mathrm{d}V - \int_\Omega \mathrm{div}(\varphi D \, \mathbf{grad}\, c) \; \mathrm{d}V + \int_\Omega D \, \mathbf{grad}\, \varphi \cdot \mathbf{grad}\, c \; \mathrm{d}V = 0 \tag{3.57}$$

The divergence theorem then permits the transformation of the second term to obtain:

$$\int_\Omega \varphi \boldsymbol{v} \cdot \mathbf{grad}\, c \; \mathrm{d}V - \int_{\partial\Omega} \varphi D \, \mathbf{grad}\, c \cdot \boldsymbol{n} \; \mathrm{d}S + \int_\Omega D \, \mathbf{grad}\, \varphi \cdot \mathbf{grad}\, c \; \mathrm{d}V = 0 \tag{3.58}$$

It should be noted at this stage that this operation has allowed the reduction of the order of derivation of c (only first derivatives are needed, and no longer the second derivatives). The other advantage resulting from this formulation is that the boundary conditions can now be taken into account very easily. In fact, the integral on the boundary can be decomposed into two sums over $\partial\Omega_N$ and $\partial\Omega_D$. The concentration c being known (given) on $\partial\Omega_D$, we can restrict the space E_φ of test functions to those that are zero on $\partial\Omega_D$: $E_{\varphi 0} = \{\varphi \in E_\varphi | \varphi = 0 \text{ on } \partial\Omega_D\}$ (the reader is referred to the bibliography for a complete justification of this step). Using the boundary condition (3.53), the ***integral*** or ***weak form***, of the posed problem is obtained:

$$\int_\Omega \varphi \boldsymbol{v} \cdot \mathbf{grad}\, c \; \mathrm{d}V + \int_{\partial\Omega_N} \varphi \alpha (c - \bar{c}_a) \; \mathrm{d}S + \int_\Omega D \, \mathbf{grad}\, \varphi \cdot \mathbf{grad}\, c \; \mathrm{d}V = 0$$

$$\forall \varphi \in E_{\varphi 0} \tag{3.59}$$

3.3.2 Discretized integral form. Galerkin method

In order to concretely solve the problem and calculate c on the domain Ω, it will be necessary to include the spatial discretization into finite elements and to choose an ensemble of test functions E_φ. For the first point, designating by c_h the discrete approximation of the continuous field c, we can write:

$$c_h(\boldsymbol{x}) = c^n \psi_n(\boldsymbol{x}) \tag{3.60}$$

the sum being over the Nn nodes of the grid. In the rest of the chapter we omit the index h, to simplify the expression. As for the discretization of the integral form (3.59), it is obtained by choosing Nn independent weighting functions (as many as the number of parameters defining the approximation c_h): $\varphi_1, \varphi_2, ..., \varphi_{Nn}$. The choice of the type of functions φ_i leads to different methods: collocation by points, by sub domains and Galerkin. We only present the latter here, which is by far the most used.

The ***Galerkin method*** consists of taking precisely the interpolation functions ψ_n of the approximation by finite elements for weighting functions. Thus if we *discretize* the weak form (3.59), applying the interpolation (3.60) and choosing $\varphi_i = \psi_i$, we obtain:

$$\int_{\Omega} \psi_i \boldsymbol{v} \cdot \mathbf{grad}(\psi_j c^j)\, dV + \int_{\partial\Omega_N} \alpha \psi_i \psi_j c^j\, dS + \int_{\Omega} D\, \mathbf{grad}(\psi_j c^j) \cdot \mathbf{grad}\, \psi_i\, dV$$

$$= \int_{\partial\Omega_N} \alpha \bar{c}_a \psi_i\, dS \tag{3.61}$$

for the nodes i for which the value c^i is not imposed. We denote the set of these degrees of freedom N^{free}, while the set of degrees of freedom for the imposed values is denoted N^{imp}.

By virtue of the linear properties of the sums and integrals, and the boundary conditions (3.54), the discrete problem to solve is then written:

$$c^j \left[\int_{\Omega} \psi_i \boldsymbol{v} \cdot \mathbf{grad}\, \psi_j\, dV + \int_{\partial\Omega_N} \alpha \psi_i \psi_j\, dS + \int_{\Omega} D\, \mathbf{grad}\, \psi_j \cdot \mathbf{grad}\, \psi_i\, dV \right]$$

$$= \int_{\partial\Omega_N} \alpha \bar{c}_a \psi_i\, dS \qquad \text{for } i \in N^{free} \tag{3.62}$$

and

$$c^i = \bar{c} \qquad \text{for } i \in N^{imp} \tag{3.63}$$

The concentration field under consideration now appears as the solution of a system of Nn equations and Nn unknowns:

$$[\boldsymbol{K}]\, \boldsymbol{c} = \boldsymbol{b} \tag{3.64}$$

in which $\boldsymbol{c}$ is the column vector having the nodal concentrations c^j, $j = 1, Nn$ for components.

The matrix $[\boldsymbol{K}]$ includes the diffusivity and advection terms. Its size is $Nn \times Nn$ and its components are given by:

$$K_{ij} = \int_{\Omega} \left[\psi_i v_k \frac{\partial \psi_j}{\partial x_k} + D \frac{\partial \psi_i}{\partial x_k} \frac{\partial \psi_j}{\partial x_k} \right] dV + \int_{\partial \Omega_N} \alpha \psi_i \psi_j \, dS \qquad \text{for } i \in N^{\text{free}} \tag{3.65}$$

and

$$K_{ij} = \delta_{ij} \qquad \text{for } i \in N^{\text{imp}} \tag{3.66}$$

k being summed from 1 to d, the spatial dimension of the problem. The generic name ***stiffness matrix***, whose origin is in the solution of mechanical equilibrium problems (sect. 3.4), is often used for the matrix $[\boldsymbol{K}]$. One can see in the present case that $[\boldsymbol{K}]$ is not symmetric, because of the term associated with advective transport. We will return to this subject in paragraph 3.6.5 for the solution of a non stationary problem; the specific treatment of the advective transport is detailed in chapter 7.

Finally, $\boldsymbol{b}$ is the right hand side vector defined by:

$$b_i = \int_{\partial \Omega_N} \alpha \bar{c}_a \psi_i \, dS \qquad \text{for } i \in N^{\text{free}} \tag{3.67}$$

and

$$b_i = \bar{c} \qquad \text{for } i \in N^{\text{imp}} \tag{3.68}$$

The method of calculation of $[\boldsymbol{K}]$ and $\boldsymbol{b}$, by decomposition for each of the elements and assembly, will be detailed in section 3.5.2. Note that if the diffusion coefficient D does not depend on the concentration c, the matrix $[\boldsymbol{K}]$ of the system will not depend on the unknown vector $\boldsymbol{c}$ and the system is therefore linear. Otherwise, the system is non linear. The techniques for solution of the system in the two cases are explained in chapter 4.

The diffusion part of matrix $[\boldsymbol{K}]$ has an interesting physical interpretation in the case of a linear triangular element: it is shown in chapter 10 that in each element, the terms of this matrix correspond to the solute flux exchanged across the medians of the triangles between two nodes.

3.3.3 The least square method

The weighted residual method is not the only one that can be used to search for a function that zeros the residual $R(c)$ on Ω. The principle of least squares, for example, consists of searching for functions c that minimize the following integral:

$$f(c) = \int_{\Omega} (R(c))^2 \, dV \tag{3.69}$$

and that respect the conditions at the boundary of Ω. This method, natural enough in appearance, suffers two major inconveniences. First, the boundary conditions are difficult to take into account, except for Dirichlet conditions (that is, $c = \bar{c}$ on $\partial\Omega = \partial\Omega_D$). Second, the order of the derivatives in R can not be reduced: in practice, this leads to high differentiability conditions on the finite element discretization. For these two reasons the method of weighted residuals discussed above is preferred.

3.3.4 Matrix presentation

We will now describe the terms appearing in (3.62) explicitly in terms of matrices. For that we denote by Ψ the line matrix composed of the Nn interpolation functions:

$$\Psi = [\psi_1(\boldsymbol{x})\ \psi_2(\boldsymbol{x})\ \ldots\ \psi_{\mathrm{Nn}}(\boldsymbol{x})] \tag{3.70}$$

For every point $\boldsymbol{x}$, the fundamental relation for interpolation (3.60) can now be written in the form of a product:

$$c(\boldsymbol{x}) = [\psi_1(\boldsymbol{x})\psi_2(\boldsymbol{x})\ldots\psi_{\mathrm{Nn}}(\boldsymbol{x})]\begin{bmatrix} c^1 \\ c^2 \\ \vdots \\ c^{\mathrm{Nn}} \end{bmatrix} = \Psi\boldsymbol{c} \tag{3.71}$$

Similarly, for each element e containing nne nodes, we can write:

$$c(\boldsymbol{x}) = [\psi_1^{\mathrm{e}}(\boldsymbol{x})\psi_2^{\mathrm{e}}(\boldsymbol{x})\ldots\psi_{\mathrm{nne}}^{\mathrm{e}}(\boldsymbol{x})]\begin{bmatrix} c^1 \\ c^2 \\ \vdots \\ c^{\mathrm{nne}} \end{bmatrix} = \Psi^{\mathrm{e}}\boldsymbol{c}^{\mathrm{e}} \tag{3.72}$$

Suppose that we want to calculate the gradient vector of the field $c(\boldsymbol{x})$. In two dimensions, this takes the form:

$$\mathbf{grad}\, c = \begin{bmatrix} \dfrac{\partial c}{\partial x_1} \\ \dfrac{\partial c}{\partial x_2} \end{bmatrix} \tag{3.73}$$

Taking into account that the derivatives of c are expressed in terms of the finite element interpolation by:

$$\frac{\partial c(\boldsymbol{x})}{\partial x_\mathrm{k}} = \frac{\partial}{\partial x_\mathrm{k}}(\psi_\mathrm{p}(\boldsymbol{x})c^\mathrm{p}) = \frac{\partial \psi_\mathrm{p}(\boldsymbol{x})}{\partial x_\mathrm{k}}c^\mathrm{p} \tag{3.74}$$

the gradient of the concentration field can be expressed, for every point in space, as a matrix product:

$$\mathbf{grad}\, c = \begin{bmatrix} \frac{\partial \psi_1}{\partial x_1} & \frac{\partial \psi_2}{\partial x_1} & \cdots & \frac{\partial \psi_\mathrm{Nn}}{\partial x_1} \\ \frac{\partial \psi_1}{\partial x_2} & \frac{\partial \psi_2}{\partial x_2} & \cdots & \frac{\partial \psi_\mathrm{Nn}}{\partial x_2} \end{bmatrix} \begin{bmatrix} c^1 \\ c^2 \\ \vdots \\ c^\mathrm{Nn} \end{bmatrix} = [\boldsymbol{B}]\boldsymbol{c} \tag{3.75}$$

Introducing the derivative column vector $\boldsymbol{\nabla}$ acting on the components of the line vector Ψ (in 2 dimensions):

$$\boldsymbol{\nabla} = \begin{bmatrix} \frac{\partial}{\partial x_1} \\ \frac{\partial}{\partial x_2} \end{bmatrix} \tag{3.76}$$

the matrix equation obtained is:

$$\mathbf{grad}\, c = \boldsymbol{\nabla} c = \boldsymbol{\nabla}\Psi(\boldsymbol{x})\boldsymbol{c} = [\boldsymbol{B}(\boldsymbol{x})]\boldsymbol{c} \tag{3.77}$$

As for the matrix expression of the weak form resulting from the Galerkin method (paragraph 3.3.2), it can be verified without difficulty that introducing the notions above in the system matrix and its right hand side vector (3.65) and (3.67), they become as follows:

$$[\boldsymbol{K}] = \int_\Omega [D[\boldsymbol{B}]^\mathrm{T}[\boldsymbol{B}] + \Psi^\mathrm{T}[\boldsymbol{B}]^\mathrm{T}\boldsymbol{v}]\,\mathrm{d}V + \int_{\partial\Omega_\mathrm{N}} \alpha\Psi^\mathrm{T}\Psi\,\mathrm{d}S \tag{3.78}$$

$$\boldsymbol{b} = \int_{\partial\Omega_\mathrm{N}} \alpha\bar{c}_\mathrm{a}\Psi^\mathrm{T}\,\mathrm{d}S \tag{3.79}$$

3.4 SOLUTION OF A VECTOR PROBLEM: MECHANICAL EQUILIBRIUM EXAMPLE

The objective of this section is to generalize the FEM formalism to problems in which the unknown is a vector field with several degrees of freedom per node, taking as an example the components of velocity or displacement in mechanics. In this case the starting point for the application of the FEM will be the principle of virtual power. We will analyze in detail a case of linear elasticity for which there exists a functional whose minimum gives the equilibrium solution, in a manner similar to the principle of virtual power.

3.4.1 Equilibrium equation and boundary conditions

In chapter 1 it was seen that in the absence of inertia and volume density forces, the local equilibrium equation is,

$$\mathbf{div}\ \boldsymbol{\sigma} = \mathbf{0} \tag{3.80}$$

where $\boldsymbol{\sigma}$ is the stress tensor. From the point of view of the application of the FEM, it is preferable to employ the equivalent equation, expressed as virtual power, as it is already in integral form. With the same assumptions, (1.49) becomes:

$$\int_{\Omega} \boldsymbol{\sigma} : \dot{\boldsymbol{\varepsilon}}^* \, \mathrm{d}V = \int_{\partial\Omega} \boldsymbol{T} \cdot \boldsymbol{v}^* \, \mathrm{d}S \qquad \text{for all } \boldsymbol{v}^* \tag{3.81}$$

or, employing the symmetry of the tensor $\boldsymbol{\sigma}$ and the relation $\dot{\boldsymbol{\varepsilon}}^* = 1/2[\mathbf{Grad}\, \boldsymbol{v}^* + (\mathbf{Grad}\, \boldsymbol{v}^*)^{\mathrm{T}}]$,

$$\int_{\Omega} \boldsymbol{\sigma} : \mathbf{Grad}\, \boldsymbol{v}^* \, \mathrm{d}V = \int_{\partial\Omega} \boldsymbol{T} \cdot \boldsymbol{v}^* \, \mathrm{d}S \qquad \text{for all } \boldsymbol{v}^* \tag{3.82}$$

where $\dot{\boldsymbol{\varepsilon}}^*$ is the tensor of virtual strain rates (associated with the virtual velocity field $\boldsymbol{v}^*$), $\boldsymbol{T}$ is the stress vector on the surface of Ω ($\boldsymbol{T} = \boldsymbol{\sigma} \cdot \boldsymbol{n}$, $\boldsymbol{n}$ being the outgoing normal vector).

The ***boundary conditions*** are principally of two types:

Imposed velocity vector (Dirichlet condition): $\boldsymbol{v} = \bar{\boldsymbol{v}}$ on $\partial\Omega_{\mathrm{D}}$ (3.83)

Imposed stress vector: $\boldsymbol{T} = \overline{\boldsymbol{T}}$ on $\partial\Omega_{\mathrm{N}}$ (3.84)

In fact, as can be seen in figure 3.14, these conditions are only rough approximations of the boundary conditions encountered in practice. In reality, the conditions are imposed *component by component* on material points.

Thus one has:

$$v_{\mathrm{k}} = \bar{v}_{\mathrm{k}} \qquad \text{in the } x_{\mathrm{k}} \text{ direction.} \tag{3.85}$$

Such is the case for the normal direction in figure 3.14, where the component of imposed velocity is that of the tool.

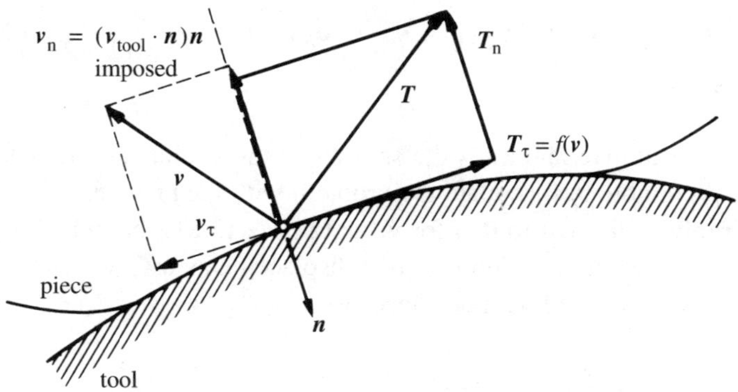

Figure 3.14 Boundary conditions for a problem in mechanics.

For the components where the velocity is not imposed, a choice exists:

$$\text{either } T_k = \overline{T}_k \text{ or } T_k = f(\boldsymbol{v}) \tag{3.86}$$

where $f(\boldsymbol{v})$ is a function of the velocity vector, which is unknown. Such is the case for the tangential component in figure 3.14, $f(\boldsymbol{v})$ coming from a friction law at the tool surface. The reader is referred to chapter 6 for a more detailed discussion of this point, as well as for a treatment of the rotations needed for the expression of the boundary conditions, as the coordinates x_k do not necessarily correspond to the laboratory coordinates.

For the continuous formulation, it is preferable to maintain the notations $\partial\Omega_D$ and $\partial\Omega_N$, to avoid needlessly encumbering the expressions. This choice has no effect on the results. However, we will see that a clarification is possible in the expression of the discretized formulation.

3.4.2 Linear elasticity example. Existence of a functional

Hooke's law for the behavior of a solid, characteristic of ***linear elasticity***, links the stress tensor $\boldsymbol{\sigma}$ to the strain tensor $\boldsymbol{\varepsilon}$ (chap. 1 equation(1.87)):

$$\sigma_{ij} = \frac{E}{1+\nu_P}\varepsilon_{ij} + \frac{E\nu_P}{(1+\nu_P)(1-2\nu_P)}\mathrm{Tr}(\boldsymbol{\varepsilon})\delta_{ij} \tag{3.87}$$

where E is Young's modulus and ν_P is Poisson's coefficient. The equation can also be expressed in tensor form:

$$\boldsymbol{\sigma} = \boldsymbol{D}^{e\ell}\boldsymbol{\varepsilon} \tag{3.88}$$

where $\boldsymbol{D}^{e\ell}$ is the elasticity tensor, of 4th order. In d dimensions, this equation can be written for the components as follows:

$$\sigma_{ij} = D^{e\ell}_{ijkl}\varepsilon_{kl} \quad \text{with} \quad D^{e\ell}_{ijkl} = \frac{E}{1+\nu_P}\delta_{ik}\delta_{jl} + \frac{E\nu_P}{(1+\nu_P)(1-2\nu_P)}\delta_{ij}\delta_{kl}$$

for $i, j, k, l = 1, d$ (3.89)

As the constitutive law links the stresses to the strains, and thus to displacements, it is preferable to formulate the problem with the ***principle of virtual work*** which is strictly equivalent to that for virtual power (3.81), but which replaces the velocity field $\boldsymbol{v}^*$ by a field of infinitesimal displacements $\delta\boldsymbol{u}^*$; which, denoting $\delta\boldsymbol{\varepsilon}^*$ the strain tensor associated with $\delta\boldsymbol{u}^*$, leads to:

$$\int_{\Omega} \boldsymbol{\sigma} : \delta\boldsymbol{\varepsilon}^* \, dV = \int_{\partial\Omega} \boldsymbol{T} \cdot \delta\boldsymbol{u}^* \, dS \qquad \text{for all } \delta\boldsymbol{u}^* \tag{3.90}$$

If $\boldsymbol{u}$ is the vector displacement for every point with respect to its initial position in Ω, without internal stress, and if $\boldsymbol{\varepsilon}$ is the strain tensor associated with $\boldsymbol{u}$ at every point, then the principle of virtual work (3.90) takes the form:

$$\int_{\Omega} \boldsymbol{D}^{e\ell}\boldsymbol{\varepsilon} : \delta\boldsymbol{\varepsilon}^* \, dV = \int_{\partial\Omega} \boldsymbol{T} \cdot \delta\boldsymbol{u}^* \, dS \qquad \text{for all } \delta\boldsymbol{u}^* \tag{3.91}$$

If we consider $\delta\boldsymbol{u}^*$ to be the variation $\delta\boldsymbol{u}$ of the real displacements $\boldsymbol{u}$, then the scalar value $\boldsymbol{D}^{e\ell}\boldsymbol{\varepsilon} : \delta\boldsymbol{\varepsilon}$ appears as the variation of the term $1/2\boldsymbol{D}^{e\ell}\boldsymbol{\varepsilon} : \boldsymbol{\varepsilon}$. The verification is simple, based on the symmetry (ij, kl) of the elasticity tensor $\boldsymbol{D}^{e\ell}$. We have:

$$\boldsymbol{D}^{e\ell}\delta\boldsymbol{\varepsilon} : \boldsymbol{\varepsilon} = D^{e\ell}_{ijkl}\delta\varepsilon_{kl}\varepsilon_{ij} = D^{e\ell}_{klij}\delta\varepsilon_{kl}\varepsilon_{ij} = D^{e\ell}_{klij}\varepsilon_{ij}\delta\varepsilon_{kl} = \boldsymbol{D}^{e\ell}\boldsymbol{\varepsilon} : \delta\boldsymbol{\varepsilon} \tag{3.92}$$

and thus:

$$\delta\left(\tfrac{1}{2}\boldsymbol{D}^{e\ell}\boldsymbol{\varepsilon} : \boldsymbol{\varepsilon}\right) = \tfrac{1}{2}\boldsymbol{D}^{e\ell}\boldsymbol{\varepsilon} : \delta\boldsymbol{\varepsilon} + \tfrac{1}{2}\boldsymbol{D}^{e\ell}\delta\boldsymbol{\varepsilon} : \boldsymbol{\varepsilon} = \boldsymbol{D}^{e\ell}\boldsymbol{\varepsilon} : \delta\boldsymbol{\varepsilon} \tag{3.93}$$

Equation (3.91) appears in these conditions as a stationary condition for the quantity W defined as:

$$W = \int_{\Omega} \tfrac{1}{2}\boldsymbol{D}^{e\ell}\boldsymbol{\varepsilon} : \boldsymbol{\varepsilon} \, dV - \int_{\partial\Omega} \boldsymbol{T} \cdot \boldsymbol{u} \, dS \tag{3.94}$$

which is the ***elastic potential energy*** and which is minimized at equilibrium. We have thus seen two equivalent expressions of the equilibrium equation for an elastic medium: first the principle of virtual work or power, and second, the variational formulation which consists of minimizing a function for the elastic potential energy.

The variational principle or the minimization of a functional (sum of integral functions over the domain and, possibly, the boundary) is shared by many problems that are called *conservative*. We will see another example in chapter 6 (section 6.5.3): the viscoplastic functional.

3.4.3 Discretization

Now we introduce finite element discretization in the principles discussed above. For the virtual and real velocity fields we have:

$$\boldsymbol{v}^* = \psi_n \boldsymbol{v}^{*n} \quad \text{and} \quad \boldsymbol{v} = \psi_n \boldsymbol{v}^n \tag{3.95}$$

where *n* varies from 1 to *Nn*, the number of nodes. The components of the tensor velocity gradient (virtual or real) can be easily expressed as functions of the components of the nodal velocity field:

$$\frac{\partial v_i^*}{\partial x_j} = \frac{\partial}{\partial x_j}(\psi_n v_i^{*n}) = \frac{\partial \psi_n}{\partial x_j} v_i^{*n} \tag{3.96}$$

where *i* and *j* vary from 1 to *d*, the spatial dimension of the problem. In general the components (i, j) of the strain rate tensor associated with $\boldsymbol{v}^*$ can be linked to the components (*n*, *k*) of the velocity field:

$$\dot{\varepsilon}_{ij}^* = \frac{1}{2}\left(\frac{\partial v_i^*}{\partial x_j} + \frac{\partial v_j^*}{\partial x_i}\right) = B_{ijnk} v_k^{*n} \tag{3.97}$$

where *k* varies from 1 to *N*, the number of unknowns per node[1]. The coefficients B_{ijnk} are given by:

$$B_{ijnk} = \frac{1}{2}\left(\frac{\partial \psi_n}{\partial x_j}\delta_{ik} + \frac{\partial \psi_n}{\partial x_i}\delta_{jk}\right) \tag{3.98}$$

Applying this discretization to the principle of virtual power (3.81):

$$\int_\Omega \sigma_{ij} B_{ijnk} v_k^{*n} \, dV = \int_{\partial\Omega} T_k \psi_n v_k^{*n} \, dS \quad \text{for all } v_k^{*n} \tag{3.99}$$

As the values v_k^{*n} can be chosen arbitrarily, the following can be deduced,

$$\int_\Omega \sigma_{ij} B_{ijnk} \, dV = \int_{\partial\Omega} T_k \psi_n \, dS \quad \text{for all } (n, k)\,. \tag{3.100}$$

1 Most often, $N = d$ (2 unknown velocity components in 2 dimensions, 3 in 3 dimensions), but it is not always true: for example, there can be three unknowns v_r, v_θ, v_z for a problem posed in a two dimensional section (*r*, *z*) of a domain.

The choice of the spatial discretization (3.95a) corresponds to a Galerkin formulation. As an exercise, the reader can verify that the result (3.100) is equivalent to a Galerkin formulation applied to each component of the local equilibrium equation **div** $\boldsymbol{\sigma} = \mathbf{0}$:

$$\int_{\Omega} \psi_{\mathrm{n}} \frac{\partial \sigma_{\mathrm{kj}}}{\partial x_{\mathrm{j}}} \, \mathrm{d}V = 0 \qquad \text{for all } (n, k) \tag{3.101}$$

Discretization of the boundary conditions. Nodal reactions

The imposition of velocity components for some nodes on the boundary $\partial\Omega$ (3.83) is expressed after discretization by:

$$v_{\mathrm{k}}^{\mathrm{n}} = \bar{v}_{\mathrm{k}}^{\mathrm{n}} \tag{3.102}$$

for certain degrees of freedom (n, k). We distinguish thus between "imposed" degrees of freedom (abbreviated *dof*) and "free", using again the notation N^{free} and N^{imp} introduced in paragraph 3.3.2 for the sets of *dof* which are "free" and "imposed", respectively. For these conditions, we can choose a discretized virtual velocity field which is zero for the "imposed" *dof*, that is, which satisfies:

$$v_{\mathrm{k}}^{*\mathrm{n}} = 0 \qquad \text{for } (n, k) \in N^{\mathrm{imp}} \tag{3.103}$$

Such a field is called ***kinematically admissible to zero***. Thus one must solve a system of (NnN) equations with as many unknowns:

$$\int_{\Omega} \sigma_{\mathrm{ij}} B_{\mathrm{ijnk}} \, \mathrm{d}V = \int_{\partial\Omega} T_{\mathrm{k}} \psi_{\mathrm{n}} \, \mathrm{d}S \qquad \text{for } (n, k) \in N^{\mathrm{free}} \tag{3.104}$$

and

$$v_{\mathrm{k}}^{\mathrm{n}} = \bar{v}_{\mathrm{k}}^{\mathrm{n}} \qquad \text{for } (n, k) \in N^{\mathrm{imp}} \tag{3.105}$$

Imagine now *the same mechanical equilibrium*, but this time obtained with *nodal forces* $\overline{\boldsymbol{F}}$ acting in the direction of the "imposed" *dof*, instead of the previously imposed velocity conditions (fig. 3.15).

The principle of virtual power in this case is written in continuous form:

$$\int_{\Omega} \boldsymbol{\sigma} : \dot{\boldsymbol{\varepsilon}}^{*} \, \mathrm{d}V - \int_{\partial\Omega} \boldsymbol{T} \cdot \boldsymbol{v}^{*} \, \mathrm{d}S - \sum_{\substack{\text{points M} \\ \text{with imposed } dof}} \overline{\boldsymbol{F}}^{\mathrm{M}} \cdot \boldsymbol{v}^{*\mathrm{M}} = 0 \tag{3.106}$$

and in discrete form:

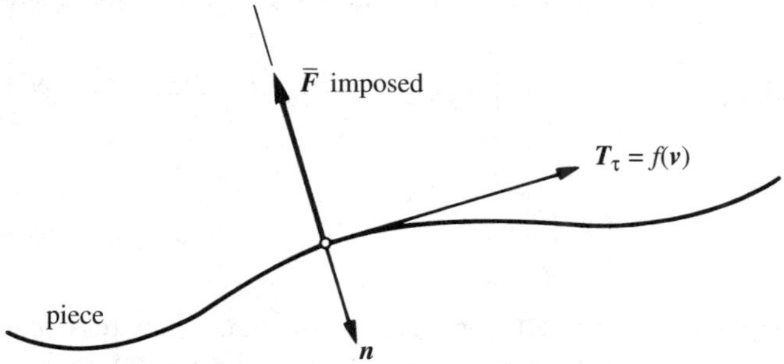

Figure 3.15 Applied normal force corresponding to the imposition of a perpendicular degree of freedom of the velocity field in figure 3.14.

$$\int_{\Omega} \sigma_{ij} B_{ijnk}\, dV - \int_{\partial\Omega} T_k \Psi_n\, dS = 0 \qquad \text{for } (n,k) \in N^{free} \tag{3.107}$$

$$\int_{\Omega} \sigma_{ij} B_{ijnk}\, dV - \int_{\partial\Omega} T_k \Psi_n\, dS - \bar{F}_k^n = 0 \qquad \text{for } (n,k) \in N^{imp} \tag{3.108}$$

We then conclude from (3.108) that the components of the *applied nodal forces* (the ***nodal reactions***) associated with the imposed degrees of freedom of velocity are in fact equal to the *residuals of the equilibrium equation according to these dofs*:

$$F_k^n = \int_{\Omega} \sigma_{ij} B_{ijnk}\, dV - \int_{\partial\Omega} T_k \Psi_n\, dS \tag{3.109}$$

Linear elasticity example

In the linear elastic case, the stresses are expressed with the constitutive law (3.89) :

$$\sigma_{ij} = D^{e\ell}_{ijpq} \varepsilon_{pq} \tag{3.110}$$

In the same way as for (3.97), the components ε_{pq} of the strain tensor can be expressed as a function of the nodal components of the displacement field, u_l^m, with the coefficients B_{pqml}:

$$\sigma_{ij} = D^{e\ell}_{ijpq} B_{pqml} u_l^m \tag{3.111}$$

Finally, we have a system of (NnN) equations to solve with as many unknowns:

$$\int_{\Omega} B_{\mathrm{ijnk}} D^{\mathrm{e}\ell}_{\mathrm{ijpq}} B_{\mathrm{pqml}} u^{\mathrm{m}}_{\mathrm{l}} \, \mathrm{d}V - \int_{\partial\Omega} T_{\mathrm{k}} \psi_{\mathrm{n}} \, \mathrm{d}S = 0 \qquad \text{for } (n, k) \in N^{\text{free}} \tag{3.112}$$

and

$$u^{\mathrm{n}}_{\mathrm{k}} = \bar{u}^{\mathrm{n}}_{\mathrm{k}} \qquad \text{for } (n, k) \in N^{\text{imp}} \tag{3.113}$$

As long as the elastic coefficients are constant and the stress vectors fixed (or can be expressed as linear functions of the displacement field), these equations are linear in the nodal unknowns $u^{\mathrm{m}}_{\mathrm{l}}$. The system can then be written:

$$K_{\mathrm{nkml}} u^{\mathrm{m}}_{\mathrm{l}} = F_{\mathrm{nk}} \qquad n, m = 1, Nn \quad \text{and} \quad k, l = 1, N \tag{3.114}$$

The system matrix, $[\boldsymbol{K}]$ – the *stiffness matrix* – has dimensions $(NnN) \times (NnN)$. When the stress vectors are fixed, the components (nk, ml) of the matrix are given by:

$$K_{\mathrm{nkml}} = \int_{\Omega} B_{\mathrm{ijnk}} D^{\mathrm{e}\ell}_{\mathrm{ijpq}} B_{\mathrm{pqml}} \, \mathrm{d}V \qquad \text{for } (n, k) \in N^{\text{free}} \tag{3.115}$$

$$K_{\mathrm{nkml}} = \delta_{\mathrm{nm}} \delta_{\mathrm{kl}} \qquad \text{for } (n, k) \in N^{\text{imp}} \tag{3.116}$$

The term on the right hand side, $\boldsymbol{F}$, is a vector of dimension (NnN), for which the components (nk) are given by:

$$F_{\mathrm{nk}} = \int_{\partial\Omega} T_{\mathrm{k}} \psi_{\mathrm{n}} \, \mathrm{d}S \qquad \text{for } (n, k) \in N^{\text{free}} \tag{3.117}$$

$$F_{\mathrm{nk}} = \bar{u}^{\mathrm{n}}_{\mathrm{k}} \qquad \text{for } (n, k) \in N^{\text{imp}} \tag{3.118}$$

It can be verified that the minimization of the functional W (3.94) with respect to the nodal unknowns, that is writing $\partial W / \partial u^{\mathrm{n}}_{\mathrm{k}} = 0$ for every pair (n, k), leads to exactly the same result.

Lastly, it is seen that the system matrix, $[\boldsymbol{K}]$, is symmetric (nk, ml). It must be noted in this context that the minimization of a functional always leads to the resolution of a symmetric system, unlike the application of the Galerkin method; which in the case where one can not define a functional, can lead to non symmetric systems such as (3.64) for example.

3.4.4 Matrix Presentation

Take the example of a two dimensional mechanics calculation (the extension to three dimensions causes no specific difficulties) in which we determine the two components of the displacement vector $\boldsymbol{u}$ at each node of the finite element discretization.

We denote by $\boldsymbol{U}$ the vector of the nodal components of the field $\boldsymbol{u}(\boldsymbol{x})$.

$$\boldsymbol{U} = \begin{bmatrix} u_1^1 \\ u_2^1 \\ u_1^2 \\ u_2^2 \\ \vdots \\ u_1^{\mathrm{Nn}} \\ u_2^{\mathrm{Nn}} \end{bmatrix} \tag{3.119}$$

We have for each of the components $u_1(\boldsymbol{x})$ and $u_2(\boldsymbol{x})$ the fundamental relation of interpolation (3.1), in matrix form:

$$\begin{bmatrix} u_1(\boldsymbol{x}) \\ u_2(\boldsymbol{x}) \end{bmatrix} = \begin{bmatrix} \psi_1(\boldsymbol{x}) & 0 & \psi_2(\boldsymbol{x}) & 0 & \ldots & \psi_{\mathrm{Nn}}(\boldsymbol{x}) & 0 \\ 0 & \psi_1(\boldsymbol{x}) & 0 & \psi_2(\boldsymbol{x}) & \ldots & 0 & \psi_{\mathrm{Nn}}(\boldsymbol{x}) \end{bmatrix} \begin{bmatrix} u_1^1 \\ u_2^1 \\ \vdots \\ u_1^{\mathrm{Nn}} \\ u_2^{\mathrm{Nn}} \end{bmatrix} \tag{3.120}$$

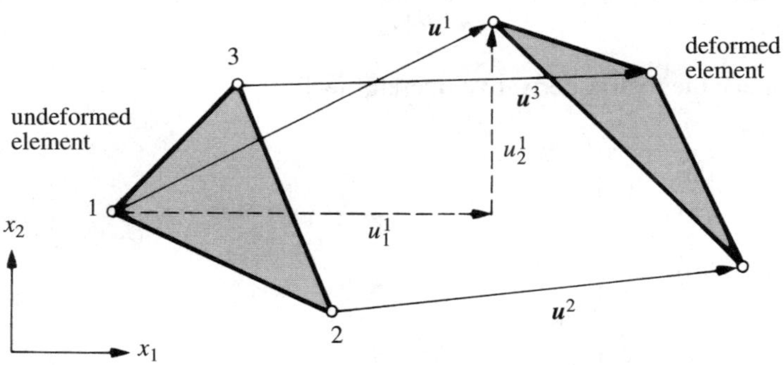

Figure 3.16 Displacements of the nodes of a linear triangular finite element.

or further condensed,

$$u(x) = [\Psi(x)]U \tag{3.121}$$

Note that this time $[\Psi]$ is a matrix of dimension $2 \times (2Nn)$, or, generally, $N \times (NNn)$. At the level of each element, we have similarly:

$$u = \begin{bmatrix} u_1(x) \\ u_2(x) \end{bmatrix} = \begin{bmatrix} \psi_1^e(x) & 0 & \psi_2^e(x) & 0 & \dots & \psi_{nne}^e(x) & 0 \\ 0 & \psi_1^e(x) & 0 & \psi_2^e(x) & \dots & 0 & \psi_{nne}^e(x) \end{bmatrix} \begin{bmatrix} u_1^1 \\ u_2^1 \\ \vdots \\ u_1^{nne} \\ u_2^{nne} \end{bmatrix}$$

$$= [\Psi^e]U^e \tag{3.122}$$

Suppose now that from the displacement field we wish to calculate the components of the strain tensor, that we give here in vector form:

$$\varepsilon(x) = \begin{bmatrix} \dfrac{\partial u_1}{\partial x_1} \\ \dfrac{\partial u_2}{\partial x_2} \\ \dfrac{\partial u_1}{\partial x_2} + \dfrac{\partial u_2}{\partial x_1} \end{bmatrix} = \begin{bmatrix} \varepsilon_{11}(x) \\ \varepsilon_{22}(x) \\ 2\varepsilon_{12}(x) \end{bmatrix} \tag{3.123}$$

The vector ε contains the three components of the strain tensor, except for an extra factor of 2 in the third component. Considering that the derivatives of u are expressed in terms of finite element interpolation:

$$\frac{\partial u_i(x)}{\partial x_k} = \frac{\partial}{\partial x_k}(\psi_p(x)u_i^p) = \frac{\partial \psi_p(x)}{\partial x_k}u_i^p \qquad (p = 1, Nn) \tag{3.124}$$

we introduce the matrix derivative operator $[\nabla]$:

$$[\nabla] = \begin{bmatrix} \dfrac{\partial}{\partial x_1} & 0 \\ 0 & \dfrac{\partial}{\partial x_2} \\ \dfrac{\partial}{\partial x_2} & \dfrac{\partial}{\partial x_1} \end{bmatrix} \tag{3.125}$$

In this manner, we have:

$$\boldsymbol{\varepsilon}(\boldsymbol{x}) = [\nabla]\boldsymbol{u}(\boldsymbol{x}) = [\nabla][\boldsymbol{\Psi}(\boldsymbol{x})]\boldsymbol{U} = [\boldsymbol{B}]\boldsymbol{U} \tag{3.126}$$

where the matrix $[\boldsymbol{B}]$, defined by the product $[\nabla]\,[\boldsymbol{\Psi}]$, is expressed as follows.

$$[\boldsymbol{B}] = [\nabla][\boldsymbol{\Psi}] = \begin{bmatrix} \dfrac{\partial\psi_1}{\partial x_1} & 0 & \dfrac{\partial\psi_2}{\partial x_1} & 0 & & \dfrac{\partial\psi_{\mathrm{Nn}}}{\partial x_1} & 0 \\ 0 & \dfrac{\partial\psi_1}{\partial x_2} & 0 & \dfrac{\partial\psi_2}{\partial x_2} & \cdots & 0 & \dfrac{\partial\psi_{\mathrm{Nn}}}{\partial x_2} \\ \dfrac{\partial\psi_1}{\partial x_2} & \dfrac{\partial\psi_1}{\partial x_1} & \dfrac{\partial\psi_2}{\partial x_2} & \dfrac{\partial\psi_2}{\partial x_1} & & \dfrac{\partial\psi_{\mathrm{Nn}}}{\partial x_2} & \dfrac{\partial\psi_{\mathrm{Nn}}}{\partial x_1} \end{bmatrix} \tag{3.127}$$

In an analogous manner, we can use formulas (3.123) through (3.126) to express the strain rate tensor $\dot{\boldsymbol{\varepsilon}}^*$ as a function of the discretized virtual velocity field, $\boldsymbol{v}^*$. Using the same vector notation,

$$\dot{\boldsymbol{\varepsilon}}^* = \begin{bmatrix} \dfrac{\partial v_1^*}{\partial x_1^*} \\ \dfrac{\partial v_2^*}{\partial x_2} \\ \dfrac{\partial v_1^*}{\partial x_2} + \dfrac{\partial v_2^*}{\partial x_1} \end{bmatrix} = \begin{bmatrix} \dot{\varepsilon}_{11}^* \\ \dot{\varepsilon}_{22}^* \\ 2\dot{\varepsilon}_{12}^* \end{bmatrix} \tag{3.128}$$

and we have:

$$\dot{\boldsymbol{\varepsilon}}^* = [\nabla]\boldsymbol{v}^* = [\nabla][\boldsymbol{\Psi}]\boldsymbol{V}^* = [\boldsymbol{B}]\boldsymbol{V}^* \tag{3.129}$$

where $\boldsymbol{V}^*$ denotes the nodal virtual velocity vector. The vectorized form of the stress tensor is given by:

$$\boldsymbol{\sigma} = \begin{bmatrix} \sigma_{11} \\ \sigma_{22} \\ \sigma_{12} \end{bmatrix} \tag{3.130}$$

Thus the volumetric strain power associated with the virtual velocity field $\boldsymbol{v}^*$ is given by the scalar product of the column vectors $\dot{\boldsymbol{\varepsilon}}^*$ and $\boldsymbol{\sigma}$:

$$\sigma_{\mathrm{ij}}\dot{\varepsilon}_{\mathrm{ij}}^* = (\dot{\boldsymbol{\varepsilon}}^*)^{\mathrm{T}}\boldsymbol{\sigma} \quad (i, j = 1, 2) \tag{3.131}$$

In this way the matrix form of the virtual power principle (3.81) – considered for the case without mass forces or inertia – is the following:

$$\int_{\Omega} (\dot{\boldsymbol{\varepsilon}}^*)^{\mathrm{T}} \boldsymbol{\sigma} \, \mathrm{d}V = \int_{\partial\Omega} ([\boldsymbol{\Psi}]\boldsymbol{V}^*)^{\mathrm{T}} \boldsymbol{T} \, \mathrm{d}S \qquad \text{for all } \boldsymbol{V}^* \tag{3.132}$$

Using (3.129), we obtain:

$$\int_{\Omega} \boldsymbol{V}^{*\mathrm{T}} [\boldsymbol{B}]^{\mathrm{T}} \boldsymbol{\sigma} \, \mathrm{d}V = \int_{\partial\Omega} \boldsymbol{V}^{*\mathrm{T}} [\boldsymbol{\Psi}]^{\mathrm{T}} \boldsymbol{T} \, \mathrm{d}S \qquad \text{for all } \boldsymbol{V}^* \tag{3.133}$$

Because this relation must be satisfied for every field $\boldsymbol{V}^*$, we have finally:

$$\int_{\Omega} [\boldsymbol{B}]^{\mathrm{T}} \boldsymbol{\sigma} \, \mathrm{d}V = \int_{\partial\Omega} [\boldsymbol{\Psi}]^{\mathrm{T}} \boldsymbol{T} \, \mathrm{d}S \tag{3.134}$$

Note that this equality is simply the discrete matrix expression of the virtual power principle, and that it is thus valid regardless of the constitutive law for the material under consideration. We will see the applications in chapter 6.

Application in the linear elastic case

If we continue now, the matrix expression of (3.87) in the context of linear elasticity and taking into account the vectorization of the stress tensor and the deformation tensor, is:

$$\boldsymbol{\sigma} = [\boldsymbol{D}^{\mathrm{e}\ell}] \boldsymbol{\varepsilon} \tag{3.135}$$

with

$$[\boldsymbol{D}^{\mathrm{e}\ell}] = \begin{bmatrix} \dfrac{E(1-\nu_{\mathrm{P}})}{(1+\nu_{\mathrm{P}})(1-2\nu_{\mathrm{P}})} & \dfrac{E\nu_{\mathrm{P}}}{(1+\nu_{\mathrm{P}})(1-2\nu_{\mathrm{P}})} & 0 \\ \dfrac{E\nu_{\mathrm{P}}}{(1+\nu_{\mathrm{P}})(1-2\nu_{\mathrm{P}})} & \dfrac{E(1-\nu_{\mathrm{P}})}{(1+\nu_{\mathrm{P}})(1-2\nu_{\mathrm{P}})} & 0 \\ 0 & 0 & \dfrac{E}{2(1+\nu_{\mathrm{P}})} \end{bmatrix} \tag{3.136}$$

$$\boldsymbol{\varepsilon} = \begin{bmatrix} \varepsilon_{11} \\ \varepsilon_{22} \\ 2\varepsilon_{12} \end{bmatrix} \qquad \boldsymbol{\sigma} = \begin{bmatrix} \sigma_{11} \\ \sigma_{22} \\ \sigma_{12} \end{bmatrix} \tag{3.137}$$

With the help of (3.126), (3.134) becomes:

$$\int_{\Omega} [\boldsymbol{B}]^{\mathrm{T}}[\boldsymbol{D}^{\mathrm{e}\ell}][\boldsymbol{B}]\boldsymbol{U}\ \mathrm{d}V = \int_{\partial\Omega} [\boldsymbol{\Psi}]^{\mathrm{T}}\boldsymbol{T}\ \mathrm{d}S \tag{3.138}$$

The vector $\boldsymbol{U}$ of nodal displacements is then a solution of the linear system

$$[\boldsymbol{K}]\boldsymbol{U} = \boldsymbol{F} \tag{3.139}$$

in which the matrix and the right hand side term are given in matrix form below, and as components by equations (3.115) and (3.117).

$$[F] = \int_{\Omega} [\boldsymbol{B}]^{\mathrm{T}}[\boldsymbol{D}^{\mathrm{e}\ell}][\boldsymbol{B}]\ \mathrm{d}V \quad \text{and} \quad [F] = \int_{\partial\Omega} [\boldsymbol{\Psi}]^{\mathrm{T}}\boldsymbol{T}\ \mathrm{d}S \tag{3.140}$$

Concluding this section on matrix notation, it should be said that it is of limited interest for the finite element program developer. In fact, the indexed notation is much closer to that which is actually programmed and leads to a more efficient vectorization. However, the matrix notation is useful for more theoretical studies of the convergence of the method, error estimation and the mesh adaptation.

3.5 IMPLEMENTATION

3.5.1 Grid and mesh notions

We call a ***grid*** the minimal description of the finite element discretization of the studied domain. It is composed of the coordinates of the nodes that make it up as well as the connectivity, that is, the manner in which the elements are formed. It is a purely geometric description.

In practice, the set of data related to the grid consists of the list of nodes and their coordinates as well as the element connection matrix which specifies each element by numbering the list of nodes of which it is composed. Figure 3.17 shows a simple example of a grid. Note that the list of nodes for each element is given counterclockwise indifferently to the first node in the list.

The ***mesh*** is a richer description than the grid of the approximation by finite elements. It contains supplementary information relative to the type of elements used, to integration, the boundaries, the boundary conditions, etc. Some of the information susceptible to being found in a mesh file are:

- The type of approximation chosen on each element (interpolation functions) as well as the type of integration (number of Gauss integration points, their position in the reference element and weights); this for the different fields approximated (interpolation is often performed differently for the velocity and pressure fields in fluid mechanics for example: cf (3.31) and (3.32)). In

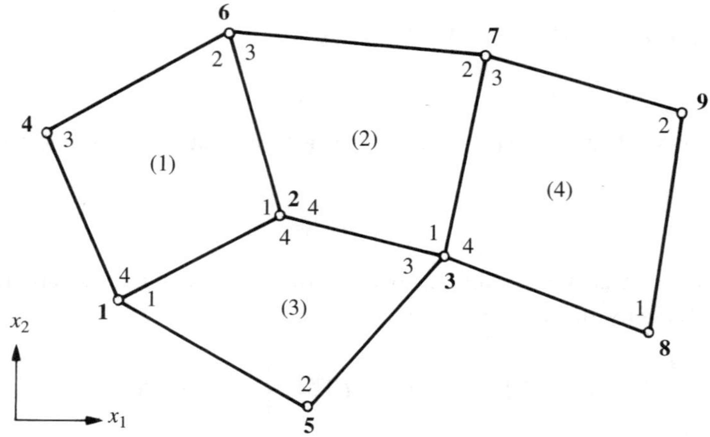

Node coordinates

1 $x_1 = .$ $x_2 = .$
2
3
⋮ ⋮ ⋮
9

connectivity

1: 2 6 4 1
2: 3 7 6 2
3: 1 5 3 2
4: 8 9 7 3

Figure 3.17 Example of a grid in 2 dimensions. The elements are linear quadrangles. The global number for each node is in bold type, the local number of the node in normal type, and the element number is in parentheses.

general, in the mesh file we use a code for each element type that refers to a library of elements which is subsequently read.

- The location of the boundary facets in 3 dimensions, or the edges in 2 dimensions. The number of the element containing the boundary facet (edge) i, as well as the local numbering of that facet (edge) in the element can also be found in the file.
- The symmetries. In 2 dimensions, the axes of symmetry are defined in the mesh file by the coordinates of a point on the axis and a direction vector, and the axis to which each boundary edge belongs can be specified. Similarly in 3 dimensions, where symmetry planes are defined by a point on the plane and the normal vector, the plane to which a boundary facet belongs can be specified.
- In the case where different types of elements are used simultaneously, the notion of a group of elements of the same type is employed. For each element, the mesh file then contains the number of the group to which it belongs.
- The same notion of groups of elements can also be used for computations mixing elements with different properties (computation with multiple materials). Similarly, it can be practical to group facets (or edges in 2 dimensions) to more easily impose boundary conditions on the domain.

The liberty of the programmer to define the data structure can be imagined. In particular, for the description of the boundaries and the boundary conditions, he will be tempted to introduce supplementary matrices into the mesh file that can help avoid repeated calculations later and thus to reduce the total computation time. This is one of the aspects of the definition of a ***data structure*** that affects the efficiency of a finite element computation. However, when two different computation programs are to be used on the same problem, the compatibility of the data structures needs to be enforced, which may require the development of interface programs for that purpose.

3.5.2 Assembly operations

Up to now we have seen that the spatial discretization of the weak formulation with finite elements, as for the finite difference method, results finally in the solution of a system of linear equations, possibly at the price of linearizing the initial problem if it is non linear (chap. 4). The matrix and the right hand side vector for this system are composed of integrals on the domain Ω or its boundary $\partial\Omega$ (cf (3.65) for example). In section 3.2 we have seen the procedure for the calculation of an integral on a finite element. The link between the local integral at the element level and the global integral is yet to be defined; this is the ***assembly operation***, which plays a central role in the implementation of the method. Its principle, which we will explain in detail below, is the following: the calculation of a global integral is subdivided into calculations of the contribution of each element, then the global problem is reconstituted to have the solution in terms of the nodal field.

Consider then a subdivision of the domain Ω into *Ne* elements each occupying a domain Ω^e. Each element possesses *nne* interpolation nodes *locally* numbered from 1 to *nne*. At each node numbered *k* locally, there is a corresponding global number v_k^e. Thus, in the example in figure 3.17, we have for the element $e = 2$:

$$v_1^2 = 3 \qquad v_2^2 = 7 \qquad v_3^2 = 6 \qquad v_4^2 = 2 \tag{3.141}$$

Consider again the stationary diffusion example, but this time without advective transport, to be more clear. The discretized weak formulation results in a global system of equations, from (3.62):

$$c^j \left[\int_{\Omega} D \,\mathbf{grad}\, \psi_j \cdot \mathbf{grad}\, \psi_i \,\mathrm{d}V + \int_{\partial\Omega_N} \alpha \psi_i \psi_j \,\mathrm{d}S \right] = \int_{\partial\Omega_N} \alpha \bar{c}_a \psi_i \,\mathrm{d}S \tag{3.142}$$

for $i = 1, Nn$

We can write each integral as a sum of integrals calculated on the different elements. Then, if the volumetric and surface contributions of the system matrix $[\boldsymbol{K}]$

are separated, we have for the volumetric part, denoted $[\boldsymbol{K}^{\text{vol}}]$:

$$K_{ij}^{vol} = \int_{\Omega} D\ \mathbf{grad}\ \psi_j \cdot \mathbf{grad}\ \psi_i\ dV = \sum_{e=1}^{Ne} \left(\int_{\Omega^e} D\ \mathbf{grad}\ \psi_j \cdot \mathbf{grad}\ \psi_i\ dV \right) = \sum_{e=1}^{Ne} K_{ij}^{e,\ vol} \tag{3.143}$$

It must be noted that in this expression, *i* and *j* are the *global* numbers for the nodes, and that the interpolation functions are extended at value zero into the elements that do not contain the nodes *i* and *j*. The term (*i*, *j*) in the integral over Ω^e is thus null if either of the two nodes *i* or *j* does not belong to the element *e*. Also, the term K_{ij}^{vol} is the sum of several element contributions if *i* and *j* are shared by several elements.

We then obtain the following strategy for the calculation of the matrix $[\boldsymbol{K}^{\text{vol}}]$. On each element Ω^e, and for each pair of local node numbers (k, k') of element *e*, we calculate:

$$K_{kk'}^{e,vol} = \int_{\Omega^e} D\ \mathbf{grad}\ \psi_{\nu_k^e} \cdot \mathbf{grad}\ \psi_{\nu_{k'}^e}\ dV = \int_{\Omega^e} D\ \mathbf{grad}\ \psi_k^e \cdot \mathbf{grad}\ \psi_{k'}^e\ dV \tag{3.144}$$

which is the term (k, k') of the elementary matrix $[\boldsymbol{K}^{e,\text{vol}}]$. We then add this term with the already existing term $(\nu_k^e, \nu_{k'}^e)$ of the global matrix $[\boldsymbol{K}^{\text{vol}}]$:

$$K_{\nu_k^e \nu_{k'}^e}^{vol} + K_{kk'}^{e,\ vol} \rightarrow K_{\nu_k^e \nu_{k'}^e}^{vol} \tag{3.145}$$

This *assembly* operation allows the building of the global matrix $[\boldsymbol{K}^{\text{vol}}]$ (of dimension $Nn \times Nn$) from the local matrices $[\boldsymbol{K}^{e,\ \text{vol}}]$ (of dimension $nne \times nne$). The matrix $[\boldsymbol{K}^{\text{vol}}]$ is then finally obtained after processing of all the elements $e = 1, Ne$.

The same strategy is obviously applied for the surface contribution and for the term on the right hand side. In the first case we have:

$$K_{ij}^{surf} = \int_{\partial\Omega_N} \alpha \psi_i \psi_j\ dS = \sum_{e=1}^{Ne} \left(\int_{\partial\Omega_N^e} \alpha \psi_i \psi_j\ dS \right) = \sum_{e=1}^{Ne} K_{ij}^{e,\ surf} \tag{3.146}$$

where $\partial\Omega_N^e$ designates the set of boundary faces of the element *e* belonging to the discretization of the surface $\partial\Omega_N$ under a Cauchy condition. If the element *e* has no boundary face its contribution to $[\boldsymbol{K}^{\text{surf}}]$ is null.

Finally, for the right hand side term, one proceeds in the same way with:

$$b_i = \int_{\partial\Omega_N} \alpha \bar{c}_a \psi_i \, dS = \sum_{e=1}^{Ne} \left(\int_{\partial\Omega_N^e} \alpha \bar{c}_a \psi_i \, dS \right) = \sum_{e=1}^{Ne} b_i^e \tag{3.147}$$

$$b_{v_k^e} + \int_{\partial\Omega_N^e} \alpha \bar{c}_a \psi_k^e \, dS \rightarrow b_{v_k^e} \tag{3.148}$$

Practical Example

Consider the very simple mesh of 3.17, composed of four elements.

The assembly of a stiffness matrix [$\boldsymbol{K}$] (for a problem with scalar field such as chemical diffusion or heat transfer) is carried out according to the following steps:

a) Construction of the local matrix [$\boldsymbol{K}^1$], on element 1.
   [$\boldsymbol{K}^1$] is a 4×4 matrix for this linear quadrangular element.
   Local nodes: 1, 2, 3, 4. Global nodes: 2, 6, 4, 1.
b) Assembly of [$\boldsymbol{K}^1$] into [$\boldsymbol{K}$]. The global matrix [$\boldsymbol{K}$] has dimension 9×9 as there are 9 nodes in total.
   Examples of the operations performed: $K_{12}^1 \rightarrow K_{26}$; $K_{13}^1 \rightarrow K_{24}$, etc.
c) Local construction of [$\boldsymbol{K}^2$].
d) Assembly of [$\boldsymbol{K}^2$] into [$\boldsymbol{K}$].
e) Local construction of [$\boldsymbol{K}^3$].
f) Assembly of [$\boldsymbol{K}^3$] into [$\boldsymbol{K}$].
g) Local construction of [$\boldsymbol{K}^4$].
   Assembly of [$\boldsymbol{K}^4$] into [$\boldsymbol{K}$].

Figure 3.18 shows examples of the updating of terms in the matrix [$\boldsymbol{K}$] performed during the assembly for elements $e = 1$ and $e = 2$.

In general, this procedure is easily programmed following the schematic of figure 3.19.

Returning to the example of figure 3.17, the global matrix [$\boldsymbol{K}$] after assembly contains the following element contributions:

$$\left.\begin{aligned} K_{11} &= K_{44}^1 + K_{11}^3 \\ K_{12} &= K_{41}^1 + K_{14}^3 \\ K_{23} &= K_{41}^2 + K_{43}^3 \\ K_{22} &= K_{11}^1 + K_{44}^2 + K_{44}^3 \\ &\text{etc.} \end{aligned}\right\} \tag{3.149}$$

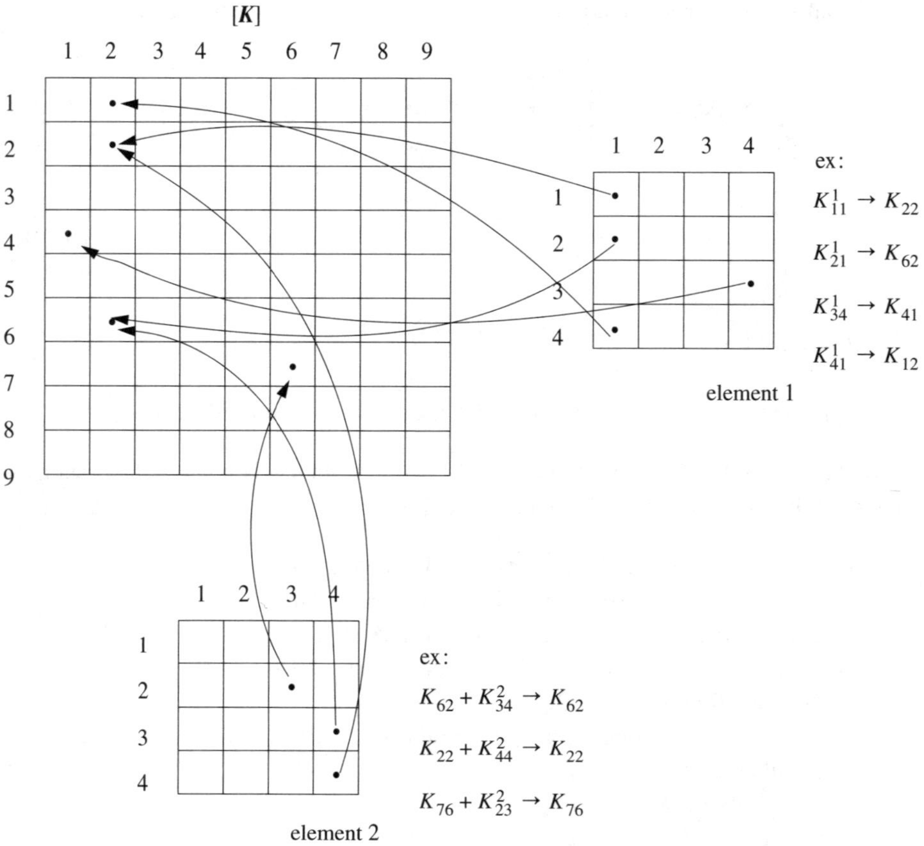

Figure 3.18 Illustration of the assembly of a stiffness matrix, in the case of an unknown *scalar* field, for the mesh of figure 3.17. Accumulation of the elementary matrix terms of the first two elements in the global matrix (only a few operations are indicated by the arrows).

Its general aspect is shown in figure 3.20.

Similarly, the assembly of the right hand side term gives, among other components:

$$\left.\begin{aligned} b_1 &= b^1_4 + b^3_1 \\ b_2 &= 0 \\ b_3 &= b^2_1 + b^3_3 + b^4_4 \\ b_4 &= b^1_3 \\ b_5 &= b^3_2 \\ &\text{etc.} \end{aligned}\right\} \tag{3.150}$$

Initialisation $[\boldsymbol{K}] = 0$
Loop over the elements e

Transfer of the necessary information: global $\rightarrow$ local
Initialization of the elementary matrix $[\boldsymbol{K}^e] = 0$

Loop over the integration points p of the reference element

Calculate the determinant of the Jacobian (real element / reference element)
Loop over the local nodes k
Loop over local nodes k'
Calculate the term kk' at point p
Accumulate in the integrals $K^e_{kk'}$
End of loop over local nodes k'
End of loop over local nodes k

End of loop over integration points p

Assembly of $[\boldsymbol{K}^e]$ into $[\boldsymbol{K}]$

End of loop over elements e

Figure 3.19 Algorithm for calculation of the matrix in a finite element problem.

	1,3	1,3	3	1	3	1			
	1,3	1,2,3	2,3	1	3	1,2	2		
	3	2,3	2,3	–	3	2	2,4	4	4
	1	1	–	1	–	1	–	–	–
$[\boldsymbol{K}]$	3	3	3	–	2,3,4	–	–	–	–
	1	1,2	2	1	–	1,2	2	–	–
		2	2,4	–	–	2	2,4	4	4
			4	–	–	–	4	4	4
			4	–	–	–	4	4	4

Figure 3.20 General aspect of the assembled stiffness matrix for the example presented in figure 3.17 (the numbers appearing in position (i, j) in the matrix are the numbers of the elements that contribute to the term K_{ij}).

3.5.3 Band structure of the matrices and the importance of the numbering

Referring to (3.142), note that all matrices in a finite element problem have a ***band structure*** around the diagonal, as shown in figure 3.21. This comes from the fact that nodes i and j must be neighbors – that is, belong to the same element – to give a non zero term to K_{ij}. As a consequence, on line i of the matrix, only zeros are found past the neighbor the furthest away from i in the sense of the numbering.

In addition, the band structure of the matrix is always symmetric: if a term K_{ij} is non zero (i.e. the nodes i and j both belong to at least one element), then the term K_{ji} is also non zero. Such matrices are called *symmetric profile* matrices. If in addition $K_{ij} = K_{ji}$ for every pair (i, j), the matrix is then symmetric.

The ***bandwidth*** of the matrix, denoted bw, for a problem with one unknown per node, is equal to $(bw_g + 1)$ where bw_g is the maximum difference between the number of two nodes in an element of the mesh. bw_g is called the *geometric bandwidth.* For example, in the case shown in figure 3.17, we have $bw_g = 6 = (9 - 3)$ for element 4. The bandwidth of the matrix is 7, as can be seen in figure 3.20.

We will see later on (chap. 4) that for a banded matrix, the time spent in factorization – which is necessary for the direct solution of a linear symmetric system of equations – is proportional to the product of the bandwidth squared times the number of unknowns. It is thus very important, for a given number of variables, to minimize the geometric bandwidth of a mesh in order to minimize the computation costs. For a given mesh topology, one must thus number the nodes in such a way that bw_g is minimized. Different algorithms have been developed for this purpose, for which we refer the reader to the bibliography at the end of the chapter. Thus, in the current case, the geometric bandwidth can be reduced from 6 to 4 with a better numbering scheme (fig. 3.22).

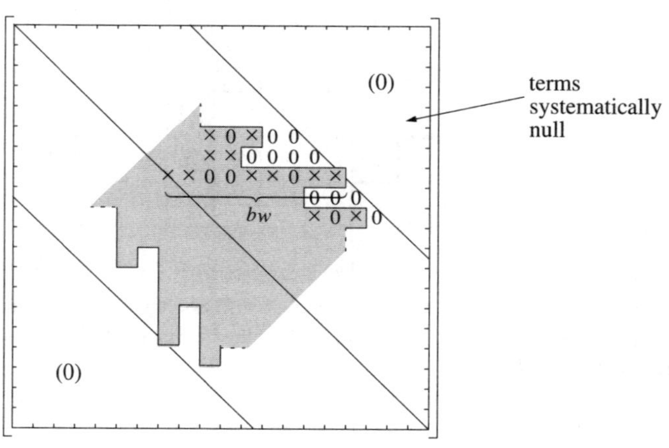

Figure 3.21 Band structure of finite element matrices (crosses symbolize non zero terms of the matrix).

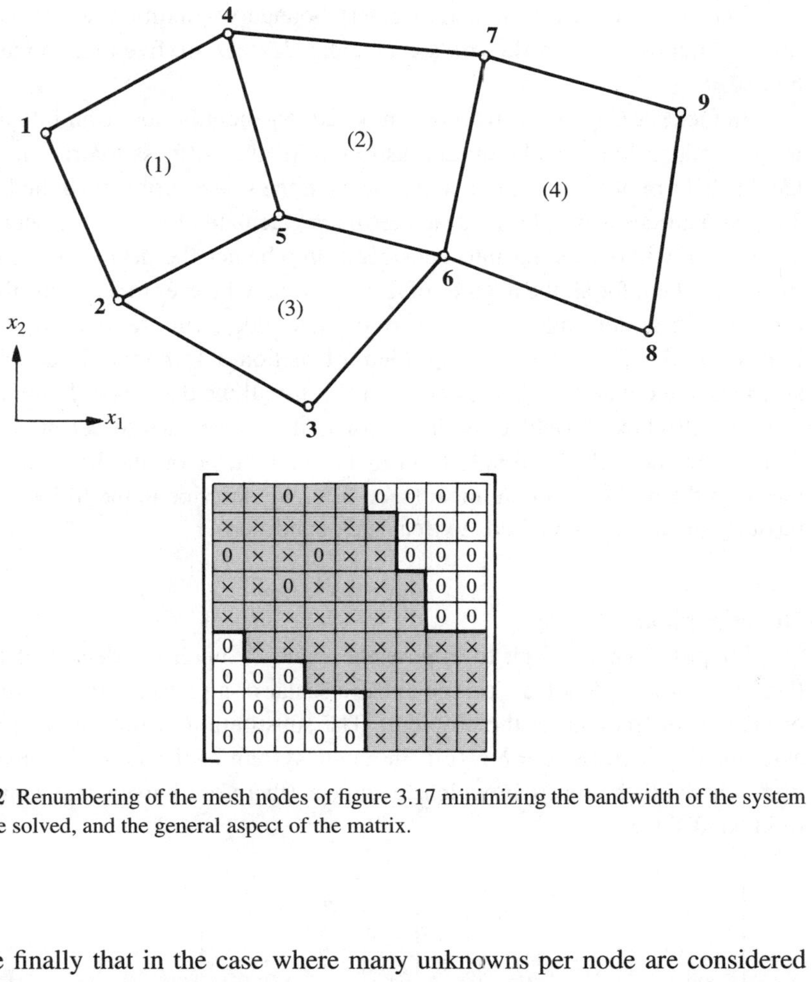

Figure 3.22 Renumbering of the mesh nodes of figure 3.17 minimizing the bandwidth of the system matrix to be solved, and the general aspect of the matrix.

Note finally that in the case where many unknowns per node are considered (solution for a vector field, or simultaneous solution of many scalar or vector fields), one tends to think in terms of nodal degrees of freedom to define the band structure of the matrix and to calculate its bandwidth. Thus if N is the number of degrees of freedom per node, the bandwidth of the matrix for the problem, expressed in degrees of freedom, is given by: $bw = N\,(bw_g + 1)$.

3.5.4 Treatment of the boundary conditions

In the previous sections 3.3 and 3.4, the mechanism for the incorporation of non-essential boundary conditions (Neumann or Cauchy) in the integral form of the equations was explained. It was illustrated for a scalar advection-diffusion problem (a given normal gradient imposed, or a function of the unknown) and for a vector problem in mechanics (a given stress vector imposed, or a stress vector as a function of the unknown velocity field).

The case of an essential (Dirichlet) boundary condition which consists of directly imposing the value of some nodal degrees of freedom, remains to be treated.

In fact, the degrees of freedom imposed is generally not taken into account in the procedures for calculation and assembly of the stiffness matrix and the right hand side term of the system to solve. Thus in the scalar problem studied in section 3.3, the matrices and right hand side term are calculated for each element, and then assembled, without taking into consideration whether the degrees of freedom are imposed. Then, for all the degrees of freedom, and whatever their status, the components of the system matrix and the right hand side term are given by (3.65) and (3.67). Similarly, for the vector problem of section 3.4, the matrix and right hand side term are defined by (3.115) and (3.117), for all the degrees of freedom.

To introduce Dirichlet conditions, it is thus necessary to act directly on the matrix and the right hand side term of the system, or on the linearized stiffness matrix if the problem is non linear (sect. 4.5). We describe in the following the two most frequently used methods to treat such conditions.

Direct method

Suppose that we want to impose the value $\bar{c}^{\mathrm{i}}$ upon the degree of freedom i (here the variable denoted c can be either a scalar or a vector, and Nn is the number of degrees of freedom of the problem). The substitution of this value in the initial assembled system $[\boldsymbol{K}]\ \boldsymbol{c} = \boldsymbol{b}$ results in a new system of the same dimension as the first, where all the terms in which $\bar{c}^{\mathrm{i}}$ appear (therefore known) are moved to the right hand side:

$$\begin{bmatrix} K_{11} & \dots 0 \dots & K_{1,\mathrm{Nn}} \\ \vdots & \vdots & \vdots \\ 0 & \dots 1 \dots & 0 \\ \vdots & \vdots & \vdots \\ K_{\mathrm{Nn},1} & \dots 0 \dots & K_{\mathrm{Nn},\mathrm{Nn}} \end{bmatrix} \begin{bmatrix} c^1 \\ \vdots \\ c^{\mathrm{i}} \\ \vdots \\ c^{\mathrm{Nn}} \end{bmatrix} = \begin{bmatrix} b_1 - K_{1\mathrm{i}}\bar{c}^{\mathrm{i}} \\ \vdots \\ \bar{c}^{\mathrm{i}} \\ \vdots \\ b_{\mathrm{Nn}} - K_{\mathrm{Nn},\mathrm{i}}\bar{c}^{\mathrm{i}} \end{bmatrix} \tag{3.151}$$

Other than the modifications to the right hand side, note those for the ith line and column of the matrix to obtain $c^{\mathrm{i}} = \bar{c}^{\mathrm{i}}$ for this degree of freedom after solution of the system. The procedure is continued this way for all the imposed values. This method is thus related to that introduced in chapter 2 for the FDM.

Penalty method

It is a bit simpler to implement because only the diagonal terms of $[\boldsymbol{K}]$ for which c is imposed are changed, and the modification of the right hand side term is easier. It consists of replacing the initial system by the following system:

$$\begin{bmatrix} K_{11} & \dots K_{1i} \dots & K_{1,\mathrm{Nn}} \\ \vdots & \vdots & \vdots \\ K_{i1} & \dots K_{ii} + \chi_p \dots & K_{i,\mathrm{Nn}} \\ \vdots & \vdots & \vdots \\ K_{\mathrm{Nn},1} & \dots K_{\mathrm{Nn},i} \dots & K_{\mathrm{Nn},\mathrm{Nn}} \end{bmatrix} \begin{bmatrix} c^1 \\ \vdots \\ c^i \\ \vdots \\ c^{\mathrm{Nn}} \end{bmatrix} = \begin{bmatrix} b_1 \\ \vdots \\ b_i + \bar{c}^i \chi_p \\ \vdots \\ b_{\mathrm{Nn}} \end{bmatrix} \tag{3.152}$$

where χ_p is "sufficiently" large with respect to the terms of the matrix $[\boldsymbol{K}]$. Take for example

$$\chi_p = 10^4 \underset{i,j=1,\mathrm{Nn}}{\mathrm{Max}} |K_{ij}| \tag{3.153}$$

The value $\bar{c}^i$ is obviously imposed on c^i with a higher and higher precision as χ_p becomes large. However, if an iterative solution of the system is employed (chap. 4), values of χ_p which are too large should be avoided as they could adversely affect the convergence rate.

In conclusion of this section on the implementation of the FEM, remember that the methods for the solution of systems of equations, linear or non linear, which are shared by the finite difference and the finite element methods, will be treated in chapter 4.

3.6 NON STATIONARY PROBLEMS

We focus in this section on detailing the solution with the finite element method of problems evolving in time. Such problems are obviously encountered very often in materials science. Consider, for example the diffusion of heat in material transformation processes, chemical diffusion at the root of segregation/homogenization phenomena in phase transformations, unsteady flows such as polymer or metal injection, etc.

This type of problem leads, as we are going to see with the example in 3.6.1, to a system of differential equations in time for which the solution requires a suitable choice of a temporal scheme, in a way similar to that seen for the finite difference method in chapter 2.

3.6.1 Example of non stationary diffusion

In the following, we again take the diffusion example already treated in chapter 2 and section 3.3. With respect to (3.52), a partial derivative with respect to time of the concentration of solute c appears when the problem is non stationary (1.74):

$$\frac{\mathrm{d}c}{\mathrm{d}t} = \frac{\partial c}{\partial t} + \boldsymbol{v} \cdot \mathbf{grad}\, c = \mathrm{div}(D\, \mathbf{grad}\, c) \tag{3.154}$$

As in the stationary case, natural boundary conditions (Neumann or Cauchy) given by (3.53) and (3.54), and essential (Dirichlet) conditions, are imposed respectively on the parts $\partial\Omega_N$ and $\partial\Omega_D$ of the domain under consideration Ω. Additionally, the temporal evolution of the concentration field requires the specification of the initial conditions $c(\boldsymbol{x}, t = 0)$ at every point in the domain (for example: uniform concentration $c = \bar{c}_0$). The domain Ω is discretized in finite elements with a mesh assumed to be fixed in space according to an Eulerian formulation (in other words, the flow "traverses" the mesh). Under these conditions, the temporal derivatives of c at the mesh nodes, which we denote $\dot{c}$ in this paragraph, are *partial* derivatives with respect to time.

As for the finite differences, the velocity field $\boldsymbol{v}$, assumed uniform, is a given of the problem. If we place ourselves in the context of simulating a process – for example the solidification of an ingot during which we would like to calculate the transport and diffusion of the alloy components – the velocity field $\boldsymbol{v}$ will no longer be constant on the domain under study, but may itself be interpolated at the mesh nodes and result from a separate mechanical computation of the flow of the metal (chap. 7).

Note finally that (3.154) is a generic equation for advection-diffusion phenomena. In particular, it can be applied to the heat transfer problem by replacing the concentration c by the temperature T and the diffusion coefficient D by the thermal diffusivity, $\kappa/\rho c_p$.

It is easy to show (section 3.3.2) that the application of a Galerkin weak formulation to (3.154) leads to the following system of equations:

$$\int_\Omega \psi_i \frac{\partial c}{\partial t}\,dV + \int_\Omega \psi_i \boldsymbol{v} \cdot \mathbf{grad}\, c\,dV + \int_\Omega D\,\mathbf{grad}\, c \cdot \mathbf{grad}\, \psi_i\,dV$$

$$= -\int_{\partial\Omega_N} \alpha\psi_i(c - \bar{c}_a)\,dS \qquad \text{for } i = 1, Nn \tag{3.155}$$

which again can be written, introducing completely nodal interpolation (3.60):

$$c^j \left[\int_\Omega D\,\mathbf{grad}\,\psi_j \cdot \mathbf{grad}\,\psi_i\,dV + \int_{\partial\Omega_N} \alpha\psi_i\psi_j\,dS + \int_\Omega \psi_i \boldsymbol{v} \cdot \mathbf{grad}\,\psi_j\,dV \right]$$

$$+\ \dot{c}^j \left[\int_\Omega \psi_i\psi_j\,dV \right] = \int_{\partial\Omega_N} \alpha\bar{c}_a\psi_i\,dS \qquad \text{for } i = 1, Nn \tag{3.156}$$

We thus arrive at the following system of differential equations:

$$[\boldsymbol{M}]\dot{\boldsymbol{c}} + [\boldsymbol{K}]\boldsymbol{c} = \boldsymbol{b} \tag{3.157}$$

in which $\boldsymbol{c}$ and $\dot{\boldsymbol{c}}$ are the column vectors having the nodal concentrations and their temporal derivatives, respectively, for components. The stiffness matrix [$\boldsymbol{K}$] and the right hand side vector $\boldsymbol{b}$ do not change with respect to the stationary formulation (cf (3.65) and (3.67)). The matrix [$\boldsymbol{M}$], called the ***mass matrix***, has the following components:

$$M_{\mathrm{ij}} = \int_{\Omega} \psi_{\mathrm{i}} \psi_{\mathrm{j}} \,\mathrm{d}V \tag{3.158}$$

This symmetric matrix has the same profile as the matrix [$\boldsymbol{K}$] and is assembled in an identical way.

Condensation of the mass matrix

For each element e the elementary mass matrix has for components in two dimensions:

$$M^{\mathrm{e}}_{\mathrm{ij}} = \int_{\Omega^{\mathrm{e}}} \psi^{\mathrm{e}}_{\mathrm{i}} \psi^{\mathrm{e}}_{\mathrm{j}} \,\mathrm{d}V = \int_{\Omega^{\mathrm{r}}} \psi^{\mathrm{r}}_{\mathrm{i}} \psi^{\mathrm{r}}_{\mathrm{j}} \det[\boldsymbol{J}^{\mathrm{e}}(\boldsymbol{\xi})] \,\mathrm{d}\xi_1 \,\mathrm{d}\xi_2 \tag{3.159}$$

For the particular case of a linear triangular element, the Jacobian matrix [$\boldsymbol{J}^{\mathrm{e}}$] is constant (section 3.2.8) and we have:

$$M^{\mathrm{e}}_{\mathrm{ij}} = \det[\boldsymbol{J}^{\mathrm{e}}] \int_0^1 \mathrm{d}\xi_1 \int_0^{1-\xi_1} \psi^{\mathrm{r}}_{\mathrm{i}} \psi^{\mathrm{r}}_{\mathrm{j}} \,\mathrm{d}\xi_2 \tag{3.160}$$

which by exact integration of the products of the interpolation functions gives (exercise 3.7.4):

$$[\boldsymbol{M}^{\mathrm{e}}] = \frac{\det[\boldsymbol{J}^{\mathrm{e}}]}{2} \begin{bmatrix} \frac{1}{6} & \frac{1}{12} & \frac{1}{12} \\ \frac{1}{12} & \frac{1}{6} & \frac{1}{12} \\ \frac{1}{12} & \frac{1}{12} & \frac{1}{6} \end{bmatrix} \tag{3.161}$$

The term appearing in front of the matrix is none other than the surface (the "mass") of the triangle as,

$$\det[\boldsymbol{J}^{\mathrm{e}}] = (x_1^2 - x_1^1)(x_2^3 - x_2^1) - (x_2^2 - x_2^1)(x_1^3 - x_1^1) \tag{3.162}$$

In the ***lumped mass matrix*** approximation, the extra-diagonal terms are summed on the diagonal, which gives then:

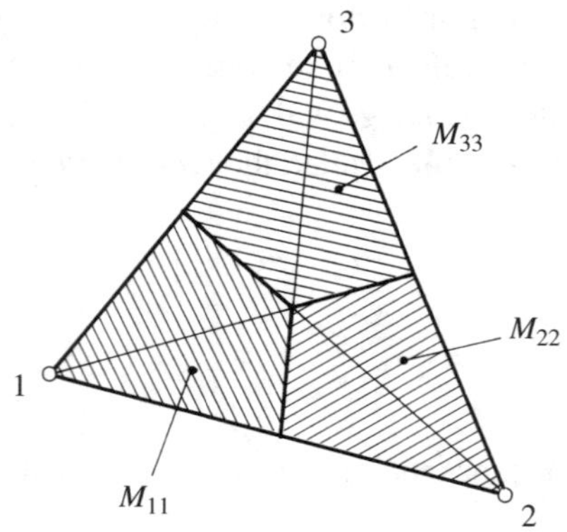

Figure 3.23 Assignment of a third of the surface of a linear triangle to the three vertex nodes, following the partition given by the medians, and corresponding to the condensation process of the mass matrix in two dimensions.

$$[\boldsymbol{M}^e] = \frac{\det[\boldsymbol{J}^e]}{2}\begin{bmatrix} \frac{1}{3} & 0 & 0 \\ 0 & \frac{1}{3} & 0 \\ 0 & 0 & \frac{1}{3} \end{bmatrix} \tag{3.163}$$

This amounts to attributing a third of the surface (mass) of the triangle to each node (fig. 3.23). It should be noted that the condensation operates in the same way for linear or quadratic quadrangular elements, but that it is not so simple for quadratic triangles, where a modification of the interpolation functions is necessary to avoid a singular condensed matrix.

The mass matrix thus being calculated, and maybe condensed, it is now a question of solving the differential system (3.157) in order to calculate the evolution of the concentration at the mesh nodes. There are many temporal integration schemes to solve this type of system. We limit our discussion to single step "*θ-method*" types, as already introduced in the context of finite differences (chap. 2).

3.6.2 Single step "θ-*method*" *type scheme*

In a manner analog to that which was done in chapter 2, assume that the solution is known up to time t. We want to calculate the new nodal concentrations $\boldsymbol{c}^{t+\Delta t}$ at time $t + \Delta t$. This type of scheme consists of performing a temporal discretization of (3.157), of the following form:

$$[\boldsymbol{M}]\frac{1}{\Delta t}(\boldsymbol{c}^{t+\Delta t} - \boldsymbol{c}^t) + [\boldsymbol{K}](\theta\boldsymbol{c}^{t+\Delta t} + (1-\theta)\boldsymbol{c}^t) = \theta\boldsymbol{b}^{t+\Delta t} + (1-\theta)\boldsymbol{b}^t \tag{3.164}$$

with

$\boldsymbol{c}^{\mathrm{t}}$	vector of concentrations at the mesh nodes at time t,
$\boldsymbol{c}^{\mathrm{t}+\Delta\mathrm{t}}$	vector of concentrations at the mesh nodes at time $t+\Delta t$,
$\boldsymbol{b}^{\mathrm{t}}$	right hand side vector at t,
$\boldsymbol{b}^{\mathrm{t}+\Delta\mathrm{t}}$	right hand side vector at $t+\Delta t$.
θ	parameter with value between 0 and 1.

Before starting the solution of the system (3.164) for $\boldsymbol{c}^{\mathrm{t}+\Delta\mathrm{t}}$, we detail three of the main schemes used in practice:

- $\theta = 0$ corresponds to the ***Euler explicit scheme***, which can be decomposed into two stages. In the first stage, the temporal derivative of $\boldsymbol{c}$ at time t is expressed, by writing (3.157) at time t:

$$\dot{\boldsymbol{c}}^{\mathrm{t}} = [\boldsymbol{M}]^{-1}(-[\boldsymbol{K}]\boldsymbol{c}^{\mathrm{t}} + \boldsymbol{b}^{\mathrm{t}}) \tag{3.165}$$

In the second stage, we integrate on the time interval with this derivative, considering it constant, to obtain $\boldsymbol{c}^{\mathrm{t}+\Delta\mathrm{t}}$:

$$\boldsymbol{c}^{\mathrm{t}+\Delta\mathrm{t}} = \boldsymbol{c}^{\mathrm{t}} + \Delta t\dot{\boldsymbol{c}}^{\mathrm{t}} \tag{3.166}$$

The combination of the two expressions above results in the following system,

$$\frac{1}{\Delta t}[\boldsymbol{M}]\boldsymbol{c}^{\mathrm{t}+\Delta\mathrm{t}} = \left(\frac{1}{\Delta t}[\boldsymbol{M}] - [\boldsymbol{K}]\right)\boldsymbol{c}^{\mathrm{t}} + \boldsymbol{b}^{\mathrm{t}} \tag{3.167}$$

which is seen to be none other than (3.164) with $\theta = 0$. The explicit scheme is called the forward marching scheme, because the derivative for the time increment is calculated from the concentrations $\boldsymbol{c}^{\mathrm{t}}$ known at time t.

- $\theta = 1$ corresponds to the ***Euler implicit scheme***, which consists this time of integrating the derivative expressed at the end of the increment over the time interval (the backward marching scheme), thus introducing the unknown $\boldsymbol{c}^{\mathrm{t}+\Delta\mathrm{t}}$:

$$\dot{\boldsymbol{c}}^{\mathrm{t}+\Delta\mathrm{t}} = [\boldsymbol{M}]^{-1}(-[\boldsymbol{K}]\boldsymbol{c}^{\mathrm{t}+\Delta\mathrm{t}} + \boldsymbol{b}^{\mathrm{t}+\Delta\mathrm{t}}) \tag{3.168}$$

$$\boldsymbol{c}^{\mathrm{t}+\Delta\mathrm{t}} = \boldsymbol{c}^{\mathrm{t}} + \Delta t\dot{\boldsymbol{c}}^{\mathrm{t}+\Delta\mathrm{t}} \tag{3.169}$$

The system to solve is thus:

$$\left(\frac{1}{\Delta t}[\boldsymbol{M}] + [\boldsymbol{K}]\right)\boldsymbol{c}^{\mathrm{t}+\Delta\mathrm{t}} = \left(\frac{1}{\Delta t}[\boldsymbol{M}]\right)\boldsymbol{c}^{\mathrm{t}} + \boldsymbol{b}^{\mathrm{t}+\Delta\mathrm{t}} \tag{3.170}$$

- $\theta = 1/2$ corresponds to the ***Crank-Nicolson scheme***, for which the temporal derivative of $\boldsymbol{c}$ over the time interval is taken equal to the arithmetic mean of the derivatives at times t and $t + \Delta t$. Then, by combination of (3.165), (3.168) and the following relation,

$$\mathbf{c}^{t+\Delta t} = \mathbf{c}^{t} + \Delta t \tfrac{1}{2}(\dot{\mathbf{c}}^{t} + \dot{\mathbf{c}}^{t+\Delta t}) \tag{3.171}$$

we obtain the system to solve:

$$\left(\frac{1}{\Delta t}[\mathbf{M}] + \tfrac{1}{2}[\mathbf{K}]\right)\mathbf{c}^{t+\Delta t} = \left(\frac{1}{\Delta t}[\mathbf{M}] - \tfrac{1}{2}[\mathbf{K}]\right)\mathbf{c}^{t} + \tfrac{1}{2}(\mathbf{b}^{t+\Delta t} + \mathbf{b}^{t}) \tag{3.172}$$

While the explicit and implicit Euler schemes have a local error which goes as the time step to the second power, this scheme has an error which goes as the third power. Globally, the Euler schemes are to first power in the time step, while the Crank-Nicolson scheme goes as the square, and is thus more accurate.

In general, for $\theta \in [0,1]$, the system to solve for $\mathbf{c}^{t+\Delta t}$ is then the following:

$$\left(\frac{1}{\Delta t}[\mathbf{M}] + \theta[\mathbf{K}]\right)\mathbf{c}^{t+\Delta t} = \left(\frac{1}{\Delta t}[\mathbf{M}] - (1-\theta)[\mathbf{K}]\right)\mathbf{c}^{t} + \theta\mathbf{b}^{t+\Delta t} + (1-\theta)\mathbf{b}^{t} \tag{3.173}$$

This linear system is to be solved by a direct or iterative method (chap. 4). It should be noted that the solution can become trivial if an explicit scheme ($\theta = 0$) and a condensed mass matrix $[\mathbf{M}]$ are used. The system matrix is then diagonal and the Nn equations become independent. The solution is obviously very economical in computation time. However, as we will see in paragraph 3.6.3, the scheme is conditionally stable.

3.6.3 Study of the stability of the scheme

The stability of the scheme comes from the properties of the homogeneous system (where $\mathbf{b} = \mathbf{0}$), which we write:

$$\mathbf{c}^{t+\Delta t} = [\mathbf{G}]\mathbf{c}^{t} \tag{3.174}$$

For the scheme to be stable, that is, for $\mathbf{c}^{t+m\Delta t}$ to be bounded as m goes to infinity, it is necessary that the maximal absolute value of the eigenvalues λ_i of the matrix $[\mathbf{G}]$ of the homogeneous system be less than 1. This maximum value, called the ***spectral radius***, is denoted $\rho([\mathbf{G}])$:

$$\rho([\mathbf{G}]) = \underset{i=1,\,Nn}{\mathrm{Max}} |\lambda_i| \tag{3.175}$$

In the case under consideration here, the expression of the matrix $[\mathbf{G}]$ is easily obtained by multiplying the two sides of the system (3.173) by $\Delta t[\mathbf{M}]^{-1}$:

$$[\boldsymbol{G}] = ([\boldsymbol{I}] + \Delta t\theta[\boldsymbol{M}]^{-1}[\boldsymbol{K}])^{-1}([\boldsymbol{I}] - \Delta t(1-\theta)[\boldsymbol{M}]^{-1}[\boldsymbol{K}]) \tag{3.176}$$

The eigenvalues of $[\boldsymbol{G}]$, λ_i, are thus deduced from the eigenvalues Λ_i of the matrix $[\boldsymbol{M}]^{-1}[\boldsymbol{K}]$ by the relationships:

$$\lambda_i = \frac{1-(1-\theta)\Delta t\Lambda_i}{1+\theta\Delta t\Lambda_i} \tag{3.177}$$

The eigenvalues Λ_i are positive by the forms of $[\boldsymbol{M}]$ and $[\boldsymbol{K}]$. Therefore, the reader can easily verify that the condition on the spectral radius of the matrix $[\boldsymbol{G}]$ ($\rho([\boldsymbol{G}] < 1$) is equivalent to:

$$1-2\theta < \frac{2}{\Delta t\Lambda_i} \quad \text{for } i = 1, Nn \tag{3.178}$$

The stability of the scheme depends then on the value chosen for θ. For $\theta < 1/2$, it is seen that the preceding equation imposes a condition on the time step; the maximum value of Δt being given by:

$$\Delta t_{\max} = \frac{2}{(1-2\theta)\rho([\boldsymbol{M}]^{-1}[\boldsymbol{K}])} = \frac{2}{(1-2\theta)\underset{i=1,Nn}{\mathrm{Max}}\Lambda_i} \tag{3.179}$$

It is said then that the scheme is ***conditionally stable***. On the other hand, the condition (3.178) is always valid for $\theta \geq 1/2$ as $1-2\theta$ is then negative. In this case, the scheme is said to be ***unconditionally stable*** (no condition on the time step). In practice, one uses preferentially the schemes with a marked implicit character ($\theta \geq 1/2$).

Of course, the results above can be generalized to problems with many degrees of freedom per node by replacing *Nn* with the global number of degrees of freedom of the mesh (*NnN*).

3.6.4 Non-linearity and non stationarity

The origins of the non linearities in the chemical or thermal diffusion problems are multiple. First of all there is the dependence of the diffusivity on the solute concentrations (or temperature, respectively). As for the thermal transfer, one adds radiation type boundary conditions as well as phase transformations.

The implementation of a semi-implicit scheme with parameter θ is expressed by the following two equations:

$$\dot{\boldsymbol{c}}^{t+\theta\Delta t} = \theta\dot{\boldsymbol{c}}^{t+\Delta t} + (1-\theta)\dot{\boldsymbol{c}}^{t} = \theta([\boldsymbol{M}]^{t+\Delta t})^{-1}(-[\boldsymbol{K}]^{t+\Delta t}\boldsymbol{c}^{t+\Delta t} + \boldsymbol{b}^{t+\Delta t}) + (1-\theta)([\boldsymbol{M}]^{t})^{-1}(-[\boldsymbol{K}]^{t}\boldsymbol{c}^{t} + \boldsymbol{b}^{t}) \tag{3.180}$$

and

$$\boldsymbol{c}^{t+\Delta t} = \boldsymbol{c}^{t} + \Delta t\dot{\boldsymbol{c}}^{t+\theta\Delta t} \tag{3.181}$$

in which the mass matrix $[\boldsymbol{M}]$ is constant if the mesh does not evolve. As for the matrix $[\boldsymbol{K}]^{t+\Delta t}$ and the right hand side term $\boldsymbol{b}^{t+\Delta t}$, they depend on the unknown $\boldsymbol{c}^{t+\Delta t}$.

Thus the solution $\boldsymbol{c}^{t+\Delta t}$ can only be *directly* calculated in the purely explicit case ($\theta = 0$), which has the stability conditions seen above.

For the case of a partially implicit scheme ($\theta > 0$), there is no direct solution. For example, one is led, for a constant mass matrix, to zero the following residual vector, a function of $\boldsymbol{c}^{t+\Delta t}$:

$$\begin{aligned}\boldsymbol{R}(\boldsymbol{c}^{t+\Delta t}) = {} & [\boldsymbol{M}](\boldsymbol{c}^{t+\Delta t} - \boldsymbol{c}^{t}) + \Delta t(\theta \boldsymbol{b}^{t+\Delta t} + (1-\theta)\boldsymbol{b}^{t} - \theta[\boldsymbol{K}]^{t+\Delta t}\boldsymbol{c}^{t+\Delta t} \\ & -(1-\theta)[\boldsymbol{K}]^{t}\boldsymbol{c}^{t})\end{aligned} \tag{3.182}$$

The iterative methods to solve this type of system of non linear equations are detailed in chapter 4.

3.6.5 Introduction to the specific treatment of the advective transport terms

Note at this stage that if we envision a Lagrangian formulation to solve (3.154), that is, if the mesh nodes move at the fluid velocity, there is then equality between the partial derivative at the nodes of the mesh and the total derivative. The (non-symmetric) term in $\boldsymbol{v} \cdot \mathbf{grad}\ \psi$ disappears in (3.156) and $\dot{\boldsymbol{c}}$ is thus the vector of total derivatives (or material derivatives) at the nodes. This type of formulation is frequently utilized to model the heat transfer in deformation mechanics, for example to simulate the non stationary processes of forming, stamping and forging of metals, thermoforming and blowing of polymers.

Conversely, the modeling of non stationary fluid flows or stationary processes of forming (rolling of metals, drawing of polymer films ...) is most often approached with an Eulerian formulation with a computation domain Ω, and thus a finite element discretization fixed in space. However, in this case, the Galerkin formulation proves to be poorly adapted when the convection is much larger than the diffusion (high Péclet numbers, chap. 2), leading to numerical oscillations in the solution. Beyond the simple reduction of the discretization spacing, which can lead to prohibitive computation time in practice, we will present the specific means to treat this problem in chapter 7.

3.7 EXERCISES

3.7.1 Prove the result (3.51) for the integration of the first derivative of a function on a linear triangle. Treat the case of a quadratic triangle with six nodes.

3.7.2 Volume calculation. We consider the volume Ω bounded by the closed surface S defined by a set of surface finite elements (triangles or quadrangles in space, for example). Using the divergence theorem, give an example of a very simple vector function for which the integral of the flux across S is equal to the measure of Ω. Discuss the case where S is not a closed surface.

3.7.3 Apply the developments made in this chapter for the chemical diffusion case to heat transfer.

3.7.4 Find the equation (3.161) for the mass matrix for a linear triangular finite element.

3.7.5 In a chemical diffusion problem solved by finite elements, the contribution of the diffusion to the stiffness matrix of an element e is given by (3.65), that is, merging the local and global node numbers for simplicity:

$$K^{\mathrm{e}}_{\mathrm{ij}} = \int_{\Omega^{\mathrm{e}}} D\,\mathbf{grad}\,\psi_{\mathrm{i}} \cdot \mathbf{grad}\,\psi_{\mathrm{j}}\,\mathrm{d}V = \int_{\Omega^{\mathrm{e}}} D\frac{\partial \psi_{\mathrm{i}}}{\partial x_{\mathrm{k}}} \cdot \frac{\partial \psi_{\mathrm{j}}}{\partial x_{\mathrm{k}}}\,\mathrm{d}V$$

where D is the coefficient of chemical diffusion. For a linear triangular element, we denote:

$$\beta^{\mathrm{i}} = y^{\mathrm{j}} - y^{\mathrm{k}} \qquad \text{and} \qquad \gamma^{\mathrm{i}} = x^{\mathrm{k}} - x^{\mathrm{j}}$$

(i, j, k) indicating a circular permutation of the three nodes at the vertices (1,2,3). Expressing the Jacobian matrix of the element as $[\boldsymbol{J}^{\mathrm{e}}]$, show that $K^{\mathrm{e}}_{\mathrm{ij}}$ can also be written:

$$K^{\mathrm{e}}_{\mathrm{ij}} = \frac{D}{2\det[\boldsymbol{J}^{\mathrm{e}}]}(\beta^{\mathrm{i}}\beta^{\mathrm{j}} + \gamma^{\mathrm{i}}\gamma^{\mathrm{j}})$$

3.8 BIBLIOGRAPHY

K. J. Bathe, *Finite Element Procedures*, Prentice Hall, Englewood Cliffs, New Jersey, 1996

P. Ciarlet, *The Finite Element Method for Elliptic Problems*, North Holland, New York, 1978

T.J.R. Hughes, *The Finite Element Method: Linear Static and Dynamic Finite Element Analysis*, McGraw-Hill, New York, 2000

R.W. Lewis, K. Morgan, H.R. Thomas, K.N. Seetharama, *The Finite Elemente Method in Heat Transfer Analysis*, J. Wiley, New York, 1996

J.N. Reddy, *An Introduction to the Finite Element Method*, Mc Graw-Hill, New York, 1985

F. Thomasset, *Implementation of Finite Element Methods for Navier-Stokes Equations*, Springer, New York, 1981

R.E. White, *An Introduction to the Finite Element Method with Application to Non-Linear Problems*, John Wiley and Sons, New York, 1985

O.C. Zienkiewicz, R.L. Taylor, *The Finite Element Method*, Mc Graw-Hill, New York, 1989

CHAPTER 4

ELEMENTS OF NUMERICAL ALGORITHMS

4.1 OBJECTIVES

- Present the main algorithms for mesh generation.
- Introduce the basic methods for the solution of linear systems and the storage of matrices in memory.
- Describe the different techniques to treat non linear problems.

As was seen in chapters 2 and 3, the finite difference and finite element numerical methods permit the transformation of a physical problem expressed in the form of conservation partial differential equations into the solution of a system of equations. The latter, either linear or non linear, lead to a set of discrete values defined at the mesh nodes and at a number of time steps. While the two preceding chapters placed the accent on the mathematical formulation that permits such a transformation, this one is focused on the more technical aspects of the implementation of such methods. As a first step, it is necessary to define the mesh on which a problem is formulated. If this step is relatively simple for the structured grids of the FDM, it is on the other hand more difficult for the unstructured meshes of the FEM or the finite volume method.

After formulation of the problem and construction of the resulting system of equations, the next step is to solve them. Many solution methods will be presented in this chapter; some consist of directly seeking the solution by the method of elimination of variables, others are iterative in nature. These two large categories of methods have implications for the storage of matrices and we will detail the principle techniques that permit their optimization.

Finally, the problems encountered in practice being most often non linear, the resulting systems of equations are also. As we will see in this chapter, these non linearities can have diverse sources: the very nature of the problem (for example, the Navier-Stokes equation is non linear in velocity), material properties that are functions of the variables (the enthalpy, for example, is strongly non linear in temperature during a phase transformation), boundary conditions (for example thermal exchanges by radiation), etc. Methods for the solution of non linear systems of equations should be considered in such situations.

4.2 METHODS FOR GENERATING MESHES

Mesh generation takes a very different appearance according to whether one considers the implementation of the finite difference or the finite element method. We have seen in chapter 2 that the FDM involves a spatial discretization tied to a coordinate system (Cartesian, cylindrical, spherical). Consequently, the discrete points must be aligned with the axes. As illustrated in figure 4.1, the spacing and the number of discrete points are prescribed by the user in each direction.

This type of grid has the advantage of being very easy to generate. However, the method suffers two major inconveniences. First, the boundary conditions can never be rigorously imposed on boundaries which are curved or slanted relative to the chosen coordinate system. Second, the correct reproduction of these boundaries, as well as of strong local gradients, necessitates a considerable local diminution of the discretization spacing which, unfortunately, propagates over the whole grid because of its structured character, thereby noticeably increasing the calculation costs. It should be noted that the great majority of finite volume methods use structured grids and thus suffer from the same handicap.

The same is not true for the FEM which allows the treatment of complex geometries with much more flexibility at a given fineness of discretization. The mesh[1] can thus be structured or unstructured. We call ***structured meshes*** all ordered networks of elements (for example a parallelepiped domain – or isomorphic with a parallelepiped – built up from parallelepiped "bricks", and in which each node can be identified by row or other axis indices for each of the coordinate axes). The grid lines are in this case more-or-less of the same form as the boundaries of the calculation domain. The mesh for a complex domain can then be considered, at the price of a decomposition of the initial domain into sub domains isomorphic with these parallelepipeds, which can be meshed separately then reconnected. Conversely,

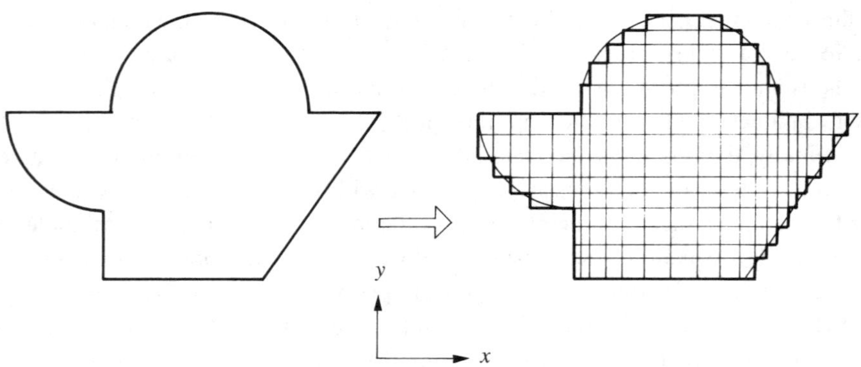

Figure 4.1 Example of spatial discretization for the finite difference method.

[1] Generic term used in this chapter, although here it only designates the discretization *grid*.

meshes that are not based on an ordered network of rows of nodes are called *unstructured*. These two types of mesh are obviously based on completely different generation methods that we will discuss in detail below.

Before that, it must be pointed out that *automatic generation* of meshes should be favored in programs intended for industrial application for the following reasons:

- Semi-automatic generation of a volume mesh from a CAD file describing the geometry, and thus the surface, of a complex three dimensional piece is a long, fastidious and therefore costly operation. If the intersection of a regular arrangement of parallelepipeds with surfaces describing the geometry is relatively easy to calculate automatically for the FDM, a volume mesh with finite elements is more complicated.
- For the case of a calculation of large deformations (the simulation of the forming of materials, for example), it is necessary to ***remesh*** the piece over time, to avoid degeneracy of the elements. These mesh regenerating operations are only conceivable if they are automated (in severe cases a simulation of forging can require a few dozen mesh regenerations!).
- The user of a finite element program is not necessarily an expert in the method and may have trouble identifying the regions that should be meshed more finely than others (for example, regions with strong gradients near interfaces, or singularities…). In addition, in a non stationary calculation, the position of these critical regions can evolve; for example a solidification front advancing inside a piece that was initially liquid. The mesh should thus be *adapted* to these difficulties over time, in order to guarantee sufficient precision in the result, and this should be done automatically. The local size of the elements of the mesh should then be driven by geometrical considerations (description of free surfaces, local penetration of the mesh in obstacles such as tools…), but also by an estimation of the numerical error attributable to the spatial discretization. We will not develop that subject in this book, but it should be known that the definition of adaptable meshes is an important research theme at this time, as it is clear that they impact the reliability and speed of the numerical calculations.

These different reasons thus motivate the development of techniques for *automatic mesh generation*. We will present first the generation techniques for structured meshes. These techniques being limited from the point of view of automation, we will show in the second part the methods for unstructured, or free, meshes, for which the algorithms are better adapted for automation.

4.2.1 Generation of structured meshes

The algorithms for the generation of structured meshes can be classed into two main categories. The first is based on the solution of partial differential equations, which can be elliptic, parabolic or hyperbolic. The second is ***algebraic methods***, based essentially on interpolation methods.

Before presenting these two techniques in detail, it should be pointed out that the meshes obtained by the solution of partial differential equations are very smooth and regular inside the domains. However, the spacing near the boundary of the domain depends on its curvature. For meshes obtained by algebraic methods, the spacing along the boundary is regular. Nonetheless, every irregularity of the boundary is propagated inside the domain.

Structured meshes generated by partial differential equations

Structured meshes for complex geometries are obtained by generation of coordinate systems that are "adjusted" to the boundaries of the domain. This is achieved by the solution of partial differential equations. The properties of the mesh are intimately dependent on the properties of these equations. Elliptic equations have a natural smoothing effect which impedes the existence of jumps or discontinuities. The solution of parabolic or hyperbolic equations is done with algorithms that march in space, which renders them very efficient for mesh generation. However, these systems are less general than elliptic systems and their use for complicated geometric forms can prove to be very difficult. In particular, when the coordinate points (and/or slopes) are specified on an entire closed boundary of the physical domain, which is frequently the case, the choice of parabolic or hyperbolic methods is not possible. In the following, we will thus present elliptic systems as an example of efficient mesh generators by solution of partial differential equations.

The method consists of generating a system of curvilinear coordinates that conform to, or are adapted to the form of the boundary. Denote Ω the restricted physical space domain that is described with Cartesian coordinates (x, y), in the two dimensional case. Let Ω_C be the regular calculation domain referred to with the curvilinear coordinates (ξ, η). Given the mapping ∂f: $\partial\Omega_C \rightarrow \partial\Omega$ which maps the boundary of Ω_C to that of Ω one-to-one; it is desired to construct a mapping f: $\Omega_C \rightarrow \Omega$ which satisfies the boundary conditions.

In the interior of Ω, we impose that (ξ, η) are harmonic functions of (x, y) and thus obtained by the Poisson equations:

$$\Delta\xi = P \quad \text{and} \quad \Delta\eta = Q \tag{4.1}$$

where the functions P and Q are known and used to control the internal construction of the mesh. With these functions the spacing of the mesh lines can be increased or reduced with respect to a point, line or surface (in 3 dimensions). In order to solve (4.1), the coordinates $x(\xi, \eta)$ and $y(\xi, \eta)$ must be prescribed on the perimeter of the physical domain. They play the role of Dirichlet conditions.

Equations such as (4.1) satisfy an extremum principle: the minimum and maximum values of (ξ, η) are attained on the boundaries. When the Poisson equations degenerate to Laplace equations ($P = Q = 0$), the coordinate lines tend to be equidistant in the absence of curvature of the boundary, due to the smoothing effect of the

Laplacian. Returning to system (4.1) with non zero functions P, Q, the spacing and orientation of the coordinate lines can be controlled (fig. 4.2).

The solution of (4.1) is most often obtained in the calculation domain Ω_C. To this end, (4.1) is transformed so that the Cartesian coordinates (x, y) are variables depending on (ξ, η), the independent variables. One obtains:

$$\begin{aligned} \alpha\frac{\partial^2 x}{\partial\xi^2} - 2\beta\frac{\partial^2 x}{\partial\xi\partial\eta} + \gamma\frac{\partial^2 x}{\partial\eta^2} + J^2\left(P\frac{\partial x}{\partial\xi} + Q\frac{\partial x}{\partial\eta}\right) &= 0 \\ \alpha\frac{\partial^2 y}{\partial\xi^2} - 2\beta\frac{\partial^2 y}{\partial\xi\partial\eta} + \gamma\frac{\partial^2 y}{\partial\eta^2} + J^2\left(P\frac{\partial y}{\partial\xi} + Q\frac{\partial y}{\partial\eta}\right) &= 0 \end{aligned} \tag{4.2}$$

with

$$\alpha = \left(\frac{\partial x}{\partial\eta}\right)^2 + \left(\frac{\partial y}{\partial\eta}\right)^2 \qquad \beta = \frac{\partial x}{\partial\xi}\frac{\partial x}{\partial\eta} + \frac{\partial y}{\partial\eta}\frac{\partial y}{\partial\xi} \qquad \gamma = \left(\frac{\partial x}{\partial\xi}\right)^2 + \left(\frac{\partial y}{\partial\xi}\right)^2 \tag{4.3}$$

$$J = \frac{\partial x}{\partial\xi}\frac{\partial y}{\partial\eta} - \frac{\partial x}{\partial\eta}\frac{\partial y}{\partial\xi} \tag{4.4}$$

The factor J is the Jacobian of the mapping. The coefficients α, β, γ are the coefficients of the metric. If $\beta = 0$, the mesh is orthogonal. However, the zeroing of β in (4.2) is insufficient to obtain the orthogonality of the mesh. It can be seen that this can lead to errors on complicated boundary forms, for example, in the neighborhood of significant concavity (fig. 4.3). This mesh generation technique owes much to Thompson who named it the method of ***boundary-fitted coordinates***.

$$P\begin{cases} < 0 \text{ for } \xi_j \quad j < i_0 \\ > 0 \text{ for } \xi_j \quad j > i_0 \end{cases} \qquad\qquad P\begin{cases} > 0 \text{ for } \xi_j \quad j < i_0 \\ < 0 \text{ for } \xi_j \quad j > i_0 \end{cases}$$

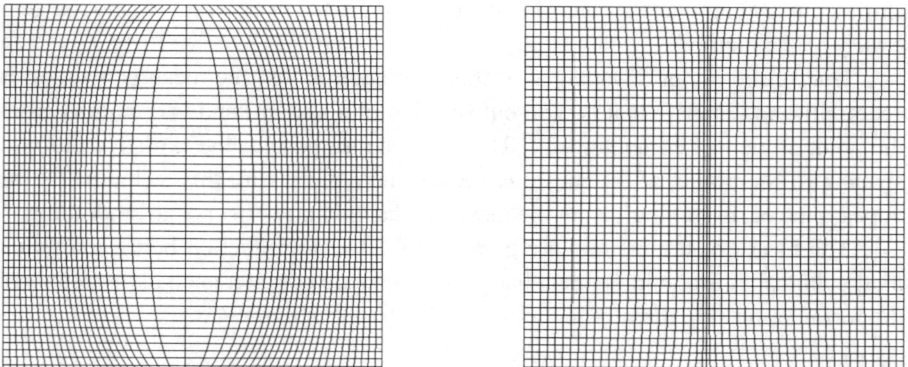

(a) i_0 (b) i_0

Figure 4.2 Example of meshes obtained by solution of partial differential equations. For the mesh on the left, $P < 0$ to the left of the centerline and $P > 0$ to the right of the centerline. The converse is true for the mesh on the right.

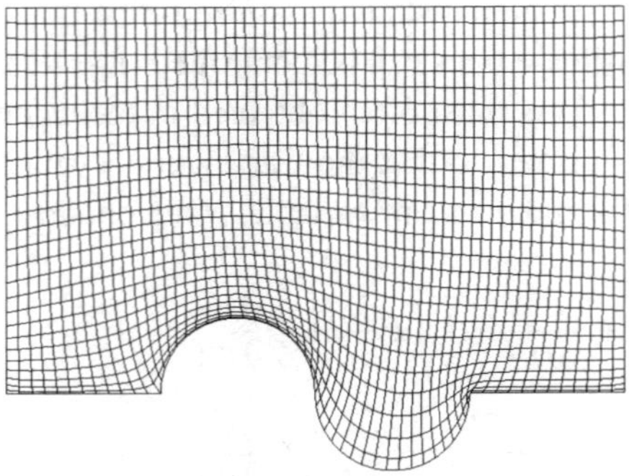

Figure 4.3 Mesh obtained by solution of partial differential equations. The mesh lines bunch up near concave boundaries and spread out from convex boundaries.

Structured meshes generated by algebraic methods

The algebraic techniques for mesh generation are based on the direct functional description of the mapping between the physical domain Ω and the calculation domain Ω_C. An example of the classic algebraic method is that of the conformal mapping in complex variable theory. Analytic expressions of the mapping can be written *in extenso* (for example, the Schwarz-Christoffel mapping which maps a domain Ω described by a polygonal contour to a half plane Ω_C). The resulting meshes are orthogonal. Nonetheless, these useful properties are counterbalanced by two major faults: the inability to specify the position of the points on the boundary and the two dimensional character of the method.

A more general class of mesh generators is based on interpolation along the boundaries and/or the intermediary surfaces in the domain. These methods can be single or multi-dimensional. In more elaborate formulations, the boundary points, slopes, curves and interior control surfaces are included in order to expand the generality of the method.

One of the major advantages of this technique is that the mesh generation is accomplished directly (that is, without solution of a system) and very rapidly from the point of view of computer time. This method also permits the explicit control of spacing of the points of the mesh. However, the danger with this technique is the appearance of discontinuities in the mesh, as the method performs no smoothing.

The basics of the method are given below. The introduction being somewhat mathematical, it is recommended that readers who are not interested in the details of the formulation skip directly to the result, illustrated in figure 4.5.

One dimensional interpolation

Given a function f of real variables, on the interval $\Omega_C \equiv [a, b]$ in Ω (a sub space of physical space $\mathbb{R}^3$). The function f is *a priori* unknown, but certain values x^i taken at n points ξ^i are known (i.e. $f(\xi^i) = x^i$ for $\xi^i \in [a, b]$, $i = 1, n$). The problem of one variable interpolation consists then of finding an interpolation function g

such that:

$$g : \Omega_C \equiv [a, b] \to \Omega \tag{4.5}$$

and

$$g(\xi^i) = x^i \qquad \text{for } i = 1, n \tag{4.6}$$

The most classical case corresponds to $\Omega \equiv \mathbb{R}$. If the variables x^i describe a two or three dimensional curve, then Ω is respectively a sub space of $\mathbb{R}^2$ (curve in a plane) or of $\mathbb{R}^3$ (curve in space). As soon as the dimension of Ω is greater than 1, $g(\xi)$ is a parametric curve and ξ is a curvilinear abscissa of this curve. The interpolant $g(\xi)$ is then a vector with d components, d being the dimension of the space. The solution of problem (4.5)–(4.6) is easily expressed in the form

$$g(\xi) = \sum_{j=1}^{n} x^j \psi_j(\xi) \tag{4.7}$$

with the cardinality condition:

$$\psi_j(\xi^i) = \delta_{ij} \tag{4.8}$$

the symbol δ_{ij} being the Kronecker delta function. The functions ψ_j form a basis for the solution of the problem (4.5)–(4.6). We know that the Lagrange basis constitutes an example of such a solution, where the ψ_j are denoted L_j and are defined as:

$$L_j(\xi) = \prod_{\substack{i=1 \\ i \neq j}}^{n} \frac{(\xi - \xi^i)}{(\xi^j - \xi^i)} \tag{4.9}$$

For $n = 2$, these functions are the linear interpolants:

$$L_1(\xi) = \frac{(\xi^2 - \xi)}{(\xi^2 - \xi^1)} \quad \text{and} \quad L_2(\xi) = \frac{(\xi - \xi^1)}{(\xi^2 - \xi^1)} \tag{4.10}$$

that we have already seen in chapter 3 introducing interpolation by finite elements ((3.9)–(3.10)).

To facilitate the following development, we introduce from the start a more compact notation:

$$g = P_\xi[f] \tag{4.11}$$

in which appears an interpolation operator P_ξ with respect to the curvilinear abscissa ξ. This operator possesses the identity property,

$$P_\xi[P_\xi] = P_\xi \tag{4.12}$$

which expresses the fact that the interpolation of the interpolant reproduces itself. Every operator that has this property (4.12) is usually called a ***projector***. The projector P_ξ is characterized by the basis functions $\psi_j(\xi)(j = 1, n)$, the expression (4.11) being formally equivalent to (4.7): g is the result of the projection of f on the basis ψ_j. Note in passing that nothing prevents the scalar (or vector) functions f and g from depending on other curvilinear coordinates than ξ.

Osculatory (or Hermite) interpolation

In some cases, it may be desirable to impose not only the local values of the unknown function f, but also some of its derivatives at the interpolation nodes. The interpolation that results is known by the name ***osculatory interpolation*** or ***Hermite interpolation***. It satisfies the following conditions:

$$g(\xi^i) = x^i \quad \text{and} \quad \frac{dg}{d\xi}(\xi^i) = d^i \quad \text{for } i = 1, n \tag{4.13}$$

It can be easily verified that, by analogy with expression (4.7), the function that is the solution of the interpolation problem is written:

$$g(\xi) = \sum_{i=1}^{n} \left(x^i \psi_i^0(\xi) + d^i \psi_i^1(\xi)\right) \tag{4.14}$$

the basis functions ψ_i^j satisfy the cardinality condition:

$$\frac{d^m}{d\xi^m} \psi_i^j(\xi^k) = \delta_{ik}\delta_{jm} \qquad j, m = 0, 1; i, k = 1, n \tag{4.15}$$

in which we have used the convention $d^0\psi_i^j/d\xi^0 = \psi_i^j$.

For the two point polynomial case, the Hermite cubic interpolation polynomials (often denoted H_i^j) are obtained, which are written:

$$\psi_1^0(s) = 1 - 3s^2 + 2s^3 \qquad \psi_2^0(s) = \psi_1^0(1-s) \tag{4.16}$$

$$\psi_1^1(s) = s(s-1)^2 \qquad \psi_2^1(s) = -\psi_1^1(1-s) \tag{4.17}$$

with

$$s = \frac{\xi - \xi^1}{\xi^2 - \xi^1} \qquad s \in [0, 1] \tag{4.18}$$

This relation is easily extended to higher order schemes. The concept will be necessary to impose orthogonality at the boundary for two and three dimensional mesh generation.

Let us group the discrete values of the function f, x^i $(i = 1, n)$ and its derivatives, d^i $(i = 1, n)$, in a single notation $\ell^i(f)$ $(i = 1, 2n)$. $\ell^i(f)$ are then the *linear functionals* of the function f: values of f or its derivatives at certain points. Each linear functional $\ell^i(f)$ is accompanied by a basis function $\psi_i(\xi)$ $(i = 1, 2n)$, the pairs (ℓ^i, ψ_i) forming an orthogonal set, such that:

$$\ell^k(\psi_i) = \delta_{ki} \qquad i, k = 1, 2n \tag{4.19}$$

It can be easily verified that the equation (4.19) is equivalent to (4.15) and, similarly, that the expression (4.11) involving the interpolation projector P_ξ, can be applied more generally than for the interpolation of point values.

We arrive at the conclusion that given a set (ℓ^i, ψ_i) $(i = 1, n)$ of any linear functionals and the associated basis functions forming an orthogonal system, the (generalized) interpolant of a function $f(\xi)$, where ξ is a curvilinear coordinate, is written:

$$g(\xi) = P_\xi[f] = \sum_{i=1}^{n} \ell^i(f)\psi_i(\xi) \tag{4.20}$$

with, using (4.19),

$$\ell^i(g) = \ell^i(f) \qquad i = 1, n \tag{4.21}$$

The expressions (4.20) and (4.21) are the basis for the development of a multi-dimensional transfinite interpolation.

Transfinite interpolation

Transfinite interpolation is a multi-dimensional interpolation method. It was developed originally for geometric modeling (Coons' surfaces). Gordon and Hall applied it to mesh generation and Eriksson introduced the osculatory concept.

Consider the two dimensional interpolation problem of a surface Ω situated in physical space $\mathbb{R}^3$. We seek the parametric equations of the surface as a function of two curvilinear coordinates u and v (to increase the clarity in the expressions below, we use here (u, v) instead of (ξ, η)):

$$x = x(u, v) \qquad y = y(u, v) \qquad z = z(u, v) \tag{4.22}$$

on the base of accessible information that is contained in two sets of linear functionals:

$$\ell_u^i(f) \qquad \ell_v^j(f) \qquad i = 1, n_u; j = 1, n_v \tag{4.23}$$

We will pay particular attention to the following points.

- The functionals ℓ_u^i and ℓ_v^j can be the point functionals as in Lagrangian interpolation, or a mixture of point and derivative functionals as in osculatory interpolation.
- The number n_u of givens along the curvilinear abscissa u is, generally, different from the number of givens along the abscissa v, n_v.
- The element f is in this case a three component vector (x, y, z), coordinates of a point on the surface Ω of $\mathbb{R}^3$.
- Let the information relative to the non discretized variable (in this case, v) of the givens of the interpolation problem, for example $\ell_u^i(f)$, be available over the continuum of points. In other words, in each of the functionals ℓ_u^i, ℓ_v^j, the variable unaffected by the functional plays the role of a parameter.

Using the results (4.20)–(4.21) of the previous paragraph, the one dimensional projectors P_u, P_v can be constructed, and one obtains:

$$P_u[f] = \sum_{i=1}^{n_u} \ell_u^i(f)\ \psi_i(u) \qquad P_v[f] = \sum_{j=1}^{n_v} \ell_v^j(f)\ \theta_j(v) \tag{4.24}$$

where $\psi_i(u)$ $(i = 1, n_u)$ and $\theta_j(v)$ $(j = 1, n_v)$ represent the orthogonal basis functions with respect to the functionals ℓ_u^i and ℓ_v^j respectively.

We now introduce the ***Boolean sum*** P^B of the operators P_u and P_v, defined by:

$$P^B = P_u + P_v - P_u[P_v] \tag{4.25}$$

From the preceding remarks, it is easy to see that the product $P_u[P_v]$ is commutative, and thus that the operator P^B is a projector:

$$P^B[P^B] = P^B \tag{4.26}$$

with

$$g(u, v) \equiv P^B[f] = \sum_{i=1}^{n_u} \ell_u^i(f)\psi_i(u) + \sum_{j=1}^{n_v} \ell_v^j(f)\theta_j(v) - \sum_{i=1}^{n_u}\sum_{j=1}^{n_v} \ell_u^i(\ell_v^j(f)\theta_j)(v)\psi_i(u) \tag{4.27}$$

Apply any functional ℓ_u^k to the two sides of (4.27). Taking into account the orthogonality relations (4.19), we have:

$$\ell_u^k(g) = \ell_u^k(f) + \sum_{j=1}^{n_v} \ell_u^k(\ell_v^j(f)\theta_j)(v) - \sum_{j=1}^{n_v} \ell_u^k(\ell_v^j(f)\theta_j)(v) \qquad k = 1, n_u \tag{4.28}$$

that is, the interpolation property:

$$\ell_u^k(g) = \ell_u^k(f) \qquad k = 1, n_u \tag{4.29}$$

Similarly, for any functional ℓ_v^k, we have the interpolation property:

$$\ell_v^k(g) = \ell_v^k(f) \qquad k = 1, n_v \tag{4.30}$$

The interpolation projector P^B thus has the property of exactly reproducing the information contained in the two continous sets of points forming the original problem.

This is precisely the property sought in the mesh problem stated above: establish a bi-directional one-to-one correspondence between the calculation domain Ω_C and the physical domain Ω, which reproduces the boundaries $\partial\Omega_C$ and $\partial\Omega$ in the continuum of points. The Boolean sum defined by (4.27), also called the ***Gordon's projector***, thus provides a convenient means to "parameterize" the interior of a domain with curvilinear boundaries.

Example

Consider an example which illustrates the preceding. We have to "mesh" the physical domain Ω shown in figure 4.4a. The boundary $\partial\Omega$ is subdivided in four segments which are numbered counterclockwise. The calculation space Ω_C is shown in figure 4.4b. Its boundary $\partial\Omega_C$ is related to $\partial\Omega$ with the following parametric equations:

$$\begin{array}{lll} \partial\Omega_1:\ x(1, v) = 1 & y(1, v) = \frac{v}{2} & -1 \le v \le 1 \\ \partial\Omega_2:\ x(u, 1) = \sqrt{2}\cos\left(\frac{\pi}{4}(2-u)\right) & y(u, 1) = \sqrt{2}\sin\left(\frac{\pi}{4}(2-u)\right) - \frac{1}{2} & -1 \le u \le 1 \\ \partial\Omega_3:\ x(-1, v) = -1 & y(-1, v) = \frac{v}{2} & -1 \le v \le 1 \\ \partial\Omega_4:\ x(u, -1) = u & y(u, -1) = -\frac{1}{2} & -1 \le u \le 1 \end{array} \tag{4.31}$$

The unknown of the problem is the two component vector $f(u, v) \equiv [x(u, v), y(u, v)]^T$ for which the Gordon projector P^B yields an approximate expression that exactly reproduces the boundary values. Each of the expressions (4.31) establishing

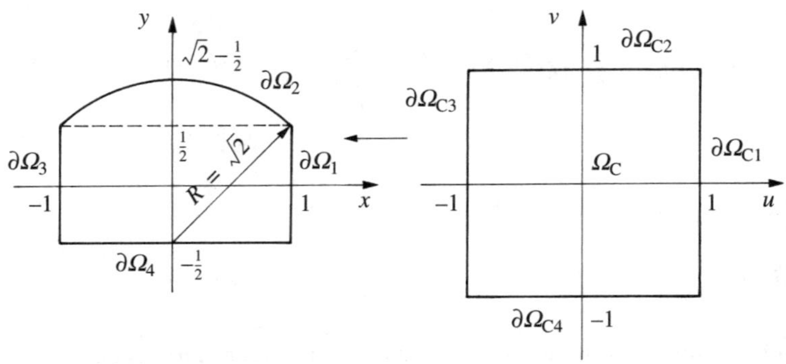

Figure 4.4 Example of the relation of a physical domain Ω, on the left, with a calculation domain Ω_C in the curvilinear coordinate space (u,v), to obtain a transfinite interpolation from Ω_C to Ω.

the correspondence $\partial\Omega_C \rightarrow \partial\Omega$ is a functional of the development (4.27). As for the basis functions, it is easily seen that they are given by the relations (4.10) with $(\xi^1, \xi^2) = (-1, 1)$.

The solution of the problem, $g(u, v) \equiv [\hat{x}(u, v), \hat{y}(u, v)]^T$ is thus obtained very simply by applying (4.27):

$$\begin{aligned} g(u, v) = \; & f(-1, v)L_1(u) + f(1, v)L_2(u) + f(u, -1)L_1(v) + f(u, 1)L_2(v) \\ & -[f(-1,-1)L_1(u)L_1(v) + f(-1, 1)L_1(u)L_2(v) \\ & + f(1,-1)L_2(u)L_1(v) + f(1, 1)L_2(u)L_2(v)] \end{aligned} \tag{4.32}$$

with

$$L_1(u) = \frac{1-u}{2} \qquad L_2(u) = \frac{1+u}{2} \tag{4.33}$$

The introduction of the expressions (4.31) in (4.32)–(4.33) yields the desired parameterization, which is written explicitly:

$$\begin{aligned} \hat{x}(u, v) &= \tfrac{1}{2}u(1-v) + \tfrac{\sqrt{2}}{2}(1+v)\cos\left(\frac{\pi}{4}(2-u)\right) \\ \hat{y}(u, v) &= \tfrac{\sqrt{2}}{2}(1+v)\sin\left(\frac{\pi}{4}(2-u)\right) - \tfrac{1}{2} \end{aligned} \tag{4.34}$$

The preceding example is based on the four given linear functionals; that is, the boundary $\partial\Omega$ of the physical domain. Note that if supplementary functionals are given, such as interior lines of the physical domain for incorporation in the mesh, or information about the interior normal derivatives along the boundary, they can be taken into account very naturally in the preceding formalism. The only difficulty of the problem is then the determination of the one dimensional basis functions

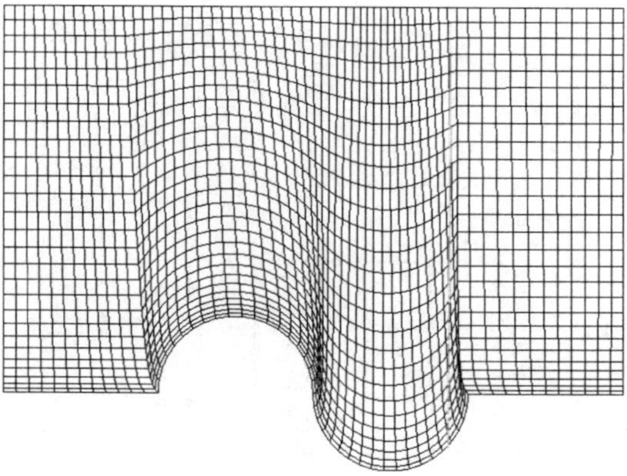

Figure 4.5 Mesh obtained by transfinite interpolation (to compare with the mesh obtained by partial differential equations in figure 4.3).

adapted to the problem, with which one proceeds to the application of (4.27) as in the example above.

In comparison to the mesh generation techniques by solution of partial differential equations, transfinite interpolation permits a better treatment of singular points as well as the orthogonality of the mesh at the boundaries of the calculation domain, as shown in figure 4.5.

Construction of complex meshes by the multi-zone method

The mesh generation techniques above can be extended to meshes with more complicated geometric domains. For that it is sufficient to decompose the domain into many *zones*, which can each be meshed separately with these techniques. Figure 4.6 shows the decomposition into three zones of a two dimensional piece assumed symmetric about the vertical axis. Eight node quadratic elements are defined in this particular case and the mappings are indicated for each zone. It goes without saying that the stitching together of the different zones requires an identical discretization of the joint boundaries.

Correctly used, this type of method allows the meshing of complex two dimensional geometries. In three dimensions, one perceives very rapidly the obvious limits of the method. Nonetheless, when it is well interfaced with a volumetric "modeler" (a CAD tool for decomposition of a domain into elementary volumes), it can still be useful for simple pieces. For more complex geometries, the effort demanded of the user to assure the compatibility conditions of the different zones quickly becomes a nightmare (it is still fairly frequent today to count in days, even in weeks, the duration of mesh generation for an industrial piece!).

In conclusion, obtaining a structured mesh in finite elements depends on the know-how of the user. He will in the end construct the mesh "by hand" with or

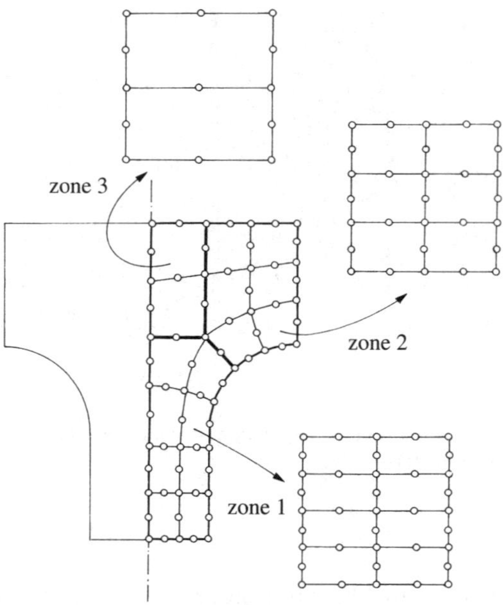

Figure 4.6 Structured mesh of a half section of an axisymmetric piece, in quadratic elements with eight nodes, by the multi-zones method (three zones here).

without the aid of a computer program. If one wants now to avoid such an operation, difficult in 2 dimensions, prohibitive in 3 dimensions, it is necessary to model it, to deduce from it an algorithm and a program to substitute for the user. That is the objective of ***automatic mesh generators*** - or ***free meshers*** – for which we will present the basic principles – in 2 dimensions – in the next paragraph.

4.2.2 Automatic mesh generation

Griding methods

Assume a closed contour, traced over a square grid. A first method consists of removing all squares strictly outside the domain delimited by the contour. One obtains a pre-mesh approximating more or less the geometry to mesh, its precision depending on the fineness of the grid. So far, the method matches the griding method necessary for the FDM. Figure 4.7 shows such a grid.

The squares strictly inside the domain are conserved, whereas the squares straddling the boundary are modified. The intersection of the boundary with the squares of the grid results in the formation of trapezoids (i), pentagons (ii) and triangles (iii). Generally the pentagons are subdivided into three triangles, and the resulting mesh is thus constituted of a mix of quadrangles and triangles.

An alternative is to keep only the squares strictly inside the contour and to generate a band of boundary elements linking the internal grid to the contour (fig. 4.8).

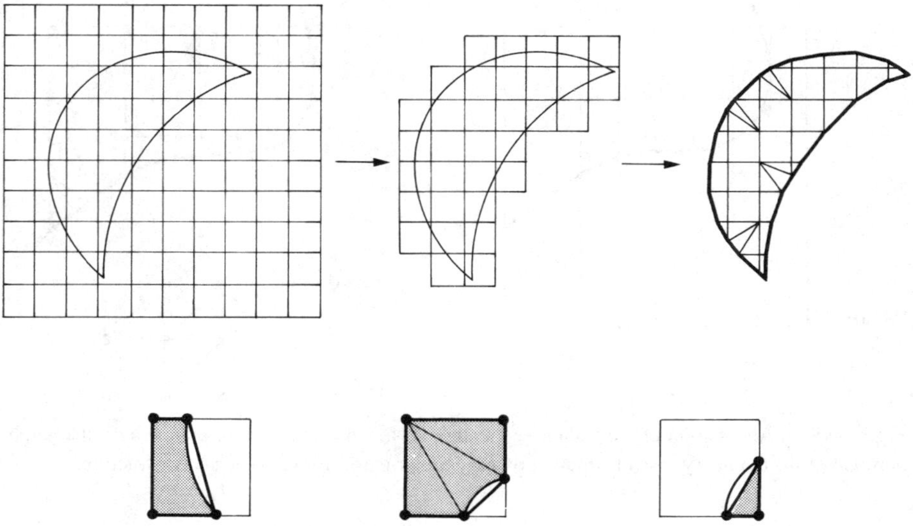

Figure 4.7 Grid method with correction of boundary elements ((i) quadrangle correction; (ii) and (iii) triangle correction).

These methods have the inconvenience of producing unbalanced meshes (structured internal part and unstructured peripheral zone) and especially of missing all detail of the contour with a dimension less than the characteristic size of the basic grid. The quadrangle tree method (*quadtree*) resolves this problem by employing a branching structure that permits geometric adaptability. This method, schematically described in figure 4.9, is as follows.

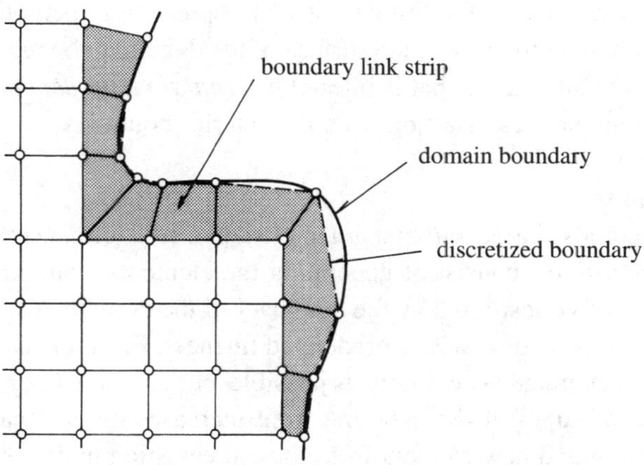

Figure 4.8 Griding method with a strip linking the boundary.

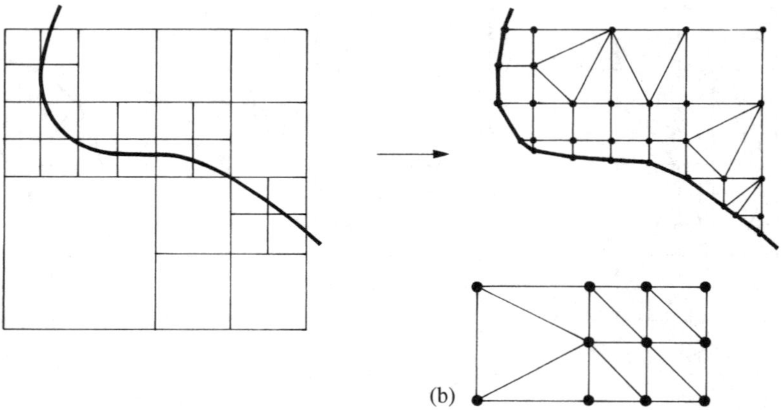

(a) (b)

Figure 4.9 a) Application of the "quadtree" method to the discretization of a domain in the neighborhood of the boundary. b) Construction of a triangular mesh from a re-balanced structure.

Starting with a relatively coarse grid, each square straddling the contour is subdivided into four squares, recursively, until the initial tolerance of the representation of the boundary is attained. The structure of the mesh obtained is then hierarchic and arborescent. In the second phase, the structure is re-balanced, imposing that the ratio of the lengths of the sides of two adjacent squares be less than or equal to two. Figure 4.9b shows the joining of a square twice the size of the adjacent squares by subdivision into three triangles. One can from there deduce a final mesh constituted of triangles or a mix of triangles and quadrangles. The method can be easily generalized in three dimensions (the *octree* method). It permits better respect of the exterior form of the domain to mesh, but at the price of a very notable increase in the number of elements (especially in 3 dimensions) when the details are not oriented with the initial grid.

The primary cause of the limitations of the preceding methods is then to not take into account the topology of the surface of the domain to be meshed. And so we present below the alternative that is offered by ***frontal methods***, which consist precisely of defining the mesh starting from the exterior boundary.

Frontal methods

These methods, based on triangular elements in 2 dimensions, and tetrahedrons in 3 dimensions, consist of generating the elements from an evolving front. The starting front is constituted by the boundary of the domain, discretized in linear segments (in 2 dimensions) with a predefined fineness. From the advancing front, a new triangle is generated as regularly as possible, either by creating a new node (the global number of edges of the evolving front increases by one; case (a) of figure 4.10), or by defining a new element from three preexisting nodes. According to the case, the latter operation can decrease by one the number of edges of the front

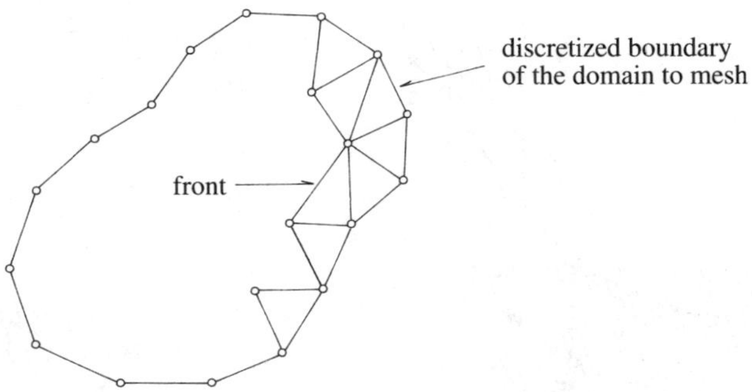

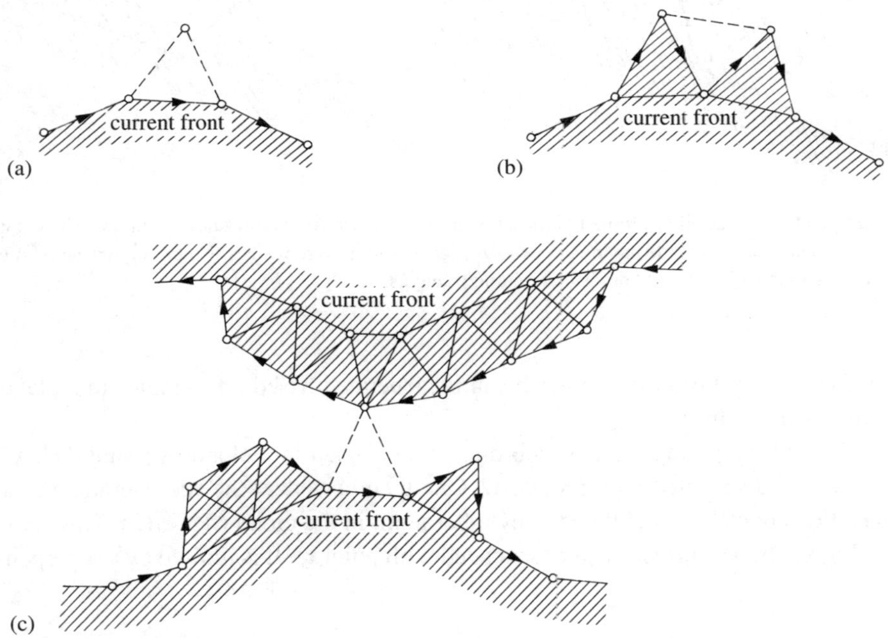

Figure 4.10 Frontal method meshing progression by generation of a new triangle a) with creation of a node, b) with creation of an edge, c) with separation of the domain. The arrows and shaded zones signal the current position of the front.

(case (b) in figure 4.10) or separate the domain, the front no longer being connected (case (c) in figure 4.10).

The difficulty obviously resides in the treatment of the collisions of the front as illustrated in figure 4.11. The nodes and the edges of the triangular elements produced on the two fronts must finish by coinciding. This generally introduces very deformed elements in the zones where the fronts join.

The Delaunay-Voronoi method

Contrary to the preceding methods, based on explicit geometric construction, the ***Delaunay-Voronoi method*** comes directly from a mathematical result, proving

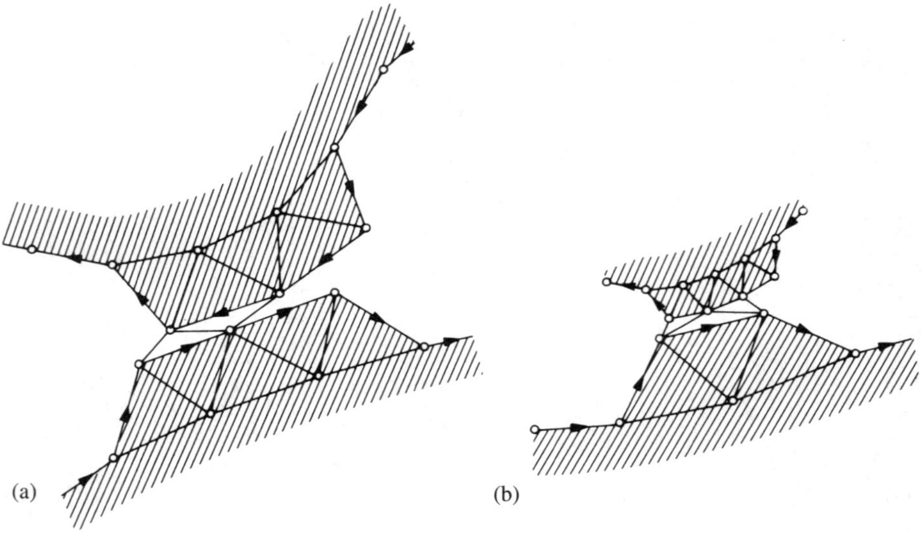

Figure 4.11 Compatibility problem during the meeting of two fronts: presence of an unmeshed gap with elements on either side having similar sizes (a) or very different sizes (b). In the two cases, it is necessary to introduce highly deformed joining triangles.

that from a cloud of points, a mesh can be generated based on triangles in a plane, on tetrahedrons in space.

Consider a point in vector coordinates $\boldsymbol{x}^1$ chosen in a cloud of points $\{\boldsymbol{x}^1, \boldsymbol{x}^2, \boldsymbol{x}^3, \ldots, \boldsymbol{x}^{\mathrm{Nn}}\}$ distributed in a domain Ω. One then considers the set of points in the plane that are closer to $\boldsymbol{x}^1$ than to any other point in the cloud (fig. 4.12). This set is a polygon: the ***Voronoi polygon*** associated with point $\boldsymbol{x}^1$. It is limited by the perpen-

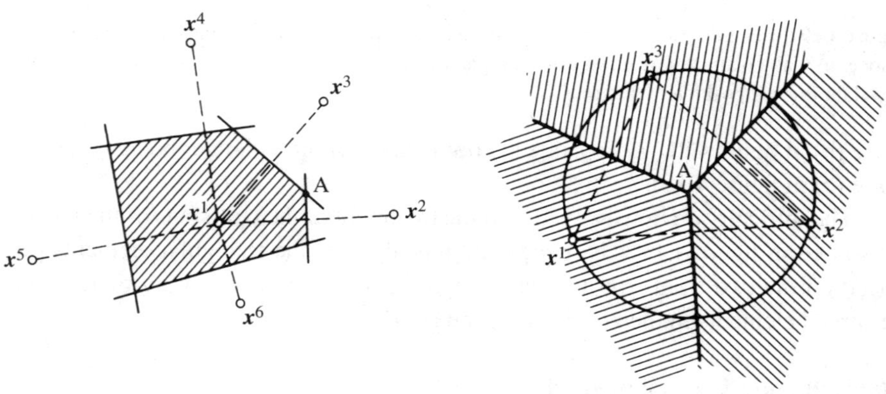

Figure 4.12 To the left, Voronoi polygon associated with point $\boldsymbol{x}^1$. To the right, the vertex A is shared with three Voronoi polygons associated with $\boldsymbol{x}^1, \boldsymbol{x}^2, \boldsymbol{x}^3$.

dicular bisectors of each of the segments $[\boldsymbol{x}^1, \boldsymbol{x}^i]$ where $\boldsymbol{x}^i$ are the points that are "neighbors" of $\boldsymbol{x}^1$ in the Voronoi sense of the term. Reciprocally, $\boldsymbol{x}^1$ is a neighbor of each of the $\boldsymbol{x}^i$ in the Voronoi sense.

Consider now a vertex A of the polygon associated with $\boldsymbol{x}^1$: it is situated at the intersection of the perpendicular bisectors of two of the segments associated with $\boldsymbol{x}^1$, for example $[\boldsymbol{x}^1, \boldsymbol{x}^2]$ and $[\boldsymbol{x}^1, \boldsymbol{x}^3]$. It follows that A is also a vertex of the polygons associated with $\boldsymbol{x}^2$ and $\boldsymbol{x}^3$. A is then the center of the circle that circumscribes the triangle formed by $\boldsymbol{x}^1$, $\boldsymbol{x}^2$ and $\boldsymbol{x}^3$. The remarkable property of the Voronoi polygons is that no other point exists in the cloud that is situated at a smaller distance from the polygon summit A than the points $\boldsymbol{x}^1$, $\boldsymbol{x}^2$ and $\boldsymbol{x}^3$. In other words, in a triangular Delaunay mesh, the circles circumscribing the triangular elements contain no other nodes of the mesh.

Another interesting property on which Delaunay's algorithm is based, is that for an arbitrary distribution of points, the probability that a polygon vertex belongs to more than three polygons is zero. As a result, the pattern produced in generating the triangles is unique. Figure 4.13 shows the Voronoi polygons defined for a cloud of points and the resulting ***Delaunay triangulation***.

The algorithm to establish the Delaunay triangulation associated with a cloud of points is thus as follows. With the addition of three auxiliary points, an initial triangle is constructed which contains all the points of the cloud; this permits the initiation of the triangulation, which will be completed by successive addition of each of the points in the cloud. Consider the triangular mesh at a given stage in its construction (fig. 4.14). The insertion of a new point from the cloud is done in the following way: current triangles for which the circumscribed circles contain the new point (the two shaded triangles in figure 4.14) are destroyed. The "hole" in the mesh that results is filled by new triangles (here: four) having as a common vertex the new point and for the other vertices the nodes delimiting the edges of the hole. This new triangulation is a Delaunay triangulation. The algorithm is repeated until all the

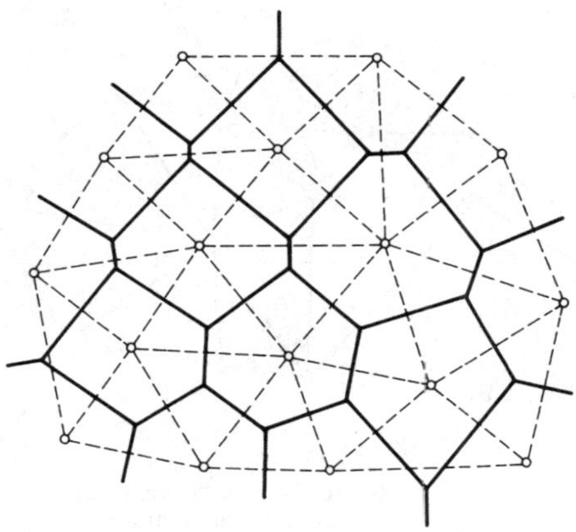

Figure 4.13 Voronoi polygons and the associated Delaunay triangulation.

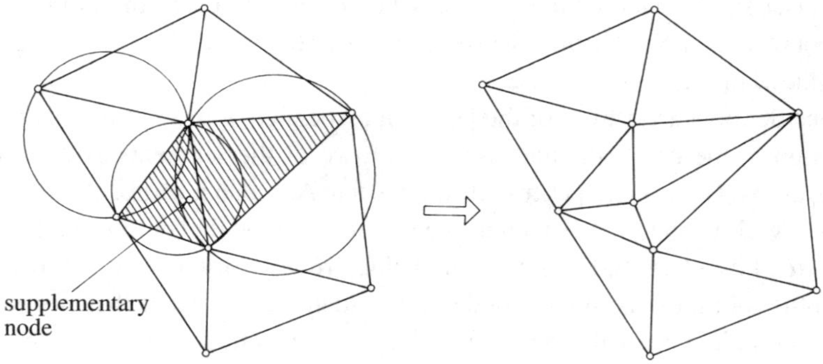

Figure 4.14 Insertion of a point in a Delaunay triangulation.

points of the cloud have been inserted. Then it only remains to destroy those triangles which have as a vertex one of the three auxiliary points chosen initially.

Suppose now that one wishes to mesh a domain Ω delimited by a boundary constituted of straight segments, for example, the gray domain in the form of an "m" in figure 4.15. The Delaunay mesh generation algorithm proceeds in the following stages.

Stage 1: meshing of the convex envelope of Ω

First, Delaunay triangulation is performed, following the procedure defined above, on the cloud of points constituted by the vertices of the boundary defining the contour of Ω. The triangulation of the convex envelope of the domain Ω is thus obtained, that is, the smallest convex polygon containing Ω. In this triangulation, shown in figure 4.15, three types of triangles appear: 1) those that are inside the domain Ω (gray triangles); 2) those outside Ω (white triangles); 3) finally those that straddle the boundary (the case of the two triangles to the right of the domain in the

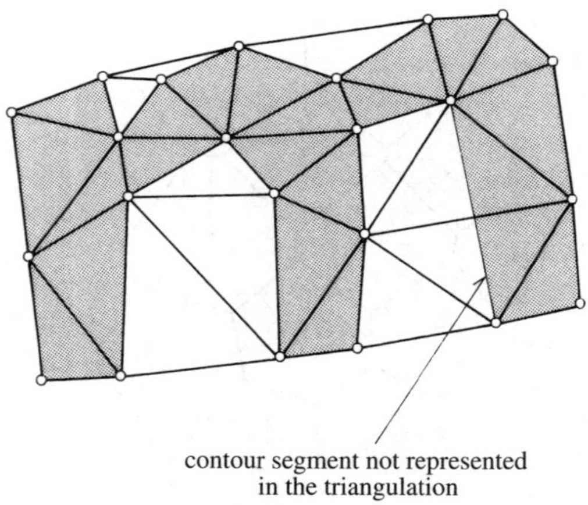

Figure 4.15 Triangulation of the convex envelope of the domain Ω.

form of an "m"). The next steps are to extract the triangles that are elements of Ω and to repair the faults that can appear.

Stage 2: extraction of the triangulation of the interior of Ω

First of all, it is necessary to insure that the contour segments are all edges of the triangulation. If that is not the case for one of the segments (example in figure 4.15), the segment is divided in two and the midpoint is inserted in the Delaunay triangulation. This dichotomous procedure is repeated on all segments where the problem occurs; the triangles of the third type, mentioned in the preceding stage, are thus eliminated. This done, the next action is to orient all the triangles having an edge on the boundary and then to propagate this orientation over the whole mesh from triangle to neighboring triangle. If the initial boundary of Ω is oriented counterclockwise, it remains only to eliminate the clockwise triangles to obtain a first mesh for Ω, without internal nodes (fig. 4.16).

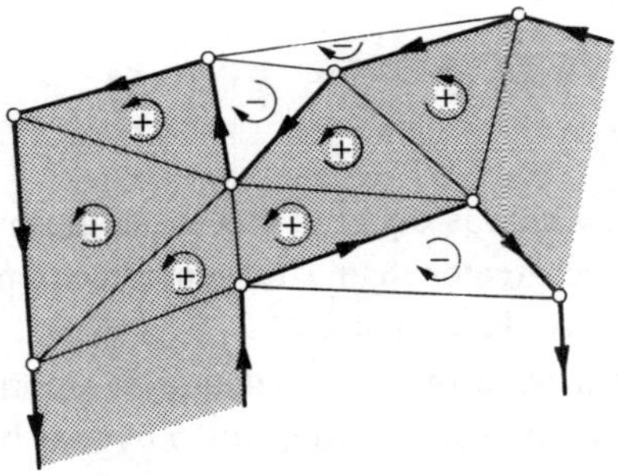

Figure 4.16 Extraction of the mesh of the interior of Ω by orientation of the elements starting from the contour. The white triangles, oriented clockwise, are eliminated.

Stage 3: creation of the internal nodes

The creation of nodes and thus internal elements should lead to triangles sufficiently regular (close to being equilateral) and of non-infinitesimal size, to insure a good compromise between precision and finite element calculation costs. It is necessary then to define before hand a ***quality*** for triangular elements. One could take for example the value Q defined by:

$$Q = \frac{4}{\sqrt{3}}\frac{S}{\bar{\ell}^2} \tag{4.35}$$

where S is the area of the triangle and $\bar{\ell}$ the average length of the edges of the triangle, the coefficient $4/\sqrt{3}$ insures the quality value $Q = 1$ for equilateral triangles.

A classic procedure for creation of internal nodes consists then of propagating the node density at the boundary into the domain. For that, a weight w^i is assigned first to each boundary node i, which is simply the arithmetic mean of the lengths of the two boundary edges to which the node belongs.

The weight serves to evaluate the size of the elements. A triangle T of the current triangulation is judged too big if

$$Q = \frac{4}{\sqrt{3}} \frac{S(T)}{(w^1 w^2 w^3)^{2/3}} > 1 \tag{4.36}$$

where $S(T)$ is the area of the triangle and w^1, w^2 and w^3 are the weights of its vertices. In this case, a new node is created inside the element and inserted in the Delaunay triangulation. This procedure is repeated until the sizes of all the triangles are judged acceptable. Many choices for the position of the new node are possible as well as the weight which is assigned to it; the following is but one solution among others:

$$\begin{aligned} \boldsymbol{x} &= \frac{1}{(w^1 + w^2 + w^3)}(w^1 \boldsymbol{x}^1 + w^2 \boldsymbol{x}^2 + w^3 \boldsymbol{x}^3) \\ w(\boldsymbol{x}) &= \alpha \frac{w^1 + w^2 + w^3}{3} \end{aligned} \tag{4.37}$$

where α is a coefficient which is equal to 1 to best approximate equilateral triangles, and which should be greater than 1 to generate triangles with growing area going towards the interior of the domain.

From a mathematical point of view, Delaunay's algorithm can be generalized to spaces in any dimension. In 3 dimensions, the polygons become polyhedrons; the triangles, tetrahedrons; and the circles, spheres. However, the extraction of a mesh from an arbitrary geometry is more difficult to solve.

Techniques for improving meshes

These techniques have as an objective the perfection of meshes, either structured or unstructured, that have been obtained with the above methods.

Centroid regularization

The most classic method of regularization, and the easiest to implement, is based on Laplace's algorithm; it consists of displacing each internal node to the geometric center of the neighboring nodes in the triangulation (the barycentric coefficients being taken equal to 1 here to obtain the centroid).

In practice, one proceeds node by node. The method is thus iterative – until stagnation, that is, until the variation of the positions of each of the nodes is sufficiently small. It requires few iterations and thus costs little in terms of calculations. It is however necessary to control some faults of the method near pronounced concavities of the boundary that can result in non conforming meshes (the orientation of the elements can become inverted).

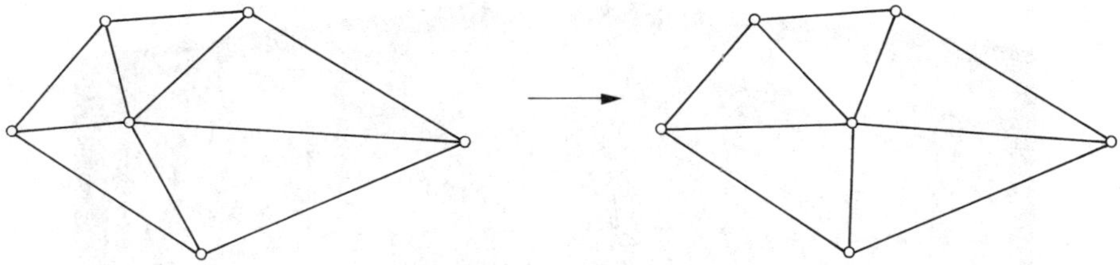

Figure 4.17 Regularization by the centroid technique.

Topological improvements

Unlike the preceding technique, this time the intervention is topological, that is, on the connectivity of the mesh. The basic idea in two dimensions consists of inverting the diagonal of the parallelogram formed by two adjacent triangles (fig. 4.18).

The topological change is made if it reduces an objective function. For example, this function can be the accumulated divergences of the two triangles from equilateral triangles:

$$\zeta = \sum_{T = T_1, T_2} (1 - Q(T))^2 \tag{4.38}$$

In practice, the inversion of the diagonals continues until there are no inversions left that are favorable. This method is often alternated with the centroid technique to perfect the meshes.

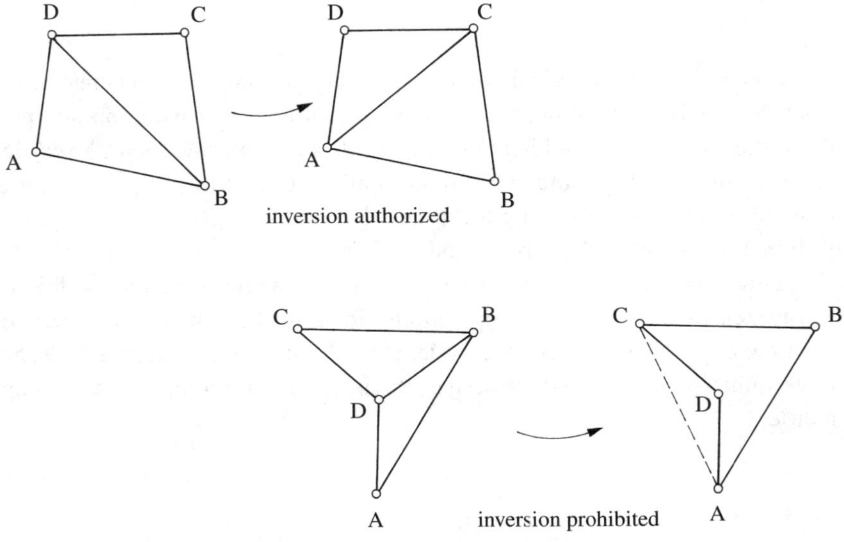

Figure 4.18 Topological improvement by inversion of the diagonals.

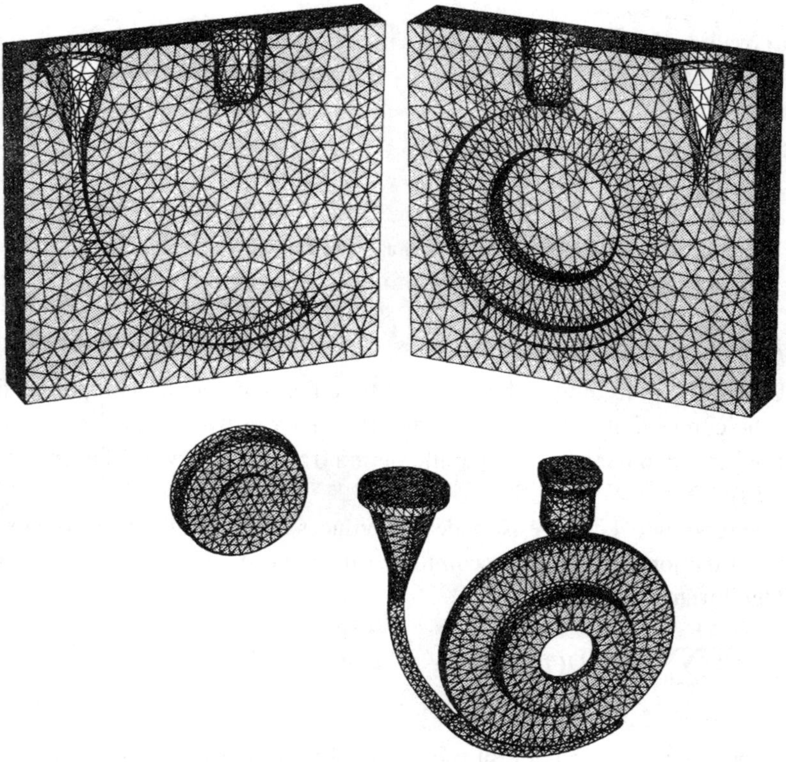

Figure 4.19 Three dimensional mesh of tetrahedral elements obtained with a free mesher of a part and of three components of a casting mold for a thermo-mechanical process analysis by finite elements (only the surface triangular elements are shown).

To conclude this introduction to the elementary notions of mesh generation, we add that there exists a multitude of possible algorithms to arrive at an acceptable result. In this domain, the architect of an algorithm is confronted with a very large choice from among what could colorfully be called "geometric recipes". The effectiveness of an automatic meshing technique depends strongly on the sequence of algorithms implemented; but also – and it is an important point that we will not develop here – on the right conception of the data structures which condition the speed of each of the elementary operations (identification of the neighbors of a node, of the edges that intersect at a node, etc.). As an example, figure 4.19 shows the three dimensional tetrahedral meshes obtained with a free mesher for complex geometries.

4.3 SOLUTION METHODS FOR LINEAR SYSTEMS

As we have seen in chapters 2 and 3, the finite difference and finite element techniques always lead to the solution of systems of linear equations of the type

$$[\boldsymbol{K}]x = f \tag{4.39}$$

where $[\boldsymbol{K}]$ has the property of being a matrix with a banded structure. We denote by n the size of the system (4.39). The two main types of methods for solving this system are, first, ***direct methods*** that are the most robust and have the advantage of giving the exact solution, within the precision of the computer, and second, ***iterative methods***, for which the solution is approximated by successive corrections. We describe below these two main types of methods.

4.3.1 Direct methods

Direct methods permit the solution of the system while avoiding the explicit calculation of the inverse matrix of $[\boldsymbol{K}]$. Among these methods, ***Gauss' method*** is the most widespread. It is based on the fact that there exists a multiplicative decomposition, or ***factorization***, of the matrix $[\boldsymbol{K}]$:

$$[\boldsymbol{K}] = [\boldsymbol{L}][\boldsymbol{U}] \tag{4.40}$$

with $[\boldsymbol{L}]$ a lower triangular matrix, for which the diagonal terms are equal to 1,
$[\boldsymbol{U}]$ an upper triangular matrix.

$$[\boldsymbol{K}] = \begin{bmatrix} 1 & & & & & (0) \\ & 1 & & & & \\ & & \ddots & & & \\ & & & 1 & & \\ & & & & 1 & \\ L_{ij} & & & & & 1 \end{bmatrix} \begin{bmatrix} U_{11} & & & U_{ij} \\ & U_{22} & & \\ & & \ddots & \\ (0) & & & U_{nn} \end{bmatrix} \tag{4.41}$$

Before returning to the determination of the coefficients of the matrices $[\boldsymbol{L}]$ and $[\boldsymbol{U}]$, we discuss the solution algorithm that results from this decomposition: the triangular form of the matrices is exploited to obtain the solution in two steps. Therefore, setting $[\boldsymbol{U}]\,\boldsymbol{x} = \boldsymbol{y}$, Gauss' method consists of two successive steps:

- First step. Forward substitution, or “descent”.
  The following system is solved:

$$[\boldsymbol{L}]\boldsymbol{y} = \boldsymbol{f} \tag{4.42}$$

  $[\boldsymbol{L}]$ being lower triangular, the first equation gives directly $y_1 = f_1$. By substitution in the second, y_2 is found, and so on until the last equation giving y_n.
- Second step. Backward substitution, or “ascent”.
  Now knowing **y**, the following system is solved:

$$[\boldsymbol{U}]\boldsymbol{x} = \boldsymbol{y} \tag{4.43}$$

  This time, $[\boldsymbol{U}]$ being upper triangular, one starts with the last equation which gives $x_n = y_n / U_{nn}$. Ascending successively through the equations, one finds by substitution the values $x_{n-1}, x_{n-2}, \dots, x_1$.

This method can be easily illustrated in the simple case of a system of 3 equations with 3 unknowns:

$$\left.\begin{aligned} K_{11}x_1 + K_{12}x_2 + K_{13}x_3 &= f_1 \\ K_{21}x_1 + K_{22}x_2 + K_{23}x_3 &= f_2 \\ K_{31}x_1 + K_{32}x_2 + K_{33}x_3 &= f_3 \end{aligned}\right\} \tag{4.44}$$

We keep the first equation, from which we take x_1 to replace in the two following equations. We obtain:

$$\left.\begin{aligned} K_{11}x_1 + K_{12}x_2 + K_{13}x_3 &= f_1 \\ \left(K_{22} - \frac{K_{21}K_{12}}{K_{11}}\right)x_2 + \left(K_{23} - \frac{K_{21}K_{13}}{K_{11}}\right)x_3 &= f_2 - \frac{K_{21}f_1}{K_{11}} \\ \left(K_{32} - \frac{K_{31}K_{12}}{K_{11}}\right)x_2 + \left(K_{33} - \frac{K_{31}K_{13}}{K_{11}}\right)x_3 &= f_3 - \frac{K_{31}f_1}{K_{11}} \end{aligned}\right\} \tag{4.45}$$

thus the new system:

$$K_{11}x_1 + K_{12}x_2 + K_{13}x_3 = f_1 \tag{4.46a}$$

$$K'_{22}x_2 + K'_{23}x_3 = f'_2 \tag{4.46b}$$

$$K'_{32}x_2 + K'_{33}x_3 = f'_3 \tag{4.46c}$$

where the K'_{ij} and f'_i appear clearly in (4.45). Performing the same operation with x_2, which is taken from the second equation (4.46b) and is replaced in the third, we obtain finally the system of equations:

$$\left.\begin{aligned} K_{11}x_1 + K_{12}x_2 + K_{13}x_3 &= f_1 \\ K'_{22}x_2 + K'_{23}x_3 &= f'_2 \\ K''_{33}x_3 &= f''_3 \end{aligned}\right\} \tag{4.47}$$

The coefficients appearing in front of the unknowns are none other than those of the matrix $[\boldsymbol{U}]$:

$$[U] = \begin{bmatrix} K_{11} & K_{12} & K_{13} \\ 0 & K'_{22} & K'_{23} \\ 0 & 0 & K''_{33} \end{bmatrix} \tag{4.48}$$

The terms on the right hand side are the components of the vector $\boldsymbol{y} = [f_1, f'_2, f''_3]^T$. The operations performed to go from (4.44) to (4.47) correspond to the solution of the first problem (4.42) : $[\boldsymbol{L}]\,\boldsymbol{y} = \boldsymbol{f}$. To get to the final values we are seeking $(x_1, x_2, \ldots, x_n)$, we have only to solve the second problem (4.43): $[\boldsymbol{U}]\,\boldsymbol{x} = \boldsymbol{y}$.

In general, for n equations, the complete algorithm including the decomposition, the descent and the back substitution is elucidated in figure 4.20. With respect to the preceding formulas, developed for the case $n = 3$, the algorithm in the figure is based on a successive decomposition method (exercise 4.6.1). The indexing convention is applied that stipulates that if the upper index of a sum has a value strictly less than that of the lower index, that sum is presumed null.

Gauss' method can be applied to full matrices as well as to banded matrices, such as those that result from the FDM and the FEM. In this context, it can be shown that this factorization is *compact* (exercise 4.6.2). It is sufficient to note in figure 4.20 that the components of [***L***] and [***U***] can all be replaced by the components [***K***] with the same indices. The band structure of matrices [***L***] and [***U***] is thus the same as that of the original matrix [***K***], which means that the software implementation of the algorithm does not require any extra memory than that to store [***K***]. Nonetheless, Gauss factorization causes *band filling*, that is, components *ij* that initially were zero, but inside the band of [***K***], become non zero in [***L***] and [***U***] in the factorization. In practice, this obliges the reservation of memory for the whole band to perform the factorization. We will return with more detail to these aspects in section 4.4.

Crout and Cholesky decompositions

The ***Crout decomposition*** is a variant of Gauss decomposition, which is written:

Factorization: $[\boldsymbol{K}] = [\boldsymbol{L}][\boldsymbol{U}]$

$U_{11} = K_{11}$
Loop over $i = 2, n$
  Loop over $j = 1, i-1$

$$L_{ij} = \left(K_{ij} - \sum_{k=1}^{j-1} L_{ik} U_{kj} \right) / U_{jj}$$

  End of loop over j
  Loop over $j = 1, i-1$

$$U_{ji} = K_{ji} - \sum_{k=1}^{j-1} L_{jk} U_{ki}$$

  End of loop over j

$$U_{ii} = K_{ii} - \sum_{k=1}^{i-1} L_{ik} U_{ki}$$

End of loop over i

Descent: $\boldsymbol{y} = [\boldsymbol{L}]^{-1} \boldsymbol{f}$

Loop over $i = 1, n$

$$y_i = f_i - \sum_{k=1}^{i-1} L_{ik} y_k$$

End of loop over i

Back substitution: $\boldsymbol{x} = [\boldsymbol{U}]^{-1} \boldsymbol{y}$

Loop over $i = n, 1$

$$x_i = \left(y_i - \sum_{k=i+1}^{n} U_{ik} x_k \right) / U_{ii}$$

End of loop over i

Figure 4.20 Solution algorithm for a linear system by Gauss' method.

$$[\boldsymbol{K}] = [\boldsymbol{L}][\boldsymbol{D}][\boldsymbol{U}'] \tag{4.49}$$

with $[\boldsymbol{L}]$ a lower triangular matrix, with diagonal terms equal to 1, $[\boldsymbol{D}]$ a diagonal matrix and $[\boldsymbol{U}']$ an upper triangular matrix, with diagonal terms equal to 1.

$$[\boldsymbol{K}] = \begin{bmatrix} 1 & & & & & & (0) \\ & 1 & & & & & \\ & & 1 & & & & \\ & & & \ddots & & & \\ & & & & 1 & & \\ & & & & & 1 & \\ L_{ij} & & & & & & 1 \end{bmatrix} \begin{bmatrix} \ddots & & & (0) \\ & \ddots & & \\ & & D_{ii} & \\ & & & \ddots \\ (0) & & & & \ddots \end{bmatrix} \begin{bmatrix} 1 & & & & & & U'_{ij} \\ & 1 & & & & & \\ & & 1 & & & & \\ & & & \ddots & & & \\ & & & & 1 & & \\ & & & & & 1 & \\ (0) & & & & & & 1 \end{bmatrix} \tag{4.50}$$

If $[\boldsymbol{K}]$ is positive definite, then the values D_{ii} of the diagonal matrix are positive and one can write:

$$[\boldsymbol{K}] = [\tilde{\boldsymbol{L}}][\tilde{\boldsymbol{U}}] \tag{4.51}$$

with

$$[\tilde{\boldsymbol{L}}] = [\boldsymbol{L}][\boldsymbol{D}]^{1/2} \quad \text{and} \quad [\tilde{\boldsymbol{U}}] = [\boldsymbol{D}]^{1/2}[\boldsymbol{U}'] \tag{4.52}$$

where the notation $[\boldsymbol{D}]^{1/2}$ designates the diagonal matrix with terms $\sqrt{D_{ii}}$. If $[\boldsymbol{K}]$ is additionally symmetric, then $[\boldsymbol{U}'] = [\boldsymbol{L}]^{\mathrm{T}}$ and the ***Cholesky decomposition*** is obtained:

$$[\boldsymbol{K}] = [\tilde{\boldsymbol{L}}][\tilde{\boldsymbol{L}}]^{\mathrm{T}} \tag{4.53}$$

Figures 4.21 and 4.22 show the Cholesky and Crout algorithms for a symmetric matrix.

$L_{11} = (K_{11})^{1/2}$
Loop over $i = 2, n$
  Loop over $j = 1, i-1$

$$L_{ij} = \left(K_{ij} - \sum_{k=1}^{j-1} L_{ik}L_{jk}\right) / L_{jj}$$

  End of loop over j

$$L_{ii} = \left(K_{ii} - \sum_{k=1}^{i-1} L_{ik}^2\right)^{1/2}$$

End of loop over i

} Factorization: $[\boldsymbol{K}] = [\tilde{\boldsymbol{L}}][\tilde{\boldsymbol{L}}]^{\mathrm{T}}$

Loop over $i = 1, n$

$$y_i = \left(f_i - \sum_{k=1}^{i-1} L_{ik}y_k\right) / L_{ii}$$

End of loop over i

} Descent: $\boldsymbol{y} = [\tilde{\boldsymbol{L}}]^{-1}\boldsymbol{f}$

Loop over $i = n, 1$

$$x_i = \left(y_i - \sum_{k=i+1}^{n} L_{ki}x_k\right) / L_{ii}$$

End of loop over i

} Back substitution: $\boldsymbol{x} = ([\tilde{\boldsymbol{L}}]^{\mathrm{T}})^{-1}\boldsymbol{y}$

Figure 4.21 Solution algorithm for a linear system by Cholesky's method for a positive definite symmetric matrix $[\boldsymbol{K}]$.

$D_{11} = K_{11}$
Loop over $i = 2, n$
  Loop over $j = 1, i-1$

$$L_{ij} = \left(K_{ij} - \sum_{k=1}^{j-1} L_{ik} L_{jk}\right) / D_{jj}$$

  End of loop over j

$$D_{ii} = K_{ii} - \sum_{k=1}^{i-1} L_{ik}^2 D_{kk}$$

End of loop over i

Factorization: $[\boldsymbol{K}] = [\boldsymbol{L}][\boldsymbol{D}][\boldsymbol{L}]^{T}$

Loop over $i = 1, n$

$$z_i = f_i - \sum_{k=1}^{i-1} L_{ik} z_k$$

End of loop over i

Descent: $\boldsymbol{z} = [\boldsymbol{L}]^{-1} \boldsymbol{f}$

Loop over $i = 1, n$
  $y_i = z_i / D_{ii}$
End of loop over i

$\boldsymbol{y} = [\boldsymbol{D}]^{-1} \boldsymbol{z}$

Loop over $i = n, 1$

$$x_i = y_i - \sum_{k=i+1}^{n} L_{ki} x_k$$

End of loop over i

Back substitution: $\boldsymbol{x} = ([\boldsymbol{L}]^{T})^{-1} \boldsymbol{y}$

Figure 4.22 Solution algorithm for a linear system by Crout's method for a symmetric matrix $[\boldsymbol{K}]$.

TDMA method for finite differences

A particular case of Gaussian elimination occurs for the finite difference method applied to structured grids, for which we have to solve multi diagonal systems. In 1 dimension, as well as with the alternating direction implicit method (ADI, chap. 2), the system is tri-diagonal. Gaussian elimination leads to Thomas' algorithm (fig. 4.23) for which the reader can easily find the formulas (exercise 4.6.3).

4.3.2 Iterative methods

A second major class of methods for solving systems of linear equations is constituted by *iterative methods*. As opposed to direct methods, iterative methods will permit the ***convergence*** towards the solution of the system, within a given precision, by successive approximations. These methods are less deterministic than the previous ones in that one can not predict *a priori* the number of iterations necessary to converge to a solution.

$U_{11} = K_{11}; y_1 = f_1$	$y_1 = f_1$
Loop over $i = 2, n$	Loop over $i = 2, n$
$U_{ii} = K_{ii} - K_{i,i-1}U_{i-1,i}/U_{i-1,i-1}$	$D_i = D_i - I_i S_{i-1}/D_{i-1}$
$y_i = f_i - K_{i,i-1}y_{i-1}/U_{i-1,i-1}$	$y_i = f_i - I_i y_{i-1}/D_{i-1}$
End of loop over i	End of loop over i
$x_n = y_n/U_{nn}$	$x_n = y_n/D_n$
Loop over $i = n-1, 1$	Loop over $i = n-1, 1$
$x_i = (y_i - K_{i,i+1}x_{i+1})/U_{ii}$	$x_i = (y_i - S_i x_{i+1})/D_i$
End of loop over i	End of loop over i

Figure 4.23 Tri-Diagonal Matrix Algorithm used in problems with the finite difference method and the alternating direction implicit method. Shown on the right are the expressions corresponding to the storage of the three "diagonals" of the matrix in the three vectors $\boldsymbol{D}$, $\boldsymbol{I}$ and $\boldsymbol{S}$ with components: $D_i = K_{ii}$ for $i = 1, n$; $I_i = K_{i,i-1}$ for $i = 2, n$ with $I_1 = 0$; $S_i = K_{i,i+1}$ for $i = 1, n-1$ with $S_n = 0$.

We turn our attention first to methods based on an additive decomposition of the matrix $[\boldsymbol{K}]$ of the system to solve: the Jacobi and Gauss-Seidel methods and their derivatives. Second, we will describe minimization methods: the gradient and conjugate gradients methods.

Iterative methods based on an additive decomposition

The basic idea of these methods is to find a simple additive decomposition (as opposed to the multiplicative decompositions that were presented in the introduction to the direct methods). Thus, if we decompose $[\boldsymbol{K}]$ in the form:

$$[\boldsymbol{K}] = [\boldsymbol{M}] - [\boldsymbol{N}] \tag{4.54}$$

we can envision the following iterative solution. Starting from an initialization $\boldsymbol{x}^{(0)}$, we calculate closer with each iteration (ν), $\boldsymbol{x}^{(\nu+1)}$ according to the scheme:

$$[\boldsymbol{M}]\boldsymbol{x}^{(\nu+1)} = [\boldsymbol{N}]\boldsymbol{x}^{(\nu)} + \boldsymbol{f} \tag{4.55}$$

$[\boldsymbol{M}]$ must obviously be invertible. We obtain:

$$\boldsymbol{x}^{(\nu+1)} = [\boldsymbol{M}]^{-1}[\boldsymbol{N}]\boldsymbol{x}^{(\nu)} + [\boldsymbol{M}]^{-1}\boldsymbol{f} \tag{4.56}$$

The matrix $[\boldsymbol{B}] = [\boldsymbol{M}]^{-1}[\boldsymbol{N}]$ is called the ***iteration matrix***. For the method to be effective, the three following conditions need to be satisfied:

- $[\boldsymbol{M}]$ must be easily invertible (in any case, more easily than $[\boldsymbol{K}]$!).
- $[\boldsymbol{M}]$ must be close to $[\boldsymbol{K}]$ (note that if $[\boldsymbol{M}] = [\boldsymbol{K}]$, the method converges in a single iteration: we are back to the direct method).

- One can prove that the method converges if the *spectral radius* of the iteration matrix, given by,

$$\rho([\boldsymbol{B}]) = \rho([\boldsymbol{M}]^{-1}[\boldsymbol{N}]) = \underset{i}{\mathrm{Max}}|\lambda_i|$$

is less than 1, the λ_i being the eigenvalues of $[\boldsymbol{B}]$.

We consider then a few of the most used methods, employing the following notations:

$$[\boldsymbol{K}] = [\boldsymbol{D}] + [\boldsymbol{L}] + [\boldsymbol{U}] \tag{4.57}$$

$[\boldsymbol{D}]$: Diagonal matrix containing the diagonal of $[\boldsymbol{K}]$,
$D_{ij} = K_{ij}$ if $i = j$, $D_{ij} = 0$ otherwise.
$[\boldsymbol{L}]$: Strictly lower triangular matrix of $[\boldsymbol{K}]$,
$L_{ij} = K_{ij}$ if $i > j$, $L_{ij} = 0$ otherwise.
$[\boldsymbol{U}]$: Strictly upper triangular matrix of $[\boldsymbol{K}]$.
$U_{ij} = K_{ij}$ if $i < j$, $U_{ij} = 0$ otherwise.

Jacobi method

It consists of taking

$$[\boldsymbol{M}] = [\boldsymbol{D}] \tag{4.58}$$

$$[\boldsymbol{N}] = -([\boldsymbol{L}] + [\boldsymbol{U}]) \tag{4.59}$$

The iterative scheme is thus defined by:

$$[\boldsymbol{D}]\boldsymbol{x}^{(\nu+1)} = -([\boldsymbol{L}] + [\boldsymbol{U}])\boldsymbol{x}^{(\nu)} + \boldsymbol{f} \tag{4.60}$$

or in components,

$$x_i^{(\nu+1)} = \frac{1}{K_{ii}}\left(f_i - \sum_{j=1\ \&\ j\neq i}^{n} K_{ij}x_j^{(\nu)}\right) \qquad \text{for } i = 1, n \tag{4.61}$$

Gauss–Seidel Method

In this case we take,

$$[\boldsymbol{M}] = [\boldsymbol{D}] + [\boldsymbol{L}] \tag{4.62}$$

$$[\boldsymbol{N}] = -[\boldsymbol{U}] \tag{4.63}$$

We obtain then:

$$([\boldsymbol{D}] + [\boldsymbol{L}])\boldsymbol{x}^{(\nu+1)} = -[\boldsymbol{U}]\boldsymbol{x}^{(\nu)} + \boldsymbol{f} \tag{4.64}$$

The matrix [***D***] + [***L***] being lower triangular (not strictly), the scheme does not require its inversion, and it can be written in terms of components:

$$x_i^{(\nu+1)} = \frac{1}{K_{ii}}\left(f_i - \sum_{j=1}^{i-1} K_{ij}x_j^{(\nu+1)} - \sum_{j=i+1}^{n} K_{ij}x_j^{(\nu)}\right) \tag{4.65}$$

It is seen that the components $x_j^{(\nu+1)}$ which appear on the right hand side have already been calculated ($j < i$). In this way, the Gauss–Seidel method is a Jacobi method for which the most recently calculated components of the unknown vector are reused to calculate the new iteration. Because of this, the method often converges more rapidly than the Jacobi method, and – second advantage – it permits an economy in storage as it does not require an additional vector for $\boldsymbol{x}$: each new value $x_i^{(\nu+1)}$ deduced from (4.65) directly replaces $x_i^{(\nu)}$ and can be used in the rest of the iteration. Figure 4.24 shows the two methods graphically for the very simple case of a system of two equations and two unknowns.

Successive Over Relaxation method (SOR)

This method consists of extrapolating the result of an iteration of the Gauss-Seidel method, multiplying the calculated correction for each component by a factor ω. One has in components:

$$x_i^{(\nu+1)} = x_i^{(\nu)} + \omega(\hat{x}_i^{(\nu+1)} - x_i^{(\nu)}) = \omega\hat{x}_i^{(\nu+1)} + (1-\omega)x_i^{(\nu)} \tag{4.66}$$

where $\hat{x}_i^{(\nu+1)}$ is the Gauss-Seidel iteration given by (4.65) and ω a positive parameter. One obtains:

$$x_i^{(\nu+1)} = \frac{\omega}{K_{ii}}\left(f_i - \sum_{j=1}^{i-1} K_{ij}x_j^{(\nu+1)} - \sum_{j=i+1}^{n} K_{ij}x_j^{(\nu)}\right) + (1-\omega)x_i^{(\nu)} \tag{4.67}$$

The multiplication of both sides by K_{ii}/ω (or D_{ii}/ω), yields the equivalent matrix expression:

$$\left(\frac{1}{\omega}[\boldsymbol{D}] + [\boldsymbol{L}]\right)\boldsymbol{x}^{(\nu+1)} = \left(\frac{(1-\omega)}{\omega}[\boldsymbol{D}] - [\boldsymbol{U}]\right)\boldsymbol{x}^{(\nu)} + \boldsymbol{f} \tag{4.68}$$

The additive decomposition corresponding to [***K***] = [***M***] – [***N***] is then written in this case:

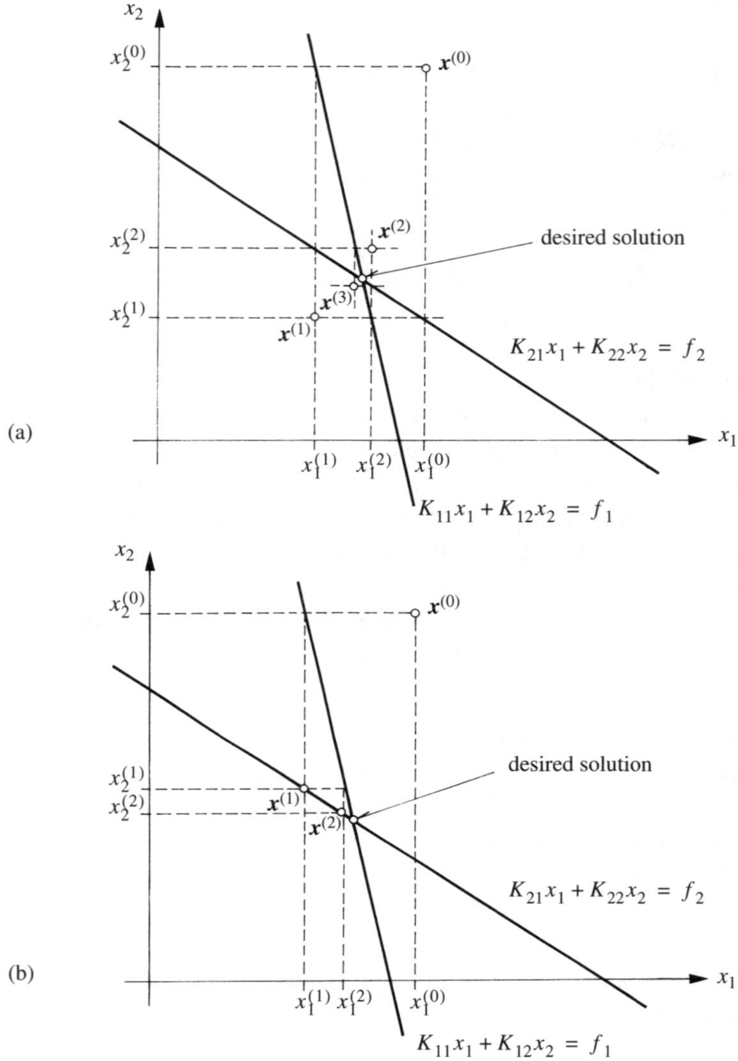

Figure 4.24 Graphical illustration of the Jacobi (a) and Gauss-Seidel (b) methods for the solution of a system of two equations with two unknowns.

$$[\boldsymbol{M}] = \frac{1}{\omega}[\boldsymbol{D}] + [\boldsymbol{L}] \quad \text{and} \quad [\boldsymbol{N}] = \frac{(1-\omega)}{\omega}[\boldsymbol{D}] - [\boldsymbol{U}] \tag{4.69}$$

The matrix $[\boldsymbol{K}]$ being symmetric positive definite, it can be shown that the condition $0 < \omega < 2$ must be satisfied for the method to converge. The inconvenience of this method is the dependence of the optimum value of the parameter ω on the problem. In any non-academic case, the search for ω_{opt} that maximizes the convergence rate is performed by trial and error. Generally, one has $\omega_{opt} > 1$ and the method is frequently called the Successive Over Relaxation method (SOR).

SSOR method (Symmetric Successive Over Relaxation)

It consists of making the SOR algorithm symmetric, with a second pass and inversion of the roles for $[\boldsymbol{L}]$ and $[\boldsymbol{U}]$. The matrices $[\boldsymbol{M}]$ and $[\boldsymbol{N}]$ are then given by:

$$[\boldsymbol{M}] = \frac{\omega}{2-\omega}\left(\frac{1}{\omega}[\boldsymbol{D}] + [\boldsymbol{L}]\right)[\boldsymbol{D}]^{-1}\left(\frac{1}{\omega}[\boldsymbol{D}] + [\boldsymbol{U}]\right) \quad \text{and} \quad [\boldsymbol{N}] = [\boldsymbol{M}] - [\boldsymbol{K}] \tag{4.70}$$

We will see later an example of the use of this method in the role of preconditioning for conjugate gradients. It should be noted that this method, although it converges more rapidly than SOR, is of little interest used alone because of its elevated calculation costs for an iteration.

Iterative minimization methods

We will now present two minimization methods based on gradients. The matrix $[\boldsymbol{K}]$ being assumed symmetric positive definite, consider the symmetric bilinear form:

$$F(\boldsymbol{x}) = \tfrac{1}{2}\boldsymbol{x}^{\mathrm{T}}[\boldsymbol{K}]\boldsymbol{x} - \boldsymbol{x}^{\mathrm{T}}\boldsymbol{f} \tag{4.71}$$

The gradient of F is given by:

$$\mathbf{grad}\, F(\boldsymbol{x}) = \frac{\partial F}{\partial \boldsymbol{x}} = [\boldsymbol{K}]\boldsymbol{x} - \boldsymbol{f} \tag{4.72}$$

Such a bilinear form is illustrated graphically in figure 4.25 for a problem with two variables (x_1, x_2), analogous to that in figure 4.24. In the (x_1, x_2) plane, the desired solution is the intersection of the two lines defined by the system of equations:

$$\begin{aligned} K_{11}x_1 + K_{12}x_2 - f_1 &= 0 \\ K_{21}x_1 + K_{22}x_2 - f_2 &= 0 \end{aligned} \tag{4.73}$$

The bilinear form F defines a surface in three dimensional space. The intersection of this surface with the plane (x_1, x_2) is a quadratic equation $F(x_1, x_2) = 0$ (fig. 4.25). At the solution we seek, the components of the gradient of F, $\partial F/\partial x_1$ and $\partial F/\partial x_2$, are null. The minimization of F, that is zeroing of its gradient (F being convex), is equivalent to the solution of the linear system $[\boldsymbol{K}]\,\boldsymbol{x} = \boldsymbol{f}$.

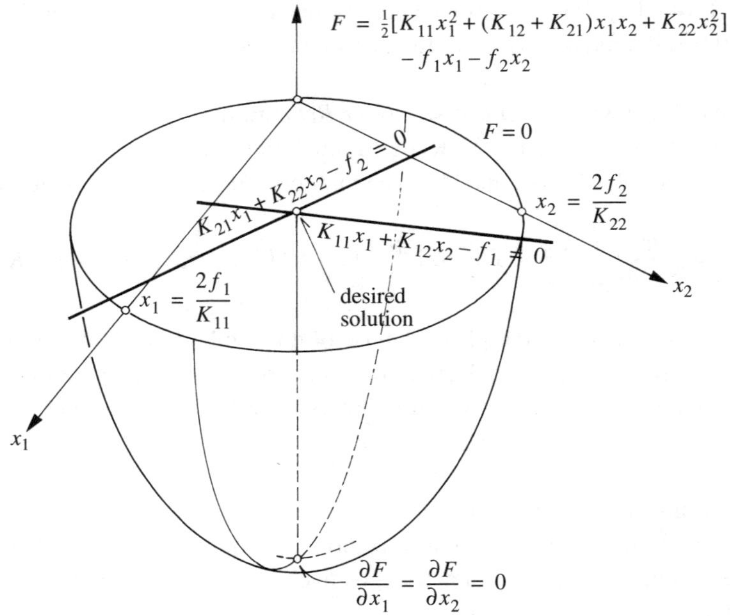

Figure 4.25 Surface defined by a symmetric linear form in two variables (x_1, x_2).

Gradient method

The method consists of minimizing F by descending along the steepest slope, as indicated by the gradient vector. This method is shown graphically in figure 4.26 in the (x_1, x_2) plane for the same two variable problem. Iso-value lines of F are indicated as well as the current gradient of $\boldsymbol{x}^{(\nu)}$.

Starting from the current point $\boldsymbol{x}^{(\nu)}$, we have:

$$\mathbf{grad}\, F(\boldsymbol{x}^{(\nu)}) = [\boldsymbol{K}]\boldsymbol{x}^{(\nu)} - \boldsymbol{f} = \boldsymbol{r}^{(\nu)} \tag{4.74}$$

The idea is then to calculate the next iteration in the form:

$$\boldsymbol{x}^{(\nu+1)} = \boldsymbol{x}^{(\nu)} - \rho \boldsymbol{r}^{(\nu)} \tag{4.75}$$

where ρ is the descent parameter, that can be optimized by minimizing $F^*(\rho) = F(\boldsymbol{x}^{(\nu)} - \rho \boldsymbol{r}^{(\nu)})$ with respect to ρ. We write then:

$$\frac{\partial F^*}{\partial \rho} = \frac{\partial F^*}{\partial \boldsymbol{x}^{(\nu+1)}} \cdot \frac{\partial(\boldsymbol{x}^{(\nu+1)})}{\partial \rho} = 0 \tag{4.76}$$

or, by virtue of (4.72), (4.74) and (4.75) :

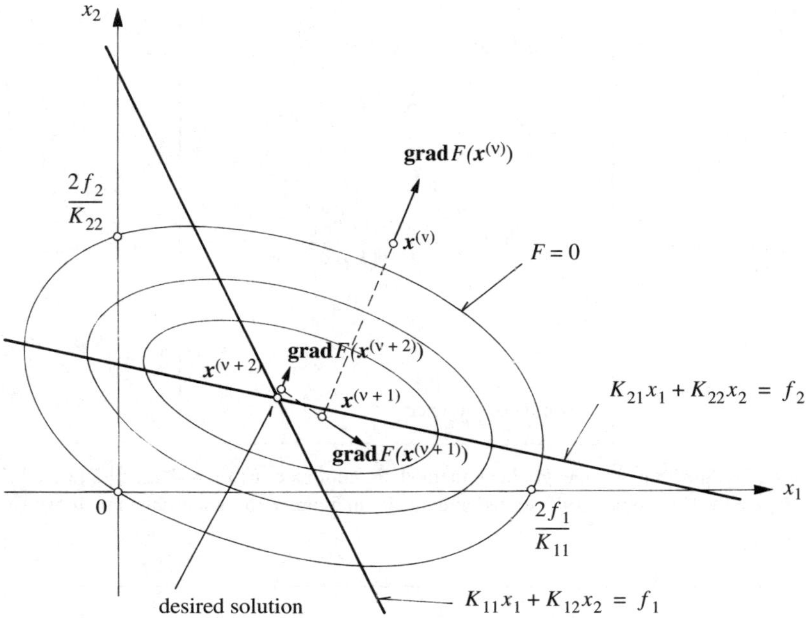

Figure 4.26 Graphical illustration of the gradient method for a symmetric bilinear form in two variables, x_1 and x_2.

$$([\boldsymbol{K}](\boldsymbol{x}^{(\nu)} - \rho\boldsymbol{r}^{(\nu)}) - \boldsymbol{f})^{\mathrm{T}}(-\boldsymbol{r}^{(\nu)}) = (\boldsymbol{r}^{(\nu)} - \rho[\boldsymbol{K}]\boldsymbol{r}^{(\nu)})^{\mathrm{T}}(-\boldsymbol{r}^{(\nu)}) = 0 \tag{4.77}$$

from which is deduced the optimal value of ρ:

$$\rho_{\mathrm{opt}} = \frac{(\boldsymbol{r}^{(\nu)})^{\mathrm{T}}\boldsymbol{r}^{(\nu)}}{(\boldsymbol{r}^{(\nu)})^{\mathrm{T}}[\boldsymbol{K}]\boldsymbol{r}^{(\nu)}} \tag{4.78}$$

The gradient method algorithm is given in figure 4.27.

Conjugate gradient method

The conjugate gradient method is a variant of the gradient method, for which the descent directions defined for successive iterations have imposed the condition of being mutually "[$\boldsymbol{K}$]-conjugate". By this terminology, we mean that the descent direction for each iteration (ν), should satisfy:

$$(\boldsymbol{d}^{(i)})^{\mathrm{T}}[\boldsymbol{K}]\boldsymbol{d}^{(\nu)} = 0 \qquad \text{for all } i < \nu \tag{4.79}$$

the $\boldsymbol{d}^{(i)}$ designate the preceding descent directions. It can be shown that this constraint guaranties the minimization of F, and thus the *solution of the system in a maximum of n iterations*, n being the size of the system. Thus, in the example illus-

Initialization: $\nu = 0$
$\boldsymbol{x}^{(0)}$ given

Iteration (ν):

$$\boldsymbol{r}^{(\nu)} = [\boldsymbol{K}]\boldsymbol{x}^{(\nu)} - \boldsymbol{f}$$

$$\rho_{\text{opt}} = \frac{(\boldsymbol{r}^{(\nu)})^{\mathrm{T}}\boldsymbol{r}^{(\nu)}}{(\boldsymbol{r}^{(\nu)})^{\mathrm{T}}[\boldsymbol{K}]\boldsymbol{r}^{(\nu)}}$$

$$\boldsymbol{x}^{(\nu+1)} = \boldsymbol{x}^{(\nu)} - \rho_{\text{opt}}\boldsymbol{r}^{(\nu)}$$

$\nu = \nu + 1$

until convergence.

Figure 4.27 Algorithm for the gradient method to minimize $F(\boldsymbol{x}) = 1/2\boldsymbol{x}^{\mathrm{T}}[\boldsymbol{K}]\boldsymbol{x} - \boldsymbol{x}^{\mathrm{T}}\boldsymbol{f}$ (thus solve $[\boldsymbol{K}]\,\boldsymbol{x} = \boldsymbol{f}$). The method is illustrated graphically in figure 4.26 for a problem with two variables, x_1 and x_2.

Initialization: $\nu = 0$
$\boldsymbol{x}^{(0)}$ given

$$\boldsymbol{r}^{(0)} = [\boldsymbol{K}]\boldsymbol{x}^{(0)} - \boldsymbol{f}$$

$\boldsymbol{d}^{(0)} = -\boldsymbol{r}^{(0)}$ (first descent direction)

Iteration (ν):

$$\alpha^{(\nu)} = \frac{(\boldsymbol{r}^{(\nu)})^{\mathrm{T}}\boldsymbol{r}^{(\nu)}}{(\boldsymbol{d}^{(\nu)})^{\mathrm{T}}[\boldsymbol{K}]\boldsymbol{d}^{(\nu)}} \qquad \text{(descent parameter optimized)}$$

$$\boldsymbol{x}^{(\nu+1)} = \boldsymbol{x}^{(\nu)} + \alpha^{(\nu)}\boldsymbol{d}^{(\nu)} \qquad \text{(descent)}$$

$$\boldsymbol{r}^{(\nu+1)} = \boldsymbol{r}^{(\nu)} + \alpha^{(\nu)}[\boldsymbol{K}]\boldsymbol{d}^{(\nu)} \qquad \text{(new residual)}$$

$$\beta^{(\nu)} = \frac{(\boldsymbol{r}^{(\nu+1)})^{\mathrm{T}}\boldsymbol{r}^{(\nu+1)}}{(\boldsymbol{r}^{(\nu)})^{\mathrm{T}}\boldsymbol{r}^{(\nu)}} \qquad \text{(parameter insuring the conjugation)}$$

$$\boldsymbol{d}^{(\nu+1)} = -\boldsymbol{r}^{(\nu+1)} + \beta^{(\nu)}\boldsymbol{d}^{(\nu)} \qquad \text{(new descent direction)}$$

$\nu = \nu + 1$

until convergence.

Figure 4.28 Conjugate gradient algorithm.

trated graphically in figure 4.26, the first iteration of the conjugate gradient is identical to that of the gradient method, but the second iteration arrives directly at the desired solution, which is not the case of the gradient method. The algorithm for the method is given in figure 4.28 (see also exercise 4.6.4).

It can be shown that the convergence rate of the conjugate gradient method depends on the ***conditioning*** of the symmetric matrix $[\boldsymbol{K}]$, defined with respect to its extreme eigenvalues:

$$\mathrm{Cond}\,[\boldsymbol{K}] = \lambda_{\max}/\lambda_{\min} \tag{4.80}$$

Often, the system matrix coming from the finite element formulation is poorly conditioned: Cond $[\boldsymbol{K}]$ is very different than 1, the ideal value, and the convergence is slow. We resort then to the *preconditioning* technique detailed below.

Preconditioned conjugate gradient method

The preconditioning consists of replacing the original system $[\boldsymbol{K}]\,\boldsymbol{x} = \boldsymbol{f}$ by a better conditioned equivalent system:

$$[\boldsymbol{C}]^{-1}[\boldsymbol{K}]\boldsymbol{x} = [\boldsymbol{C}]^{-1}\boldsymbol{f} \tag{4.81}$$

The preconditioning matrix $[\boldsymbol{C}]$ should be symmetric, invertible and chosen such that the conditioning of $[\boldsymbol{C}]^{-1}\,[\boldsymbol{K}]$, the new system matrix, is this time much closer to 1. In theory, the best choice is obviously $[\boldsymbol{C}] = [\boldsymbol{K}]$, but this is equivalent to a direct solution! In practice, $[\boldsymbol{C}]$ must be close to $[\boldsymbol{K}]$ while still being easily invertible. In particular, one can find inspiration in the iterative methods based on additive decomposition of $[\boldsymbol{K}]$ to choose $[\boldsymbol{C}]$: we give two examples below. But there exist other preconditioners based on incomplete Cholesky decompositions that we will not detail here.

- Diagonal preconditioning. It is the simplest choice:

  $$[\boldsymbol{C}] = \mathrm{Diag}[\boldsymbol{K}] = [\boldsymbol{D}] \tag{4.82}$$

  Often enough, this preconditioning is sufficient to obtain at low cost a very significant acceleration of the convergence rate of the conjugate gradient method.

- SSOR preconditioning. In this case, for matrix $[\boldsymbol{C}]$ one takes the matrix $[\boldsymbol{M}]$ defined by (4.70). In fact, to be able to apply the conjugate gradient method on this new system, it is first necessary to make the system symmetric (as the product of the symmetric matrices $[\boldsymbol{C}]^{-1}\,[\boldsymbol{K}]$ is not necessarily symmetric). $[\boldsymbol{C}]^{-1}$ being symmetric, it can be decomposed by Cholesky's method (4.54) which we denote

  $$[\boldsymbol{C}]^{-1} = [\boldsymbol{C}]^{-1/2}([\boldsymbol{C}]^{-1/2})^{\mathrm{T}} \tag{4.83}$$

  The equivalent symmetric system to the initial system is then:

  $$([\boldsymbol{C}]^{-1/2})^{\mathrm{T}}[\boldsymbol{K}][\boldsymbol{C}]^{-1/2}\boldsymbol{x}' = ([\boldsymbol{C}]^{-1/2})^{\mathrm{T}}\boldsymbol{f} \tag{4.84}$$

with

$$[\boldsymbol{C}]^{-1/2}\boldsymbol{x}' = \boldsymbol{x} \tag{4.85}$$

If we apply the conjugate gradient algorithm previously described to this system, we obtain the preconditioned conjugate gradient algorithm, detailed in figure 4.29.

Initializations: $\nu = 0$

$\boldsymbol{x}^{(0)}$ given

$\boldsymbol{r}^{(0)} = [\boldsymbol{K}]\boldsymbol{x}^{(0)} - \boldsymbol{f}$

$\boldsymbol{d}^{(0)} = -[\boldsymbol{C}]^{-1}\boldsymbol{r}^{(0)}$

$\boldsymbol{m}^{(0)} = -\boldsymbol{d}^{(0)}$

Iterations on (ν):

$$\alpha^{(\nu)} = \frac{(\boldsymbol{r}^{(\nu)})^{\mathrm{T}}\boldsymbol{m}^{(\nu)}}{(\boldsymbol{d}^{(\nu)})^{\mathrm{T}}[\boldsymbol{K}]\boldsymbol{d}^{(\nu)}}$$

$$\boldsymbol{x}^{(\nu+1)} = \boldsymbol{x}^{(\nu)} + \alpha^{(\nu)}\boldsymbol{d}^{(\nu)}$$

$$\boldsymbol{r}^{(\nu+1)} = \boldsymbol{r}^{(\nu)} + \alpha^{(\nu)}[\boldsymbol{K}]\boldsymbol{d}^{(\nu)}$$

$$\boldsymbol{m}^{(\nu+1)} = [\boldsymbol{C}]^{-1}\boldsymbol{r}^{(\nu+1)}$$

(only supplementary calculation with respect to the classic conjugate gradient, fig. 4.28)

$$\beta^{(\nu)} = \frac{(\boldsymbol{r}^{(\nu+1)})^{\mathrm{T}}\boldsymbol{m}^{(\nu+1)}}{(\boldsymbol{r}^{(\nu)})^{\mathrm{T}}\boldsymbol{m}^{(\nu)}}$$

$$\boldsymbol{d}^{(\nu+1)} = -\boldsymbol{m}^{(\nu+1)} + \beta^{(\nu)}\boldsymbol{d}^{(\nu)}$$

$\nu = \nu + 1$

until convergence.

Figure 4.29 Preconditioned conjugate gradient algorithm.

It should be noted that it is not necessary to form $[\boldsymbol{C}]^{-1/2}$ (only $[\boldsymbol{C}]$ needs to be known), and that the only supplementary calculation to perform per iteration is the solution of $[\boldsymbol{C}]\,\boldsymbol{m}^{(\nu+1)} = \boldsymbol{r}^{(\nu+1)}$.

Multi-grid methods

A remark to conclude this Section consecrated to iterative solution methods; there exists another type: *multi-grid methods*. They are based on the utilization of several meshes (or grids) of increasing fineness. Very schematically, the principle consists of attaining the solution of the problem posed on the initial grid, denoted the fine grid, using successive solutions performed on coarser grids. These methods thus use operators for interpolation (transition from a coarse grid to a fine grid), sampling (transition from a fine grid to a coarse grid) and smoothing (constituted of a few iterations of one of the classic iterative methods presented above). The corrections effected on the coarse grids are calculated by direct or iterative methods. Each of them is much less costly than a solution performed on the initial fine grid; and one thus constructs, by cycles between the different grid levels, a particularly effective iteration method when the different operators insure good convergence. Being outside the scope of this book, these methods will not be detailed; nevertheless, they are very often used for finite differences or finite elements on structured grids. The grids with increasing fineness are then simply deduced and the operators are also easily obtained. Such is not the case for unstructured grids, for which the implementation is clearly more difficult; but can be very effective.

We have thus seen in review the principle solution algorithms for linear systems. Before discussing the choice among methods, it is appropriate to detail the storage methods for the systems to be solved. The capacity of a computer being finite, the efficiency of a solution algorithm is measured by the number of elementary operations that are entailed; and also by the memory space necessary to perform those calculations.

4.4 STORAGE OF MATRICES IN MEMORY

As we have seen in the preceding chapters, the matrices that result from a FDM or FEM formulation are essentially sparse: they contain a large number of zeros. More precisely, the number of non zero terms in these matrices for a scalar problem with a symmetric matrix is given by the product $Nn(nvn + 1)$ where Nn is the number of nodes in the mesh, and nvn the average number of “neighbors”. It is out of the question, of course, to store all the Nn^2 terms of the matrix. Consider simply a two dimensional problem for a square divided in 100×100 points for the FDM: the resulting matrix will have 10^8 terms, although there are in fact only 50,000 non zero terms! One sees then that it is important to optimize the storage of such matrices.

4.4.1 Storage of banded matrices: skyline storage and band storage

A first method exploits the band structure of the system matrix, storing only the components contained in the ***profile*** of the matrix. The profile of a matrix $[\boldsymbol{K}]$ is given by the minimal values of the index of the first non zero coefficient of the lines, stated otherwise, by the set of $\lambda(i)$, $i = 1, n$ such that:

$$\lambda(i) = \text{Min}\{j|\ 1 \leq j \leq i \text{ and } K_{ij} \neq 0\} \tag{4.86}$$

Figure 4.30 illustrates the profile graphically. It should be noted in passing that the profile of the matrices resulting from the finite element method, or from the

$\lambda(1) = 1$	K_{11}	K_{12}	K_{13}	K_{14}	0	0	0	0	0
$\lambda(2) = 1$	K_{21}	K_{22}	K_{23}	K_{24}	K_{25}	K_{26}	0	0	0
$\lambda(3) = 1$	K_{31}	K_{32}	K_{33}	0	0	K_{36}	K_{37}	K_{38}	0
$\lambda(4) = 1$	K_{41}	K_{42}	0	K_{44}	K_{45}	0	0	0	0
$\lambda(5) = 2$	0	K_{52}	0	K_{54}	K_{55}	K_{56}	0	0	0
$\lambda(6) = 2$	0	K_{62}	K_{63}	0	K_{65}	K_{66}	K_{67}	0	K_{69}
$\lambda(7) = 3$	0	0	K_{73}	0	0	K_{76}	K_{77}	K_{78}	K_{79}
$\lambda(8) = 3$	0	0	K_{83}	0	0	0	K_{87}	K_{88}	0
$\lambda(9) = 6$	0	0	0	0	0	K_{96}	K_{97}	0	K_{99}

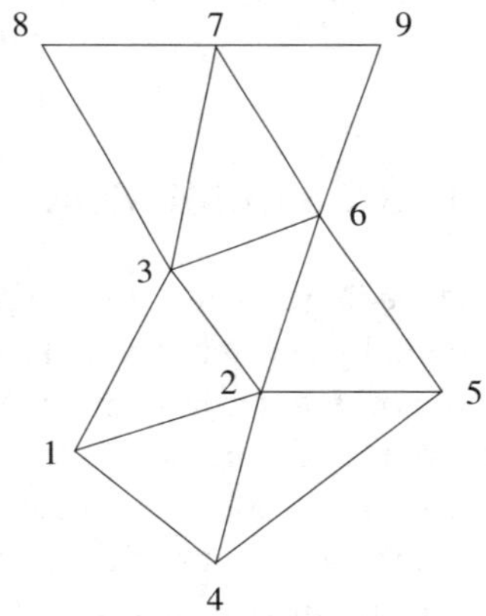

Figure 4.30 The profile of a matrix to solve a diffusion problem on a finite element mesh composed of 8 linear triangle elements and 9 nodes is shown. The K_{ij} are the non zero components of the matrix. The boxed components are those for which the storage is necessary for a direct solution.

finite difference method, is by nature symmetric: the profile of the transposed matrix $[\boldsymbol{K}]^T$ is the same as that of $[\boldsymbol{K}]$.

A determining element is that when one uses a direct solution method, Gauss' method for example, the profile of the lower and upper triangular matrices used in the factorization (4.40) is the same as that of the matrix to be inverted. However, as was seen in Section 4.3.1, during this factorization $[\boldsymbol{K}] = [\boldsymbol{L}]\,[\boldsymbol{U}]$, the coefficients L_{ij} and U_{ij} situated in the interior of the profile of the matrix are not necessarily zero, even if the K_{ij} were (for example $L_{74} \neq 0$ although $K_{74} = 0$ in the example of figure 4.30). With these considerations, the memory space occupied by the components of $[\boldsymbol{L}]$ and $[\boldsymbol{U}]$ is the same as that required for $[\boldsymbol{K}]$ assuming that the zero values of $[\boldsymbol{K}]$ in the interior of the profile are also stored. One stores in memory the components in the profile, that is the boxed components in figure 4.30. The storage of matrix $[\boldsymbol{K}]$ then requires n_s memory addresses:

$$n_s = \sum_{i=1}^{n} [2(i - \lambda(i)) + 1] \tag{4.87}$$

Knowing that the profile of the matrix is symmetric, the storage of the coefficients K_{ij} in a single vector can be done simply by indicating the values of $\lambda(i)$, or, which amounts to the same, the position $\mu(i)$ of the diagonal coefficients K_{ii} (marked with an asterisk in figure 4.31).

The position $\mu(i)$ is given by:

$$\mu(i) = \sum_{k=1}^{i} [2(k - \lambda(k)) + 1] = \mu(i-1) + 2(i - \lambda(i)) + 1 \tag{4.88}$$

for $i = 2, n$ $\quad (\mu(1) = 1)$

In the case of a symmetric matrix $[\boldsymbol{K}]$, Gauss' method is replaced by Crout's or Cholesky's:

$$[\boldsymbol{K}] = [\boldsymbol{L}][\boldsymbol{D}][\boldsymbol{L}]^T \quad \text{or} \quad [\boldsymbol{K}] = [\tilde{\boldsymbol{L}}][\tilde{\boldsymbol{L}}]^T \tag{4.89}$$

The storage of the symmetric components is thus superfluous and it is sufficient to store the matrix in a vector conforming to the scheme in figure 4.32.

K_{11}	K_{21}	K_{12}	K_{22}	K_{31}	K_{32}	K_{13}	K_{23}	K_{33}	K_{41}	K_{42}	K_{43}	K_{14}	K_{24}	K_{34}	K_{44}
*			*					*			0			0	*

K_{52}	K_{53}	K_{54}	K_{25}	K_{35}	K_{45}	K_{55}	K_{62}	K_{63}	K_{64}	K_{65}	K_{26}	K_{36}	K_{46}	K_{56}	K_{66}
	0			0		*			0				0		*

K_{73}	K_{74}	K_{75}	K_{76}	K_{37}	K_{47}	K_{57}	K_{67}	K_{77}	K_{83}	K_{84}	K_{85}	K_{86}	K_{87}	K_{38}	K_{48}
	0	0			0	0		*		0	0	0			0

K_{58}	K_{68}	K_{78}	K_{88}	K_{96}	K_{97}	K_{98}	K_{69}	K_{79}	K_{89}	K_{99}
0	0		*			0			0	*

Figure 4.31 Storage of the components of a banded matrix (the same matrix as in figure 4.30) in a vector. The asterisks mark the diagonal components and the values equal to zero are indicated.

K_{11}	K_{21}	K_{22}	K_{31}	K_{32}	K_{33}	K_{41}	K_{42}	K_{43}	K_{44}	K_{52}	K_{53}	K_{54}	K_{55}	K_{62}	K_{63}
*		*			*			$\ddot{0}$	*		$\ddot{0}$		*		

K_{64}	K_{65}	K_{66}	K_{73}	K_{74}	K_{75}	K_{76}	K_{77}	K_{83}	K_{84}	K_{85}	K_{86}	K_{87}	K_{88}	K_{96}	K_{97}
$\ddot{0}$		*		$\ddot{0}$	$\ddot{0}$		*		$\ddot{0}$	$\ddot{0}$	$\ddot{0}$		*		

K_{98}	K_{99}
$\ddot{0}$	*

Figure 4.32 Storage of a symmetric matrix (the same matrix as in figure 4.30) in a vector.

The number of stored components is reduced in this case to:

$$n_s = \sum_{i=1}^{n} [i - \lambda(i) + 1] \tag{4.90}$$

and the position of the diagonal components, K_{ii}, is given by:

$$\mu(i) = \sum_{k=1}^{i} [k - \lambda(k) + 1] = \mu(i-1) + i - \lambda(i) + 1 \tag{4.91}$$

for $i = 2, n$ $\quad(\mu(1) = 1)$

The storage described above is called ***skyline***. One of the principle consequences of its use for the FEM is the modification of the assembly procedure: when accumulating the elementary contributions into a term K_{ij}, the accumulation occurs in component number $\mu(i) - (i - j)$ of the vector (for a symmetric matrix, when $i > j$).

Note that one often uses a form of storage a bit simpler to manipulate but less economical: ***band storage***. This method consists of calculating the ***bandwidth*** of the matrix, *bw*, which is none other than the largest dimension of the profile seen in the matrix:

$$bw = \underset{i = 1, n}{\text{Max}} (i - \lambda(i) + 1) \tag{4.92}$$

For example, in figure 4.30, $bw = 6$. All the terms included in the band of width *bw* around the diagonal are stored. Figure 4.33 shows an illustration of this for the case of a symmetric matrix. In this case, the number of components stored is equal to the product ($bw\ n$) if for programming reasons one also stores the null terms situated "outside" of the matrix.

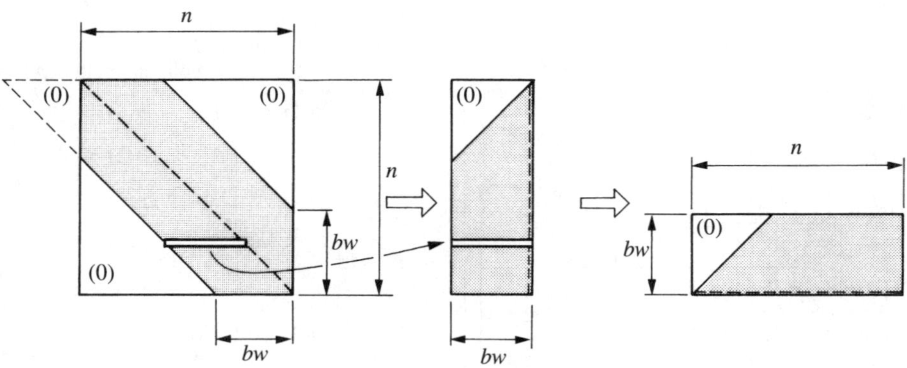

Figure 4.33 Band storage of a symmetric matrix. Only the lower part is stored. Note that according to the language used, it may be preferable to store the transpose of this reduced matrix, of dimension $bw \times n$, for programming efficiency in the solver: it is more rapid to operate on variables with neighboring addresses in memory. The dashed line shows the position of the diagonal components of the matrix.

4.4.2 Compact storage

We have seen in the preceding paragraph that economies of matrix storage were necessarily limited by the utilization of direct solution methods. Although in this case the profile of the initial matrix is retained in the transformation into triangular matrices, the profile is filled out during Gaussian or Crout elimination; the initially zero valued components become non zero. However, this limitation disappears when iterative solution methods are used. One can then envision the storage of only the components K_{ij} of the system matrix $[\boldsymbol{K}]$ that are non zero. This is called ***compact storage*** or condensed storage (also called Morse storage).

The principle of this type of storage is illustrated in figure 4.34; it consists of scanning the matrix $[\boldsymbol{K}]$ line by line, keeping only the non zero components of K_{ij} that are then arranged in a vector $\boldsymbol{T}$. Two *pointers* are employed to manage the correspondence between $[\boldsymbol{K}]$ and $\boldsymbol{T}$:

- a first pointer, already employed in skyline storage, $\mu(i)$, $i = 1, n$ which gives for each line i of the matrix $[\boldsymbol{K}]$ the address of the diagonal term K_{ii} in the vector $\boldsymbol{T}$ (cf (4.88) and (4.91)). The vector $\boldsymbol{\mu}$ is called the ***line pointer***.
- a second pointer, $\boldsymbol{cp}(k)$ gives the column number in $[\boldsymbol{K}]$ of each term T_k of the vector $\boldsymbol{T}$. The vector $\boldsymbol{cp}$ has the size of the vector $\boldsymbol{T}$, *a priori* unknown (number of non zero components) and is called the ***column pointer***.

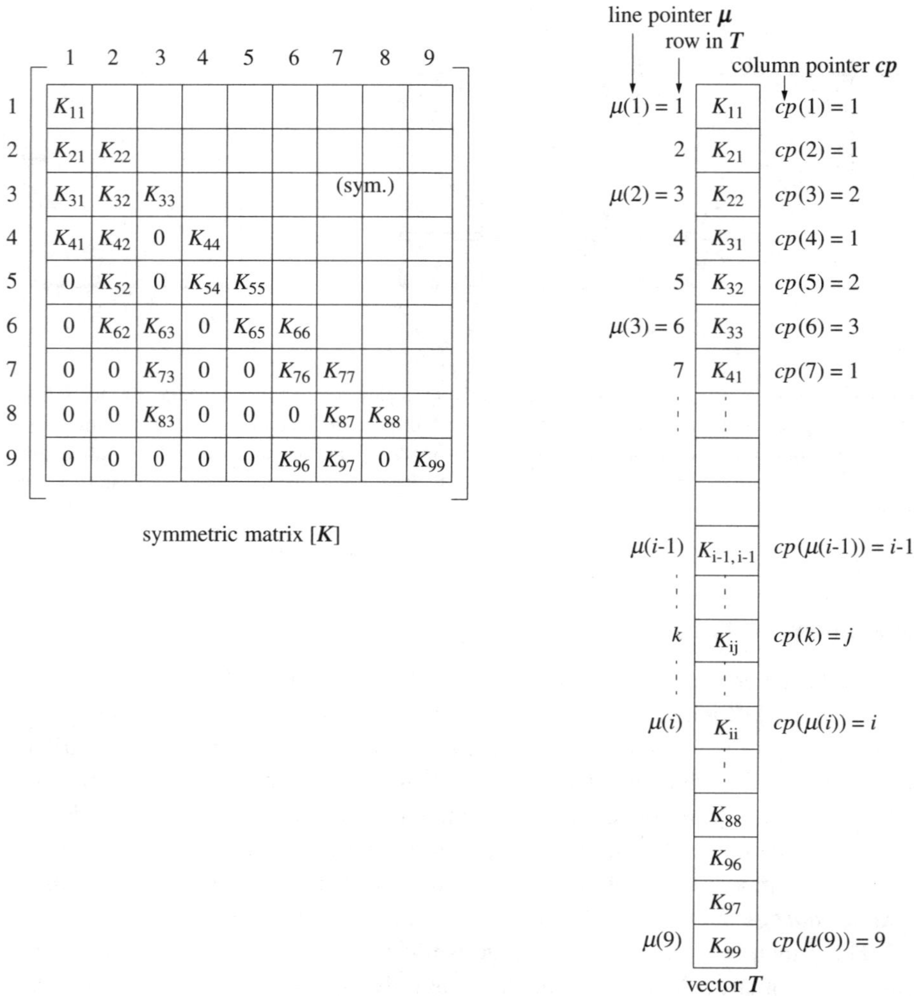

Figure 4.34 Example of the application of compact storage to the case of the symmetric matrix of figure 4.30. Referencing is with line and column pointers.

4.4.3 Consequences of the storage mode on the assembly and solution procedures

With respect to the assembly and solution procedures described earlier in this text, it is clear that the choice of the matrix storage method for a finite element or finite difference problem will have some consequences.

Consequences on the assembly procedure

The assembly of a contribution to a term K_{ij} of the global stiffness matrix [$\boldsymbol{K}$] (sect. 2.3.6 for the FDM and sect. 3.5.2 for the FEM) should be concretely realized (i.e. "programmed") by an accumulation in the vector $\boldsymbol{T}$ where the components of $[\boldsymbol{K}]$ are stored.

Take the example of band storage for a symmetric matrix (fig. 4.33). The vector $\boldsymbol{T}$ is constituted of columns of the reduced matrix (size $bw \times n$), that results from the extraction of the lower band of the matrix $[\boldsymbol{K}]$. The reader can verify that the addition of a local contribution to a coefficient K_{ij} amounts to adding the contribution to the component $(i\,bw + j - i)$ of vector $\boldsymbol{T}$.

In the case of Morse storage (fig. 4.34), the assembly of the diagonal terms K_{ii} of the global matrix is trivial: one sums in the component $\mu(i)$ of the vector $\boldsymbol{T}$. However, for a non diagonal term K_{ij} of the matrix, it is necessary to determine the address k, within the interval $]\mu(i-1), \mu(i)[$, such that $cp(k) = j$. One then adds the local value to T_k.

Consequences for the solution

Similar modifications affect the solution algorithms.

Thus, for the case of band storage used in direct methods, the Gauss, Crout and Cholesky algorithms, given in figures 4.20 to 4.22, must be modified. As in the preceding, the operations on the components K_{ij}, L_{ij}, and U_{ij}, must be carried out on the corresponding components of the storage vector $\boldsymbol{T}$.

For compact storage, very often used with iterative solvers, the scalar product and the matrix-vector product are used intensively (cf (4.40)). For example, we give in figure 4.35 the algorithm operating on the components of $\boldsymbol{T}$, in the case of a symmetric matrix (exercise 4.6.5).

```
Loop over i = 1, n
    r(i) = 0
End of loop over i
μ(0) = 0

Loop over i = 1, n

    Loop over j = μ(i – 1) + 1, μ(i)
        r(i) = r(i) + T(j) u(cp(j))
    End of loop over j

    Loop over j = μ(i – 1) + 1, μ(i)
        r(cp(j)) = r(cp(j)) + T(j) u(i)
    End of loop over j

End of loop over i
```

Figure 4.35 Calculation algorithm for the product of a symmetric matrix $[\boldsymbol{K}]$, of dimension n, stored in compact form in a vector $\boldsymbol{T}$, and a vector $\boldsymbol{u}$ with the result stored in the vector $\boldsymbol{r}$.

4.4.4 The choice of an algorithm for the solution of linear systems

As we have already underlined above, the efficiency of a solver depends on the number of elementary operations and the memory space necessary to attain the solution.

Calculation cost

It is clearly easier to estimate for direct methods, for which it is possible to exactly evaluate the number of elementary operations required (i.e. additions, multiplications, divisions, square roots).

For a symmetric positive definite matrix system, stored in the form of a band – as is very frequently the case in finite elements and finite differences – Crout's method has the best performance. If n denotes the number of equations to solve and bw the band width of the matrix, the number of operations, and thus the calculation time, varies as the product ($n\ bw^2$); Cholesky's method requires n square roots in addition.

Take for example the regular linear quadrangle meshes in 2 dimensions, or linear hexahedrons in 3 dimensions, with the nodes numbered line-by-line and plane-by-plane respectively (a "lexicographic" ordering), on which mechanical finite element calculations are performed, and for which there are in general as many unknowns per node as spatial dimensions (the components of a velocity or displacement field).

- In 2 dimensions, if Nn denotes the number of nodes, then the number of equations (or of unknowns) is $n = 2Nn$. The number of nodes in each dimension is (approximately) $\sqrt{Nn}$ and thus the bandwidth is equal to $2\sqrt{Nn}$. An estimation of the calculation time is given by:

$$t_{\text{calc}} = O(2Nn[2(Nn)^{1/2}]^2) = O(Nn^2) = O(n^2) \tag{4.93}$$

- In 3 dimensions, $n = 3Nn$, $bw = 3(Nn)^{2/3}$ and thus:

$$t_{\text{calc}} = O(3Nn[3(Nn)^{2/3}]^2) = O(Nn^{7/3}) = O(n^{7/3}) \tag{4.94}$$

The calculation time thus varies very rapidly as a function of the number of nodes Nn of the discretization: a division by two of the spatial step in each dimension increases the calculation time a factor of 16 in 2 dimensions, and a factor of 128 in 3 dimensions! Also, for any mesh, the advantage of minimizing the band width of the matrix by adopting the optimal numbering of the mesh nodes is evident.

As for the iterative methods, one can only calculate the number of elementary operations *per iteration* and the global cost of the calculation will be a function of the convergence rate of each method, that is, the number of iterations required for a given precision. For the conjugate gradient method, the calculation time per iteration varies as follows:

$$t_{calc/iter} = O(n^{4/3}) \tag{4.95}$$

As for the convergence rate, it depends on the conditioning of the system matrix (thus on the type of problem being treated), as we have seen above.

Cost of storage in memory

The direct solvers seen above conserve the profile of the system matrix and permit band or skyline storage. In the case of a symmetric matrix calculated on a regular cubic mesh, band storage requires the following memory space (expressed in number of variables, each requiring 8 bytes).

- In 2 dimensions:

$$\text{Sto}_{\text{band}} = bwn = 2(Nn)^{1/2}2Nn = O((Nn)^{3/2}) = O(n^{3/2}) \tag{4.96}$$

- In 3 dimensions:

$$\text{Sto}_{\text{band}} = bwn = 3(Nn)^{2/3}3Nn = O((Nn)^{5/3}) = O(n^{5/3}) \tag{4.97}$$

The memory utilization can of course be reduced with skyline storage or, even further, using compact storage. If *nvn* is the number of neighboring nodes averaged over all the nodes of the mesh (8 for a regular mesh of linear quadrangles, 26 for a regular mesh of linear hexahedrons), the storage space needed is as follows.

- In 2 dimensions:

$$\text{Sto}_{\text{compact}} = 2(nvn + 1)n \approx 18n \approx 36Nn \tag{4.98}$$

- In 3 dimensions:

$$\text{Sto}_{\text{compact}} = 3(nvn + 1)n \approx 81n \approx 243Nn \tag{4.99}$$

As an example, the above formulas give for a 10,000 node mesh (100×100) in 2 dimensions: $\text{Sto}_{\text{band}} = 4{,}080{,}000$ while $\text{Sto}_{\text{compact}} = 360{,}000$, and thus the difference is a factor of around 11.

For a mesh with 27,000 nodes ($30 \times 30 \times 30$) in 3 dimensions, $\text{Sto}_{\text{band}} = 226 \times 10^6$ and $\text{Sto}_{\text{compact}} = 6{,}516{,}000$, and thus a ratio of about 35; in addition, note explicitly that the first storage mode requires no less than 1.8×10^9 bytes (1.8 GB = 1,800 MB).

In conclusion, one could say that the iterative solution methods are the only choice possible for the solution of very large systems in three spatial dimensions, despite their inconveniences already mentioned. It is thus essential to master the convergence rates by choosing well adapted preconditioners.

For intermediate sized problems, the choice remains open between iterative and direct methods. Note here that poor conditioning can also affect the result of the latter (exercise 4.6.6). When it's a case of solving many systems of equations having the same matrix, and only the right hand side is different, the direct methods become particularly interesting as the matrix factorization performed to solve the first system can clearly be reused for the others, and thus the solution requires only the short calculation time necessary for a "descent-back substitution".

Finally, for more modest calculations, direct methods are the most reliable and efficient.

4.5 NON LINEAR PROBLEMS

There are multiple sources of the non linearities in the formulation of the problems.

- The *very nature of the problem* can be a source of non-linearity. Terms that are non linear with respect to the unknowns can appear in the partial differential conservation equations. Thus, for large deformations, the components of the Green-Lagrange strain tensor are given by (chap. 1):

$$\varepsilon_{ij} = \frac{1}{2}\left(\frac{\partial u_i}{\partial x_j} + \frac{\partial u_j}{\partial x_i} + \sum_{k=1}^{3} \frac{\partial u_k}{\partial x_i}\frac{\partial u_k}{\partial x_j}\right) \tag{4.100}$$

  the last term generates the nonlinearity. Similarly, in fluid flow, the acceleration term [**Grad** $\boldsymbol{v}$] $\boldsymbol{v}$ of the Navier-Stokes equation introduces a non-linearity for the calculation of the unknown velocity field. Finally, the enthalpy-temperature relation is strongly non linear for the case of a phase change (chap. 5).

- The *physical parameters of the model* can vary with the unknowns of the problem. In linear elasticity, one assumes that Young's modulus and Poisson's coefficient are independent of the transformation of the material. The assumption is no longer valid for non linear elasticity (behavior of rubber for example), and *a fortiori* for elasto-plastic or elasto-viscoplastic comportment for which the history of the material determines the physical properties at a given time. Non linearity can also be found in thermal problems, when the heat capacity or the conductivity of a material are functions of the temperature itself.

- The *boundary conditions* can cause non linearity: the radiation of a black body in thermal problems is a typical case for which the fourth power of an unknown (the temperature) appears. It is the same for friction laws in mechanics.

- Finally, the *time discretization* of a non stationary transformation can cause the non linearity. For example, in the study of the inflation of a polymer membrane (blowing of bottles or of hollow bodies), one can write, at each time increment, the mechanical equilibrium of the membrane surface at the end of the increment; the equilibrium equation being written in the form of an integral. The integration domain is then the membrane configuration at the end of the increment: it depends on the displacement during the increment which is an unknown of the problem, thus the non linearity.

Nonetheless, the non linear character of the problem to be solved does not introduce fundamental differences in the formulation employed, and the finite element or finite difference methods lead to a discrete formulation of the non linear problems which can be written in the following form,

$$\boldsymbol{R}(\boldsymbol{x}) = [\boldsymbol{K}(\boldsymbol{x})]\boldsymbol{x} - \boldsymbol{f}(\boldsymbol{x}) = \boldsymbol{0} \tag{4.101}$$

in which $\boldsymbol{x}$ is the column vector having as components the nodal unknowns of the problem. The vectors $\boldsymbol{R}$ (global residual, or first derivative of a functional with respect to the unknowns) and $\boldsymbol{f}$ have the same dimension as $\boldsymbol{x}$, which we denote n.

The non linearity thus turns the system matrix (4.101) and the right hand side vector resulting from the boundary conditions, into functions depending on the unknown $\boldsymbol{x}$. To solve this non linear equation in $\boldsymbol{x}$, that is, to find $\boldsymbol{x}$ which zeros the residual vector $\boldsymbol{R}$, one is led to use iterative methods, and thus to implement an algorithm of the type shown in figure 4.36.

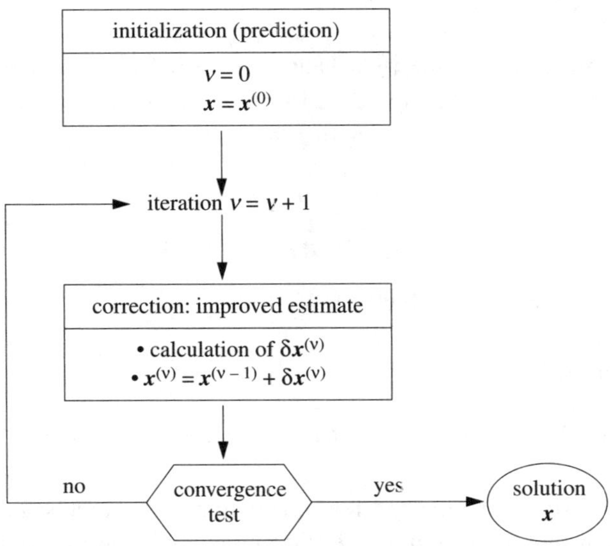

Figure 4.36 Iterative solution of a non linear problem.

Following an initialization phase, the successive corrections imposed on the unknown variable are destined to drive the residual vector as close as possible to zero. One then says that there is convergence (to within a certain precision) towards the solution.

In practice, the majority of algorithms lead to the solution of a system of linear equations at each iteration: one has *linearized* the original problem. The algorithm choice depends on many factors:

- the risk of divergence (possibly acceptable in a laboratory program, but intolerable in an industrial code),
- the rate of convergence (which determines the calculation time),
- the degree and type of non linearities of the problem,
- the uniqueness or not of the solution (local extrema problem),
- the possibility of using an approximate method.

There are a great number of iterative methods: substitution method, Newton-Raphson method, non linear conjugate gradient method, BFGS method (Broyden-Fletcher-Goldfarb-Shanno), etc. In the limits of this chapter, we restrict the explanation to the substitution and Newton-Raphson methods.

4.5.1 The substitution method or Picard's method

The method consists of linearizing the system of equations (4.101) by solving the following linear system at each iteration (ν):

$$[\boldsymbol{K}(\boldsymbol{x}^{(\nu-1)})]\boldsymbol{x}^{(\nu)} = \boldsymbol{f}(\boldsymbol{x}^{(\nu-1)}) \tag{4.102}$$

The matrix $[\boldsymbol{K}(\boldsymbol{x})]$ and the right hand side term $\boldsymbol{f}(\boldsymbol{x})$ are made "explicit" by using the value of $\boldsymbol{x}$ from the preceding iteration. The correction obtained at each iteration can easily be expressed by noting that:

$$\begin{aligned}\boldsymbol{R}(\boldsymbol{x}^{(\nu-1)}) &= [\boldsymbol{K}(\boldsymbol{x}^{(\nu-1)})]\boldsymbol{x}^{(\nu-1)} - \boldsymbol{f}(\boldsymbol{x}^{(\nu-1)})\\ &= [\boldsymbol{K}(\boldsymbol{x}^{(\nu-1)})]\boldsymbol{x}^{(\nu-1)} - [\boldsymbol{K}(\boldsymbol{x}^{(\nu-1)})]\boldsymbol{x}^{(\nu)}\\ &= [\boldsymbol{K}(\boldsymbol{x}^{(\nu-1)})](\boldsymbol{x}^{(\nu-1)} - \boldsymbol{x}^{(\nu)})\end{aligned} \tag{4.103}$$

One has thus:

$$\boldsymbol{x}^{(\nu)} = \boldsymbol{x}^{(\nu-1)} + \delta\boldsymbol{x}^{(\nu)} \quad \text{with} \quad [\boldsymbol{K}(\boldsymbol{x}^{(\nu-1)})]\delta\boldsymbol{x}^{(\nu)} = -\boldsymbol{R}(\boldsymbol{x}^{(\nu-1)}) \tag{4.104}$$

Figure 4.37 shows the method for an equation with single scalar unknown in the case of a constant right hand side. For strongly non linear functions, the convergence is very dependent on the choice of the initial value. The advantage of the

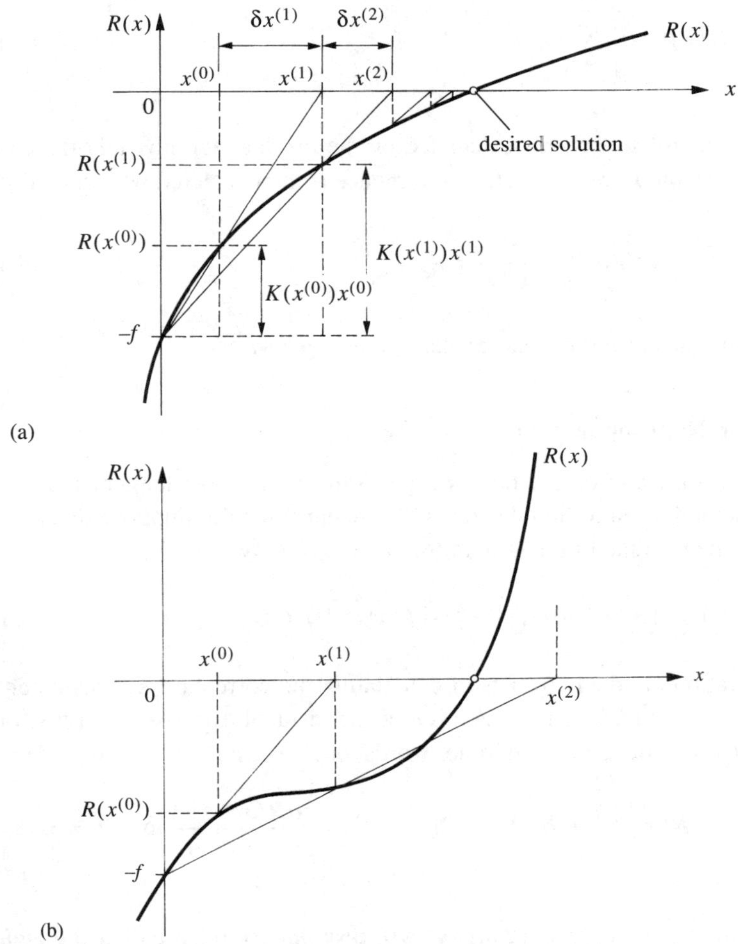

Figure 4.37 Substitution or Picard's method. Graphical illustration in the simple case of a scalar function of one variable and a constant right hand side: a) convergence, b) divergence. See also exercise 4.6.7.

residual method is its simplicity of implementation: for each iteration it is sufficient to recalculate (to assemble) the matrix $[\boldsymbol{K}(\boldsymbol{x})]$ and to solve the system (4.102).

Definition of convergence

The convergence of the method can be evaluated by calculating the norm of the residual vector $\boldsymbol{R}$ at each iteration. One can use the so-called ***max norm***:

$$|\boldsymbol{R}| = \operatorname*{Sup}_{i = 1, n} (|R_i|) \tag{4.105}$$

or the ***Euclidean norm***:

$$|\boldsymbol{R}| = \sqrt{\boldsymbol{R}\cdot\boldsymbol{R}} = \left[\sum_{i=1}^{n}(R_i)^2\right]^{1/2} \qquad (4.106)$$

In practice, relative norms calculated on the residual and possibly the corrections to the variable $\boldsymbol{x}$ are used. The convergence is then declared when:

$$\frac{|\boldsymbol{R}^{(\nu)}|}{|\boldsymbol{R}^{(0)}|} < \varepsilon_r \quad \text{and/or} \quad \frac{|\delta\boldsymbol{x}^{(\nu)}|}{|\boldsymbol{x}^{(\nu)}|} < \varepsilon_c \qquad (4.107)$$

ε_r and ε_c being the convergence tolerances given *a priori.*

4.5.2 Newton-Raphson method

The Newton-Raphson method is a generalization of the tangent method, or Newton's method, used to find the zeros of a scalar function. Suppose that at iteration $(\nu - 1)$, the residual of the system (4.101) is given by:

$$\boldsymbol{R}(\boldsymbol{x}^{(\nu-1)}) = [\boldsymbol{K}(\boldsymbol{x}^{(\nu-1)})]\boldsymbol{x}^{(\nu-1)} - \boldsymbol{f}(\boldsymbol{x}^{(\nu-1)}) \neq \boldsymbol{0} \qquad (4.108)$$

The iteration (ν) then consists of calculating the correction $\delta\boldsymbol{x}^{(\nu)}$ as being the growth of $\boldsymbol{x}^{(\nu-1)}$ that results in the zeroing of the residual expressed as a first-order Taylor expansion around $\boldsymbol{x}^{(\nu-1)}$. In other words, one writes:

$$\boldsymbol{R}(\boldsymbol{x}^{(\nu)}) = \boldsymbol{R}(\boldsymbol{x}^{(\nu-1)} + \delta\boldsymbol{x}^{(\nu)}) \approx \boldsymbol{R}(\boldsymbol{x}^{(\nu-1)}) + \left[\frac{\partial\boldsymbol{R}(\boldsymbol{x}^{(\nu-1)})}{\partial\boldsymbol{x}}\right]\delta\boldsymbol{x}^{(\nu)} = \boldsymbol{0} \qquad (4.109)$$

The matrix $[\partial\boldsymbol{R}/\partial\boldsymbol{x}]$ is the ***tangent stiffness matrix*** (also called the ***rigidity matrix***, or more generally the Jacobian) of the non linear system. As components:

$$R_i(\boldsymbol{x}^{(\nu-1)}) + \sum_{j=1}^{n}\left[\frac{\partial R_i(\boldsymbol{x}^{(\nu-1)})}{\partial x_j}\right]\delta x_j^{(\nu)} = 0 \qquad (4.110)$$

for $i = 1, n$

The correction is then obtained by:

$$\delta\boldsymbol{x}^{(\nu)} = -\left[\frac{\partial\boldsymbol{R}(\boldsymbol{x}^{(\nu-1)})}{\partial\boldsymbol{x}}\right]^{-1}\boldsymbol{R}(\boldsymbol{x}^{(\nu-1)}) \qquad (4.111)$$

with:

$$\boldsymbol{x}^{(\nu)} = \boldsymbol{x}^{(\nu-1)} + \delta\boldsymbol{x}^{(\nu)} \qquad (4.112)$$

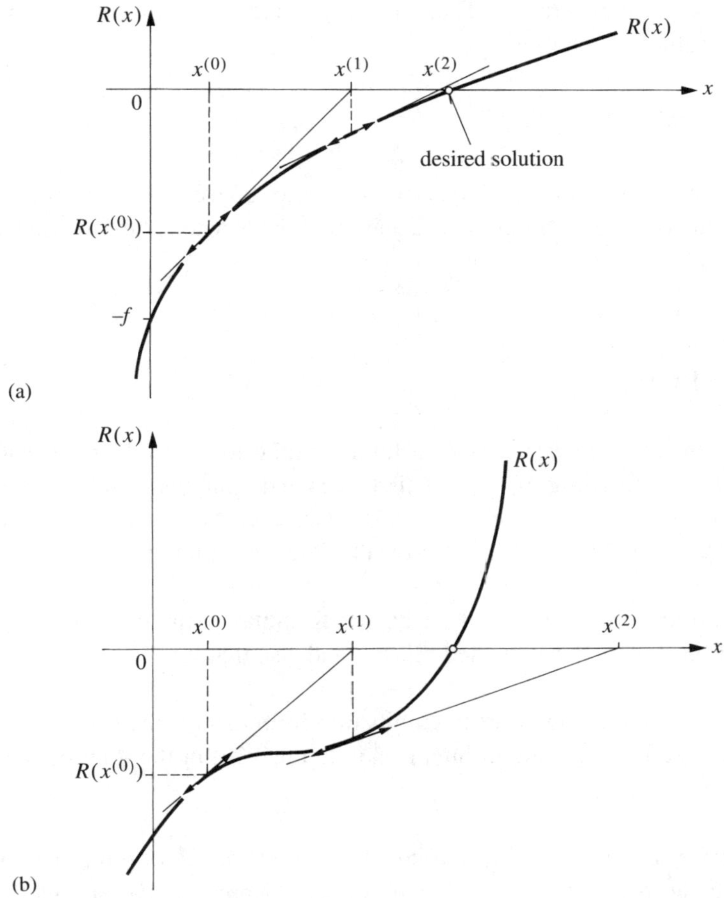

Figure 4.38 Newton-Raphson method. Graphical illustration for the simple case of a scalar function of one unknown and a constant right hand side: a) convergence, b) divergence. See also exercise 4.6.7.

The non linearity of $\boldsymbol{R}$ is responsible for $\boldsymbol{R}(\boldsymbol{x}^{(\nu)})$ not being null: it is therefore necessary to proceed with other corrections of the same type until convergence. When the Newton-Raphson method converges, the convergence is quadratic, that is, the error at iteration (ν) is bounded by the square of the error at iteration $(\nu - 1)$. In other words, the method yields a sequence of iterations that doubles the number of significant digits at each step. Picard's method only converges linearly.

As with the substitution method, *the convergence is not guaranteed.* A technique to make the algorithm more reliable is *under-incrementing* or line search: it consists of verifying that the correction calculated actually results in a diminution of the norm of the residual $\boldsymbol{R}$. If that is not the case, only a part of the correction is employed. In general, the correction is divided by a constant factor at each under-incrementing iteration. Thus, at the pth under-incrementation, dividing by two, for example, each time:

$$\boldsymbol{x}^{(\nu)} = \boldsymbol{x}^{(\nu-1)} + (\tfrac{1}{2})^{p}\delta\boldsymbol{x}^{(\nu)} \tag{4.113}$$

This technique, although it does not guarantee convergence, in practice permits the handling of severe nonlinearities, and gives the Newton-Raphson method a real robustness.

4.6 EXERCISES

4.6.1 Given $[\boldsymbol{K}]$ a $n \times n$ matrix of a linear system to solve. Assume that the sub matrix of $[\boldsymbol{K}]$ consisting of its $i-1$ first lines and columns could be decomposed into a product $[\boldsymbol{L}][\boldsymbol{U}]$, of the form (4.41). Continue the decomposition to obtain line i of $[\boldsymbol{L}]$ and column i of $[\boldsymbol{U}]$. Derive thus the algorithm in figure 4.20.

4.6.2 Starting from the Gauss factorization formulas in figure 4.20, show that the profile of the matrix to be inverted fills up; and also that it is conserved.

4.6.3 Starting from the Gaussian elimination formulas for the solution of a linear system, derive Thomas's algorithm given in figure 4.23 for the solution of a tridiagonal system.

4.6.4 Derive the expression of the conjugate gradients algorithm, given in figure 4.28. One can proceed in the following manner. Knowing that one seeks $\boldsymbol{x}^{(\nu+1)}$ in the form $\boldsymbol{x}^{(\nu+1)} = \boldsymbol{x}^{(\nu)} + \alpha^{(\nu)}\boldsymbol{d}^{(\nu)}$, show that one must necessarily have $\alpha^{(\nu)} = -(\boldsymbol{d}^{(\nu)})^{\mathrm{T}}\boldsymbol{r}^{(\nu)}/((\boldsymbol{d}^{(\nu)})^{\mathrm{T}}[\boldsymbol{K}]\boldsymbol{d}^{(\nu)})$. Knowing that one seeks the new descent direction in the form $\boldsymbol{d}^{(\nu+1)} = -\boldsymbol{r}^{(\nu+1)} + \sum_{\mu=0}^{\nu}\beta^{(\mu)}\boldsymbol{d}^{(\mu)}$, show that in order for the direction to be $[\boldsymbol{K}]$-conjugate with all the previous, one must have $\beta^{(\mu)} = ((\boldsymbol{r}^{(\nu+1)})^{\mathrm{T}}[\boldsymbol{K}]\boldsymbol{d}^{(\mu)})/((\boldsymbol{d}^{(\mu)})^{\mathrm{T}}[\boldsymbol{K}]\boldsymbol{d}^{(\mu)})$ and deduce then that $\beta^{(\mu)} = 0$ for $\mu \leq \nu-1$ and that $\beta^{(\nu)}$ and $\alpha^{(\nu)}$ have the expressions given in figure 4.28.

4.6.5 Establish the algorithm for the calculation of the product of a symmetric matrix $[\boldsymbol{K}]$, of dimension n, stored in compact form in a vector $\boldsymbol{T}$, by a vector $\boldsymbol{u}$.

4.6.6 Given an arbitrary 2×2 system $[\boldsymbol{A}]\boldsymbol{x} = \boldsymbol{b}$, with the vector $\boldsymbol{x} = [1, 1]^{\mathrm{T}}$ as a solution. Perturb the values of the right hand side with an arbitrary relative variation (for example 0.1%) and examine the relative variation of the solution $\boldsymbol{x}$. Repeat this operation choosing varying conditioning numbers of matrix $[\boldsymbol{A}]$ and show that the amplitude of the errors in the solution grow larger and larger as the conditioning of the matrix gets worse (i.e. Cond $[\boldsymbol{A}]$ large).

Noting $|[A]| = \underset{x \neq 0}{\mathrm{Max}} |[A]x|/|x|$, and given that $\mathrm{Cond}\,[A] = |[A]||[A]^{-1}|$, show that: $|\Delta x|/|x| \leq \mathrm{Cond}\,[A]|\Delta b|/|b|$.

4.6.7 Take the graphical illustration of the Picard and the Newton-Raphson algorithms for the case of a non constant scalar function on the right hand side. One first sketches *R*(x) and *f*(x) before explaining the functioning of the two methods.

4.7 BIBLIOGRAPHY

O. AXELSSON, "Solution of Linear Systems of Equations: Iterative Methods", *Lecture Notes in Mathematics*, vol. 572, Spinger Verlag, Berlin, New York, 1976.

B. CARNAHAN, H.A. LUTHER, J.O. WILKES, *Applied Numerical Methods*, John Wiley and Sons, New York, 1969.

P. CIARLET, *Introduction à l'Analyse Numérique Matricielle et à l'Optimisation*, Collection Mathématiques Appliquées pour la Maîtrise, Masson, Paris, 1982.

T. COUPEZ, H. DIGONNET, R. DUCLOUX, "Parallel Meshing and Remeshing", Applied Mathematical Modelling, vol. 25, pp. 153–175, 2000.

H. EDELSBRUNNER, *Geometry and Topology for Mesh Generation*, Cambridge University Press, Cambridge, 2001.

L.E. ERIKSSON, "Practical Three-Dimensional Mesh Generation Using Transfinite Interpolation", *SIAM J. Sci. Stat. Comput.*, vol. 6, pp. 712–740, 1985.

P.J. FREY, P.L. GEORGE, *Mesh Generation. Application to Finite Elements,* Hermes Science Pub., Oxford, Paris, 2000.

G.H. GOLUB, C.F. VAN LOAN, *Matrix Computations*, Johns Hopkins Univ. Press, Baltimore, 1996.

W.J. GORDON, C. HALL, "Construction of Curvilinear Coordinate Systems and Applications to Mesh Generation", *Int. J. Num. Meth. Eng.*, vol. 7, pp. 461–477, 1973.

P. LASCAUX, R. THEODORE, *Analyse Numérique Matricielle Appliquée à l'Art de l'Ingénieur*, Masson, Paris, 1986.

J.K. REID, *On the Method of Conjugate Gradients for Solution of Large Sparse Systems of Linear Equations*, Academic Press, San Diego, California, 1971.

J.F. THOMPSON, *Numerical Grid Generation. Foundation and Applications*, North-Holland, New York, 1985.

D.M. YOUNG, *Iterative Solution of Large Linear Systems*, Academic Press, San Diego, California, 1971.

CHAPTER 5

PHASE TRANSFORMATIONS

5.1 OBJECTIVES

- Review the concepts of phase diagrams and the equilibrium conditions that apply at an interface between two phases.
- Introduce the different scales of a process and the local transformation path concept.
- Derive the average continuity equations for two phases.
- Describe the convection and radiation contributions that intervene in heat transfer phenomena.
- Present the numerical methods for tracking an interface and those related to the average equations.
- Apply these methods to the modelling of processes and the calculation of microstructures.

Phase transformations are of course one of the key aspects of materials science and technology. The following are all problems involving phase transformations: the processing of iron in a blast furnace or of aluminum in an electrolysis cell starting from their respective minerals, the decarburizing of steels, the solidification of metal alloys and their solid state transformations, the production by a flux technique of high melting point single crystals, the precipitation of ceramic powders starting from aqueous solutions, the growth of a film at the interface between two materials (diffusion couple), corrosion or oxidation reactions on a surface, etc. The temperature at which these transformations take place plays – along with the equilibrium phase diagrams – an essential role, but convective movements of a liquid phase or the gas pressure during a reaction can also be important. Moreover, phase diagrams only give the equilibrium conditions at the interface between two phases; the chemical diffusion of species, much slower than thermal diffusion, will affect the formation of microstructures and thus introduce a new scale between the process dimensions and the interatomic distance (chap. 1). The purpose of this chapter is not to cover all the aspects of the modeling of phase transformations; but, rather to illustrate – incompletely – the different procedures that can be applied in this very diverse and complex domain.

5.2 STATE EQUATIONS

5.2.1 Phase diagrams

The equilibrium conditions for N_p phases ν_i ($i = 1, N_p$), with N_c chemical constituents A_j ($j = 1, N_c$), are given by thermodynamics. For example, in the aluminum-silicon system, for which the phase diagram is shown in figure 5.1, we will have, $\nu_1 = \ell$, the liquid phase of the mixture (Al + Si), $\nu_2 = \alpha$, the cubic face centered phase (cfc) of aluminum, and $\nu_3 = \beta$, the diamond structure of silicon. There are two chemical constituents in this case: A_1 = Al and A_2 = Si.

The thermodynamic equilibrium conditions stipulate that:

$$p^{\nu_1} = p^{\nu_2} = p^{\nu_3} = \ldots = p \tag{5.1a}$$

$$T^{\nu_1} = T^{\nu_2} = T^{\nu_3} = \ldots = T \tag{5.1b}$$

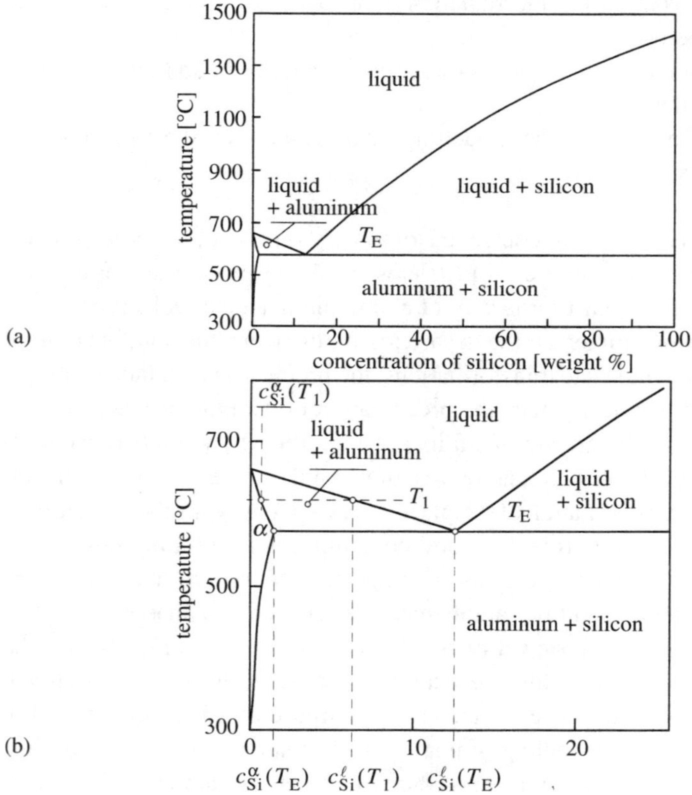

Figure 5.1 Phase diagram of the aluminum-silicon system (a) and enlargement of the zone rich in aluminum (b) showing the equilibrium of two phases at a temperature T_1, and of three phases at the eutectic temperature T_E.

$$\mu_{A_j}^{\nu_1} = \mu_{A_j}^{\nu_2} = \ldots = \mu_{A_j}^{\nu_{N_p}} = \mu_{A_j} \qquad \text{for } j = 1, N_c \tag{5.1c}$$

In other words, the pressure, p^{ν_i}, the temperature, T^{ν_i}, as well as the chemical potentials, $\mu_{A_j}^{\nu_i}$, of each of the elements A_j must be equal for all the phases, ν_i, present. If only one phase is present, there are $(N_c + 1)$ degrees of freedom for the system: the pressure, the temperature and the $(N_c - 1)$ chemical concentrations, c_j, of the solute elements (defined per unit volume in chapter 1). One of the elements effectively plays the role of solvent as we have the equation: $c_1 + c_2 + \ldots + c_{N_c} = 1$. When N_p phases are present, the number of variables describing the system is equal to $(N_p (N_c - 1) + 2)$, that is, the $(N_c - 1)$ concentrations in each of the N_p phases, the temperature and the pressure. However, equations (5.1c) introduce $(N_p - 1)$ constraints for each of the chemical species A_j present in the system, thus, $N_c (N_p - 1)$ equations. The number of degrees of freedom of the system is then equal to $(N_c + 2 - N_p)$, including the pressure (or to $(N_c + 1 - N_p)$ if p is fixed). This is ***Gibbs' phase rule***.

For a binary system such as Al-Si (fig. 5.1), we have then 2 degrees of freedom at fixed pressure when a single phase is present: the temperature and the concentration c_{Si}. That is the case in the zones marked "α" or "ℓ" for which only a single phase is present. If two phases are present, for example the liquid and the phase α, only one degree of freedom remains. The temperature being fixed at T_1, the concentrations $c_{Si}^{\alpha}(T_1)$ and $c_{Si}^{\ell}(T_1)$ in the phases α and ℓ are given by the solidus and the liquidus of the alloy, respectively (fig. 5.1). Three zones of coexistence of two phases can be distinguished in the Al-Si diagram. If the three phases α, β and ℓ are present, the system no longer has any degree of freedom: the temperature, T_E, and the concentrations of silicon in the three phases, $c_{Si}^{\alpha}(T_E)$, $c_{Si}^{\beta}(T_E)$ and $c_{Si}^{\ell}(T_E)$ are fixed for a given pressure. This state clearly corresponds to the eutectic point visible in figure 5.1. For more details on phase diagrams, the reader is referred to Porter and Easterling in the bibliography.

5.2.2 Out-of-equilibrium phenomena

As material transformation processes take place over a finite length of time, the phases present are rarely in equilibrium! Therefore, two types of deviation from the equilibrium given by a phase diagram are distinguished:

- In the first case, the chemical species and/or the heat do not have sufficient time to diffuse and thus to even out the gradients of solute and temperature that are inevitably created in a phase transformation. In this case, only *the interfaces* between the phases are considered to be effectively in equilibrium. Figure 5.2a shows the case of growth of a pure single crystal at velocity $v^{s/\ell}$: only the interface is at the melting temperature of the material, T_f, and a thermal gradient exists in the two phases. The case of a solidifying aluminum-silicon alloy is shown in figure 5.2b: the interface being at the temperature T_1, the concentrations at the interface

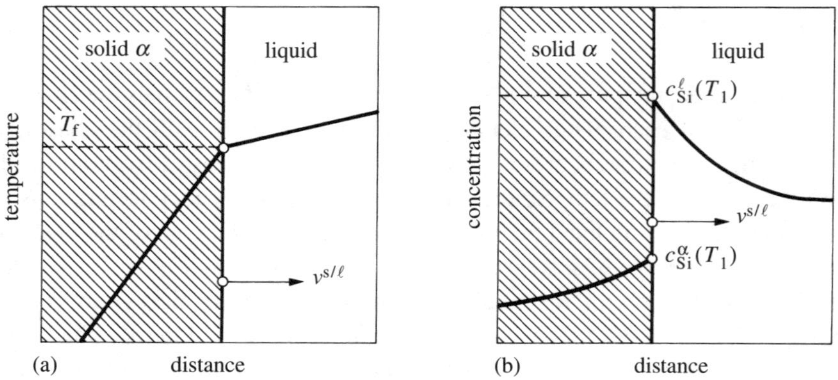

Figure 5.2 Thermal equilibrium of an interface between two phases, solid and liquid, for a solidifying pure material (a). Chemical equilibrium of an interface between two phases, α and ℓ for a binary system (b).

are still given by $c^{\alpha}_{Si}(T_1)$ and $c^{\ell}_{Si}(T_1)$ of the phase diagram (fig. 5.1), but one notes the existence of chemical gradients in the two phases, α and liquid, in addition to the thermal gradients already mentioned.

- The interfaces themselves can depart from the equilibrium given by the phase diagram for three reasons. The interface surface energy γ, between the phases, modifies the energy balance, and thus the equilibrium, when the characteristic size of the phases is small. Thus, for a pure material, it is known that the melting point, T^{r}_{f}, of a small sphere of radius R is lowered with respect to the temperature, T^{∞}_{f}, of the bulk material (infinite radius). The temperature difference, ΔT_r, associated with this contribution is called ***curvature undercooling***, with the value:

$$\Delta T_r = T^{\infty}_{f} - T^{r}_{f} = \Gamma \Xi = \frac{\gamma T^{\infty}_{f}}{\rho L_f} \frac{2}{R} \tag{5.2}$$

where Γ is the Gibbs-Thomson coefficient, ρL_f the specific latent heat of fusion and $\Xi = 2R^{-1}$ the mean curvature of the interface. For the case of a non spherical interface separating the two materials, the local mean curvature Ξ is defined as $(R_1^{-1} + R_2^{-1})$, where R_1 and R_2 are the principle radii of curvature. For an alloy, it is as if the portion of the phase diagram describing the equilibrium between α and ℓ were displaced by ΔT_r (fig. 5.3). This curvature undercooling is exactly that associated with the energy barrier that must be overcome when a new solid seed of radius R is created in a liquid (nucleation).

For the precipitation and growth of *coherent* phases in a solid phase, that is, of phases having almost the same crystalline structure as that of the matrix from which they come, the concept of interface energy has no real significance. However, the

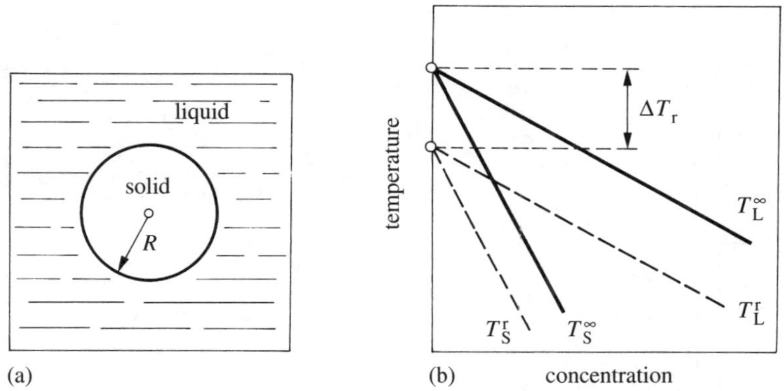

Figure 5.3 Equilibrium between a solid sphere of radius R and the liquid (a) and influence on the phase diagram (b).

local *stresses* induced by the small parameter differences of the crystalline structure become the cause of a deviation from equilibrium. Such is notably the case for the spinodal decomposition of an alloy.

The two other causes of a deviation from equilibrium at an interface are, first, the ***kinetics of attachment of atoms*** *or* ***molecules*** and, second, ***solute trapping***. The first contribution is associated with the rearrangement of the atoms or molecules as they pass from the structure of one phase to the other. The kinetic undercooling will then be greater the more the structures present differ (and thus have elevated transformation entropies). The second contribution is linked to the redistribution of the solute elements accross the moving interface of an alloy. Such deviations from equilibrium are important for molecules with complex forms (for example, polymers), but also for metallic alloys at high transformation rates.

5.2.3 Simple models of phase transformations

Two extreme cases can be presented for the occurrence of a phase transformation; they are illustrated in figure 5.4. In the first case (fig. 5.4(a)), the interface between the liquid and solid phases of a pure material that solidifies in a Bridgman furnace is relatively "smooth", that is, its geometric characteristics are on the scale of the size of the process. Here, the curvature of the interface is comparable with the radius of the specimen and is the result of thermal exchanges between the hot and cold zones of the furnace by conduction, radiation and convection in the liquid. In such a situation, the modeling by finite elements or by finite differences can take into account the discrete regions corresponding to the two phases. These evolve with time: the method must use an *evolving* mesh with ***conservation equations at the interfaces,*** which will be deduced in the following section.

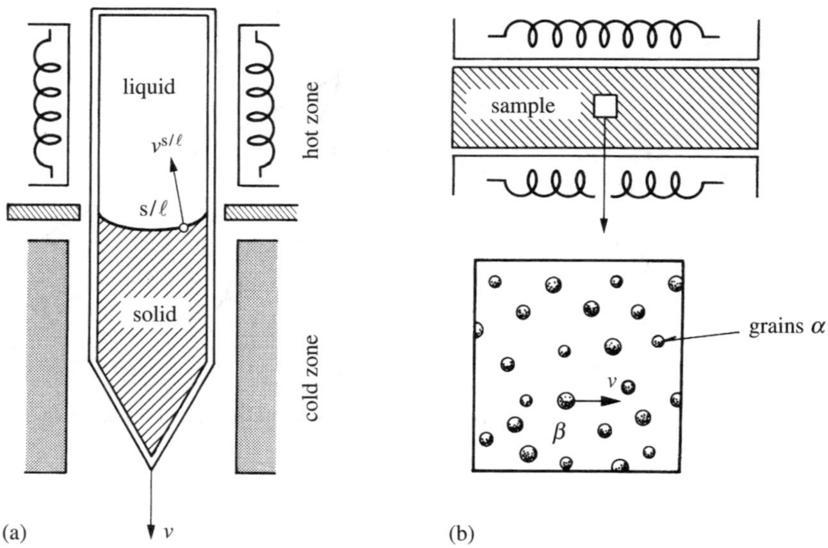

Figure 5.4 Examples of a smooth interface of a pure crystal solidifying in a Bridgman furnace (a) and of an interface fragmented into numerous spherical grains nucleating in the primary phase volume (b).

Figure 5.4(b) illustrates the situation of a transformation governed by the nucleation of the second phase α in the volume of the primary phase β (transformation $\beta \rightarrow \alpha$). In this simple case, the curvature of the interface is dictated by the radius of the grains which can be considerably smaller than the characteristic dimension of the domain. Since it is impossible to solve the continuity equations at the scale of the process or of the part with a mesh that would also be finer than the size of the grains, one turns to ***average continuity equations***, in which appears the concept of a ***phase fraction***. Then, in the case of figure 5.4(b) where the grains of the phase α appear by random nucleation in the volume of homogeneous phase β, the volume fraction f_α of the transformed phase is given by the well known Avrami equation (exercise 5.6.1):

$$f_\alpha(t) = 1 - \exp[-f_\alpha^0(t)] \tag{5.3}$$

where the exponential factor takes into account the impingement of grains. The fraction appearing in the exponential, $f_\alpha^0(t)$, is called the ***extended fraction*** of the transformed phase and can be calculated by neglecting the impingement of grains. The grains remain spherical during their growth if the rate at which the seed interface advances is isotropic, and if they do not collide. The extended fraction $f_\alpha^0(t)$ is thus given in this case by the convolution:

$$f_{\alpha}^{0}(t) = \int_{0}^{t} \frac{4\pi}{3}(R(\tau, t))^3 \, dn = \int_{0}^{t} \frac{4\pi}{3}(R(\tau, t))^3 \dot{n}(\tau) \, d\tau \tag{5.4}$$

where dn is the number of grains formed per unit volume between times τ and $\tau + d\tau$, and $R(\tau, t)$ is the radius of the grains that nucleate at instant τ and are observed at instant t. dn can be expressed as $\dot{n}(\tau)\, d\tau$, where $\dot{n}(\tau)$ is the nucleation rate at instant τ. In the case where the growth rate, v, of the grain interface is a function only of the local undercooling, ΔT, of the sample (deviation from equilibrium), this radius is given by the equation:

$$R(\tau, t) = \int_{\tau}^{t} v(\Delta T(u)) \, du + R(\tau, \tau) \tag{5.5}$$

where $R(\tau, \tau)$ is the initial radius of the seeds at the moment of their nucleation (often much smaller than $R(\tau, t)$ and thus generally neglected). In the isothermal situation (ΔT = constant), it can easily be shown that $f_{\alpha}^{0}(t)$ varies as t^3 if all the grains nucleate at the same time, while a variation as t^4 is deduced for a nucleation rate $\dot{n}$ that is constant (exercise 5.6.2). For situations where the morphology of the grains is more complex and/or nucleation is not randomly distributed in the volume, one turns to empirical Avrami laws:

$$f_{\alpha}(t) = 1 - \exp[-bt^{q}] \tag{5.6}$$

where b and q are adjustable parameters, functions of the temperature and deduced from isothermal measurements. The phase fractions $f_{\alpha}(t)$ so obtained are transcribed in the form of time-temperature-transformation (TTT) diagrams such as that shown in figure 5.5. Along a horizontal (isothermal transformation), the first point encountered corresponds to the starting time of the transformation (usually $f_{\alpha} = 1\%$) and the second to the end ($f_{\alpha} = 99\%$). This type of diagram or equation (5.6) can be used for anisothermal transformations by employing a superposition principle (sect. 5.5.3).

5.2.4 Interface conservation equations

In the following, we will assume that the interfaces are at equilibrium, and thus that the concentrations and the temperature at the interface are given by the corresponding phase diagram (we also assume fixed pressure). Then in the neighborhood of the interface, we can establish the conservation of mass, solute concentration, energy or momentum. The equations so obtained will largely drive the evolution of the interface, whose position is *a priori* unknown.

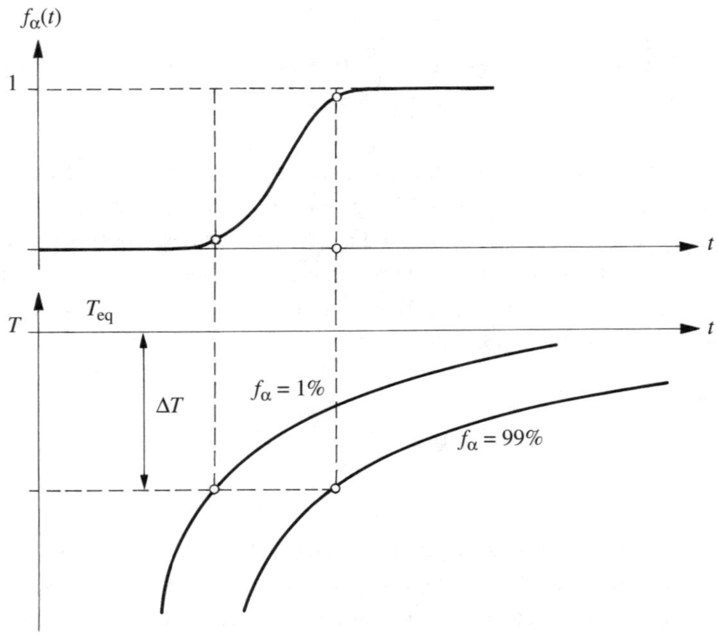

Figure 5.5 Evolution of the transformed fraction $f_\alpha(t)$ at a given temperature and associated TTT diagram.

General equation

Consider the situation illustrated in figure 5.6, of an interface between two phases α and β moving at velocity $\boldsymbol{v}^{\alpha/\beta}$ in a given, fixed reference frame. $\boldsymbol{n}^{\alpha/\beta}$ is the surface normal of the interface pointing out of α (thus going into β). The effective velocities of the material measured in the fixed reference frame are $\boldsymbol{v}^\alpha$ and $\boldsymbol{v}^\beta$ for each of the phases α and β, respectively. In a reference frame that translates with the interface, the general conservation equation obtained in chapter 1 can be written for each phase:

$$\frac{\partial}{\partial t}(\rho\xi)^\nu + \mathrm{div}[(\rho\xi)^\nu(\boldsymbol{v}^\nu - \boldsymbol{v}^{\alpha/\beta})] + \mathrm{div}\,\boldsymbol{j}_\xi^\nu = \dot{Q}_\xi^\nu \quad \text{for } \nu = \alpha, \beta \tag{5.7}$$

where $(\boldsymbol{v}^\nu - \boldsymbol{v}^{\alpha/\beta})$ is none other than the relative velocity in this moving reference frame of the material in phase ν. The four terms appearing in this equation correspond to the local temporal variation of the value ξ, its transport, diffusion and the source term respectively. The value ξ can correspond to the material itself ($\xi = 1$), to the concentration per unit mass of a chemical species i, ($\xi = w_i$), to the enthalpy ($\xi = H$) or to one component of the velocity along an axis x_j ($\xi = v_j$).

We integrate this equation for a small "flat" volume element, Ω, around the interface (fig. 5.6). Using the divergence theorem, we obtain:

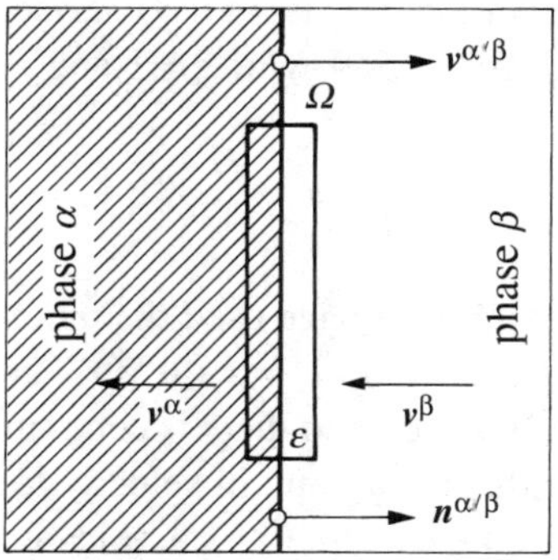

Figure 5.6 Motion of an interface and motion of the material.

$$\frac{\mathrm{d}}{\mathrm{d}t}\int_{\Omega}(\rho\xi)^{\mathrm{v}}\,\mathrm{d}V + \int_{\partial\Omega}[(\rho\xi)^{\mathrm{v}}(\boldsymbol{v}^{\mathrm{v}} - \boldsymbol{v}^{\alpha/\beta})]\cdot\boldsymbol{n}\,\mathrm{d}S + \int_{\partial\Omega}\boldsymbol{j}_{\xi}^{\mathrm{v}}\cdot\boldsymbol{n}\,\mathrm{d}S = \int_{\Omega}\dot{Q}_{\xi}^{\mathrm{v}}\,\mathrm{d}V \quad (5.8)$$

where $\boldsymbol{n}$ is the outgoing normal of the domain Ω. The first integral and the integral of the source term are zero as the thickness, ε, of the chosen volume element goes to zero. Similarly, in this limit, the contributions of the surface integrals are only from the surfaces parallel to the interface. Noting that $\boldsymbol{n} \equiv \boldsymbol{n}^{\alpha/\beta}$ and $\boldsymbol{n} \equiv -\boldsymbol{n}^{\alpha/\beta}$ for the portions of $\partial\Omega$ contained in β and α, respectively, we have finally:

$$\begin{aligned}
&\{(\rho\xi)(\boldsymbol{v} - \boldsymbol{v}^{\alpha/\beta}) + \boldsymbol{j}_{\xi}\}_{\alpha}^{\beta}\cdot\boldsymbol{n}^{\alpha/\beta}\\
&\quad = \{(\rho\xi)\boldsymbol{v} + \boldsymbol{j}_{\xi}\}_{\alpha}^{\beta}\cdot\boldsymbol{n}^{\alpha/\beta} - \boldsymbol{v}^{\alpha/\beta}\cdot\boldsymbol{n}^{\alpha/\beta}\{\rho\xi\}_{\alpha}^{\beta}\\
&\quad = [(\rho\xi)^{\beta}(\boldsymbol{v}^{\beta} - \boldsymbol{v}^{\alpha/\beta}) + \boldsymbol{j}_{\xi}^{\beta}]\cdot\boldsymbol{n}^{\alpha/\beta} - [(\rho\xi)^{\alpha}(\boldsymbol{v}^{\alpha} - \boldsymbol{v}^{\alpha/\beta}) + \boldsymbol{j}_{\xi}^{\alpha}]\cdot\boldsymbol{n}^{\alpha/\beta} = 0
\end{aligned} \quad (5.9)$$

where the notation $\{A\}_{\alpha}^{\beta} = A^{\beta} - A^{\alpha}$ has been adopted to represent the jump of a value A across the interface between α and β. This general conservation equation will be applied specifically for each of the continuity equations derived in chapter 1.

Conservation of mass

In the case where the two phases present, α and β, have the same density, the interface between them can advance without global displacement of the material. However, in most cases, the densities are not equal. For example, most solid bodies, with the exception of water and some semiconductors, such as germanium or silicon, have a higher density for the solid than for the liquid. During solidification, liquid must be displaced towards the interface to compensate for the loss of volume (or the increase in specific mass).

To establish the conservation of mass at the interface α/β for the general case, it is sufficient to apply equation (5.9) with $\xi = 1$ and $\boldsymbol{j}_\xi = 0$ (no mass diffusion). Then:

$$v_n^\beta \rho^\beta - v_n^\alpha \rho^\alpha = v_n^{\alpha/\beta}(\rho^\beta - \rho^\alpha) \tag{5.10}$$

To simplify the expression, we have used the index "n" to denote the normal velocity components along $\boldsymbol{n}^{\alpha/\beta}$ (thus $v_n = \boldsymbol{v} \cdot \boldsymbol{n}^{\alpha/\beta}$). For the particular case where $\rho^\alpha = \rho^\beta$, the term on the right is null and there is continuity of the normal velocities at the interface. Otherwise there is a jump in velocity to compensate the change in specific mass. Taking again the example of solidification, we can assume that the solid phase does not move ($\boldsymbol{v}^\alpha = 0$) and then:

$$v_n^\ell = v_n^{s/\ell}\left[1 - \frac{\rho^s}{\rho^l}\right] \tag{5.11}$$

Thus, if $\rho^s > \rho^\ell$, the velocity of the liquid is opposite to the motion of the interface: material is drawn to the solid during solidification.

Conservation of solute

The conservation of solute at an interface is illustrated in figure 5.2(b) for solidification. Setting $\xi = w$ and $\boldsymbol{j}_\xi = -D\,\mathbf{grad}\,c$, D being the chemical diffusion coefficient, and noting that $c^\nu = \rho^\nu w^\nu$ in each phase (we have omitted the index "i" for the number of the element of solute considered), we obtain from equation (5.9):

$$v_n^\beta c^{\beta*} - v_n^\alpha c^{\alpha*} - v_n^{\alpha/\beta}(c^{\beta*} - c^{\alpha*}) = D^\beta\left[\frac{\partial c}{\partial n}\right]^{\beta*} - D^\alpha\left[\frac{\partial c}{\partial n}\right]^{\alpha*} \tag{5.12}$$

To signify that the concentrations are those at the interface, we have added the index "*". The gradients of the concentration field appearing in the term on the right are also taken at the interface, in the direction of $\boldsymbol{n}^{\alpha/\beta}$. In the case where the solute transport associated with the velocities v_n^β and v_n^α can be neglected ($v_n^\beta = v_n^\alpha = 0$), equation (5.12) establishes that the velocity of the interface is given by the solute flux jump divided by the concentration jump, with a minus sign:

$$v_n^{\alpha/\beta} = -\frac{D^\beta[\partial c/\partial n]^{\beta*} - D^\alpha[\partial c/\partial n]^{\alpha*}}{(c^{\beta*} - c^{\alpha*})} \tag{5.13}$$

Conservation of heat

Similarly as with the conservation of solute, it is easy to show that:

$$v_n^\beta(\rho H)^{\beta*} - v_n^\alpha(\rho H)^{\alpha*} - v_n^{\alpha/\beta}[(\rho H)^{\beta*} - (\rho H)^{\alpha*}] = \kappa^\beta\left[\frac{\partial T}{\partial n}\right]^{\beta*} - \kappa^\alpha\left[\frac{\partial T}{\partial n}\right]^{\alpha*} \tag{5.14}$$

where κ is the thermal conductivity. For the case where $v_n^\beta = v_n^\alpha = 0$, the velocity of the interface is given by the difference of the heat flux in the two phases divided by the volumetric enthalpy jump. The latter is none other than the transformation enthalpy per unit volume, $(\rho L)^{\alpha/\beta}$, for example the specific latent heat of fusion in the case of figure 5.2(a). We have then:

$$v_n^{\alpha/\beta} = -\frac{\kappa^\beta[\partial T/\partial n]^{\beta*} - \kappa^\alpha[\partial T/\partial n]^{\alpha*}}{(\rho L)^{\alpha/\beta}} \tag{5.15}$$

This explains why, in the situation of controlled solidification shown in figure 5.2(a), the gradient in the liquid is weaker than in the solid if the thermal conductivities of the two phases are close: the latent heat liberated by the phase transformation must be drawn off by the solid.

REMARK

In the case of a binary alloy, it is seen that the velocity of the interface is given by two conditions: solute conservation (5.13) and heat conservation (5.15). The concentrations at the interface are of course related to the temperature, T^*, at the interface. Nevertheless, we have two conditions for the velocity of a single interface in a binary alloy; the two relations are not necessarily compatible. It is this incompatibility that is in fact responsible in alloys for the destabilization of a plane front (solutal undercooling) and thus for the diversity of microstructures encountered in materials science.

Conservation of momentum

Choosing a coordinate system where x is perpendicular to the interface α/β, equation (5.9) combined with the conservation of mass (5.10) gives, after calculation for the normal component of momentum (exercise 5.6.3):

$$\rho^\beta[(v_n^\beta)^2 - (v_n^{\alpha/\beta})^2] - \rho^\alpha[(v_n^\alpha)^2 - (v_n^{\alpha/\beta})^2] = \sigma_n^\beta - \sigma_n^\alpha \tag{5.16}$$

In the case where the densities of the two phases are equal, the term to the left is zero and the equation is reduced to the continuity of the normal stresses at the interface: $\sigma_n^\beta = \sigma_n^\alpha$. For the two tangential components, equation (5.9) gives:

$$\rho^\beta v_\tau^\beta[v_n^\beta - v_n^{\alpha/\beta}] - \rho^\alpha v_\tau^\alpha[v_n^\alpha - v_n^{\alpha/\beta}] = \sigma_\tau^\beta - \sigma_\tau^\alpha \tag{5.17}$$

5.2.5 Domain composed of two phases

Take again the very simple case of a domain Ω initially containing a phase β of a pure material which is transformed over time into phase α. It could be, for example, the crystallization of a pure material from the liquid phase (fig. 5.7). We assume for simplicity that the two phases have the same density and that, in addition, there

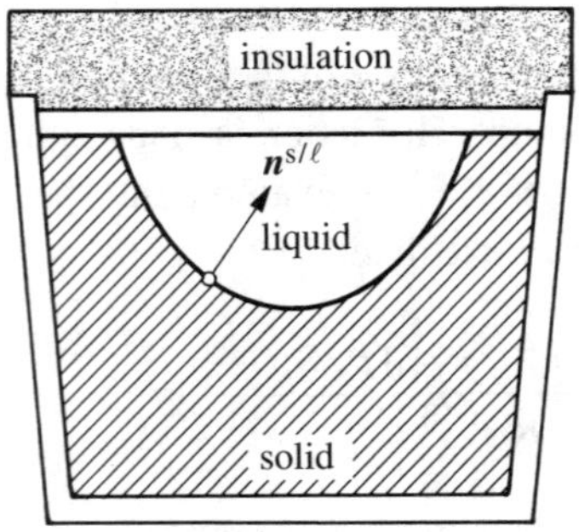

Figure 5.7 Solidification of a pure metal in a mold (planar front).

is no advection. In this case, the problem can be reduced to one of thermal exchange by conduction in the two phases, with however, the difficulty of a free interface, α/β, moving in time. Inside each of the sub-domains, Ω^α and Ω^β, characterizing the phases α and β, the heat equation can be written (chap. 1):

$$(\rho c_p)^\nu \frac{\partial T}{\partial t} - \mathrm{div}(\kappa^\nu \,\mathbf{grad}\, T) = 0 \qquad \text{for } \nu = \alpha, \beta \tag{5.18}$$

with the boundary conditions on the boundary $\partial\Omega$ and defined initial conditions.

At the free interface α/β, *two conditions* must be specified for the problem to be well posed. First, the equilibrium condition (if equilibrium is assumed!):

$$T(\boldsymbol{x}, t) = T^{\alpha/\beta} \qquad \text{for } \boldsymbol{x} \in \partial\Omega^{\alpha/\beta} \tag{5.19}$$

where $T^{\alpha/\beta}$ is the equilibrium temperature between the two phases (for example the melting point in the case of crystallization). The second condition to apply to the moving interface α/β is given by the heat conservation derived above (5.15).

Taking into account the undercooling related to the curvature of this interface (5.2), equation (5.19) becomes:

$$T(\boldsymbol{x}, t) = T^{\alpha/\beta}\left[1 - \frac{\gamma^{\alpha/\beta}}{(\rho L)^{\alpha/\beta}} \Xi^{\alpha/\beta}\right] \qquad \text{for } \boldsymbol{x} \in \partial\Omega^{\alpha/\beta} \tag{5.20}$$

The curvature, $\Xi^{\alpha/\beta}$, is measured by the change of orientation of the normal $\boldsymbol{n}^{\alpha/\beta}$ pointing toward the exterior of the phase α (the curvature is negative in figure 5.7). The value $d_\gamma = \gamma^{\alpha/\beta}/(\rho L)^{\alpha/\beta}$ has the dimensions of a length and, being proportional to the surface tension, it has the character of a ***capillarity length***. Although the value ($d_\gamma\, \Xi^{\alpha/\beta}$) is negligible for the case in figure 5.7, it plays a very important role in the formation of microstructures.

In the case of a binary alloy with two phases present, the problem is similar: the heat equation and the solute conservation equation are to be solved in each phase with the heat and solute conservation equations at the interface α/β described in the previous section, and with the equilibrium conditions at the interface given by the equality of the temperatures and of the chemical potentials (5.1).

5.2.6 Representative volume to take average values

As we have already mentioned, the diffusion of solute is considerably slower than the diffusion of heat in many materials: the non dimensional Lewis number defined as the ratio of the thermal and chemical diffusion coefficients ($Le = \alpha/D$ with $\alpha = \kappa/(\rho c_p)$, see Chapter 10) is on the order of 10^4 for a liquid metal (10^7 for a solid metal). As a result, the interface between two phases α and β is rarely smooth and regular at the scale of a process or a treated part as indicated in figure 5.7, manifesting instead a very fine and complex morphology. We have seen the case of a simple Avrami transformation where the interface was initially fragmented in a large number of grains by a nucleation mechanism (fig. 5.4(b)). This "fragmentation" of the interface between the phases α and β can also be the result of a morphological growth instability: the formation of ***dendrites*** during the solidification of alloys (fig. 5.8) or of Widmanstätten plates in solid state transformations. The coupled growth of lamellae or fibers of two phases α and β starting from a homogeneous phase γ (eutectic or eutectoid) is also a form of fragmentation of the interfaces α/β, α/γ and β/γ frequently encountered in phase transformations.

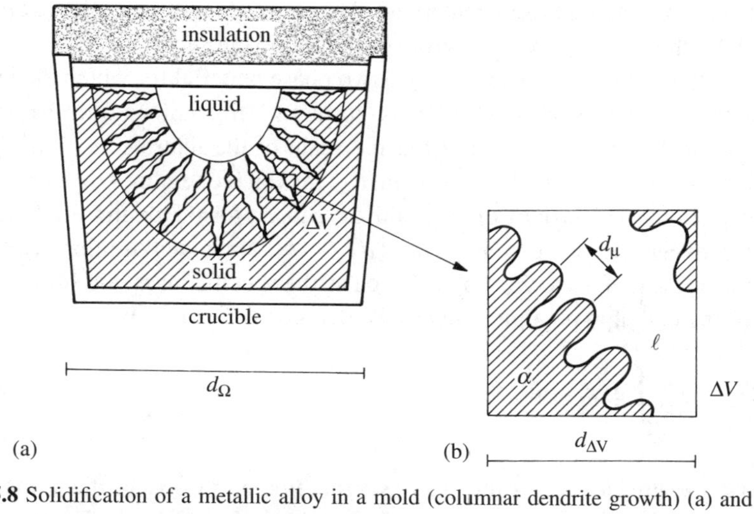

Figure 5.8 Solidification of a metallic alloy in a mold (columnar dendrite growth) (a) and volume element used for averaging the continuity equations (b).

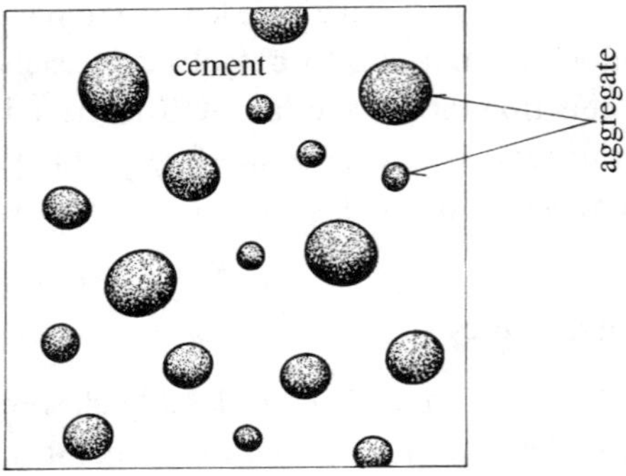

Figure 5.9 Characteristic volume element for concrete.

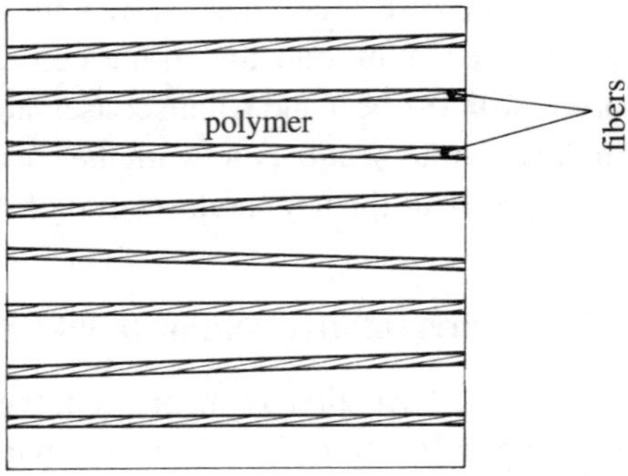

Figure 5.10 Characteristic volume element for a polymer reinforced with oriented fibers.

The material heterogeneities are of course not necessarily related to phase transformations. Think of concrete, for example, composed of solid granules in a cement matrix (fig. 5.9), or of kevlar fibers in a polymer matrix (fig. 5.10), and more generally of all composite materials for which two or more phases are intentionally mixed.

In such situations, solving the continuity equations separately in the two phases at the scale of a process (typically 0.1–1 m) but with the spatial resolution of the microstructure (typically of the order of a few μm) is unfortunately not realistic! It would be the same for calculation of the deformation of a bridge (tens of meters) with a resolution at the scale of aggregates (centimeters or even millimeters)! Because of this, when very different scales are present in a heterogeneous material, ***averaged*** continuity equations are used (one also speaks of ***homogenization theory*** in mechanics of heterogeneous materials). The principle will be explained immediately below; the equations will be given in the following section.

Consider a volume element, ΔV, in a two phase material for which the interface between the two phases present, $\partial\Omega^{\alpha/\beta}$, is "tortuous" (fig. 5.8(b)). The characteristic dimension of the microstructure, d_μ, at the scale of the interface is much smaller than the characteristic size of the domain, d_Ω, in which one hopes to solve the continuity equations. The size of the domain ΔV chosen for the continuity balances is situated between the two extremes, $d_\mu < d_{\Delta V} < d_\Omega$. Now suppose that we take the average inside ΔV of a specific value ξ such as the specific mass, the solute concentration or the enthalpy per unit volume. We denote this value $\langle\xi\rangle$:

$$\langle\xi\rangle = \frac{1}{\Delta V}\int_{\Delta V} \xi(\boldsymbol{x}, t)\ \mathrm{d}V \tag{5.21}$$

How does this average value vary as a function of the size of the domain $d_{\Delta V}$? Such a variation is schematically illustrated in figure 5.11. When ΔV is small, typi-

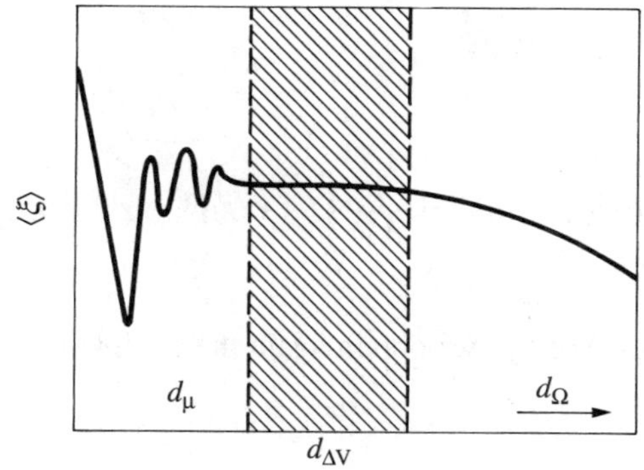

Figure 5.11 Influence of the volume element size, $d_{\Delta V}$, used to calculate the average of a value $\langle\xi\rangle$.

cally of the order of d_μ or even smaller, $\langle\xi\rangle$ may fluctuate enormously. According to whether the domain "falls" principally in phase α or β, the average value of ξ is very different. When the size of the domain increases, this average value tends to stabilize as statistically sufficient fractions of the two phases are taken into account (shaded zone in figure 5.11). One then speaks of a ***representative volume element***. However, if the volume becomes too large, variations are again observed, but this time related to heterogeneities at the scale of the process. In the following, we always place ourselves in the case of a representative volume neither "too small" so that the fluctuations at the scale of the microstructure are averaged, nor "too large" with respect to the macroscopic variations of the value at the scale of the domain.

We define the following values:

- The volume fraction, f_α, of a phase α inside ΔV:

$$f_\alpha = \frac{1}{\Delta V}\int_{\Delta V} \chi^\alpha(\boldsymbol{x}, t)\, \mathrm{d}V = \frac{\Delta V^\alpha}{\Delta V} \tag{5.22}$$

where χ^α is the distribution function of the phase α:

$$\chi^\alpha(\boldsymbol{x}, t) = \begin{cases} = 0 \text{ if } \boldsymbol{x} \notin \Omega^\alpha \\ = 1 \text{ if } \boldsymbol{x} \in \Omega^\alpha \end{cases} \tag{5.23}$$

ΔV^α is the volume of the sub-domain defined by $\Omega^\alpha \cap \Delta V$. An identical expression can be defined for the fraction, f_β, of the phase β, where the corresponding distribution function, χ^β, is complementary to χ^α. For the two phase system chosen here, the relation $\Delta V = \Delta V^\alpha + \Delta V^\beta$, or $f_\alpha + f_\beta = 1$, is of course satisfied.

- The ***average values***, $\langle\xi^\alpha\rangle$ and $\langle\xi^\beta\rangle$, of a value ξ weighted over the sub-domains ΔV^α and ΔV^β :

$$\langle \xi^{\alpha} \rangle = \frac{1}{\Delta V} \int_{\Delta V} \xi(\boldsymbol{x}, t) \chi^{\alpha}(\boldsymbol{x}, t) \, \mathrm{d}V \tag{5.24a}$$

$$\langle \xi^{\beta} \rangle = \frac{1}{\Delta V} \int_{\Delta V} \xi(\boldsymbol{x}, t) \chi^{\beta}(\boldsymbol{x}, t) \, \mathrm{d}V = \frac{1}{\Delta V} \int_{\Delta V} \xi(\boldsymbol{x}, t)[1 - \chi^{\alpha}(\boldsymbol{x}, t)] \, \mathrm{d}V \tag{5.24b}$$

With the definition (5.21) of $\langle \xi \rangle$, we obtain again the complementary relation:

$$\langle \xi \rangle = \langle \xi^{\alpha} \rangle + \langle \xi^{\beta} \rangle \tag{5.25}$$

- The *intrinsic* average values, $\langle \xi \rangle^{\alpha}$ and $\langle \xi \rangle^{\beta}$, specified in each sub-domain ΔV^{α} or ΔV^{β}:

$$\langle \xi \rangle^{\alpha} = \frac{1}{\Delta V^{\alpha}} \int_{\Delta V} \xi(\boldsymbol{x}, t) \chi^{\alpha}(\boldsymbol{x}, t) \, \mathrm{d}V = \frac{\Delta V}{\Delta V^{\alpha}} \langle \xi^{\alpha} \rangle = \frac{\langle \xi^{\alpha} \rangle}{f_{\alpha}} \tag{5.26a}$$

$$\langle \xi \rangle^{\beta} = \frac{1}{\Delta V^{\beta}} \int_{\Delta V} \xi(\boldsymbol{x}, t) \chi^{\beta}(\boldsymbol{x}, t) \, \mathrm{d}V = \frac{\Delta V}{\Delta V^{\beta}} \langle \xi^{\beta} \rangle = \frac{\langle \xi^{\beta} \rangle}{f_{\beta}} \tag{5.26b}$$

which are equal to the average values divided by the corresponding phase fractions.

One must also be able to take the average of the temporal and spatial derivatives of the value ξ appearing in the continuity equations.

Temporal derivative of ξ

By definition of the volumetric fraction of the phase α (5.22), the temporal derivative of f_{α}, in the "local" sense of the element ΔV, is given by:

$$\frac{\partial f_{\alpha}}{\partial t} = \frac{1}{\Delta V} \int_{\Delta V} \frac{\partial \chi^{\alpha}(\boldsymbol{x}, t)}{\partial t} \, \mathrm{d}V = \frac{1}{\Delta V} \frac{\mathrm{d}(\Delta V^{\alpha})}{\mathrm{d}t} \tag{5.27}$$

The distribution function of the phase α is such that its temporal derivative is zero everywhere except at the interface α/β. More precisely:

$$\frac{\partial \chi^{\alpha}(\boldsymbol{x}, t)}{\partial t} = \boldsymbol{v}^{\alpha/\beta} \cdot \boldsymbol{n}^{\alpha/\beta} \delta(\boldsymbol{x} - \boldsymbol{x}^{\alpha/\beta}) = v_{\mathrm{n}}^{\alpha/\beta} \delta(\boldsymbol{x} - \boldsymbol{x}^{\alpha/\beta}) \tag{5.28}$$

where, as for the equations at the interfaces, $\boldsymbol{v}^{\alpha/\beta}$ is the local velocity of the interface, $\boldsymbol{n}^{\alpha/\beta}$ the normal pointing to the exterior of the domain Ω^{α} and $v_{\mathrm{n}}^{\alpha/\beta}$ the normal component of the velocity (fig. 5.12). $\delta(\boldsymbol{x} - \boldsymbol{x}^{\alpha/\beta})$ is the Dirac function cen-

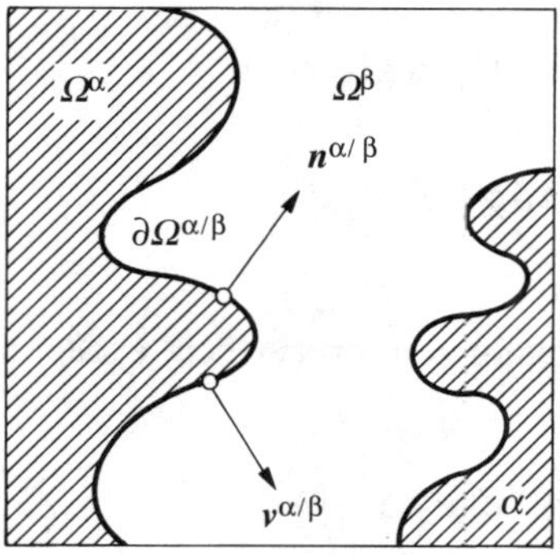

Figure 5.12 Volume element used to average the conservation equations.

tered at the points $x^{\alpha/\beta}$ of the interface: it equals ∞ if x is situated on the interface (singularity) and 0 otherwise. By the properties of this function, (5.27) is equivalent to:

$$\frac{\partial f_\alpha}{\partial t} = \frac{1}{\Delta V}\int_{\Delta V}\frac{\partial \chi^\alpha(x,t)}{\partial t}\,\mathrm{d}V = \frac{1}{\Delta V}\int_{\partial\Omega^{\alpha/\beta}} v_n^{\alpha/\beta}\,\mathrm{d}S \tag{5.29}$$

The integral of the normal velocity on the surface of the interface α/β corresponds to the temporal evolution of the domain ΔV^α. Using (5.29), it is thus easy to average the temporal derivative of any value ξ:

$$\frac{\partial\langle\xi^\alpha\rangle}{\partial t} = \frac{1}{\Delta V}\int_{\Delta V}\frac{\partial\xi(x,t)}{\partial t}\chi^\alpha(x,t)\,\mathrm{d}V + \frac{1}{\Delta V}\int_{\Delta V}\xi(x,t)\frac{\partial\chi^\alpha(x,t)}{\partial t}\,\mathrm{d}V$$

$$= \langle\frac{\partial\xi^\alpha}{\partial t}\rangle + \frac{1}{\Delta V}\int_{\partial\Omega^{\alpha/\beta}}\xi^{\alpha*}v_n^{\alpha/\beta}\,\mathrm{d}S = \langle\frac{\partial\xi^\alpha}{\partial t}\rangle + \langle\xi^\alpha v_n^{\alpha/\beta}\rangle^* \tag{5.30}$$

The same brackets "⟨.⟩" have been used in the equation above to denote an average taken on the interface α/β, while "*", as in section 5.2.4, designates values taken at the interface. In other words, the temporal derivative of the average of the value ξ is equal to the average of its temporal derivative plus an average on the interface of the product of $\xi^{\alpha*}$, the value taken at the interface in the phase α, and the normal component of the interface velocity $v_n^{\alpha/\beta}$.

Spatial derivative of ξ

Similarly, the gradient of f_α is given by:

$$\mathbf{grad}\, f_\alpha = \frac{1}{\Delta V} \int_{\Delta V} \mathbf{grad}\, \chi^\alpha(\boldsymbol{x}, t)\, \mathrm{d}V \tag{5.31}$$

the gradient of χ^α being again zero everywhere except at the interface α/β. More precisely:

$$\mathbf{grad}\, \chi^\alpha(\boldsymbol{x}, t) = -\boldsymbol{n}^{\alpha/\beta}\delta(\boldsymbol{x} - \boldsymbol{x}^{\alpha/\beta}) \tag{5.32}$$

The integration of this function leads again to the reduction of a volume integral (5.31) to an integral on the interface α/β:

$$\mathbf{grad}\, f_\alpha = -\frac{1}{\Delta V} \int_{\partial\Omega^{\alpha/\beta}} \boldsymbol{n}^{\alpha/\beta}\, \mathrm{d}S \tag{5.33}$$

Proceeding as for the temporal derivative of an average value, the gradient of the average of a value ξ can be expressed:

$$\mathbf{grad}\, \langle \xi^\alpha \rangle = \langle \mathbf{grad}\, \xi^\alpha \rangle - \frac{1}{\Delta V} \int_{\partial\Omega^{\alpha/\beta}} \xi^{\alpha *} \boldsymbol{n}^{\alpha/\beta}\, \mathrm{d}S = \langle \mathbf{grad}\, \xi^\alpha \rangle - \langle \xi^\alpha \boldsymbol{n}^{\alpha/\beta} \rangle^* \tag{5.34}$$

It is thus equal to the average of the gradient less an average on the interface of the product of $\xi^{\alpha *}$ and the normal vector at the interface, $\boldsymbol{n}^{\alpha/\beta}$. The same type of relation is obtained with the divergence of an averaged vector value:

$$\begin{aligned} \mathrm{div}\, \langle \boldsymbol{\xi}^\alpha \rangle &= \langle \mathrm{div}\, \boldsymbol{\xi}^\alpha \rangle - \frac{1}{\Delta V} \int_{\partial\Omega^{\alpha/\beta}} \boldsymbol{\xi}^{\alpha *} \cdot \boldsymbol{n}^{\alpha/\beta}\, \mathrm{d}S = \langle \mathrm{div}\, \boldsymbol{\xi}^\alpha \rangle - \langle \boldsymbol{\xi}^\alpha \cdot \boldsymbol{n}^{\alpha/\beta} \rangle^* \\ &= \langle \mathrm{div}\, \boldsymbol{\xi}^\alpha \rangle - \langle \xi_\mathrm{n}^\alpha \rangle^* \end{aligned} \tag{5.35}$$

where ξ_n^α is the normal component along $\boldsymbol{n}^{\alpha/\beta}$ of the vector value $\boldsymbol{\xi}$.

5.2.7 Average continuity equations

With the aid of the general relations obtained in the preceding section, we will now derive the average equations for a system constituted of two phases α and β evolving in time. The conservation equations at the interface obtained in section 5.2.4 will naturally appear in these average equations.

Mass conservation equation

The equation of conservation of mass averaged over two phases is the simplest to derive as it includes neither diffusive nor source terms. Locally, the equation is written (chap. 1):

$$\frac{\partial \rho}{\partial t} + \operatorname{div}(\rho \boldsymbol{v}) = 0 \tag{5.36}$$

where ρ is the density and $\boldsymbol{v}$ the velocity of the medium under consideration. The average of this equation over the volume $\Delta V^{\alpha} \equiv \Omega^{\alpha} \cap \Delta V$ is obtained by multiplication by the distribution function of the phase α, $\chi^{\alpha}(\boldsymbol{x}, t)$, and integration over ΔV:

$$\frac{1}{\Delta V}\int_{\Delta V} \frac{\partial \rho}{\partial t}\chi^{\alpha}\, dV + \frac{1}{\Delta V}\int_{\Delta V} \operatorname{div}(\rho \boldsymbol{v})\chi^{\alpha}\, dV = 0 \tag{5.37}$$

This equation can be transformed according to relations (5.30) and (5.35) previously deduced:

$$\frac{\partial \langle \rho^{\alpha} \rangle}{\partial t} - \langle \rho^{\alpha} v_n^{\alpha/\beta} \rangle^* + \operatorname{div}\langle \rho^{\alpha} \boldsymbol{v}^{\alpha} \rangle + \langle \rho^{\alpha} v_n^{\alpha} \rangle^* = 0 \tag{5.38}$$

Similarly, the average over the phase β in the element of volume ΔV leads to the relation:

$$\frac{\partial \langle \rho^{\beta} \rangle}{\partial t} + \langle \rho^{\beta} v_n^{\alpha/\beta} \rangle^* + \operatorname{div}\langle \rho^{\beta} \boldsymbol{v}^{\beta} \rangle - \langle \rho^{\beta} v_n^{\beta} \rangle^* = 0 \tag{5.39}$$

Note that the signs appearing in front of the averages on the interface change according to the convention for the direction of the normal $\boldsymbol{n}^{\alpha/\beta}$ at the interface (outgoing from the domain Ω^{α} and entering into Ω^{β}). Adding the averages obtained on the two sub-domains:

$$\frac{\partial}{\partial t}[\langle \rho^{\alpha} \rangle + \langle \rho^{\beta} \rangle] + \operatorname{div}[\langle \rho^{\alpha} \boldsymbol{v}^{\alpha} \rangle + \langle \rho^{\beta} \boldsymbol{v}^{\beta} \rangle] + \langle (\rho^{\beta} - \rho^{\alpha}) v_n^{\alpha/\beta} + \rho^{\alpha} v_n^{\alpha} - \rho^{\beta} v_n^{\beta} \rangle^* = 0 \tag{5.40}$$

As can be seen, the term averaged over the interface and appearing inside the brackets "$\langle\,\rangle^*$" is none other than the conservation of mass at the interface derived in section 5.2.4 (5.10). It is thus null. The averages appearing in the other terms can be replaced by the intrinsic averages, using $\langle \xi^{\alpha} \rangle = f_{\alpha} \langle \xi \rangle^{\alpha}$ (5.26). Finally:

$$\frac{\partial \langle \rho \rangle}{\partial t} + \operatorname{div}\langle \rho \boldsymbol{v} \rangle = 0 \tag{5.41}$$

with the usual definition of the average values:

$$\langle \rho \rangle = \langle \rho^{\alpha} \rangle + \langle \rho^{\beta} \rangle = f_{\alpha} \langle \rho \rangle^{\alpha} + f_{\beta} \langle \rho \rangle^{\beta} = \frac{1}{\Delta V}\int_{\Delta V} (\rho \chi^{\alpha} + \rho \chi^{\beta})\, dV = \frac{1}{\Delta V}\int_{\Delta V} \rho\, dV \tag{5.42}$$

$$\langle \rho v \rangle = f_\alpha \langle \rho v \rangle^\alpha + f_\beta \langle \rho v \rangle^\beta = \frac{1}{\Delta V} \int_{\Delta V} (\rho v \chi^\alpha + \rho v \chi^\beta)\, \mathrm{d}V$$

$$= \frac{1}{\Delta V} \int_{\Delta V} \rho v \, \mathrm{d}V \tag{5.43}$$

The average mass continuity equation thus has exactly the same form as the equation for a single phase system written in chapter 1 with the mass density, ρ, and the momentum density, ρv, being replaced by the average values defined in (5.42) and (5.43). The physical signification of the averaged equation (5.41) is simple enough: supposing that the two phases, α and β, do not have the same mass density but that both are constant ($\rho^\alpha = \mathrm{cnst}$ and $\rho^\beta = \mathrm{cnst}$, $\rho^\alpha \neq \rho^\beta$), the two phase media is no longer "incompressible" in the sense that its average mass density $\langle \rho \rangle$ is no longer constant. Therefore movements of material are necessary ($\rho v \neq 0$) to accommodate the variations of mass density related to phase transformations.

A simple case is that of a solidifying medium (a liquid phase, $\beta \equiv \ell$, and a solid phase $\alpha \equiv s$). Assuming that only the liquid moves, and that the mass densities of the solid and liquid phases are constant ($\rho^s = \mathrm{cnst}$ and $\rho^\ell = \mathrm{cnst}$, $\rho^s \neq \rho^\ell$), the equation of conservation of mass becomes, after division by ρ^ℓ:

$$\left[\frac{\rho^s}{\rho^\ell} - 1\right] \frac{\partial f_s}{\partial t} + \mathrm{div}(f_\ell \langle v \rangle^\ell) = 0 \tag{5.44}$$

The first term represents the ***shrinkage upon solidification***: as $\rho^s > \rho^\ell$ for most metals and alloys, this term is positive if the solid fraction f_s increases (solidification). In this case, the liquid must move to feed the growth front, in the opposite direction of the front velocity: this liquid flux is represented by the second term of (5.44) (negative divergence). In the case of a solid-liquid mixture where the solid is fixed (fig. 5.13), the liquid has to flow between the solid particles, which can be considered to be a *porous medium*. The flow in such a medium is often modeled by Darcy's law, which will be derived in more detail during the averaging of the momentum:

$$\langle v \rangle = f_\ell \langle v \rangle^\ell = -\frac{K(f_s, d_\mu)}{\mu} (\mathbf{grad}\, p - \rho^\ell \boldsymbol{g}) \tag{5.45}$$

The velocity $\langle v \rangle$ is the macroscopic velocity of the fluid, that is, the average real velocity of the fluid between the solid particles, $\langle v \rangle^\ell$, weighted by the liquid fraction, f_ℓ. In the context of the analysis of water flows in porous media performed by Darcy to calculate the flux of the fountains of Dijon, $\langle v \rangle$ is also called the ***effective velocity***. $K(f_s, d_\mu)$ is the ***permeability*** of the medium, which strongly depends on the solid fraction, f_s, and the spacing between the solid particles, d_μ (characteristic size of the microstructure). μ is the dynamic viscosity, p the pressure and $\boldsymbol{g}$ the

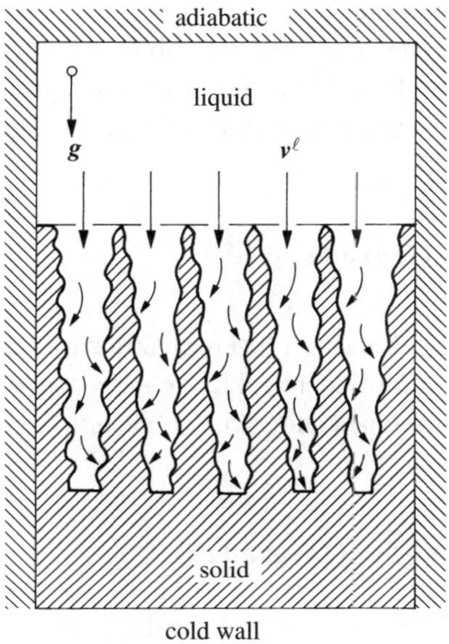

Figure 5.13 Liquid flow in a two phase region composed of dendrites.

gravitational vector. Note that the real velocity of the liquid $\langle v \rangle^\ell$ could have been used in Darcy's equation, with a consequent modification of the definition of the permeability. The insertion of (5.45) into the conservation of mass leads to an equation expressed in pressure:

$$\left[\frac{\rho^s}{\rho^\ell} - 1\right]\frac{\partial f_s}{\partial t} - \operatorname{div}\left[\frac{K}{\mu}(\mathbf{grad}\, p - \rho^\ell g)\right] = 0 \tag{5.46}$$

Such an equation allows the calculation of the pressure drop in the liquid medium inside the porous solid, as long as:

- the shrinkage term is known, for example by a thermal calculation giving $f_s(t)$,
- a model of the permeability $K(f_s, d_\mu)$ is prescribed (we will return later to this value for the averaged momentum equation),
- the boundary conditions are specified. At the boundary separating the two-phase medium from the liquid, the pressure could be specified, for example, as being given by the metallostatic pressure (Dirichlet condition). At the end of the two-phase zone, a eutectic precipitation between the dendrites imposes a condition of type (5.11) for the normal velocity of the fluid (Neumann condition for the normal component of the pressure gradient).

Solute conservation equation

The same approach can be applied to establish the equation for the solute of an alloy averaged over two phases. For that, it is sufficient to take the local conservation equation with c representing the volumetric concentration ($c = \rho w$, see chapter 1):

$$\frac{\partial c}{\partial t} + \mathrm{div}\,(c\boldsymbol{v}) - \mathrm{div}\,(D\,\mathbf{grad}\,c) = 0 \tag{5.47}$$

where the internal chemical reaction (source) term has been omitted for simplicity. The first two terms have a form identical to the mass conservation equation, and the last term can again be developed using (5.35). Then, for the average over the phase α:

$$\frac{\partial\langle c^\alpha\rangle}{\partial t} + \mathrm{div}\,\langle c^\alpha \boldsymbol{v}^\alpha\rangle - \mathrm{div}\,\langle D^\alpha\,\mathbf{grad}\,c^\alpha\rangle$$
$$- \langle c^\alpha v_n^{\alpha/\beta}\rangle^* + \langle c^\alpha v_n^\alpha\rangle^* - \langle D^\alpha\,\mathbf{grad}\,c^\alpha \cdot \boldsymbol{n}\rangle^* = 0 \tag{5.48}$$

A similar relation can be obtained for the average over the sub-domain ΔV^β. The sum of the contributions over the two sub-domains ΔV^α and ΔV^β gives finally:

$$\frac{\partial\langle c\rangle}{\partial t} + \mathrm{div}\,\langle c\boldsymbol{v}\rangle - \mathrm{div}\,\langle D\,\mathbf{grad}\,c\rangle$$
$$+ \left\langle (c^\beta - c^\alpha)v_n^{\alpha/\beta} - c^\beta v_n^\beta + c^\alpha v_n^\alpha + D^\beta\frac{\partial c^\beta}{\partial n} - D^\alpha\frac{\partial c^\alpha}{\partial n}\right\rangle^* = 0 \tag{5.49}$$

where the notation $\partial c/\partial n$ has been employed to represent the normal derivative of the concentration gradient in each of the phases. The surface term appearing again between the brackets "$\langle\ \rangle^*$" is identical to the solute conservation equation obtained for the moving interface (5.12). The solute conservation equation has thus again the same form as that written for a single phase domain:

$$\frac{\partial\langle c\rangle}{\partial t} + \mathrm{div}\,\langle c\boldsymbol{v}\rangle - \mathrm{div}\,\langle D\,\mathbf{grad}\,c\rangle = 0 \tag{5.50}$$

with the definition of the averaged values:

$$\langle c\rangle = \langle c^\alpha\rangle + \langle c^\beta\rangle = f_\alpha\langle c\rangle^\alpha + f_\beta\langle c\rangle^\beta = \frac{1}{\Delta V}\int_{\Delta V} c\;\mathrm{d}V \tag{5.51}$$

$$\langle c\boldsymbol{v}\rangle = f_\alpha\langle c\boldsymbol{v}\rangle^\alpha + f_\beta\langle c\boldsymbol{v}\rangle^\beta = \frac{1}{\Delta V}\int_{\Delta V} c\boldsymbol{v}\;\mathrm{d}V \tag{5.52}$$

$$\langle D\,\mathbf{grad}\,c\rangle = f_\alpha\langle D\,\mathbf{grad}\,c\rangle^\alpha + f_\beta\langle D\,\mathbf{grad}\,c\rangle^\beta = \frac{1}{\Delta V}\int_{\Delta V} D\,\mathbf{grad}\,c\;\mathrm{d}V \tag{5.53}$$

Before describing a few particular cases relative to the average solute conservation, it is first appropriate to make an important remark about the last term of (5.50). This term – as also the transport term – intervenes in solutions of the averaged continuity equation at the process scale. However, the diffusion of chemical species at this scale is very slow (typically 10^5 – 10^8 seconds on the scale of a centimeter!). It thus appears reasonable to neglect this term at the macroscopic scale. As for the transport term, it is only important if a liquid phase is present or in the case of large deformations. If the transport and diffusion terms in (5.50) are null or negligible, the local solute conservation equation is that of a ***closed system***: the average concentration $\langle c\rangle$ is constant and equal to the nominal composition of the alloy, c_0. Finally, it should be noted that this solute equation, along with the phase diagram, affects the *transformation path* of the material, that is, the trajectory in time of a volume element ΔV in the phase diagram. A few particular cases are briefly cited below.

- If the diffusion at the scale of the small volume element ΔV is sufficiently rapid in the two phases, the average intrinsic local concentrations are then equal to those at the interface ($\langle c\rangle^\nu \simeq c^{\nu *} = c^\nu$, $\nu = \alpha, \beta$), the latter being imposed by the phase diagram. With this assumption, the well known ***lever rule*** is obtained for phase transformations:

$$f_\alpha c^\alpha + f_\beta c^\beta = f_\alpha c^\alpha + (1-f_\alpha)c^\beta = c_0 \tag{5.54}$$

where c_0 is the nominal composition of the alloy. In the case of a very simple binary phase diagram (fig. 5.14), the equilibrium concentrations c^α and c^β are given as functions of the temperature by the corresponding equilibrium lines. The local temperature of the two phases being the same (equilibrium), the following linear relations can be written:

$$T = T^{\alpha/\beta}(c^\alpha) = T^{\alpha/\beta}_{(0)} + m^\alpha c^\alpha = T^{\beta/\alpha}(c^\beta) = T^{\alpha/\beta}_{(0)} + m^\beta c^\beta \tag{5.55}$$

The lines $T^{\alpha/\beta}(c^\alpha)$ and $T^{\beta/\alpha}(c^\beta)$, with slopes m^α and m^β, respectively, are the solidus and liquidus in the case of an equilibrium between a solid phase and the liquid, the temperature $T^{\alpha/\beta}_{(0)}$ being the equilibrium temperature of the two phases α and β of the pure material (the melting temperature for solidification). In the case of a simple linear diagram (fig. 5.14), the partition coefficient, $k^{\alpha/\beta}$, between the two phases is defined indiscriminately as the ratio of either the concentrations or of the slopes:

$$k^{\alpha/\beta} = \frac{c^\alpha}{c^\beta} = \frac{m^\beta}{m^\alpha} \tag{5.56}$$

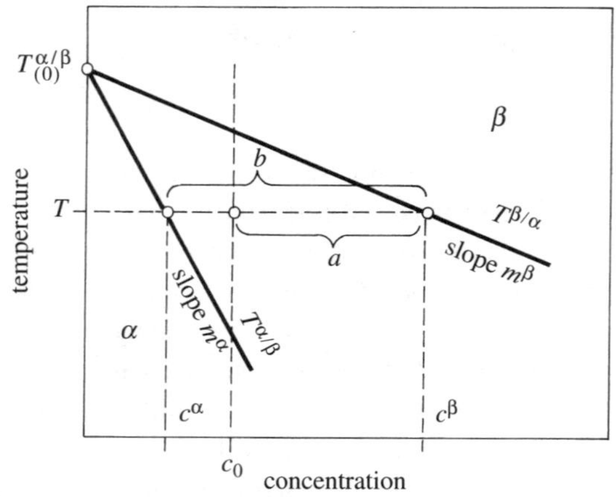

Figure 5.14 Ideal phase diagram showing the principle of the lever rule.

Equation (5.54) then becomes:

$$f_\alpha = \frac{c^\beta - c_0}{c^\beta - c^\alpha} = \frac{c^\beta - c_0}{c^\beta(1 - k^{\alpha/\beta})} = \frac{1}{1 - k^{\alpha/\beta}} \frac{T - T^{\beta/\alpha}(c_0)}{T - T^{\alpha/\beta}_{(0)}} \tag{5.57}$$

For $c^\beta = c_0$, the fraction f_α is zero while $f_\alpha = 1$ for $c^\beta = c_0/k^{\alpha/\beta}$. Between these two extreme values, the transformed fraction f_α is the ratio between the two segments a and b of figure 5.14, from which the name: lever rule.

- Another situation often encountered in phase transformations is that of a phase β in which the diffusion is rapid at the scale d_μ ($D^\beta \simeq \infty$) and a phase α for which there is practically no diffusion ($D^\alpha \simeq 0$). This approximation, named ***Scheil*** or ***Scheil-Gulliver***, describes relatively well the solidification path of the majority of alloys, with the exception however of the iron-carbon system, for which the diffusion of carbon in the solid state is significant. In this case, the concentration in phase β, here, the liquid, is always such that: $\langle c \rangle^\beta \simeq c^{\beta *} = c^\beta$, while the average concentration of phase α is given by (fig. 5.15):

$$\langle c^\alpha \rangle = \frac{1}{\Delta V} \int_{\Delta V} c^\alpha(\boldsymbol{x}) \chi^\alpha \, \mathrm{d}V = \frac{1}{\Delta V} \int_{\Delta V^\alpha} c^\alpha(\boldsymbol{x}) \, \mathrm{d}V^\alpha = \int_0^{f_\alpha(t)} c^\alpha(f) \, \mathrm{d}f \tag{5.58}$$

Although the concentration profile is "fixed" in phase α, the average concentration $\langle c^\alpha \rangle$ depends on time as the domain ΔV^α varies during the transformation. This dependence appears explicitly in the upper limit of integration when the integral over dV is replaced by an integral over the volumetric fraction df (df = d$V^\alpha/\Delta V$ and $f_\alpha(t) = \Delta V^\alpha/\Delta V$). The concentration at the interface in the phase α appears in the time derivative of this equation:

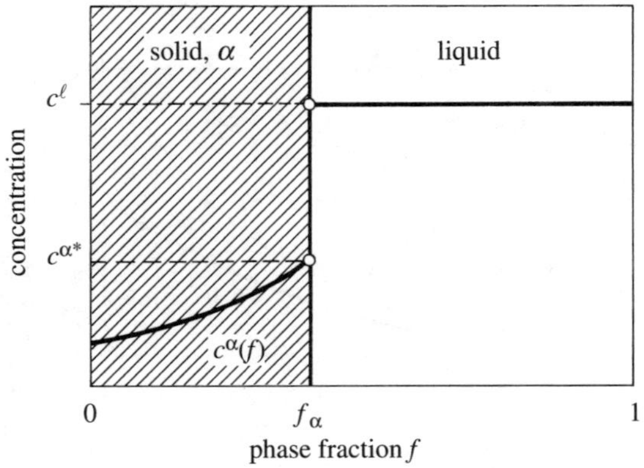

Figure 5.15 Concentration profile during solidification of an alloy according to Scheil's model.

$$\frac{\partial\langle c^\alpha\rangle}{\partial t} = c^\alpha(f_\alpha)\frac{\mathrm{d}f_\alpha}{\mathrm{d}t} = c^{\alpha*}\frac{\mathrm{d}f_\alpha}{\mathrm{d}t} \tag{5.59}$$

Inserting this relation in the macroscopic conservation equation with neither transport nor diffusion (5.50), taking into account that c^β is uniform and using the equilibrium condition at the interface (5.56):

$$c^{\alpha*}\frac{\mathrm{d}f_\alpha}{\mathrm{d}t} + \frac{\mathrm{d}}{\mathrm{d}t}[c^\beta(1-f_\alpha)] = -\frac{\mathrm{d}f_\alpha}{\mathrm{d}t}[c^\beta(1-k^{\alpha/\beta})] + (1-f_\alpha)\frac{\mathrm{d}c^\beta}{\mathrm{d}t} = 0 \tag{5.60}$$

(Note that the temporal derivatives are not partial derivatives at the level of the small, closed local volume elements used for the average). The integration of this equation with the initial conditions $c^\beta(f_\alpha = 0) = c_0$ gives Scheil's relation:

$$f_\alpha = 1 - \left[\frac{c^\beta}{c_0}\right]^{1/(k^{\alpha/\beta}-1)} = 1 - \left[\frac{T - T_{(0)}^{\alpha/\beta}}{T^{\beta/\alpha}(c_0) - T_{(0)}^{\alpha/\beta}}\right]^{1/(k^{\alpha/\beta}-1)} \tag{5.61}$$

The transformation path thus obtained is compared with that calculated with the lever rule in figure 5.16 for an aluminum-7wt% silicon alloy. It should be noted that Scheil's model (5.61) is no longer for an equilibrium transformation path – even if the interface is at equilibrium – as the diffusion in the solid is not complete. Therefore, the transformed fraction $f_\alpha = 1$ is only attained for $c^\beta \to \infty$. Before attaining this unphysical limit, there will be precipitation of another phase γ, for example of silicon in the case of the selected alloy (eutectic reaction).

- The solute conservation equations for a closed system with partial diffusion in one or two phases must generally be solved numerically. Analytic solutions have been derived for some particular cases in one spatial dimension (f_α can be replaced by a spatial coordinate).

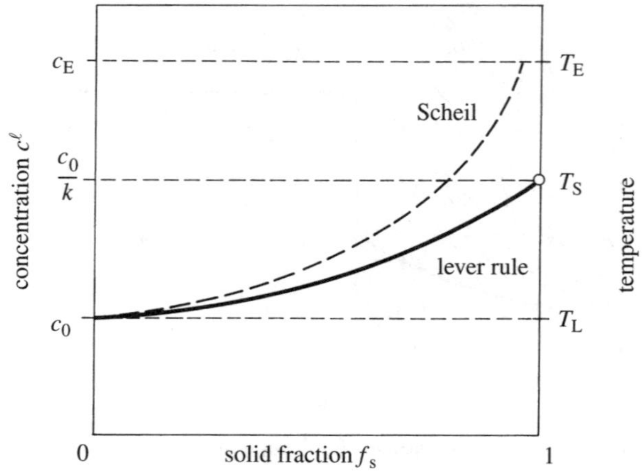

Figure 5.16 Liquid concentration profiles, $c^\ell(f_s)$, obtained with the lever rule and Scheil's model for solidification of an alloy.

Heat conservation equation

The equation of conservation of heat (chap. 1) without a source term becomes, with Fick's second law:

$$\frac{\partial(\rho H)}{\partial t} + \mathrm{div}\,(\rho H \boldsymbol{v}) - \mathrm{div}\,(\kappa\, \mathbf{grad}\, T) = 0 \tag{5.62}$$

Having exactly the same form as the equation of solute conservation (5.47), the averaging of the two phases α and β leads to a similar expression:

$$\frac{\partial\langle\rho H\rangle}{\partial t} + \mathrm{div}\,\langle\rho H \boldsymbol{v}\rangle - \mathrm{div}\,\langle\kappa\, \mathbf{grad}\, T\rangle$$
$$+ \left\langle(\rho^\beta H^\beta - \rho^\alpha H^\alpha) v_n^{\alpha/\beta} - \rho^\beta H^\beta v_n^\beta + \rho^\alpha H^\alpha v_n^\alpha + \kappa^\beta \frac{\partial T^\beta}{\partial n} - \kappa^\alpha \frac{\partial T^\alpha}{\partial n}\right\rangle^* = 0 \tag{5.63}$$

The term appearing between the brackets "$\langle\ \rangle^*$" for the moving interface evaluates to zero again (5.14), and the averaged heat conservation equation takes the form:

$$\frac{\partial\langle\rho H\rangle}{\partial t} + \mathrm{div}\,\langle\rho H \boldsymbol{v}\rangle - \mathrm{div}\,\langle\kappa\, \mathbf{grad}\, T\rangle = 0 \tag{5.64}$$

with the definitions of the averaged values:

$$\langle\rho H\rangle = \langle(\rho H)^\alpha\rangle + \langle(\rho H)^\beta\rangle = f_\alpha\langle\rho H\rangle^\alpha + f_\beta\langle\rho H\rangle^\beta = \frac{1}{\Delta V}\int_{\Delta V} \rho H \,\mathrm{d}V \tag{5.65}$$

$$\langle \rho H v \rangle = f_\alpha \langle \rho H v \rangle^\alpha + f_\beta \langle \rho H v \rangle^\beta = \frac{1}{\Delta V} \int_{\Delta V} \rho H v \, dV \tag{5.66}$$

$$\langle \kappa \, \mathbf{grad}\, T \rangle = f_\alpha \langle \kappa \, \mathbf{grad}\, T \rangle^\alpha + f_\beta \langle \kappa \, \mathbf{grad}\, T \rangle^\beta = \frac{1}{\Delta V} \int_{\Delta V} \kappa \, \mathbf{grad}\, T \, dV \tag{5.67}$$

Although having the same form as the average solute equation, the average heat equation nonetheless presents important differences in the physical implications. First of all, the "macroscopic" diffusion term cannot generally be neglected at the scale of a process or a piece. The local temperature (in the sense of ΔV) in the two phases being the same (equilibrium condition), (5.67) is replaced by:

$$\langle \kappa \, \mathbf{grad}\, T \rangle = \langle \kappa \rangle \, \mathbf{grad}\, T \qquad \text{with } \langle \kappa \rangle = f_\alpha \langle \kappa \rangle^\alpha + f_\beta \langle \kappa \rangle^\beta \tag{5.68}$$

where $\langle \kappa \rangle$ is none other than the thermal conductivity averaged over the phases.

The averaged volumetric enthalpy, $\langle \rho H \rangle$, can be expressed in terms of the specific heats of each of the phases and the latent heat of the transformation $(\rho L)^{\alpha/\beta}$:

$$\langle \rho H \rangle = f_\alpha \int_0^T (\rho c_p)^\alpha \, d\theta + f_\beta \left[\int_0^T (\rho c_p)^\beta \, d\theta + (\rho L)^{\alpha/\beta} \right] \tag{5.69}$$

This average volumetric enthalpy is a function of the transformation path via the volumetric phase fraction $f_\alpha(c^\beta)$ or $f_\alpha(T)$ (for example (5.61) for Scheil's model). It is shown in figure 5.17(a) again for the case of an aluminum-7wt% silicon alloy. The liberation of latent heat near the liquidus and during the eutectic reaction is clearly apparent in this figure.

Defining the average specific heat per unit of volume, $\langle \rho c_p \rangle$ by:

$$\int_0^T \langle \rho c_p \rangle \, d\theta = f_\alpha \int_0^T (\rho c_p)^\alpha \, d\theta + f_\beta \int_0^T (\rho c_p)^\beta \, d\theta \tag{5.70}$$

and assuming that it does not depend explicitly on time, the temporal variation of the average volumetric enthalpy becomes:

$$\frac{\partial \langle \rho H \rangle}{\partial t} = \langle \rho c_p \rangle \frac{\partial T}{\partial t} - (\rho L)^{\alpha/\beta} \frac{\partial f_\alpha}{\partial t} \tag{5.71}$$

Two contributions appear: the first is related to a variation of the average specific heat over the two phases, while the second can be assimilated with a source term for heat, related to the evolution of the latent heat release during the transfor-

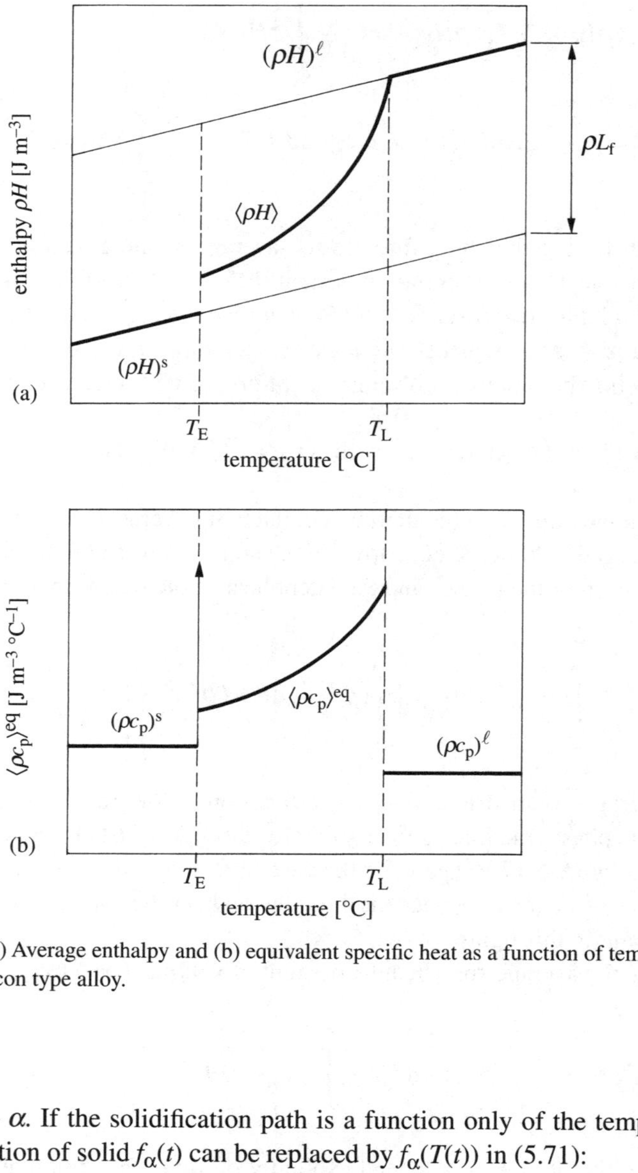

Figure 5.17 (a) Average enthalpy and (b) equivalent specific heat as a function of temperature for an aluminum-silicon type alloy.

mation $\beta \rightarrow \alpha$. If the solidification path is a function only of the temperature, the volume fraction of solid $f_\alpha(t)$ can be replaced by $f_\alpha(T(t))$ in (5.71):

$$\frac{\partial \langle \rho H \rangle}{\partial t} = \langle \rho c_p \rangle \frac{\partial T}{\partial t} - (\rho L)^{\alpha/\beta} \frac{\mathrm{d} f_\alpha}{\mathrm{d} T} \frac{\partial T}{\partial t} = \langle \rho c_p \rangle^{eq} \frac{\partial T}{\partial t} \tag{5.72}$$

The ***equivalent average specific heat***, $\langle \rho c_p \rangle^{eq}$, defined as follows:

$$\langle \rho c_p \rangle^{eq} = \langle \rho c_p \rangle - (\rho L)^{\alpha/\beta} \frac{\mathrm{d} f_\alpha}{\mathrm{d} T} \tag{5.73}$$

includes the contributions of the specific heat and the latent heat. It allows the expression of the average enthalpy variations as a function of those of the temperature alone. This simplicity of formulation eases the treatment of phase transformations in commercial programs that solve the heat diffusion equation. It has, however, a few problems as shown in figure 5.17(b): by its definition, the equivalent specific heat manifests very strong variations near the start of a transformation (for example near the liquidus), or even singularities during an isothermal transformation (for example on the eutectic plateau). These very strong non-linearities of the equivalent specific heat can make the numerical scheme non conservative. We will return to these aspects in section 5.4.

The average heat equation can then be written as follows:

$$\langle \rho c_{\mathrm{p}} \rangle \frac{\partial T}{\partial t} - (\rho L)^{\alpha/\beta} \frac{\partial f_{\alpha}}{\partial t} + \operatorname{div} \langle \rho H \boldsymbol{v} \rangle - \operatorname{div}(\langle \kappa \rangle \, \mathbf{grad}\, T) = 0 \tag{5.74}$$

As for the conservation of solute, the term for heat transport by convection $\langle \rho H \boldsymbol{v} \rangle$ only intervenes if at least one phase is moving: this is the case for a liquid phase or for large deformations of a solid. Two limiting cases can be envisioned: i) the two phases are transported at the same velocity ($\boldsymbol{v}^{\alpha} = \boldsymbol{v}^{\beta}$); ii) one of the phases is immobile in the chosen reference frame (for example $\boldsymbol{v}^{\alpha} = 0$).

- For the case where $\boldsymbol{v}^{\alpha} = \boldsymbol{v}^{\beta} = \boldsymbol{v}$, all the enthalpy of the material is transported with this velocity, including the latent heat of transformation. The corresponding term can be put in the form:

$$\operatorname{div} \langle \rho H \boldsymbol{v} \rangle = \operatorname{div}[\boldsymbol{v} \langle \rho H \rangle] \tag{5.75}$$

where $\langle \rho H \rangle$ is given by (5.69). If, in addition, the densities of the phases α and β are equal and constant ($\operatorname{div} \boldsymbol{v} = 0$), this term is equal to:

$$\operatorname{div} \langle \rho H \boldsymbol{v} \rangle = \rho \boldsymbol{v} \cdot \mathbf{grad}\, \langle H \rangle \tag{5.76}$$

- For the case where $\boldsymbol{v}^{\alpha} = 0$ and $\rho^{\alpha} = \rho^{\beta} = \rho = \text{constant}$, the conservation of mass gives $\operatorname{div}\langle \boldsymbol{v} \rangle = \operatorname{div}(f_{\beta} \boldsymbol{v}^{\beta}) = 0$. With the definition of H^{β} given by the term in square brackets in equation (5.69), the transport term in the average heat equation can then be written for these assumptions:

$$\operatorname{div} \langle \rho H \boldsymbol{v} \rangle = \langle \boldsymbol{v} \rangle \cdot \mathbf{grad}\, (\rho H)^{\beta} = (\rho c_{\mathrm{p}})^{\beta} f_{\beta} \langle \boldsymbol{v} \rangle^{\beta} \cdot \mathbf{grad} T \tag{5.77}$$

In this case, only the specific heat of the phase in motion β is transported with the average velocity (or flux) of the fluid $\langle \boldsymbol{v} \rangle = f_{\beta} \langle \boldsymbol{v} \rangle^{\beta}$ (we have assumed that the latent heat of the transformation appearing in (5.69) is constant as well as c_{p}^{β}).

Momentum conservation equation

This average equation is only interesting if one of the two phases is in motion relative to the other. For example, this situation can be encountered for solid particles in suspension in a liquid, for a vapor phase coexisting with the liquid or for two liquid phases either miscible or inmiscible. Many specialized texts having been written for flows in *two phase media*, we restrict ourselves here to the treatment of the case of a fixed solid phase α in the presence of a liquid phase β: this is the approximation of a ***packed bed***. Unlike the previously derived average equations for which the behavioral laws of the two phases were similar (for example Fick's first law applied to each phase with different diffusion coefficients), very different behavioral laws for the two media appear in the conservation equation for momentum. To simplify, we assume that the solid is non deformable and the liquid Newtonian. In addition, we assume that $\rho^s = \rho^\ell = \rho$ (we use the indices "s" and "ℓ" to distinguish the solid and liquid phases, respectively). Using the general relations deduced in section 5.2.6, the following contributions appear in the Navier-Stokes equation along the x axis (1.102) averaged over the liquid phase ($\beta \equiv \ell$):

- $$\left\langle \frac{\partial(\rho^\ell v_x^\ell)}{\partial t} \right\rangle = \frac{\partial \langle \rho^\ell v_x^\ell \rangle}{\partial t} + \langle \rho^\ell v_x^\ell v_n^{s/\ell} \rangle^* \tag{5.78}$$

(The normal is taken as entering the liquid). The solid being immobile and the densities of the phases being equal, we deduce from the mass equation (5.10) that $v_n^\ell = 0$ at the solid-liquid interface. Also, with the assumption of continuity of the tangential shear stresses in the two phases, the momentum balance at the interface (5.17) implies that $v_\tau^\ell = v_\tau^s = 0$. This means that the fluid velocity is zero at every point of the solid-liquid interface (no slip condition). The surface average term appearing in (5.78) is thus null.

- $$\langle \mathrm{div}\,(\rho^\ell v_x^\ell \boldsymbol{v}^\ell) \rangle = \mathrm{div}\,\langle \rho^\ell v_x^\ell \boldsymbol{v}^\ell \rangle - \langle \rho^\ell v_x^\ell \boldsymbol{v}^\ell \cdot \boldsymbol{n} \rangle^* \tag{5.79}$$

The last term is also null for the same reasons.

- Assuming constant viscosity, the viscous dissipation term is averaged as follows:

$$\begin{aligned} \mu\langle \mathrm{div}\,\mathbf{grad}\, v_x^\ell \rangle &= \mu\,\mathrm{div}\,\langle \mathbf{grad}\, v_x^\ell \rangle - \mu \langle \mathbf{grad}\, v_x^\ell \cdot \boldsymbol{n} \rangle^* \\ &= \mu\,\mathrm{div}\,\mathbf{grad}\,\langle v_x^\ell \rangle - \mu\,\mathrm{div}\,\langle v_x^\ell \boldsymbol{n} \rangle^* - \mu \langle \mathbf{grad}\, v_x^\ell \cdot \boldsymbol{n} \rangle^* \end{aligned} \tag{5.80}$$

Note that the general expressions obtained for the spatial derivative averages were used in two places. The term $\mu\,\mathrm{div}\,\langle v_x^\ell \boldsymbol{n} \rangle^*$ of this equation is again null, because of the non slip assumption for the liquid in contact with the solid. However, the last term $F_x^\mu = \mu \langle \mathbf{grad}\, v_x^\ell \cdot \boldsymbol{n} \rangle^*$ is not zero as we will see further on, and represents the ***drag*** exerted by the solid on the fluid flow through shear forces.

- The average of the volume forces yields simply:

$$\langle \rho^\ell g_x \rangle = \rho^\ell f_\ell g_x \tag{5.81}$$

- More care must be taken with the average of the pressure term. We can write:

$$\langle \frac{\partial p^\ell}{\partial x} \rangle = \frac{\partial \langle p^\ell \rangle}{\partial x} - \langle p^\ell n_x \rangle^* = \frac{\partial (f_\ell \langle p \rangle^\ell)}{\partial x} - \langle p^\ell n_x \rangle^* = f_\ell \frac{\partial \langle p \rangle^\ell}{\partial x} + F_x^p \tag{5.82}$$

The term F_x^p also represents a friction force of the solid on the fluid flow, but via the pressure variation (this term is called "form drag").

Regrouping all the contributions, using relation (5.26) relating an average to an intrinsic average and taking ρ^ℓ to be constant, we finally obtain:

$$\frac{\partial (\rho^\ell f_\ell \langle v_x \rangle^\ell)}{\partial t} + \operatorname{div}(\rho^\ell f_\ell \langle v_x \boldsymbol{v} \rangle^\ell) - \mu \Delta (f_\ell \langle v_x \rangle^\ell) + F_x^f = \rho^\ell f_\ell g_x - f_\ell \frac{\partial \langle p \rangle^\ell}{\partial x} \tag{5.83}$$

with

$$F_x^f = (F_x^\mu + F_x^p) = \mu \langle \mathbf{grad}\, v_x^\ell \cdot \boldsymbol{n} \rangle^* + \langle p \rangle^\ell \frac{\partial f_\ell}{\partial x} - \langle p^\ell n_x \rangle^* \tag{5.84}$$

We are now going to calculate the friction term F_x^p in an extremely simple situation in order to show that it can be reduced to an equation such as Darcy's, characterizing a fluid flow in porous media (5.45). We take a volume composed of a solid with cylindrical holes of radius R parallel to the direction x, and in which the fluid can flow (fig. 5.18). The liquid fraction is given by: $f_\ell = n\pi R^2$, where n is the surface density of holes in a transverse section of the flow. In each of these holes, we assume a stationary flow with a no slip condition on the walls (Hagen-Poiseuille flow). In this case, f_ℓ is independent of x and n_x is zero. Only the friction term F_x^μ remains. For a Newtonian fluid, the velocity profile inside each hole is given by (exercise 5.6.5):

$$v_x^\ell(r) = v_{max}\left[1 - \left(\frac{r}{R}\right)^2\right] = -\frac{\partial p^\ell}{\partial x}\frac{R^2}{4\mu}\left[1 - \left(\frac{r}{R}\right)^2\right] \tag{5.85}$$

where v_{max} is the maximum velocity at the center of the tube.

It is easy to show that the friction term F_x^μ appearing in (5.84) is given in this case by:

$$F_x^\mu = \mu \langle \frac{\partial v_x^\ell}{\partial n} \rangle^* = -\frac{\mu}{\Delta V} \int_{\partial \Omega^{s/\ell}} \left[\frac{\partial v_x^\ell}{\partial r}\right]_{r = R} dS = -\frac{\mu f_\ell}{\Delta V^\ell} \int_{\partial \Omega^{s/\ell}} \left[\frac{\partial v_x^\ell}{\partial r}\right]_{r = R} dS$$

$$= -f_\ell \frac{\partial p^\ell}{\partial x} \tag{5.86}$$

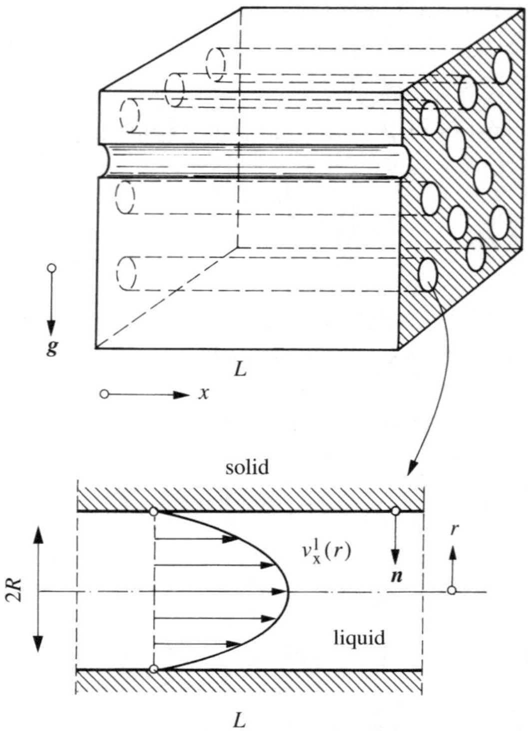

Figure 5.18 Flow in a porous medium constituted of equal diameter holes aligned with the x axis.

where ΔV is a characteristic volume element given by L/n, L being a length for the tube, and ΔV^ℓ the corresponding fluid volume ($\Delta V^\ell = \pi R^2 L$ and $\Delta S^\ell = 2\pi RL$). The minus sign comes from the direction of r opposite that of the normal entering into the liquid. It is seen that this term is none other than the pressure variation term appearing on the right hand side of equation (5.83). Replacing the pressure gradient in (5.86) with (5.85) by employing the intrinsic average velocity of the fluid v_x^ℓ inside a tube (we have in fact $\langle v_x \rangle^\ell$, equal to $v_{max}/2$, but the brackets indicating the average have been omitted to simplify the expression). We have then:

$$F_x^\mu = -f_\ell \frac{\partial p^\ell}{\partial x} = f_\ell \frac{8\mu v_x^\ell}{R^2} \tag{5.87}$$

with

$$v_x^\ell = \langle v_x \rangle^\ell = \frac{1}{\pi R^2} \int_0^R v_x^\ell(r) 2\pi r \, dr \tag{5.88}$$

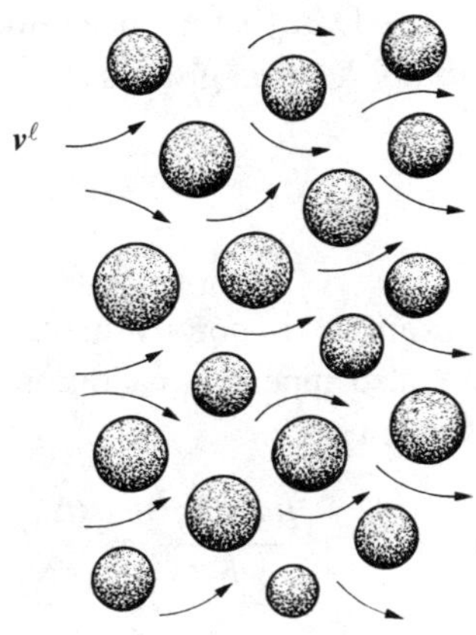

Figure 5.19 Flow in a tortuous porous medium.

For a system in which the liquid path is much more *tortuous* (fig. 5.19), an equivalent hydrodynamic radius, R_{eq}, is defined with respect to the effective section in which the liquid can flow and the wetted perimeter. For each cylinder of figure 5.18, one has then:

$$R_{eq} = \frac{\pi R^2}{2\pi R} = \frac{R}{2} \tag{5.89}$$

and

$$F_x^\mu = f_\ell \frac{2\mu v_x^\ell}{R_{eq}^2} \tag{5.90}$$

The equivalent hydrodynamic radius can also be expressed as the ratio of the effective volume of the liquid, ΔV^ℓ, to the solid-liquid interface area, $\Delta S^{s/\ell}$. Dividing by the volume element ΔV:

$$R_{eq} = \frac{\Delta V^\ell / \Delta V}{\Delta S^{s/\ell}/\Delta V} = \frac{f_\ell}{(\Delta S^{s/\ell}/\Delta V^s)(\Delta V^s/\Delta V)} = \frac{f_\ell}{(\Delta S^{s/\ell}/\Delta V^s) f_s}$$
$$= \frac{f_\ell}{d_\mu^{-1}(1-f_\ell)} = \frac{d_\mu f_\ell}{(1-f_\ell)} \tag{5.91}$$

The value d_μ is a characteristic value of the porous solid defined as the ratio of its surface area in contact with the fluid over its volume. Inserting (5.91) in (5.90):

$$F_x^\mu = f_\ell \frac{2\mu v_x^\ell}{R_{eq}^2} = f_l \frac{2(1-f_\ell)^2}{d_\mu^2 (f_\ell)^2} \mu v_x^\ell = -f_\ell \frac{\partial p^\ell}{\partial x} \tag{5.92}$$

We thus end up with a law like Darcy's for a one dimensional flow (5.45) without a gravity term, the permeability K being given by:

$$K(f_\ell, \mathrm{d}_\mu) = \frac{d_\mu^2 (f_\ell)^3}{2(1-f_\ell)^2} \tag{5.93}$$

In a general case, for example a porous medium consisting of spheres (fig. 5.19), the friction force F_x^p also appears. In addition, the flow may not be laminar and K is more generally written as:

$$K = A\frac{d_\mu^2 (f_\ell)^3}{(1-f_\ell)^2} \quad \text{and} \quad F_x^f = \frac{\mu v_x^\ell (f_\ell)^2}{K} = \frac{\mu v_x f_\ell}{K} \tag{5.94}$$

where A is a parameter and where the averaged velocity $\boldsymbol{v} = f_\ell \boldsymbol{v}^\ell$ has been introduced (actually $\langle v_x \rangle = f_\ell \langle v_x \rangle^\ell$). Going back to the notation where brackets are used for intrinsic average values, the average equation (5.83) of the momentum in the x direction becomes:

$$\frac{\partial(\rho^\ell \langle v_x \rangle)}{\partial t} + \frac{1}{f_\ell}\mathrm{div}(\rho^\ell \langle v_x \rangle \langle \boldsymbol{v} \rangle) - \mu\Delta\langle v_x \rangle + \frac{\mu \langle v_x \rangle f_\ell}{K} = \rho^\ell f_\ell g_x - f_\ell \frac{\partial p^\ell}{\partial x} \tag{5.95}$$

(Note that the gradients of f_ℓ have been neglected in the Stokes term as f_ℓ came from the divergence). It is seen that in the limit $f_\ell = 1$, the permeability K becomes infinite and we end up thus with the Navier-Stokes equation (1.102) established in chapter 1. If, however, the permeability becomes very small, the friction term becomes dominant on the left hand side of the equation (5.95) and we end up with Darcy's law (5.45) with a gravity term.

5.3 INITIAL AND BOUNDARY CONDITIONS

5.3.1 Generalities

Although the concentrations, or the pressure, play an important role in phase transformations, they are primarily the result of a *thermal history*, $T(t)$, of the material. The temperature variations can be imposed on a part placed in a furnace, in a homogenization or precipitation treatment, for example, or naturally introduced during a process. In some cases, *concentration variations* can be imposed at the material surface to induce transformations. Such is the case for nitriding or carburizing treatment of iron alloys.

From here on, this section dedicated to initial and boundary conditions will be essentially focused on ***thermal transfer*** at a free surface. When justified, the parallel with chemical transfers will be mentioned. The initial conditions will not be the object of a special section as they are either uniform and trivial (for example $T(\boldsymbol{x}, t = 0) = T_0$), or non uniform. In the latter case, they can come from an experimental analysis (for example the distribution of concentrations and phases in a metalographic cut before homogenization treatment) or be fixed by an earlier calculation (for example, inhomogeneous temperature distribution in an injected polymer deduced from a simulation of the filling of the mold).

It was seen in chapter 1 that the thermal boundary conditions for a domain could be of three types:

- *Dirichlet condition* (or essential condition): the surface temperature of the domain is imposed, $T(\boldsymbol{x}, t) = \overline{T}_a$ for $\boldsymbol{x} \in \partial\Omega$. Note that this imposed temperature can depend on time and position.
- *Neumann condition* (or natural condition): the heat flux, q_T, at the surface of the domain is imposed:

$$\kappa\frac{\partial T}{\partial n} = q_T \qquad \text{for } \boldsymbol{x} \in \partial\Omega \tag{5.96}$$

  (q_T is negative by convention if heat is exiting the domain). The particular case of null flux corresponds to an *adiabatic* condition or a *homogeneous Neumann* condition.
- *Cauchy condition* (or mixed natural condition): the heat flux at the surface of the domain is imposed via a thermal transfer coefficient, h:

$$\kappa\frac{\partial T}{\partial n} = -h(T - \overline{T}_a) \qquad \text{for } \boldsymbol{x} \in \partial\Omega \tag{5.97}$$

Convection and radiation are the principle physical mechanisms for thermal transfer and they are briefly presented in the following sections. For more details, the reader is referred to the much more complete books mentioned at the end of this chapter.

5.3.2 Transfer by forced convection

Consider first a horizontal plate over which a liquid or a gas flows (fig. 5.20). The fluid velocity far from the surface, v_∞, is approximately parallel to the surface and is assumed zero in contact with the surface (non slip condition). The region in which the velocity profile changes from 0 to v_∞ is called the ***boundary layer***. Its thickness, δ_v, increases with distance as indicated in figure 5.20. The non dimensional Reynolds number (chap. 10) that characterizes this flow is defined as follows:

$$Re_x = \frac{v_\infty x}{\nu} \tag{5.98}$$

where ν is the kinematic viscosity. The number is given with an index to indicate that it is a function of the position x in the flow direction. For a critical Reynolds

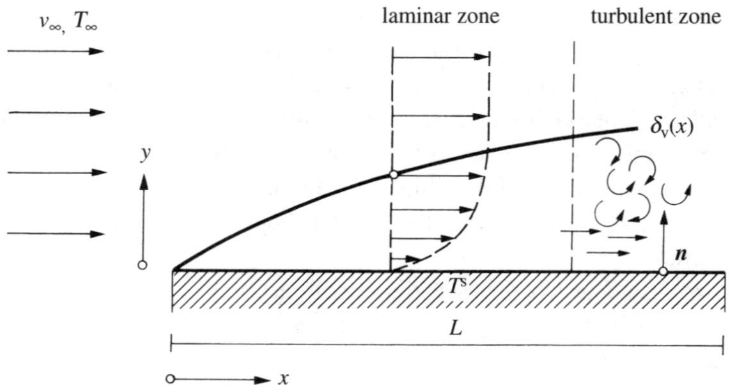

Figure 5.20 Development of a boundary layer in a flow parallel to a solid surface.

number between 10^5 and 10^6, the flow becomes turbulent: the velocity initially parallel to the surface develops a chaotic character. For increasing values of x along the surface, the flux can change from a laminar flow to a turbulent flow, as shown in figure 5.20 (chap. 7).

In the same way, if the surface of the solid does not have the same temperature as the fluid, a *thermal boundary layer* appears, with thickness δ_T growing with x. At the surface, the temperatures of the material and the fluid must be equal and the heat flux must be continuous. By extension of (5.97), this flux has the value:

$$\kappa^s \left[\frac{\partial T}{\partial n}\right]^s = -h(T^s - T_\infty) = \kappa^f \left[\frac{\partial T}{\partial n}\right]^f \quad \text{for } \boldsymbol{x} \in \partial\Omega \tag{5.99}$$

the indices "*s*" and "*f*" designate the solid medium (the material) and the fluid (gas or liquid) respectively. The temperature of the fluid, T_∞, is measured far from the interface while the temperature T^s is that of the solid surface, which is assumed imposed and uniform. The normal derivatives of the temperature profile are taken on each side of the interface, the normal $\boldsymbol{n}$ being defined as exiting the material (in the direction y in figure 5.20). As it is common in the theory of transport phenomena to introduce non dimensional numbers, the values of temperature, space, and velocity are normalized:

$$\tilde{T}(y) = \frac{T(y) - T^s}{T_\infty - T^s},\ \tilde{x} = \frac{x}{L},\ \tilde{y} = \frac{y}{L},\ \tilde{v}_x = \frac{v_x}{v_\infty},\ \tilde{v}_y = \frac{v_y}{v_\infty} \tag{5.100}$$

where L is the characteristic length of the surface along the flow. Equation (5.99) becomes:

$$\frac{h(x)L}{\kappa^f} = Nu_x = \left[\frac{\partial \tilde{T}}{\partial \tilde{y}}\right]^f \quad \text{for } \boldsymbol{x} \in \partial\Omega \tag{5.101}$$

The coefficient h being a function of distance, x, along the flow, the non dimensional ***Nusselt number***, Nu_x, is also. This number more-or-less measures the non dimensional temperature gradient, perpendicular to the surface, and thus the local thermal transfer coefficient. To find an expression for Nu_x, it is necessary to return to the continuity equations (again!) for an incompressible fluid (constant pressure approximation), in stationary two dimensional flow. From which (chap. 1):

Mass $$\frac{\partial v_x}{\partial x} + \frac{\partial v_y}{\partial y} = 0 \qquad (5.102)$$

x momentum $$v_x \frac{\partial v_x}{\partial x} + v_y \frac{\partial v_x}{\partial y} = \nu \frac{\partial^2 v_x}{\partial y^2} \qquad (5.103)$$

Heat $$v_x \frac{\partial T}{\partial x} + v_y \frac{\partial T}{\partial y} = \alpha \frac{\partial^2 T}{\partial y^2} \qquad (5.104)$$

Note that the assumptions $v_x \gg v_y$, $(\partial T/\partial y) \gg (\partial T/\partial x)$ and $(\partial v_x/\partial y) \gg (\partial v_x/\partial x)$ have been made (the values vary much more rapidly in the direction perpendicular to the surface than in the direction of the flow). Adopting the normalizations defined in (5.100), these equations can be written:

Mass $$\frac{\partial \tilde{v}_x}{\partial \tilde{x}} + \frac{\partial \tilde{v}_y}{\partial \tilde{y}} = 0 \qquad (5.105)$$

x momentum $$\tilde{v}_x \frac{\partial \tilde{v}_x}{\partial \tilde{x}} + \tilde{v}_y \frac{\partial \tilde{v}_x}{\partial \tilde{y}} = \frac{\nu}{L v_\infty} \frac{\partial^2 \tilde{v}_x}{\partial \tilde{y}^2} = \frac{1}{Re_L} \frac{\partial^2 \tilde{v}_x}{\partial \tilde{y}^2} \qquad (5.106)$$

Heat $$\tilde{v}_x \frac{\partial \tilde{T}}{\partial \tilde{x}} + \tilde{v}_y \frac{\partial \tilde{T}}{\partial \tilde{y}} = \frac{\alpha}{L v_\infty} \frac{\partial^2 \tilde{T}}{\partial \tilde{y}^2} = \frac{1}{Pr Re_L} \frac{\partial^2 \tilde{T}}{\partial \tilde{y}^2} \qquad (5.107)$$

The non dimensional Reynolds number, Re_L, for the full length of the surface appears naturally as well as the *Prandtl* number, Pr, defined as the ratio of the kinematic viscosity and the thermal diffusivity of the fluid:

$$Pr = \frac{\nu}{\alpha} = \frac{\mu c_p^f}{\kappa^f} \qquad (5.108)$$

With the boundary conditions corresponding to those of figure 5.20, equations (5.105–5.107) can be solved to obtain the normal derivative of the normalized temperature at the surface and thus the local heat transfer coefficient, $h(x)$, or the Nusselt number, Nu_x (5.101). This number will be a function of the position coordinate, $\tilde{x}$, and the non dimensional Reynolds and Prandtl numbers, Re_L and Pr, appearing in the continuity equations. The Nusselt number, $\langle Nu \rangle$, can be defined to measure the heat transfer coefficient, $\langle h \rangle$, averaged over the length of the surface of exchange as:

$$\langle Nu \rangle = \frac{\langle h \rangle L}{\kappa^f} = \frac{1}{\kappa^f}\int_0^L h(x)\,\mathrm{d}x = \frac{1}{L}\int_0^L Nu_x\,\mathrm{d}x \tag{5.109}$$

and will only be a function of the Reynolds and Prandtl numbers. The particular case of a laminar flow over a plate (fig. 5.20) gives:

$$\langle Nu \rangle = 0.664 Re_L^{1/2} Pr^{1/3} \tag{5.110}$$

More generally (partially turbulent flow, complex form of the surface of exchange), thermal exchanges are characterized by a relation of the type:

$$\langle Nu \rangle = A Re_L^m Pr^n \tag{5.111}$$

where A, m and n are parameters. Exchanges of a chemical nature between a surface and fluid circulation take a similar form, replacing the Nusselt and Prandtl numbers by the *Sherwood* and *Schmidt* numbers (chap. 10). Expressions of the same type can also be used for situations where the surface temperature, T^s, is no longer assumed to be constant along the flow direction.

Finally, it should be noted that the case in figure 5.20 corresponds to an *external* flow around a material, that is, a situation where the boundary layers in the fluid can develop freely in the direction perpendicular to the surface. In the case of an *internal* flow, circulation of a fluid inside a cooled mold for example, the boundary layers will be superimposed after a characteristic distance from the entrance (fig. 5.21). As long as $\delta_v \ll R$ and $\delta_T \ll R$, where R is the radius of the pipe, the flow can be treated as an external flow. When the distance x along the pipe is sufficiently large, the flow is said to be ***fully developed***: the velocity profile of a Newto-

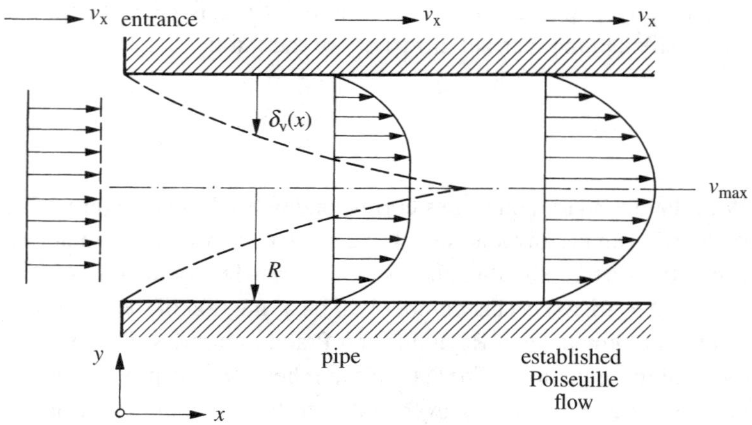

Figure 5.21 Development of the flow in a pipe up to the establishment of a Hagen-Poiseuille flow.

nian fluid in a cylindrical pipe is then that of Hagen-Poiseuille flow (5.85) (fig. 5.18), as long as it remains laminar. If it is turbulent, the profile is flatter in the large fraction of the center section, with higher velocity gradients near the walls.

5.3.3 Transfer by natural convection

Parts are often cooled *naturally* in air; in this case, the air flow is itself created by the temperature variations – and thus those of specific mass – of the gaseous medium induced by the surface of the part. This situation is much more complex as the continuity equations for energy and momentum are *coupled*. Notably, there are many situations for the orientation of the surface normal with respect to gravity which have to be considered (fig. 5.22). For the case of a vertical plate, the density

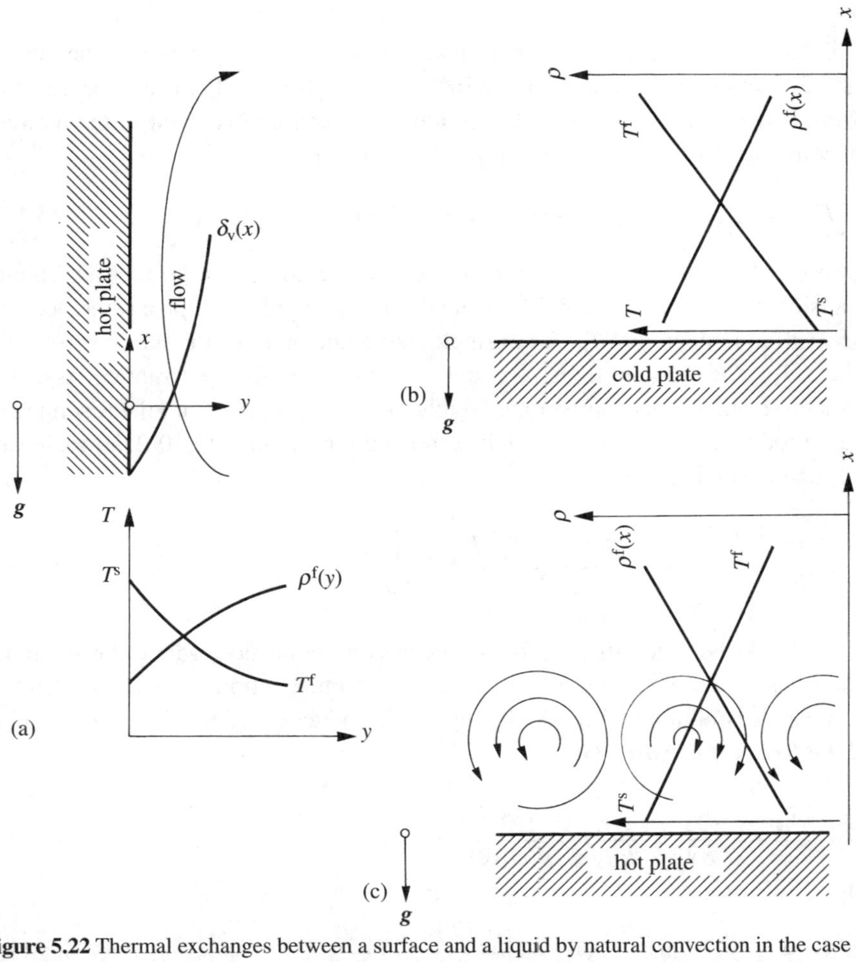

Figure 5.22 Thermal exchanges between a surface and a liquid by natural convection in the case of a horizontal density gradient (a), stable vertical gradient (b), and an unstable vertical gradient (c).

gradients are induced in the horizontal direction. In case (a) in figure 5.22, the plate is assumed hotter than the fluid, which flows practically parallel to the surface, from bottom to top; the fluid layers neighboring the plate are hotter, and thus less dense, than those further away. A clockwise motion tends to develop. The problem is similar to that treated in the preceding section as we will see. For vertical density gradients, there are two cases to distinguish according to the sign of $\mathbf{grad}\,\rho^{\mathrm{f}} \cdot \boldsymbol{g}$, where $\boldsymbol{g}$ is the gravity vector. If $\mathbf{grad}\,\rho^{\mathrm{f}} \cdot \boldsymbol{g} > 0$, the denser layers of the fluid are found below with respect to gravity: they are ***stratified*** and the fluid is stagnant (fig. 5.22(b)). Such is the case for the upper surface of a part cooler than the fluid, or a lower surface warmer than the fluid. For $\mathbf{grad}\,\rho^{\mathrm{f}} \cdot \boldsymbol{g} < 0$, the situation becomes complicated as the less dense layers are below and the resulting buoyancy force tends to create a vertical flux. Unlike the situation for a horizontal density gradient, this flow can have a complex form and leads to ***Rayleigh-Bénard instabilities*** (the formation of many convection cells favoring thermal exchanges from bottom to top, figure 5.22(c)).

In the simple case of a vertical plate with the x axis parallel to the surface (fig. 5.22, situation (a)), the continuity equations for the fluid are the same as before. However the gravitational term must be taken into account in the momentum equation along x. This term can be written (1.71):

$$F_{\mathrm{v,x}} = \rho g_{\mathrm{x}} = -\rho_0 g[1 - \beta_{\mathrm{T}}(T(\boldsymbol{x}, t) - T_0)] \tag{5.112}$$

where g is the modulus (positive) of the gravity vector, β_{T} is the thermal volume expansion coefficient of the fluid (positive for most fluids) and ρ_0 a reference specific mass at temperature T_0, for example the temperature of the fluid far from the surface, T_∞. The constant term, $\rho_0\, g$, only contributes to the hydrostatic pressure of the medium and is thus compensated by the pressure gradient with the assumptions here. Introducing this force term in the x momentum equation (5.106), after division by ρ_{o} and normalization:

$$\tilde{v}_{\mathrm{x}}\frac{\partial \tilde{v}_{\mathrm{x}}}{\partial \tilde{x}} + \tilde{v}_{\mathrm{y}}\frac{\partial \tilde{v}_{\mathrm{x}}}{\partial \tilde{y}} = \frac{g\beta_{\mathrm{T}}(T^{\mathrm{s}} - T_\infty)L}{v_0^2}\tilde{T} + \frac{\nu}{v_0 L}\frac{\partial^2 \tilde{v}_{\mathrm{x}}}{\partial \tilde{y}^2} \tag{5.113}$$

It should be noted that, unlike a forced convection flow where the reference velocity is given far from the surface, the normalization velocity v_0 here is unknown. It is eliminated in (5.113) by the introduction of the Reynolds number and the ***Grashof number***, Gr:

$$\tilde{v}_{\mathrm{x}}\frac{\partial \tilde{v}_{\mathrm{x}}}{\partial \tilde{x}} + \tilde{v}_{\mathrm{y}}\frac{\partial \tilde{v}_{\mathrm{x}}}{\partial \tilde{y}} = \frac{Gr}{Re^2}\tilde{T} + \frac{1}{Re}\frac{\partial^2 \tilde{v}_{\mathrm{x}}}{\partial \tilde{y}^2} \tag{5.114}$$

with

$$Re = \frac{L v_0}{\nu} \quad \text{and} \quad Gr = \frac{g\beta_{\mathrm{T}}(T^{\mathrm{s}} - T_\infty)L^3}{\nu^2} \tag{5.115}$$

The Grashof number measures the ratio of the gravitational and viscous forces. For the case where $Gr/Re^2 << 1$, the gravitational term is of little importance and forced convection dominates: the thermal exchanges characterized by the Nusselt number are functions of the Reynolds and Prandtl numbers as before. In the opposite case ($Gr/Re^2 >> 1$), natural convection dominates and Nu depends essentially on the Grashof and Prandtl numbers, or the Rayleigh number, which is the product of the two ($Ra = Gr\,Pr$). Note the third power of the characteristic dimension L appearing in the Grashof number: the natural convection will increase rapidly with the size of the domain that contains the fluid.

5.3.4 Transfer by boiling/condensation

Thermal transfer by phase transformations of a fluid, generally water, can be extremely effective. To convince oneself, it is sufficient to compare the latent heat of vaporization and the specific heat of water: around $2.25 \cdot 10^6$ J kg^{-1} compared to $4.2 \cdot 10^3$ J kg^{-1} K^{-1}. Thus the energy necessary to vaporize a gram of hot water is five times greater than that required to raise its temperature from 0 to 100 K! However, vapor phase formation can also be very bad for heat exchange as the thermal conductivity of this medium is very low (typically 0.025 W m^{-1} K^{-1} for water vapor, compared to 0.68 W m^{-1} K^{-1} for liquid at 100 °C). To insure the best heat exchange, a vapor phase should ideally be formed in contact with the surface, but it also must be rapidly removed to avoid the formation of a gaseous film which is practically insulating.

The curve in figure 5.23 shows the complexity of such exchanges: it gives the flux of heat extracted, q_T, (absolute value) from a hot wire in a water bath at the boiling point as a function of the temperature difference, ΔT_s, between the wire and the

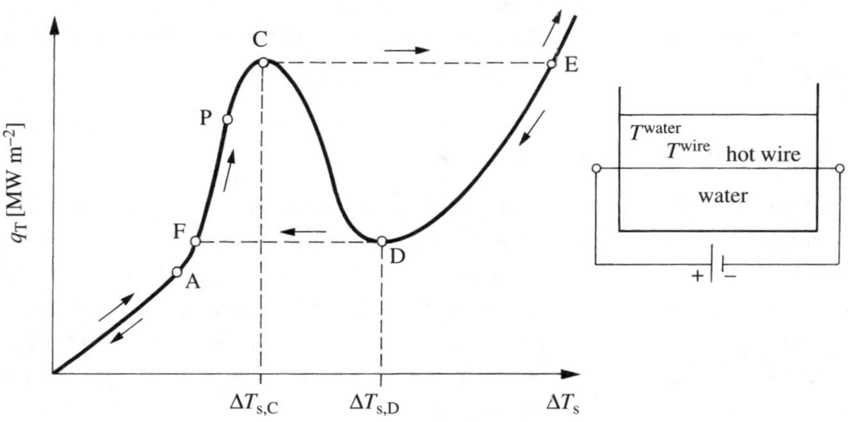

Figure 5.23 Heat flux extracted from a hot wire in a water bath at the boiling temperature as a function of the temperature difference of the wire $\Delta T_s = T^{wire} - T^{water}$.

bath (experiment performed by Nukiyama). By controlling the *temperature* of the wire, the solid curve can be obtained. For $\Delta T_s = 0$ K, there is no heat exchange. Progressively increasing ΔT_s, the heat given off by the wire is first extracted by natural convection of the liquid without formation of a vapor phase. The liquid in direct contact with the wire needs a certain *heat increment* to overcome the nucleation barrier associated with the creation of bubbles (a term analogous to the curvature undercooling described for phase transformations). Beyond point A where *boiling starts*, bubbles form and are rapidly entrained, which considerably increases the extracted heat flux. At point P, the slope of the curve $q_T(\Delta T_s)$, and thus the heat transfer coefficient h between the wire and the bath, is at a maximum. The vapor bubbles are no longer isolated, and vapor columns form. At point C, the extracted heat flux is at a maximum: beyond $\Delta T_{s,C}$, the vapor film formed at the surface at the wire becomes more and more continuous, thus reducing the thermal exchange. For $\Delta T_s = \Delta T_{s,D}$, the vapor film surrounding the wire is continuous and the extracted flux is minimized. Past this value, the flux increases again slowly by conduction and thermal radiation inside the continuous vapor film.

In Nukiyama's original experiment, rather than the temperature, the flux q_T was controlled by the electrical power dissipated in the wire. In this case, the curve obtained is slightly different: increasing q_T beyond point C, the operating point jumps abruptly to point E (path indicated by the dashed line) taking the wire to a temperature that can possibly result in its destruction. Lowering the value of the electrical power dissipated in the wire, the system follows the upper branch of the curve up to point D, and below this point jumps abruptly to point F (abrupt interruption of the vapor film).

The curve in figure 5.23 corresponds to a well controlled thermal exchange by boiling-condensation in which convection of the liquid and vapor are not forced. In this case, the maximum heat flux extracted by the fluid is of the order of 10^6 Wm^{-2}. Forced convection of a liquid near its boiling point allows going well beyond this limit (values of $35 \cdot 10^6$ Wm^{-2} have been reported) by favoring the nucleation of bubbles in a turbulent liquid, and at the same time, avoiding the formation of a continuous film.

5.3.5 Transfer by radiation

Radiation transfer, which of course has no equivalent for solute transfer, plays a very important role in most materials processes. Under vacuum, it remains the only mode of energy transfer to the exterior. As it is a very complex subject – the proof is the voluminous and excellent book by Siegel and Howell consecrated to this theme – this section will be limited to the case of a ***diffuse gray body***. This approximation is actually the one most often used in practice. It is equivalent to making the following restrictive assumptions:

- The material is opaque, that is, an incident wave is absorbed in a very short distance, of the order of the wavelength of light (~μm);

- The *surface* is *diffuse* and *gray*, that is, it neither emits nor absorbs radiation in a way that depends on the orientation or the wavelength;
- The *incident radiation* on the surface is *diffuse*, that is, its intensity does not depend on the angle from which it arrives.

For these conditions, the surface of a body at temperature T (in Kelvin) emits an energy flux (radiation, $\mathrm{Wm^{-2}}$) given by the Stefan-Boltzmann law:

$$q_T^e = \sigma_B \varepsilon_r T^4 \tag{5.116}$$

where σ_B is the Stefan-Boltzmann constant ($\sigma_B = 5.67 \cdot 10^{-8}\ \mathrm{Wm^{-2}\,K^{-4}}$) and ε_r is the relative ***emissivity*** of the material ($0 < \varepsilon_r < 1$). Note that this flux is radiated by the surface element in a hemisphere located above it. For a so-called ***black body***, that is, one that totally absorbs all incident radiation, $\varepsilon_r = 1$. The flux, $\mathrm{d}q_T^e$, emitted per unit surface in a solid angle $\mathrm{d}\Omega = \sin\theta \mathrm{d}\theta \mathrm{d}\phi$ is not simply proportional to $\mathrm{d}\Omega$ because the angle at which the radiating surface is viewed varies as $\cos\theta$ (fig. 5.24(a)). It is thus given by the relation:

$$\mathrm{d}q_T^e = \sigma_B \varepsilon_r T^4 \frac{\cos\theta \sin\theta \mathrm{d}\theta \mathrm{d}\phi}{\int_0^{2\pi} \mathrm{d}\phi \int_0^{\pi/2} \mathrm{d}\theta \sin\theta \cos\theta} = \sigma_B \varepsilon_r T^4 \frac{\cos\theta \sin\theta \mathrm{d}\theta \mathrm{d}\phi}{\pi} \tag{5.117}$$

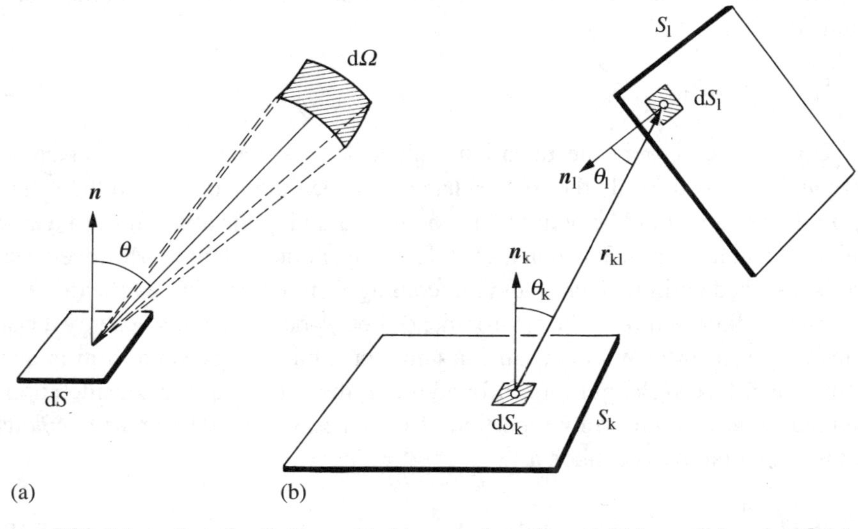

Figure 5.24 Schematics of radiation emitted by a surface element $\mathrm{d}S$ in a solid angle $\mathrm{d}\Omega$ (a) and mutual radiation between two surfaces S_k and S_l, each having a uniform temperature (b).

It can be noted that the integral of this flux over the hemisphere correctly yields (5.116). Consequently, the radiative power expressed in Watts, P^{e}_{kl}, which is emitted by a finite sized surface, S_k, at uniform temperature T_k, and received at surface S_l, (fig. 5.24(b)) is equal to:

$$P^{e}_{kl} = \int_{S_k} dS_k \int_{S_l} dq^{e}_{T} = \frac{\sigma_B \varepsilon_{r,k} T^4_k}{\pi} \int_{S_k}\int_{S_l} \frac{\cos\theta_k \cos\theta_l}{r^2_{kl}} \, dS_k dS_l \tag{5.118}$$

The solid angle $d\Omega = \sin\theta d\theta d\phi$ appearing in the expression of dq^{e}_{T} was replaced by $r^{-2}_{kl}\, dS_l \cos\theta_l$, where $dS_l \cos\theta_l$ is the projection of the surface element dS_l on the direction $\boldsymbol{r}_{kl}$ linking the two surface elements dS_k and dS_l. The angles θ_k and θ_l are defined as:

$$\cos\theta_k = \boldsymbol{n}_k \cdot \boldsymbol{r}_{kl} \qquad \text{and} \qquad \cos\theta_l = -\boldsymbol{n}_l \cdot \boldsymbol{r}_{kl} \tag{5.119}$$

where $\boldsymbol{n}_k$ and $\boldsymbol{n}_l$ are the two outgoing normals of the surfaces S_k and S_l. The ***view factor***, F_{kl}, between the two facets S_k and S_l defined as:

$$F_{kl} = \frac{1}{S_k} \int_{S_k}\int_{S_l} \frac{\cos\theta_k \cos\theta_l}{\pi r^2_{kl}} \, dS_k dS_l \tag{5.120}$$

allows (5.118) to be written:

$$P^{e}_{kl} = \sigma_B \varepsilon_{r,k} T^4_k S_k F_{kl} \tag{5.121}$$

Given the symmetric form of (5.120), the following ***reciprocity relation*** can be obtained:

$$S_k F_{kl} = S_l F_{lk} \tag{5.122}$$

Up until here, only the radiation emitted by a surface has been taken into account: if it was the only thermal exchange between the material and the exterior medium, the part would be obliged to cool until attaining the absolute zero temperature (0 K), temperature at which it would no longer emit radiation. However, a surface is also bathed in incident radiation coming from other surfaces (belonging to the same medium and/or to the environment). Let q^{i}_{T} be the radiative energy density incident on a surface. With the same assumptions, this energy is uniform in angle (diffuse) and in wavelength (gray). In addition, the surface under consideration is assumed to *absorb* uniformly a portion of this energy. The ***absorption coefficient*** for this radiation, α_r, is defined as:

$$\alpha_r = \frac{q^{a}_{T}}{q^{i}_{T}} \tag{5.123}$$

where q_T^a is the power absorbed per unit surface. By definition, the black body is such that $\alpha_r = \varepsilon_r = 1$. A gray diffusing surface is such that $\alpha_r = \varepsilon_r < 1$. Under these conditions, the energy balance for the incident radiation flux is written:

$$q_T^i = q_T^a + q_T^r = \alpha_r q_T^i + (1 - \alpha_r) q_T^i = \varepsilon_r q_T^i + (1 - \varepsilon_r) q_T^i \tag{5.124}$$

q_T^r is the ***reflected*** power (or more precisely, rediffused) per unit surface. Summing this power with that radiated by the material (5.116), the total flux, q_T^s, exiting the facet is obtained:

$$q_T^s = q_T^e + q_T^r = \sigma_B \varepsilon_r T^4 + (1 - \varepsilon_r) q_T^i \tag{5.125}$$

This exiting flux q_T^s is also commonly called ***radiosity***. As for the energy balance for the facet, it is given by the radiative power absorbed, q_T^a, minus the power radiated:

$$q_T = q_T^a - q_T^e = \varepsilon_r q_T^i - \sigma_B \varepsilon_r T^4 \tag{5.126}$$

Now take the balance over the set of facets forming an *enclosure* (fig. 5.25), each of them assumed planar and having a uniform temperature. For such an enclosure, it is easy to show starting from (5.120) that the following ***closure relation*** is satisfied (exercise 5.6.6) :

$$\sum_{l \neq k} F_{kl} = 1 \tag{5.127}$$

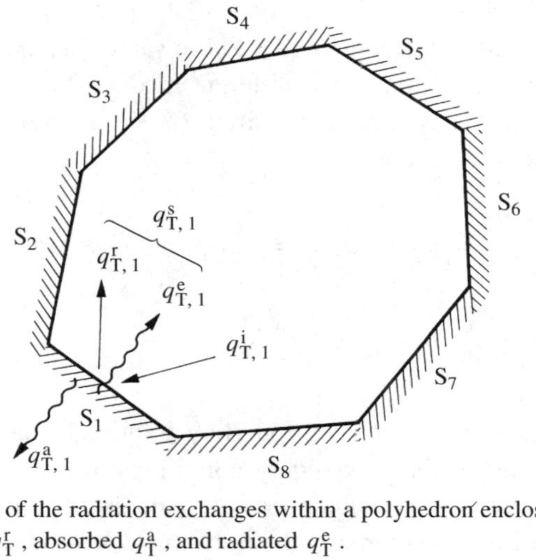

Figure 5.25 Schematics of the radiation exchanges within a polyhedron enclosure showing the flux incident q_T^i, rediffused q_T^r , absorbed q_T^a , and radiated q_T^e .

The incident flux, $q^{i}_{T,k}$, on a facet S_k is the sum of the outgoing flux of all the facets multiplied by the corresponding view factors. The power incident on the facet, P^{i}_{k}, is thus given by:

$$P^{i}_{k} = S_k q^{i}_{T,k} = \sum_{l \neq k} F_{lk} S_l q^{s}_{T,l} = S_k \sum_{l \neq k} F_{kl} q^{s}_{T,l} \tag{5.128}$$

Note that the reciprocity relation (5.122) has been used to write the last equality. Using (5.125), the outgoing power of the same facet has the value:

$$\begin{aligned} P^{s}_{k} = S_k q^{s}_{T,k} &= S_k[\sigma_B \varepsilon_{r,k} T^4_k + (1-\varepsilon_{r,k}) q^{i}_{T,k}] \\ &= S_k\left[\sigma_B \varepsilon_{r,k} T^4_k + (1-\varepsilon_{r,k}) \sum_{l \neq k} F_{kl} q^{s}_{T,l}\right] \end{aligned} \tag{5.129}$$

Dividing by S_k, a system of N equations for the N outgoing fluxes of the facets, $q^{s}_{T,k}$ is obtained:

$$\sum_{l} A_{kl} q^{s}_{T,l} = b_k \qquad \text{for } k = 1, N \tag{5.130}$$

or

$$[\boldsymbol{A}]\boldsymbol{q}^{s}_{T} = \boldsymbol{b} \tag{5.131}$$

with

$$A_{kl} = \delta_{kl} - (1-\varepsilon_{r,k}) F_{kl} \qquad \text{and} \qquad b_k = \sigma_B \varepsilon_{r,k} T^4_k \tag{5.132}$$

δ_{kl} is the Kronecker delta function. The temperatures of the facets known at a given instant, and thus the terms b_k, with the system of equations (5.130) the values $q^{s}_{T,k}$ can be found by inverting the matrix $[\boldsymbol{A}]$. These values can be used for the solution of the heat diffusion equation in the different media. The flux $q_{T,k}$ appearing for each of the facets in the Neumann condition (5.96) is given by equation (5.126). Using (5.128), it finally becomes:

$$\begin{aligned} \kappa\left[\frac{\partial T}{\partial n}\right]_{S_k} &= q_{T,k} = q^{a}_{T,k} - q^{e}_{T,k} = \varepsilon_{r,k} q^{i}_{T,k} - \sigma_B \varepsilon_{r,k} T^4_k \\ &= \varepsilon_{r,k}\left[\sum_{l} F_{kl} q^{s}_{T,l} - \sigma_B T^4_k\right] \end{aligned} \tag{5.133}$$

where the heat fluxes $q^{s}_{T,j}$ have been found by solution of the system of equations (5.130). With the aid of these boundary conditions, the new temperatures in the media, and thus also at the boundaries, are calculated with a standard method (FDM, FVM or FEM). In summary, the ***net radiation method*** consists of decomposing each time step into two steps: 1) calculation of the radiation flux using (5.131) and the known temperatures of the surface; 2) calculation of the diffusion in the materials with the boundary conditions (5.133) to find the new temperatures at every point.

The case of black facets

The case of facets all belonging to black bodies is very simple. Setting $\varepsilon_{r,\,k} = 1$ for all the facets, the matrix $[\boldsymbol{A}]$ is diagonal and consequently $q^s_{T,\,k} = \sigma_B T^4_k$. Inserting these values in (5.133) and using the closure relation (5.127), we have:

$$\kappa\left[\frac{\partial T}{\partial n}\right]_{S_k} = q_{T,\,k} = -\sigma_B \sum_l F_{kl}[T^4_k - T^4_l] \tag{5.134}$$

This relation is none other than a sum over all the facets of the radiative energies weighted by the corresponding view factors.

5.4 NUMERICAL TREATMENT

This section presents the formulation in finite differences/volumes or finite elements of the continuity equations encountered in phase transformations. To simplify the presentation, the convective transport term will be assumed null and attention will be focused on the heat diffusion equation with phase transformations. As the average continuity equations have the advantage of being solvable for a single domain invariant in time, they will be presented first. The numerical methods for the tracking of an evolving front between two phases will then be mentioned.

5.4.1 Average methods

For the case where the velocity field is null and where chemical diffusion at the microscopic scale is neglected (5.50), it was seen that the local transformation path of a material was that of a small *closed* volume. The lever rule or Scheil's model established for a binary alloy can be generalized for multicomponent systems. Such models of ***microsegregation*** of alloy elements between the various phases aim to predict the transformation path, that is the trajectory through the phase diagram $\{c^{v_i}_{A_j}(t)\}$ where v_i designates the phases and A_j the solute elements, as well as the fractions of the different phases, $\{f_{v_i}(t)\}$, as a function of the local thermal history $T(t)$. The two extreme cases which are represented by the lever rule and Scheil's model have shown that the path was actually independent of the thermal history: time can thus be omitted and the transformation path can be expressed in this case as $\{c^{v_i}_{A_j}(T)\}$ and $\{f_{v_i}(T)\}$.

The knowledge of the different phase fractions, $\{f_{v_i}(T)\}$, of their intrinsic thermal properties and the corresponding latent heat of transformation allows the deduction of the average enthalpy of the material $\langle\rho H\rangle$ as a function of temperature. As for the example already presented of solidification of an aluminum-silicon alloy (fig. 5.16), we assume in the following that only two phases, α and β, are present in the domain, the latent heat of transformation being given by $L^{\alpha/\beta}$. Omitting for simplicity of expression the "$\langle\ \rangle$" for the average values, the heat equation governing the phase transformation is given by (5.74):

$$\frac{\partial(\rho H)}{\partial t} - \mathrm{div}(\kappa\,\mathbf{grad}\,T) = 0 \tag{5.135}$$

or:

$$\rho c_{\mathrm{p}}^{\mathrm{eq}}\frac{\partial T}{\partial t} - \mathrm{div}(\kappa\,\mathbf{grad}\,T) = 0 \quad \text{with} \quad \rho c_{\mathrm{p}}^{\mathrm{eq}} = \rho c_{\mathrm{p}} - \rho L^{\alpha/\beta}\frac{\partial f_\alpha}{\partial T} \tag{5.136}$$

We have assumed that the phase fraction, f_α, was a function only of T. The formulations in finite volumes (FVM) or in finite elements (FEM) of the latter equation are completely analogous to those which were seen for chemical diffusion in chapters 2 and 3. It leads in the two cases to a system of equations:

$$[\boldsymbol{C}(T)]\frac{\boldsymbol{T}^{\mathrm{t}+\Delta\mathrm{t}} - \boldsymbol{T}^{\mathrm{t}}}{\Delta t} + [\boldsymbol{K}(T)]\boldsymbol{T}^{\theta} = \boldsymbol{b}(T) \tag{5.137}$$

with the following definitions:

- $\boldsymbol{T}^{\theta}$ is the vector of the temperatures at the mesh nodes, weighted values between the explicit values, $\boldsymbol{T}^{\mathrm{t}}$, known at time t and the implicit values, $\boldsymbol{T}^{\mathrm{t}+\Delta\mathrm{t}}$, unknown, at time $t + \Delta t$:

$$\boldsymbol{T}^{\theta} = (1-\theta)\boldsymbol{T}^{\mathrm{t}} + \theta\boldsymbol{T}^{\mathrm{t}+\Delta\mathrm{t}} \quad \text{with} \quad 0 \le \theta \le 1 \tag{5.138}$$

- $[\boldsymbol{C}(T)]$ is the equivalent specific heat matrix, which strongly depends on the temperature via the transformation latent heat (fig. 5.17(b)). Analogous to the mass matrix defined in chapters 2 and 3, this matrix takes the form shown in table 5.1 for the FVM and FEM.
- $[\boldsymbol{K}(T)]$ is the rigidity matrix (or here, the diffusion matrix), which can also depend on the temperature via the thermal conductivity, $\kappa(T)$.
- $\boldsymbol{b}(T)$ is the term corresponding to the boundary conditions, which also can depend on the temperature (for example in radiative heat exchanges).

The heat equation (5.135) in which the enthalpy appears explicitly, can be formulated in an analogous manner, yielding:

$$[\boldsymbol{M}]\frac{(\rho\boldsymbol{H})^{\mathrm{t}+\Delta\mathrm{t}} - (\rho\boldsymbol{H})^{\mathrm{t}}}{\Delta t} + [\boldsymbol{K}(T)]\boldsymbol{T}^{\theta} = \boldsymbol{b}(T) \tag{5.139}$$

Table 5.1 Finite volume and finite element expressions of the terms appearing in equations (5.137) and (5.139) and the mixed type boundary conditions.

FVM (local numbering)	**FEM** (global numbering)
$C_{CC} = \rho c_p^{eq} \Delta x \Delta y$	$C_{ij} = \int_\Omega \rho c_p^{eq} \psi_i \psi_j \, dV$
$M_{CC} = \Delta x \Delta y$	$M_{ij} = \int_\Omega \psi_i \psi_j \, dV$
$K_{CE} = -\kappa_{CE}\dfrac{\Delta y}{\Delta x_E},\ K_{CW} = -\kappa_{CW}\dfrac{\Delta y}{\Delta x_W}$ $K_{CN} = -\kappa_{CN}\dfrac{\Delta x}{\Delta y_N},\ K_{CS} = -\kappa_{CS}\dfrac{\Delta x}{\Delta y_S}$ $K_{CC} = -(K_{CE} + K_{CN} + K_{CW} + K_{CS}) +$ boundary term if neighbors are absent and h fixed	$K_{ij} = \int_\Omega \kappa(T) \mathbf{grad}\, \psi_i \cdot \mathbf{grad}\, \psi_j \, dV$ $+ \int_{\partial\Omega_N} h \psi_i \psi_j \, dS$
$\kappa_{CJ} = \kappa(\overline{T}_{CJ})$ and $\overline{T}_{CJ} = \dfrac{T_C + T_J}{2}$ $J = E, W, S, N$	
$b_C = h\overline{T}_a \Delta y$ (or $h\overline{T}_a \Delta x$) for a missing neighbor E, W (or N, S)	$b_i = \int_{\partial\Omega_N} h \psi_i \overline{T}_a \, dS$

where the mass matrix, $[\boldsymbol{M}]$, the expression of which is also given in table 5.1, is independent of temperature. It is now a matter of solving these non linear equations.

Explicit scheme ($\theta = 0$)

It was shown in chapters 2 and 3 for the purely diffusive problem without phase transformations, that this temporal scheme is subject to a stability condition, such that the diffusion length associated with the time step Δt must be smaller than the smallest grid element. Aside from this question, the explicit scheme corresponding to equations (5.137) and (5.139) poses no particular problems. With the method called the ***equivalent specific heat*** (5.137), the matrix $[\boldsymbol{C}]^t$ is calculated as a function of $\boldsymbol{T}^t$ and is thus known. By simple inversion of the latter, the variations of the temperature at the nodes, $\Delta \boldsymbol{T} = \boldsymbol{T}^{t+\Delta t} - \boldsymbol{T}^t$, can be calculated with the aid of the relation:

$$\Delta \boldsymbol{T} = \boldsymbol{T}^{t+\Delta t} - \boldsymbol{T}^t = ([\boldsymbol{C}]^t)^{-1}\{-[\boldsymbol{K}]^t \boldsymbol{T}^t + \boldsymbol{b}\}\Delta t \tag{5.140}$$

The term inside the braces is also known as it is evaluated at time t. As the matrix $[\boldsymbol{C}]^t$ evolves in time, it must be inverted at each time step. The small values of Δt imposed by the Fourier criterion require that this inversion be rapid, which will be the case if the matrix is *diagonal*. This is naturally the case with the FVM, while for the FEM it is necessary to use the condensed form of this matrix (chap. 3). Finally, it should be noted that because of the singularities appearing in the equiva-

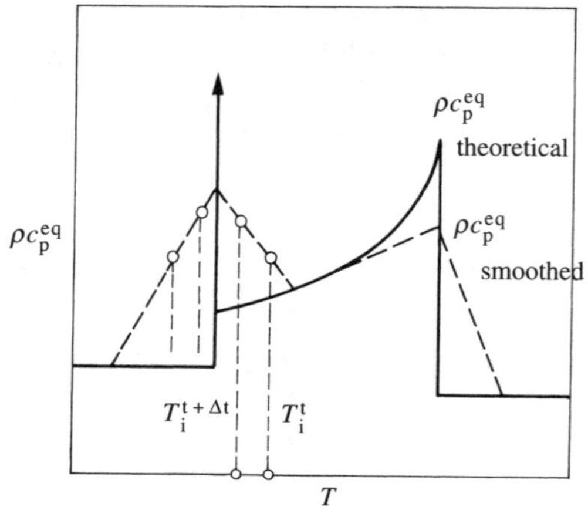

Figure 5.26 Equivalent specific heat $\rho c_p^{eq}(T)$ of an alloy and redistribution of the singularities and jumps for the employment of a numerical method.

lent specific heat ρc_p^{eq}, this method is not *conservative* in energy. For example, the Dirac peak associated with the latent heat liberated during an isothermal reaction (fig. 5.17) might be totally omitted! A palliative is to redistribute such a singularity over a sufficiently large temperature interval so as to be able to cover it with many small time steps (fig. 5.26), while at the same time taking care to conserve energy during the operation (the integral of ρc_p^{eq} over the temperature must remain constant).

From this point of view, the so-called ***enthalpy method*** based on (5.139) is much more rigorous. It consists of calculating the variation of enthalpy $\Delta(\rho \boldsymbol{H})$ by inversion of the matrix $[\boldsymbol{M}]$:

$$\Delta(\rho \boldsymbol{H}) = (\rho \boldsymbol{H})^{t+\Delta t} - (\rho \boldsymbol{H})^t = [\boldsymbol{M}]^{-1}\{-[\boldsymbol{K}]^t \boldsymbol{T}^t + \boldsymbol{b}\}\Delta t \qquad (5.141)$$

Once the $\Delta(\rho H)_i$ or the new volumetric enthalpies, $(\rho H)_i^{t+\Delta t}$, are known at each of the nodes i, the corresponding temperature, $T_i^{t+\Delta t}$, is updated with the inverse relation of $\rho H(T)$, that is: $T_i^{t+\Delta t} = g((\rho H)_i^{t+\Delta t})$ where $g(\rho H)$ is the inverse function of $\rho H(T)$ (fig. 5.27). It can be shown that this method is *conservative*, that is, that it conserves the global energy of the system. As are all explicit methods, it is subject to a Fourier stability criteria.

Mixed scheme $(\theta \neq 0)$

Temporal schemes with $0 < \theta < 1/2$ are of little interest: they are conditionally stable but require at each time step the inversion of the matrix $[\boldsymbol{K}]$, which is not diagonal by nature. The Crank-Nicolson mixed scheme corresponding to $\theta = 1/2$ is stable and offers a temporal precision of higher order than that of the fully implicit Euler scheme ($\theta = 1$). To simplify the equations below, it is nevertheless this

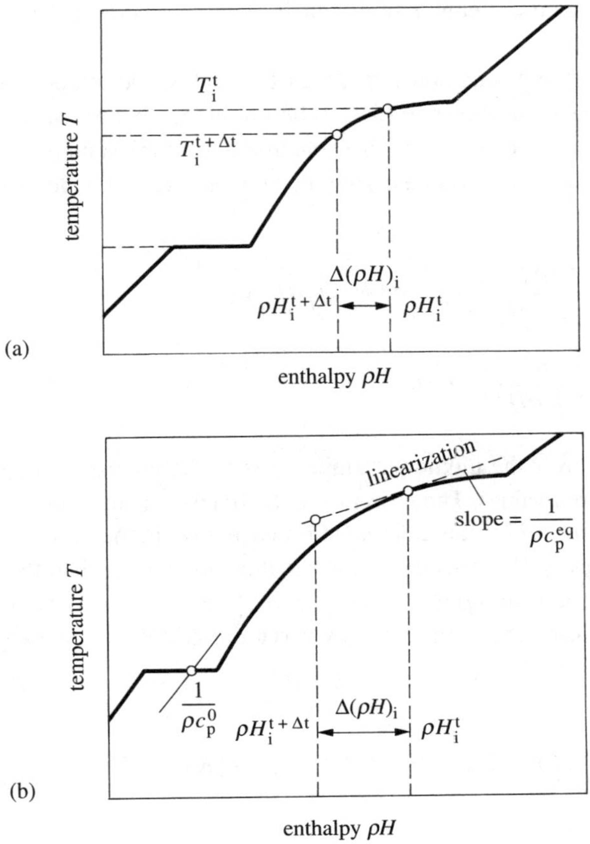

Figure 5.27 Principle of the enthalpy method for an explicit scheme (a) and an implicit scheme (b).

scheme that will be considered (for more details on mixed schemes, the reader is referred to chapters 2 or 3).

Starting again from (5.137) based on an equivalent specific heat, such a scheme permits the new temperatures to be found:

$$\boldsymbol{T}^{t+\Delta t} = \left\{\frac{1}{\Delta t}[\boldsymbol{C}]^{t+\Delta t} + [\boldsymbol{K}]^{t+\Delta t}\right\}^{-1}\left\{\boldsymbol{b} + \frac{1}{\Delta t}[\boldsymbol{C}]^{t+\Delta t}\boldsymbol{T}^{t}\right\} = ([\boldsymbol{A}]^{t+\Delta t})^{-1}\boldsymbol{b}' \tag{5.142}$$

Although the non diagonal matrix $[\boldsymbol{A}]$ must be inverted at each time step, the size of the time step is limited only by the desired *precision* here. However, with large time steps, the problem of the non conservation of energy passing through phase transformations becomes more difficult. As the matrices must be estimated at time $t + \Delta t$ with unknown temperatures, an iterative method becomes necessary (chap. 4).

Because the enthalpy method introduces two fields, the temperature and the enthalpy, it is necessary to linearize (5.139) before being able to use an implicit or mixed scheme. Among the many possible schemes, the simplest consists of keeping the enthalpy as the variable and linearizing the temperature at a node using a Taylor expansion:

$$\begin{aligned}\tilde{T}_i^{t+\Delta t} &= T_i^t + \left(\frac{dT}{d(\rho H)}\right)_i^t ((\rho H)_i^{t+\Delta t} - (\rho H)_i^t) + \ldots \\ &= T_i^t + \left(\frac{dT}{d(\rho H)}\right)_i^t \Delta(\rho H)_i^t + \ldots \end{aligned} \tag{5.143}$$

(A "~" was placed over the unknown temperature to distinguish this approximate value from the desired value). The term inside the brackets is the slope of the curve $g(\rho H)$ shown in figure 5.27b: it is calculated for each node in the mesh at the current value of the enthalpy, ρH_i^t. This expression is thus none other than the inverse of the equivalent specific heat, $(\rho c_p^{eq})^{-1}$ taken at ρH_i^t. By keeping only the first two terms of the expansion (5.143), the enthalpy variations $\Delta(\rho \boldsymbol{H})$ are given by the solution of the system:

$$\Delta(\rho \boldsymbol{H}) = \left\{\frac{1}{\Delta t}[\boldsymbol{M}] + [\boldsymbol{K}]^{t+\Delta t}[\rho \boldsymbol{c}_p^{eq}]^{-1}\right\}^{-1} \{\boldsymbol{b} - [\boldsymbol{K}]^{t+\Delta t}\boldsymbol{T}^t\} \tag{5.144}$$

As the thermal conductivity of most materials varies relatively little with temperature, the matrix $[\boldsymbol{K}]$ can be estimated at time t, and then equation (5.144) gives directly the enthalpy variations at the nodes and thus the new enthalpies $(\rho H)_i^{t+\Delta t}$. Note that the corresponding linearized temperatures $\tilde{T}_i^{t+\Delta t}$, calculated with (5.143), are not on the curve $T = g(\rho H)$ defined by the solidification path (fig. 5.27b). A simple method is then to perform an *isoenthalpic* recovery, that is, to take for the new temperature $T_i^{t+\Delta t}$ the value defined by $g(\rho H_i^{t+\Delta t})$. Such a scheme, which evidently conserves the global energy of the system, tends to introduce *numerical diffusion* as did the upwind schemes encountered for diffusion-convection problems (chap. 2). This is manifested by a phase transformation zone more spread out in space when the time step is large. A solution would be to reduce the time step or to introduce iterations at each time step.

It should be noted that the value of $(\rho c_p^{eq})^{-1}$ introduced in the linearization scheme (5.143) becomes null when the enthalpy ρH_i^t is on the isothermal plateau of a phase transformation (fig. 5.27b). For the node under consideration, this locally reduces the linearization scheme to $\tilde{T}_i^{t+\Delta t} = T_i^t$, that is, to an explicit scheme! Although this does not make the scheme globally unstable, it creates local oscillations and prevents the iterations from converging. The same type of phenomena can occur at the start of a transformation when the latent heat released is high (or (ρc_p^{eq}) large). In such situations, the linearization scheme (5.143) can be replaced by a relation of the type:

$$\tilde{T}_i^{t+\Delta t} = T_i^t + \frac{1}{\rho c_p^0}((\rho H)_i^{t+\Delta t} - (\rho H)_i^t) \qquad (5.145)$$

where ρc_p^0 is a fixed maximal value (fig. 5.27b). Although the method remains globally conservative in energy, it introduces even more numerical diffusion.

Finally, it is worth mentioning that the enthalpy scheme, although conservative in energy, induces "wiggles" in the cooling curves when the temperature interval associated with the transformation $\alpha \rightarrow \beta$ is of the order of, or smaller than, the temperature difference between two points in the mesh. This is especially evident in one dimensional problems. The amplitude of these wiggles is proportional to the latent heat release and to the size of the mesh elements: they are the result of the average of the transformation over each mesh element. Other linearization schemes between the temperature and the enthalpy have been introduced to reduce these numerical problems.

To conclude, it must be said that the methods based on an equivalent specific heat were mostly used in the past as they could be introduced without difficulty into any commercial thermal calculation program: it was sufficient that the program could take into account a specific heat as a function of temperature. However, as it is difficult to insure conservation of energy, these methods have largely been replaced by the enthalpy method.

5.4.2 Front tracking methods

The implementation of so-called *front tracking* methods is much more delicate, notably for two or three dimensional problems. Nevertheless, they are very useful for isothermal phase transformations (solidification of a pure body for example) for which the macroscopic interface is smooth (fig. 5.7). We assume in this case that the phase transformation $\alpha \rightarrow \beta$ is completely controlled by the temperature. A very well adapted technique for this type of problem is the *boundary element method,* since only the moving frontier is meshed. As the method has not been covered in this book, we only present:

- the coordinate transformation methods, valid with the FVM, FDM or FEM,
- the methods for an evolving mesh, specific to the FEM.

Coordinate transformations

To illustrate this method, consider the simple problem of a transformation in one spatial dimension, x: the Bridgman growth of a monocrystal for example (fig. 5.4a). The interface plane being at a position $x^*(t)$ which evolves with time, the problem consists of solving the heat equation in each of the subdomains α and β, with the continuity conditions on the moving interface, (5.15) and (5.19). The idea of the method is to transform the two evolving one-dimensional subdomains, Ω^α and Ω^β, into *fixed* subdomains (fig. 5.28). To that end, the following coordinate transformations are applied:

$$x \in \Omega^\alpha \equiv [0, x^*(t)] \rightarrow \xi \in [0, 1] \quad \text{with} \quad \xi(x, t) = \frac{x}{x^*(t)} \tag{5.146}$$

$$x \in \Omega^\beta \equiv [x^*(t), L] \rightarrow \eta \in [0, 1] \quad \text{with} \quad \eta(x, t) = \frac{x - x^*(t)}{L - x^*(t)} \tag{5.147}$$

The temperature field in each of the phases, $T(x, t)$, then becomes $T^\alpha(\xi(x, t), t)$ and $T^\beta(\eta(x, t), t)$ in each subdomain. As the first derivative in time and the second derivative in x appear in the heat equation (5.18) in one dimension, the following rules for the derivative are applied in the subdomain Ω^α :

$$\left(\frac{\partial T}{\partial x}\right)_t = \left(\frac{\partial T^\alpha}{\partial \xi}\right)_t \left(\frac{\partial \xi}{\partial x}\right)_t = \frac{1}{x^*}\left(\frac{\partial T^\alpha}{\partial \xi}\right)_t \tag{5.148}$$

$$\left(\frac{\partial^2 T}{\partial x^2}\right)_t = \frac{1}{(x^*)^2}\left(\frac{\partial^2 T^\alpha}{\partial \xi^2}\right)_t \tag{5.149}$$

$$\begin{aligned}\left(\frac{\partial T}{\partial t}\right)_x &= \left(\frac{\partial T^\alpha}{\partial t}\right)_\xi + \left(\frac{\partial T^\alpha}{\partial \xi}\right)_t \left(\frac{\partial \xi}{\partial t}\right)_x = \left(\frac{\partial T^\alpha}{\partial t}\right)_\xi - \frac{x}{(x^*)^2}\frac{\mathrm{d}x^*}{\mathrm{d}t}\left(\frac{\partial T^\alpha}{\partial \xi}\right)_t \\ &= \left(\frac{\partial T^\alpha}{\partial t}\right)_\xi - \frac{v_\xi}{x^*}\left(\frac{\partial T^\alpha}{\partial \xi}\right)_t\end{aligned} \tag{5.150}$$

The velocity of a particular point of the domain $v_\xi = v^* \xi$, where v^* is the velocity of the interface, has been introduced to express the stretching over time of the domain Ω^α. For $\xi = 0$, v_ξ is zero, while at the interface $v_{\xi=1} = v^*$. Similar relations are obtained for the domain $\Omega^\beta \equiv [x^*, L]$ if the velocity $v_\eta = v^*(1 - \eta)$ is introduced. The heat equation in each of the subdomains becomes, assuming constant thermal properties:

$$\frac{\alpha^\alpha}{(x^*)^2}\frac{\partial^2 T^\alpha}{\partial \xi^2} + \frac{v_\xi}{x^*}\frac{\partial T^\alpha}{\partial \xi} - \frac{\partial T^\alpha}{\partial t} = 0 \qquad \xi \in [0, 1] \tag{5.151}$$

$$\frac{\alpha^\beta}{(L - x^*)^2}\frac{\partial^2 T^\beta}{\partial \eta^2} + \frac{v_\eta}{L - x^*}\frac{\partial T^\beta}{\partial \eta} - \frac{\partial T^\beta}{\partial t} = 0 \qquad \eta \in [0, 1] \tag{5.152}$$

where α^α and α^β are the thermal diffusivities of the two phases α and β. It is seen that the formulation of the heat equation in the fixed domains [0, 1] naturally introduces transport terms of the type $\boldsymbol{v} \cdot$ **grad** T, as if the material moved locally at a velocity $-v_\xi$ or $-v_\eta$. Such equations can be solved in each subdomain, e.g., with the finite difference method having nodes evenly spaced in ξ or η (fig. 5.28), as long as the boundary conditions are given. The boundary conditions on the outside bound-

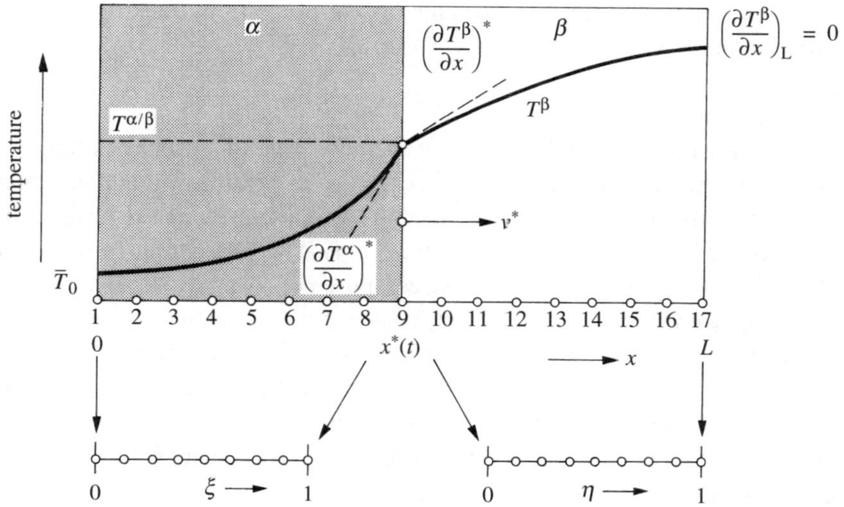

Figure 5.28 Coordinate transformation (or Landau transformation) to track an interface whose position $x^*(t)$ is variable: the intervals $[0, x^*(t)]$ of the domain α and $[x^*(t), L]$ of the domain β are transformed on a fixed interval [0, 1].

aries (at $\xi = 0$ $(x = 0)$ and $\eta = 1$ $(x = L)$) are standard: in fig. 5.28, a Dirichlet condition, $T = \bar{T}_0$, has been imposed at $x = 0$ while the heat flux has been assumed null at $x = L$ (homogeneous Neumann condition). The temperature being imposed at the interface α/β (for example, if thermodynamic equilibrium is assumed, $T = T^{\alpha/\beta}$), the velocity v^* appears as an unknown in the flux condition (5.15). Assuming that the node shared between the two subdomains is node N ($N = 9$ in figure 5.28), the flux continuity condition (5.15) can be expressed in finite differences:

$$v^* = \frac{-1}{\rho L^{\alpha/\beta}}\left[\kappa^\beta \frac{T_{N+1} - T^{\alpha/\beta}}{\Delta x_\beta} - \kappa^\alpha \frac{T^{\alpha/\beta} - T_{N-1}}{\Delta x_\alpha}\right] \tag{5.153}$$

which is an equation linking v^* and the temperatures of the two nodes next to the interface, T_{N-1} and T_{N+1}, Δx_α and Δx_β are the mesh parameters in each of the subdomains. Once the problem has been assembled for the two subdomains, the vector of unknowns will then be constituted of the temperatures at the nodes of domain α, the velocity v^* of the interface, and the temperatures of the nodes of domain β; the equilibrium temperature of the node on the interface appears on the right hand side of the equations for nodes $N - 1$, N and $N + 1$.

This coordinate transformation method can be adapted to two or even three dimensions, but with the condition that the interface retains a relatively simple morphology. Note that mixed derivatives generally appear in the expression for the Laplacian of the temperature field, which can complicate the finite difference formulation. This inconvenience does not appear in the front tracking method for an evolving mesh with finite elements.

Evolving mesh

Except for the boundary element method, which has not been treated in this book, the finite element method is, because of its flexibility, the most suitable for tracking a moving interface in two or three spatial dimensions. Consider then two domains Ω^{α} and Ω^{β} evolving in time during a phase transformation $\alpha \to \beta$ (fig. 5.29). At a given time t, the mesh composed of Nn nodes, for which the coordinates are given by the vectors $\{\boldsymbol{x}^{k}(t)\}$, is such that a certain number of the nodes fall on the interface $\partial\Omega^{\alpha/\beta}$. These nodes being assumed to track the interface through its movement, the overall mesh can *evolve,* that is, the coordinates of the nodes are functions of time. The interpolation functions (chap. 3) are then such that $\psi_i(\boldsymbol{x}, \boldsymbol{x}^1(t), \boldsymbol{x}^2(t), \ldots, \boldsymbol{x}^{Nn}(t))$, where $\boldsymbol{x}$ is the spatial coordinate vector of a point. One can then write:

$$d\psi_i = \mathbf{grad}\, \psi_i \cdot d\boldsymbol{x} + \sum_{k=1}^{Nn} \mathbf{grad}_k\, \psi_i \cdot d\boldsymbol{x}^k \tag{5.154}$$

where $\mathbf{grad}_k\, \psi_i$ is the derivative of the function ψ_i with respect to the coordinates $\boldsymbol{x}^k$ of a node. Moreover, the isoparametric case (chap. 3) implies that:

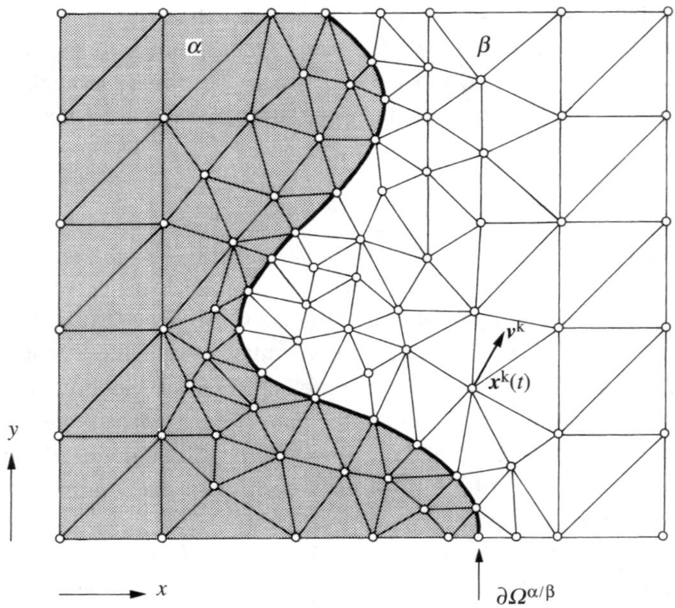

Figure 5.29 Triangle mesh of a rectangular two phase domain with nodes coinciding with the interface $\partial\Omega^{\alpha/\beta}$.

$$x = \sum_{k=1}^{Nn} x^k \psi_k(\xi) \tag{5.155}$$

where ξ are spatial variables that allow the definition of each mesh element (triangle, quadrangle) from a reference element. To simplify the mathematical expression, the element index has been omitted. For the situation where the points x simply follow the spatial evolution of the mesh, ξ is constant and so are ψ_i and all the ψ_ks. Differentiating (5.155) and inserting dx in (5.154), one obtains:

$$\mathrm{d}\psi_i = 0 = \sum_{k=1}^{Nn} (\mathbf{grad}_k\, \psi_i + \psi_k\, \mathbf{grad}\, \psi_i) \cdot \mathrm{d}x^k \tag{5.156}$$

As this relation is valid regardless of the displacement of the mesh nodes, we have finally:

$$\mathbf{grad}_k\, \psi_i = -\psi_k\, \mathbf{grad}\, \psi_i \tag{5.157a}$$

or, as components

$$\frac{\partial \psi_i}{\partial x_j^k} = -\psi_k \frac{\partial \psi_i}{\partial x_j} \tag{5.157b}$$

Start now from the heat equation in a subdomain Ω^α and apply the Galerkin formalism (the test functions are the same as those used for the development of the temperature field). Using the equation just above, we have for the temporal derivative:

$$\int_{\Omega^\alpha} \rho c_p \frac{\partial T}{\partial t} \psi_j\, \mathrm{d}V = \int_{\Omega^\alpha} \rho c_p \sum_{i=1}^{Nn} \frac{\partial (T^i \psi_i(x, x^1(t), x^2(t), \ldots, x^{Nn}(t)))}{\partial t} \psi_j\, \mathrm{d}V$$

$$= \sum_{i=1}^{Nn} \frac{\mathrm{d}T^i}{\mathrm{d}t} \int_{\Omega^\alpha} \rho c_p \psi_i \psi_j\, \mathrm{d}V + \sum_{i=1}^{Nn} T^i \int_{\Omega^\alpha} \rho c_p \psi_j \sum_{k=1}^{Nn} \mathbf{grad}_k\, \psi_i \frac{\mathrm{d}x^k}{\mathrm{d}t}\, \mathrm{d}V$$

$$= \sum_{i=1}^{Nn} \frac{\mathrm{d}T^i}{\mathrm{d}t} \int_{\Omega^\alpha} \rho c_p \psi_i \psi_j\, \mathrm{d}V - \sum_{i=1}^{Nn} T^i \int_{\Omega^\alpha} \rho c_p \psi_j\, \mathbf{grad}\, \psi_i \sum_{k=1}^{Nn} \psi_k \frac{\mathrm{d}x^k}{\mathrm{d}t}\, \mathrm{d}V$$

$$= \sum_{i=1}^{Nn} C_{ji} \frac{\mathrm{d}T^i}{\mathrm{d}t} - \sum_{i=1}^{Nn} V_{ji}^{e} T^i \tag{5.158}$$

with the following definitions:

$$C_{\text{ji}} = \int_{\Omega^\alpha} \rho c_{\text{p}} \psi_{\text{i}} \psi_{\text{j}} \, \text{d}V \tag{5.159}$$

and

$$V^{\text{e}}_{\text{ji}} = \int_{\Omega^\alpha} \rho c_{\text{p}} \psi_{\text{j}} \boldsymbol{v}^{\text{e}} \cdot \textbf{grad}\, \psi_{\text{i}} \, \text{d}V \quad \text{with } \boldsymbol{v}^{\text{e}} = \sum_{\text{k}=1}^{\text{Nn}} \psi_{\text{k}} \boldsymbol{v}^{\text{k}} = \sum_{\text{k}=1}^{\text{Nn}} \psi_{\text{k}} \frac{\text{d}\boldsymbol{x}^{\text{k}}}{\text{d}t} \tag{5.160}$$

As for the coordinate transformation method, an ***energy transport*** term appears with an entraining velocity $-\boldsymbol{v}^{\text{e}}$ given by interpolation of the displacement $\boldsymbol{v}^{\text{k}}$ of the mesh nodes. The diffusion term not being affected by the displacement of the nodes, the rigidity matrix $[\boldsymbol{K}]$ retains the same form as that described in chapter 3 (also table 5.1). The heat diffusion equation in a phase thus takes the form:

$$[\boldsymbol{C}]\frac{\text{d}\boldsymbol{T}}{\text{d}t} - [\boldsymbol{V}^{\text{e}}]\boldsymbol{T} + [\boldsymbol{K}]\boldsymbol{T} = \boldsymbol{b} \tag{5.161}$$

It is thus possible to follow the displacement of the mesh nodes, notably those belonging to the interface $\partial\Omega^{\alpha/\beta}$, via the introduction of a transport term (5.160). A similar equation will be obtained for the domain Ω^β. The displacement of the interface nodes is imposed by a flux balance condition (5.15), coupled with an equilibrium type condition (5.19), while that for the nodes on the exterior boundary of Ω (assumed immobile) is null. However, it must be noted that equation (5.15) only specifies the normal component of the velocity. Two situations can then be observed for this thermally controlled phase transformation (the latent heat $L^{\alpha/\beta}$ is assumed positive):

- $v_{\text{n}}^* > 0$ and $(\partial T/\partial n)^{\beta *} > 0$: the phase α grows at the expense of phase β in a positive temperature gradient (fig. 5.30(a)). Note that for such a situation – called ***constrained growth*** – the heat flux extracted from α, q_{T}^{α}, must be greater than that arriving from β, q_{T}^{β} (conservation equation (5.15)). In this case, the interface α/β is stable. Any convex perturbation of the interface tends to augment the gradient in front of the interface and thus to induce a higher heat flux from the domain β: the velocity will automatically decrease according to (5.15). The numerical method implemented to move the nodes at the interface in such a situation will retain its smooth character.
- $v_{\text{n}}^* > 0$ and $(\partial T/\partial n)^{\beta *} < 0$: the phase α grows at the expense of phase β in a negative temperature gradient (fig. 5.30(b)). In this case – called ***undercooled growth*** – the interface α/β is unstable: a convex perturbation of the interface due to a local increase of v_{n}^* again increases the thermal gradient (absolute value) and thus $|q_{\text{T}}^{\beta}|$, which permits the interface to advance still more rapidly. In this case, the numerical method will not retain the smooth character of the interface. The morphological instabilities thus created, reflecting the unstable physical nature of the interface, will be directly influ-

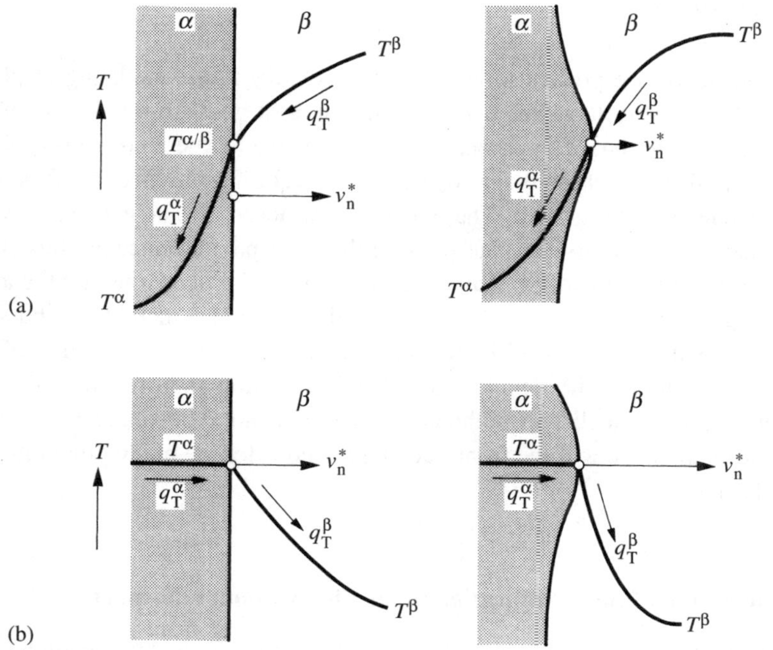

Figure 5.30 Stable (a) and unstable (b) situations of an interface α/β of a pure material growing in positive (a) and negative (b) thermal gradients, respectively, with $v_n^* > 0$ (***n*** is the outgoing normal from α). The latent heat of transformation $(\rho L)^{\alpha/\beta}$ is assumed positive.

enced by the mesh. In other words, the characteristic dimensions of the destabilized interface $\partial\Omega^{\alpha/\beta}$ will be linked to the fineness of the mesh elements. If it is desired to obtain an interface morphology that is conditioned by the physical phenomena and not by the mesh, it is necessary to introduce the interface curvature term (5.20) which tends to stabilize the interface.

Note that if v_n^* is negative (transformation $\alpha \rightarrow \beta$, as $\boldsymbol{n}$ is the outgoing normal of α), the stable and unstable situations of fig. 5.30 are also inverted. Finally, it should be noted that while the heat conservation equation at the interface (5.15) allows the displacement of the interface nodes, many strategies can be adopted for the displacement of the volume nodes:

- They can remain fixed ($\boldsymbol{x}^k$ = constant). Unless new elements are introduced and removed in the neighborhood of the interface, this method will be limited to small displacements.
- The position of the volume nodes can be recalculated at each time step with appropriate interpolations, and possibly also mesh regeneration to retain sufficient spatial resolution (chap. 4). For more details, the reader can consult the review article by Lynch in the bibliography.

5.5 EXAMPLES

In this section we present a few examples relative to the modeling of phase transformations, a very large subject by definition, as it goes from the scale of the microstructure up to that of the process and can involve all the conservation equations to say nothing of phase diagrams and out-of-equilibrium phenomena associated with nucleation and growth! The object is thus not to give an exhaustive view of the subject, but rather to illustrate some of the concepts introduced in this chapter. We will first present two applications at the scale of a process that use the average equations, then a coupling with out-of-equilibrium phenomena for solid state transformations described by TTT diagrams. Another example of coupling of the conservation equations with detailed models of nucleation and growth is given at the end of chapter 9 (stochastic methods) and will thus not be treated here. Finally, to finish the chapter, we will see front tracking methods for the prediction of microstructure formation.

5.5.1 Calculation of the solidification of an alloy without transport

Neglecting the diffusion of solute at the macroscopic scale, and assuming that there is no advective transport, we can obtain a unique relation between the transformed fraction (or the average enthalpy) and the temperature with the lever rule (or Scheil's model) as seen in section 5.2.7. Once this relation has been obtained, phase transformations are governed by thermal transfer, that is, by (5.64) or (5.74) with $v = 0$. The only problem then is the strong non-linearity between enthalpy and temperature, for which the enthalpy method is the most appropriate.

To illustrate such a method, we consider a three dimensional part of rather complex shape: a turbine blade produced by investment casting (fig. 5.31). This process consists first of forming a part in wax by an injection method. The part is covered in successive layers of ceramic powder with a binder so as to form a shell. This shell is then thermally consolidated by sintering the ceramic grains after having removed the wax. The mold remains; packed in insulating layers, it is preheated and then placed under vacuum. The nickel base metal alloy (superalloy) is then cast in the mold. From here on solidification is controlled by the initial conditions (temperatures of the mold and metal) and the cooling parameters: largely radiation and thermal contact, for example, with a water cooled copper chill. Thus are obtained blades constituted of equiaxed grains (fig. 5.31(a)), columnar grains (fig. 5.31(b)), or even a single grain (single crystal blades), (fig. 5.31(c)) depending on the solidification conditions.

Figure 5.32(a) shows the three dimensional mesh of a similar blade intended for a jet engine, a mesh that can be obtained directly from a CAD file describing its geometry. Next, the mesh elements necessary for the description of the thermal exchanges in the ceramic mold, the insulation and the liquid metal runners and ris-

Figure 5.31 Superalloy turbine blades obtained from investment casting showing equiaxed (a), columnar (b) and single crystal (c) structures (Courtesy of Sulzer Innotec).

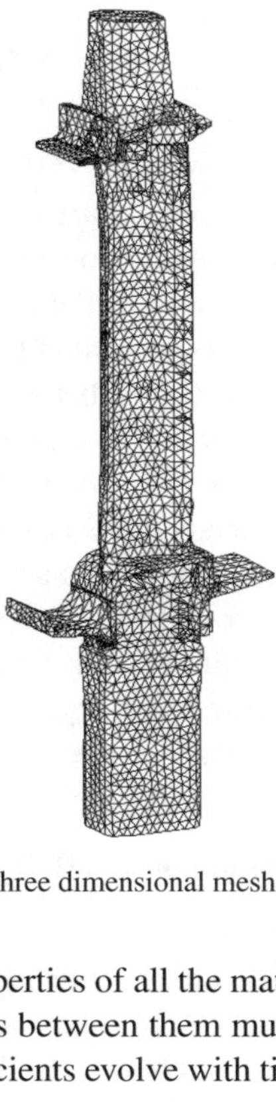

Figure 5.32 Three dimensional mesh of a turbine blade.

ers are added. Besides the properties of all the materials (metal, ceramic, insulation, cooler), the thermal exchanges between them must be specified, which is not easy. In fact, the heat transfer coefficients evolve with time not only because the radiation

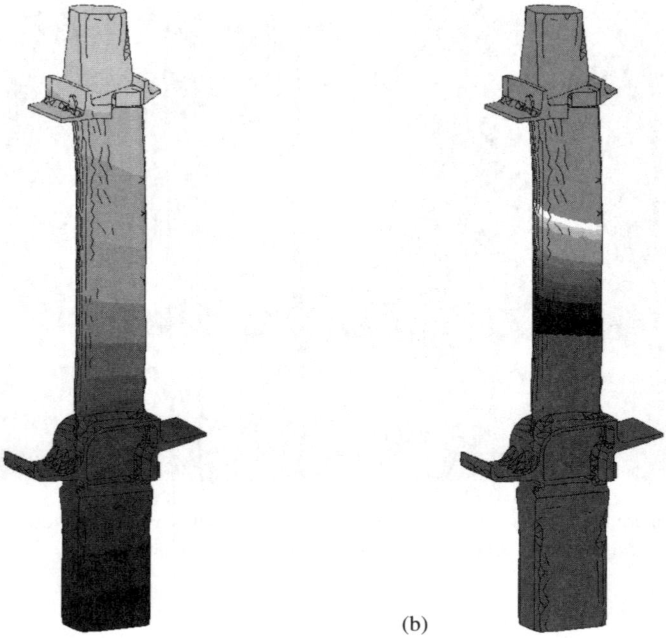

(a) (b)

Figure 5.33 Temperature map (a) and liquid fraction (b) during solidification of the blade shown in 5.32.

terms are to the fourth power of the temperature (sect. 5.3.5), but also because the contact between the various media (conduction) is directly influenced by thermal contractions. For semi-porous media such as the mold or the insulation, the radiation transfers between grains or fibers also modifies their apparent conductivity. Finally, the radiation of the external surfaces, between each other and the surroundings, requires a preliminary calculation of the view factors. If the mold moves with respect to a furnace (controlled solidification), these view factors also evolve over time. One sees, even in a case where there is only one conservation equation involved – heat – that the modeling is far from being simple, and often requires the calibration with certain parameters from instrumented experiments.

Figure 5.33(a) shows a temperature map in a blade during solidification while the zone composed of solid dendrites and liquid ($0 < f_\ell < 1$) is shown in figure 5.33(b) by the liquid iso-fraction lines.

5.5.2 Calculation of the solidification of an alloy with transport

With a convective transport term, the situation is more complicated than that presented in the previous section: solute can locally enter in the small volume element used to perform the averages. Therefore, the relation $\rho H(T)$ is no longer unique and depends on convective movements. To illustrate this situation, we take

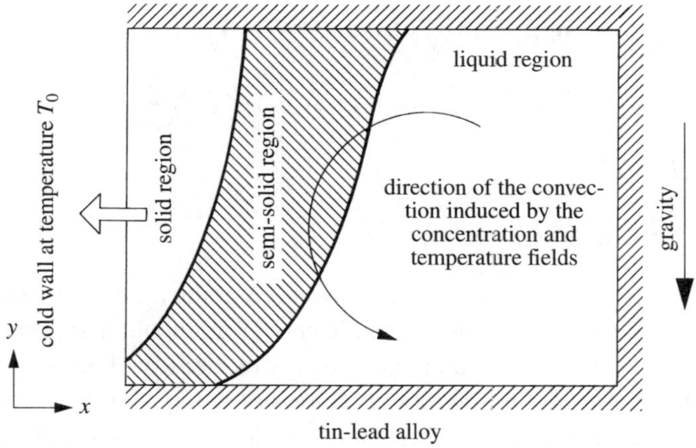

Figure 5.34 Schematics of a rectangular cavity used to solidify a tin-5wt% lead alloy.

the case of a rectangular domain initially containing a tin-5wt% lead alloy (fig. 5.34). Everything is initially at a uniform temperature above the liquidus of the alloy. At time $t = 0$, the vertical surface on the left is cooled, entraining the progressive solidification of the alloy, while the three other surfaces are assumed adiabatically insulated.

Without convective motion, the isotherms would be perfectly vertical. However, the colder, and thus denser, layers to the left of the cavity tend to entrain the liquid in a counterclockwise motion. This effect is reinforced by the fact that as lead is partly rejected into the liquid by the tin dendrites, the resulting liquid in the mushy region (containing solid and liquid) is heavier. There is then a natural convection phenomenon reinforced by solute convection (they are called double diffusion phenomena). Note that the convection caused by solute gradients would have been reversed with respect to the thermally induced convection for the case of a lead alloy containing a small amount of tin. Assuming that the solid is fixed, the average equations necessary to calculate such a situation in the incompressible case ($\rho^\ell = \rho^s = \rho_0$) are:

$$\frac{\partial(\rho_0 H)}{\partial t} + \rho_0 c_p \boldsymbol{v} \cdot \mathbf{grad}\, T - \mathrm{div} \kappa \mathbf{grad}\, T = 0 \tag{5.162}$$

$$\frac{\partial c}{\partial t} + \boldsymbol{v} \cdot \mathbf{grad}\, c^\ell = 0 \tag{5.163}$$

$$\mathrm{div}\, \boldsymbol{v} = 0 \tag{5.164}$$

$$\rho_0 \frac{\partial v_x}{\partial t} + \frac{\rho_0}{f_\ell} \boldsymbol{v} \cdot \mathbf{grad}\, v_x - \mu \Delta v_x + \frac{\mu v_x f_\ell}{K} = -f_\ell \frac{\partial p}{\partial x} \tag{5.165}$$

$$\rho_0 \frac{\partial v_y}{\partial t} + \frac{\rho_0}{f_\ell} \boldsymbol{v} \cdot \mathbf{grad}\, v_y - \mu \Delta v_y + \frac{\mu v_y f_\ell}{K}$$
$$= -\rho_0 (1 - \beta_T (T - T_0) - \beta_c (c^\ell - c_0^\ell)) f_\ell g - f_\ell \frac{\partial p}{\partial y} \tag{5.166}$$

$$H = c_p T + L f_\ell \tag{5.167}$$

$$c = c^\ell f_\ell + c^s f_s = c^\ell f_\ell + k c^\ell (1 - f_\ell) \tag{5.168}$$

The brackets for the average values have been omitted here for simplicity. β_T and β_c are the volumetric expansion coefficients of the liquid associated with the temperature and concentration variations and ρ_0 is the density of the alloy taken at a reference temperature T_0 and liquid concentration c_0^ℓ (eq. (1.71)). The density is thus assumed constant except in the volumetric force term of the Navier-Stokes equations, the term which drives the natural convection; this is the ***Boussinesq's approximation***. Note that all the material properties (viscosity μ, specific heat c_p, etc.) have been assumed to be constant and known, and gravity is in the $-y$ direction. The permeability term, K, is given by an expression of type (5.93) as a function of the liquid fraction, f_ℓ, and the characteristic distance of the microstructures, d_μ. The latter value can be calculated from a coarsening model and the local solidification time. The unknown fields appearing in equations (5.162–5.168) are then: the enthalpy, $H(\boldsymbol{x}, t)$, the temperature, $T(\boldsymbol{x}, t)$, the average local concentration, $c(\boldsymbol{x}, t)$, the liquid concentration, $c^\ell(\boldsymbol{x}, t)$, the liquid volume fraction, $f_\ell(\boldsymbol{x}, t)$, the two components of the average velocity field, $v_x = f_\ell v_x^\ell(\boldsymbol{x}, t)$ and $v_y = f_\ell v_y^\ell(\boldsymbol{x}, t)$, and the pressure, $p(\boldsymbol{x}, t)$; altogether, 8 scalar fields. Yet, we only have 7 equations: we lack the relation linking the temperature and the concentration field in the zone where the two phases, solid and liquid, coexist, that is:

- If $f_\ell = 1$ (liquid) or $f_\ell = 0$ (solid), f_ℓ is no longer an unknown in the problem, and T is independent of c^ℓ,
- In the two phase zone, where f_ℓ is effectively a variable $(0 < f_\ell < 1)$, $T = T_f + m^\ell c^\ell$ (5.55), where T_f is the melting point of the pure body and m^ℓ the slope of the liquidus (fig. 5.14).

It is seen that the phenomena are closely coupled: the temperature variations induce solidification, thus the segregation phenomena at the scale of the microstructure (microsegregation). These two phenomena entrain liquid movement by natural thermal and solutal convection, the solute transport changes the average local concentration, c, leading to a segregation at the scale of the domain or of the process (macrosegregation). In order to solve equations (5.162–5.168), an implicit scheme and a finite element formulation in enthalpy have been adopted (sect. 5.4). A pressure penalization (chap. 6 and 7) has been used to replace the pressure appearing in the Navier-Stokes equations with the conservation of mass (5.164):

$$\operatorname{div} \boldsymbol{v} = -\frac{1}{\chi_p''} \tag{5.169}$$

where χ_p'' is a penalization parameter (typically $\varepsilon_p = 10^{10}$). The conservation equations have been partially "decoupled", that is, "foreign" fields in any equation (for example, the temperature field in the Navier-Stokes equations or the velocity field in the solute equation) are replaced by the last known values. One thus advances in time successively resolving the heat equation with the velocity field from the preceding step, then the solute equation and finally the pressure penalized Navier-Stokes equations with the new temperature and concentration fields.

Figure 5.35 shows the result of such a calculation one minute after having started the cooling of the rectangular domain from the left boundary. The different fields shown are the temperature, the solid fraction, the streamlines with indication of the maximum velocity in the liquid and the relative concentration variations of lead (0% = nominal concentration). As can be seen, the densest liquid layers are of course those that contain the most lead: they tend to descend in the cavity, which causes counterclockwise convective motion. This motion is seen in the form of the isotherms, which have larger spacing in the lower portion of the domain. Note that as the temperature is relatively uniform in the cavity, the solid fraction field, $f_s = (1 - f_\ell)$, has practically the same form as that of the average local concentration in the two phase zone.

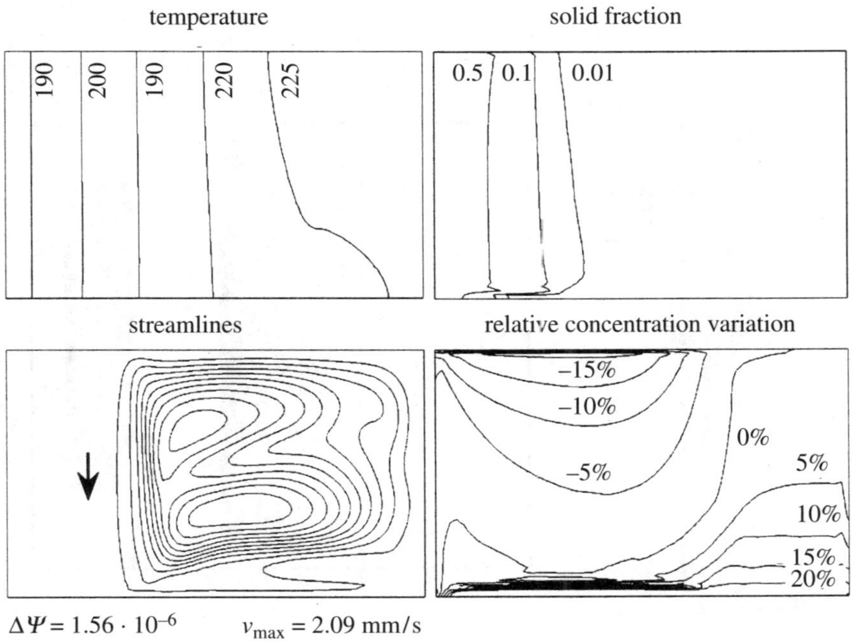

Figure 5.35 Temperature, solid fraction, streamlines, and relative concentration variation maps for a tin-5wt% lead alloy solidified in the cavity of figure 5.34.

5.5.3 Calculation of phase transformations with the help of TTT diagrams

The two preceding examples assume local equilibrium at the interface. We show below a calculation that takes into account the transformation kinetics – thus deviations from equilibrium – of a ferrito-pearlitic type alloy, whose surface is hardened by induction heating (fig. 5.36). The principle of the process is the following: an alternating electric current in an induction coil induces a current, $\boldsymbol{j}_E$, in the cylin-

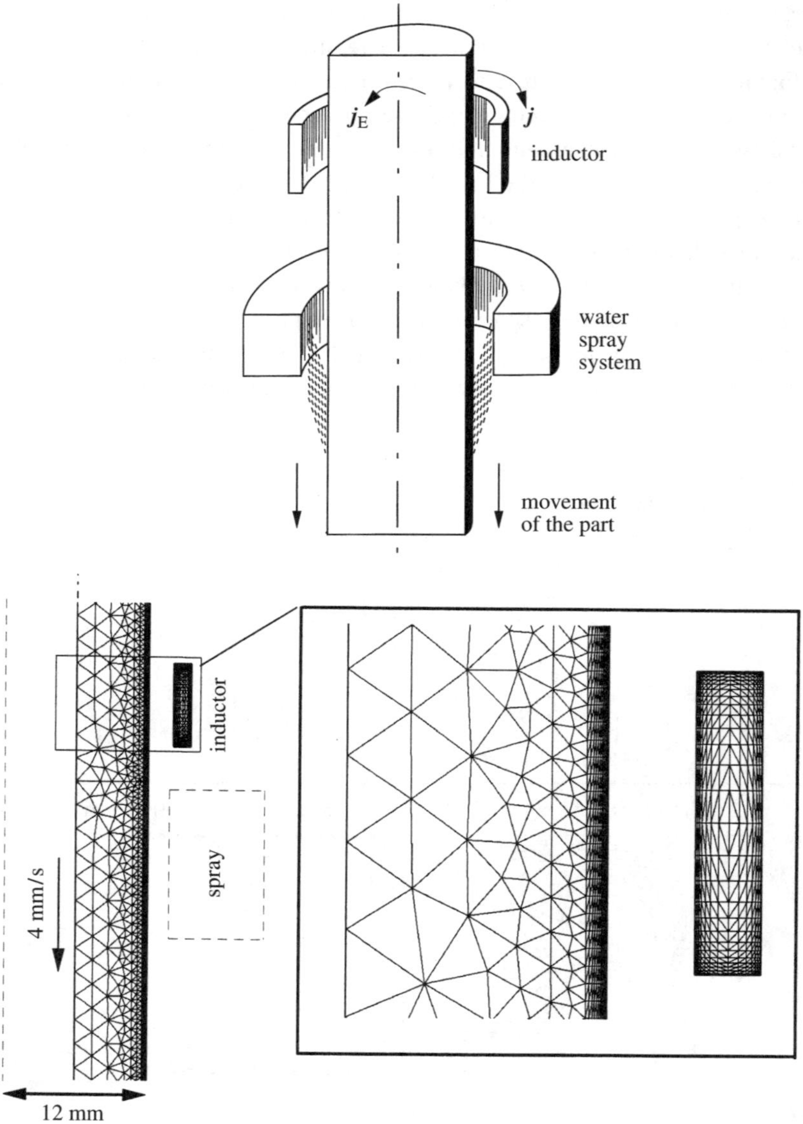

(a)

(b)

Figure 5.36 Schematics of the induction heating surface hardening process (a) and associated mesh for the inductor and the cylinder being treated (b).

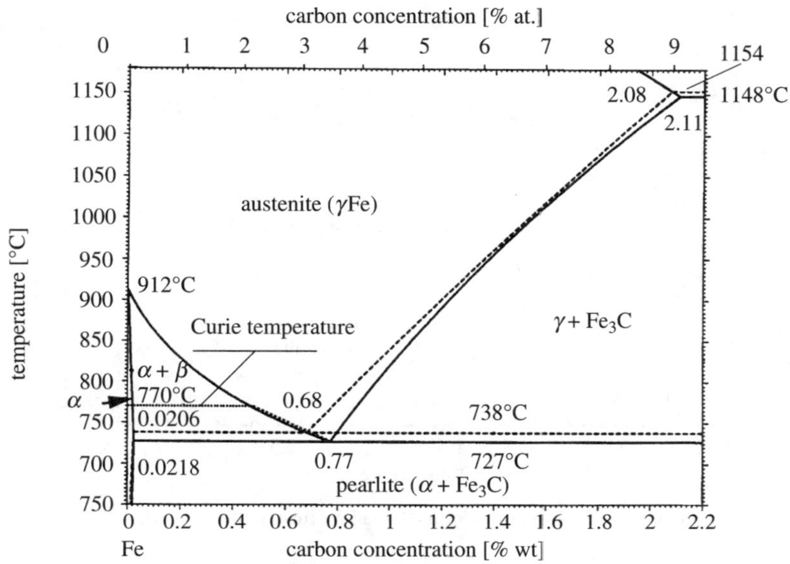

Figure 5.37 Iron-carbon phase diagram for solid state transformations.

der placed in the center. This current is localized principally in a thin skin at the surface, the thickness of which varies as the square root of the period of the alternating current (skin depth effect). The induced current provokes a rapid local heating in the cylinder by the Joule effect (1.81), and a surface layer of the alloy can thus be transformed into austenite (phase diagram in figure 5.37). As the cylinder moves at constant velocity in the coil, the heated zone is then cooled in a water spray, which induces a transformation of the austenized structure into various phases: ferrite, pearlite, bainite, martensite. The kinetics of these transformations during cooling of the austenite are shown in figure 5.38 in the form of a TTT diagram (sect. 5.2.3). Such a diagram is read along horizontal lines (isothermal transformations): the start and end of each transformation is indicated by two times (for example, transformation of austenite (A) into ferrite (F), pearlite (P) or bainite (B)). The martensitic transformation being massive and non diffusive, only the temperature at the start of the transformation, M_s, is indicated. In addition to values from the literature, experimental points measured by dilatometry have been added.

As TTT diagrams can in principle only be used for isothermal transformations, we fall back on an *additive* principle to predict the transformation kinetics of an arbitrary thermal history $T(t)$. This principle is shown in figure 5.39. The cooling curve is first cut up into a number of isothermal steps corresponding to time steps. At the start of each time step $(j + 1)$, the transformed fraction, for example, that of ferrite f_α^j, is known from the previous time step (j). The TTT diagram corresponding to this new temperature level gives a new relation $f_\alpha(t, T^{j+1})$: it is thus necessary to introduce a *virtual time*, θ_j, for which: $f_\alpha(\theta_j, T^{j+1}) = f_\alpha^j$. Once this time is

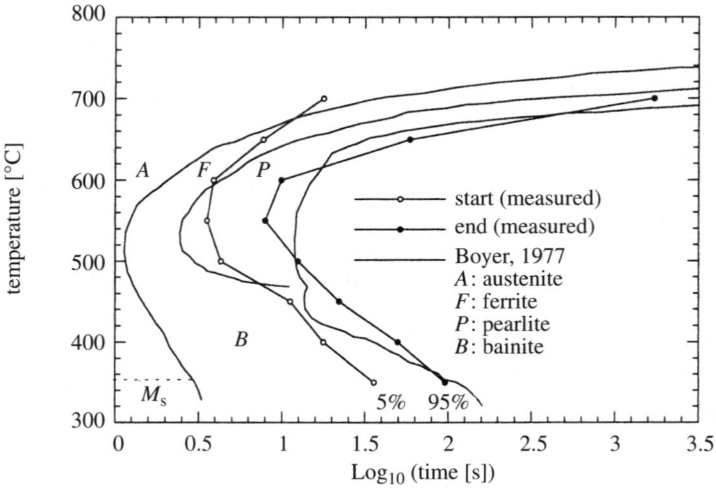

Figure 5.38 TTT diagram for a Ck45 alloy (iron-0.49wt% carbon).

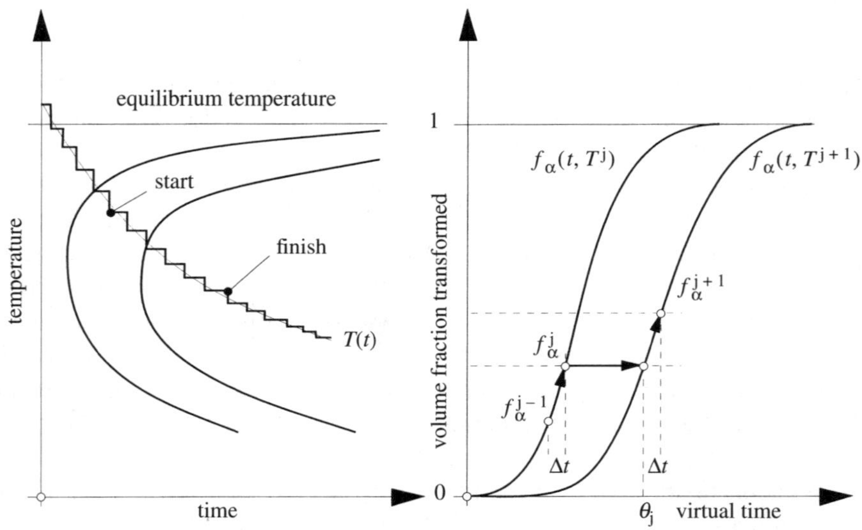

Figure 5.39 Additive principle for the use of TTT diagrams in non isothermal situations.

known, it is sufficient to calculate the new transformed fraction after the time step $f_\alpha^{j+1} = f_\alpha(\theta_j + \Delta t, T^{j+1})$.

Figure 5.40 shows the result of such a calculation after a stationary thermal condition in the cylinder has been established. Because of the symmetry, only half the cylinder was modeled; the axis of the cylinder is on the left in the figure. The rectangular section of the inductor in the upper part and the spray zone are seen in

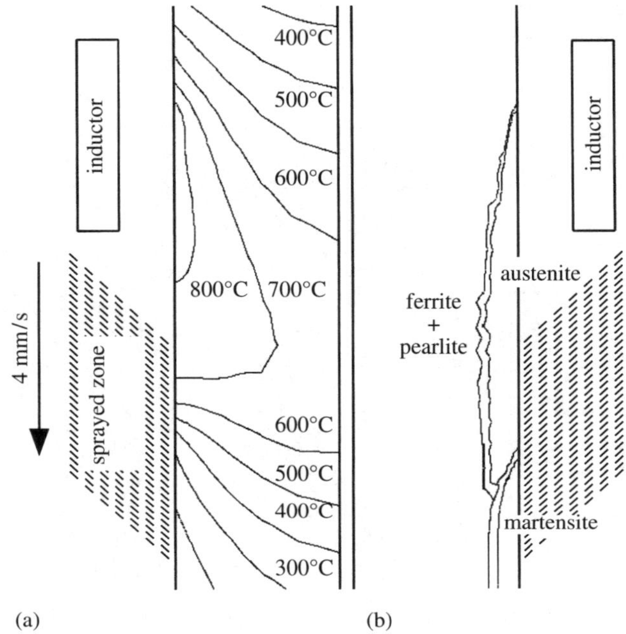

Figure 5.40 Maps of temperature (a) and phase distribution (b) during the induction heating hardening process.

the figure. The thermal transfer coefficient in that zone was deduced by an inverse method (chapter 9). The figure on the left shows the isotherms while that on the right shows the austenitic and martensitic zones during treatment. Note that a particular numerical technique had to be implemented for the electromagnetic calculations in the part, the inductor and in air in order to obtain the induced currents. A finite element formulation was adopted for the solution of Maxwell's equations in the part and in the induction coil, while the coupling between the two media was obtained with Green's functions discretized by the boundary element method. With this technique, meshing of the air can be avoided, thus permitting the free motion of the inductor/spray system with respect to the cylinder. The mesh used is shown in figure 5.36, while the measured and calculated temperature curves for various points of the cylinder are compared in figure 5.41.

5.5.4 Calculation of microstructures with a front tracking method

We discussed in section 5.4 the instabilities of an interface during the growth of a pure body in a negative temperature gradient (fig. 5.30(b)). Figure 5.42 shows the destabilization of the solidification front of a transparent organic succinonitrile-4wt% acetone compound. The experiment carried out under an optical micro-

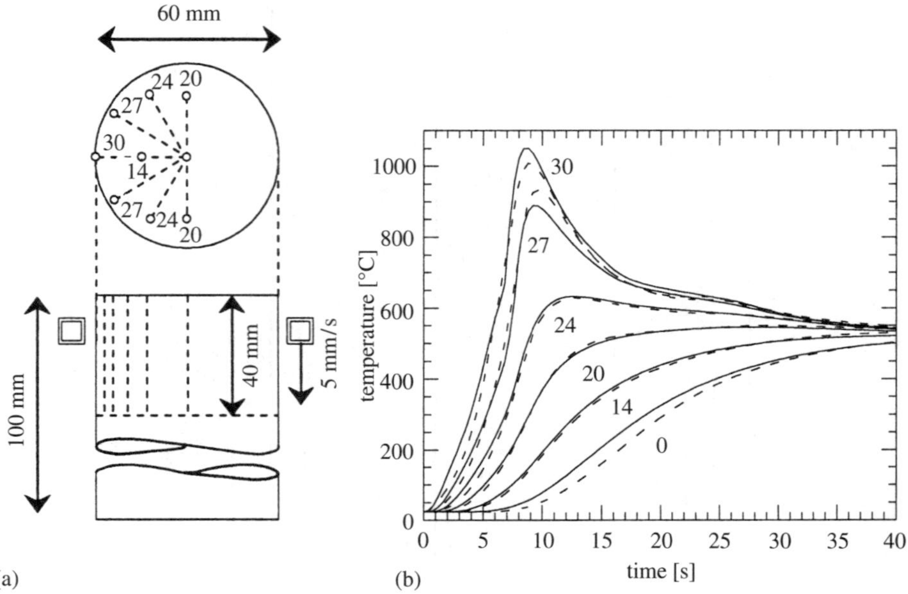

Figure 5.41 Measured and calculated temperature curves at various points of a small cylinder heated by induction without cooling. The variations of heating rate are due to the change in the magnetic permeability of the material as it passes through the Curie point.

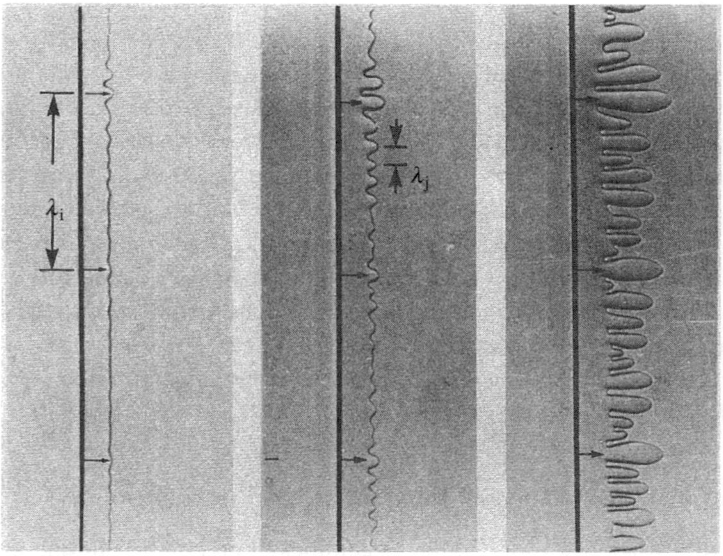

Figure 5.42 Destabilization of a growth front in succinonitrile-4wt% acetone for the sudden increase in the growth velocity from 0.06 μm s^{-1} to 3.4 μm s^{-1} (Courtesy of R. Trivedi).

scope is analogous to Bridgman growth as shown in figure 5.4(a): the compound between two parallel glass plates moves at a constant velocity in a thermal gradient fixed by hot and cold sources placed on either side of the microscope objective. Attaining a stationary state, the solidification front advances at the same velocity (to the right in figure 5.42) as that of the motion of the cell (to the left): thus to the observer it appears to be immobile. Without solute rejection, the growth front would be stable, as it is in a positive thermal gradient.

However, as the solid rejects solute ahead of the interface, (fig. 5.2(b)), there exists a critical velocity beyond which the front becomes unstable (***constitutional undercooling criterion***). Therefore, in figure 5.42 a change of the velocity from 0.06 μm s^{-1} to 3.4 μm s^{-1} causes cells to appear which become more and more pronounced over time. Waiting long enough, these cells give birth to dendrites.

Figure 5.43 is a simulation by finite elements of the destabilization of a planar front for an alloy. On the left, the deformed mesh permits the direct visualization of the position of the interface: it was obtained by coordinate transformation of a regular mesh. The figure on the right shows the concentration profiles in the liquid (above) and solid (below). In this situation, the calculation of solute diffusion was performed in a reference frame "attached to the microscope", which causes an

(a) (b)

Figure 5.43 Simulation of cell formation: quadrangle mesh obtained by coordinate transformation (a) and concentration field (b) (After K. Tsiveriotis *et al.*).

(a) (b)

Figure 5.44 Front tracking FEM method with remeshing for a pure material solidified from an undercooled liquid bath: (a) progressive destabilization of the interface; (b) mesh for the last position shown in (a) (after J.M. Sullivan *et al.*).

advective transport term to appear in the equations. The curvature term for the interface (5.2) was of course taken into account, such that the phase diagram is displaced by a curvature undercooling $\Delta T_r = \Gamma \Xi$ (fig. 5.3(b)) where Γ is the Gibbs-Thomson coefficient and Ξ the local curvature of the interface. The latter can be directly calculated at every node of the interface from the angles formed by the edges connecting neighboring nodes of the interface. Note that the thermal field was assumed known in this case.

Figure 5.44 shows the instabilities that can be created at the interface of a pure material solidified from an undercooled liquid (negative thermal gradient). The entire domain is initially at a temperature lower than the melting point. At $t = 0$, solidification starts on the left boundary. This situation is thus analogous to that shown in figure 5.30(b). In this case, kinetic undercooling, proportional to the normal velocity v_n^*, has been added to the curvature undercooling term (5.2) to express the equilibrium condition at the solid-liquid interface. As can be seen in the mesh, a node displacement method has been adopted with a remeshing technique to retain a certain fineness at the interface. The displacement of the nodes at the interface is known through the thermal balance (5.15); that of the volume nodes was calculated from … an elasticity calculation! This procedure thus permits the stretching of the mesh in a progressive manner. Note that the result of this calculation was obtained for a small, closed cavity, thus without an advection term in the heat equation.

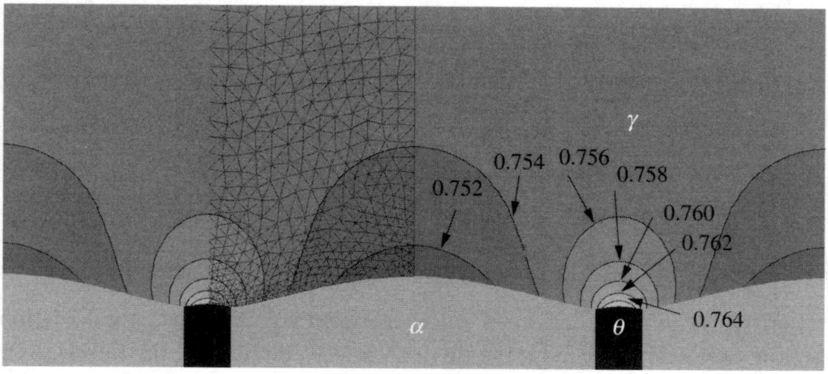

Figure 5.45 Concentration map and mesh of the computational domain used to calculate the pearlite → austenite transformation (0.5 μm lamellar spacing, 1K difference from equilibrium).

The same finite elements technique with front tracking and mesh regeneration was used to model the reaustenization of the pearlitic zones of carbon steel at the microstructure scale during induction heating (fig. 5.36). For that, the initial eutectoid structure was assumed periodic with alternating layers of ferrite α and cementite θ (Fe_3C), the respective fractions of which are given by the equilibrium phase diagram (fig. 5.37). The austenic phase γ nucleates at the grain boundaries between the pearlitic and ferritic zones, and then grows at the expense of the phases α and θ. Here, the mechanism is the inverse of a eutectic or eutectoid growth, with however, a major difference: the spacing of the layers in this case is fixed by the initial conditions. The carbon concentration in α being practically zero, and the phase θ being stoichiometric, the diffusion calculation is limited to the phase γ. The situation is shown in figure 5.45 with the finite element mesh used for the diffusion calculation. This mesh evolves over time as the nodes on the interfaces γ/α and γ/θ move towards the bottom. The velocity of the latter is determined by the flux condition (5.13) in which the concentrations at the interfaces, $(c^{\gamma/\alpha})^*$ and $(c^{\gamma/\theta})^*$, take into account the contribution of the curvature. The temperature of the domain being uniform and known, the contribution of curvature (5.2) can be effectively converted into a term for the difference from the equilibrium of the concentrations at the interfaces (fig. 5.3(b)). After interpolation of the velocities at all the mesh nodes, the solute diffusion equation analogous to (5.161) is solved with an implicit scheme. A new distribution of concentration and the position of the new interface are thus deduced and one then advances to the next time step. Note that when the mesh becomes too deformed, a remeshing procedure is performed.

The concentration map and the form of the interfaces γ/α and γ/θ in figure 5.45 correspond to a steady state for a domain temperature 1 K above the eutectoid temperature. In order to better visualize the periodic nature of the results, they have been reproduced by two symmetric reflections. The closed system conditions used for the lateral edges of the domain (center of the layers) can be noted. It can also be seen that the interfaces γ/α and γ/θ are not isoconcentration surfaces: at the triple point $\gamma/\alpha/\theta$, the curvature contribution is maximal and the concentration variations in the phase γ are very small. Inversely, at the centers of the layers, the curvatures are weak, but the concentration variations are maximal (0.765% and 0.751% here).

5.5.5 Calculation of microstructures with a pseudo front tracking method

For two or three dimensional situations most solute diffusion calculations do not explicitly take into account the flux continuity condition at the interface (5.13). As illustrated in the previous section, it is in fact difficult to track the front. A simple and original technique has recently been developed to overcome the difficulties of front tracking while taking solute conservation at the interface into account. This method, based on the finite volume method, is shown in figure 5.46 for a regular hexagonal mesh (squares would also have worked). In this calculation, the temperature is assumed uniform (or at least known) at the microstructure scale and known at all times. The transformations of ferrito-pearlitic steel under heating are to be described, but this time for the ferritic zones remaining after the pearlitic transformation. Two phases are initially present: the ferritic phase α and the austenitic phase γ coming from the regions that were pearlitic ($\alpha + \theta$). It is assumed that the dissolving of the very fine flakes of pearlite ($\alpha + \theta \rightarrow \gamma$) has already occurred according to the model presented in section 5.5.4, thus transforming the pearlite colonies into austenite zones relatively rich in carbon.

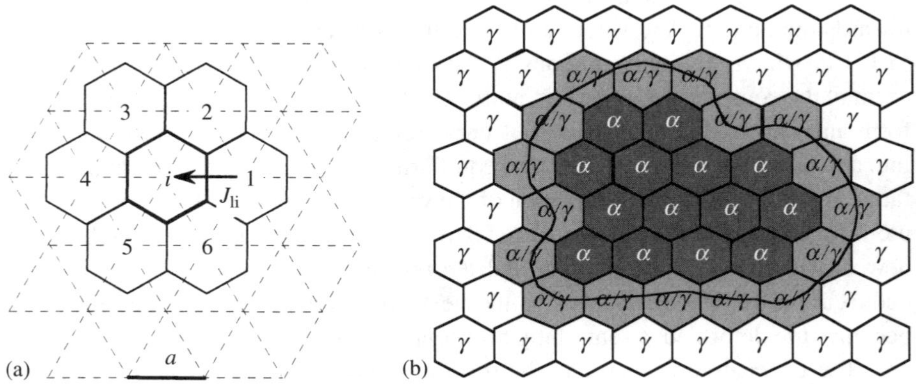

Figure 5.46 Regular hexagonal mesh used to solve the solute diffusion equation with a finite volume method (a) and cell types introduced to take into account the phases α, γ and the solute balance at the interface α/γ (b).

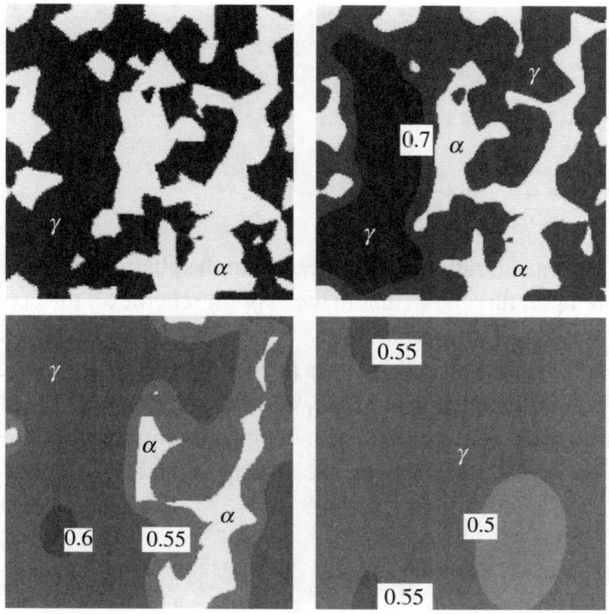

Figure 5.47 Calculated carbon concentration distribution and distribution of the phases in an iron-0.49wt% carbon alloy during 10 K s^{-1} heating starting from the eutectoid temperature: $t = 0$ (a), 2.5 s (b), 5 s (c) and 8 s (d).

The method thus consists of taking an initial microstructure composed of hexagonal cells α and γ. These zones are shown as light and dark respectively in figure 5.47(a). The initial proportions of these cells are given by the lever rule at the eutectoid temperature and their distribution is given by a Voronoi type algorithm (chap. 4) starting from randomly distributed nucleation centers. The pseudo front tracking method then consists of introducing a layer of α/γ interface cells between cells of different types. These interface cells have a phase fraction $f_\alpha = 1$ initially. The diffusion equations in the phases α and γ are then solved using an explicit finite volumes method. If a is the hexagonal grid parameter, it is easy to show that the solute balance for each volume can be reduced to:

$$V\frac{\Delta c_{\mathrm{i}}}{\Delta t} = \sum_{\mathrm{k}} SJ_{\mathrm{ki}} \tag{5.170}$$

where $V = \sqrt{3}/2a^2$ is the volume (the area in two dimensions) of the cell and $S = a/\sqrt{3}$ is the surface area (the length) across which the solute flux J_{ki} passes. The sum is over the six nearest neighbors k of cell i. If the two cells i and k belong to the same phase ν ($\nu = \alpha$ or γ), this flux is simply given by:

$$J_{\mathrm{ki}} = D^{\mathrm{v}} \frac{c_{\mathrm{k}} - c_{\mathrm{i}}}{a} \tag{5.171}$$

whereas if k is an interface cell:

$$J_{\mathrm{ki}} = D^{\mathrm{v}} \frac{c_{\mathrm{k}}^{\mathrm{v}*} - c_{\mathrm{i}}}{a} \tag{5.172}$$

where $c_{\mathrm{k}}^{\mathrm{v}*}$ is the concentration at the interface in the phase ν given by the phase diagram and the temperature. Note that in the explicit scheme all the concentrations are taken at time step t and are thus known.

For the cells α and γ, (5.170) allows the new concentrations c_{i} to be found. For the interface cells α/γ, the variation Δc_{i} corresponds to the average concentration weighted over the two phases. The new average concentration, c_{i}, thus found is then transformed into a phase fraction $f_{\alpha,\,\mathrm{i}}$ according to the lever rule for each cell:

$$f_{\alpha,\,\mathrm{i}} = \frac{c_{\mathrm{i}}^{\gamma*} - c_{\mathrm{i}}}{c_{\mathrm{i}}^{\gamma*} - c_{\mathrm{i}}^{\alpha*}} \tag{5.173}$$

When $f_{\alpha,\,\mathrm{i}} = 0$, the interface cell becomes a γ cell and new interface cells are introduced in its immediate neighborhood to separate it from the α cells.

The concentration maps of figure 5.47 show the austenitization of a Fe-0.49wt% C steel during continuous heating at 10 K s^{-1}. The different gray levels correspond to the iso-concentration lines. In (a), the initial microstructure clearly shows the distribution of phases α (ferritic steel zones) and γ (old pearlitic zones) at the eutectoid temperature. Figures (b), (c) and (d) correspond to the times 2.5, 5 and 8 seconds (i.e., 25, 50 and 80 °C above the eutectoid temperature). The dimension of the zone shown is 12.5 μm square. As can be seen, the steel is totally austenitized and practically homogeneous in concentration after 8 s when it is heated at 10 K s^{-1}.

In figure 5.47, the type of situation modeled is stable and so is the pseudo front tracking method: it corresponds to the reduction of the α zones by diffusion towards the exterior (γ zones). In an inverse situation (growth of α by diffusion into γ), the physical situation – and the numerical method – become unstable. For example, a calculation of solidification starting from a seed will create the dendrite in figure 5.48: as can be seen, the method reproduces the unstable nature of the physical situation. However, the direction of the growth of the dendrites follows exactly the axes of the hexagonal grid and the radius of curvature of the tip is directly related to the size of the elements. A real dendrite is the result of two competing contributions: diffusion, which tends to destabilize the interface, and curvature, which acts in the opposite direction. The latter contribution can be introduced in the pseudo front tracking method: the curvature is given by the divergence of the normal at the interface $\boldsymbol{n}^{s/\ell}$, which can be calculated with the normalized gradient of the solid fraction, $\mathbf{grad}\, f_{\mathrm{s}} |\mathbf{grad}\, f_{\mathrm{s}}|^{-1}$ (exercise 5.6.7). However, the solid fraction is rather

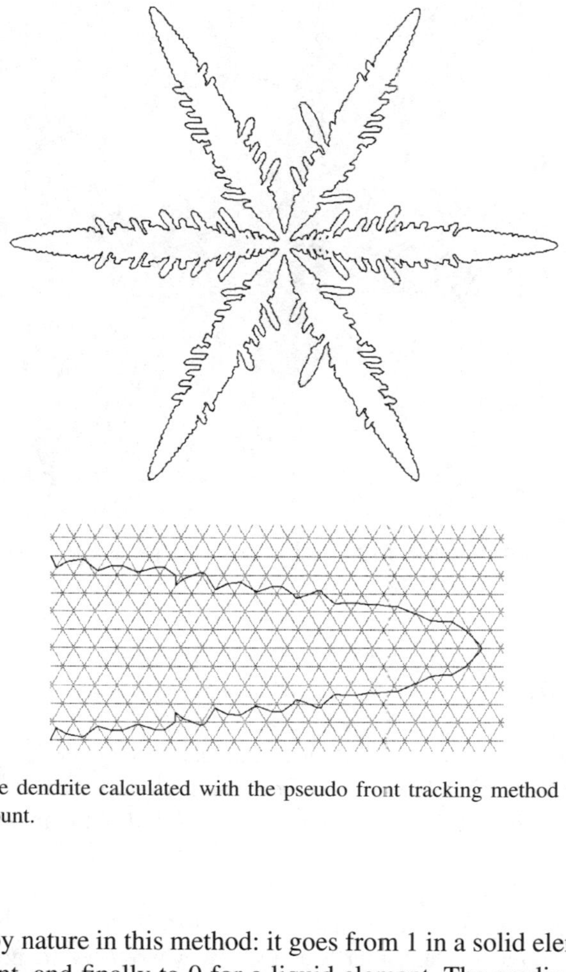

Figure 5.48 Solute dendrite calculated with the pseudo front tracking method without taking the curvature into account.

discontinuous by nature in this method: it goes from 1 in a solid element, to f_s in an interface element, and finally to 0 for a liquid element. The gradient is thus rather poorly defined and the curvature estimation is not very precise. The *phase field* method, in which the Laplacian of the solid fraction also appears, diffuses the solid-liquid interface over many elements and thus allows much more precision in the calculation of the curvature contribution. Although inevitably introducing anisotropies linked to the mesh, this method allows the simulation of the formation of dendrites for which the growth is not necessarily aligned with the grid axes (fig. 5.49). The interested reader is referred to the books in the bibliography.

Figure 5.49 Solute dendrite calculated with a phase field method.

5.6 EXERCISES

5.6.1. Assuming a random distribution of nucleation centers for grains of α in a matrix β, prove Avrami's relation (5.3). For that, take into account the fact that the probability for a point arbitrarily chosen in space to fall in the matrix β is given by $(1 - f_\alpha)$.

5.6.2. Using (5.4) and (5.5), show that $f_\alpha^0(t) = 4\pi/3v^3Nt^3$ for the instantaneous nucleation case and a constant interface velocity, where N is the grain density, and that $f_\alpha^0(t) = \pi/3v^3It^4$ if the seeds appear continuously with a constant nucleation rate, I.

5.6.3. Derive (5.16) for the continuity of the normal stresses at the interface between two media.

5.6.4. Consider a material solidifying in a directional and stationary manner ($f_s(x, t) = f_s(x - v_T t)$, where v_T is the velocity of the advancing isotherms), and for which the density is such that ρ^s = constant and ρ^ℓ = constant. Show, using (5.46), that the microscopic velocity of the liquid between the solid particles is constant and given at every point of the region by:

$$v_x^\ell = -\left[\frac{\rho^s}{\rho^\ell} - 1\right] v_T$$

How does the effective liquid flux vary? Show that in this case the pressure gradient in the liquid is given by:

$$\frac{\mathrm{d}p}{\mathrm{d}x} = \frac{\mu}{K}\left[\frac{\rho^{\mathrm{s}}}{\rho^{\ell}} - 1\right] f_{\ell} v_{\mathrm{T}}$$

What does it correspond to? What could be the consequences?

5.6.5. Starting from the Navier-Stokes equation (1.103) for an incompressible fluid and the Laplacian expression in cylindrical coordinates (chap. 10), deduce (5.85) describing the distribution of the steady state velocity $v_{\mathrm{x}}(r)$ in a pipe.

5.6.6. Prove the closure relation (5.127) for a set of internal facets of a closed domain. What happens in the case of a calculation domain that is not closed?

5.6.7. Prove that the curvature Ξ of a solid/liquid interface such as that used in the pseudo front tracking method is given by:

$$\Xi = \mathrm{div}\left\{\frac{\mathbf{grad}\, f^{\mathrm{s}}}{|\mathbf{grad}\, f^{\mathrm{s}}|}\right\}$$

5.7 BIBLIOGRAPHY

R. B. BIRD, W. E. STEWART and E. N. LIGHTFOOT, *Transport Phenomena,* J. Wiley & Sons, New York, 1960.

F. P. INCROPERA and D. P. DE WITT , *Fundamentals of Heat and Mass Transfer,* J. Wiley & Sons, New York, 2nd Edition, 1985.

A. JACOT, M. RAPPAZ, "A Two-Dimensional Diffusion Model for the Prediction of Phase Transformations : Application to Reaustenitisation and Homogenisation of Hypoeutectoid Fe-C Steels", *Acta Mater.* vol. 45 (1997) 575–85.

A. JACOT, M. RAPPAZ and R. REED, "Modelling of Reaustenitisation from the Pearlite Structure in Steel", *Acta Mater.* vol. 46 (1998) 3949.

D. R. LYNCH, "Unified Approach to Simulation on Deforming Elements with Application to Phase Change Problems", *J. Computational Physics* 47 (1982) 387-411.

D. A. PORTER and K. E. EASTERLING, *Phase Transformations in Metals and Alloys,* Van Nostrand Rheinhold, Berkshire, Great Britain, 1984.

R. SIEGEL and J. R. HOWELL, *Thermal Radiation Heat Transfer,* Mc Graw Hill, New York, 1981.

J. M. SULLIVAN, Jr and D. R. LYNCH, *Grid Generation for Dendritic Growth Simulations on Deforming Elements*, in *Numerical Methods in Thermal Problems*, Eds. R. W. Lewis, K. Morgan and W. G. Habashi (Pineridge Press, Cambridge, 1987) p. 5.

J. SZEKELY, *Fluid Flow Phenomena in Metals Processing,* Academic Press, New York, 1979.

K. TSIVERIOTIS and R. A. BROWN, "Solution of Free-Boundary Problems using Finite-Element/Newton Methods and Locally Refined Grids: Application to Analysis of Solidification Microstructure", *Int. Journal Numer. Methods Fluids*, vol. 16, 1993.

J. WARREN, W. J. BOETTINGER, "Prediction of Dendritic Growth and Microsegregation Patterns in a Binary Alloy using the Phase-Field Method", *Acta Metall. Mater.*, vol. 43, (1995) 689.

CHAPTER 6

DEFORMATION OF SOLIDS

6.1 OBJECTIVES

- Introduce the concepts of irreversible deformations of continuous solid media: the plasticity criterion and the law of plastic flow.
- Review the constitutive equations for elastoplastic, elasto-viscoplastic and pure viscoplastic materials.
- Present the unilateral contact conditions and the main friction laws.
- Introduce the principal characteristics of the numerical treatment of the equilibrium equations.

In chapter 1, the notions of elastic deformation and the ideally viscous medium (Newtonian fluid) were presented. However, most materials, either solid or liquid, very rarely follow such idealized behavior. Chapter 7 will treat non Newtonian fluids such as molten polymers.

In this chapter we center attention on irreversible deformations of solid state metals and polymers. We introduce the main constitutive laws, insisting on the notion of plasticity: the property of some materials to undergo permanent deformation without breaking, with the primary application being forming.

We subsequently give a view of the boundary conditions characteristic of a plastic deformation problem, presenting the modeling of contact and friction.

We then discuss the numerical treatment of the equations to be solved and show how the finite element method, associated with an incremental procedure, allows the modeling of complex forming processes.

6.2 CONSTITUTIVE EQUATIONS

6.2.1 Elastoplasticity

Plastic behavior: one dimensional phenomenological study

The behavior of a cold metallic test sample, with an applied force in simple tension (fig. 6.1), shows us all the necessary elements for the comprehension of the plasticity phenomenon common to most metals when deformed at low temperature and to some solid state polymers.

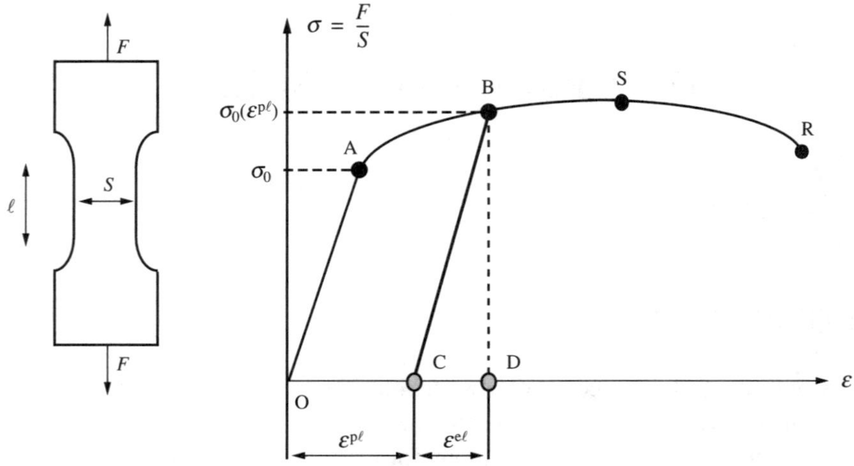

Figure 6.1 Uniaxial tension of a metallic test sample.

The sample undergoes an imposed global strain, noted ε, for which the stress $\sigma(\varepsilon)$ is recorded; this stress is equal to F/S, F being the measured force, and S the varying area of the transverse section of the sample. Different regions of the curve $\sigma(\varepsilon)$ (fig. 6.1), corresponding to different behaviors, can be distinguished.

Between the points O and A, the curve is a straight line, the slope of which is Young's modulus, or the elastic modulus. We have then the simple relation:

$$\sigma = E\varepsilon \tag{6.1}$$

which is the *linear elastic constitutive law*, seen already in chapter 1. The essential characteristic of the deformation is then its *reversibility:* when the load is released (thus $\sigma = 0$), the sample returns to its initial length ℓ_0 giving $\varepsilon = 0$ at the end.

Conversely, beyond point A, that is past the stress σ_0 (***plasticity threshold***), *irreversible* or *plastic* deformation occurs: when the sample is freed from a point such as B, it returns along BC. It is seen that the sample does not return to its original length ℓ_0 and that there exists a residual plastic strain denoted $\varepsilon^{p\ell}$. Thus the strain at point B is decomposed into an elastic part $\varepsilon^{e\ell}$ and an irreversible plastic part:

$$\varepsilon = \varepsilon^{e\ell} + \varepsilon^{p\ell} \tag{6.2}$$

Note that the unloading path of the material after release is represented by a straight line (BC), the slope of which is Young's modulus E. The unloading is elastic; if a new elongation is imposed on the sample from the state represented by the point C, the path followed is again the line CB. New plastic strain occurs beyond point B. The plasticity threshold beyond which plastic strain is created, also called the ***flow stress***, has changed with respect to the initial threshold σ_0. In general, the flow stress depends on the plastic strain already experienced by the material according to an increasing function $\sigma_0(\varepsilon^{p\ell})$; this is the phenomenon of ***work hardening***. In the opposite case (decrease of the flow stress with $\varepsilon^{p\ell}$), we have ***softening***.

Beyond the point S, the cross sectional area of the test sample no longer remains uniform. The tensile force passes through a maximum and an instability is observed leading to the local diminution of the section: the ***necking*** phenomenon. The stress calculated by the ratio of the tensile force to the section calculated assuming homogeneous strain decreases regularly until rupture, shown by the curve SR in figure 6.1. A more precise analysis in the section which is reduced by necking shows that the real stress state is no longer one dimensional. The flow stress that can be deduced continues in fact to increase, as long as the material does not undergo softening.

In conclusion, we should retain the ideas that plasticity is irreversible and incremental by nature. Contrary to elasticity, there is no longer a one-to-one relation between a state of stress and a state of strain. To correctly describe the plastic behavior of a material it is necessary to take its history into account.

Plasticity criterion for the multi axial case

We pose the problem in the following manner: given a stress tensor $\boldsymbol{\sigma}$ at a point in the material, it is necessary to decide if the state of the stress corresponds to the elastic domain, or if it is in the plastic domain. To solve this problem we employ a mathematical function *f*, called the ***plasticity criterion***, which possesses the following characteristics:

$$\left.\begin{array}{ll} f(\boldsymbol{\sigma}) < 0 & \text{characteristic of a local elastic state,} \\ f(\boldsymbol{\sigma}) = 0 & \text{corresponds to a state of local plastic flow,} \\ f(\boldsymbol{\sigma}) > 0 & \text{is considered as unphysical and is thus prohibited.} \end{array}\right\} \quad (6.3)$$

In stress space, we call the surface of the equation $f(\boldsymbol{\sigma}) = 0$ the ***plasticity surface***, or ***load surface***, the ***elastic domain*** being defined by $f(\boldsymbol{\sigma}) < 0$. We must then give the form of the function *f*. For that, we employ some experimental observations, but also simplifying assumptions.

Isotropic assumption

This simplifying – yet nonetheless very often acceptable – assumption insures the invariance of *f* regardless of the coordinates chosen for the expression of the stresses, no spatial direction being preferred (for example, by the morphological or crystallographic texture if the material is polycrystalline). In particular, the tensor $\boldsymbol{\sigma}$ being symmetric, the principal coordinates of the stresses can be used to express *f*:

$$f(\sigma_{\text{I}}, \sigma_{\text{II}}, \sigma_{\text{III}}) \leq 0 \quad (6.4)$$

where σ_{I}, σ_{II}, σ_{III}, are the principal stresses (eigenvalues of $\boldsymbol{\sigma}$). Moreover, the assumption requires that a permutation of two arguments in (6.4) does not change the value of the function (f is symmetric with respect to σ_{I}, σ_{II}, σ_{III}) :

$$f(\sigma_{\mathrm{II}}, \sigma_{\mathrm{I}}, \sigma_{\mathrm{III}}) = f(\sigma_{\mathrm{I}}, \sigma_{\mathrm{II}}, \sigma_{\mathrm{III}}), \text{ etc.} \tag{6.5}$$

Assumption of equivalence of tension and compression

This assumption means that for every state of stress:

$$f(-\sigma_{\mathrm{I}}, -\sigma_{\mathrm{II}}, -\sigma_{\mathrm{III}}) = f(\sigma_{\mathrm{I}}, \sigma_{\mathrm{II}}, \sigma_{\mathrm{III}}) \tag{6.6}$$

Note that most metals do not verify this simplifying assumption. In fact, after a first plastic strain in uniaxial tension, if the load is reversed, the tension is released and replaced with a uniaxial compression, and the material will undergo plastic strain at a value of compression stress lower than the flow stress in tension (***Bauschinger effect***). We will return later to this assumption.

From (6.6), the volume in principal stress space defined by (6.4) (elastic domain + plasticity surface) is symmetric with respect to the origin. In addition, its intersection with each of the principal planes is symmetric with respect to the bisectors of each of these planes (6.5).

Assumption of independence from the hydrostatic pressure

It is an experimental reality for metallic materials and solid state polymers that can be verified by, for example, an ***oedometric test*** (fig. 6.2). This test consists of submitting a sample to a hydrostatic pressure applied by a piston in a rigid container.

The ambient pressure affects neither the appearance of plastic strain, nor the plastic flow itself. This can be generalized by the superposition of a purely hydrostatic state of stress on the stress state defined by the tensor $\boldsymbol{\sigma}$. Thus for any pressure P:

$$f(\sigma_{\mathrm{I}} + P, \sigma_{\mathrm{II}} + P, \sigma_{\mathrm{III}} + P) = f(\sigma_{\mathrm{I}}, \sigma_{\mathrm{II}}, \sigma_{\mathrm{III}}) \tag{6.7}$$

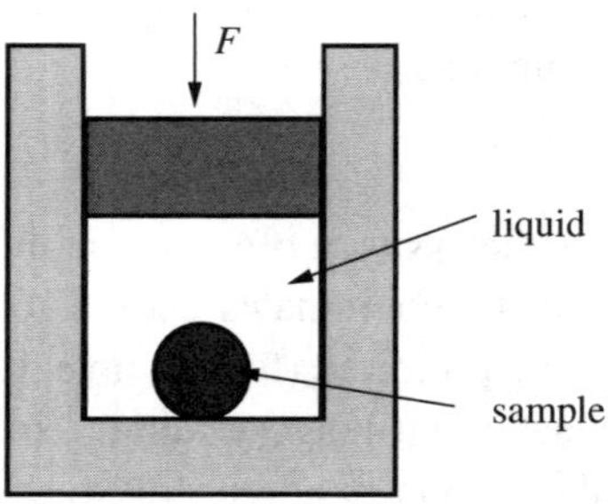

Figure 6.2 Oedometric test demonstrating the impossibility of plastifying a metal by increasing the hydrostatic pressure.

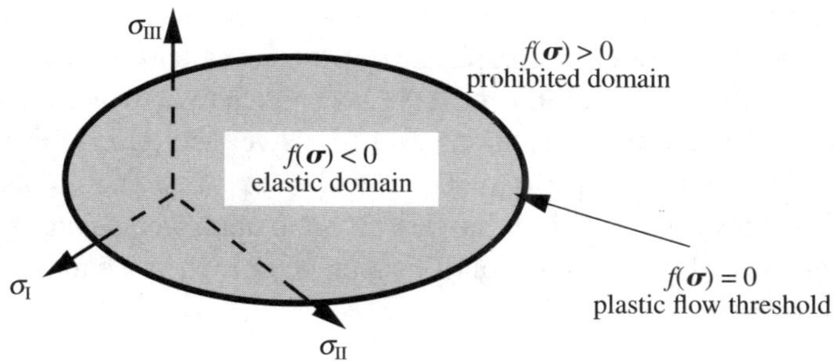

Figure 6.3 Sketch of the elastic domain and the plasticity surface.

This relation implies that the plasticity surface is a cylinder in the space of principal stresses with the axis parallel to (1,1,1). Moreover, taking P to be the *hydrostatic pressure p associated with the stress tensor* $\boldsymbol{\sigma}$ ($p = -1/3 \,\mathrm{Tr}\boldsymbol{\sigma}$), one concludes that it should be possible to express the plasticity criterion as a function only of the principal components of the deviatoric stress tensor $\boldsymbol{s} = \boldsymbol{\sigma} + p\boldsymbol{I}$ (chap. 1; $\boldsymbol{s}$ is a tensor with a trace equal to zero, and the same principal directions as $\boldsymbol{\sigma}$):

$$f(s_I, s_{II}, s_{III}) = f(\sigma_I, \sigma_{II}, \sigma_{III}) \tag{6.8}$$

Assumption of convexity

As we will see later on, the maximum work principle implies that the plasticity surface is convex. The elastic domain and the plasticity surface represented by equations (6.3) can then be sketched as in figure 6.3.

Tresca's plasticity criterion

It seems that this was the first plasticity criterion that was proposed, more than a century ago. Tresca postulated that plastification occurs at a point in the material when the maximal shear stress at the point, τ_{max}, reaches a critical value, τ_0. This assumption has its origin in the microscopic plastic phenomena of the sliding of the most dense atomic planes in certain directions with respect to each other (we will return to this point further on). The plasticity criterion for this assumption is then written :

$$f = \tau_{max} - \tau_0 \tag{6.9}$$

where τ_0 is the shear limit, a characteristic of the material under consideration. Analysis of Mohr's circle shows clearly that the maximal shear τ_{max}, attained on two particularly oriented facets, is equal to half the maximum difference of the principal stresses. The expression (6.9) of the criterion is then equivalent (within a factor of 2) to:

$$f = \underset{I, J = 1,3}{\mathrm{Max}} |\sigma_I - \sigma_J| - 2\tau_0 \tag{6.10}$$

In other words, plastification occurs when the maximum difference between two of the three principal components of the stress tensor attains a critical value equal to double the shear limit τ_0. It is easy enough to convince oneself that the criterion satisfies the four conditions evoked above. Moreover, for a uniaxial stress state where $\sigma_I \neq 0$ and $\sigma_{II} = \sigma_{III} = 0$, one has $f = \sigma_I - 2\tau_0$. The variable τ_0 introduced in (6.9) is thus equal to half of the flow stress in uniaxial tension, σ_0. Tresca's criterion signifies – by (6.9) – that plastification occurs when the maximal shear attains the critical value $\sigma_0/2$.

Representation in the space of principal stresses

We have seen that the load surface must be a cylinder. For the case of Tresca's criterion, the base of the cylinder is a regular hexagon defined in the plane perpendicular to the direction (1,1,1), with sides of $\sqrt{2/3}\,\sigma_0$ (fig. 6.4).

von Mises plasticity criterion

von Mises proposed a criterion close to Tresca's, but with a form that is mathematically simpler to manipulate. It is a cylinder of revolution circumscribed about the hexagonal cylinder of Tresca's criterion.

Representation in the space of principal stresses

The function for von Mises' criterion is written:

$$f = (\sigma_I - \sigma_{II})^2 + (\sigma_{II} - \sigma_{III})^2 + (\sigma_{III} - \sigma_I)^2 - 2\sigma_0^2 \tag{6.11}$$

It is immediately verified that σ_0 in (6.11) is the flow stress in uniaxial tension by taking $\sigma_I \neq 0$ and the other components equal to zero.

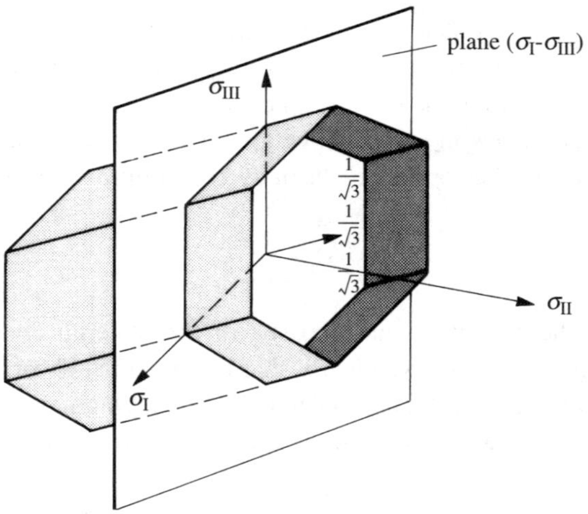

Figure 6.4 Tresca's criterion in principal stress space.

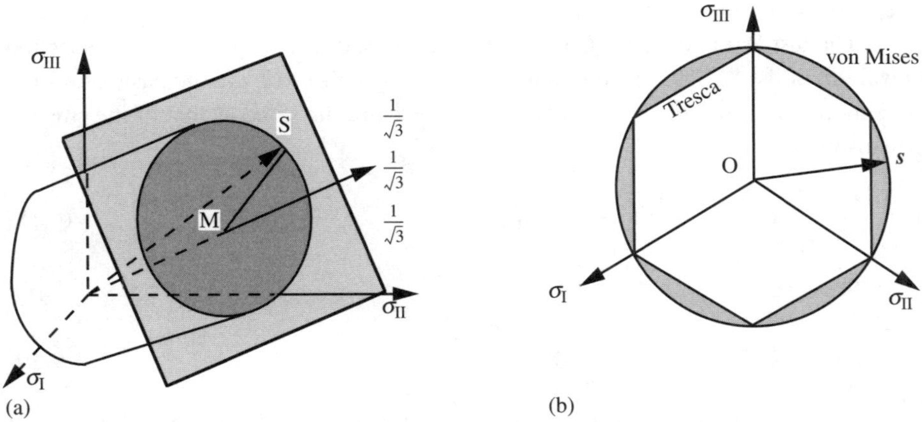

Figure 6.5 (a) von Mises plasticity surface in principal stress space and the octahedric plane; (b) comparison of the criteria of von Mises and Tresca in the deviatoric plane.

Using the deviatoric stress $\mathbf{s}$, in (6.11), another form of the criterion is obtained, noting that $\sigma_I - \sigma_J = s_I - s_J$ (for $I, J = 1, 3$) and that $\mathbf{s}$ has a trace equal to zero:

$$f' = \tfrac{1}{3} f = s_I^2 + s_{II}^2 + s_{III}^2 - \tfrac{2}{3}\sigma_0^2 \tag{6.12}$$

Given the points S and M representative of stress state and its hydrostatic part, respectively (fig. 6.5(a)); one has:

$$\mathrm{OM} = \begin{bmatrix} -p \\ -p \\ -p \end{bmatrix} \quad \mathrm{OS} = \begin{bmatrix} \sigma_I \\ \sigma_{II} \\ \sigma_{III} \end{bmatrix} \quad \mathrm{MS} = \begin{bmatrix} \sigma_I + p \\ \sigma_{II} + p \\ \sigma_{III} + p \end{bmatrix} = \begin{bmatrix} s_I \\ s_{II} \\ s_{III} \end{bmatrix} \tag{6.13}$$

The vector MS is orthogonal to the axis of the plasticity criterion (OM · MS = 0). The plane orthogonal to this axis containing the points M and S is called the ***octahedric plane***. The parallel plane passing through the origin of the coordinates is called the ***deviatoric plane***. It is seen that:

$$\mathrm{MS}^2 = s_I^2 + s_{II}^2 + s_{III}^2 \tag{6.14}$$

which shows that the von Mises plasticity surface is a cylinder with a circular base, oriented along the axis (1,1,1) of the principal stress coordinates, and with a radius $\sqrt{2/3}\,\sigma_0$. The intersection of this cylinder with the deviatoric plane is thus a circle, which is circumscribed about a regular hexagon representing Tresca's criterion (fig. 6.5(b)).

Representation in arbitrary coordinates

The last form obtained (6.12) is particularly seducing as it involves the second invariant of the deviatoric stress tensor, classically denoted J_2. In the basis of principal coordinates of the deviatoric stresses (which are the same as that of the stresses themselves):

$$J_2 = \frac{1}{2}\mathrm{Tr}(s^2) = \frac{1}{2}(s_{\mathrm{I}}^2 + s_{\mathrm{II}}^2 + s_{\mathrm{III}}^2) \tag{6.15}$$

thus,

$$f'' = \frac{3}{2}f' = \frac{1}{2}f = 3J_2 - \sigma_0^2 \tag{6.16}$$

It is thus possible to express the criterion function in arbitrary coordinates (not necessarily the principal axes) using the invariance of J_2. In any orthonormal Cartesian coordinate system (using the summation convention for repeated indices and the symmetry of the deviatoric tensor):

$$J_2 = \frac{1}{2}(s^2)_{\mathrm{ii}} = \frac{1}{2}s_{\mathrm{ij}}s_{\mathrm{ji}} = \frac{1}{2}s_{\mathrm{ij}}s_{\mathrm{ij}} = \frac{1}{2}\boldsymbol{s} : \boldsymbol{s} \tag{6.17}$$

The ***von Mises equivalent stress*** σ_{eq} is thus defined:

$$\sigma_{\mathrm{eq}} = \sqrt{\frac{3}{2}s_{\mathrm{ij}}s_{\mathrm{ij}}} = \sqrt{\frac{3}{2}\boldsymbol{s} : \boldsymbol{s}} \tag{6.18}$$

and the criterion function is expressed in the form:

$$f'' = \sigma_{\mathrm{eq}}^2 - \sigma_0^2 \tag{6.19}$$

or with σ_{eq} and σ_0 being positive,

$$f''' = \sigma_{\mathrm{eq}} - \sigma_0 \tag{6.20}$$

One then grasps the concept of the plasticity criterion which consists of reducing a multiaxial stress state in the material characterized by a tensor, to a simple *scalar equivalent* that when compared with the plasticity threshold of the material in uniaxial stress is sufficient to know whether there is plastification, and thus appearance of irreversible strain.

In simple shear, the stress tensor has only one non zero component (one of the off diagonal components, for example: $\sigma_{xz} = \sigma_{zx} = \tau \neq 0$). Then $\sigma_{\mathrm{eq}} = \sqrt{3}\,\tau$, and one deduces from (6.20) that the maximum shear at plastification is $\tau_{\max} = \sigma_0/\sqrt{3}$ (compared to $\sigma_0/2$ for Tresca's criterion).

The reader can practice by proving, starting from (6.16) and (6.17), that in non principal directions (x, y, z), von Mises' criterion can be expressed directly from the components of the stress tensor:

$$f = (\sigma_{xx} - \sigma_{yy})^2 + (\sigma_{yy} - \sigma_{zz})^2 + (\sigma_{zz} - \sigma_{xx})^2 + 6\sigma_{xy}^2 + 6\sigma_{yz}^2 + 6\sigma_{zx}^2 - 2\sigma_0^2 \tag{6.21}$$

Comparison of Tresca's and von Mises' plasticity criteria in a simple case: plane stress state

Consider a plane stress state, characterized in principal axes by a tensor of components:

$$\boldsymbol{\sigma} = \begin{bmatrix} \sigma_I & 0 & 0 \\ 0 & \sigma_{II} & 0 \\ 0 & 0 & 0 \end{bmatrix} \tag{6.22}$$

The elastic domains ($f < 0$) of Tresca's and von Mises' criteria are given by the equations (applying (6.10) and (6.11), respectively):

$$\text{Tresca: } |\sigma_I - \sigma_{II}| < \sigma_0 \quad \text{and} \quad |\sigma_I| < \sigma_0 \quad \text{and} \quad |\sigma_{II}| < \sigma_0 \tag{6.23}$$

$$\text{von Mises: } \sigma_I^2 + \sigma_{II}^2 - \sigma_I \sigma_{II} < \sigma_0^2 \tag{6.24}$$

The inequalities (6.23) define the interior of a hexagon in the plane (σ_I, σ_{II}) (fig. 6.6): the intersection of Tresca's hexagonal cylinder (fig. 6.4) and the plane $\sigma_{III} = 0$. The inequality (6.24) delimits the interior of an ellipse, the intersection of von Mises' circular cylinder and the same plane.

The difference between the two criteria can be seen, notably in pure shear (two principal stresses of the same modulus but opposite sign), where the values $\sigma_0/\sqrt{3}$ for von Mises and $\sigma_0/2$ for Tresca are again found.

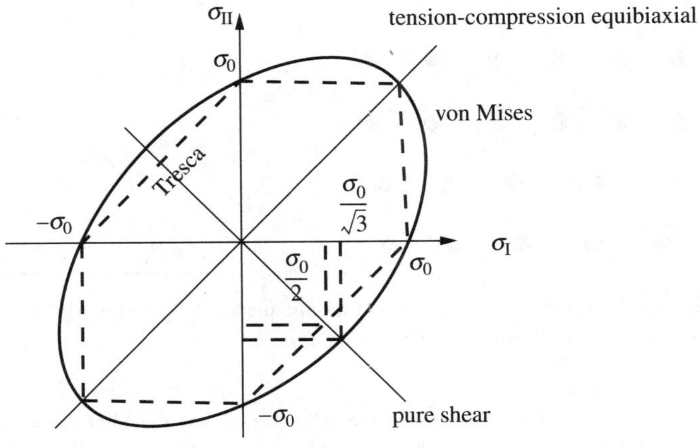

Figure 6.6 Tresca's and von Mises' criteria for plane stresses.

The law of plastic flow: maximum work principle

Having introduced the notion of the plasticity criterion, and remembering the incremental nature of plastic strain, we now examine the plastic flow rule, that is the form of the plastic strain rate tensor, $\dot{\boldsymbol{\varepsilon}}^{p\ell}$. The tensor $\dot{\boldsymbol{\varepsilon}}^{p\ell}$ is the irreversible part of the strain rate tensor $\dot{\boldsymbol{\varepsilon}}$ (the symmetric part of the velocity gradient tensor, also denoted $\boldsymbol{d}$, introduced in chapter 1, Sect. 1.2.5). The multi-dimensional and incremental generalization of (6.2) is given by the following tensor relation:

$$\dot{\boldsymbol{\varepsilon}} = \boldsymbol{d} = \dot{\boldsymbol{\varepsilon}}^{e\ell} + \dot{\boldsymbol{\varepsilon}}^{p\ell} \quad \text{or in components,} \quad \dot{\varepsilon}_{ij} = \dot{\varepsilon}^{e\ell}_{ij} + \dot{\varepsilon}^{p\ell}_{ij} \tag{6.25}$$

As we are dealing with *instantaneous* plasticity, it should be noted that time simply plays the role of an arbitrary chronological scale. The following developments could as well have been derived from a relation for the infinitesimal tensor increments $d\boldsymbol{\varepsilon} = d\boldsymbol{\varepsilon}^{e\ell} + d\boldsymbol{\varepsilon}^{p\ell}$.

In the case of metal alloys, the strain generally involves an aggregation of crystalline grains. The plastic strain corresponds to the permanent strain of the grains. It occurs by slip of atomic planes (in general the densest planes) with respect to one another in the direction of an atomic ordering; the combination of a plane and a direction is called a ***slip system***[1]. These elementary slips result from the propagation of crystal lattice defects (dislocations) and explain how *plastic strain does not cause any relative volume change*. The condition for the activation of a slip system is given by Schmid's law, as explained in figure 6.7.

The slip system $s = (\pi, \boldsymbol{u})$ is active when the vector $\boldsymbol{\tau}_s$, projection in the slip direction $\boldsymbol{u}$ of the stress vector $\boldsymbol{T}$ acting on the slip plane π, attains the modulus of a critical value τ_{cs}. The vector $\boldsymbol{\tau}_s$ is called ***resolved shear*** in the slip system, τ_{cs} being the ***critical resolved shear***, a characteristic value of the slip system under consideration.

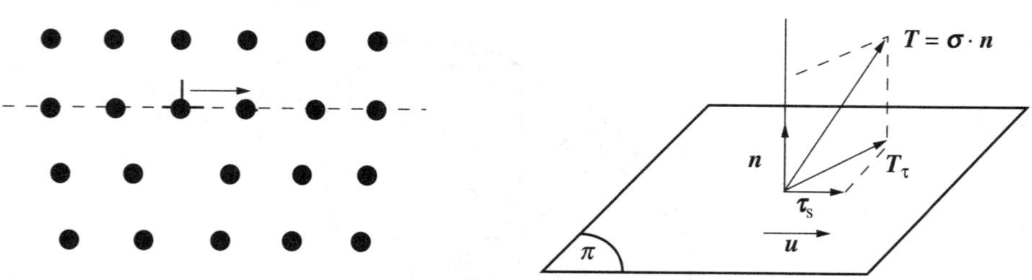

Figure 6.7 Sketch of a relative slip of atomic planes, by propagation of the dislocations, defects in the crystal lattice. Illustration of Schmid's law.

1 In the following development, we will not enter into the details of dislocation theory, for which the reader is referred to the books cited in the bibliography. The goal pursued here is to give the reader an overview of the physical justifications of the macroscopic laws of plasticity.

For a single crystal of given crystallographic structure, there are thus a finite number, N, of slip systems. For example, for metals with a cubic face centered structure (cfc), one has the combinations of the dense planes $\{111\}$ and the directions $\langle 110 \rangle$ (so $N = 12$ possible systems, combinations of the four planes of the family $\{111\}$ and the three directions of the family $\langle 110 \rangle$[2]). A plastic strain, characterized by a tensor $\dot{\boldsymbol{\varepsilon}}^{p\ell}$, results from the activation of A systems among the N possible. One measures thus a slip velocity, $\boldsymbol{v}_s$, for these systems. Assuming for simplicity that they are the first A,

$$\text{for} \quad 1 \le s \le A \quad \text{we have} \quad |\boldsymbol{\tau}_s| = \tau_{cs} \quad \boldsymbol{v}_s \ne \mathbf{0} \tag{6.26}$$

$$\text{for} \quad A < s \le N \quad \text{we have} \quad |\boldsymbol{\tau}_s| < \tau_{cs} \quad \boldsymbol{v}_s = \mathbf{0} \tag{6.27}$$

Now consider a virtual stress state, $\boldsymbol{\sigma}^*$, that respects Schmid's law for all the slip systems (a stress state called ***plastically admissible***). We have thus $|\boldsymbol{\tau}_s^*| \le \tau_{cs}$ for every system s. We compare the strain power dissipated in plastic flow by the stress field that actually exists in the single crystal, $\boldsymbol{\sigma}$, and that by the virtual stress field $\boldsymbol{\sigma}^*$. In the first case, we have (the index s is not summed in the integral) :

$$\int_\Omega \boldsymbol{\sigma} : \dot{\boldsymbol{\varepsilon}}^{p\ell}\, dV = \sum_{s=1}^{N} \int_{\pi(s)} \boldsymbol{\tau}_s \cdot \boldsymbol{v}_s\, dS = \sum_{s=1}^{A} \int_{\pi(s)} \boldsymbol{\tau}_s \cdot \boldsymbol{v}_s\, dS + \sum_{s=A+1}^{N} \int_{\pi(s)} \boldsymbol{\tau}_s \cdot \mathbf{0}\, dS$$

$$= \sum_{s=1}^{A} \tau_{cs} |\boldsymbol{v}_s|\, dS \tag{6.28}$$

whereas for $\boldsymbol{\sigma}^*$, we have:

$$\int_\Omega \boldsymbol{\sigma}^* : \dot{\boldsymbol{\varepsilon}}^{p\ell}\, dV = \sum_{s=1}^{A} \int_{\pi(s)} \boldsymbol{\tau}_s^* \cdot \boldsymbol{v}_s\, dS + \sum_{s=A+1}^{N} \int_{\pi(s)} \boldsymbol{\tau}_s^* \cdot \mathbf{0}\, dS \le \sum_{s=1}^{A} \int_{\pi(s)} \tau_{cs} |\boldsymbol{v}_s|\, dS \tag{6.29}$$

We deduce finally the ***principle of maximum work***: for a given plastic strain rate $\dot{\boldsymbol{\varepsilon}}^{p\ell}$, among all the plastically admissible stress states, it is the real stress that dissipates the highest plastic strain power. This is expressed by:

$$\forall \boldsymbol{\sigma}^* | f(\boldsymbol{\sigma}^*) \le 0, \quad \sigma^*_{ij} \dot{\varepsilon}^{p\ell}_{ij} \le \sigma_{ij} \dot{\varepsilon}^{p\ell}_{ij} \quad \text{or} \quad \boldsymbol{\sigma}^* : \dot{\boldsymbol{\varepsilon}}^{p\ell} \le \boldsymbol{\sigma} : \dot{\boldsymbol{\varepsilon}}^{p\ell} \tag{6.30}$$

Provable for a single crystal, this expression should be considered as a principle for polycrystalline materials. The principle of maximum work, also called ***Hill's***

2 For the case of cfc crystals, all the slip systems are crystallographically equivalent and the critical resolved shear has the same value for all of them.

principle, is well verified for dense metals. The form of the plastic flow rules can thus be easily deduced as we are going to see.

Consider a stress state $\boldsymbol{\sigma}$ verifying the plasticity criterion, that is, such that $f(\boldsymbol{\sigma}) = 0$, and the plastic strain rate tensor $\dot{\boldsymbol{\varepsilon}}^{p\ell}$ which is associated with it. For a material obeying the principle of maximum work, we have for every other stress state $\boldsymbol{\sigma}^*$ such that $f(\boldsymbol{\sigma}^*) \leq 0$:

$$(\boldsymbol{\sigma}^* - \boldsymbol{\sigma}) : \dot{\boldsymbol{\varepsilon}}^{p\ell} \leq 0 \quad \text{or as components,} \quad (\sigma^*_{ij} - \sigma_{ij})\dot{\varepsilon}^{p\ell}_{ij} \leq 0 \tag{6.31}$$

The double contracted product of the two tensors can define a scalar product between tensors. In a two dimensional case such as that in figure 6.8 (stress state in a principal plane, for example), it is relatively easy to see that relation (6.31) insures that the angle between the vector of components of $\dot{\boldsymbol{\varepsilon}}^{p\ell}$ and the vector of components of $\boldsymbol{\sigma}^* - \boldsymbol{\sigma}$ is obtuse (a right angle at best). This implies that $\dot{\boldsymbol{\varepsilon}}^{p\ell}$ is necessarily directed towards the outside perpendicular of the plasticity surface. If that were not the case, it would be very easy to find a $\boldsymbol{\sigma}^*$ such that the angle in question was acute, and the principle of maximum work would be violated.

The reader can find this result for the general case (scalar product in $\mathbb{R}^6$) by assuming a counter example and performing a first-order expansion of f in the neighborhood of $\boldsymbol{\sigma}$. The result is that $\dot{\boldsymbol{\varepsilon}}^{p\ell}$ must be proportional to the exterior normal (tensor $\partial f/\partial \boldsymbol{\sigma}$) of the plasticity surface:

$$\exists \lambda^{p\ell} > 0 \quad \text{such that} \quad \dot{\boldsymbol{\varepsilon}}^{p\ell} = \lambda^{p\ell} \frac{\partial f}{\partial \boldsymbol{\sigma}} \tag{6.32}$$

or as components,

$$\dot{\varepsilon}^{p\ell}_{ij} = \lambda^{p\ell} \frac{\partial f}{\partial \sigma_{ij}} \tag{6.33}$$

Relation (6.33) is called the ***plastic flow rule***, defined by the ***normal law***. One also says that the *plastic flow rule is associated* with the plasticity criterion (under-

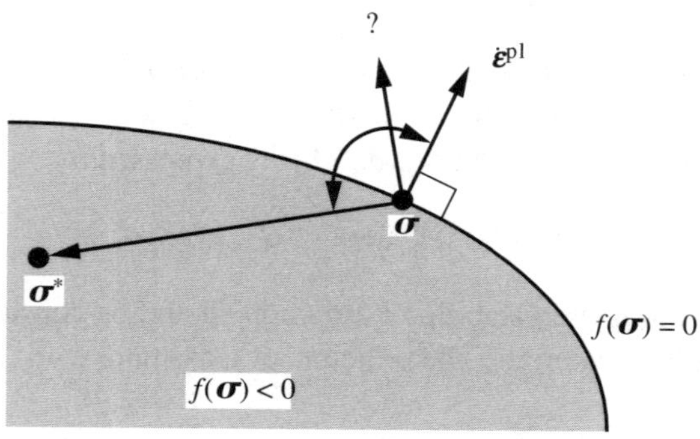

Figure 6.8 Normal law resulting from the principle of maximum work.

stood: by the principle of maximum work). The coefficient $\lambda^{p\ell}$ is named the ***plastic multiplier***. Finally another direct consequence of the principle of maximum work is the necessary convexity of the plasticity surface. Repeating the construction of figure 6.8, it is easy to show that a local concavity around $\boldsymbol{\sigma}$ would allow the violation of the principle.

Plastic flow rule associated with the von Mises' criterion

Relation (6.33) requires that f be differentiable with respect to the stress tensor components, which is not the case for Tresca's criterion. The plasticity surface sketched in figure 6.4 exhibits edges on this surface, along which the normal to f is undefined; it will be the same for $\dot{\boldsymbol{\varepsilon}}^{p\ell}$. It is a fundamental difficulty in the practical application of Tresca's criterion for the calculation of plastic flows. Therefore we only describe here the plastic flow rule associated with von Mises' criterion. If we take it in the form (6.20) and apply the plastic flow rule (6.33), we obtain:

$$\dot{\varepsilon}^{p\ell}_{ij} = \lambda^{p\ell}\frac{\partial\sigma_{eq}}{\partial\sigma_{ij}} = \lambda^{p\ell}\frac{1}{2\sigma_{eq}}3s_{kl}\frac{\partial s_{kl}}{\partial\sigma_{ij}} \tag{6.34}$$

but,

$$s_{kl} = \sigma_{kl} - \tfrac{1}{3}\,\mathrm{Tr}\,\boldsymbol{\sigma}\,\delta_{kl} \tag{6.35}$$

and thus,

$$\frac{\partial s_{kl}}{\partial\sigma_{ij}} = \delta_{ki}\delta_{jl} - \tfrac{1}{3}\delta_{ij}\delta_{kl} \tag{6.36}$$

The trace of $\boldsymbol{s}$ being equal to zero, relation (6.34) takes the form:

$$\dot{\varepsilon}^{p\ell}_{ij} = \frac{3\lambda^{p\ell}}{2\sigma_{eq}}s_{ij} \tag{6.37}$$

Note at this stage that $\boldsymbol{s}$ verifies the plasticity criterion (there is flow). Then $\sigma_{eq} = \sigma_0$ and we have:

$$\sigma_0^2 = \tfrac{3}{2}s_{ij}s_{ij} = \tfrac{3}{2}\left(\frac{2\sigma_0}{3\lambda^{p\ell}}\right)^2\dot{\varepsilon}^{p\ell}_{ij}\dot{\varepsilon}^{p\ell}_{ij} \tag{6.38}$$

Defining the ***equivalent plastic strain rate*** by:

$$\dot{\varepsilon}^{p\ell}_{eq} = \sqrt{\tfrac{2}{3}\dot{\varepsilon}^{p\ell}_{ij}\dot{\varepsilon}^{p\ell}_{ij}} \tag{6.39}$$

we have, with (6.38), $\lambda^{p\ell} = \dot{\varepsilon}^{p\ell}_{eq}$ and the flow rule (6.37) is written:

$$\dot{\varepsilon}^{p\ell}_{ij} = \frac{3\dot{\varepsilon}^{p\ell}_{eq}}{2\sigma_0}s_{ij} \tag{6.40}$$

The proportionality of $\dot{\boldsymbol{\varepsilon}}^{p\ell}$ to $\boldsymbol{s}$ allows the verification of the *incompressibility of plastic strain* ($\mathrm{Tr}\ \dot{\boldsymbol{\varepsilon}}^{p\ell} = 0$).

Power dissipated by plastic strain

Consider again a material obeying von Mises' criterion; the power dissipated *per unit volume* by the plastic strain is equal to:

$$\dot{w}^{p\ell} = \sigma_{ij}\dot{\varepsilon}^{p\ell}_{ij} = (s_{ij} - p\delta_{ij})\dot{\varepsilon}^{p\ell}_{ij} = s_{ij}\dot{\varepsilon}^{p\ell}_{ij} \tag{6.41}$$

in which the incompressibility condition for plastic strain is used. Employing (6.40) and (6.39):

$$\dot{w}^{p\ell} = \frac{2\sigma_0}{3\dot{\varepsilon}^{p\ell}_{eq}}\dot{\varepsilon}^{p\ell}_{ij}\dot{\varepsilon}^{p\ell}_{ij} = \sigma_0\dot{\varepsilon}^{p\ell}_{eq} \tag{6.42}$$

The power dissipated by plastic strain per unit volume is then very easily expressed for the multiaxial case using scalar invariants: the flow stress and the equivalent plastic strain rate.

Introduction to the notion of work hardening: isotropic and kinematic models

At the start of this chapter, we introduced the notion of work hardening through the dependence of the flow stress threshold σ_0 on the plastic strain accumulated by the material in its history. In the case of multiaxial stresses, σ_0 will depend on the equivalent plastic strain:

$$\varepsilon^{p\ell}_{eq} = \int_0^t \dot{\varepsilon}^{p\ell}_{eq}(t')\, dt' \tag{6.43}$$

This first approach to work hardening consists of dilating the elastic domain (the domain defined by $f < 0$) conserving the isotropic character of f: this is the ***isotropic work hardening*** model (fig. 6.9).

Practically, the criterion function is written:

$$f(\boldsymbol{\sigma}, \varepsilon^{p\ell}_{eq}) = \sigma_{eq} - \sigma_0(\varepsilon^{p\ell}_{eq}) \le 0 \tag{6.44}$$

where the law for the evolution of σ_0 is determined with a simple rheological test such as a uniaxial tension or compression test. As examples, we give a few classic models for isotropic work hardening:

Hollomon's law or power law $\quad \sigma_0 = \sigma_{0H}(\varepsilon^{p\ell}_{eq})^{n_H} \quad$ (6.45)

Ludwik's law: $\quad \sigma_0 = \sigma_{00L} + \sigma_{0L}(\varepsilon^{p\ell}_{eq})^{n_L} \quad$ (6.46)

Krupkowski's law: $\quad \sigma_0 = \sigma_{0K}(\varepsilon^{p\ell}_{eq} + \varepsilon_K)^{n_K} \quad$ (6.47)

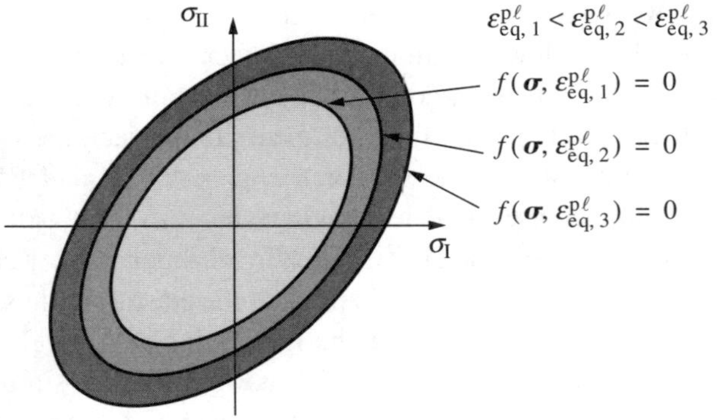

Figure 6.9 Diagram of isotropic work hardening of a von Mises material, for a planar stress state, in principal stress space.

The reality can however be noticeably different. For example, the Bauschinger effect, already evoked in paragraph 6.2.1, corresponding to a softening of the material when the load is reversed. Then in tension-compression, the work hardening occurring in the tension phase does not carry over to the value of the flow stress in compression (fig. 6.10(a)). This phenomenon is modeled with the introduction of a translation of the elasticity domain in stress space by the value of an internal stress tensor $\boldsymbol{\chi}$, (fig. 6.10(b)). A ***kinematic work hardening*** model is then defined, the criterion function being written (denoting by $(.)^{dev}$ the deviatoric part of a tensor):

$$f(\boldsymbol{\sigma}-\boldsymbol{\chi}) = \sigma_{eq}(\boldsymbol{\sigma}-\boldsymbol{\chi}) - \sigma_0 = \sqrt{\tfrac{3}{2}(\boldsymbol{\sigma}-\boldsymbol{\chi})^{dev} : (\boldsymbol{\sigma}-\boldsymbol{\chi})^{dev}} - \sigma_0 \leq 0 \qquad (6.48)$$

The difficulty is in the modeling of the evolution of the internal stress tensor $\boldsymbol{\chi}$, which is outside the scope of this book.

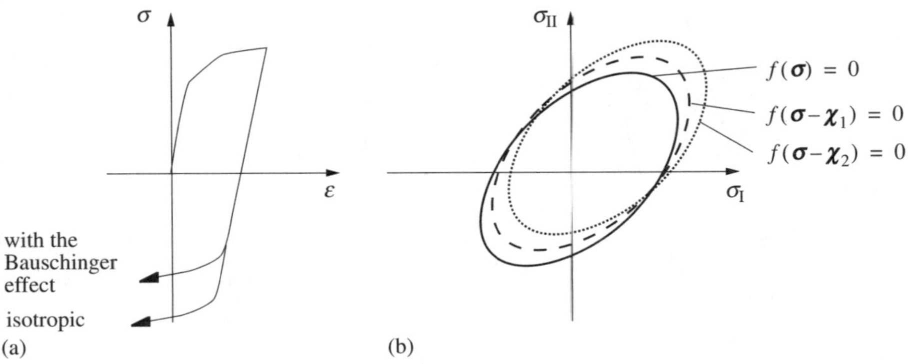

Figure 6.10 Bauschinger effect in uniaxial tension-compression (a) and kinematic work hardening model (b).

Prandtl-Reuss elastoplastic constitutive law

The plastic flow equations having been established, consider again a typical measurement of a tension-compression curve from a tension test (fig. 6.11).

When there is plastic flow, for example between the two points A and B, we note by comparison of the similar triangles AA'A" and BB'B" that not only the plastic strain increases, which is obvious, but also the elastic strain. Then, when the flow stress is attained, the flow is in reality elastoplastic. As can be seen in the figure – typical for a cold metal – the increase of the elastic strain is much weaker than that of the plastic strain because of the high value of Young's modulus. One is thus tempted to approximate the material as rigid-plastic: simply ignoring the elastic strain; the material is considered undeformable as long as $f(\sigma) < 0$, and $\dot{\varepsilon} = \dot{\varepsilon}^{p\ell}$ when $f(\sigma) = 0$. However, it is more rigorous to take the elastoplasticity into account, especially when the material is not entirely plastified, for the calculation of the stresses that develop in this case.

In the following, we consider elastoplastic flow assuming the elastic strain rate to be low with respect to the plastic strain rate; thus it can be shown that this justifies the additive decomposition of the strain rate tensor already given in (6.25):

$$\dot{\boldsymbol{\varepsilon}} = \dot{\boldsymbol{\varepsilon}}^{e\ell} + \dot{\boldsymbol{\varepsilon}}^{p\ell} \tag{6.49}$$

The multiaxial elastoplastic behavior of a metal obeying von Mises' criterion and for which the work hardening is isotropic can be summarized in the following manner:

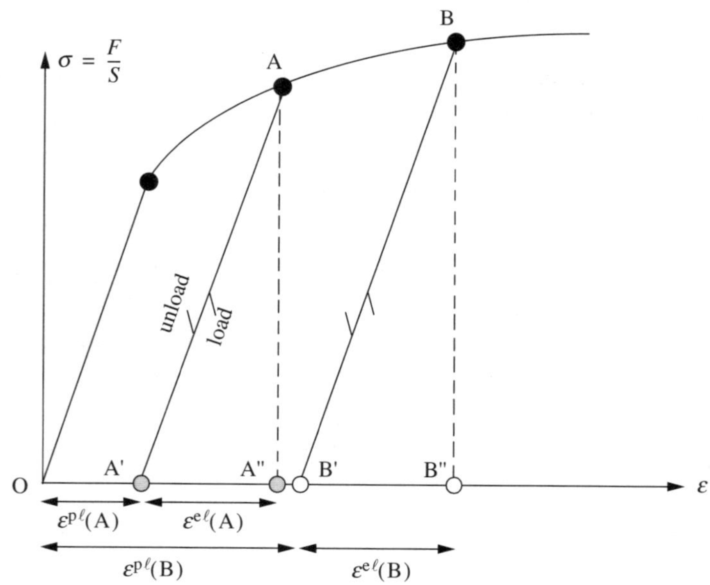

Figure 6.11 Uniaxial elastoplastic tension (sketch).

- As long as the equivalent stress σ_{eq} is less than the flow stress $\sigma_0(\varepsilon_{eq}^{p\ell})$, the behavior is elastic and governed by Hooke's law (chap. 1):

$$\boldsymbol{\sigma} = \boldsymbol{D}^{e\ell}\boldsymbol{\varepsilon}^{e\ell} \tag{6.50}$$

$\boldsymbol{D}^{e\ell}$ being the **elasticity tensor** (4th order).

- When the equivalent stress attains σ_0, the strain contains a plastic part, and the plastic part of the strain rate follows the normal law.

The elastoplastic flow equations (called the ***Prandtl-Reuss equations***) are then:

$$\left.\begin{aligned} \dot{\boldsymbol{\varepsilon}} &= \dot{\boldsymbol{\varepsilon}}^{e\ell} + \dot{\boldsymbol{\varepsilon}}^{p\ell} \\ \dot{\boldsymbol{\varepsilon}}^{e\ell} &= (\boldsymbol{D}^{e\ell})^{-1}\dot{\boldsymbol{\sigma}} \\ \dot{\boldsymbol{\varepsilon}}^{p\ell} &= \lambda^{p\ell}\frac{\partial f}{\partial \boldsymbol{\sigma}} = \frac{3}{2}\frac{\dot{\varepsilon}_{eq}^{p\ell}}{\sigma_0}\boldsymbol{s} \\ f(\boldsymbol{\sigma}, \varepsilon_{eq}^{p\ell}) &= 0 \end{aligned}\right\} \tag{6.51}$$

The different situations which can occur: elastic loading, elastoplastic loading and elastic unloading, are illustrated in figure 6.12 and the corresponding equations are summarized in table 6.1.

Objective derivation in the Prandtl-Reuss equations

In general, a constitutive equation should satisfy a few fundamental principles, notably that of material invariance or of "objectivity". In other words, it should be independent of the Galilean reference frame in which it is expressed, or independent of the observer. To illustrate this notion in an intuitive manner, take an object of

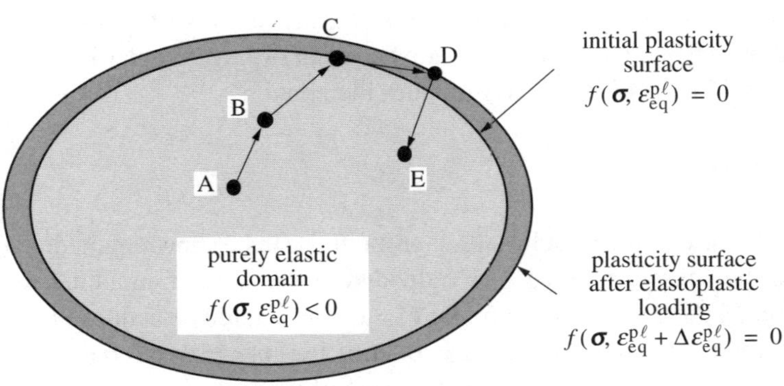

Figure 6.12 The different paths of elastoplastic strain.

Table 6.1

	$\dot{\varepsilon}^{e\ell}$	$\dot{\varepsilon}^{p\ell}$	$f(\boldsymbol{\sigma}, \varepsilon_{eq}^{p\ell})$	$\frac{\partial f}{\partial \boldsymbol{\sigma}} : \dot{\boldsymbol{\sigma}}$
Path AB (or BC) (elastic load)	$(\boldsymbol{D}^{e\ell})^{-1}\dot{\boldsymbol{\sigma}}$	$\boldsymbol{0}$	< 0	
Path CD (elastoplastic load)	$(\boldsymbol{D}^{e\ell})^{-1}\dot{\boldsymbol{\sigma}}$	$\lambda^{p\ell}\frac{\partial f}{\partial \boldsymbol{\sigma}}$	0	≥ 0
Path DE (elastic unloading)	$(\boldsymbol{D}^{e\ell})^{-1}\dot{\boldsymbol{\sigma}}$	$\boldsymbol{0}$	0	< 0

known mass, for example a kilogram, and suspend it with a given spring. The spring will stretch a certain amount due to the weight, say a centimeter. Then the object and spring are placed on a horizontal disk and the spring is attached to the center of the disk. Rotation of the disk commences at constant angular velocity such that the spring stretches again by one centimeter. Two observers (one attached to the laboratory reference frame, the other attached to the rotating disk) agree that the centripetal force, that is, the force necessary to prevent the object from being launched, is exactly 9.81 N. Clearly, the mechanical response of the spring is not affected by the rigid body movement.

In the same way, it is required that the response of the material to its strain be invariant under a change of reference frame. This is the ***principle of objectivity***. A reference frame change is written:

$$\begin{aligned} \boldsymbol{x}^* &= \boldsymbol{c}(t) + \boldsymbol{Q}(t) \cdot \boldsymbol{x} \\ t^* &= t + a \end{aligned} \tag{6.52}$$

where t and $\boldsymbol{x}$ are the time and the position, respectively, in the old reference frame, and t^* and $\boldsymbol{x}^*$ are the time and position in the new reference frame; the constant a being a real number. The first relation expresses that the position in the new reference frame results from the relative movement of the two frames, composed of a translation $\boldsymbol{c}(t)$ and a rigid body rotation with an orthogonal matrix $\boldsymbol{Q}(t)$. The stress tensor is said to be ***objective*** if it satisfies the relation:

$$\boldsymbol{\sigma}^* = \boldsymbol{Q}(t)\boldsymbol{\sigma}\boldsymbol{Q}(t)^{\mathrm{T}} \tag{6.53}$$

In the case of the Cauchy stress tensor, this can be shown immediately by writing that the stress vector (the force divided by the area of an elementary surface) expressed in the new reference frame $\boldsymbol{T}^* = \boldsymbol{\sigma}^* \cdot \boldsymbol{n}^*$ should be deduced from the stress vector expressed in the old frame $\boldsymbol{T} = \boldsymbol{\sigma} \cdot \boldsymbol{n}$ by the change of reference frames resulting from the rotation, $\boldsymbol{T}^* = \boldsymbol{Q}(t) \cdot \boldsymbol{T}$. The general conditions of objectivity are the following:

$$\alpha = \alpha^* \qquad \text{for an objective scalar}$$

$$\boldsymbol{b}^* = \boldsymbol{Q}(t) \cdot \boldsymbol{b} \qquad \text{for an objective vector}$$

$$\boldsymbol{\Sigma}^* = \boldsymbol{Q}(t)\boldsymbol{\Sigma}\boldsymbol{Q}(t)^{\mathrm{T}} \qquad \text{for an objective tensor}$$

From (6.52), it can easily be shown that a material vector joining two points is objective. Inversely, according to intuition, the velocity vector of a material point is not objective, because the derivative of (6.52) gives:

$$\boldsymbol{v}^* = \frac{\mathrm{d}\boldsymbol{x}^*}{\mathrm{d}t^*} = \frac{\mathrm{d}\boldsymbol{x}^*}{\mathrm{d}t} = \dot{\boldsymbol{c}}(t) + \dot{\boldsymbol{Q}}(t) \cdot \boldsymbol{x} + \boldsymbol{Q}(t) \cdot \boldsymbol{v} \tag{6.54}$$

We have seen in chapter 1, (1.23), that for any point M, the velocity gradient tensor is the sum of the strain rate tensor $\dot{\boldsymbol{\varepsilon}}$ (symmetric part of $\mathbf{Grad}\, \boldsymbol{v} = \partial \boldsymbol{v} / \partial \boldsymbol{x}$, also denoted $\boldsymbol{d}$) and the rotation rate tensor[3] $\boldsymbol{\omega}$ (anti-symmetric part). Physically, this decomposition signifies that the local velocity field at M is the sum of a bulk translation (null gradient tensor), of a pure strain (gradient tensor $\dot{\boldsymbol{\varepsilon}}$) and a pure rotation (gradient tensor $\boldsymbol{\omega}$). It is easy to show, differentiating (6.54), that the strain rate tensor $\dot{\boldsymbol{\varepsilon}}$ is objective while the rotation rate tensor is not.

It can thus be verified that the power dissipated $\boldsymbol{\sigma} : \dot{\boldsymbol{\varepsilon}}$ is an objective scalar, independent of the reference frame. Moreover, all the scalar quantities expressed by tensor invariants are objective scalars as they remain constant during a change of reference frame. This is notably the case for the equivalent strain rate $\dot{\varepsilon}^{\mathrm{p}\ell}_{\mathrm{eq}}$ and, by extension, its integration in time: the equivalent strain $\varepsilon^{\mathrm{p}\ell}_{\mathrm{eq}}$.

From (6.52), the following results can also be shown:

- The gradient tensor of the transformation $\boldsymbol{F}$ is not objective: $\boldsymbol{F}^* = \boldsymbol{Q}(t)\boldsymbol{F}$.
- The right Cauchy-Green tensor $\boldsymbol{C} = \boldsymbol{F}^{\mathrm{T}}\boldsymbol{F}$ is not objective, because it is invariant: $\boldsymbol{C}^* = \boldsymbol{C}$.
- Consequently, the same is true for the Green-Lagrange tensor $\boldsymbol{\varepsilon} = 1/2(\boldsymbol{C} - \boldsymbol{I})$.

Return now to the Prandtl-Reuss constitutive law: is it expressed in the same manner regardless of the reference frame? In view of the different remarks above and the nature of the plasticity criterion function, it is the case for the four equations (6.51), excepting the second which includes the material derivative of the stress tensor. In general, taking the derivative with respect to time of (6.53), it is shown that the material derivative of an objective tensor is not an objective tensor. For the case of the Cauchy stress tensor, one can convince oneself intuitively taking the example of a material domain undergoing a pure rotation $\boldsymbol{R}$ over a time interval Δt. Although the total derivative of the stress tensor is zero in the laboratory frame (the strain being null), it is seen that after the rotation the state of stress has evolved and that the tensor is in fact given by: $\boldsymbol{\sigma} + \Delta\boldsymbol{\sigma} = \boldsymbol{R}\boldsymbol{\sigma}\boldsymbol{R}^{\mathrm{T}}$.

---

3 To convince oneself that $\boldsymbol{\omega}$ really corresponds to a rotation, the reader can verify that the vector $\boldsymbol{\omega}\,\mathbf{d}\boldsymbol{x}$ has the same components as the vector product $\boldsymbol{\Omega} \otimes \mathbf{d}\boldsymbol{x}$ where $\boldsymbol{\Omega}$ is the vector of components $[\omega_{32}\omega_{13}\omega_{21}]^{\mathrm{T}}$.

The relation (6.51b) should thus be expressed in an objective way. The strain rate tensor being objective, the derivative of the stress tensor used in this relation then should also be objective. Then (6.51b) will have the same form in every reference frame. For example, one could take Jaumann's derivative:

$$\overset{J}{\boldsymbol{\sigma}} = \dot{\boldsymbol{\sigma}} - \boldsymbol{\omega}\boldsymbol{\sigma} - \boldsymbol{\sigma}\boldsymbol{\omega}^{\mathrm{T}} \tag{6.55}$$

or the Oldroyd convective derivative:

$$\overset{\nabla}{\boldsymbol{\sigma}} = \dot{\boldsymbol{\sigma}} - [\mathbf{Grad}\,\boldsymbol{v}]\boldsymbol{\sigma} - \boldsymbol{\sigma}[\mathbf{Grad}\,\boldsymbol{v}]^{\mathrm{T}} \tag{6.56}$$

but this is only one possible choice among other objective derivatives, the choice should be guided by physical considerations.

An alternative solution very often used in practice to deal with large rotations consists of integrating the constitutive law in the frame turning with the material at the rotation rate $\boldsymbol{\omega}$ (corotational frame) in the same manner that one would have done with the total derivative. A stress tensor expressed in a corotational frame $(\boldsymbol{\sigma} + \Delta\boldsymbol{\sigma})_{\mathrm{R}}$ is obtained. This tensor is then brought back to the fixed observation frame performing:

$$\boldsymbol{\sigma} + \Delta\boldsymbol{\sigma} = \boldsymbol{R}(\boldsymbol{\sigma} + \Delta\boldsymbol{\sigma})_{\mathrm{R}}\boldsymbol{R}^{\mathrm{T}} \tag{6.57}$$

$\boldsymbol{R}$ being the rotation given by $\boldsymbol{R} = \Delta t\boldsymbol{\omega}$.

6.2.2 Elasto-viscoplasticity

So far, we have considered that the behavior of the material was independent of the strain rate. This assumption is relatively well verified for cold metals, at temperatures $T < 1/3\ T_{\mathrm{m}}$, where T_{m} is the melting point (for temperatures expressed in Kelvin). It is less well verified for higher temperatures, as illustrated in figure 6.13 for which a clear dependence exists for the stress state as a function of the strain rate.

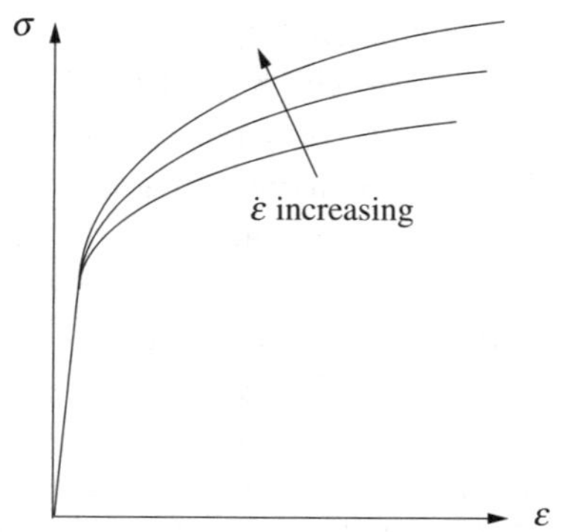

Figure 6.13 Schematic representation of a uniaxial tension test for different strain rates.

Norton was the first to give a one dimensional approximation of this phenomenon, with what is commonly called a ***power law***:

$$\sigma = k\dot{\varepsilon}^{m} \tag{6.58}$$

The exponent m is called the ***strain rate sensitivity coefficient***. The behavior of a *viscous fluid* flowing in response to a stress (even if this stress is weak) can be recognized in this relation (chap. 7). In other words, there is not a flow stress threshold. A particular case is a ***Newtonian fluid***, for which $m = 1$. This relation can be generalized to a behavior in which one takes account of the presence of a stress threshold σ_0, under which the behavior is elastic, and above which the irreversible portion of the strain – which is named the ***viscoplastic strain rate*** $\dot{\varepsilon}^{vp}$ – is now linked to the *difference* between σ and σ_0 by the power law:

$$\sigma = \sigma_0 + k(\dot{\varepsilon}^{vp})^{m} \tag{6.59}$$

In this context, one speaks of σ_0 as the ***static plasticity threshold***, the difference $\sigma - \sigma_0$ being called ***extra-stress***. It should be noted that unlike elastoplasticity, the stress can now exceed the plasticity threshold. When the stress state is multiaxial, relation (6.59) can be generalized by replacing σ and $\dot{\varepsilon}^{vp}$ by the scalar invariants in the sense of von Mises:

$$\sigma_{eq} = \sigma_0 + k(\dot{\varepsilon}^{vp}_{eq})^{m} \quad \text{with} \quad \dot{\varepsilon}^{vp}_{eq} = \sqrt{\tfrac{2}{3}\dot{\varepsilon}^{vp}_{ij}\dot{\varepsilon}^{vp}_{ij}} \tag{6.60}$$

Figure 6.14 shows the various approaches to viscoplastic behavior and the comparison with elastoplastic behavior.

It can be seen in the figure that the notion of a threshold is especially sensitive in the domain of weak strain rates, whereas for more rapid strains (materials form-

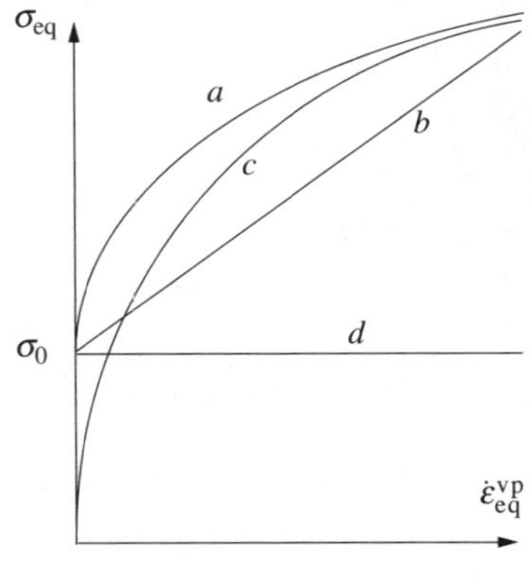

a: $\sigma_{eq} = \sigma_0 + k(\dot{\varepsilon}^{vp}_{eq})^{m}$ Norton with threshold

b: $\sigma_{eq} = \sigma_0 + k\dot{\varepsilon}^{vp}_{eq}$ Bingham (with threshold, but a linear law)

c: $\sigma_{eq} = k(\dot{\varepsilon}^{vp}_{eq})^{m}$ Norton without threshold

d: $\sigma_{eq} = \sigma_0$ von Mises (elastoplastic)

Figure 6.14 Relations between the equivalent stress and the equivalent plastic (or viscoplastic) strain rate.

ing), the behavior can be modeled equivalently by laws with or without a threshold.

The tensor relations of elasto-viscoplastic flow were formalized by Perzyna, in a form analogous to that already encountered for elastoplastic flow:

$$\dot{\boldsymbol{\varepsilon}} = \dot{\boldsymbol{\varepsilon}}^{\mathrm{e}\ell} + \dot{\boldsymbol{\varepsilon}}^{\mathrm{vp}} \tag{6.61a}$$

$$\dot{\boldsymbol{\varepsilon}}^{\mathrm{e}\ell} = (\boldsymbol{D}^{\mathrm{e}\ell})^{-1}\dot{\boldsymbol{\sigma}} \tag{6.61b}$$

$$\dot{\boldsymbol{\varepsilon}}^{\mathrm{vp}} = \frac{\sqrt{3}}{2\sigma_{\mathrm{eq}}}\left\langle\frac{\sigma_{\mathrm{eq}}-\sigma_0}{\sqrt{3}K}\right\rangle^{1/\mathrm{m}}\boldsymbol{s} \tag{6.61c}$$

The parameter K, characteristic of the material, is called the ***viscoplastic consistency***. The brackets[4] in the last expression reflect the presence of the static plasticity threshold using the following convention:

$$\langle x\rangle = x \ \text{ if } \ x \geq 0 \quad \text{and} \quad \langle x\rangle = 0 \ \text{ if } \ x < 0$$

or

$$\langle x\rangle = \tfrac{1}{2}(x + |x|) \tag{6.62}$$

In stress space, viscoplasticity with a threshold is shown in figure 6.15. Remark in particular that in an elasto-viscoplastic flow situation, the unloading from a given stress state is first elasto-viscoplastic until the interception of the static

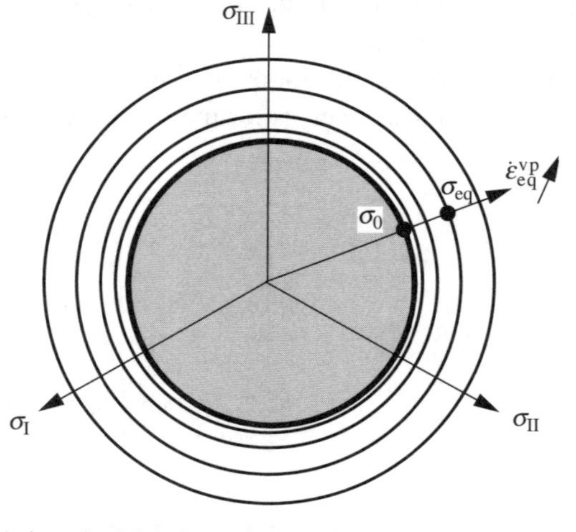

Figure 6.15 In elasto-viscoplasticity, the equivalent viscoplastic strain rate $\dot{\varepsilon}^{\mathrm{vp}}_{\mathrm{eq}}$ increases, according to a power law with coefficient $1/m$, with the difference $\sigma_{\mathrm{eq}} - \sigma_0$, which expresses the difference of the stress state from the plasticity surface. The graphical representation is shown here in the deviatoric plane.

[4] Here we have retained the notation most often used in the literature, not to be confused with the spatial average, also denoted in this book with angle brackets.

plasticity surface, and from there it continues in a purely elastic manner (contrary to the elastoplastic model, for which the unloading is purely elastic in its totality).

If the equivalent viscoplastic strain rate is calculated (6.60b) from the flow rule (6.61c), it is easily shown that:

$$\sigma_{eq} = \sigma_0 + K(\sqrt{3})^{m+1}(\dot{\varepsilon}_{eq}^{vp})^m \tag{6.63}$$

This relation, in turn inserted into (6.61c) leads to:

$$\dot{\boldsymbol{\varepsilon}}^{vp} = \frac{3\dot{\varepsilon}_{eq}^{vp}}{2\sigma_{eq}}\boldsymbol{s} \tag{6.64}$$

which is von Mises' flow rule, already found for elastoplasticity (6.40).

The phenomena of work hardening can be expressed in many ways. Isotropic work hardening can be modeled with the increasing functions $\sigma_0(\varepsilon_{eq}^{vp})$ and $K(\varepsilon_{eq}^{vp})$, (ε_{eq}^{vp} being the equivalent plastic strain calculated from formula (6.43), replacing the indices "$p\ell$" by "vp") whereas kinematic work hardening leads to substituting $\sigma_{eq}(\boldsymbol{\sigma} - \boldsymbol{\chi})$ for $\sigma_{eq}(\boldsymbol{\sigma})$ in (6.61c).

6.2.3 Elasto-viscoplasticity without a threshold and pure viscoplasticity

In the case of metals deformed while hot ($T > 2/3\ T_m$, in K), it is usually difficult to demonstrate a static plasticity threshold, very weak stresses generating deformation by creep. One tends then to take $\sigma_0 = 0$ in the viscoplastic flow rule (6.61c), thereby obtaining the elasto-viscoplastic flow equations without threshold:

$$\left.\begin{aligned} \dot{\boldsymbol{\varepsilon}} &= \dot{\boldsymbol{\varepsilon}}^{e\ell} + \dot{\boldsymbol{\varepsilon}}^{vp} \\ \dot{\boldsymbol{\varepsilon}}^{e\ell} &= (\boldsymbol{D}^{e\ell})^{-1}\dot{\boldsymbol{\sigma}} \\ \dot{\boldsymbol{\varepsilon}}^{vp} &= \frac{\sqrt{3}}{2\sigma_{eq}}\left(\frac{\sigma_{eq}}{\sqrt{3}K}\right)^{1/m}\boldsymbol{s} = \frac{3\dot{\varepsilon}_{eq}^{vp}}{2\sigma_{eq}}\boldsymbol{s} \end{aligned}\right\} \tag{6.65}$$

The third equation can be inverted into the form:

$$\boldsymbol{s} = 2K(\sqrt{3}\dot{\varepsilon}_{eq}^{vp})^{m-1}\dot{\boldsymbol{\varepsilon}}^{vp} \tag{6.66}$$

If in addition, the elastic strain component is neglected ($\dot{\boldsymbol{\varepsilon}}^{e\ell} = \boldsymbol{0}$ and $\dot{\boldsymbol{\varepsilon}} = \dot{\boldsymbol{\varepsilon}}^{vp}$), the constitutive equations reduce to the pure viscoplastic constitutive law, called the ***Norton-Hoff law***:

$$\boldsymbol{s} = 2K(\sqrt{3}\dot{\varepsilon}_{eq})^{m-1}\dot{\boldsymbol{\varepsilon}} \tag{6.67}$$

The constitutive law for an incompressible ***generalized Newtonian fluid***, with non linear viscosity $\mu = K(\sqrt{3}\dot{\varepsilon}_{eq})^{m-1}$ can be recognized here (the limiting case of a Newtonian fluid corresponds to $m = 1$; see also chap. 7). The deviatoric stress tensor thus depends on a ***viscoplastic potential***, function of the strain rates:

$$s = \frac{\partial \varphi}{\partial \dot{\boldsymbol{\varepsilon}}} \quad \text{with} \quad \varphi = \frac{K}{m+1}(\sqrt{3}\dot{\varepsilon}_{eq})^{m+1} \tag{6.68}$$

which gives, by analogy with (6.58):

$$\sigma_{eq} = k\dot{\varepsilon}_{eq}{}^{m} = K(\sqrt{3})^{m+1}\dot{\varepsilon}_{eq}{}^{m} \tag{6.69}$$

Note that the Norton-Hoff law is also much used for modeling of molten polymer flows, when the elastic effects encountered in these materials are ignored. The viscoelastic equations of polymers are qualitatively close to the elasto-viscoplastic behavior without threshold that we have presented: they will be detailed in chapter 7.

6.3 BOUNDARY CONDITIONS

6.3.1 The unilateral contact conditions

We will now examine the conditions of mechanical contact between two bodies on either side of a contact interface. In the context of solid deformation, these two bodies could be a part that is to be formed and the forming tool, or more generally, two materials interacting in a process (for example a cast part that cools in its mold, two mechanical components sliding one along the other…). Figure 6.16 will allow us to correctly pose the problem of contact between these two bodies. If $\partial\Omega_c$ is the contact surface (macroscopic) between these two; then for a point M in $\partial\Omega_c$ two material points are defined, M_1 and M_2, belonging to bodies 1 and 2 respec-

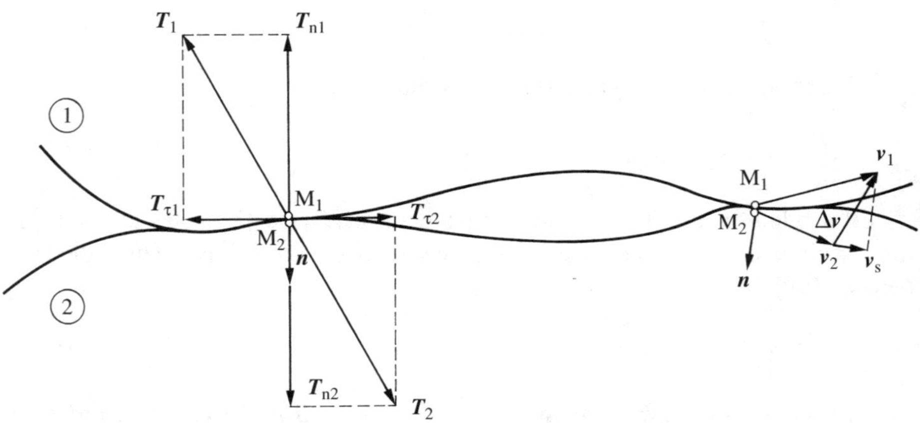

Figure 6.16 Schematic of the contact and principal kinematic and mechanical variables.

tively, and coinciding at time t with the point M. The normal vector to $\partial\Omega_c$ at M is denoted $\boldsymbol{n}$ in the outgoing direction with respect to body 1.

The stress states represented by the tensors $\boldsymbol{\sigma}_1$ and $\boldsymbol{\sigma}_2$ exist at M_1 and M_2. The stress vectors on the surface, T_1 and T_2 – opposites by the principle of action and reaction – are given by (chap. 1):

$$\boldsymbol{T}_1 = \boldsymbol{\sigma}_1 \cdot \boldsymbol{n} \quad \text{and} \quad \boldsymbol{T}_2 = \boldsymbol{\sigma}_2 \cdot (-\boldsymbol{n}) = -\boldsymbol{T}_1 \tag{6.70}$$

Each stress vector $\boldsymbol{T}$ is decomposed into normal and tangential components for each body (the indices are left off for clarity).

- The ***normal stress vector*** is defined by:

$$\boldsymbol{T}_{\text{n}} = (\boldsymbol{T} \cdot \boldsymbol{n})\boldsymbol{n} = T_{\text{n}}\boldsymbol{n} \tag{6.71}$$

  The scalar product $T_{\text{n}} = \boldsymbol{T} \cdot \boldsymbol{n}$ is called the ***normal stress*** and denoted σ_{n}. If there is contact, the stress state is necessarily compressive and thus $\sigma_{\text{n}} \leq 0$. Also, $p_{\text{c}} = -\sigma_{\text{n}}$ ($p_{\text{c}} \geq 0$ if contact) is called the ***contact pressure***, a pressure that must not be confused with the hydrostatic pressure p associated with the stress tensor. In the case when the contact is broken, we assume that the surfaces become free surfaces, atmospheric pressure being negligible with respect to the other stresses. In this case, $\boldsymbol{T} = \boldsymbol{\sigma} \,.\, \boldsymbol{n} = \boldsymbol{0}$ and thus $\sigma_{\text{n}} = 0$.
- The ***tangential stress vector*** is defined by:

$$\boldsymbol{T}_{\tau} = \boldsymbol{T} - (\boldsymbol{T} \cdot \boldsymbol{n})\boldsymbol{n} \tag{6.72}$$

  The vector $\boldsymbol{T}_{\tau}$ is orthogonal to $\boldsymbol{n}$ and thus contained in the tangent plane. Its norm T_{τ} is the ***friction tangential stress*** – or ***shear***.

From a kinematic point of view, one defines the velocity vectors $\boldsymbol{v}_1$ and $\boldsymbol{v}_2$ of M_1 and M_2, and the ***relative velocity vector*** (here for body 1 with respect to body 2), $\Delta\boldsymbol{v} = \boldsymbol{v}_1 - \boldsymbol{v}_2$, whose tangential component is the ***relative slip velocity vector*** $\boldsymbol{v}_{\text{g}}$:

$$\boldsymbol{v}_{\text{s}} = \Delta\boldsymbol{v} - (\Delta\boldsymbol{v} \cdot \boldsymbol{n})\boldsymbol{n} \tag{6.73}$$

The two bodies can not penetrate each other, so it is clear that:

$$\Delta\boldsymbol{v} \cdot \boldsymbol{n} \leq 0 \tag{6.74}$$

The equality $\Delta\boldsymbol{v} \cdot \boldsymbol{n} = 0$ signifies the continuation of contact at point M between bodies 1 and 2, with possibly a relative slip of one with respect to the other (in the case where $\boldsymbol{v}_{\text{s}}$ is non zero), while $\Delta\boldsymbol{v} \cdot \boldsymbol{n} < 0$ is synonymous with the separation of the two bodies. This possibility of loss of contact defines the notion of ***unilateral contact***; it is in this context that we place ourselves from here on. The kinematic and mechanical equations of unilateral contact reduce to a form named for ***Signorini***:

$$\left.\begin{array}{l}\sigma_n \leq 0 \\ \Delta \boldsymbol{v} \cdot \boldsymbol{n} \leq 0 \\ (\Delta \boldsymbol{v} \cdot \boldsymbol{n})\sigma_n = 0\end{array}\right\} \qquad (6.75)$$

The third relation comes from $\sigma_n = 0$ (loss of contact), or from $\Delta \boldsymbol{v} \cdot \boldsymbol{n} = 0$ (continued contact without interpenetration).

6.3.2 The principal friction laws

A ***friction law*** is a relation that permits the calculation of the tangential stress vector as a function of the very numerous contact parameters, among which:

- the thermomechanical variables σ_n, $\boldsymbol{v}_s$, T (temperature);
- the rheology of bodies 1 and 2;
- their "surface state" (roughness, chemical composition, surface energy);
- in lubricated cases, the rheology of the lubricant;
- the time, through the evolution of one of the above variables.

It is of course impossible to include the influence of all these parameters in a single mathematical formula. In practice, one tries to express the friction law as a function of the most important variables according to the context; the other variables, being hidden, determine the values of the non dimensional ***friction coefficients*** which will appear in the law. We limit ourselves here to the presentation of three friction models representing the essential part of the practical applications. The reader is referred to the books of the bibliography for further information on contact friction.

Coulomb's law

According to this model, it is the contact pressure p_c which with a friction coefficient μ_f, determines the value of a tangential stress threshold below which no relative slip appears. This threshold is called the ***slip threshold***:

$$\tau_c = \mu_f p_c \qquad (\mu_f \geq 0) \qquad (6.76)$$

$$T_\tau < \tau_c \Rightarrow \boldsymbol{v}_s = \boldsymbol{0} \qquad (6.77)$$

The most common assumption is that of ***isotropic friction***, which consists of assuming that the threshold τ_c does not depend on the direction of relative slip velocity $\boldsymbol{v}_s$ in the tangent plane. In addition, when this slip threshold is attained, the friction shear vector is parallel (and by nature opposed) to the relative slip velocity:

$$T_\tau = \tau_c \Rightarrow \exists \lambda_f \geq 0 \quad \text{such that} \quad \boldsymbol{v}_s = -\lambda_f \boldsymbol{T}_\tau \qquad (6.78)$$

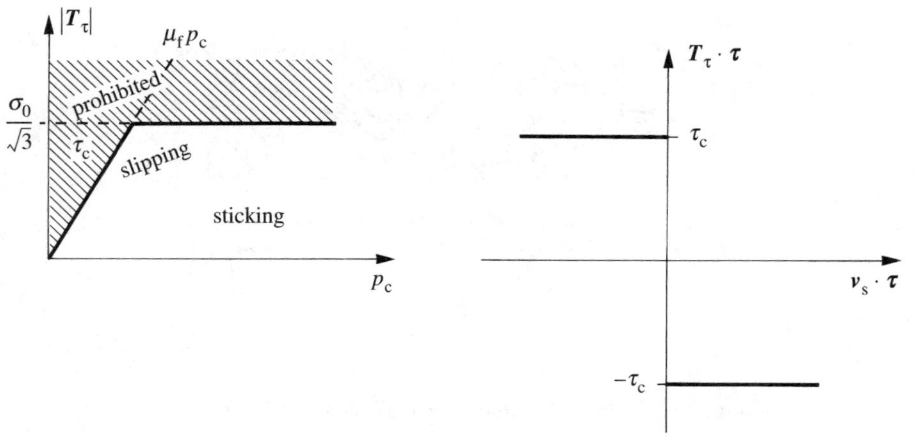

Figure 6.17 Representation of Coulomb's friction law. The vector $\boldsymbol{\tau}$ is a unitary tangent vector at contact, parallel to the slip velocity.

Moreover, the friction shear is compelled to maintain the same modulus, equal to τ_c. One then has:

$$T_\tau > \tau_c \qquad \text{impossible}^5 \tag{6.79}$$

From (6.78) the explicit friction shear expression is deduced as a function of the relative slip velocity, *when the latter is not null* (fig. 6.17):

$$\boldsymbol{T}_\tau = -\mu_f p_c \frac{1}{|\boldsymbol{v}_s|} \boldsymbol{v}_s \tag{6.80}$$

The physical interpretation of Coulomb's law is the following. The growth of the shear threshold with p_c is physically due to the increase in the real contact surface at the microscopic level of roughness (fig. 6.18). The normal force $\boldsymbol{F}_n$ is applied on the real contact surface S_{rc} (smaller than the apparent contact surface S_c) and has the modulus:

$$F_n = p_{rc} S_{rc} \tag{6.81}$$

where p_{rc} is a real contact pressure, averaged over S_{rc}.

Taking into account the geometry of the contact at a microscopic scale, this pressure can be seen as an indentation pressure which can be shown by plasticity theory to have the value:

$$p_{rc} \approx 3\sigma_0 \tag{6.82}$$

5 Here is seen clearly the analogy with the volume plastic behavior of metals, for which the threshold is also a criterion.

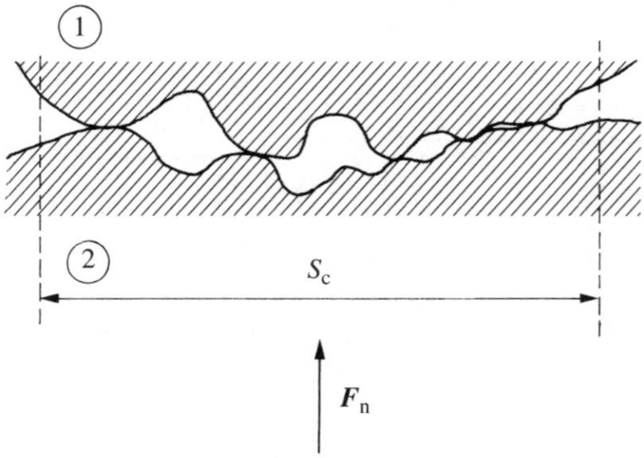

Figure 6.18 Physical interpretation of Coulomb's friction law.

σ_0 being the flow stress of the softest material. As for the tangential force $\boldsymbol{F}_\tau$, its modulus can be written:

$$F_\tau = T_{\tau r} S_{rc} \tag{6.83}$$

when $T_{\tau r}$ is an averaged real shear stress, which can be related to the maximal shear stress τ_{max} of the softest material, for a von Mises body, for example:

$$T_{\tau r} = \tau_{max} = \sigma_0 / \sqrt{3} \tag{6.84}$$

The ratio of the tangential forces to the normal forces is thus constant:

$$\mu_f = \frac{F_\tau}{F_n} = \frac{T_{\tau r}}{p_{rc}} \approx \frac{1}{3\sqrt{3}} \approx 0.19 \tag{6.85}$$

This physical interpretation corresponds well to the order of magnitude of the friction coefficients observed in practice. Note however that local work hardening, surface energies and the real geometry which can clearly have a large influence on the value of μ_f, have not been taken into account here.

If one of the bodies has an incompressible plastic behavior, its plasticity criterion limits the shear to $T_\tau \leq \tau_{max}$. This limit is generally integrated into Coulomb's law, again for a von Mises body, by posing:

$$\tau_c = \mathrm{Min}\left(\mu_f p_c, \frac{\sigma_0}{\sqrt{3}}\right) \tag{6.86}$$

Tresca's law

Here we assume that the slip threshold τ_c is a fraction $\overline{m}$ – friction coefficient – of the maximum shear τ_{max} of the softest material. For a von Mises body, we have the following relations:

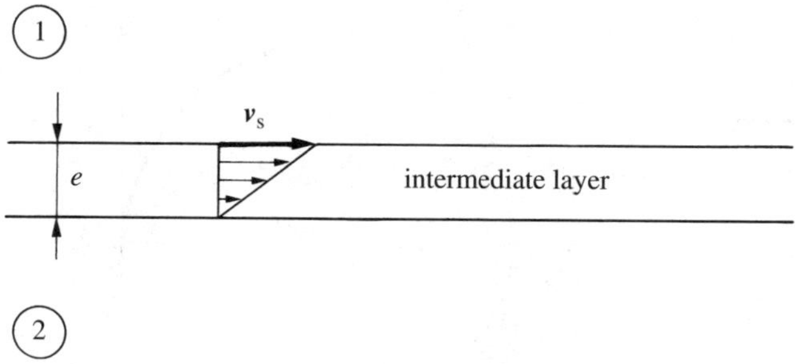

Figure 6.19 Physical interpretation of Tresca's friction law (boundary layer model). The relative velocity profile with respect to body 2 is shown in the boundary layer of thickness *e*.

$$\left.\begin{array}{lll} \tau_c = \bar{m}\sigma_0/\sqrt{3} & & \bar{m} \in [0,1] \\ T_\tau < \bar{m}\sigma_0/\sqrt{3} & \Rightarrow & \boldsymbol{v}_s = \mathbf{0} \\ T_\tau = \bar{m}\sigma_0/\sqrt{3} & \Rightarrow & \exists \lambda_f \geq 0 \quad \text{such that} \quad \boldsymbol{v}_s = -\lambda_f \boldsymbol{T}_\tau \\ T_\tau > \bar{m}\sigma_0/\sqrt{3} & & \text{impossible} \end{array}\right\} \tag{6.87}$$

When the relative slip velocity $\boldsymbol{v}_s$ is non zero, the tangential friction stress vector is expressed:

$$\boldsymbol{T}_\tau = -\bar{m}\frac{\sigma_0}{\sqrt{3}}\frac{1}{|\boldsymbol{v}_s|}\boldsymbol{v}_s \tag{6.88}$$

Consider the following physical interpretation for this law. It is as if a boundary layer existed at the interface of the two bodies. This boundary layer is composed of a rigid plastic material, with flow stress $\bar{m}\sigma_0$ (maximum shear $\bar{m}\sigma_0/\sqrt{3}$), that is sheared between the two bodies. This situation is illustrated in figure 6.19; the graphical interpretation of equations (6.87) is given in figure 6.20(a).

Viscoplastic type law

The preceding Coulomb and Tresca laws simultaneously manifest the notions of a threshold and a criterion: once the slip threshold is attained, the shear is compelled to have constant modulus, equal to τ_c. This characterizes the plastic type of laws. In rheology, we have seen earlier that the more flexible viscoplastic laws eliminate the notion of a prohibited domain ($T_\tau > \tau_c$); in such a case, the stress above the threshold becomes higher and higher as the strain rate increases. We can construct the analogy for the friction laws:

$$\begin{array}{l} T_\tau < \tau_c \Rightarrow \boldsymbol{v}_s = \mathbf{0} \\ T_\tau \geq \tau_c \Rightarrow T_\tau - \tau_c = \operatorname{Min}\left(\alpha_f \dfrac{\sigma_0}{\sqrt{3}}\left(\dfrac{|\boldsymbol{v}_s|}{U}\right)^{p_f}, \dfrac{\sigma_0}{\sqrt{3}}\right) \end{array} \tag{6.89}$$

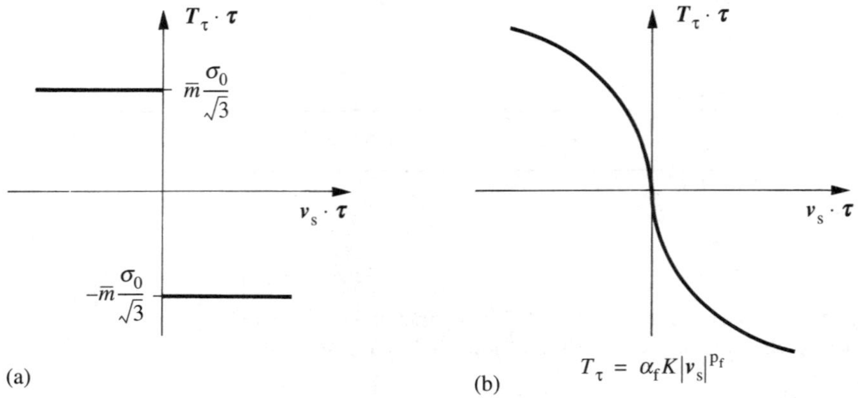

Figure 6.20 Boundary layer type friction laws: (a) Tresca's law; (b) Norton's viscoplastic law.

where α_f and p_f are the coefficients characterizing the friction, the velocity U and $\sigma_0/\sqrt{3}$ being employed to make the coefficient α_f non dimensional. As for the threshold τ_c, it can be given by Coulomb's or Tresca's law.

A particular case is Norton's friction law (fig. 6.20(b)), thus named by analogy with the power law for creep. It is a law without a threshold ($\tau_c = 0$), used for pure viscoplastic bodies, for which the shears are not *a priori* limited:

$$T_\tau = \alpha_f K|\boldsymbol{v}_s|^{p_f} \quad \text{or, in vector form} \quad \boldsymbol{T}_\tau = -\alpha_f K|\boldsymbol{v}_s|^{p_f - 1}\boldsymbol{v}_s \tag{6.90}$$

The accompanying physical assumption is again of a boundary layer. Consider a boundary layer of thickness e, composed of the softest of the bodies (consistency K, strain rate sensitivity coefficient m) sheared at the interface. In this layer, the component $n\tau$ of the strain rate tensor (shear) is:

$$\dot{\varepsilon}_{n\tau} = \frac{|\boldsymbol{v}_s|}{2e} \tag{6.91}$$

which, by the Norton-Hoff viscoplastic law (6.67), gives a deviatoric shear component equal to:

$$s_{n\tau} = K\left(\frac{|\boldsymbol{v}_s|}{e}\right)^m \tag{6.92}$$

By comparison with the proposed law (6.90a), this boundary layer model leads to $p_f = m$ and to the interpretation of the coefficient α_f in terms of boundary layer thickness, according to the relation $\alpha_f = (1/e)^m$. The coefficient α_f is thus dimensional in this case. Note that the microscopic characteristics of the contact or the presence of a third body (a lubricant for example) can explain the differences with respect to these theoretical values.

Finally, note that in a viscoplastic type friction law, the tangential friction stress vector $\boldsymbol{T}_\tau$ appears as being derived from a viscoplastic friction potential φ_f (in an analogous manner to the deviatoric stress tensor, (6.68)):

$$\boldsymbol{T}_\tau = -\frac{\partial \varphi_\mathrm{f}}{\partial \boldsymbol{v}} \quad \text{with} \quad \varphi_\mathrm{f} = \frac{\alpha_\mathrm{f} K}{p_\mathrm{f}+1}|\boldsymbol{v}_\mathrm{g}|^{p_\mathrm{f}+1} \tag{6.93}$$

Regularization of the friction laws

The discontinuities of the threshold friction laws pose serious numerical problems. Take Coulomb's law for example. When $\boldsymbol{T}_\tau$ grows, at a given contact pressure p_c, the relative slip velocity $|\boldsymbol{v}_\mathrm{s}|$ starts at zero then discontinuously assumes a value which is – even worse – undetermined. Inversely, if $|\boldsymbol{v}_\mathrm{s}|$ is imposed, $\boldsymbol{T}_\tau$ is undetermined for $|\boldsymbol{v}_\mathrm{s}| = 0$, and then takes a value $\mu_\mathrm{f} p_\mathrm{c}$ (possibly with a discontinuity) when the slip commences. To remove these difficulties, the friction laws are regularized by replacing the discontinuous relation $\boldsymbol{T}_\tau(|\boldsymbol{v}_\mathrm{s}|)$ with a continuous smoothed function that can be differentiated at the origin and which goes asymptotically towards the exact solution (fig. 6.21). For example we can write the following expression, the regularization variable for Coulomb's law $v_{\mathrm{s}0}$ being sufficiently small with respect to the characteristic velocities of the process:

$$\boldsymbol{T}_\tau = -\tau_\mathrm{c}\frac{1}{\sqrt{v_\mathrm{s}^2 + v_{\mathrm{s}0}^2}}\boldsymbol{v}_\mathrm{s} \tag{6.94}$$

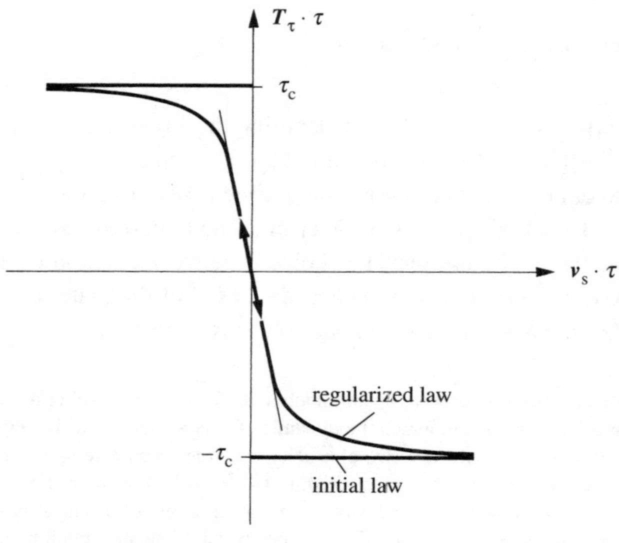

Figure 6.21 Regularization of Coulomb's friction law.

6.4 NUMERICAL TREATMENT

6.4.1 Temporal discretization for the analysis of non stationary processes

We will now describe the principle of analysis of a deformation problem. To fix an idea, we consider a non stationary forming process. We will use the terms "part" and "tool". We are thus in the category of non stationary processes of forging and stamping as opposed to rolling or drawing which are stationary by nature (excepting the transitional phases of engaging and disengaging the material). The numerical approach to stationary processes will be detailed in chapter 7 (sect. 7.5). To start, we also make the assumption that the tools are not deformable.

A non stationary process is generally analyzed by a decomposition into a sequence of time ***increments***, possibly of variable size, Δt, but relatively small in general compared to the duration of the process. In a repeating fashion, we suppose that for each time interval $[t, t + \Delta t]$, the state of the system is known at t and we desire to calculate its new state at $t + \Delta t$.

We have already given an example of a non stationary numerical approach for the solution of the diffusion equations (chemical or thermal) in chapters 2 and 3. The approach is different in the context of the analysis of deformations: in this case the domain for which a solution is obtained evolves in time, and thus the choice of a temporal discretization and a domain updating mode must precede the finite element spatial discretization of the equations.

In general, at each time increment, we will have to solve the equilibrium equations in integral form, that is, the equation for the principle of virtual power. In this case – as very often for forming, where inertia and gravity can be neglected – we have (Sect. 1.2.6):

$$\int_{\Omega} \boldsymbol{\sigma} : \dot{\boldsymbol{\varepsilon}}(\boldsymbol{v}^*)\, \mathrm{d}V - \int_{\partial\Omega} \boldsymbol{T} \cdot \boldsymbol{v}^*\, \mathrm{d}S = 0 \tag{6.95}$$

for every virtual velocity field $\boldsymbol{v}^*$, $\boldsymbol{\sigma}$ denoting the stress tensor and $\boldsymbol{T}$ the stress vector on the surface $\partial\Omega$ of the domain Ω. According to the case, the integration domain can either be the configuration at time t, Ω^t, or that at the end of the increment $\Omega^{t+\Delta t}$. Thus the necessity to determine the temporal integration scheme used for updating the configuration. The choice is between two approaches:

- The *explicit approach*[6]: the coordinates $\boldsymbol{x}^t$ of the material particles at t (that is Ω^t) are assumed known, and (6.95) is solved on Ω^t in terms of the veloc-

6 Not to be confused with what is commonly called the purely explicit formulation, frequently used to solve problems in which the dynamic effects associated with inertial forces are dominant. The solution is thus in terms of acceleration, the coordinate updating resulting from two successive temporal integrations according to an explicit scheme. Formulations solving for velocity, such as those described above, are named implicit formulations in this context (even though it will be seen with (6.96) that they can be based on an explicit integration of the velocity field!)

ity field at time t, $\boldsymbol{v}^{\mathrm{t}}$, the updating scheme being:

$$\boldsymbol{x}^{\mathrm{t}+\Delta\mathrm{t}} = \boldsymbol{x}^{\mathrm{t}} + \Delta t\, \boldsymbol{v}^{\mathrm{t}} \tag{6.96}$$

- The *implicit approach*: the coordinates $\boldsymbol{x}^{\mathrm{t}}$ of the material particles at t (that is Ω^{t}) are assumed known, and (6.95) is solved on $\Omega^{\mathrm{t}+\Delta\mathrm{t}}$ in terms of the velocity field at time $t + \Delta t$, $\boldsymbol{v}^{\mathrm{t}+\Delta\mathrm{t}}$, the updating scheme being:

$$\boldsymbol{x}^{\mathrm{t}+\Delta\mathrm{t}} = \boldsymbol{x}^{\mathrm{t}} + \Delta t\, \boldsymbol{v}^{\mathrm{t}+\Delta\mathrm{t}} \tag{6.97}$$

It is seen that the implicit approach introduces a geometric non-linearity as the integrals are calculated on a configuration that depends on the unknown variable, in this case the velocity field at the end of the increment.

Lets look a bit in detail at these strategies; how is the equilibrium equation (6.95) formulated with only the velocity field as an unknown? For that, it is necessary to express each of the terms as a function of that unknown.

The external stress vectors $\boldsymbol{T}$ are either linked directly to $\boldsymbol{v}$ (such is the case for example, of the friction stresses, through the friction law), or indirectly linked to $\boldsymbol{v}$. For example, the pressure of a fluid – assumed perfect – applied to the surface of the part is written for the implicit approach, $\boldsymbol{T}^{\mathrm{t}+\Delta\mathrm{t}} = -p_{\mathrm{a}}\boldsymbol{n}^{\mathrm{t}+\Delta\mathrm{t}}$, where $\boldsymbol{n}^{\mathrm{t}+\Delta\mathrm{t}}$ is the surface normal, depending on $\boldsymbol{v}$ through the scheme (6.97) and p_{a}, the pressure of the fluid.

As for the stress tensor $\boldsymbol{\sigma}$, it has to be linked to the velocity field through the constitutive law; $\boldsymbol{\sigma}$ depends on the strain rate tensor $\dot{\boldsymbol{\varepsilon}}$, and thus on $\boldsymbol{v}$. This obviously amounts to another source of non-linearity along with the friction. In the following we give examples of the treatment of the friction and constitutive laws.

The above remarks can be used in the context of a problem that one would solve in terms of the incremental displacement variable $\boldsymbol{u}$ between the two configurations instead of the velocity. In either case, a single equation is obtained, non linear in general, that can be written generically:

$$R(\boldsymbol{v}) = 0 \tag{6.98}$$

$\boldsymbol{v}$ being either the velocity field or the displacement, depending on the case. The solution of this equation is obtained with different techniques: first of all the finite element method, that permits the spatial discretization, and then the solution methods for non linear equations, such as the Newton-Raphson method presented in chapter 4.

6.4.2 Finite element spatial discretization

Now we introduce the spatial discretization of the equilibrium equation with the finite element method (chap. 3). For that, the part is meshed with finite elements, and the following interpolation of the velocity field is introduced (valid for the real velocity field and for virtual velocity fields),

$$v = \psi_n v^n \tag{6.99}$$

The sum is over the *Nn* nodes of the mesh. This gives for the strain rate tensor:

$$\dot{\varepsilon}_{ij} = \frac{1}{2}\left(\frac{\partial v_i}{\partial x_j} + \frac{\partial v_j}{\partial x_i}\right) = B_{ijnk} v_k^n \text{ with } B_{ijnk} = \frac{1}{2}\left(\frac{\partial \psi_n}{\partial x_j}\delta_{ik} + \frac{\partial \psi_n}{\partial x_i}\delta_{jk}\right) \tag{6.100}$$

In these expressions the index *k* varies from 1 to *N* (number of components of the velocity field).

Consider the virtual velocity fields discretized according to (6.99) and kinematically admissible to zero (chap. 3, Sect. 3.4). Equation (6.95) is thus transformed into a system of (*Nn N*) equations for which the unknowns are the nodal components of the velocity field v_k^n:

$$\int_{\Omega} \sigma_{ij} B_{ijnk} \, dV = \int_{\partial\Omega} T_k \psi_n \, dS \qquad \text{for } (n, k) \in N^{free} \tag{6.101}$$

$$v_k^n = \bar{v}_k^n \qquad \text{for } (n, k) \in N^{imp} \tag{6.102}$$

where N^{imp} and N^{free} denote respectively the sets of imposed and free degrees of freedom of the problem.

The volume integral, on the left hand side of (6.101) is calculated by assembly of all the different elements of the mesh of the domain Ω. For each of these elements, the integral is numerically calculated by Gauss' method, which means that the stresses σ_{ij}, and the coefficients B_{ijnk} must be calculated at the Gauss points of the element (chap. 3 and 10).

The surface integral, on the right hand side, is calculated by assembly of the different edges (in 2 dimensions) or facets (in 3 dimensions) of the boundary of the mesh of Ω. The integral is numerically calculated by Gauss' method on each of the edges or facets, which means that the interpolation functions ψ_n, and the external stress vectors, $\boldsymbol{T}$, must be calculated at the Gauss points of the edge or facet.

The stresses and external stress vectors, possibly depending on the velocity field (or the displacement), are locally calculated at each Gauss point as functions of the velocity vector (or displacement vector) interpolated in each element or facet treated. As we have seen in chapter 3 (Sect. 3.4.3), for linear elastic behavior, a linear system of equations results, the unknowns being the components of the displacement vector at the nodes of the mesh.

Such is not the case when non linear constitutive or friction laws are employed, the non-linearity being greater when they are used in association with an implicit temporal integration scheme. In this case a system of non linear equations results, which can be written generically denoting by $\boldsymbol{V}$ the vector of the nodal components of the velocity field:

$$\boldsymbol{R}(\boldsymbol{V}) = \boldsymbol{0} \tag{6.103}$$

Or as components,

$$R_k^n(V) = \int_{\Omega} \sigma_{ij} B_{ijnk} \, dV - \int_{\partial\Omega} T_k \psi_n \, dS = 0 \qquad \text{for } (n, k) \in N^{\text{free}} \tag{6.104}$$

$$v_k^n = \bar{v}_k^n \qquad \text{for } (n, k) \in N^{\text{imp}} \tag{6.105}$$

The vector $\boldsymbol{R}$ is called the ***residual vector*** of the discretized equilibrium equation.

6.4.3 Solution of non linear problems

Different iterative solution methods for non linear systems such as (6.103) were presented in chapter 4. We will show below the application in detail of the Newton-Raphson method in the context of mechanical solutions.

Assume then that at the start of the iteration (ν), the residual of our system (6.103) is given by:

$$\boldsymbol{R}(\boldsymbol{V}^{(\nu-1)}) \neq \boldsymbol{0} \tag{6.106}$$

The iteration (ν) thus consists (chap. 4) of calculating the correction $\delta \boldsymbol{V}^{(\nu)}$ of $\boldsymbol{V}^{(\nu-1)}$ by solving the following linear system:

$$\left[\frac{\partial \boldsymbol{R}(\boldsymbol{V}^{(\nu-1)})}{\partial \boldsymbol{V}}\right] \delta \boldsymbol{V}^{(\nu)} = -\boldsymbol{R}(\boldsymbol{V}^{(\nu-1)}) \tag{6.107}$$

in which $[\partial \boldsymbol{R}/\partial \boldsymbol{V}]$ is the ***tangent stiffness matrix*** of the non linear system. In components, for each pair (n, k), we have:

$$\left[\frac{\partial R_k^n}{\partial v_l^m}\right]^{(\nu-1)} \delta v_l^{m(\nu)} = -R_k^n(\boldsymbol{V}^{(\nu-1)}) \tag{6.108}$$

The new estimate $\boldsymbol{V}^{(\nu)}$ is obtained by:

$$\boldsymbol{V}^{(\nu)} = \boldsymbol{V}^{(\nu-1)} + \delta \boldsymbol{V}^{(\nu)} \tag{6.109}$$

The velocity corrections (or the displacement corrections) are then iteratively calculated until a convergence criterion is satisfied. Different criteria can be defined, for example:

$$|\boldsymbol{R}^{(\nu)}| < \varepsilon_r \qquad \text{with} \qquad |\boldsymbol{R}| = \Big(\sup_{(n,k)\in N^{free}} |R_k^n|\Big) \Big/ \Big(\sup_{(n,k)\in N^{imp}} |R_k^n|\Big) \tag{6.110}$$

ε_r being a tolerance fixed a priori (typically 10^{-4}) relative to the essential conditions fixed at the boundary. This criterion offers the advantage of normalizing the differences from equilibrium with respect to the nodal reactions (remember that R_k^n are the components of the applied reaction forces according to the "imposed" degrees of freedom, chap. 3, sect. 3.4).

Application: expression of the stiffness matrix

The calculation of the tangent stiffness matrix appearing in (6.108) necessitates the differentiation of the stress tensor with respect to the components of the velocity field, which is written:

$$\frac{\partial \sigma_{ij}}{\partial v_l^m} = \frac{\partial \sigma_{ij}}{\partial \dot{\varepsilon}_{gh}} \frac{\partial \dot{\varepsilon}_{gh}}{\partial v_l^m} = \frac{\partial \sigma_{ij}}{\partial \dot{\varepsilon}_{gh}} B_{ghml} \tag{6.111}$$

After differentiation of (6.104):

$$\frac{\partial R_k^n}{\partial v_l^m} = \int_{\Omega} B_{ghml} \frac{\partial \sigma_{ij}}{\partial \dot{\varepsilon}_{gh}} B_{ijnk} \, dV + \int_{\Omega} \sigma_{ij} \frac{\partial B_{ijnk}}{\partial v_l^m} \, dV - \int_{\partial\Omega} \frac{\partial T_k}{\partial v_l^m} \psi_n \, dS \tag{6.112}$$

For an explicit scheme (6.96), the second integral disappears, the coefficients B being associated with the configuration Ω^t and not varying with the unknown velocity field $\boldsymbol{v}^t$. Conversely, as the integrals are calculated on $\Omega^{t+\Delta t}$, for the implicit scheme (6.97), rigorous interpretation of (6.112) would require terms for the differentiation of the domain: those coming from the differentiation with respect to the velocity components of the Jacobian of the mapping from the real elements to the reference elements. Nevertheless, these terms can often be neglected when the incremental displacements are weak.

Finally, the first integral in (6.112) highlights the importance of the ***tangent constitutive modulus*** $\partial\boldsymbol{\sigma}/\partial\dot{\boldsymbol{\varepsilon}}$. This modulus must be calculated during the solution of the constitutive law. Its precision obviously conditions the convergence of the Newton-Raphson method.

6.4.4 Treatment of contact

Now we must incorporate the contact boundary conditions described above in the spatial and temporal discretization of the equations of mechanical equilibrium. As a first approximation, each tool is considered non deformable, its surface being partitioned into linear discrete elements, (segments in 2 dimensions, triangles or quadrangles in 3 dimensions). The global surface of the tooling is thus defined at instant t by a set of linear equations $g_e^t(\boldsymbol{x}) = 0$ (e being the index for the discretization elements) that we write in global notation $g^t(\boldsymbol{x}) = 0$. By convention, we denote by $\boldsymbol{n}$ the normal (defined for each element of the discretization) oriented towards the interior of the tools (defined by $g^t(\boldsymbol{x}) < 0$). Referring to section 6.3.1, we deduce that a point M of the part, with coordinates $\boldsymbol{x}$ at instant t, is in contact if:

$$g^t(\boldsymbol{x}) = 0 \quad \text{and} \quad \sigma_n < 0 \tag{6.113}$$

the future evolution of this point being governed by Signorini's equations (6.75) for which we repeat the physical interpretation:

$$\sigma_n < 0 \quad \text{and} \quad \Delta\boldsymbol{v} \cdot \boldsymbol{n} = 0 \quad \text{point remaining in contact} \tag{6.114}$$

$$\sigma_n = 0 \quad \text{and} \quad \Delta\boldsymbol{v} \cdot \boldsymbol{n} < 0 \quad \text{point losing contact} \tag{6.115}$$

Assume then that the set of points in contact is known at instant t, the start of an increment. Two problems emerge. How is it to be insured that the points in contact verify the conditions at the end of the increment? And how are the modifications of the surface in contact to be managed (some points lose contact, others come into contact)? A starting assumption is that each point remains in the same state (in contact or not) during the increment, and the changes of state (entry into contact, loss of contact) only occur at the end of the increment (or at the start of the next, which amounts to the same). A set of boundary conditions thus being determined throughout the increment, the incremental displacements are calculated, then the new positions of the points, with their stresses. One then tests conditions (6.113) to determine the state of contact at the end of the increment.

Boundary conditions applicable at a contact point

In general, because of the discrete nature of the function g, it is difficult to impose that a point in contact with a tool exactly satisfies the condition $g^{t+\Delta t}(\boldsymbol{x} + \Delta t\, \boldsymbol{v}) = 0$ at the end of the increment. One thus tends to apply preferentially the condition $\Delta\boldsymbol{v} \cdot \boldsymbol{n} = 0$. In this case, the solution only imposes that the relative vector velocity of point M of the part at position $\boldsymbol{x}$ be tangent to the surface of the tool, which amounts to saying that its relative displacement between t and $t + \Delta t$ should be tangent to the tool in $\boldsymbol{x}$. If the tool is curved, the point M leaves the geometric surface of the tool (fig. 6.22) and two types of situation can appear.

- *Locally concave tool*: the point of the part penetrates the tool and must be projected onto the surface (an operation that causes loss of volume, all the more if the time step is large).
- *Locally convex tool*: the point of the part moves away from the tool, even if $\Delta\boldsymbol{v} \cdot \boldsymbol{n} = 0$. It is therefore necessary to test the normal stress at the end of the

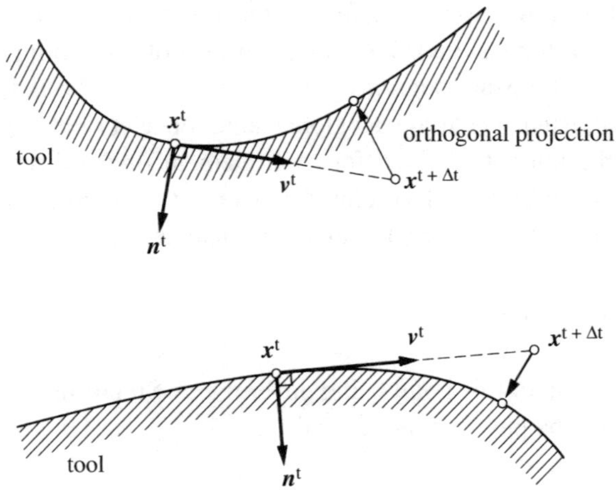

Figure 6.22 Boundary condition applied to nodes in contact.

increment: if $\sigma_n^{t+\Delta t} < 0$, the point remains in contact and its projection on the surface causes an increase of the volume; if $\sigma_n^{t+\Delta t} \geq 0$, the point has left the tool surface, and there is no reason to modify its new position.

Updating of the state of contact

In order for a point to be considered in contact, it should fulfill a geometric condition ($g^t(\boldsymbol{x}) = 0$) and a stress condition ($\sigma_n < 0$). It is thus important that the position and the normal stress be known simultaneously. This is not obvious *a priori* and has important consequences on the contact algorithm, according to whether the approach is explicit or implicit.

- *Explicit approach:* one assumes the state of contact in $\boldsymbol{x}^t$, then calculates the velocity field $\boldsymbol{v}^t$ which allows the updating of the positions (6.96). It is thus only at this moment that σ_n^t can be known in $\boldsymbol{x}^t$ and the contact assumption made at instant t verified! There is a lag of an increment between the "geometric" management of the contact (position of the point) and its "mechanical" management (normal stress).
  Thus if a positive normal stress is determined at this stage, two strategies are possible: either the solution is validated, but the point must be free of all conditions $\Delta \boldsymbol{v} \cdot \boldsymbol{n} = 0$ for the next increment; or the calculation is redone for this increment freeing the point from the condition. A good solution could be to perform the calculation of σ_n^t at each Newton-Raphson iteration, and to adjust the assumption on the contact state, which finally insures the calculation of $\boldsymbol{v}^t$ on $\boldsymbol{x}^t$ with the correct stress $\sigma_n(\boldsymbol{x}, t)$. However, the convergence is slowed.
- *Implicit approach:* in this case, $\boldsymbol{v}^{t+\Delta t}$ is calculated directly and thus the stresses at $t + \Delta t$ and $\sigma_n(\boldsymbol{x} + \Delta\boldsymbol{x}, t + \Delta t)$. Then the positions, velocities and

stresses are all determined simultaneously at $t + \Delta t$. In the same way, the contact condition $\Delta \boldsymbol{v}^{t+\Delta t} \cdot \boldsymbol{n}^{t+\Delta t} = 0$ can be updated during the Newton-Raphson iterations for $\sigma_n(\boldsymbol{x} + \Delta \boldsymbol{x}, t + \Delta t)$.

The treatment of new points entering into contact is simpler. It corresponds to the fact that a point M such that $g^t(\boldsymbol{x}) > 0$ (strictly outside the tool) verifies the condition at the end of the increment:

$$g^{t+\Delta t}(\boldsymbol{x} + \Delta t\, \boldsymbol{v}) \leq 0 \tag{6.116}$$

Of course, the evaluation of the function g, as well as the calculations of the projections on the surface, require procedures adapted to the type of discretization of the tool surface, with minimal computation time, but these details will not be discussed here.

We will now review the two methods that take into account the condition $\Delta \boldsymbol{v} \cdot \boldsymbol{n} = 0$ in the global formulation.

Direct nodal method

The solution of the discretized equilibrium equations (6.103) should be performed taking into account the following boundary conditions:

$$\boldsymbol{v}^m \cdot \boldsymbol{n} = \boldsymbol{v}_{tool} \cdot \boldsymbol{n} \quad \text{for every node } m \text{ in contact } (m = 1, Nn) \tag{6.117}$$

where $\boldsymbol{v}_{tool}$ is the velocity vector of the tool, assumed known, and where it should be noted that $\boldsymbol{n}$ can be chosen to be defined outgoing from the mesh of the part or from the discretization of the tool. This relation can be put in the general form of the imposition of the velocity degrees of freedom (6.105) under the condition that it be written in a local reference (($\boldsymbol{n}$, $\boldsymbol{\tau}$) in 2 dimensions, $\boldsymbol{\tau}$ being a unitary tangent vector; ($\boldsymbol{n}$, $\boldsymbol{\tau}_1$, $\boldsymbol{\tau}_2$) in 3 dimensions). For every node m in contact, a change of variables is performed for its degrees of freedom: in 3 dimensions, $(v_x^m, v_y^m, v_z^m) \rightarrow (v_n^m, v_{\tau_1}^m, v_{\tau_2}^m)$. This only involves simple manipulations of the residual vector and the stiffness matrix. Then after solution of the global system, we obtain the velocity vector solution at node m in the local reference associated with this node. It only remains the inverse change of variables to express it in the initial fixed reference.

The test for updating the contact condition is difficult to implement when it is based on the sign of the normal stress. This supposes a real calculation of the stresses (which, by the way is not always necessary) and which in addition is at the integration points of the boundary surfaces or nodes, which assumes a supplementary "smoothing" and thus an error source. The examination of the *nodal reactions* is however more reliable. We point out here that they are given for every "imposed" degree of freedom (m, k), by (Chap. 3, Sect. 3.4.3):

$$F_k^m = \int_\Omega \sigma_{ij} B_{ijmk}\, dV - \int_{\partial\Omega} T_k \Psi_m\, dS \tag{6.118}$$

The loss of contact is thus triggered (zeroing of the condition $\Delta \boldsymbol{v}^{\mathrm{m}} \cdot \boldsymbol{n} = 0$) when $F_{\mathrm{k}}^{\mathrm{m}}$ is positive, k being the normal direction.

Penalty method

The method consists of penalizing the velocity fields having a penetrating normal component. For that, all penetrating normal velocities induce a proportional repulsive normal force. However, no impenetrability condition of the type $\Delta \boldsymbol{v} \cdot \boldsymbol{n} = 0$ is applied. We write then:

$$\int_{\Omega} \boldsymbol{\sigma} : \dot{\boldsymbol{\varepsilon}}^* \, \mathrm{d}V - \int_{\partial\Omega} \boldsymbol{T} \cdot \boldsymbol{v}^* \, \mathrm{d}S + \int_{\partial\Omega_{\mathrm{c}}} \chi_{\mathrm{p}} \langle (\boldsymbol{v} - \boldsymbol{v}_{\mathrm{tool}}) \cdot \boldsymbol{n} \rangle \boldsymbol{n} \cdot \boldsymbol{v}^* \, \mathrm{d}S = 0 \tag{6.119}$$

the coefficient χ_{p} being the penalty coefficient. The bracket, whose definition is given in (6.62), expresses the fact that the penalty only applies to penetrating fields. The third integral could well be over the entire boundary $\partial\Omega$, but in practice it is restricted to the "penalizable" part of the boundary $\partial\Omega_{\mathrm{c}}$, in the neighborhood of contact with the tools (i.e. for which the points verify $g(\boldsymbol{x}) \approx 0$). It is seen by comparison with the general expression of the principle of virtual power that the normal stress vector and the normal stress on $\partial\Omega_{\mathrm{c}}$ are expressed respectively by:

$$\boldsymbol{T}_{\mathrm{n}} = -\chi_{\mathrm{p}} \langle (\boldsymbol{v} - \boldsymbol{v}_{\mathrm{tool}}) \cdot \boldsymbol{n} \rangle \boldsymbol{n} \qquad \sigma_{\mathrm{n}} = -\chi_{\mathrm{p}} \langle (\boldsymbol{v} - \boldsymbol{v}_{\mathrm{tool}}) \cdot \boldsymbol{n} \rangle \tag{6.120}$$

When there is actually contact ($g(\boldsymbol{x}) < 0$) the vector $\boldsymbol{T}$ in the second integral now represents only the imposed tangential component (or function of the velocity field) via a friction law. The finite element discretization gives for every degree of freedom (m, k):

$$R_{\mathrm{k}}^{\mathrm{m}} = \int_{\Omega} \sigma_{\mathrm{ij}} B_{\mathrm{ijmk}} \, \mathrm{d}V - \int_{\partial\Omega} T_{\mathrm{k}} \psi_{\mathrm{m}} \, \mathrm{d}S + \int_{\partial\Omega_{\mathrm{c}}} \chi_{\mathrm{p}} \langle (\psi_{\mathrm{r}} \boldsymbol{v}^{\mathrm{r}} - \boldsymbol{v}_{\mathrm{tool}}) \cdot \boldsymbol{n} \rangle n_{\mathrm{k}} \psi_{\mathrm{m}} \, \mathrm{d}S = 0 \tag{6.121}$$

Note that this expression of the third integral, which we denote $R_{\mathrm{k}}^{(\mathrm{pen})\mathrm{m}}$, implies that the algorithm that recognizes contact (calculation of the function g and the normal vector $\boldsymbol{n}$) be performed at the integration points of the boundary facets or edges. The contribution of the penalty term to the system stiffness matrix is then deduced:

$$\frac{\partial R_{\mathrm{k}}^{(\mathrm{pen})\mathrm{m}}}{\partial v_{\mathrm{l}}^{\mathrm{q}}} = \int_{\partial\Omega_{\mathrm{c}}} \chi_{\mathrm{p}} H((\psi_{\mathrm{r}} \boldsymbol{v}^{\mathrm{r}} - \boldsymbol{v}_{\mathrm{tool}}) \cdot \boldsymbol{n}) \psi_{\mathrm{q}} n_{\mathrm{l}} n_{\mathrm{k}} \psi_{\mathrm{m}} \, \mathrm{d}S \tag{6.122}$$

where H is the Heaviside function ($H(x) = 1$ if $x \geq 0$; $H(x) = 0$ otherwise).

Note that a more rudimentary penalization method is often employed: nodal penalization, in contrast with integral penalization presented above. Then the penalty forces defined directly at the mesh nodes are employed:

$$\int_{\Omega} \boldsymbol{\sigma} : \dot{\boldsymbol{\varepsilon}}^* \, \mathrm{d}V - \int_{\partial\Omega} \boldsymbol{T} \cdot \boldsymbol{v}^* \, \mathrm{d}S + \sum_{r \in \partial\Omega_c} S_r \chi'_p \langle (\boldsymbol{v}^r - \boldsymbol{v}_{tool}) \cdot \boldsymbol{n}^r \rangle \boldsymbol{n}^r \cdot \boldsymbol{v}^{*r} = 0 \quad (6.123)$$

where S_r is a "control surface" attached to node *r*, which gives for a node *m* belonging to $\partial\Omega_c$ (penalizable):

$$R_k^m = \int_{\Omega} \sigma_{ij} B_{ijmk} \, \mathrm{d}V - \int_{\partial\Omega} T_k \psi_m \, \mathrm{d}S + S_m \chi'_p \langle (\boldsymbol{v}^m - \boldsymbol{v}_{tool}) \cdot \boldsymbol{n}^m \rangle n_k^m = 0 \quad (6.124)$$

and a diagonal term this time (on the nodal indices) in the stiffness matrix (to compare with (6.122)) :

$$\frac{\partial R_k^{(pen)m}}{\partial v_l^q} = S_m \chi'_p H((\boldsymbol{v}^m - \boldsymbol{v}_{tool}) \cdot \boldsymbol{n}^m) n_l^m n_k^m \delta_{mq} \quad (6.125)$$

The advantage of the penalty method is obviously its simplicity. Nevertheless, the major difficulty comes from the choice of the penalty coefficient: the larger it is, the less penetration is allowed, but values too high affect the conditioning of the matrices and the convergence of the solution algorithms. The optimum thus depends on the problem treated. It is even apparent that some problems with a large distribution of normal stresses can not be correctly treated with this method without adapting *local* values of the penalty coefficient. The method of Lagrange multipliers, more accurate but more costly, is often preferred. We will not develop that method here.

Generalization to the treatment of contact between many deformable bodies

To solve a problem with many deformable bodies means to solve the equilibrium equation (6.95) in each body accompanied by the boundary conditions (6.75) imposed by the presence of the other bodies. There exist two principal strategies, for which we will only discuss the principles, without going into detail.

- A *global calculation* consisting of considering the set of solids in contact as a unique domain on which the set of equations is solved. The possibilities for spatial discretization and imposition of the contact conditions are then multiple: meshes of the different bodies coinciding with contact elements (thin elements at the interfaces); meshes possibly not coincident with a penalty method or a Lagrange multiplier method.
- An *alternating solution* on each body by turns, taking for boundary conditions the forces applied by the neighboring bodies, or the normal velocities to respect because of their presence. Such a solution would be iterative, stopping when convergence is attained on each of the bodies. The meshes can

thus be non coinciding, so that each of the systems to be solved is of significantly smaller size than that resulting from a global calculation.

6.4.5 Treatment of large deformations: remeshing

Triggering of remeshing

The incremental update procedure for the position of the mesh nodes according to (6.96) or (6.97), leads inevitably in the case of large deformations to ***degeneracy of the elements***: the mapping between the real element and the reference element loses its bi-directional character and the Jacobian matrix can no longer be inverted. It is thus necessary to generate a new mesh, based on the existing mesh, in order to continue the calculations. Actually, this case is an extreme situation that should be avoided. In practice, mesh regeneration is «triggered» earlier, in the following conditions.

- The *local penetration* into the tools is excessive: an element intersects the interior of the tools present, and the intersected surface or volume is greater than a critical value (fig. 6.23(a)).
- The *local curvature* is excessive at the boundary of the mesh. The evaluation will differ according to the element type employed and of course the dimension of the problem (fig. 6.23(b)). Exceeding a maximal value entrains the regeneration of the mesh.
- *Periodic mesh regeneration*, triggered as a preventative measure; the period can be expressed as a number of increments or in time.
- An *error estimate* due to the spatial discretization that is judged too high. Methods exist for estimating this error (the most common of which are based on the analysis of "jumps" of the gradients of variables calculated in

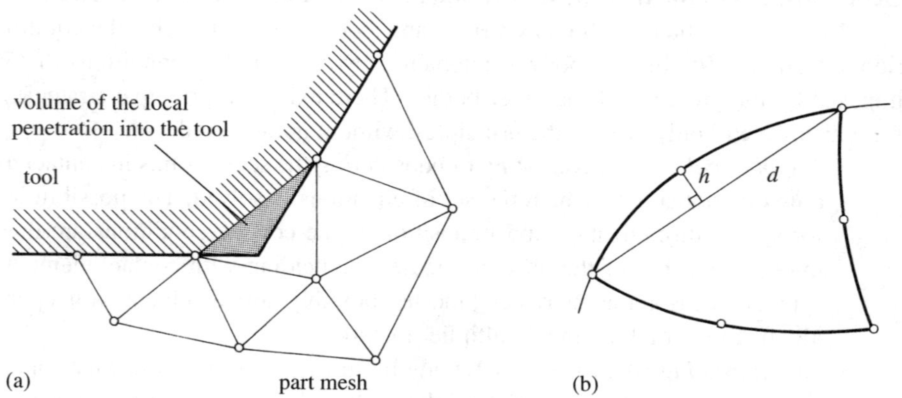

Figure 6.23 Illustration of the criteria of local penetration (a) and of local curvature (b) for the triggering of remeshing. The ratio *h*/*d* is an indicator of the curvature of an edge of a quadratic triangular element.

neighboring elements). The finite element theory allows the deduction of the desirable size of the elements in the new mesh. The mesh regeneration is thus driven by the error estimator and has the quality of being an adaptive remeshing, but this theme will not be developed here.

Execution of the remeshing

Mesh regeneration is generally performed in two stages. First, the surface is treated, then secondly, the internal nodes are generated.

Surface remeshing

One way of proceeding is to systematically divide all the boundary segments or facets (fig. 6.24), thereby creating an over discretization of this boundary. These over discretization points are projected onto the surface of the tools in the zones where penetration was detected (fig. 6.24). In general, they are repositioned on the boundary to satisfy certain conditions defined *a priori*: given progression of the distance between successive nodes, respect of the curvature of the boundary and of the volume of the part, minimization of the penetration, etc. This repositioning, as well as the elimination of nodes judged to be useless, is iteratively performed running along the boundary analyzing the neighborhood of each point, edge or facet of the over discretization.

Another technique, associated with preventative mesh regeneration triggered with a relatively high frequency, consists not of systematically over discretizing, but rather to proceed by local improvements (changing the connection of the elements, creation or removal of nodes), in the neighborhood of the surface nodes. We will see an example in chapter 7, section 7.6.

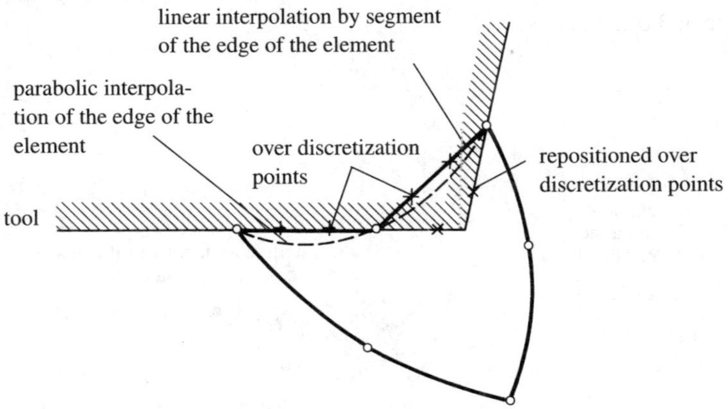

Figure 6.24 Over discretization and adaptation to contact in case of local penetration of the mesh in the tool.

Volume remeshing

The new surface having been defined, one proceeds to the creation of internal nodes. This can be performed according to the meshing methods presented in chapter 4, section 4.2. The size of the internal mesh elements is thus directly induced from the size of the boundary edges or facets.

Another strategy consists of proceeding by local improvements, according to the same principles as for the surface.

Transport of the variables after remeshing

After generation of a new mesh, according to the type of calculation performed, it may be necessary to *transport* various types of variables from the old mesh to the new. The variables can be scalar fields (for example, the temperature, the equivalent plastic deformation), vector fields (velocity) or tensor fields (stresses). In the case of vector and tensor fields the components are generally transported separately, leading back to the problem of transport of a scalar field. We distinguish nonetheless two sub cases, according to whether the field is defined at the nodes or at the integration points of the mesh.

Transport of a nodal field

Such is the case of the temperature, the concentration or the components of the velocity field. Given a scalar field denoted f, it is a matter of calculating at each node of the new mesh the value $f_{(T)}$ (value after transport) as a function of the values of f at the nodes of the old mesh.

The method most generally employed is that of direct interpolation, shown in figure 6.25, which is decomposed into three phases:

- *Localization of the element e of the old mesh containing the new node*. This analysis is easy in the case of linear triangular elements in 2 dimensions, but can be more complicated in the case of quadratic or quadrangular elements, or in 3 dimensions.

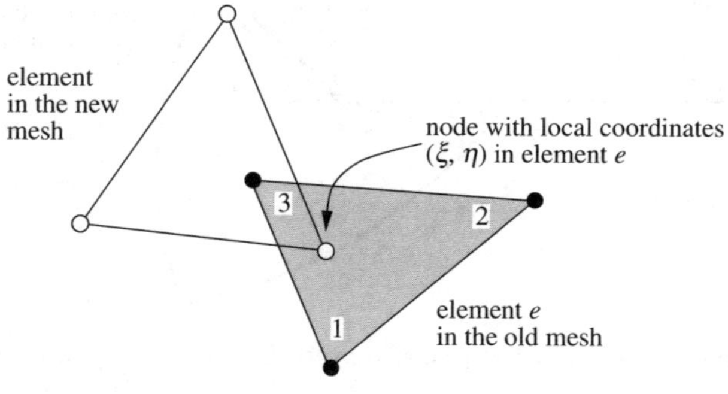

Figure 6.25 Nodal transport by direct interpolation.

- *Calculation of the local coordinates* (ξ, η) *of the node in element e*. In 2 dimensions, a system of two equations with two unknowns must be solved (three equations with three unknowns in 3 dimensions) for which the order varies as the order of the interpolation functions used:

$$x = \psi_i(\xi, \eta)x^i \tag{6.126}$$
$$y = \psi_i(\xi, \eta)y^i$$

where (x, y) are the coordinates of the analyzed node, the sum being over the *nne* nodes of the element *e* (to simplify, we have identified the local and global numberings here).

- *Interpolation in element e*, to calculate the transported value, $f_{(T)}$, according to the formula:

$$f_{(T)} = \psi_i(\xi, \eta)f^i \tag{6.127}$$

Transport of a field defined at the integration points

Such is the case for example of the components of the stress tensor or the equivalent plastic strain. We show two methods here.

The first is a ***transport by proximity***. It consists for each integration point of the new mesh of identifying which is the closest integration point among those of the old mesh. The value of f at this point is then directly moved to the new integration point (fig. 6.26).

The second method, that we will describe below, consists of a ***transport by direct interpolation***, preceded by a ***nodal smoothing***. As a first step, the field f is "smoothed" at the nodes of the old mesh. A nodal field $\tilde{f}$ being obtained, one proceeds, in a second step, to direct interpolation of the values of $\tilde{f}$ at the new integration point, as explained above for the transport of a nodal variable.

The first step of nodal smoothing can be performed by the least squares method, which consists of minimizing the functional Ξ, defined as follows, on the

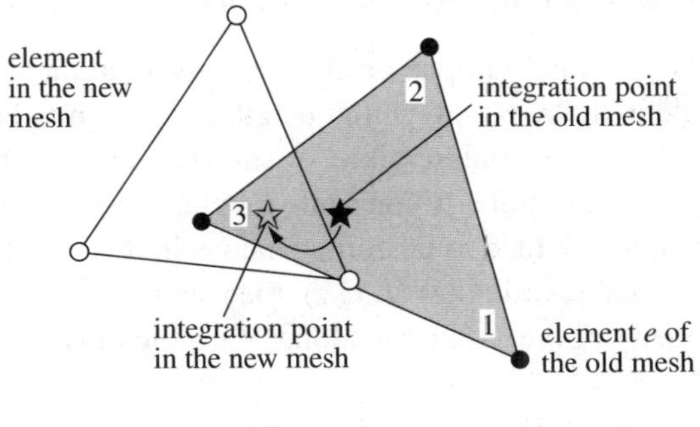

Figure 6.26 Transport of a field defined at the integration points.

old mesh, with respect to the nodal unknowns $\tilde{f}^{\mathrm{i}}$.

$$\Xi = \int_{\Omega} (\psi_{\mathrm{j}}\tilde{f}^{\mathrm{j}} - f)^2 \, \mathrm{d}V \tag{6.128}$$

where the sum is over the total number of nodes (*Nn*) of the old mesh. In other words, the following problem is to be solved:

$$\frac{\partial \Xi}{\partial \tilde{f}^{\mathrm{i}}} = 0 \qquad \text{for all } i = 1, Nn \tag{6.129}$$

A system of *Nn* linear equations in *Nn* unknowns is thus obtained:

$$\int_{\Omega} \psi_{\mathrm{i}}\psi_{\mathrm{j}}\tilde{f}^{\mathrm{j}} \, \mathrm{d}V = \int_{\Omega} \psi_{\mathrm{i}} f \, \mathrm{d}V \tag{6.130}$$

which can be expressed in matrix form $[\boldsymbol{M}]\,\tilde{\boldsymbol{f}} = \boldsymbol{b}$ where $\tilde{\boldsymbol{f}}$ is the vector of nodal unknowns and where the matrix (mass matrix) and the right hand side have the components:

$$M_{\mathrm{ij}} = \int_{\Omega} \psi_{\mathrm{i}}\psi_{\mathrm{j}} \, \mathrm{d}V \qquad b_{\mathrm{i}} = \int_{\Omega} \psi_{\mathrm{i}} f \, \mathrm{d}V \tag{6.131}$$

As has already been mentioned, the calculation of the values transported $f_{(\mathrm{T})}$ to the integration points of the new mesh is achieved by direct interpolation of the values of $\tilde{f}$ at these points (according to formula (6.127) replacing f by $\tilde{f}$).

6.5 EXAMPLES

6.5.1 An academic example: uniaxial tension of a test sample

Consider a cylindrical sample, or more precisely, an idealization of the central part of the sample in figure 6.1. Its initial length is ℓ_0, its initial radius R_0 (fig. 6.27). The material is assumed incompressible, or stated otherwise, the volume variations due to elasticity are negligible. A constant stretching velocity v_0 is imposed on the sample. The strain is assumed to be homogeneous in the sample.

Using cylindrical coordinates (r, θ, z), the rotational symmetry of the problem and the homogeneity of the strain rate along z are such that:

$$v_{\mathrm{r}} = v_{\mathrm{r}}(r) \qquad v_{\theta} = 0 \qquad v_{\mathrm{z}} = \frac{z}{\ell} v_0 \tag{6.132}$$

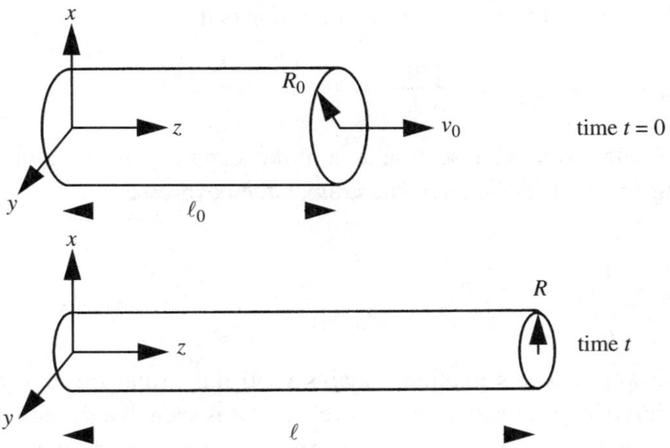

Figure 6.27 Stretching of a cylindrical sample.

From incompressibility, we have the following relation (see chapter10 for the expression of the divergence in cylindrical coordinates):

$$\operatorname{div} \boldsymbol{v} = \operatorname{Tr} \dot{\boldsymbol{\varepsilon}} = \dot{\varepsilon}_{rr} + \dot{\varepsilon}_{\theta\theta} + \dot{\varepsilon}_{zz} = \frac{\partial v_r}{\partial r} + \frac{v_r}{r} + \frac{v_0}{\ell} = 0 \tag{6.133}$$

It should be noted that in this expression, $\partial v_r/\partial r$ is identical to $\mathrm{d}v_r/\mathrm{d}r$ as v_r depends only on r. The integration can be easily performed by integrating the homogeneous equation without the left hand side and using the method of variation of the constant. It can also be seen that:

$$\frac{\mathrm{d}v_r}{\mathrm{d}r} + \frac{v_r}{r} = \frac{1}{r}\frac{\mathrm{d}}{\mathrm{d}r}(rv_r) \tag{6.134}$$

One thus obtains:

$$v_r = -\frac{v_0}{2\ell}r + \frac{A}{r} \qquad \text{with } A = \text{cnst} \tag{6.135}$$

The condition $v_r = 0$ on the axis of the sample imposes that the integration constant be null: $A = 0$. From that the expression of the strain rate tensor (diagonal) is deduced (chap. 10):

$$\dot{\boldsymbol{\varepsilon}} = \begin{bmatrix} -\frac{v_0}{2\ell} & 0 & 0 \\ 0 & -\frac{v_0}{2\ell} & 0 \\ 0 & 0 & \frac{v_0}{\ell} \end{bmatrix} \tag{6.136}$$

The equivalent or generalized strain rate is then:

$$\dot{\varepsilon}_{eq} = \frac{v_0}{\ell} = \frac{1}{\ell}\frac{d\ell}{dt} = -\frac{1}{S}\frac{dS}{dt} \tag{6.137}$$

where S is the transverse section area of the sample (uniform along z), the product $S\ell$ being constant. The equivalent strain is then expressed:

$$\varepsilon_{eq} = \int_0^t \dot{\varepsilon}_{eq}\, du = \ln\frac{\ell}{\ell_0} \tag{6.138}$$

We note that this relation remains valid for a time varying traction velocity, $v_0(t)$. Moreover, at constant traction velocity, it is seen that the equivalent strain rate decreases with time. To maintain the strain rate constant during a traction test, the velocity of the jack must be constantly increased (exponentially). The conclusions are of course the opposite for the case of a compression test.

Stability of the plastic deformation

We turn our interest now to the localization of the strain, characteristic of the necking phenomenon that leads to the destruction of the sample. The purpose of this section is to make apparent the influence of the rheological coefficients on the initiation of this phenomenon. We follow Hart's analysis of the instability in uniaxial tension (fig. 6.28).

The stability of the deformation can be expressed in the following way: for each section z of the sample, the rate of decrease of the transverse section area $S(z, t)$ should vary in a manner proportional to the surface area, so that the smallest sections are reduced more rapidly than the larger sections. Such an evolution leads to a delocalization (a "diffusion") of a possible narrowing of the section. Expressed in equations, this can be written as a doubly logarithmic derivative and denoting $\dot{S} = \partial S/\partial t$:

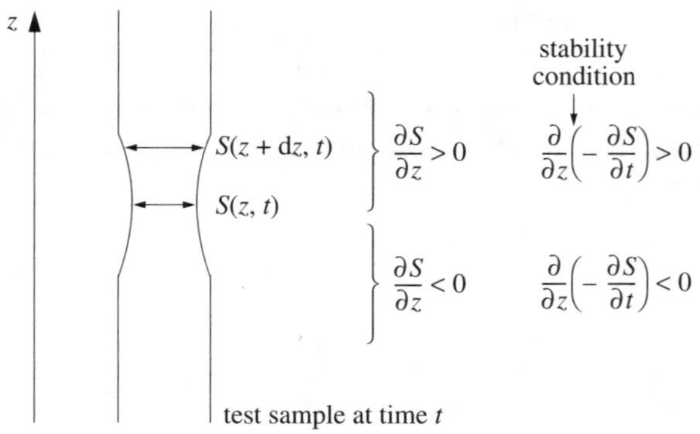

Figure 6.28 Hart's analysis of the instability in uniaxial tension. Stability criterion in a section z anywhere in the sample.

$$\frac{\dfrac{\partial \ln(-\dot{S})}{\partial z}}{\dfrac{\partial \ln S}{\partial z}} > 0 \tag{6.139}$$

At the same time, the equilibrium of the sample implies that the product $F = S\sigma_{eq}$ does not depend on z (assumption of uniaxial stress). Thus:

$$\frac{\partial \ln \sigma_{eq}}{\partial z} + \frac{\partial \ln S}{\partial z} = 0 \tag{6.140}$$

Assume that the material follows a constitutive law of the type $\sigma_{eq} = \sigma_{eq}(\dot{\varepsilon}_{eq}, \varepsilon_{eq})$, in which the plastic and total strain rates are mixed for simplicity (the same for the strains). The strain rate sensitivity coefficient, m (already seen above, (6.60)), and the work hardening coefficient, γ, are thus defined by:

$$m = \frac{\partial \ln \sigma_{eq}}{\partial \ln \dot{\varepsilon}_{eq}} = \frac{\dot{\varepsilon}_{eq}}{\sigma_{eq}} \frac{\partial \sigma_{eq}}{\partial \dot{\varepsilon}_{eq}} \qquad \gamma = \frac{\partial \ln \sigma_{eq}}{\partial \varepsilon_{eq}} = \frac{1}{\sigma_{eq}} \frac{\partial \sigma_{eq}}{\partial \varepsilon_{eq}} \tag{6.141}$$

The latter coefficient γ is linked to the more common work hardening coefficient $n = \partial \ln \sigma_{eq} / \partial \ln \varepsilon_{eq}$ by the expression $\gamma = n/\varepsilon_{eq}$. After (6.141):

$$\frac{\partial \ln \sigma_{eq}}{\partial z} = \frac{1}{\sigma_{eq}}\left(\frac{\partial \sigma_{eq}}{\partial \varepsilon_{eq}} \frac{\partial \varepsilon_{eq}}{\partial z} + \frac{\partial \sigma_{eq}}{\partial \dot{\varepsilon}_{eq}} \frac{\partial \dot{\varepsilon}_{eq}}{\partial z}\right) = \gamma \frac{\partial \varepsilon_{eq}}{\partial z} + m \frac{\partial \ln \dot{\varepsilon}_{eq}}{\partial z} \tag{6.142}$$

Writing relation (6.137) at the *local* level:

$$\dot{\varepsilon}_{eq}(z) = -\frac{\dot{S}(z)}{S(z)} \tag{6.143a}$$

and thus

$$\frac{\partial \varepsilon_{eq}}{\partial z} = -\frac{1}{S} \frac{\partial S}{\partial z} \tag{6.143b}$$

and

$$\frac{\partial \dot{\varepsilon}_{eq}}{\partial z} = \frac{1}{S^2} \frac{\partial S}{\partial z} \dot{S} - \frac{1}{S} \frac{\partial \dot{S}}{\partial z} \tag{6.143c}$$

which gives in (6.142):

$$\frac{\partial \ln \sigma_{eq}}{\partial z} = -(m + \gamma) \frac{\partial \ln S}{\partial z} + m \frac{1}{\dot{S}} \frac{\partial \dot{S}}{\partial z} \tag{6.144}$$

From which, taking into account (6.140):

$$(m + \gamma - 1) \frac{\partial \ln S}{\partial z} = m \frac{\partial \ln(-\dot{S})}{\partial z} \tag{6.145}$$

and Hart's stability criterion (6.139) thus takes the form:

$$\frac{(m+\gamma-1)}{m}>0$$

or, m being positive,

$$m+\gamma-1>0 \tag{6.146}$$

The stability of plastic deformation is thus directly linked to the rheology coefficients of the material. More generally and more concretely, this criterion shows that the material has two means of stabilizing its plastic deformation: first by work hardening (the flow stress in the already deformed regions increases, and it becomes the turn of the neighboring regions to undergo deformation), and by sensitivity to the velocity (the flow stress in the regions deformed the most rapidly increases and their strain slows, again passing the strain to the neighboring regions). We will study the following two particular cases:

- Materials insensitive to the strain rate ($m \approx 0$), as for example cold metallic alloys. Considere's criterion is found in this case: the deformation is stable as long as $\gamma > 1$, or $\varepsilon_{eq} < n$.
- Materials with very weak work hardening ($\gamma \approx 0$). The deformation is only stable for $m > 1$. This condition is thus never attained for most materials. Nonetheless, it can be shown that starting from $m > 0.5$, the propagation of the initiation of necking is very slow. Such is the case of molten polymers or metallic alloys that are called ***superplastic*** ($0.5 < m < 0.8$). The stability condition is most closely approximated by materials with a Newtonian behavior ($m = 1$) such as molten glass. This property was exploited by the master glassmakers of Murano to stretch mixtures of glass melts through very large deformations.

To finish this analysis, we will obtain the relation between the appearance of the instability and the maximum tensile force. This maximum corresponds to:

$$\frac{\partial F}{\partial \varepsilon_{eq}} = 0 \quad \text{or} \quad \frac{\partial \ln F}{\partial \varepsilon_{eq}} = 0 \tag{6.147}$$

with

$$\frac{\partial \ln F}{\partial \varepsilon_{eq}} = \frac{\partial \ln S}{\partial \varepsilon_{eq}} + \frac{\partial \ln \sigma_{eq}}{\partial \varepsilon_{eq}} + \frac{\partial \ln \sigma_{eq}}{\partial \dot{\varepsilon}_{eq}}\frac{\partial \dot{\varepsilon}_{eq}}{\partial \varepsilon_{eq}} \tag{6.148}$$

The first partial derivative on the right hand side is (–1) according to (6.143b). As for the derivative $\partial \dot{\varepsilon}_{eq}/\partial \varepsilon_{eq}$, it is clearly null for a test at constant strain rate, as for a test at constant pulling velocity v_0, we have, using (6.137) and (6.138):

$$\frac{\partial \dot{\varepsilon}_{eq}}{\partial \varepsilon_{eq}} = \frac{\partial}{\partial \varepsilon_{eq}}\left(\frac{v_0}{\ell}\right) = \frac{\partial}{\partial \varepsilon_{eq}}\left(\frac{v_0}{\ell_0}e^{-\varepsilon_{eq}}\right) = -\dot{\varepsilon}_{eq} \tag{6.149}$$

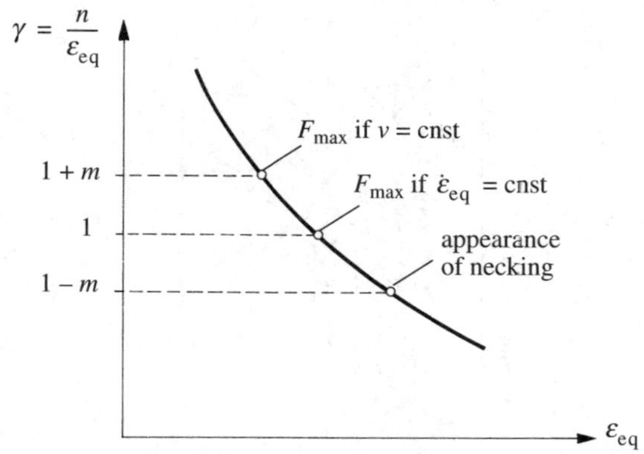

Figure 6.29 Relation between the maximum tensile force and the appearance of instability as expressed by Hart's criterion.

and thus the maximum force is obtained in this case when:

$$\frac{\partial \ln F}{\partial \varepsilon_{eq}} = 0 = -1 + \gamma - m \tag{6.150}$$

For a test at constant strain rate, the maximum force is obtained for $\gamma = 1$, or $n = \varepsilon_{eq}$ (we have again Considere's criterion). In general, as $\gamma = n/\varepsilon_{eq}$ decreases during the test, Hart's instability criterion, expressed on the areas, is obtained after the maximum tensile force (fig. 6.29).

Implementation of a finite element model

This section presents an illustration of the preceding results using a finite element simulation of a tension test. Figure 6.30 shows the initial geometry of the axisymmetric sample, as well as its finite element mesh in quadratic triangles.

Because of the axisymmetry, and by exploiting the horizontal symmetry plane, only a quarter of the sample section is calculated. The top surface is assumed to be in a condition of bilateral contact (i.e. separation prohibited) and attached to a tool whose vertical velocity applies the tension.

The chosen constitutive law is of the pure viscoplastic type, of the form:

$$\sigma_{eq} = K(\sqrt{3})^{m+1} \varepsilon_{eq}^{n} \dot{\varepsilon}_{eq}^{m} \tag{6.151}$$

Figure 6.31 shows clearly the influence of the rheology coefficients n and m in the resistance to necking. For a work hardening material which is sensitive to the strain rate, a diffuse necking is obtained, whereas the necking is more and more marked as n and m decrease, the influence of these two coefficients acting together. One notes that the calculations of samples with necking require automatic mesh regeneration in order to avoid the degeneracy of the elements.

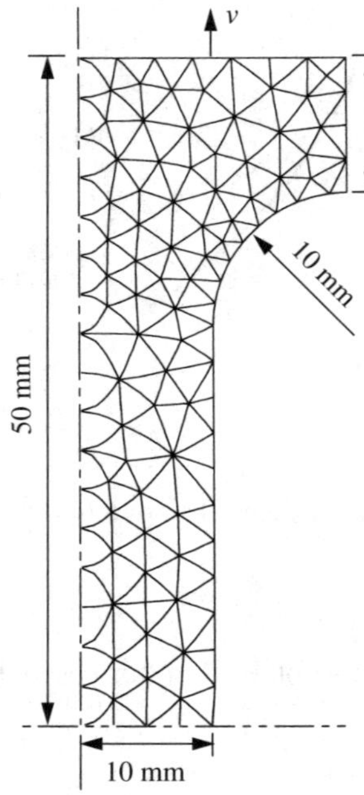

Figure 6.30 Finite element simulation of a uniaxial tension test on an axisymmetric sample. Initial mesh and boundary conditions.

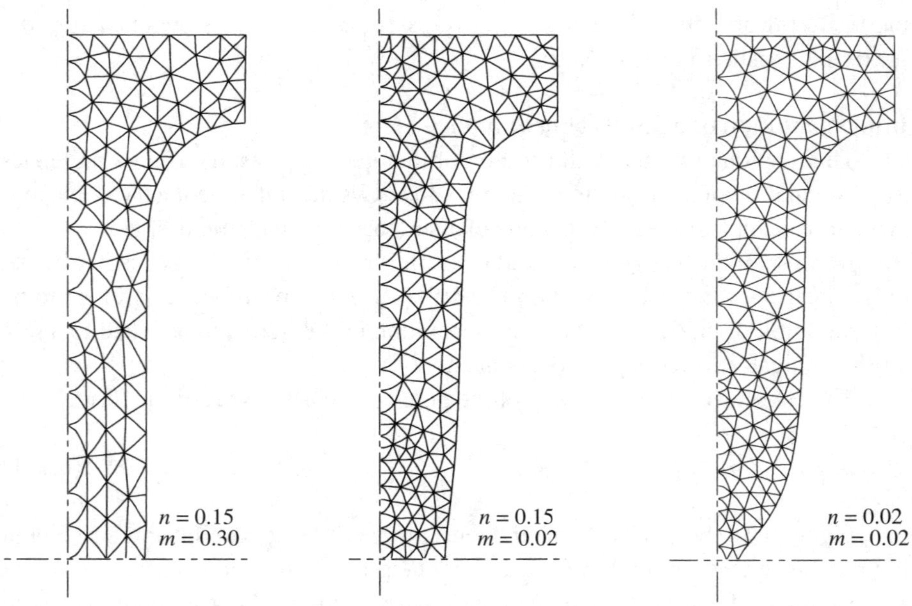

Figure 6.31 Three calculated configurations for the same imposed displacement of the top section and three sets of rheology coefficients n and m (6.151). Work hardening (coefficient n) and strain rate sensitivity (coefficient m) stabilize the plastic deformation.

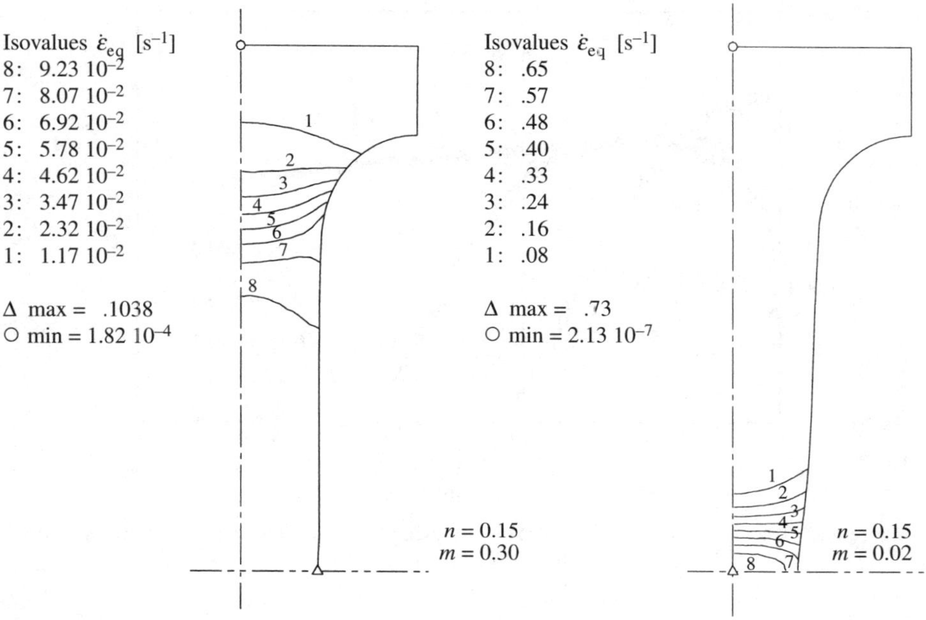

Figure 6.32 Iso-value maps of the strain rate in two cases, without and with necking.

In each of the three tests, the traction velocity is identical and calculated to insure a constant strain rate $\dot{\varepsilon}_c = 0.1 \text{ s}^{-1}$ for the assumption of homogeneous strain of the useful part of the sample. The sample having an initial length $\ell_0 = 30$ mm, we have, by integration of (6.137):

$$\ell = \ell_0 \exp(\dot{\varepsilon}_c t) \tag{6.152}$$

$$v = \frac{d\ell}{dt} = \ell_0 \dot{\varepsilon}_c \exp(\dot{\varepsilon}_c t) \tag{6.153}$$

Figure 6.32 shows a strain rate map associated with the necking, giving an illustration of the localization of the strain. The nominal value of 0.1 s^{-1} is maintained during the test without necking whereas a very noticeable increase is seen in the neighborhood of the necking, the rest of the central part of the sample undergoing practically no further strain.

Finally the plot of the traction force as a function of the displacement of the top of the sample in figure 6.33, is in agreement with Hart's analysis; the maximum force precedes the appearance of the necking and is found at $\varepsilon_{eq} = n$, that is, for a displacement of:

$$\Delta\ell = \ell_0(\exp n - 1) \tag{6.154}$$

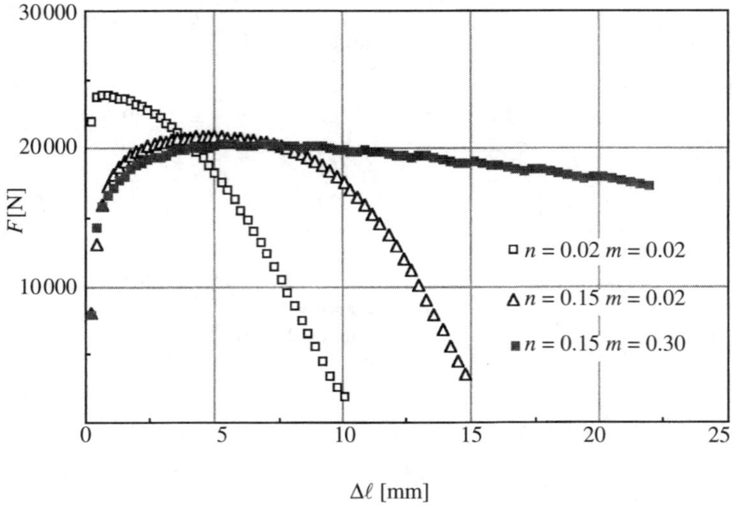

Figure 6.33 Traction force as a function of the imposed displacement for three simulations of tests.

6.5.2 Thermoforming of polymers

Processes and behavior of thermoformed polymers

This forming process consists of the deformation of a preheated polymer sheet with gaseous pressure or possibly a punch to force it into the form of a mold (fig. 6.34). It is applied with a large number of polymers, mostly amorphous thermoplastics (example: polystyrene) or semicrystalline thermoplastics (example: polyethylene), in the manufacturing of packaging in the food industry, or of parts for various technologies (interior components in the automotive industry, tanks for refrigerating devices, bathtubs ...).

One of the main problems in the implementation of a manufacturing line is to obtain thicknesses in the final part conforming to the specification. Numerical simulation permits the testing of different operating conditions (polymer type, tool tem-

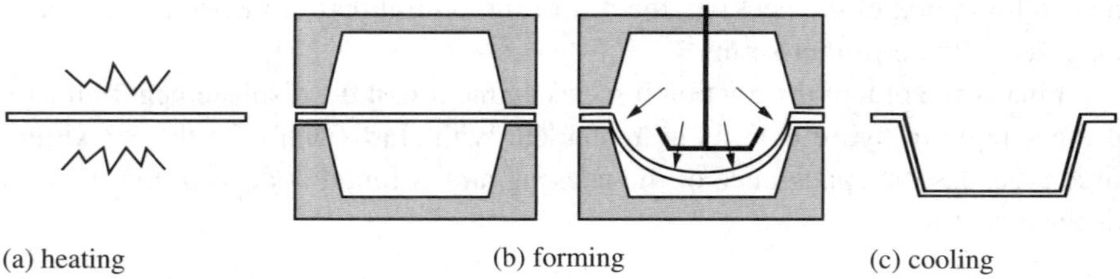

Figure 6.34 Process for thermoforming of polymers (schematic).

perature, initial temperature of the sheet, applied pressure, etc.) and thus to rapidly optimize the process at least cost. We describe below the principal characteristic of a finite element simulation model.

The polymers are thermoformed at high temperature, in a domain where their flow stress is low, that is, above the glass transition temperature for the amorphous polymers, and near the melting temperature for the semi-crystalline polymers. The behavior of such polymers can be modeled using the G'Sell and Jonas law, the expression of which is given below:

$$\sigma_{eq} = K_p(T)[1 - \exp(-w\varepsilon_{eq})]\exp(h\varepsilon_{eq}^2)(\dot{\varepsilon}_{eq})^m \tag{6.155}$$

This constitutive law should be considered as a particular case of Norton-Hoff's viscoplastic law without a threshold (6.69). The term in brackets gives zero flow stress at zero strain, and increases rapidly with the strain to take into account the viscoelastic effects in monotonic loading (that is, without unloading). The second exponential term corresponds to the hardening of the semi-crystalline polymers at the largest strains. This law then applies to a large variety of polymers in a large range of temperatures; two examples are shown in figure 6.35.

The coefficients K_p, w, h and m can be determined by uniaxial traction tests. For an example, the values measured for a polystyrene are given in table 6.2.

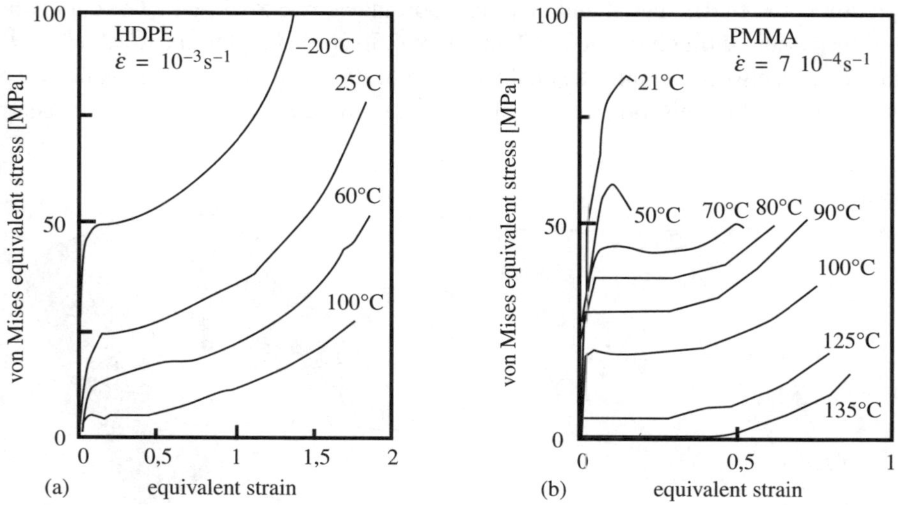

Figure 6.35 Example of behavior in uniaxial traction of two polymers as a function of temperature: (a) a semi-crystalline (high density polyethylene); (b): an amorphous polymer (polymethylmetacrylate).

Table 6.2 Rheology parameters of a polystyrene for the G'Sell and Jonas law.

T[°C]	w	m	h	K_p [MPa s^m]
20	3.7	0.1	0.5	133
50	3.7	0.1	0.5	92
90	3.7	0.1	0.5	44
100	3.7	0.1	0.5	27
110	3.7	0.32	0.46	2.7
120	3.7	0.46	0.47	2.6
130	3.7	0.56	0.23	2.2
140	3.7	0.68	0.13	1.9

Mechanical model: the membrane assumption and use of material curvilinear coordinates

The low thickness of thermoformed sheets, with respect to the radii of curvature encountered, permits the treatment of their deformation as that of geometric surfaces. This amounts to neglecting the power dissipated by the bending and shear forces in the thickness by comparison to those dissipated by the expansion forces. This is what is covered by the "*membrane assumption*". In other words, the stress state is considered to be planar, in the plane tangent to the average surface of the sheet at each point.

This tangent plane is easily defined with the ***Lagrangian*** or ***curvilinear coordinates***: a material point M is identified by two curvilinear coordinates θ^1 and θ^2 which, for example, are its initial coordinates x_1 and x_2 in the undeformed configuration of the sheet (fig. 6.36). Then, at every instant t, there exists a bidirectional relation between the spatial and material coordinates $\boldsymbol{x}^t = \boldsymbol{x}^t(\theta^1, \theta^2)$. The curves θ^1 and θ^2 passing through the point M^t in figure 6.36 correspond to fixed values of θ^2 and θ^1 respectively. An infinitesimal variation of $d\theta^i$ ($i = 1, 2$) is accompanied by a variation $d\boldsymbol{x}^t$ of the position vector, tangent to the curve θ^i. It follows that the vector

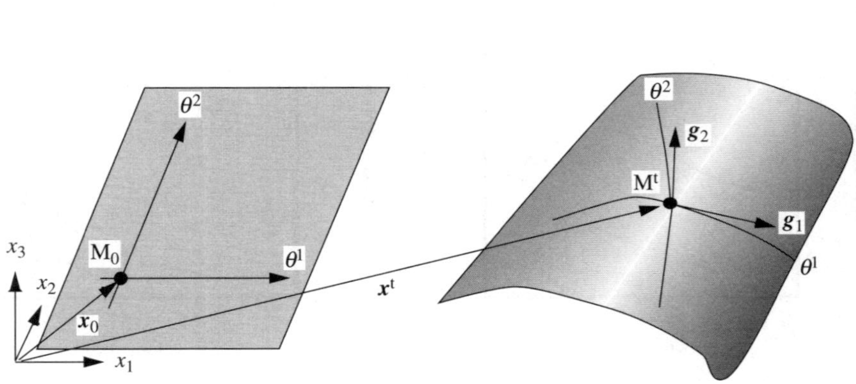

$t = 0$: initial configuration t: deformed configuration

Figure 6.36 Curvilinear coordinates.

$$g_i = \frac{\partial x^t}{\partial \theta^i} \tag{6.156}$$

is tangent to the curve θ^i at point M^t. The sheet thus has a tangent basis at every point M^t defined by:

$$g_1 = \frac{\partial x^t}{\partial \theta^1} \qquad g_2 = \frac{\partial x^t}{\partial \theta^2} \qquad g_3 = \frac{1}{|g_1 \otimes g_2|} g_1 \otimes g_2 \tag{6.157}$$

From this tangent basis, for which the vectors are neither orthogonal nor unitary (excepting the third which is orthogonal to the first two and normalized), a reciprocal basis can be defined, denoted (g^1, g^2, g^3) and defined by:

$$g_i \cdot g^j = \delta_i^j \qquad \text{for every pair } (i, j) \text{ of indices varying from 1 to 3,} \tag{6.158}$$

where δ_i^j is the Kronecker delta function for which we clearly distinguish the two indices to simplify the sums to be seen later. The vectors g^1 and g^2 are thus normal to the curves θ^2 and θ^1 respectively; the vectors g^3 and g_3 are identical. If we denote by g_{ij} the components of the tangent vectors, g_i, in the reciprocal basis, that is,

$$g_i = g_{ij} g^j \qquad (i, j = 1, 3) \tag{6.159}$$

we then deduce from (6.158):

$$g_i \cdot g_j = g_{ik} g^k \cdot g_j = g_{ik} \delta_j^k = g_{ij} \tag{6.160}$$

The g_{ij} are the components of the ***metric tensor***. We can also introduce the dual decomposition:

$$g^i = g^{ij} g_j \qquad (i, j = 1, 3) \tag{6.161}$$

from which:

$$g^{ij} = g^i \cdot g^j \qquad g^{ik} g_{kj} = \delta_j^i \qquad (i, j, k = 1, 3) \tag{6.162}$$

Every vector v or tensor T can also be expressed in the two bases, introducing different types of components: the ***covariant components*** (lower indices), and the ***contravariant components*** (upper indices), and the mixed components for the tensors:

$$v = v^i g_i = v_i g^i \tag{6.163}$$

$$T = T^{ij} g_i \times g_j = T_{ij} g^i \times g^j = T_j^i g_i \times g^j \tag{6.164}$$

The relations between the different types of components are deduced from the preceding relations:

$$v^{i} = g^{ij}v_{j} \qquad v_{i} = g_{ij}v^{j} \tag{6.165}$$

$$T^{ij} = g^{im}T^{j}_{m} = g^{im}g^{jn}T_{mn} \qquad T_{ij} = g_{im}T^{m}_{j} = g_{im}g_{jn}T^{mn} \tag{6.166}$$

The ***covariant derivative*** is defined for the contravariant component v^i of the vector $\boldsymbol{v}$ in the tangent basis of vectors $\boldsymbol{g}_i$ (respectively, for the covariant component v_i in the reciprocal basis of vectors $\boldsymbol{g}^i$) as being the component in this basis of the partial derivative of the vector $\boldsymbol{v}$ with respect to θ^i:

$$\frac{\partial \boldsymbol{v}}{\partial \theta^{j}} = v^{i}_{|j}\boldsymbol{g}_{i} = v_{i|j}\boldsymbol{g}^{i} \tag{6.167}$$

We then have:

$$v^{i}_{|j} = \frac{\partial \boldsymbol{v}}{\partial \theta^{j}} \cdot \boldsymbol{g}^{i} \qquad v^{i}_{|j} = g^{ik}v_{k|j} \tag{6.168}$$

This formalism being established, we now examine the expression of the deformation of the sheet. At point M, we denote by $\boldsymbol{G}_i$ and $\boldsymbol{g}_i$ the tangent vectors in the initial and deformed configurations of the sheet, respectively. G_{ij} and g_{ij} are the covariant components of the associated metric tensors. Having seen in chapter 1 (Sect. 1.2.5) that the deformations are defined by the variations of the scalar product between material vectors at M, we consider two vectors defined in the neighborhood of M by:

$$\mathrm{d}\boldsymbol{x}_{0} = \mathrm{d}\theta^{i}\boldsymbol{G}_{i} \qquad \mathrm{d}\boldsymbol{x}'_{0} = \mathrm{d}\theta'^{i}\boldsymbol{G}_{i} \tag{6.169}$$

and which become at time t:

$$\mathrm{d}\boldsymbol{x} = \mathrm{d}\theta^{i}\boldsymbol{g}_{i} \qquad \mathrm{d}\boldsymbol{x}' = \mathrm{d}\theta'^{i}\boldsymbol{g}_{i} \tag{6.170}$$

The variation of the scalar product is then:

$$\begin{aligned} \mathrm{d}\boldsymbol{x}\cdot\mathrm{d}\boldsymbol{x}' - \mathrm{d}\boldsymbol{x}_{0}\cdot\mathrm{d}\boldsymbol{x}'_{0} &= \mathrm{d}\theta^{i}\boldsymbol{g}_{i}\cdot\mathrm{d}\theta'^{j}\boldsymbol{g}_{j} - \mathrm{d}\theta^{i}\boldsymbol{G}_{i}\cdot\mathrm{d}\theta'^{j}\boldsymbol{G}_{j} \\ &= (g_{ij} - G_{ij})\,\mathrm{d}\theta^{i}\,\mathrm{d}\theta'^{j} \end{aligned} \tag{6.171}$$

With relations (1.33) and (1.35), the Green-Lagrange strain tensor, $\boldsymbol{\varepsilon}$, is defined by:

$$\mathrm{d}\boldsymbol{x}\cdot\mathrm{d}\boldsymbol{x}' - \mathrm{d}\boldsymbol{x}_{0}\cdot\mathrm{d}\boldsymbol{x}'_{0} = 2\,\mathrm{d}\boldsymbol{x}_{0}\cdot(\boldsymbol{\varepsilon}\cdot\mathrm{d}\boldsymbol{x}'_{0}) \tag{6.172}$$

One thus deduces that $\boldsymbol{\varepsilon}$ has the covariant components:

$$\varepsilon_{ij} = \tfrac{1}{2}(g_{ij} - G_{ij}) \tag{6.173}$$

In other words, the strains are given directly by the variations of the metric tensor. If, in addition, one considers the displacement $\boldsymbol{u} = \boldsymbol{x}^t - \boldsymbol{x}_0$ that the point M undergoes, one can also write:

$$\boldsymbol{g}_i = \frac{\partial \boldsymbol{x}^t}{\partial \theta^i} = \frac{\partial}{\partial \theta^i}(\boldsymbol{x}_0 + \boldsymbol{u}) = \boldsymbol{G}_i + u^m_{|i}\boldsymbol{G}_m = (\delta^m_i + u^m_{|i})\boldsymbol{G}_m \tag{6.174}$$

and from this relation, another expression for $\boldsymbol{\varepsilon}$ is deduced after a few calculations, as a function of the covariant derivatives of the displacement field, which is to be compared to the expression (1.35) (the partial derivatives are replaced by the covariant derivatives):

$$\varepsilon_{ij} = \tfrac{1}{2}(u_{i|j} + u_{j|i} + u_{m|i}u^m_{|j}) \tag{6.175}$$

The same can be shown for the velocities:

$$\dot{\varepsilon}_{ij} = \tfrac{1}{2}(v_{i|j} + v_{j|i}) \tag{6.176}$$

The equilibrium of the deformed sheet is expressed by the principle of virtual power, written here:

$$\forall \boldsymbol{v}^* \quad \int_{\Omega} \boldsymbol{\sigma}: \dot{\boldsymbol{\varepsilon}}^* e \,\mathrm{d}S - \int_{\Omega_p} p_a \boldsymbol{g}_3 \cdot \boldsymbol{v}^* \,\mathrm{d}S - \int_{\Omega_c} \boldsymbol{T}_\tau \cdot \boldsymbol{v}^* \,\mathrm{d}S = 0 \tag{6.177}$$

where the volume element is expressed in the form of the product $e\mathrm{d}S$, e being the local thickness of the sheet and $\mathrm{d}S$ the surface element. $\boldsymbol{T}_\tau$ denotes the friction stress vector in the regions Ω_c of the sheet in contact with the tools and p_a is the pressure difference between the two sides of the sheet, applied on the region Ω_p of the deformed surface Ω.

The virtual deformation power term has a particular expression due to the curvilinear coordinates. Conforming to the general definition of the doubly contracted product of two tensors of order 2, we have:

$$\boldsymbol{\sigma}: \dot{\boldsymbol{\varepsilon}}^* = \mathrm{Tr}(\boldsymbol{\sigma}^T \dot{\boldsymbol{\varepsilon}}^*) \tag{6.178}$$

The trace of a tensor is given in curvilinear coordinates by the sum of its *mixed* components (so as to be independent of the basis in which it is expressed). The mixed components of a product of two tensors being obtained from the product of their mixed components ($(\boldsymbol{AB})^i_j = A^i_k B^k_j$), one thus has in this case using also (6.166), the symmetry of the two tensors, and (6.176):

$$\boldsymbol{\sigma}: \dot{\boldsymbol{\varepsilon}}^* = (\boldsymbol{\sigma}^{\mathrm{T}}\dot{\boldsymbol{\varepsilon}}^*)^{\mathrm{i}}_{\mathrm{i}} = (\boldsymbol{\sigma}^{\mathrm{T}})^{\mathrm{i}}_{\mathrm{j}}\dot{\varepsilon}^{*\mathrm{j}}_{\mathrm{i}} = (\boldsymbol{\sigma}^{\mathrm{T}})^{\mathrm{ik}}g_{\mathrm{kj}}\dot{\varepsilon}^{*\mathrm{j}}_{\mathrm{i}} = \sigma^{\mathrm{ik}}\dot{\varepsilon}^*_{\mathrm{ik}} = \sigma^{\mathrm{ij}}v^*_{\mathrm{i|j}} \tag{6.179}$$

Finally, the principle of virtual power is expressed:

$$\forall \boldsymbol{v}^* \int_{\Omega} \sigma^{\mathrm{ij}} v^*_{\mathrm{i|j}} e \,\mathrm{d}S - \int_{\Omega_{\mathrm{p}}} p_{\mathrm{a}} \boldsymbol{g}_3 \cdot \boldsymbol{v}^* \,\mathrm{d}S - \int_{\Omega_{\mathrm{c}}} \boldsymbol{T}_{\tau} \cdot \boldsymbol{v}^* \,\mathrm{d}S = 0 \tag{6.180}$$

In accordance with the membrane assumption, the sum in the first term of (6.180) is limited to variation of the indices i and j from 1 to 2. In practice the stress tensor is stored in the form of vector with three components: σ^{11}, σ^{22} and σ^{12}.

Temporal discretization

An implicit integration scheme (6.97) is used for the time steps. The equilibrium equation (6.180) then must be expressed as a function only of the unknown $\boldsymbol{v}^{\mathrm{t}+\Delta\mathrm{t}}$. We will not treat the constitutive law in detail here; however, we note that by using the vector forms of $\dot{\boldsymbol{\varepsilon}}$ and $\boldsymbol{\sigma}$ given in chapter 3 (Sect. 3.4.4), it is possible to express the tensor flow rule derived from (6.155), in the form:

$$\dot{\boldsymbol{\varepsilon}} = \frac{1}{k}\dot{\varepsilon}_{\mathrm{eq}}^{(1-\mathrm{m})}[\boldsymbol{A}]\boldsymbol{\sigma} \tag{6.181}$$

in which k is the power law coefficient deduced from the comparison of (6.155) and (6.69), and $[\boldsymbol{A}]$ is a 3×3 matrix with components which are functions of the components g_{ij} of the metric tensor $\boldsymbol{g}$.

The incremental strains between two configurations, at times t and $t + \Delta t$, can be expressed with the aid of relation (6.173):

$$\Delta\varepsilon_{\mathrm{ij}} = \tfrac{1}{2}(g^{\mathrm{t}+\Delta\mathrm{t}}_{\mathrm{ij}} - g^{\mathrm{t}}_{\mathrm{ij}}) \tag{6.182}$$

and using an implicit temporal integration scheme for the strain rates:

$$\Delta\boldsymbol{\varepsilon} = \Delta t\,\dot{\boldsymbol{\varepsilon}}^{\mathrm{t}+\Delta\mathrm{t}} \tag{6.183}$$

The combination of (6.181) written at time $t + \Delta t$, (6.182) and (6.183) allows the expression of $\boldsymbol{\sigma}^{\mathrm{t}+\Delta\mathrm{t}}$ as a function of $\boldsymbol{g}^{\mathrm{t}+\Delta\mathrm{t}}$, and thus of $\boldsymbol{v}^{\mathrm{t}+\Delta\mathrm{t}}$ (which is the objective) using (6.97) and (6.157).

A Coulomb friction law is employed. In this case, the contact pressure p_{c} is the gas pressure applied to the sheet, and the friction stress is expressed at time $t + \Delta t$, including the regularization (cf (6.80) and (6.94)):

$$\boldsymbol{T}^{\mathrm{t}+\Delta\mathrm{t}}_{\tau} = -\mu_{\mathrm{f}} p^{\mathrm{t}+\Delta\mathrm{t}}_{\mathrm{a}} \frac{1}{\sqrt{(\boldsymbol{v}^{\mathrm{t}+\Delta\mathrm{t}}_{\mathrm{s}})^2 + v^2_{\mathrm{s0}}}} \boldsymbol{v}^{\mathrm{t}+\Delta\mathrm{t}}_{\mathrm{s}} \tag{6.184}$$

Finally, the local thickness at time $t + \Delta t$ can be calculated, knowing $\boldsymbol{v}^{t+\Delta t}$, and using the conservation of volume equation – the material being assumed incompressible. In curvilinear coordinates, the surface element dS is written:

$$dS = \sqrt{\det \boldsymbol{g}}\, d\theta^1\, d\theta^2 \qquad (6.185)$$

from which, the thickness is deduced, writing $(e dS)^t = (e ds)^{t+\Delta t}$:

$$e^{t+\Delta t} = e^t \sqrt{\det \boldsymbol{g}^t / \det \boldsymbol{g}^{t+\Delta t}} \qquad (6.186)$$

And lastly, with the velocity field given at $t + \Delta t$, $\boldsymbol{v}^{t+\Delta t}$, it is possible to determine $\boldsymbol{x}^{t+\Delta t}$, $\boldsymbol{\sigma}^{t+\Delta t}$, $e^{t+\Delta t}$ and $\boldsymbol{T}_{\tau}^{t+\Delta t}$. The equilibrium equation (6.180) written at $t + \Delta t$ can thus be formulated as a non linear equation with as a unique unknown, the velocity field $\boldsymbol{v}^{t+\Delta t}$.

Spatial discretization and solution: finite element method

The sheet is discretized in linear triangular elements with three nodes and a single Gauss point (the stresses, strains and thickness are constant in each element). Using the notations introduced in chapter 3:

$$\boldsymbol{v} = [\boldsymbol{\Psi}]\boldsymbol{V} \qquad \dot{\boldsymbol{\varepsilon}} = [\boldsymbol{B}]\boldsymbol{V} \qquad (6.187)$$

where $\boldsymbol{V}$ is the vector of nodal components of the velocity field, $[\boldsymbol{\Psi}]$ the interpolation matrix and $[\boldsymbol{B}]$ the discrete derivative operator. The discretized form of the equilibrium equation is:

$$\boldsymbol{R}(\boldsymbol{V}^{t+\Delta t}) = \int_{\Omega^{t+\Delta t}} ([\boldsymbol{B}]^{t+\Delta t})^T \boldsymbol{\sigma}^{t+\Delta t} e^{t+\Delta t}\, dS - \int_{\Omega_p^{t+\Delta t}} p_a^{t+\Delta t} ([\boldsymbol{\Psi}]^{t+\Delta t})^T \boldsymbol{g}_3^{t+\Delta t}\, dS$$
$$- \int_{\Omega_c^{t+\Delta t}} ([\boldsymbol{\Psi}]^{t+\Delta t})^T \boldsymbol{T}_{\tau}^{t+\Delta t}\, dS = \boldsymbol{0} \qquad (6.188)$$

Then, this non linear system for the unknown $\boldsymbol{V}^{t+\Delta t}$, is solved at each time step by the Newton-Raphson method. After solution, the configuration is updated applying the implicit scheme (6.97) at each node of the mesh.

A contact algorithm is used to calculate the relative position of each mesh node with respect to the tool surfaces, also discretized with triangles. In the case where the updated position of a node, $\boldsymbol{x}^{t+\Delta t}$, is found inside a tool, the algorithm calculates the normal velocity to impose in the following step so that at its end, the node returns to the surface of the tool. A null velocity in the direction normal to the tool is imposed for nodes in contact with the tools.

Treatment of thermal coupling

Given the thinness of the sheets, the short duration of the process (a few seconds at most) and the weak thermal diffusivity of the polymers, it can be assumed that the heat transfer is essentially one dimensional in the thickness of the sheets. Therefore, if s denotes the coordinate in the direction normal to the tangent plane of the sheet $s \in [-e/2, +e/2]$), the one dimensional heat transfer equation is written:

$$\rho c_\mathrm{p} \frac{\mathrm{d}T}{\mathrm{d}t} = \frac{\partial}{\partial s}\left(\kappa \frac{\partial T}{\partial s}\right) + \boldsymbol{\sigma} : \dot{\boldsymbol{\varepsilon}} \tag{6.189}$$

where ρ denotes the density, c_p the mass specific heat at constant pressure and κ the thermal conductivity. The mechanical dissipation term $\boldsymbol{\sigma} : \dot{\boldsymbol{\varepsilon}}$, deduced from the mechanical solution, is a volumetric source term in the thickness of the sheet. The boundary condition on the sheet at the ambient air interface is the following:

$$-\kappa \frac{\partial T}{\partial n} = h(T - T_\mathrm{a}) \tag{6.190}$$

where h is the convection coefficient with air (temperature T_a). As the thermoforming tools are in general metallic (high thermal diffusivity with respect to polymers), and have regulated temperatures, this temperature is imposed at the sheet-tool interface. The initial temperature of the sheet is assumed known, possibly obtained from a preliminary numerical simulation of the heating phase.

Equation (6.189) is discretized in space and time at each integration point of the triangle elements constituting the average mesh of the surface of the sheet (fig. 6.37). A one dimensional quadratic finite element method (3 nodes per element) with a Crank-Nicolson temporal scheme is used (chap. 3, sect. 3.6).

Knowing the temperature profile at each integration point, the thermal mechanical coupling is performed according to the algorithm summarized in figure 6.38.

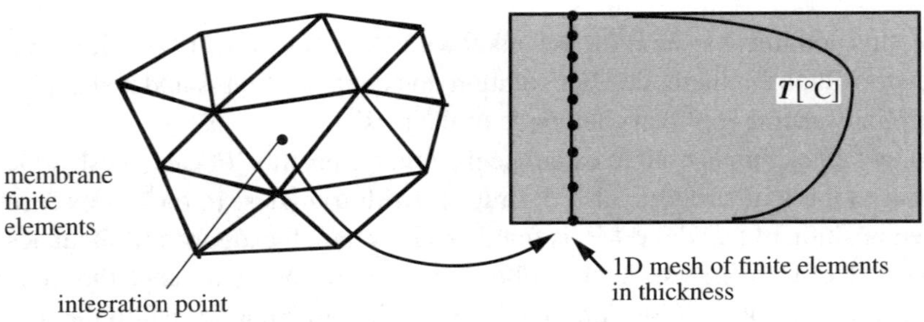

Figure 6.37 Thermal coupling.

A *Thermal solution*

- loop over all the finite elements of the membrane
  - at the center of the element, 1D FEM calculation of the new temperature field $T^{t+\Delta t}$ in the thickness. The source term is given by the mechanical solution at the preceding time step.
  - at each integration point *ip* of the 1D mesh, updating of the temperature dependent rheology coefficients: $K_p(T^{t+\Delta t})$, $m(T^{t+\Delta t})$ and $h(T^{t+\Delta t})$.
- end of the loop over all the finite elements of the membrane

B *Mechanical solution*

B1 calculation of the equilibrium residual $\boldsymbol{R}(\boldsymbol{V}^{t+\Delta t})$

- loop over all the finite elements of the membrane
  - at the center of the element, loop over the integration points *ip* of the 1D mesh
    - solution of the constitutive equation: calculation of $\boldsymbol{\sigma}^{t+\Delta t}(ip)$ and the tangent modulus $\partial\boldsymbol{\sigma}^{t+\Delta t}/\partial\Delta\boldsymbol{\varepsilon}(ip)$, using the values of the coefficients $K_p(T^{t+\Delta t})$, $m(T^{t+\Delta t})$ and $h(T^{t+\Delta t})$ at the point *ip*.
  - calculation of the average values in the thickness $<\boldsymbol{\sigma}^{t+\Delta t}>$ and $<\partial\boldsymbol{\sigma}^{t+\Delta t}/\partial\Delta\boldsymbol{\varepsilon}>$ and sum in the residual vector $\boldsymbol{R}$ and in the stiffness matrix $\partial\boldsymbol{R}/\partial\boldsymbol{V}$.
- end of the loop over all the finite elements of the membrane

B2 Newton-Raphson iterative solution of $\boldsymbol{R}(\boldsymbol{V}^{t+\Delta t}) = \boldsymbol{0}$

Figure 6.38 Thermal mechanical coupling algorithm executed at each time step.

Comparison of experiments and calculations for axisymmetric forming

Thermoforming experiments have been performed with polystyrene; the mold geometry is shown in figure 6.39. The forming parameters are the following: initial sheet thickness, 1.1 mm; initial temperature, 130 °C; aluminum mold temperature, 20 °C; linearly varying pressure in time (0.6 MPa s^{-1}).

Since the problem is axisymmetric, only a narrow angular section was meshed, with 160 triangles. The rheology data from table 6.2 have been used, with the following values for the thermal parameters: $\rho c_p = 1.92 \times 10^6$ J m^{-3} K^{-1}; $\kappa = 0.2$ W m^{-1} K^{-1}, $h = 33$ W m^{-2} K^{-1}. Coulomb friction with coefficient $\mu_f = 0.4$ characterizes the sheet-mold contact. It is seen in figure 6.39 that the thickness profile predicted by the calculation is in good agreement with the experiment.

Going beyond this first result, it is interesting to analyze the influence of the thermal coupling. In figure 6.40 the evolution of the thermal profile in the thickness of the sheet is shown for the central point. During the first phase of free blowing, the sheet undergoes weak heating due to the plastic dissipation, as well as a weak cooling by convection with the ambient air. The deformation is thus quasi-isothermal and the sheet thins rapidly. After contact with the mold the polymer cools much more rapidly, and significant thermal gradients appear in the thickness. The average flow stress increases progressively, making the sheet more and more difficult to deform locally. After 0.25 s, the deformation is blocked; slipping and thinning of

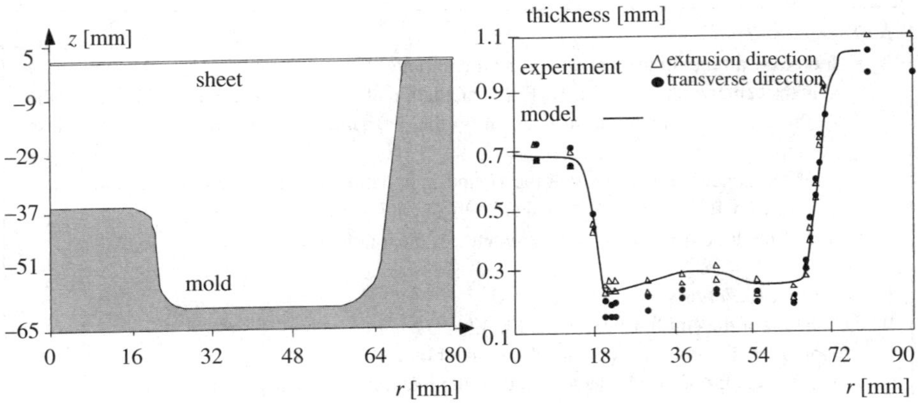

Figure 6.39 Geometry of the axisymmetric mold and comparison of the calculated and experimental thickness profiles. For the latter, the minimal and maximal values are indicated for each radius. A certain anisotropy is seen in the sheet. This anisotropy was not taken into account in the model, and as the rheology coefficients were obtained in the direction of the extrusion, the model predictions should be compared with the measurements obtained in the same direction.

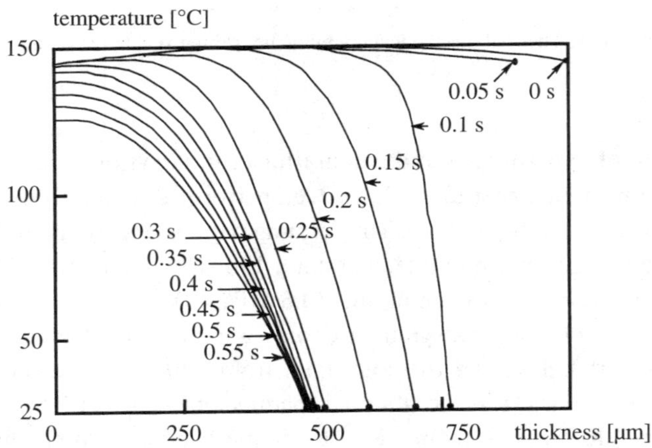

Figure 6.40 Evolution of the temperature profile in the thickness of the sheet at the center point in axisymmetric thermoforming.

the sheet are no longer possible. The sheet is fixed thereafter, and continues to cool in contact with the mold.

Figure 6.41 shows the thickness distribution obtained with and without thermal coupling for the same friction coefficient, as well as with the assumption of sticking contact. It is seen that the result of the coupled calculation is close to that obtained with sticking contact. Therefore, the latter assumption is employed in numerous calculation programs, even though for solid state polymers it does not correspond to

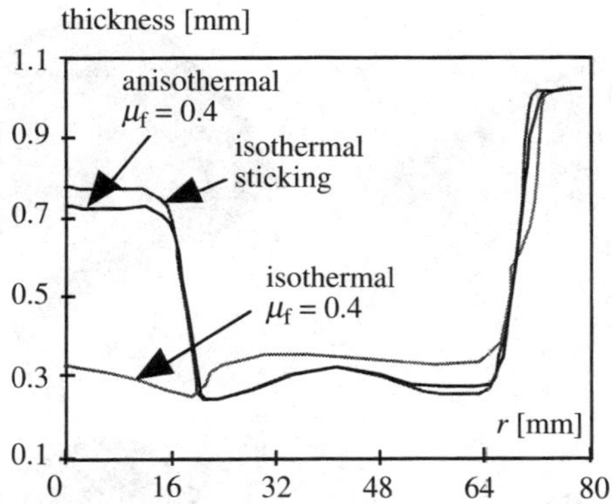

Figure 6.41 Calculated thickness distribution showing the effects of thermal coupling and the contact conditions.

any physical reality. On the other hand, the coupled formulation including both thermal and mechanical aspects allows the distinction of the friction and thermal effects. It also opens the door for better control of the thickness distributions through control of the temperatures and the nature of the thermoforming tools.

Application to an industrial part

The thermoforming of a polypropylene tray for the food industry has been simulated. The initial thickness of the sheet is 0.475 mm. The operating parameters are less well known than in the previous example, but can be taken as follows: initial temperature of the sheet, 150 °C; regulated temperature of the aluminum mold, 5 °C; linearly increasing pressure in time 1.4 MPa s^{-1}; air temperature 20 °C.

A quarter of the symmetric real part was calculated with 7591 triangles, the mold surface being discretized with 6463 triangles. Figure 6.42 shows a few successive configurations of the sheet.

Figure 6.43 shows the deformed mesh at the end of blowing along with the cuts along which the thickness of the real piece was measured. The comparison between the calculated and measured thickness is shown in figure 6.44 for one of these cuts.

6.5.3 Hot forging of metals

Generalities about the process

Forging consists of pounding metal blocks to impose a complex geometric form by plastic deformation. There are two main types of forging to distinguish.

- ***Free forging***, performed with tools that are generally of simple form, by applying a large number of blows.

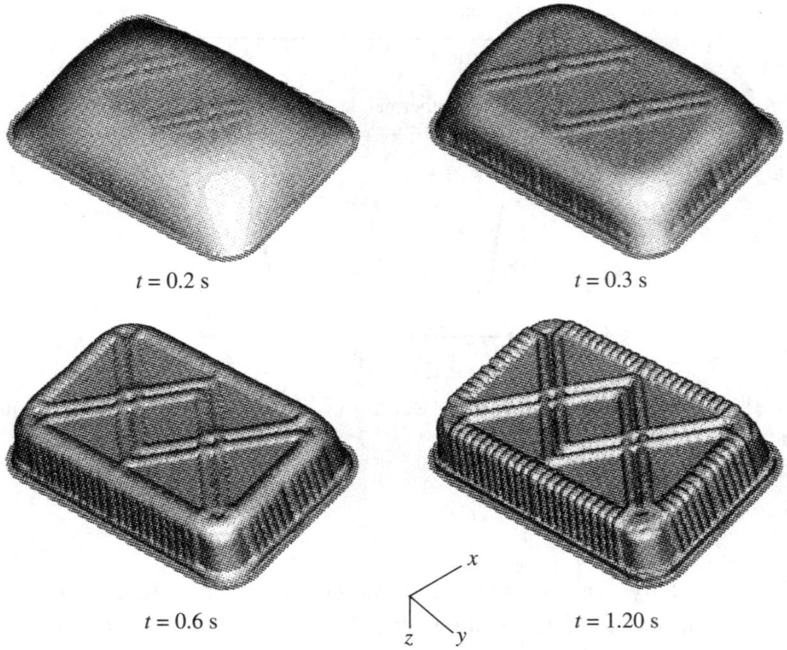

Figure 6.42 Evolution of the form of the sheet during the forming process.

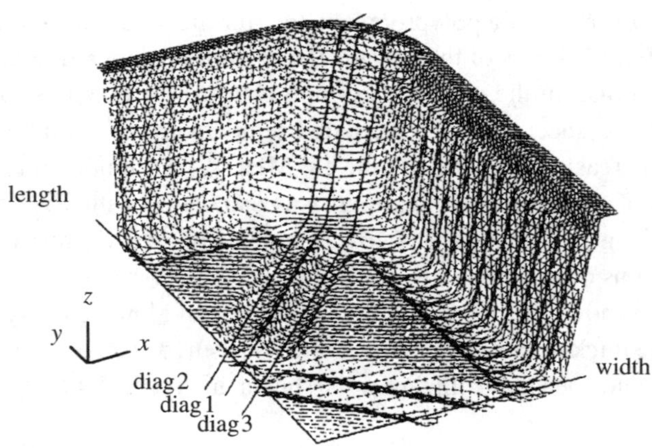

Figure 6.43 Final deformed mesh and cuts for the final thickness measurements.

- **Stamping**, which consists of obtaining the desired form with few actions during which the metal deformation can be very large. The lateral flashes which appear during the process induce an increase of pressure in the workpiece and are used to control the shaping of the material (fig. 6.45).

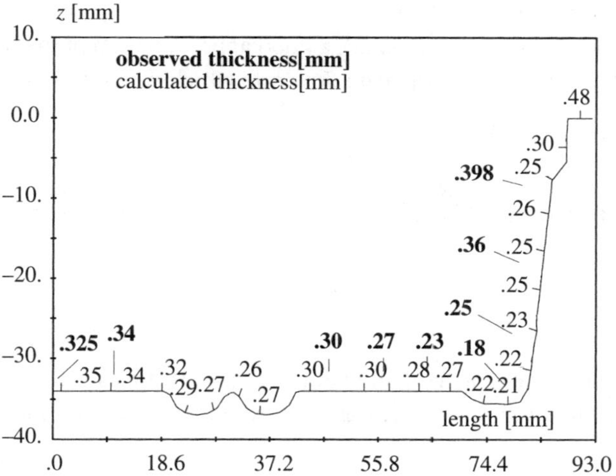

Figure 6.44 Comparison between calculated and measured thickness.

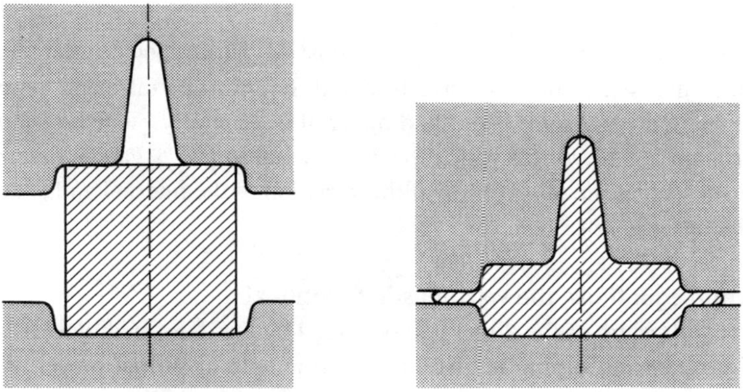

Figure 6.45 Schematic illustration of the forging process.

Metal forging can be performed in various temperature domains:

- cold (that is, the ambient temperature for common metals);
- hot, for temperatures above 50% of the melting temperature in Kelvin, the elevated temperature allowing reduction of the forces applied and an increase in the ductility of the metal;
- more rarely, at intermediate temperature ("warm"), at a temperature around 30% of the melting temperature.

In the scope of this book, we are interested primarily in hot forging. The block is thus preheated. The tools are generally cooler, and large thermal exchanges exist between the block and the tools.

Behavior of hot forged metals

As mentioned above, (Sect. 6.2.3), a good approximation of the behavior of hot metals is the Norton-Hoff viscoplastic flow rule which we present again here:

$$s = 2K(\sqrt{3}\dot{\varepsilon}_{eq})^{m-1}\dot{\boldsymbol{\varepsilon}} \tag{6.191}$$

Note right away that this relation assumes the incompressibility of the material (s is deviatoric) thus:

$$\operatorname{div} \boldsymbol{v} = \operatorname{Tr} \dot{\boldsymbol{\varepsilon}} = 0 \tag{6.192}$$

In this basic model, the hardening due to plastic strain as well as the increase in ductility associated with the increased temperature can be taken into account with the relations:

$$K = K_0(\varepsilon_{eq} + \varepsilon_0)^n \exp(-\beta T) \tag{6.193}$$

$$m = m_0 + m_1 T \tag{6.194}$$

The coefficients K_0, ε_0, n, β, m_0, m_1, must be determined by elementary tests (traction, compression, torsion) performed at different strain rates and temperatures.

The friction between the tool and material is generally modeled by the viscoplastic friction law (6.90) for which the coefficients α_f and p_f can be determined by simple tests such as the observation of the variation of the internal and external radii during the upsetting of rings.

Equilibrium equations, the viscoplastic functional

The equilibrium of the domain Ω occupied by the deformed block (the "part") at any time t during forging is given by the principle of virtual power. We denote $\partial\Omega_v$ the part of the boundary where the vector velocity is imposed, $\partial\Omega_T$ the part where the stress vector is imposed, and $\partial\Omega_c$ the region subject to a contact condition with slip and friction. Then, with (6.95):

$$\int_{\Omega} \boldsymbol{\sigma} : \dot{\boldsymbol{\varepsilon}}^* \, dV - \int_{\partial\Omega_v} \boldsymbol{T} \cdot \boldsymbol{v}^* \, dS - \int_{\partial\Omega_T} \bar{\boldsymbol{T}} \cdot \boldsymbol{v}^* \, dS - \int_{\partial\Omega_c} \boldsymbol{T} \cdot \boldsymbol{v}^* \, dS = 0 \tag{6.195}$$

Note that this expression neglects gravity and inertia, which is justifiable in the context of forging (except for the inertial terms in the case of very high tool velocities, as for drop hammers, a case which we exclude here). Also note that in practice, $\partial\Omega_v$ and $\partial\Omega_T$ do not have real significance, the boundary conditions either being applied by the tools (thus a contact zone $\partial\Omega_c$), or being free surfaces ($\bar{\boldsymbol{T}} = \boldsymbol{0}$ on $\partial\Omega_T$, neglecting atmospheric pressure). These integrals will nonetheless be retained for the generality of the presentation.

The restriction of the virtual velocity fields to those that are "kinematically admissible to zero" (that is null where the velocities are applied) allows the zeroing of the integral over $\partial\Omega_v$ and reduces the last to include only the tangential friction shear. Expanding the stress tensor $\boldsymbol{\sigma}$, we obtain:

$$\int_{\Omega} 2K(\sqrt{3}\dot{\varepsilon}_{eq})^{m-1}\dot{\boldsymbol{\varepsilon}}:\dot{\boldsymbol{\varepsilon}}^*\,dV - \int_{\Omega} p\boldsymbol{I}:\dot{\boldsymbol{\varepsilon}}^*\,dV - \int_{\partial\Omega_T} \overline{\boldsymbol{T}}\cdot\boldsymbol{v}^*\,dS - \int_{\partial\Omega_c} \boldsymbol{T}_\tau\cdot\boldsymbol{v}^*\,dS = 0 \tag{6.196}$$

Again restricting the space of the virtual velocity fields to incompressible fields (that is, having zero divergence and thus such that $\boldsymbol{I}:\dot{\boldsymbol{\varepsilon}}^* = \mathrm{Tr}\,\dot{\boldsymbol{\varepsilon}}^* = \mathrm{div}\,\boldsymbol{v}^* = 0$; called ***plastically admissible***), we see that (6.196) is simply the stationary condition for the velocity field functional, Φ, defined by:

$$\Phi(\boldsymbol{v}) = \int_{\Omega} \varphi(\boldsymbol{v})\,dV - \int_{\partial\Omega_T} \overline{\boldsymbol{T}}\cdot\boldsymbol{v}\,dS + \int_{\partial\Omega_c} \varphi_f(\boldsymbol{v})\,dS \tag{6.197}$$

where φ and φ_f are the viscoplastic potentials defined by (6.68) and (6.93). We have:

$$\boldsymbol{s} = 2K(\sqrt{3}\dot{\varepsilon}_{eq})^{m-1}\dot{\boldsymbol{\varepsilon}} = \frac{\partial\varphi(\boldsymbol{v})}{\partial\boldsymbol{v}} \quad \text{and} \quad \boldsymbol{T}_\tau = -\alpha_f K\left(\frac{|\boldsymbol{v}_s|}{V_{ref}}\right)^{p_f}\frac{1}{|\boldsymbol{v}_s|}\boldsymbol{v}_s = -\frac{\partial\varphi_f(\boldsymbol{v})}{\partial\boldsymbol{v}} \tag{6.198}$$

Then the relation (6.188), written for all incompressible and kinematically admissible to zero $\boldsymbol{v}^*$, is equivalent to the *stationarity of Φ over the space of velocity fields that are kinematically and plastically admissible* (if $\boldsymbol{v} + d\boldsymbol{v}$ is kinematically admissible, then $d\boldsymbol{v}$ is kinematically admissible to zero). One can show that Φ is convex and that the solution corresponds to a minimum of Φ. Therefore, the solution of the equilibrium equation at every time step can be considered equivalently by either solution of the principle of virtual power, or by minimization of the velocity field functional. In the two cases, two additional constraints must be added, the incompressibility from (6.192) and the respect of the boundary conditions imposed on the velocity field.

Numerical Model

Penalty method

In this first method, the initial problem is "slightly" modified by allowing weak compressibility of the material:

$$p = -\chi_p''\,\mathrm{div}\,\boldsymbol{v} \tag{6.199}$$

in which χ_p'' is a positive constant with an arbitrary large value: thus a slight compression causes a large increase in the hydrostatic pressure (slight dilatation causes a large diminution). The introduction of this compressibility relation in the principle of virtual power gives:

$$\int_{\Omega} \boldsymbol{s} : \dot{\boldsymbol{\varepsilon}}^* \, \mathrm{d}V + \int_{\Omega} \chi_p'' \, \mathrm{div}\, \boldsymbol{v} \, \mathrm{div}\, \boldsymbol{v}^* \, \mathrm{d}V - \int_{\partial\Omega_T} \overline{\boldsymbol{T}} \cdot \boldsymbol{v}^* \, \mathrm{d}S - \int_{\partial\Omega_c} \boldsymbol{T}_\tau \cdot \boldsymbol{v}^* \, \mathrm{d}S = 0 \tag{6.200}$$

It is thus seen that the principle of virtual power, in full generality, is equivalent to the minimization of the following functional:

$$\Phi_p(\boldsymbol{v}) = \Phi(\boldsymbol{v}) + \int_{\Omega} \frac{\chi_p''}{2} (\mathrm{div}\, \boldsymbol{v})^2 \, \mathrm{d}V \tag{6.201}$$

This time it is performed over the space of kinematically admissible velocity fields (the incompressibility restriction has disappeared). The last term appears as a *penalization* term: velocity fields with non zero divergence are "punished".

The penalized formulation is clearly convenient as it involves only the velocity field as an unknown. However, it leads to errors in pressure and also to very poorly conditioned systems of equations (a very large spectrum of eigenvalues) resulting in the failure of iterative solution methods. For these reasons, one prefers the Lagrange multiplier method, for which we present the principle and the finite element discretization in the following.

Lagrange multiplier method

In this second method, the problem is solved with variable velocity and pressure, requiring that the solution ($\boldsymbol{v}$, p) satisfy the following two equations along with the velocity boundary conditions:

$$\int_{\Omega} 2K(\sqrt{3}\dot{\bar{\varepsilon}}_{eq})^{m-1} \dot{\boldsymbol{\varepsilon}} : \dot{\boldsymbol{\varepsilon}}^* \, \mathrm{d}V - \int_{\Omega} p \, \mathrm{div}\, \boldsymbol{v}^* \, \mathrm{d}V - \int_{\partial\Omega_T} \overline{\boldsymbol{T}} \cdot \boldsymbol{v}^* \, \mathrm{d}S - \int_{\partial\Omega_c} \boldsymbol{T}_\tau \cdot \boldsymbol{v}^* \, \mathrm{d}S = 0$$

$$\int_{\Omega} p^* \, \mathrm{div}\, \boldsymbol{v} \, \mathrm{d}V = 0 \tag{6.202}$$

for every field $\boldsymbol{v}^*$ kinematically admissible to zero and for every field p^*. These two equations constitute the stationarity condition of the following functional (Lagrangian), in the space of kinematically admissible velocity fields and pressure fields:

$$\Phi_L(v, p) = \Phi(v) - \int_\Omega p \,\mathrm{div}\, v \, \mathrm{d}V \tag{6.203}$$

The pressure p thus appears as a *Lagrange multiplier* of the incompressibility condition.

For the finite element formulation, the velocity and pressure fields are assumed discretized differently according to the relations:

$$v = \sum_{k=1}^{Nnv} \psi_k^v v^k \quad \text{and} \quad p = \sum_{m=1}^{Nnp} \psi_m^p p^m \tag{6.204}$$

where *Nnv* and *Nnp* are the number of interpolation nodes used for the velocity and the pressure respectively, ψ_k^v and ψ_m^p being the corresponding interpolation functions associated with nodes k and m respectively. Inserting these discretizations in (6.202) and using (6.100) for the differentiation of the velocity field:

$$\int_\Omega s_{ij}(v) B_{ijnk} \, \mathrm{d}V - \int_\Omega \psi_m^p p^m \frac{\partial \psi_n^v}{\partial x_k} \, \mathrm{d}V - \int_{\partial\Omega_T} \bar{T}_k \psi_n^v \, \mathrm{d}S - \int_{\partial\Omega_c} T_{\tau k}(v) \psi_n^v \, \mathrm{d}S = 0$$

for $n = 1, Nnv$ and $k = 1, 3$

(6.205)

$$\int_\Omega \psi_l^p \frac{\partial \psi_n^v}{\partial x_k} v_k^n \, \mathrm{d}V = 0$$

for $l = 1, Nnp$

Denoting by $\boldsymbol{V}$ and $\boldsymbol{P}$ the vectors containing the nodal velocity and pressure unknowns ($\boldsymbol{V}$ has ($N\,Nnv$) components, N being the number of components of the velocity field, and $\boldsymbol{P}$ has *Nnp* components), equations (6.205) can be written in condensed form:

$$\begin{aligned} \boldsymbol{R}_v(\boldsymbol{V}, \boldsymbol{P}) &= \boldsymbol{0} \\ \boldsymbol{R}_p(\boldsymbol{V}) &= \boldsymbol{0} \end{aligned} \tag{6.206}$$

It can be shown that the two interpolation types, in velocity and pressure, can not be chosen independently; compatibility conditions exist which we will not discuss here that guarantee the uniqueness of the solution and its convergence to the solution of the continuous problem as the discretization size goes to zero (see chap. 7 Sect. 7.4.2). Figure 6.46 gives a few examples of compatible finite elements.

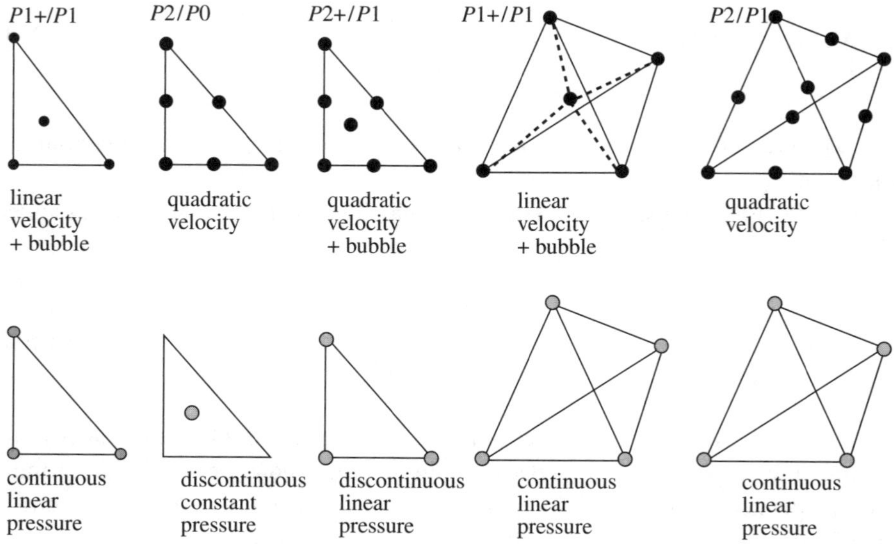

Figure 6.46 Examples of compatible finite elements in 2 and 3 dimensions.

Note that some elements have additional degrees of freedom in the center of the element for the velocity field interpolation. The associated interpolation function equals 1 at this central node and 0 at the surface of the element: it is commonly called a "bubble" function. This function can be 3rd order on a 2 dimensional element (element $P2+/P1$) or linear, defined for each of the three sub triangles of a triangular element in two dimensions, or for each of the four sub tetrahedra of a tetrahedral element in three dimensions (element $P1+/P1$). These additional degrees of freedom being strictly internal for each element, and thus not connected with the other nodes that belong to the element, they can be eliminated at the level of each element during the system assembly (6.206). However, we will not enter into these very technical details, which are outside the scope of this book.

In general, when the exponents of the constitutive and friction laws are other than 1 (non Newtonian), problem (6.206) is non linear. It can be solved by the Newton-Raphson method. Developing $\boldsymbol{R}_v$ and $\boldsymbol{R}_p$ to first-order in V and P, one deduces that at iteration (ν), the corrections of the velocity and pressure vectors must satisfy the following system:

$$\begin{bmatrix} \dfrac{\partial \boldsymbol{R}_v^{(\nu-1)}}{\partial \boldsymbol{V}} & \dfrac{\partial \boldsymbol{R}_v^{(\nu-1)}}{\partial \boldsymbol{P}} \\ \dfrac{\partial \boldsymbol{R}_p^{(\nu-1)}}{\partial \boldsymbol{V}} & 0 \end{bmatrix} \begin{bmatrix} \delta \boldsymbol{V}^{(\nu)} \\ \delta \boldsymbol{P}^{(\nu)} \end{bmatrix} = \begin{bmatrix} -\boldsymbol{R}_v(\boldsymbol{V}^{(\nu-1)}, \boldsymbol{P}^{(\nu-1)}) \\ -\boldsymbol{R}_p(\boldsymbol{V}^{(\nu-1)}) \end{bmatrix} \tag{6.207}$$

This linear system, whose symmetry follows from (6.205), is solved by a direct or iterative method (chap. 4) and the next iteration is determined by:

$$\boldsymbol{V}^{(\nu)} = \boldsymbol{V}^{(\nu-1)} + \delta\boldsymbol{V}^{(\nu)} \qquad (6.208)$$

$$\boldsymbol{P}^{(\nu)} = \boldsymbol{P}^{(\nu-1)} + \delta\boldsymbol{P}^{(\nu)}$$

The process is continued until convergence. One can then update the mesh configuration according to a purely explicit temporal integration scheme. Having solved the system of equations (6.206) on the configuration Ω^t at the start of the time step $[t, t + \Delta t]$, the updating of the nodal coordinates $\boldsymbol{X}$ is simply given by:

$$\boldsymbol{X}^{t+\Delta t} = \boldsymbol{X}^t + \Delta t \boldsymbol{V}^t \qquad (6.209)$$

The general organization of such a calculation is given in figure 6.47.

Example: simulation of hot forging of an industrial part

The example is for the hot forging of a steel part, a steering knuckle for an automobile suspension. Figure 6.48 shows the preform and the upper and lower tools in the initial configuration. The upper tool moves vertically, the lower remains stationary. The preform is obtained from a preliminary rolling phase.

The problem is solved in velocity and pressure using the tetrahedral finite element $P1+/P1$, conforming to the algorithm given in figure 6.47. The surfaces of the tools are discretized in triangles and contact is imposed by a nodal penalization method. Figures 6.49 through 6.51 show the progress of the forging. As can be seen,

Initializations: load the initial mesh
load the tool discretization
load the general data for the calculation (material characteristics ...)

Loop over the time steps

- mechanical solution ⇒ $(\boldsymbol{V}^t, \boldsymbol{P}^t)$
- update $\boldsymbol{X}^{t+\Delta t} = \boldsymbol{X}^t + \Delta t\, \boldsymbol{V}^t$
- analyze contact between the part and the tools
- possible mesh regeneration
- transport of the different variables from the old to the new mesh
- solution of the heat equation ⇒ $\boldsymbol{T}^{t+\Delta t}$
- update of the values of the material parameters (K, m ...) as functions of $\boldsymbol{T}$
- update the different variables: $\boldsymbol{X}^t \Leftarrow \boldsymbol{X}^{t+\Delta t}$; $\boldsymbol{T}^t \Leftarrow \boldsymbol{T}^{t+\Delta t}$
- test for end of the simulation
- possible storage of the results

end of loop over the time steps

Figure 6.47 general organization of the calculations.

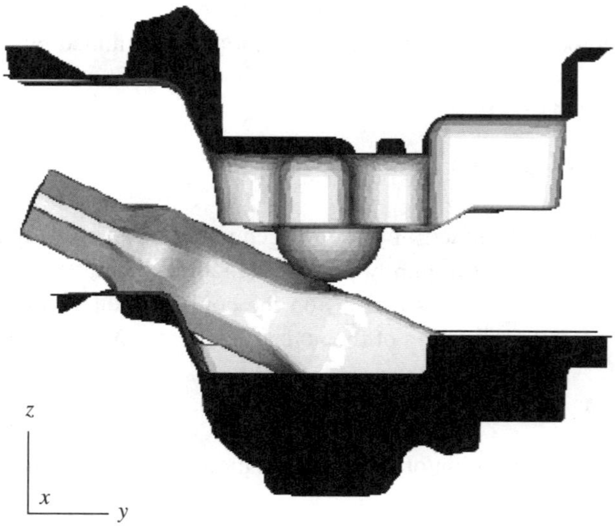

Figure 6.48 Finite element simulation of hot forging of an automobile steering knuckle in steel.

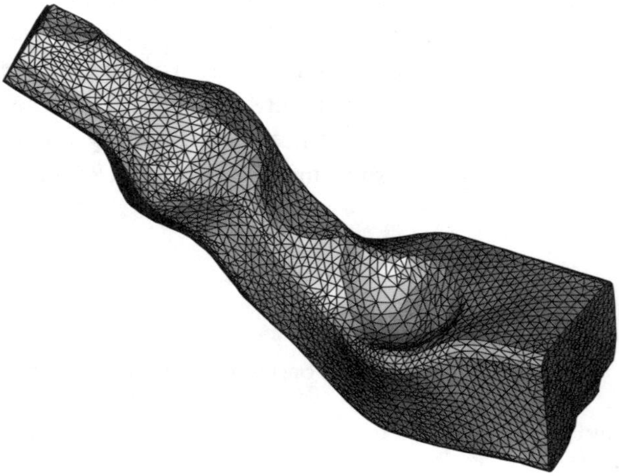

Figure 6.49 Intermediate configuration during forging.

the deformations that the material undergoes are very large and require numerous utilizations of the automatic remeshing procedure. In addition, the complex form of the part makes a fine mesh necessary. Thus, in the last phases of the simulation, the mesh has around 130 000 elements. The calculation time is then in the neighborhood of 8 weeks on one processor of a Sun 4000 computer. It can be reduced by performing parallel calculations on multiple processors. Despite the difficult implementation, such a simulation permitted the solution of an actual problem of material

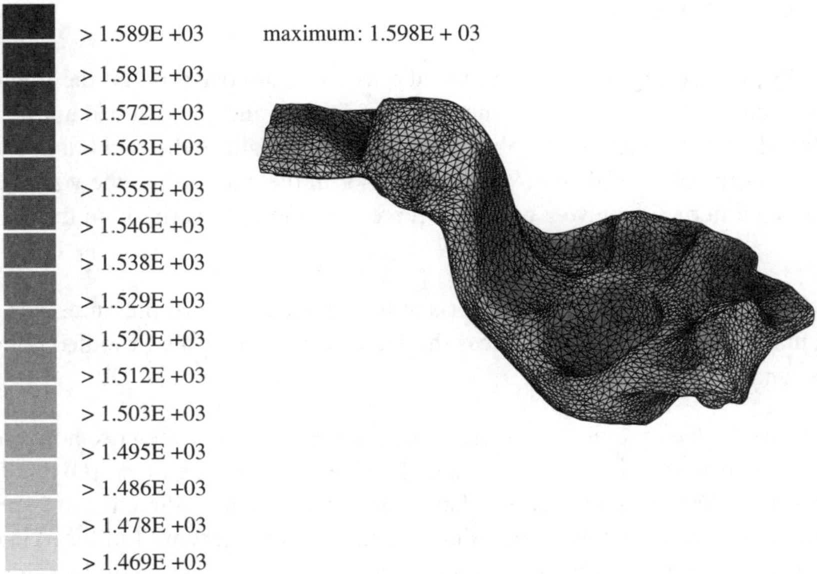

Figure 6.50 Intermediate configuration during forging. Temperature distribution [K] in the part.

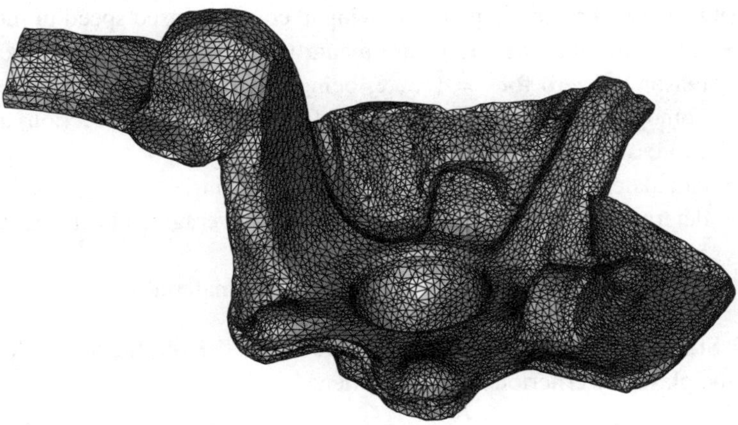

Figure 6.51 Final configuration.

folding: between two solutions consisting of lengthening the preform or modifying the tool, the simulation allowed the correct choice in good time and for a very competitive cost with respect to a full scale test.

6.6 EXERCISES

6.6.1 Given a rectangular plate in biaxial tension in the plane (x, y): the traction forces in x and y have F and G for modulii respectively, and are assumed uniformly distributed on the edges of the plate. Denote L the length, ℓ the width and e the thickness of the plate. Calculate the stress tensor in the plate. Write the von Mises and Tresca criteria, and give a graphical representation of the criteria in the (F, G) plane.

6.6.2 Write the von Mises criterion for a thin tube under internal pressure, then for a thin tube in traction-torsion. Express the Tresca criterion for a thin walled hollow sphere, under internal pressure.

6.6.3 Consider the upsetting of a rectangular parallelepiped between two horizontal planes (dimensions: $2a \times 2b \times h$, centered at the origin of the plane (Oxy)). The lower plane is fixed while the upper plane moves in the minus z direction at a speed of modulus V_0. Assuming that the strain rate tensor is homogeneous (uniform) in the material, and that the material is incompressible, give the expression of the velocity field and of $\dot{\boldsymbol{\varepsilon}}$.

6.6.4 Given a bar of rigid-plastic von Mises material of length L, rectangular section (width $2a$, height $2h$). Assume that $L >> h$ and $L >> a$. Study the crushing of the bar between two horizontal planes moving at equal opposed speed of modulus V:

- assuming that the strains are planar (the components $\dot{\varepsilon}_{i3}$ of the strain rate tensor are zero for $i = 1, 3$, x_3 being the axial direction of the bar) find a homogeneous planar strain rate field (uniform in every section) and give the expression of the velocity field;
- calculate the power dissipated by plastic strain;
- determine the forging force as well as the average contact pressure on each face;
- determine the stress tensor for a von Mises material.

6.6.5 Starting from (6.33) and (6.35), show that the plastic flow rule associated with the plasticity criterion f can be written:

$$\dot{\boldsymbol{\varepsilon}}^{p\ell} = \lambda^{p\ell}\frac{\partial f}{\partial s}$$

6.6.6 Hill proposed a plasticity criterion in 1948 taking into account the anisotropy of the material. Assuming the orthogonality of the principal axes of anisotropy (OX, OY and OZ), this criterion is written:

$$f = F(\sigma_{YY} - \sigma_{ZZ})^2 + G(\sigma_{ZZ} - \sigma_{XX})^2 + H(\sigma_{XX} - \sigma_{YY})^2 + 2L\sigma_{YZ}^2 + 2M\sigma_{ZX}^2 + 2N\sigma_{XY}^2 - 1$$

Show that each of the three coefficients F, G, and H can be interpreted as functions of the flow stress of the material in simple traction or compression in the three directions of anisotropy while the coefficients L, M, and N can be interpreted as functions of limiting shear stress in each of the anisotropy planes. Determine the plastic flow rule associated with this plasticity criterion.

6.6.7 Contrary to dense metals, porous metals or metals in powder form can undergo plastic strain due to hydrostatic pressure. One then defines the plasticity criterion in the following way: $f(\boldsymbol{\sigma}) = f(p, q) \leq 0$, where p is the hydrostatic pressure and q the equivalent stress in the sense of von Mises. Show that if a flow rule is associated with such a criteria, then the orientation of the normal vector at the plasticity surface, with components $(\partial f/\partial p, \partial f/\partial q)$ expresses the densification or the de-densification of the metal (neglecting the elastic contributions to the strain).

Apply to the following criteria:

- Green's criterion $\quad f = \sqrt{aq^2 + 9bp^2} - \sigma_0$
- modified Cam-clay criterion $\quad f = q^2 + M^2(p - P_0)(p - P_1)$

in which a, b, M, P_0 and P_1 are the characteristic parameters of the material. Sketch the plasticity surfaces in the plane (p,q) and the associated normals.

6.6.8 Prove formulas (6.122) and (6.125) for the contribution to the tangent matrix of the contact penalization terms. Derive the expression of the linear system (6.130) for the least squares smoothing method.

6.6.9 Show the result (6.197) giving the expression of the viscoplastic functional.

6.6.10 Derive the expression (6.205) of the weak form of the equilibrium and incompressibility equations discretized in finite elements. Calculate the detailed expressions of the tangent matrix and the right hand side term of the linear system resulting form the Newton-Raphson method (6.207).

6.7 BIBLIOGRAPHY

P. Baque, E. Felder, J. Hyafil, Y. D'Escatha, *Mise en Forme des Métaux, Calcul par la plasticité, Dunod*, Paris, 1974.

N. Bay, "Friction stress and normal stress in bulk metal forming processes", *J. Mech. Working Technol.*, vol. 14, 1987, p. 203-223.

M. Bellet, J.L. Chenot, L. Fourment, E. Massoni, P. Montmitonnet, Séminaire de Plasticité - éléments finis et mise en forme des métaux, Ecole des Mines de Paris, volumes 1-3, 1994.

M. Bellet, M.H. Vantal, B. Monasse, "Thermal effects in the numerical simulation of the thermoforming of multilayered polymer sheets", *Int. Polymer Processing*, vol. 13, 1998, pp. 299-308.

T. Belytschko, W.K. Liu, B. Moran, *Nonlinear Finite Elements for Continua and Structures*, Wiley, New York, 2001.

J.L. CHENOT, T. COUPEZ, L. FOURMENT, R. DUCLOUX, "3-D finite element simulation of the forging process: scientific developments and industrial applications", Proc. ICTP, Int. Conf. on Technology of Plasticity, T. Altan editor, pp. 465-474, 1996.

A. CURNIER, *Méthodes numériques en mécanique des solides*, Presses polytechniques et universitaires romandes, Lausanne, 1993.

D. FRANCOIS, A. PINEAU, A. ZAOUI, *Comportement Mécanique des Matériaux. Traité des Nouvelles Technologies*, Série Mécanique, Hermès, Paris, 1991.

J.M. HAUDIN, F. MONTHEILLET, *Notions fondamentales sur les matériaux, déformation plastique et essais mécaniques*, SNPMD Editor, Paris, 1989.

R. HILL, The Mathematical Theory of Plasticity, The Oxford Engineering Science Series, Clarendon Press, Oxford, 1967.

A.S. KHAN, S. HUANG, *Continuum theory of Plasticity*, John Wiley and Sons, New York, 1995.

U.F. KOCKS, A.S. ARGON, M.F. ASHBY, *Thermodynamics and Kinetics of Slip*, Pergamon Press, Oxford, 1975.

J. LEMAITRE, J.L. CHABOCHE, *Mécanique des matériaux solides*, Dunod, Paris, 1988.

J. MANDEL, *Propriétés Mécaniques des Matériaux - Rhéologie et Plasticité*, Eyrolles, Paris, 1978.

P. MONTMITONNET, "Lois de frottement et déformation plastique", *Matériaux et Techniques*, No. 1-3, pp. 8-21, 1993.

J.T. ODEN, E.B. PIRES, "Non local and non linear friction laws and variational principles for contact problems in elasticity". *J. Appl. Mech.*, vol. 50, 1983, p. 67-76.

R.H. WAGONER, J.L. CHENOT, *Fundamentals of metal forming*, John Wiley, New York, 1996.

CHAPTER 7

INCOMPRESSIBLE FLUID FLOW

7.1 OBJECTIVES

- Establish the mathematical model of the problem being studied. In particular, we will introduce the concept of the constitutive equation for a fluid, that is, the relation linking the stress tensor and the velocity gradients. The initial conditions for dynamic cases and the boundary conditions will also be discussed.
- Discuss the discretization methods for the numerical treatment of the mathematical model. Concentrating on the "velocity-pressure" formulation which is the most appropriate for the general treatment of three dimensional problems, we will present the finite difference and finite element discretization methods. As the advection terms are often dominant in these equations, we will be particularly interested in the upwind schemes.
- Propose a few applications as examples in order to illustrate the implementation of the proposed methods.

In many of the forming processes of materials, the material exists in the form of a "fluid". Examples are molten metals in the steel industry, casting, the production of crystalline semiconductors in Czochralski or Bridgman furnaces, molten glass in float glass processes, polymers, plastics for the fabrication of films, and fibers, paints, food processing …

This non exhaustive list shows the extent of the industrial applications covered by the title and suggests the diversity, richness and also the difficulty of the physical problems at hand. We will restrict ourselves for simplicity to the case of incompressible fluids. This excludes all the applications of external aerodynamics (as encountered in aeronautics: airplanes, the space shuttle) and internal aerodynamics (combustion in motors, flows in gas turbines).

However, our view focused on incompressible flows in the industrial world is not simplistic, as the palette is vast, and "worth the detour".

7.2 CONSTITUTIVE EQUATIONS

7.2.1 General considerations

The theory of mechanics of continuous media allows the description of the constitutive laws that describe the deformations of fluid medium. These laws, in combination with the general conservation principles introduced in chapter 1, form the system of partial differential equations, which are equal in number to the number of unknowns of the system.

The assumption of incompressibility being made for the rest of this chapter (density ρ = constant), the conservation of mass (1.8) imposes:

$$\operatorname{div} \boldsymbol{v} = 0 \tag{7.1}$$

The velocity field, $\boldsymbol{v}$, of an incompressible flow must therefore be divergence-free. For the notations of this chapter, we denote the components of the velocity vector $\boldsymbol{v}$ by u, v, w, when we are using rectangular Cartesian coordinate system (x, y, z).

Viscous incompressible fluids can be divided into two main classes: Newtonian viscous fluids, and non Newtonian fluids. In order to distinguish them, one can perform the following experiment. The fluid is placed between two parallel plates (see figure 1.12). The lower plate is fixed, while the upper moves in its plane at a constant velocity U. The flow obtained is known as planar Couette flow.

This experimental apparatus can be used as a viscometer. It permits the measurement of the shear stress σ_{xy} as a function of the shear rate $(\partial u/\partial y)$, denoted $\dot{\gamma}$. If the plot obtained is linear (curve (a) in figure 7.1), the dynamic viscosity μ is constant and the fluid is Newtonian, (this case shown in figure 1.12). In the case where $\mu = \mu(\dot{\gamma})$ (curve b), the fluid is non Newtonian, and its viscosity depends on the shear rate.

A second experiment consists of flowing the fluid in a vertical cylindrical tube and observing the form of the jet at the orifice. The section of the jet for a Newto-

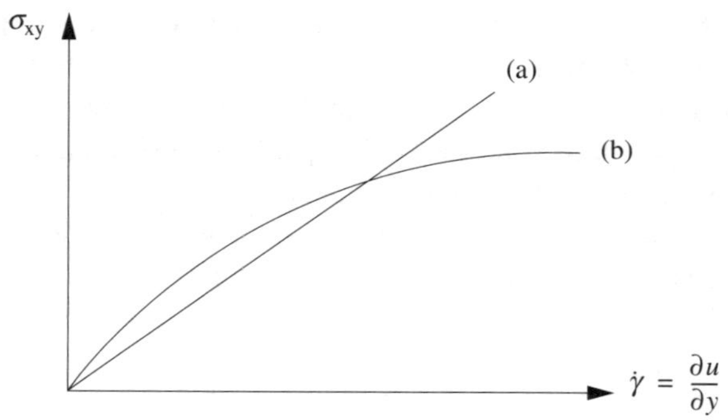

Figure 7.1 Experimental measurement of the stress σ_{xy} as a function of the shear rate. (a) Newtonian fluid; (b) non Newtonian fluid

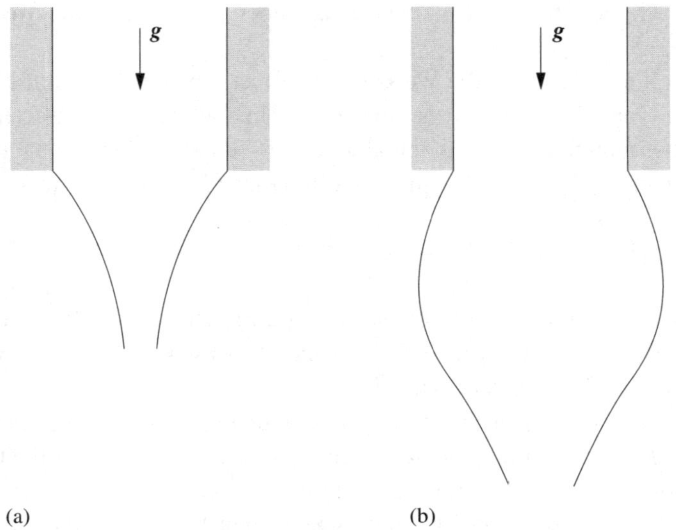

Figure 7.2 Jet at the exit of a cylindrical tube: (a) Newtonian fluid; (b) non Newtonian fluid.

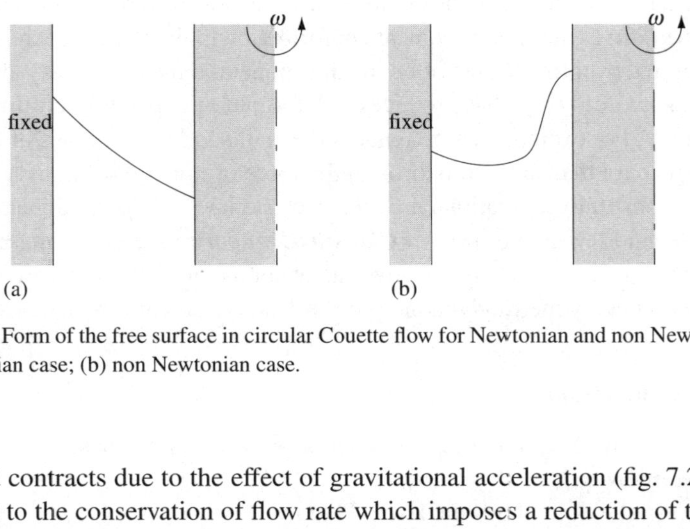

Figure 7.3 Form of the free surface in circular Couette flow for Newtonian and non Newtonian flows (a) Newtonian case; (b) non Newtonian case.

nian fluid contracts due to the effect of gravitational acceleration (fig. 7.2(a)). This conforms to the conservation of flow rate which imposes a reduction of the section to compensate the increase of the average velocity. Conversely, the non Newtonian fluid expands at the exit of the tube: the "swelling effect" (fig. 7.2(b)); in certain cases the jet section can be twice that of the tube. Note that this phenomenon also exists for Newtonian fluids, but it is much less pronounced.

Examine now the form of the free surface of a fluid between two concentric vertical cylinders, the inner cylinder rotating at a constant angular velocity ω and the outer cylinder fixed. The Newtonian fluid is pushed toward the outer cylinder due to the centrifugal force (fig. 7.3(a)) and the form of the free surface is that of a paraboloid of revolution, whereas the non Newtonian fluid collects and rises along the interior cylinder (fig. 7.3(b)). This latter effect is known as the ***Weissenberg effect***.

The swelling effect and the Weissenberg effect, observed for non Newtonian fluids, are the manifestations of the non linear phenomena that characterize these fluids. Another manifestation is the production of secondary flows in the transverse section of a rectangular pipe. Defining the difference of the normal stresses by:

$$N_1 = \sigma_{xx} - \sigma_{yy} \quad \text{and} \quad N_2 = \sigma_{yy} - \sigma_{zz} \tag{7.2}$$

it is seen that the secondary flows, as the swelling and Weissenberg effects, are caused by the non equilibrium of the normal stresses $N_1 \neq 0$ and $N_2 \neq 0$, while $N_1 = N_2 = 0$ for the Newtonian case.

Certain non Newtonian fluids also show memory effects. These are the ***viscoelastic fluids*** for which the stresses at the present time depend on the history of the deformation of the fluid. Under a brief and intense force, they react as elastic solid materials while over very long time scales, their behavior is that of a viscous fluid. Between these two extremes, one encounters viscoelastic behavior. The assumption of ***fading memory*** allows the attribution of more weight to recent events than to earlier ones. The viscoelastic fluids are characterized by the phenomenon of ***stress relaxation***: to maintain a constant deformation in the fluid, a decreasing force must be applied over time (most often, in an exponential manner). They also present the phenomenon of ***creep*** that produces a deformation depending on time under a constant force. For example, creep renders the stained glass windows of cathedrals thicker at the base than at the top through the force of gravity acting over centuries. Finally the constitutive equations must respect the laws of thermodynamics, especially the second law in the form of ***Clausius-Duhem's inequality***: the latter takes into account the irreversibility of physical phenomena at the macroscopic scale, expressing that every thermodynamic process leads to an entropy increase.

7.2.2 Newtonian fluids

The constitutive equation of a ***Newtonian fluid*** is written based on the following considerations:

- At rest, the fluid obeys the laws of statics.
- The equation of fluid behavior is objective, that is independent of the Galilean reference frame in which it is expressed, independent also of the observer (see chapter 6, Sect. 6.2.1).
- The stress tensor $\boldsymbol{\sigma}$ is a function of the velocity gradient tensor. More specifically, the tensor $\boldsymbol{\sigma}$ is an isotropic function of $\boldsymbol{d}$, the strain rate tensor, the rotation rate tensor $\boldsymbol{\omega}$ not being objective.
- The fluid is an isotropic medium, that is, independent of the orientation of the coordinate system axes.

With these assumptions, the objective equation of incompressible Newtonian viscous fluid behavior (see chapter 1) is:

$$\boldsymbol{\sigma} = -p\boldsymbol{I} + 2\mu\boldsymbol{d} \tag{7.3}$$

The tensor $\boldsymbol{I}$ is the identity tensor, and the tensor $\boldsymbol{d}$ is defined by the relation (1.22a):

$$\boldsymbol{d} = \tfrac{1}{2}[\mathbf{Grad}\,\boldsymbol{v} + (\mathbf{Grad}\,\boldsymbol{v})^{\mathrm{T}}] \tag{7.4}$$

Inserting (7.4) in (1.20) and making the assumption of constant dynamic viscosity, μ, the ***Navier-Stokes equations*** are obtained:

$$\rho\frac{\mathrm{d}\boldsymbol{v}}{\mathrm{d}t} = -\,\mathbf{grad}\,p + \mu\Delta\boldsymbol{v} + \boldsymbol{F}_{\mathrm{v}} \tag{7.5}$$

with the material derivative d/dt defined by the relation (1.41):

$$\frac{\mathrm{d}\boldsymbol{v}}{\mathrm{d}t} = \frac{\partial\boldsymbol{v}}{\partial t} + (\mathbf{Grad}\,\boldsymbol{v})\cdot\boldsymbol{v} \tag{7.6}$$

and the Laplacian operator Δ:

$$\Delta\boldsymbol{v} = \mathbf{div}(\mathbf{Grad}\,\boldsymbol{v})$$

Rewriting equation (7.5) in non dimensional form introduces the Reynolds number. For that, one chooses well the scales used to non dimensionalize the variables, that is a reference length L for space, an inertial time L/U and a characteristic velocity U. The variables become $\boldsymbol{x}' = \boldsymbol{x}/L$, $t' = Ut/L$, $\boldsymbol{v}' = \boldsymbol{v}/U$, $p' = (p - p_0)/(\rho U^2)$, $\boldsymbol{F}_{\mathrm{v}}' = \boldsymbol{F}_{\mathrm{v}}/(\rho U^2/L)$. To simplify the notation, the primes are dropped, and the non dimensional equation is obtained:

$$\frac{\mathrm{d}\boldsymbol{v}}{\mathrm{d}t} = -\,\mathbf{grad}\,p + (Re)^{-1}\Delta\boldsymbol{v} + \boldsymbol{F}_{\mathrm{v}} \tag{7.7}$$

The ***Reynolds number*** is defined by the relation:

$$Re = \frac{\rho UL}{\mu} = \frac{UL}{\nu} \tag{7.8}$$

in which $\nu = \mu/\rho$ is the kinematic viscosity. The latter is expressed in $\mathrm{m^2\,s^{-1}}$. The Reynolds number compares the importance of inertia UL to the effects of viscosity, characterized by ν.

A limiting form of the Navier-Stokes equations is obtained when $Re \to \infty$. Then:

$$\frac{\mathrm{d}\boldsymbol{v}}{\mathrm{d}t} = -\,\mathbf{grad}\,p + \boldsymbol{F}_{\mathrm{v}} \tag{7.9}$$

This last equation is the Euler equation; it describes the dynamics of a perfect (inviscid) incompressible fluid. It can be applied, for example, in the study of internal flows in hydraulic machinery (Francis, Kaplan turbines, etc.) or external flows around wings (airplanes, the space shuttle, ...) or a car body. These flows have characteristic Reynolds numbers of the order of millions, and the Euler equation is a valid model when the effects of turbulence are neglected.

If one wants to examine the case of slow flows (creeping flows), $Re \to 0$, the previous non dimensionalization – which comes from aerodynamic and hydrodynamic applications – is no longer appropriate. It is necessary to relate the characteristic scales to the dominant physical phenomenon, i.e., the viscosity.

Defining now $t' = \nu t/L^2$ and $p' = (p - p_0)/(\mu U/L)$, neglecting the volume forces and ignoring the primes:

$$\frac{\partial \boldsymbol{v}}{\partial t} + Re(\mathbf{Grad}\, \boldsymbol{v}) \cdot \boldsymbol{v} = -\mathbf{grad}\, p + \Delta \boldsymbol{v} \tag{7.10}$$

The Reynolds number zero limit gives the non dimensional ***Stokes equation***:

$$\frac{\partial \boldsymbol{v}}{\partial t} = -\mathbf{grad}\, p + \Delta \boldsymbol{v} \tag{7.11}$$

The Stokes equation (7.11) is linear (that is, that every combination of solutions is itself a solution) and parabolic, whereas the Euler equation (7.9) is non linear, and hyperbolic with an elliptic term coming from the pressure gradient. Finally the Navier-Stokes equation (7.7) is parabolic of the advection-diffusion type. Recall that the type of partial differential equation characterizes its nature. Therefore, referring to mathematical models of physics, a Laplace equation (for example, stationary diffusion) is elliptic, whereas the transient diffusion equation is parabolic. Finally, a wave equation is hyperbolic.

The flows corresponding to low or zero Reynolds numbers are laminar flows. For example a slow flow where the Reynolds number is of order unity is that of molten glass in a melting furnace or that of the injection of semi-solid metal. The velocity of glass at the liquid surface is around $5 \cdot 10^{-2}$ m s^{-1}, the characteristic length is that of the depth of the glass melt, around 1m, and the kinematic viscosity of glass is around $5 \cdot 10^{-2}$ m^2 s^{-1}. For comparison, $\nu_{air} = 10^{-5}$ m^2 s^{-1} and $\nu_{water} = 10^{-6}$ m^2 s^{-1}.

Laminar flows are simple (and very often stationary), and the physics associated with them is principally determined by the boundary conditions of the problem, that is, by the geometric configuration. The fluid particles mix little and follow regular trajectories such that the fluid layers flow over each other.

As the Reynolds number increases, a value is attained, called the critical Reynolds number, at which a transition occurs. The physics becomes more complicated with non stationary effects, hysteresis and complex spatial structures.

Increasing the Reynolds number, other transitions or bifurcations are attained which, bit by bit, take us to the turbulent state. The latter is characterized by the following observations:

- the flow is fully three dimensional;
- it is time dependent with a marked random character;
- it is dissipative; in other words, the viscous terms of the Navier-Stokes equations can not be neglected, even though the Reynolds number can be very high;
- the flow is diffusive; it shows strong mixing which is exploited in some processes to more rapidly homogenize the material properties or to obtain a mixture of several species.

The passage from the laminar state to the turbulent state through a succession of bifurcations is a spectral transition. It is well illustrated by Couette flow between two concentric cylinders. The inner cylinder of radius R_1 rotates at the angular velocity ω and the outer cylinder of radius R_2 is fixed. Redefine the Reynolds number by relation:

$$Re = \frac{\omega R_1 (R_2 - R_1)}{\nu} \tag{7.12}$$

and fix the ratio of the radii at 0.87, which corresponds to the experimental conditions reported in the literature by Coles (1965).

For $Re < 41.5$, the stationary laminar state is described by the azimuthal velocity (exercise 7.1):

$$v_\theta = Ar + \frac{B}{r} \tag{7.13}$$

with $A = \omega R_1^2/(R_2^2 - R_1^2)$ and $B = \omega R_1^2 R_2^2/(R_2^2 - R_1^2)$.

For $Re = 42$, the first transition appears, to Taylor vortices, which, in a constant longitudinal section $\theta =$ constant, are seen in counter rotating pairs (fig. 7.4).

Increasing the inner cylinder velocity, and for $Re = 66$, a second transition is reached such that the Taylor vortices become deformed in the circumferential direction. Later transitions modify the number of vortex pairs in the vertical direction and the wave number of the azimuthal deformations.

In certain cases the transition can be brutal. A typical example is that of flow in a circular pipe where, for a Reynolds number of the order of 2000 based on the velocity on the axis and the diameter of the pipe, the transition is instantaneous from the laminar to the turbulent state.

To describe turbulent flows, the Navier-Stokes equations can be used as is. One then performs ***direct numerical simulation*** (DNS). However, for reasons related to the computation capacities of current computers, DNS is limited to Reynolds num-

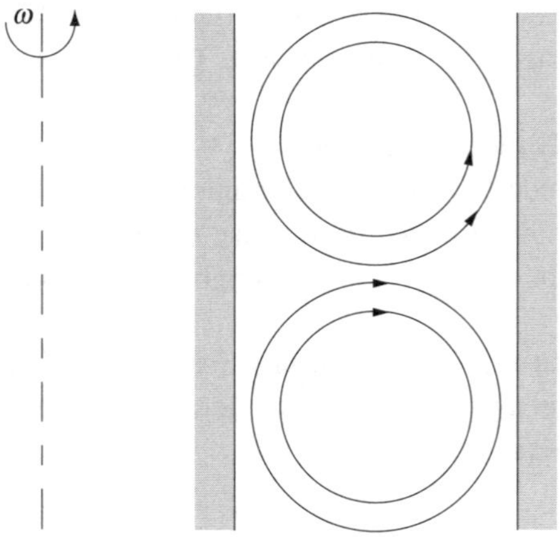

Figure 7.4 Taylor vortices.

bers of the order of a few thousand (around 20 000). Thus one most often turns to Reynolds modeling of turbulent flow which is presented in section 7.2.4.

7.2.3 Non Newtonian fluids

The development of the constitutive equations for non Newtonian fluids relies on mathematical considerations which are outside the introductory scope of this book. For more details the reader is referred to the books cited in the bibliography (Crochet *et al.* 1984, Joseph 1990). We thus restrict ourselves to the presentation of a few constitutive equations of the "partial differential equation" type.

Decompose the stress tensor into an arbitrary pressure field and an extra-stress tensor that we denote in agreement with the majority of the books on rheology, $\boldsymbol{T}_{\mathrm{v}}$. The trace of $\boldsymbol{T}_{\mathrm{v}}$ is non zero (contrary to (1.98)):

$$\boldsymbol{\sigma} = -p\boldsymbol{I} + \boldsymbol{T}_{\mathrm{v}} \tag{7.14}$$

This extra-stress tensor is itself decomposed into a viscoelastic component $\boldsymbol{T}_{\mathrm{v1}}$ and another purely viscous component:

$$\boldsymbol{T}_{\mathrm{v}} = \boldsymbol{T}_{\mathrm{v1}} + 2\mu_0\boldsymbol{d} \tag{7.15}$$

with μ_0 the "Newtonian" dynamic viscosity. This latter viscosity is that of the Newtonian solvent in which the polymer may be diluted.

The constitutive equation for $\boldsymbol{T}_{\mathrm{v1}}$ is obtained by invoking the objectivity (6.53) of the tensor $\boldsymbol{T}_{\mathrm{v1}}$, the fading memory (that which gives the more weight to events in the recent past than to the more distant past) and the tensor relations linked to the

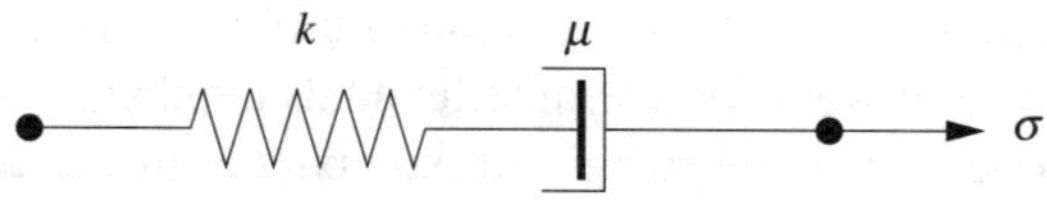

Figure 7.5 Maxwell's model in one dimension.

history of the strain gradients:

$$\boldsymbol{T}_{v1} + \lambda_1 \overset{\triangledown}{\boldsymbol{T}}_{v1} = 2\mu_1 \boldsymbol{d} \tag{7.16}$$

where λ_1 and μ_1 represent respectively the relaxation time and the viscosity of the viscoelastic component of the over stress. $\overset{\triangledown}{\boldsymbol{T}}_{v1}$ represents a derivative of the tensor $\boldsymbol{T}_{v1}$ that we will discuss below. This relation is the generalization to three spatial dimensions of Maxwell's one dimensional viscoelastic model which consists of a spring with constant k for the elastic behavior and a dashpot of viscosity μ such that $\lambda = \mu / k$ for the viscous behavior (Figure 7.5).

The stress-strain relations are $\sigma = k\,\varepsilon_1$ for the spring and $\sigma = \mu\dot{\varepsilon}_2$ for the dashpot. The same force applied in the model, σ, produces a displacement σ/k in the spring and a velocity σ/μ in the dashpot. The total velocity is the sum

$$\dot{\varepsilon} = \dot{\varepsilon}_1 + \dot{\varepsilon}_2 = \frac{\dot{\sigma}}{k} + \frac{\sigma}{\mu} \tag{7.17}$$

One can then write the differential equation

$$\sigma + \lambda\dot{\sigma} = \mu\dot{\varepsilon}, \quad \lambda = \frac{\mu}{k} \tag{7.18}$$

As for the constitutive equation for $\boldsymbol{T}_{v1}$, one would like to generalize the relation (7.18), for three spatial dimensions in the form

$$\boldsymbol{T}_{v1} + \lambda_1 \frac{\mathrm{d}\boldsymbol{T}_{v1}}{\mathrm{d}t} = 2\mu_1 \boldsymbol{d} \tag{7.19}$$

Unfortunately, (7.19) is not valid, as the total derivative of the stress tensor is not objective. The introduction of the ***upper-convected derivative***, denoted $\overset{\triangledown}{\boldsymbol{T}}_{v1}$, resolves this difficulty. It is given by the equation:

$$\overset{\triangledown}{\boldsymbol{T}}_{v1} = \frac{\mathrm{d}\boldsymbol{T}_{v1}}{\mathrm{d}t} - (\mathbf{Grad}\,\boldsymbol{v})\boldsymbol{T}_{v1} - \boldsymbol{T}_{v1}(\mathbf{Grad}\,\boldsymbol{v})^{\mathrm{T}} \tag{7.20}$$

The upper-convected derivative is thus an objective form of the temporal derivative. The relations (7.15), (7.16) and (7.20) are those of an Oldroyd-B fluid. If the viscosity μ_0 is zero, we obtain the Maxwell-B fluid model. Note that equation (7.20) is hyperbolic, which complicates the solution of problems with non Newtonian fluids, even for the case of slow flows with a Reynolds number near zero.

As these models are cumbersome to manipulate, in the case of viscometric flows – where the fluid particle undergoes a history characterized by constant stretching – one can limit oneself to simpler constitutive relations as for example that of a fluid of second order:

$$\boldsymbol{T}_{v1} = 2\mu\boldsymbol{d} - 2\psi_{1,0}\overset{\nabla}{\boldsymbol{d}} + 4\psi_{2,0}\boldsymbol{dd} \qquad (7.21)$$

In this last equation, the coefficients $\psi_{1,0}$ and $\psi_{2,0}$ are defined as follows:

$$\psi_{1,0} = \frac{N_1}{\dot{\gamma}^2} \qquad (7.22)$$

$$\psi_{2,0} = \frac{N_2}{\dot{\gamma}^2} \qquad (7.23)$$

$$\dot{\gamma} = \sqrt{2d_{ij}d_{ij}} \qquad (7.24)$$

the subscript zero in (7.22) and (7.23) indicates that the shear rate $\dot{\gamma}$ tends to have small values; recall also that N_1 and N_2 are the differences of the normal stresses (7.2). Equation (7.24) shows that the shear rate is the second invariant of the tensor $\boldsymbol{d}$. Note that $\dot{\gamma}$ is equal to $\sqrt{3}\dot{\varepsilon}_{eq}$ introduced in chapter 6, section 6.2.

In a number of practical applications, one turns to the generalized Newtonian fluid model for which $\boldsymbol{T}_v$ is a function of $\boldsymbol{d}$:

$$\boldsymbol{T}_v = \boldsymbol{T}_v(\boldsymbol{d}) \qquad (7.25)$$

The most well known model is that of the Newtonian fluid (eq. (7.6)):

$$\boldsymbol{T}_v = 2\mu\,\boldsymbol{d} \qquad (7.26)$$

For a generalized Newtonian fluid:

$$\mu = \mu(\dot{\gamma}) \qquad (7.27)$$

a commonly used relation being the ***power law***:

$$\mu(\dot{\gamma}) = K\,\dot{\gamma}^{m-1} \qquad (7.28)$$

with K the consistency and m the power law index ($m = 1$ for the Newtonian case). For many polymers, m is between 0.3 and 0.6. Fluids for which m is less than unity are ***pseudoplastic*** (or viscoplastic); those for which m is greater than unity are the

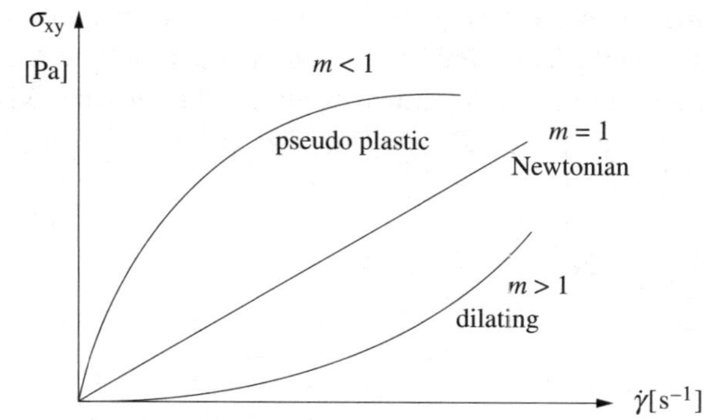

Figure 7.6 Evolution of the shear stress as a function of the shear rate.

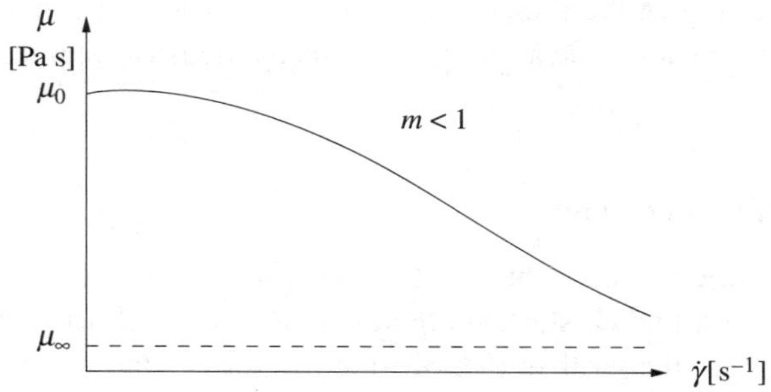

Figure 7.7 Variation of the viscosity as a function of the shear rate for a pseudoplastic fluid.

"dilating" fluids. Figure 7.6 shows the evolution of the shear stress in a planar Couette flow as a function of the shear rate for various values of m. Figure 7.7 shows the variation of the viscosity as a function of the shear rate for the pseudoplastic fluid: it is seen that it diminishes as $\dot{\gamma}$ increases; in the literature one speaks of *shear thinning* viscosity. In the opposite case, where the viscosity μ increases while $\dot{\gamma}$ also increases, one has shear thickening viscosity[1]. The power law does not allow the description of a Newtonian plateau; in addition it produces an infinite viscosity for zero shear rate. In industrial applications it is replaced by ***Bird-Carreau's law***:

$$\mu - \mu_\infty = (\mu - \mu_0)[1 + (\lambda\dot{\gamma})^2]^{(m-1)/2} \tag{7.29}$$

where μ_0 is the viscosity of the Newtonian region at low shear rate, μ_∞ the viscosity of the second Newtonian zone at very high shear rate, λ being a characteristic time of the material.

1 One should pay attention to avoid a false interpretation of the expressions used (dilating, thickening, thinning) which, although common, are nonetheless incorrect in the context of incompressible fluids.

It is customary to make the equations of non Newtonian fluids non dimensional by reference to a length L, a velocity U and to the inertial time L/U. The pressure and the extra-stresses are non dimensionalized through the viscosity, that is $p' = (p - p_0)/(\mu U/L)$ and $\boldsymbol{T}_v' = \boldsymbol{T}_v/(\mu U/L)$. The constitutive equation (7.19) becomes:

$$\boldsymbol{T}_{v1}' + We\overset{\nabla}{\boldsymbol{T}}{}_{v1}' = 2\boldsymbol{d}' \tag{7.30}$$

with

$$We = \lambda_1 \frac{U}{L} \tag{7.31}$$

This last number is the ***Weissenberg number***. It compares the relaxation time of the fluid, λ_1, to the characteristic time of the flow given by L/U. The higher its value, the more hyperbolic the character of (7.30). The Weissenberg number can take values between unity and a few dozen in many industrial applications involving molten polymers.

7.2.4 Turbulence modeling

The working or forming of metallic materials often involves a phase during which they are in a molten state. As their viscosity is small (in some cases of the order of 100 times smaller than that of water, as for crystalline semiconductors or molten tin), the characteristic Reynolds numbers are often very high and the flows are turbulent. As finely detailed information of the geometric structure of the flow and its temporal dynamics is not desired, for fully developed turbulence (large Reynolds numbers of the order of one million), the statistical Reynolds averaging approach is employed: it consists of decomposing all the quantities into time averaged values and fluctuations. The velocity field is the sum of the two quantities:

$$\boldsymbol{v} = \bar{\boldsymbol{v}} + \boldsymbol{v}' \tag{7.32}$$

where $\bar{\boldsymbol{v}}$ is the average velocity and $\boldsymbol{v}'$ the fluctuation around the average. An analogous definition is used for the pressure. The average velocity field at a point with position $\boldsymbol{x}$ is defined by a temporal average over the interval T:

$$\bar{\boldsymbol{v}} = \frac{1}{T} \int_{t-T/2}^{t+T/2} \boldsymbol{v}(\tau)\, d\tau \tag{7.33}$$

With this definition of the average and relation (7.32), it is easy to show that the average of the fluctuations $\bar{\boldsymbol{v}}'$ is zero. Applying the averaging process to equations (7.1) and (7.5), the ***Reynolds averaged Navier-Stokes*** (RANS) ***equations*** are obtained, neglecting the volume forces:

$$\mathrm{div}\,\bar{v} = 0 \tag{7.34}$$

$$\rho\frac{\bar{\mathrm{d}}}{\mathrm{d}t}\bar{v} = \mathbf{div}(\bar{\boldsymbol{\sigma}} + \boldsymbol{R}) \tag{7.35}$$

with

$$\bar{\boldsymbol{\sigma}} = -\bar{p}\boldsymbol{I} + 2\mu\bar{\boldsymbol{d}} \tag{7.36}$$

$$\bar{\boldsymbol{d}} = \tfrac{1}{2}[\mathbf{Grad}\,\bar{v} + (\mathbf{Grad}\,\bar{v})^{\mathrm{T}}] \tag{7.37}$$

The average material derivative now involves the average velocity field in the transport term

$$\frac{\bar{\mathrm{d}}}{\mathrm{d}t}\bar{v} = \frac{\partial\bar{v}}{\partial t} + (\mathbf{Grad}\,\bar{v})\cdot\bar{v} \tag{7.38}$$

It can be shown (see exercise 7.2) that the average of the non linear term $(\mathbf{Grad}\,\boldsymbol{v})\cdot\boldsymbol{v}$ written in the form $\mathbf{div}(\boldsymbol{v}\times\boldsymbol{v})$ produces a contribution from the average velocity field $(\mathbf{Grad}\,\bar{v})\cdot\bar{v}$ and a new term $\mathbf{div}(\overline{\boldsymbol{v}'\times\boldsymbol{v}'})$. We denote with the symbol $-\boldsymbol{R}/\rho$ the quantity thus obtained inside the divergence operator. The symmetric turbulent stress tensor, called the ***Reynolds tensor***, $\boldsymbol{R}$, thus has as components:

$$R_{\mathrm{ij}} = -\rho\overline{v_{\mathrm{i}}'v_{\mathrm{j}}'} \tag{7.39}$$

By analogy with $\bar{\boldsymbol{\sigma}}$, it is interpreted as the tensor providing the turbulent contribution to the average flow. It is seen that (7.34) and (7.35) provide four equations for ten unknowns: the average pressure, the three components of the average velocity field and the six components of the Reynolds stress. We thus must close the system in order that it becomes solvable. If the equations for the dynamics of R_{ij} are written, triple correlations of the type $\overline{v_{\mathrm{i}}'v_{\mathrm{j}}'v_{\mathrm{k}}'}$ appear and introduce more new unknowns than equations. We are confronted with what is called a diverging hierarchy in mathematics. It is the famous closure problem of turbulence modeling.

The current state of the art turns to a closure with two equations that involve the two supplementary quantities that are the ***turbulent kinetic energy*** per unit mass k, defined by the relation (as always, we sum over repeated indices)

$$k = \tfrac{1}{2}\overline{v_{\mathrm{i}}'v_{\mathrm{i}}'} \tag{7.40}$$

and the ***average dissipation rate*** of this energy

$$\varepsilon = \nu\overline{\frac{\partial v_{\mathrm{i}}'}{\partial x_{\mathrm{j}}}\frac{\partial v_{\mathrm{i}}'}{\partial x_{\mathrm{j}}}} \tag{7.41}$$

Following the reasoning of continuum mechanics, the closure is obtained expressing the Reynolds stress tensor (second order correlation) as a function of the average velocity field (first-order correlation) by a constitutive equation:

$$\boldsymbol{R} = -\tfrac{2}{3}\rho k\boldsymbol{I} + 2\mu_T\bar{\boldsymbol{d}} \tag{7.42}$$

in which μ_T represents the ***turbulent viscosity*** introduced for the first time by Boussinesq in 1877. We must however note that this reasoning by analogy is not totally justified as the turbulence is not a material characteristic, but rather a qualitative property of a flow. When (7.42) is inserted in (7.35) – as we will do in section 7.4.5 – the isotropic term of (7.42) combines with the average pressure to give a modified pressure, while the term characterized by the turbulent viscosity combines with the molecular viscosity term. The turbulence modeling thus has the effect of increasing the local viscosity (which increases the robustness of the numerical methods implemented) with the expectation that the non constant turbulent viscosity will be sufficiently anisotropic to represent the entire complexity of the physics.

By a dimensional analysis argument, the turbulent viscosity (units kg m^{-1} s^{-1}) is written in the form:

$$\mu_T = C_\mu \rho \frac{k^2}{\varepsilon} \tag{7.43}$$

where C_μ is an empirical constant evaluated at 0.09 on the basis of experimental results obtained in simple flows.

To obtain the equation for k, one proceeds in the following manner. The average momentum equation (7.35) is subtracted from the Navier-Stokes (7.5) equation with $\boldsymbol{F}_v = \boldsymbol{0}$ and the equations (in each of the spatial directions $i = 1,3$) that govern the dynamics of the velocity fluctuations are obtained:

$$\frac{\partial v_i'}{\partial t} + \frac{\partial}{\partial x_j}(\bar{v}_j v_i') = -\frac{\partial}{\partial x_j}(\bar{v}_i v_j') - \frac{\partial}{\partial x_j}(v_i' v_j' - \overline{v_i' v_j'}) - \frac{1}{\rho}\frac{\partial p'}{\partial x_i} + \nu\frac{\partial^2 v_i'}{\partial x_j^2} \tag{7.44}$$

This equation is then multiplied by v_i' and averaged. After a few algebraic manipulations:

$$\frac{\mathrm{d}}{\mathrm{d}t}k = -\overline{v_i' v_j'}\frac{\partial \bar{v}_i}{\partial x_j} - \nu\overline{\frac{\partial v_i'}{\partial x_j}\frac{\partial v_i'}{\partial x_j}} - \frac{\partial}{\partial x_j}J_j \tag{7.45}$$

$$J_j = \frac{1}{\rho}\overline{p' v_j'}\delta_{ij} + \tfrac{1}{2}\overline{v_i' v_i' v_j'} - \nu\frac{\partial k}{\partial x_j} \tag{7.46}$$

We have generated new correlations by this process. In order to be able to close the system of equations, the assumption that consists of replacing the pressure-velocity correlation and the triple correlation by a diffusion term for the kinetic energy is introduced:

$$-\left(\frac{1}{\rho}\overline{p'v'_j}\delta_{ij} + \tfrac{1}{2}\overline{v'_iv'_iv'_j}\right) = \frac{\nu_T}{\sigma_k}\frac{\partial k}{\partial x_j} \tag{7.47}$$

where $\nu_T = \mu_T/\rho$ and σ_k is a constant. The term expressing the interaction of the Reynolds stress and the gradient of the average velocity becomes:

$$-\overline{v'_iv'_j}\frac{\partial \bar{v}_i}{\partial x_j} = 2\nu_T\bar{d}_{ij}\bar{d}_{ij} \tag{7.48}$$

To write relation (7.48), use was made of (7.39) and (7.42) (see also exercise 7.3). The equation in k is written:

$$\frac{\bar{\mathrm{d}}}{\mathrm{d}t}k = \mathrm{div}\left(\left(\nu + \frac{\nu_T}{\sigma_k}\right)\mathbf{grad}\,k\right) + 2\nu_T\bar{\boldsymbol{d}} : \bar{\boldsymbol{d}} - \varepsilon \tag{7.49}$$

The equation for ε is more difficult to model. However, one can consider that it is obtained from (7.49) by formally "multiplying" it by the quantity ε/k. Then:

$$\frac{\bar{\mathrm{d}}}{\mathrm{d}t}\varepsilon = \mathrm{div}\left(\left(\nu + \frac{\nu_T}{\sigma_\varepsilon}\right)\mathbf{grad}\,\varepsilon\right) + 2\nu_T\bar{\boldsymbol{d}} : \bar{\boldsymbol{d}}C_{\varepsilon_1}\frac{\varepsilon}{k} - \frac{\varepsilon^2}{k}C_{\varepsilon_2} \tag{7.50}$$

where the empirical constants take the following values: $C_{\varepsilon_1} = 1.44$; $C_{\varepsilon_2} = 1.92$; $\sigma_k = 1.0$; $\sigma_\varepsilon = 1.3$.

It should be noted that over the last decade, nonlinear $k - \varepsilon$ models have been proposed in order to incorporate non equilibrium of the turbulent normal stresses, in the same way that the differences of the stresses N_1 and N_2 do not cancel for non Newtonian fluids. These non linear models can predict the generation of secondary turbulent flows in the transverse sections of non cylindrical pipes, whereas the classic linear models erroneously predict a longitudinal turbulent flow with no secondary flows. Speziale (1991) proposed writing the Reynolds stress tensor with quadratic terms in $\bar{\boldsymbol{d}}$:

$$\frac{1}{\rho}R_{ij} = -\tfrac{2}{3}k\delta_{ij} + 2C_\mu\frac{k^2}{\varepsilon}\bar{d}_{ij} + 4C_DC_\mu^2\frac{k^3}{\varepsilon^2}(\bar{d}_{ik}\bar{d}_{kj} - \tfrac{1}{3}\bar{d}_{mn}\bar{d}_{mn}\delta_{ij})$$
$$+ 4C_EC_\mu^2\frac{k^3}{\varepsilon^2}(\overset{\nabla}{\bar{d}}_{ij} - \tfrac{1}{3}\overset{\nabla}{\bar{d}}_{mm}\delta_{ij}) \tag{7.51}$$

with $C_D = C_E = 1.68$. The definition of the tensor $\overset{\nabla}{\bar{\boldsymbol{d}}}$ is identical to that of the extra-stress (7.20) of the non Newtonian fluids. It is seen that there exists a strong conceptual link between the modeling of turbulence and of the mechanics of non Newtonian fluids. The comparison of relations (7.21) and (7.51) shows the profound analogy between the two approaches which employ quadratic terms in the strain rate tensors.

7.3 BOUNDARY AND INITIAL CONDITIONS

7.3.1 Initial conditions

In Newtonian fluid mechanics, the fluid can be imposed to be initially at rest, with zero velocity and pressure:

$$\boldsymbol{v}_0 = \boldsymbol{0} \quad \text{and} \quad p_0 = 0 \tag{7.52}$$

Regardless of the physical situation, the initial velocity field must be incompressible:

$$\operatorname{div} \boldsymbol{v}_0 = 0 \tag{7.53}$$

This latter constraint may require an auxiliary calculation that we will present later in the context of the numerical treatment of Stokes flow. Remember that in an incompressible fluid, the pressure is known except possibly for a constant. The constant is fixed by imposing a reference pressure at a point or by imposing an average pressure over the calculation domain.

For non Newtonian fluids, the extra stresses are initially zero:

$$\boldsymbol{T}_\mathrm{v} = \boldsymbol{0} \tag{7.54}$$

Finally, for turbulent flow, the Reynolds stress tensor is assumed initially zero:

$$\boldsymbol{R} = \boldsymbol{0} \tag{7.55}$$

The initial conditions can be considered as the first stage of an iterative process which yields a stationary solution at convergence, or as the primer for an evolving calculation that does not necessarily have a stationary state.

7.3.2 Boundary conditions

Solid walls

For the classic case of Newtonian fluids, we have seen the non slip condition at a wall (1.123), which can be generalized

$$\boldsymbol{v} = \boldsymbol{v}_\mathrm{w} \tag{7.56}$$

where $\boldsymbol{v}_\mathrm{w}$ represents the imposed velocity of the wall.

The calculation domains are not necessarily confined. To show how to establish the boundary conditions of a simple problem, consider the two dimensional flow over the backward facing step in figure 7.8. Here, $\boldsymbol{v} = (u, v)$. On the walls AF and BCDE, $u = v = 0$ is imposed. For the entry section AB, we assume that devel-

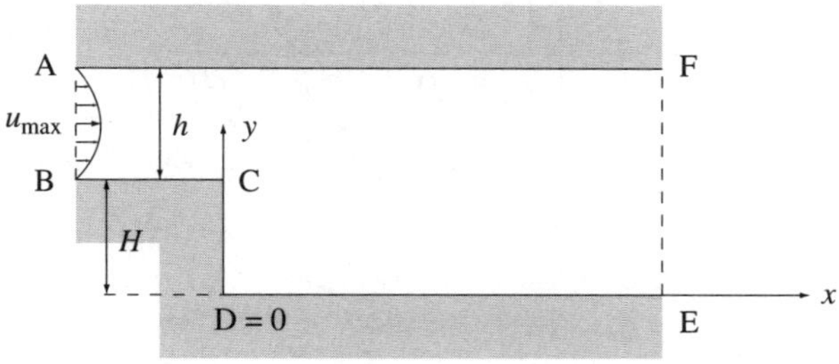

Figure 7.8 The two dimensional problem of a backward facing step.

oped flow arrives from an infinitely long pipe and that there is a Poiseuille velocity profile in y (see exercise 7.4):

$$u = \frac{4}{h^2}(h + H - y)(y - H)u_{\max} \qquad \text{with } v = 0 \tag{7.57}$$

At the output section EF, the fluid is permitted to exit at its velocity, that is, the fluid is not pulled with an imposed velocity to make it exit the domain. This amounts to imposing a zero normal stress component (1.126a):

$$\sigma_{xx} = -p + 2\mu\frac{\partial u}{\partial x} = 0 \tag{7.58}$$

and with (1.126b):

$$\frac{\partial u}{\partial y} + \frac{\partial v}{\partial x} = 0 \tag{7.59}$$

Often, equation (7.59) is replaced with $v = 0$; which assumes that the output section EF is sufficiently far from the step so that the flow is again developed ($u = u(y)$). Then, $\partial u/\partial x = 0$ and $p = 0$, by (7.58).

The case of non Newtonian fluids is more difficult. The hyperbolic nature of the constitutive equation requires the imposition of the viscoelastic extra-stress at the entry sections. On solid walls the non slip condition is imposed (7.56). However, some polymers or polymer solutions show behavior at the walls that is better approximated by the assumption that the fluid slips over the wall. Then a valid slip law is written in the form:

$$\boldsymbol{T}_\tau = -a|\boldsymbol{v}_s|^{p_f - 1}\boldsymbol{v}_s \tag{7.60}$$

where $\boldsymbol{T}_\tau$ is the shear stress vector and $\boldsymbol{v}_s$ the tangential slip velocity at the wall. This law is analogous to the viscoplastic friction law (6.89) used for metals. The coefficients a and p_f depend on the fluid and the roughness of the wall. For example, for the extrusion of PVC tubes, $a = 10^5$ kg m^{-2} s^{-1} and $p_f = 0.2$.

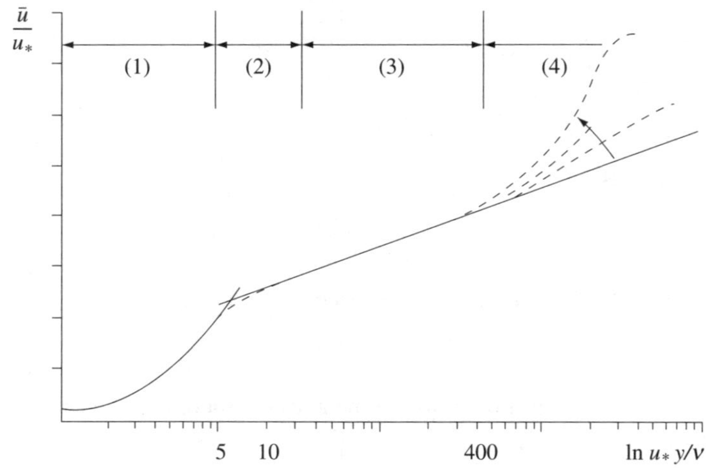

Figure 7.9 Velocity profile in a turbulent flow sheared at a wall.

For turbulent flows, the non slip condition at a wall implies that the local Reynolds number is small. Taking the flow between two plane parallel plates as an example (Tennekes and Lumley, 1972, Ryhming, 1991), four zones can be distinguished (figure 7.9):

- the laminar viscous boundary sub layer where the molecular viscosity is preponderant (1),
- the transition layer where the viscous effects are comparable to the turbulent effects (2),
- the turbulent boundary layer where the molecular viscosity is negligible with respect to the turbulent viscosity (3),
- the central zone (4).

When the Reynolds number is high, the first two layers are very thin. To give an idea, take the example of flow between two plane parallel plates separated by 20 cm and a Reynolds number of 10^5. In these conditions, the turbulent boundary layer starts at 4 mm from the wall. In order to avoid having to place many mesh points near the wall to resolve the behavior of the viscous sub layer and the boundary layer, the boundary conditions are imposed at start of the turbulent boundary layer rather than at the wall directly. This method employs "wall functions". It can be shown, assuming constant shear in the boundary layer equal to that at the wall, that the velocity profile in a turbulent boundary layer follows a logarithmic law (see Ryhming):

$$\frac{\bar{u}}{u_*} = \frac{1}{\kappa}\left(\ln\frac{u_* y}{\nu} - \ln\beta\right) \tag{7.61}$$

where the friction velocity u_* is defined by the relation:

$$|\boldsymbol{T}_\tau| = \rho u_*^2 \tag{7.62}$$

The value $(u_* y/\nu)$ defines the distance to the wall; it is denoted y^+. The von Karman constant κ equals 0.41 and β is of the order 10^{-1}. To obtain (7.61), it has been assumed that the shear stress is constant in the boundary layer equal to the shear stress at the wall $\boldsymbol{T}_\tau$. With (7.61) and (7.62), k and ε can be evaluated in the turbulent boundary layer ($30 < y^+ < 100$). After a few simple calculations taking into account (7.43) and (7.61) and assuming equilibrium of the turbulent production and the dissipation, k is found to be constant:

$$k = \frac{u_*^2}{\sqrt{C_\mu}} \tag{7.63}$$

while

$$\varepsilon = \frac{u_*^3}{\kappa y} \tag{7.64}$$

These values are thus imposed at the walls for the calculations of k and ε (Dirichlet conditions) with u_* obtained with (7.62). At the fluid entry, profiles are given for k and ε obtained by solving a one dimensional problem; at the fluid exit, homogenous Neumann conditions are traditionally employed, of the type:

$$\frac{\partial k}{\partial n} = \frac{\partial \varepsilon}{\partial n} = 0 \tag{7.65}$$

Interface

An interface is a zone in space where two immiscible fluids are in direct contact. In the case of planar (two dimensional) flow, the interface is a curved line in the plane whereas for a general three dimensional flow, it is a surface.

An example of an interface comes from the production of photographic paper. This process, "coating flow", is used to deposit many layers of fluids containing silver salts on a sheet of paper passing at high velocity (fig. 7.10). These multiple layers are the basis for common color photos. The fluid layers are obtained from vertical supply tubes that open at the end to air. The viscous fluids – containing all the chemical products necessary for the reaction with light that results in a photo – flow from the tubes slowly down an inclined surface. Two supply tubes are shown in figure 7.10, producing an interface where the two layers flow downstream, one over the other. At contact with air another surface appears, known as a ***free surface***.

Consider figure 7.11. The interface condition expresses the mechanical equilibrium of the contact forces applied by fluids I and II on either side of the interface.

We have then

$$\boldsymbol{T}_{\mathrm{I}} + \boldsymbol{T}_{\mathrm{II}} = \boldsymbol{0} \tag{7.66}$$

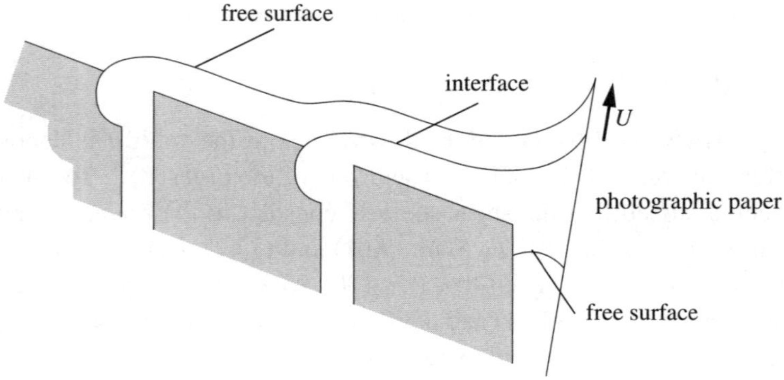

Figure 7.10 Coating flow.

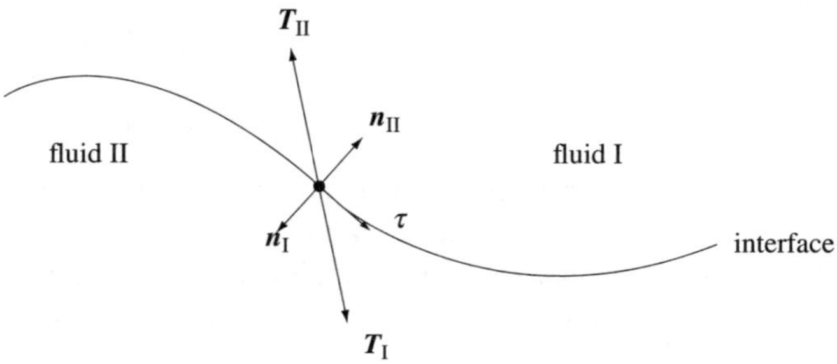

Figure 7.11 Treatment of an interface.

with the definition (1.26) for T. With (7.66) and the equality $\boldsymbol{n}_I = -\boldsymbol{n}_{II}$, we obtain

$$\boldsymbol{\sigma}_I \cdot \boldsymbol{n}_I = \boldsymbol{\sigma}_{II} \cdot \boldsymbol{n}_I \tag{7.67}$$

The interface conditions are projected on the normal $\boldsymbol{n}_I$ and the tangent vector $\boldsymbol{\tau}$ giving

$$(\boldsymbol{\sigma}_I \cdot \boldsymbol{n}_I) \cdot \boldsymbol{n}_I = (\boldsymbol{\sigma}_{II} \cdot \boldsymbol{n}_I) \cdot \boldsymbol{n}_I \tag{7.68}$$

$$(\boldsymbol{\sigma}_I \cdot \boldsymbol{n}_I) \cdot \boldsymbol{\tau} = (\boldsymbol{\sigma}_{II} \cdot \boldsymbol{n}_I) \cdot \boldsymbol{\tau} \tag{7.69}$$

It is clear that the fluids in contact can be Newtonian or non Newtonian.

Laminar free surface

A very interesting particular case is that of a free surface (see figure 7.12) where a laminarly flowing viscous fluid is in contact with a perfect fluid: air, assumed to be at rest.

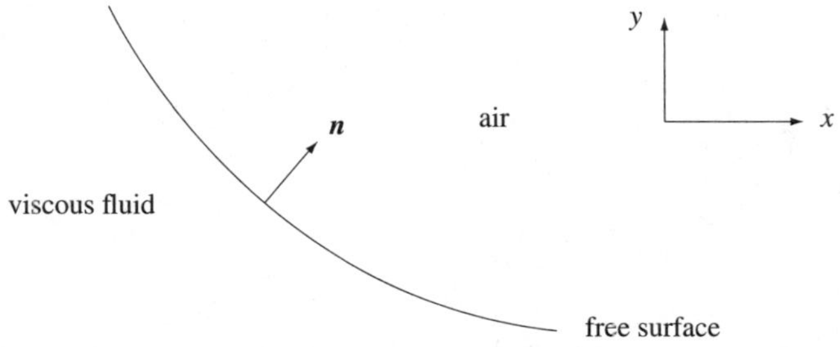

Figure 7.12 Treatment of a free surface

Taking $\boldsymbol{n}$, the outgoing normal from the viscous fluid, as a reference, we have, $\boldsymbol{\sigma}_{\mathrm{II}} = -p_{\mathrm{air}}\boldsymbol{I}$, where p_{air} is the pressure in the air, considered as a perfect fluid without viscosity. The fluid index I is omitted for simplicity. The conditions (7.68) and (7.69) thus become

$$(\boldsymbol{\sigma} \cdot \boldsymbol{n}) \cdot \boldsymbol{n} = -p_{\mathrm{air}}(\boldsymbol{n} \cdot \boldsymbol{n}) = -p_{\mathrm{air}} \tag{7.70}$$

$$(\boldsymbol{\sigma} \cdot \boldsymbol{n}) \cdot \boldsymbol{\tau} = 0 \tag{7.71}$$

The latter relation indicates that the tangential component of the contact force at the free surface must be zero, as the air (non viscous) is incapable of taking a shear force. This assumes that surface tension phenomena are neglected; they are treated later (see also chapter 1). Writing equations (7.70) and (7.71) for a planar problem, denoting by p_{viscous} the pressure in the viscous medium, to distinguish it from that in the outside medium:

$$-p_{\mathrm{viscous}} + 2\mu\frac{\partial u}{\partial x}n_{\mathrm{x}}^2 + 2\mu\left(\frac{\partial u}{\partial y} + \frac{\partial v}{\partial x}\right)n_{\mathrm{x}}n_{\mathrm{y}} + 2\mu\frac{\partial v}{\partial y}n_{\mathrm{y}}^2 = -p_{\mathrm{air}} \tag{7.72}$$

$$2\mu n_{\mathrm{x}}n_{\mathrm{y}}\left(\frac{\partial u}{\partial x} - \frac{\partial v}{\partial y}\right) + \mu\left(\frac{\partial u}{\partial y} + \frac{\partial v}{\partial x}\right)(n_{\mathrm{y}}^2 - n_{\mathrm{x}}^2) = 0 \tag{7.73}$$

Note that free surface flows are particularly difficult, as they imply a knowledge of the boundary of the integration domain, which is itself a solution of the problem through the normal components appearing in relations (7.72) and (7.73). We will indicate next how the situation can be simplified if the surface is a uniform function.

It should first be mentioned that the case of a turbulent free surface is more difficult to treat than a laminar free surface. The modeling of the Reynolds stress tensor in the viscous fluid and the turbulence in air must be taken into account. With

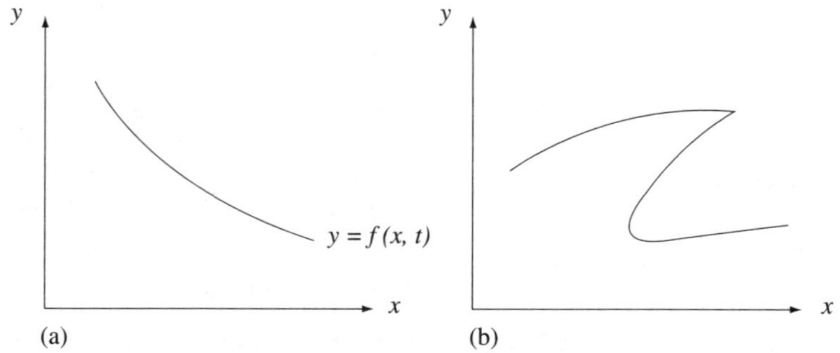

Figure 7.13 Free surface: (a) uniform representation by equation (7.74); (b) multiform.

that, the influence of wind, for example, can be obtained (calculation of boat hulls in waves) taking into account the turbulent Reynolds stress induced by the (turbulent) shear due to the wind.

If the free surface can be explicitly represented by the equation

$$F(x, y, t) = y - f(x, t) = 0 \tag{7.74}$$

– which implies that the free surface is uniform, and excludes multiform functions such as a wave which breaks on a gently sloped beach (fig. 7.13) – then the components of $\boldsymbol{n}$ can be written as

$$n_x = \frac{1}{(1+f'^2)^{1/2}} \quad \text{and} \quad n_y = \frac{-f'}{(1+f'^2)^{1/2}} \tag{7.75}$$

where f' denotes the partial derivative of f with respect to x. In addition, in the case of a transient free surface, some algorithms profit from the kinematic constraint expressing that the free surface remains a material surface composed of the same fluid particles over time. Therefore, using (7.74)

$$\frac{\mathrm{d}F}{\mathrm{d}t} = 0 \quad \text{that is,} \quad \frac{\partial f}{\partial t} + u\frac{\partial f}{\partial x} = v \tag{7.76}$$

When the surface tension effects are important (see chapter 1), they are taken into account in the boundary condition projected onto the normal (7.72) which is written with the surface tension coefficient γ

$$(\boldsymbol{\sigma} \cdot \boldsymbol{n}) \cdot \boldsymbol{n} = -p_{air} - \gamma\left(\frac{1}{R_1} + \frac{1}{R_2}\right) \tag{7.77}$$

In the last relation, R_1 and R_2 are the principal radii of curvature of the surface. If the flow is planar, $R_1 = \infty$ and $1/R_2 = -f''/(1+f'^2)^{3/2}$. The quantity inside the parentheses multiplying the surface tension in equation (7.77), is equal to 2 Ξ, where Ξ is the average curvature of the surface. Its sign depends on the concavity (–) or the convexity (+) of the surface.

7.4 NUMERICAL TREATMENT OF THE NAVIER-STOKES PROBLEM

7.4.1 Formulation in primitive variables

The Navier-Stokes equations for an incompressible fluid are derived from (1.103) and (1.8):

$$\operatorname{div} \boldsymbol{v} = 0 \tag{7.78}$$

$$\rho \frac{\mathrm{d}\boldsymbol{v}}{\mathrm{d}t} = -\mathbf{grad}\, p + \mu \Delta \boldsymbol{v} + \boldsymbol{F}_{\mathrm{v}} \tag{7.79}$$

The problem (7.78)–(7.79) is a problem under the constraint (7.78); this latter condition implies the pressure field that insures the incompressibility. One notices that there is no dynamic (non stationary) equation linked to the pressure calculation. Its existence would simplify the approach, as the same reasoning would apply to (7.78) and (7.79). We will see in the following how to calculate the pressure in order to insure a value of the velocity divergence coherent with the discretization errors due to the numerical approximation. It is in fact necessary to ask oneself what is the signification of imposing zero divergence on a computer that performs its calculations with floating point representation. There is very little probability that this divergence will be identically zero (0.0). Therefore, it is accepted that it is of the order of 10^{-2} or 10^{-3}, as a function of the global number of mesh points in the calculation being performed.

The pressure calculation can be avoided by resorting to the vorticity stream function formulation (in two dimensions) or the vorticity potential vector (in three dimensions). The vorticity $\boldsymbol{\omega}$ being defined by

$$\boldsymbol{\omega} = \mathbf{curl}\, \boldsymbol{v} \tag{7.80}$$

the application of the operator **curl** to (7.79) causes the pressure term to disappear as **curl** (**grad** p) = **0**, and gives the dynamic vorticity equation. Unfortunately, this procedure displaces the problem to the generation of vorticity at the walls. For this reason it is by far preferable to work with the velocity-pressure formulation (7.78)–(7.79), called the primitive variable formulation, because it is more general and covers all applications including free surface problems.

A first calculation strategy consists of treating the equations (7.78)-(7.79) in a coupled manner. For a two dimensional problem, this approach is excellent. Conversely, for three spatial dimensions, the bottleneck of the method comes from the solution of an algebraic system of large size and large band width if a direct solution method is used. The best choice in this case is to resort to iterative solution methods.

A second strategy implements "***splitting***" of the Navier-Stokes operator by a decoupled calculation of the velocities and the pressure. To this end, a Poisson equation is generated for the pressure by taking the divergence of (7.79) under the condition (7.78). As they are linear, the partial derivative with respect to time and the Laplacian operator commute with the divergence operator. No contribution comes from (7.78). The divergence of the pressure gradient generates the Laplacian; finally, the non linear term yields a source term for pressure in the Poisson equation also including the volume force term

$$\Delta p = \mathrm{div}(-(\rho(\mathbf{Grad}\,\boldsymbol{v})\cdot\boldsymbol{v})+\boldsymbol{F}_{\mathrm{v}}) \tag{7.81}$$

This second order equation can only be solved with the imposition of boundary conditions. These are obtained classically by applying a Neumann condition to the normal component of the Navier-Stokes equation:

$$\frac{\partial p}{\partial n} = \mathbf{grad}\,p\cdot\boldsymbol{n} = \boldsymbol{n}\cdot\left(\boldsymbol{F}_{\mathrm{v}}+\mu\Delta\boldsymbol{v}-\rho(\mathbf{Grad}\,\boldsymbol{v})\cdot\boldsymbol{v}-\rho\frac{\partial\boldsymbol{v}}{\partial t}\right) \tag{7.82}$$

In order to not confront all the difficulties at the same time, we will first study the Stokes problem, in which the inertial effects are neglected. This problem has the advantage of being linear and yet showing all the calculation difficulties of the pressure in primitive variables.

7.4.2 The stationary Stokes problem

The finite difference method

We seek to solve the coupled equations

$$\mathrm{div}\,\boldsymbol{v} = 0 \tag{7.83}$$

$$-\mathbf{grad}\,p+\mu\Delta\boldsymbol{v}+\boldsymbol{F}_{\mathrm{v}} = \mathbf{0} \tag{7.84}$$

In order to facilitate the presentation of the most classic discretizations, we refer first to the finite difference method. This will be performed in two dimensions with a regular grid, the three dimensional extension being obvious. We denote by d the spatial dimension of the problem ($d = 2$ or 3).

The collocation method puts the discretized velocities and pressures at the center of the calculation cells (fig. 7.14 a). Identifying the center of the cells with the index i in the x direction and by the index j in the y direction, equations (7.83) and (7.84) approximated by the second order centered schemes yield the discrete equations (see chap. 2)

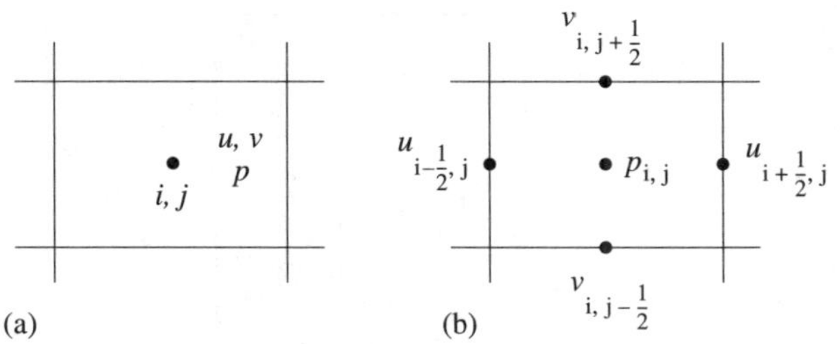

Figure 7.14 Grids for the velocity-pressure formulation: (a) collocated, (b) staggered

$$\frac{u_{i+1,j} - u_{i-1,j}}{2\Delta x} + \frac{v_{i,j+1} - v_{i,j-1}}{2\Delta y} = 0 \tag{7.85}$$

$$-\left(\frac{p_{i+1,j} - p_{i-1,j}}{2\Delta x}\right) + \mu\Delta_{i,j}u_{i,j} + F_{x,i,j} = 0 \tag{7.86}$$

$$-\left(\frac{p_{i,j+1} - p_{i,j-1}}{2\Delta y}\right) + \mu\Delta_{i,j}v_{i,j} + F_{y,i,j} = 0 \tag{7.87}$$

where the discrete Laplacian operator is obtained by second order finite differences:

$$\Delta_{i,j}u_{i,j} = \frac{u_{i+1,j} - 2u_{i,j} + u_{i-1,j}}{\Delta x^2} + \frac{u_{i,j+1} - 2u_{i,j} + u_{i,j-1}}{\Delta y^2} \tag{7.88}$$

The system of equations (7.85)–(7.87), assembled over the whole discretization, can be solved by a standard direct or iterative method (see chapter 4). However, the centered collocated discretization suffers from a well known deficiency: the appearance of a ***spurious pressure mode***. It is in fact an artifact of the numerical approximation, which is manifested by wiggles in the pressure field. This is the spurious "checker board" mode; as on a checker board, the pressure alternates between a positive and a negative value as the grid is regularly scanned in the two spatial directions (fig.7.15). The mode originates from the fact that the pressure calculation for the nodes with even $i+j$ is totally decoupled from that for the nodes with odd $i+j$. To convince oneself, it is sufficient to write the "divergence" operator for the right hand side of (7.81) with a second order centered approximation. For a calculation of the pressure in a node with even $i+j$, the velocities involved in the evaluation of the discrete divergence are located on the nodes with odd $i+j$ (inversely, the pressure nodes with odd $i+j$ are calculated from velocities at nodes with even $i+j$). There are thus two decoupled pressure networks, for which the pressure constants (which set the reference pressure in an incompressible calculation) can be completely different, causing the oscillations. The spurious modes do not affect the velocity field calculation, as it can easily be seen that they produce

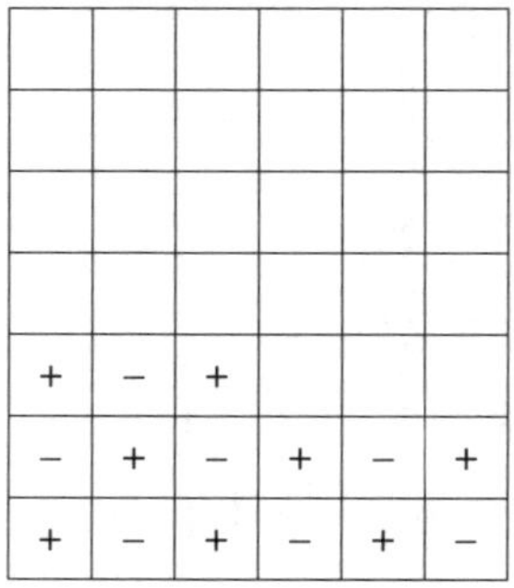

Figure 7.15 The "checker board" spurious pressure mode.

zero pressure gradient. A means to eliminate the "checker" mode is to introduce a dissipation term such as a biharmonic operator that will smooth the oscillations. This smoothing is particularly important for free surface problems where the pressure is a very sensitive variable, as well as in applications where it is necessary to evaluate the forces exerted by the fluid on the wall.

In order to avoid the presence of a spurious mode, a ***staggered grid*** can be employed (fig. 7.14 b) where the pressure is centered, the horizontal component of the velocity is defined on the left and right faces of the cell while the vertical component is attached to the upper and lower faces. The staggered grid was proposed by Harlow and Welch in 1965 for the ***MAC method*** (Marker And Cell) which was intended for numerical simulation of free surface flows. This method is very compact in the sense that the discrete velocities are distributed closely around the pressure nodes. It yields compact approximations that involve very close neighboring points. The staggered grid has spatial convergence properties that make it an excellent choice. The discretized equations are written for a regular grid with the pressures in the center, indexed by i and j:

$$\frac{u_{i+1/2,j} - u_{i-1/2,j}}{\Delta x} + \frac{v_{i,j+1/2} - v_{i,j-1/2}}{\Delta y} = 0 \tag{7.89}$$

$$-\left(\frac{p_{i+1,j} - p_{i,j}}{\Delta x}\right) + \mu \Delta_{i+1/2,j} u_{i+1/2,j} + F_{x,i+1/2,j} = 0 \tag{7.90}$$

$$-\left(\frac{p_{i,j+1} - p_{i,j}}{\Delta y}\right) + \mu \Delta_{i,j+1/2} v_{i,j+1/2} + F_{y,i,j+1/2} = 0 \tag{7.91}$$

The discrete Laplacian $\Delta_{i+1/2,j}\, u_{i+1/2,j}$ is defined by analogy with relation (7.88). The generalization of the staggered grid in curvilinear coordinates is more elaborate than that of the collocated mesh. This is why the latter, despite its inconveniences, is preferred for engineering problems with complex geometry.

The finite element method

The finite element method is traditionally based on the Galerkin weighted residuals method (sect. 3.3). For problems posed with homogeneous Dirichlet kinematic boundary conditions, the weak formulation of the problem (7.83)–(7.84) is established by the scalar product of (7.84) with a test function $\boldsymbol{w}$ which is zero at the boundaries, and of (7.83) with a scalar function q that is square integrable. Integration by parts is performed using Green's formula, such that

$$-\mu\int_{\Omega}\Delta\boldsymbol{v}\cdot\boldsymbol{w}\,\mathrm{d}V = -\mu\int_{\Omega}\frac{\partial^2 v_\mathrm{i}}{\partial x_\mathrm{j}\partial x_\mathrm{j}}w_\mathrm{i}\,\mathrm{d}V =$$

$$= -\mu\int_{\partial\Omega}w_\mathrm{i}\frac{\partial v_\mathrm{i}}{\partial x_\mathrm{j}}n_\mathrm{j}\,\mathrm{d}S + \mu\int_{\Omega}\frac{\partial v_\mathrm{i}}{\partial x_\mathrm{j}}\frac{\partial w_\mathrm{i}}{\partial x_\mathrm{j}}\,\mathrm{d}V = \mu\int_{\Omega}\frac{\partial v_\mathrm{i}}{\partial x_\mathrm{j}}\frac{\partial w_\mathrm{i}}{\partial x_\mathrm{j}}\,\mathrm{d}V$$

and

$$\int_{\Omega}(\mathbf{grad}p)\cdot\boldsymbol{w}\,\mathrm{d}V = \int_{\Omega}\frac{\partial p}{\partial x_\mathrm{i}}w_\mathrm{i}\,\mathrm{d}V = \int_{\partial\Omega}pn_\mathrm{i}w_\mathrm{i}\,\mathrm{d}S - \int_{\Omega}p\frac{\partial w_\mathrm{i}}{\partial x_\mathrm{i}}\,\mathrm{d}V = -\int_{\Omega}p\frac{\partial w_\mathrm{i}}{\partial x_\mathrm{i}}\,\mathrm{d}V$$

More compactly, for the stationary Stokes equation (7.83)–(7.84)

$$\mu\int_{\Omega}(\mathbf{Grad}\,\boldsymbol{v}):(\mathbf{Grad}\,\boldsymbol{w})\,\mathrm{d}V - \int_{\Omega}p\,\mathrm{div}\,\boldsymbol{w}\,\mathrm{d}V = \int_{\Omega}\boldsymbol{F}_\mathrm{v}\cdot\boldsymbol{w}\,\mathrm{d}V \tag{7.92}$$

$$-\int_{\Omega}q\,\mathrm{div}\,\boldsymbol{v}\,\mathrm{d}V = 0 \tag{7.93}$$

Note the presence in equation (7.92) of a classic elliptic term related to the viscous dissipation and a pressure term. The latter can be considered as a constraint due to the incompressibility where the pressure plays the role of a Lagrange multiplier of the condition (7.83). It can be shown that the weak form (7.92)–(7.93) of the Stokes problem corresponds to a ***saddle point problem***, such that the energy functional modified by the pressure term is minimized for the velocity solution, and maximized for the pressure solution. The name saddle point for the problem comes from the geometric characteristics of the functional, which is convex for the minimization of the velocity and concave for the maximization of the pressure, offering thus the image of a saddle with double curvature.

However, a certain number of Newtonian fluid mechanical problems are posed with boundary conditions expressed in terms of stresses, as for example, in free surface flows ((7.70)–(7.71)), or in the exit section of the calculation domain ((7.58)–(7.59)). We will see later that for non Newtonian fluids, it is also easier to reason in terms of stress as a variable of the problem. In this case, the stationary continuous Stokes problem is written in the following form

$$\mathrm{div}\,\boldsymbol{v} = 0 \tag{7.94}$$

$$\mathbf{div}\,\boldsymbol{\sigma} + \boldsymbol{F}_{\mathrm{v}} = 0 \tag{7.95}$$

The weak formulation is obtained as before by multiplying (7.95) by a test function $\boldsymbol{w}$ that does not go to zero at the boundary, and (7.94) by a square integrable scalar function q. The continuity equation yields a relation identical to (7.93). It also leads to the following equality

$$\int_{\Omega}\left(\frac{\partial\sigma_{\mathrm{ij}}}{\partial x_{\mathrm{j}}} + F_{\mathrm{vi}}\right)w_{\mathrm{i}}\ \mathrm{d}V = \int_{\Omega}\frac{\partial}{\partial x_{\mathrm{j}}}(\sigma_{\mathrm{ij}}w_{\mathrm{i}})\ \mathrm{d}V - \int_{\Omega}\left(\sigma_{\mathrm{ij}}\frac{\partial w_{\mathrm{i}}}{\partial x_{\mathrm{j}}} - F_{\mathrm{vi}}w_{\mathrm{i}}\right)\mathrm{d}V = 0$$

With the divergence theorem, the first term of the right hand side is transformed to obtain

$$\int_{\partial\Omega}\sigma_{\mathrm{ij}}w_{\mathrm{i}}n_{\mathrm{j}}\ \mathrm{d}S - \int_{\Omega}\left(\sigma_{\mathrm{ij}}\frac{\partial w_{\mathrm{i}}}{\partial x_{\mathrm{j}}} - F_{\mathrm{vi}}w_{\mathrm{i}}\right)\mathrm{d}V = 0$$

or

$$\int_{\Omega}\boldsymbol{\sigma} : \mathbf{Grad}\,\boldsymbol{w}\ \mathrm{d}V = \int_{\Omega}\boldsymbol{F}_{\mathrm{v}}\cdot\boldsymbol{w}\ \mathrm{d}V + \int_{\partial\Omega}\overline{\boldsymbol{T}}\cdot\boldsymbol{w}\ \mathrm{d}S \tag{7.96}$$

where the symbol $\overline{\boldsymbol{T}}$ denotes the imposed stress vector at the boundary. Relation (7.96) expresses the equivalence between the weak form of the mechanical equilibrium equation in the sense of the principle of virtual power, and the Galerkin formulation as was seen in chapters 1 and 3 (sect. 3.4).

The domain Ω is covered with a finite element mesh. In order to discretize the integral equations (7.92)–(7.93), the velocities and pressures are respectively approximated by the relations (the index h indicates a discrete version)

$$\boldsymbol{v}_{\mathrm{h}}(\boldsymbol{x}) = \sum_{\mathrm{j}=1}^{\mathrm{Nnv}}\boldsymbol{v}^{\mathrm{j}}\psi_{\mathrm{j}}^{\mathrm{v}}(\boldsymbol{x}), \quad p_{\mathrm{h}}(\boldsymbol{x}) = \sum_{\mathrm{j}=1}^{\mathrm{Nnp}}p^{\mathrm{j}}\psi_{\mathrm{j}}^{\mathrm{p}}(\boldsymbol{x}) \tag{7.97}$$

where the velocities and pressures are defined at *Nnv* and *Nnp* nodes, respectively. Recall that the local velocity and pressure approximations were given by equations (3.31) and (3.32), respectively. The Galerkin formulation is obtained by writing (7.92)–(7.93) in the form

$$\mu \int_{\Omega_h} (\mathbf{Grad}\, \boldsymbol{v}_h) : (\mathbf{Grad}\, \boldsymbol{w}_h)\, dV - \int_{\Omega_h} p_h \operatorname{div} \boldsymbol{w}_h\, dV = \int_{\Omega_h} \boldsymbol{F}_v \cdot \boldsymbol{w}_h\, dV \tag{7.98}$$

$$-\int_{\Omega_h} q_h \operatorname{div} \boldsymbol{v}_h\, dV = 0 \tag{7.99}$$

where Ω_h is the finite element mesh; in two dimensions, $\boldsymbol{w}_h = (\psi_j^v(\boldsymbol{x}), \psi_j^v(\boldsymbol{x}))$, $j = 1, Nnv$ and $q_h = \psi_j^p(\boldsymbol{x})$, $j = 1, Nnp$. Careful examination of relation (7.98) shows that if the polynomial order of the velocity interpolant is arbitrarily chosen, the order of the pressure interpolant is automatically equal to one less than that of the velocity. Take the example of an interpolant of second order for the velocities. Then the integrand (**Grad** $\boldsymbol{v}_h$): (**Grad** $\boldsymbol{w}_h$) is also of second order. Desiring that p_h div $\boldsymbol{w}_h$ also be approximated by a second order polynomial, implies the choice of first-order for p_h, as div $\boldsymbol{w}_h$ is itself of first-order. This shows that the velocity and pressure interpolants can not be chosen at liberty. The discretizations must satisfy a compatibility condition, to which we will return below.

The matrix representation of the problem (7.98)–(7.99) is as follows for a two dimensional problem

$$\begin{bmatrix} \boldsymbol{A} & \boldsymbol{0} & -\boldsymbol{D}_1^T \\ \boldsymbol{0} & \boldsymbol{A} & -\boldsymbol{D}_2^T \\ -\boldsymbol{D}_1 & -\boldsymbol{D}_2 & \boldsymbol{0} \end{bmatrix} \begin{bmatrix} \boldsymbol{v}_{1,h} \\ \boldsymbol{v}_{2,h} \\ \boldsymbol{p}_h \end{bmatrix} = \begin{bmatrix} \boldsymbol{f}_1 \\ \boldsymbol{f}_2 \\ \boldsymbol{0} \end{bmatrix} \tag{7.100}$$

In equation (7.100), $\boldsymbol{v}_{1,h}$ is the global vector of the nodal values of the velocity component in the x direction; $\boldsymbol{v}_{2,h}$ is defined analogously for the y direction. The matrix $[\boldsymbol{A}]$ has the components

$$A_{mn} = \mu \int_{\Omega_h} \mathbf{grad}\, \psi_m^v \cdot \mathbf{grad}\, \psi_n^v\, dV, \quad m, n = 1, Nnv \tag{7.101}$$

while the matrices $[\boldsymbol{D}_1]$ and $[\boldsymbol{D}_2]$ are obtained with the relations

$$D_{1,mn} = \int_{\Omega_h} \psi_m^p \frac{\partial \psi_n^v}{\partial x}\, dV, \quad m = 1, Nnp, n = 1, Nnv \tag{7.102}$$

$$D_{2,mn} = \int_{\Omega_h} \psi_m^p \frac{\partial \psi_n^v}{\partial y}\, dV, \quad m = 1, Nnp, n = 1, Nnv \tag{7.103}$$

Note that the discrete gradients matrices are the transpose of those appearing in the discrete divergence. The discretization of the complete equation (7.96), with a stress term on the boundary yields analogous expressions, slightly more complex, to the matrix [A]. The source terms f_1 and f_2 are obtained from the product of the mass matrix [M], having components

$$M_{\mathrm{mn}} = \int_{\Omega_{\mathrm{h}}} \psi_{\mathrm{m}}^{\mathrm{v}} \psi_{\mathrm{n}}^{\mathrm{v}} \, \mathrm{d}V, \quad m, n = 1, Nnv \tag{7.104}$$

and the global vector of the nodal values of the vector F_{v} in the x and y directions, respectively.

To facilitate the rest of the presentation, the system (7.100) is rewritten generalized to d spatial dimensions in the form

$$\begin{pmatrix} A & -D^{\mathrm{T}} \\ -D & 0 \end{pmatrix} \begin{pmatrix} V \\ P \end{pmatrix} = \begin{pmatrix} f \\ 0 \end{pmatrix} \tag{7.105}$$

where [A] is a matrix composed of d positive definite diagonal blocks A of order Nnv, while [D] is rectangular of size $Nnp \times (d\ Nnv)$. The vector V is the global vector of nodal velocity components, and P is the global vector of nodal pressures.

For an incompressible fluid, in the absence of a free surface, the pressure is known within a constant which is fixed at a well chosen point in the mesh, generally at the boundary of the domain, or by imposing that the average pressure over the domain be such that

$$\int_{\Omega} p \, \mathrm{d}V = 0 \tag{7.106}$$

In order to be able to solve the matrix Stokes problem (7.105), it is necessary that it be regular. For that, the presence of spurious modes that produce null pressure gradients must be avoided. Therefore, the functional spaces – the approximations (7.97) – must be chosen in a compatible way. The theoretical studies refer to the work of the mathematicians F. Brezzi and I. Babuška. This compatibility condition is also known by the name ***inf-sup condition***. To respect it, the local form functions which are the global form functions $\psi_{\mathrm{j}}^{\mathrm{v}}$ and $\psi_{\mathrm{j}}^{\mathrm{p}}$, restricted to the element under consideration, can not be taken arbitrarily. We present below two simple elements that are often used; the books by Thomasset (1981), Girault-Raviart (1986), Gresho et al. (1998) and Pironneau (1988) offer a complete presentation of the finite element methods applied to the Navier-Stokes equations.

The first element commonly used is a triangle (fig. 7.16) on which the velocities are represented by a second order complete polynomial (fig. 3.9) and the pressures by a first-order complete polynomial (fig. 3.8). There are thus six nodal values for the velocities and three for the pressures. This element has the name $P2/P1$. The second element is a quadrangle (fig. 7.17) where the velocities are approximated by

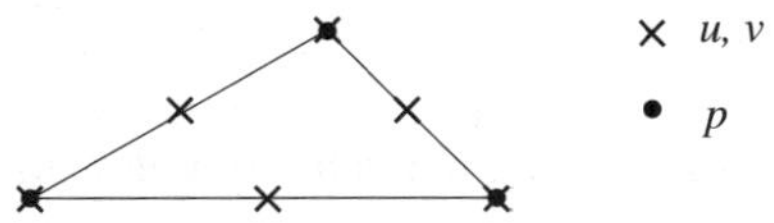

Figure 7.16 Triangular velocity-pressure element.

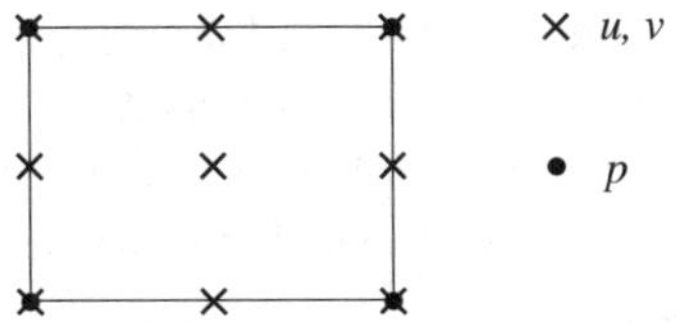

Figure 7.17 Quadrangular velocity-pressure element.

quadratic interpolants (fig. 3.12) and the pressures by linear interpolants (fig. 3.10). It is called $Q2/Q1$. It should be noted that the quadrangular element with linear velocities and constant pressure ($Q1/Q0$) does not satisfy the inf-sup condition, and has a checker board type spurious mode. However, filtering methods exist that permit the recovery of a non oscillating pressure field. The interested reader should refer to the article by Sani et al. (1981) given in the bibliography for this particular point outside the scope of this book.

The generalization of these elements to three spatial dimensions yields a tetrahedron with 10 nodal velocity values (30 components) and 4 for the pressures, and a parallelepiped with 27 velocity nodes and 8 pressure nodes. The latter element has 81 velocity unknowns and 8 pressure unknowns. It is thus very expensive in terms of memory space and computer time. That is one reason why the element with linear velocity and constant pressure is so popular for three dimensional applications despite the spurious pressure mode deficiency; it involves only 24 velocity and one pressure unknowns.

It can be shown that the triangular $P2/P1$ and quadrangular $Q2/Q1$ elements are globally of second order, which means that their spatial convergence rate is of order $O(h^2)$ where h is the spatial discretization size of the mesh.

The matrix system (7.105) can be solved by a direct algorithm such as Gauss elimination. However, the bandwidth of the system is such that direct solution induces a prohibitive cost for three dimensional problems. That is why modern algorithms use iterative solvers. In order to develop such an algorithm, we present ***Uzawa's method***. Multiplying the first relation of (7.105) by $[\boldsymbol{D}][\boldsymbol{A}]^{-1}$ and taking into account the second

$$[\boldsymbol{S}]\boldsymbol{P} = -[\boldsymbol{D}][\boldsymbol{A}]^{-1}\boldsymbol{f} \tag{7.107}$$

where $\boldsymbol{S}$ is the Uzawa operator defined by the equation

$$[\boldsymbol{S}] = [\boldsymbol{D}][\boldsymbol{A}]^{-1}[\boldsymbol{D}]^{\mathrm{T}} \tag{7.108}$$

Knowing $\boldsymbol{P}$ with (7.107), the velocity field can be evaluated with the relation

$$[\boldsymbol{A}]\boldsymbol{V} = \boldsymbol{f} + [\boldsymbol{D}]^{\mathrm{T}}\boldsymbol{P} \tag{7.109}$$

Note that $[\boldsymbol{S}]$ corresponds to a "zero order" operator, as it sequentially calculates a gradient $[\boldsymbol{D}]^{\mathrm{T}}$, the inverse of a Laplacian $[\boldsymbol{A}]^{-1}$ and a divergence $[\boldsymbol{D}]$. The inverse $[\boldsymbol{A}]^{-1}$ is obviously not calculated, as it would be a full matrix requiring a lot of memory space. Rather, the Laplace equations are solved, for which a preconditioned conjugate gradient technique (for example, by the diagonal) is a judicious choice (see chapter 4).

7.4.3 The non stationary Stokes problem

Now consider the coupled equations

$$\mathrm{div}\, \boldsymbol{v} = 0 \tag{7.110}$$

$$\rho \frac{\partial \boldsymbol{v}}{\partial t} = -\,\mathbf{grad}\, p + \mu \Delta \boldsymbol{v} + \boldsymbol{F}_{\mathrm{v}} \tag{7.111}$$

with the condition of null velocity on the border $\partial\Omega$ for the simplicity of the presentation. The initial condition is given by (7.52) and must respect the incompressibility condition (7.53). Note in passing that the linear equation (7.111) is parabolic and that in view of the diffusive damping character of the viscous term, its solutions can only decrease in time. It is none other than relation (7.11) written in dimensional variables.

We will present in the following a few integration algorithms that represent the state of the art. This will be done in the context of a finite element spatial discretization which is more satisfying from the conceptual viewpoint.

The non stationary Uzawa algorithm

The non stationary problem (7.110)–(7.111) is more complicated than the stationary case. We apply a first-order implicit Euler scheme for the temporal discretization. The following finite element formulation results:

$$\rho \int_{\Omega_{\mathrm{h}}} \frac{\boldsymbol{v}_{\mathrm{h}}^{\mathrm{t}+\Delta\mathrm{t}} - \boldsymbol{v}_{\mathrm{h}}^{\mathrm{t}}}{\Delta t} \boldsymbol{w}_{\mathrm{h}}\, \mathrm{d}V + \mu \int_{\Omega_{\mathrm{h}}} \mathbf{Grad}\, \boldsymbol{v}_{\mathrm{h}}^{\mathrm{t}+\Delta\mathrm{t}} : \mathbf{Grad}\, \boldsymbol{w}_{\mathrm{h}}\, \mathrm{d}V - \int_{\Omega_{\mathrm{h}}} p_{\mathrm{h}}^{\mathrm{t}+\Delta\mathrm{t}} \mathrm{div}\, \boldsymbol{w}_{\mathrm{h}}\, \mathrm{d}V$$

$$= \int_{\Omega_{\mathrm{h}}} \boldsymbol{F}_{\mathrm{v}}^{\mathrm{t}+\Delta\mathrm{t}} \cdot \boldsymbol{w}_{\mathrm{h}}\, \mathrm{d}V \tag{7.112}$$

$$-\int_{\Omega_h} q_h \operatorname{div} v_h^{t+\Delta t} \, dV = 0 \tag{7.113}$$

In matrix form, equations (7.112)–(7.113) are written for a two dimensional problem

$$\begin{bmatrix} \boldsymbol{H} & \boldsymbol{0} & -\boldsymbol{D}_1^{\mathrm{T}} \\ \boldsymbol{0} & \boldsymbol{H} & -\boldsymbol{D}_2^{\mathrm{T}} \\ -\boldsymbol{D}_1 & -\boldsymbol{D}_2 & \boldsymbol{0} \end{bmatrix} \begin{bmatrix} \boldsymbol{v}_{1,h}^{t+\Delta t} \\ \boldsymbol{v}_{2,h}^{t+\Delta t} \\ \boldsymbol{p}_h^{t+\Delta t} \end{bmatrix} = \begin{bmatrix} \boldsymbol{f}_1^{t+\Delta t} \\ \boldsymbol{f}_2^{t+\Delta t} \\ \boldsymbol{0} \end{bmatrix} \tag{7.114}$$

In this last relation, the matrix $[H]$ is defined by the equation

$$[\boldsymbol{H}] = \frac{\rho}{\Delta t}[\boldsymbol{M}] + [\boldsymbol{A}] \tag{7.115}$$

In (7.114), the explicit term $\rho/\Delta t\ [\boldsymbol{M}]\ \boldsymbol{V}^t$ has been transferred to the terms on the right hand side $\boldsymbol{f}_1^{t+\Delta t}$ and $\boldsymbol{f}_2^{t+\Delta t}$.

For d spatial dimensions, with a notation analogous to that of equation (7.105)

$$\begin{pmatrix} \boldsymbol{H} & -\boldsymbol{D}^{\mathrm{T}} \\ -\boldsymbol{D} & \boldsymbol{0} \end{pmatrix} \begin{pmatrix} \boldsymbol{V}^{t+\Delta t} \\ \boldsymbol{P}^{t+\Delta t} \end{pmatrix} = \begin{pmatrix} \boldsymbol{f}^{t+\Delta t} \\ \boldsymbol{0} \end{pmatrix} \tag{7.116}$$

Multiplying the first equation of (7.104) by $[\boldsymbol{D}][\boldsymbol{H}]^{-1}$

$$[\boldsymbol{S}]\boldsymbol{P}^{t+\Delta t} = -[\boldsymbol{D}][\boldsymbol{H}]^{-1}\boldsymbol{f}^{t+\Delta t} \tag{7.117}$$

$$[\boldsymbol{H}]\boldsymbol{V}^{t+\Delta t} - [\boldsymbol{D}]^{\mathrm{T}}\boldsymbol{P}^{t+\Delta t} = \boldsymbol{f}^{t+\Delta t} \tag{7.118}$$

where the non stationary Uzawa operator $[S]$ is defined by

$$[\boldsymbol{S}] = [\boldsymbol{D}][\boldsymbol{H}]^{-1}[\boldsymbol{D}]^{\mathrm{T}} \tag{7.119}$$

Examining relations (7.117) and (7.118), it is seen that after having obtained the pressure by solving (7.117), the pressure gradient in (7.118) can be evaluated and the equation – of Helmholtz type – solved for the velocity field.

For large values of $\mu\Delta t$, (recall that μ is incorporated in the definition of $[\boldsymbol{A}]$), the non stationary Uzawa operator goes to the stationary operator (ignoring the constants). For small values of $\mu\Delta t$, the situation is somewhat different, as $[S]$ goes to a pseudo-Laplacian $[E]$ defined by

$$[\boldsymbol{E}] = [\boldsymbol{D}][\boldsymbol{M}]^{-1}[\boldsymbol{D}]^{\mathrm{T}} \tag{7.120}$$

If the mass matrix is condensed to a diagonal – which is performed by summing all the coefficients on the same row of the matrix, assigning the result to the diagonal coefficient and zeroing the rest (see chapter 3) – equation (7.120) yields a Poisson equation for the pressure which serves as a preconditioner for the Uzawa operation. It should be noted that the pseudo-Laplacian $[\boldsymbol{E}]$ takes into account the velocity boundary conditions through $[\boldsymbol{M}]^{-1}$.

Block projection and LU factorization methods

Modern calculation algorithms tend to offer spatial and temporal errors of the same order of magnitude. As the non stationary Uzawa algorithm is temporally of first-order and spatially of second order, we propose a second order temporal method in this sub section. We choose an implicit Euler method of second order (see exercise 7.7.5). The acceleration term is then

$$\rho \int_{\Omega_h} \frac{3/2\boldsymbol{v}_h^{t+\Delta t} - 2\boldsymbol{v}_h^{t} + 1/2\boldsymbol{v}_h^{t-\Delta t}}{\Delta t} \tag{7.121}$$

while the other terms of equations (7.112) and (7.113) remain identical. Setting $\beta_2 = 3/2$ and taking into account (7.121), the global system matrix is written

$$\begin{pmatrix} \boldsymbol{H} & -\boldsymbol{D}^{T} \\ -\boldsymbol{D} & \boldsymbol{0} \end{pmatrix} \begin{pmatrix} \boldsymbol{V}^{t+\Delta t} \\ \boldsymbol{P}^{t+\Delta t} \end{pmatrix} = \begin{pmatrix} \boldsymbol{f}^{t+\Delta t} \\ \boldsymbol{0} \end{pmatrix} \tag{7.122}$$

with the definition of $[\boldsymbol{H}]$ given by the relation

$$[\boldsymbol{H}] = \beta_2 \frac{\rho}{\Delta t}[\boldsymbol{M}] + [\boldsymbol{A}] \tag{7.123}$$

and a modification of the right hand side $\boldsymbol{f}^{t+\Delta t}$ taking into account the explicit terms at times t and $t - \Delta t$.

The following presentation of the projection methods is performed on the matrix equation (7.122) which already incorporates the discrete boundary conditions of the problem. This permits the correct pressure boundary condition to be found *a posteriori* for the continuous equations (7.82), but also to propose the generalized *LU* factorization by blocks for the system

$$\begin{pmatrix} \boldsymbol{H} & -\boldsymbol{HQD}^{T} \\ -\boldsymbol{D} & \boldsymbol{0} \end{pmatrix} \begin{pmatrix} \boldsymbol{V}^{t+\Delta t} \\ \boldsymbol{P}^{t+\Delta t} \end{pmatrix} = \begin{pmatrix} \boldsymbol{f}^{t+\Delta t} \\ \boldsymbol{0} \end{pmatrix} \tag{7.124}$$

in which the choice of the arbitrary matrix $[\boldsymbol{Q}]$ determines the ***projection method***. Of course, $[\boldsymbol{Q}]$ must be close to $[\boldsymbol{H}]^{-1}$ to control the splitting error, defined as that produced when the momentum equations are not solved coupled to the incompressibility constraint. This splitting error shows up essentially in the global temporal order of the scheme when the intermediate steps are eliminated. Concretely it is seen that this splitting error results in first-order temporal schemes, even if each step of the split process is of second order.

Recall that the *LU* factorization is none other than Gaussian elimination (see chapter 4). The *LU* factorization by blocks of (7.124) is performed in two stages:

$$\begin{pmatrix} \boldsymbol{H} & \boldsymbol{0} \\ -\boldsymbol{D} & -\boldsymbol{D}\boldsymbol{Q}\boldsymbol{D}^{\mathrm{T}} \end{pmatrix} \begin{pmatrix} \boldsymbol{V}^{*} \\ \boldsymbol{P}^{\mathrm{t}+\Delta\mathrm{t}} \end{pmatrix} = \begin{pmatrix} \boldsymbol{f}^{\mathrm{t}+\Delta\mathrm{t}} \\ \boldsymbol{0} \end{pmatrix} \tag{7.125}$$

and

$$\begin{pmatrix} \boldsymbol{I} & -\boldsymbol{Q}\boldsymbol{D}^{\mathrm{T}} \\ \boldsymbol{0} & \boldsymbol{I} \end{pmatrix} \begin{pmatrix} \boldsymbol{V}^{\mathrm{t}+\Delta\mathrm{t}} \\ \boldsymbol{P}^{\mathrm{t}+\Delta\mathrm{t}} \end{pmatrix} = \begin{pmatrix} \boldsymbol{V}^{*} \\ \boldsymbol{P}^{\mathrm{t}+\Delta\mathrm{t}} \end{pmatrix} \tag{7.126}$$

It is seen that the intermediate velocity field $\boldsymbol{V}^{*}$ is not incompressible. The second stage (7.126) consists of projecting the field $\boldsymbol{V}^{*}$ with non zero divergence on the null divergence field $\boldsymbol{V}^{\mathrm{t}+\Delta\mathrm{t}}$. It is for this reason that the method is known as the projection method. The factorization being carried out on the discrete system including the boundary conditions, splitting errors of the operator that might have been induced by pressure boundary conditions are avoided. We follow the presentation of J. B. Perot (1993).

The first choice of $[\boldsymbol{Q}]$, the most obvious, is

$$[\boldsymbol{Q}] = [\boldsymbol{H}]^{-1} \tag{7.127}$$

In this case, the factored equations (7.125) and (7.126) are equivalent to the original system (7.124), and no decoupling error appears. This method is none other than the Uzawa algorithm, the performance of which depends on the efficiency of the calculation of the inverse of $[\boldsymbol{H}]$.

The second choice consists of taking

$$[\boldsymbol{Q}] = \frac{\Delta t}{\rho\beta_2}[\boldsymbol{M}]^{-1} \tag{7.128}$$

We are thus in the context of what is called the ***fractional step method***. With (7.125) and (7.126), we easily verify that the splitting error, introduced by the fact that $[\boldsymbol{Q}]$ is only an approximation of $[\boldsymbol{H}]^{-1}$ is obtained by the relation

$$-[\boldsymbol{D}]^{\mathrm{T}}\boldsymbol{P}^{\mathrm{t}+\Delta\mathrm{t}} - (-[\boldsymbol{H}][\boldsymbol{Q}][\boldsymbol{D}]^{\mathrm{T}}\boldsymbol{P}^{\mathrm{t}+\Delta\mathrm{t}}) = \frac{\Delta t}{\rho\beta_2}[\boldsymbol{A}][\boldsymbol{M}]^{-1}[\boldsymbol{D}]^{\mathrm{T}}\boldsymbol{P}^{\mathrm{t}+\Delta\mathrm{t}} = O(\Delta t) \tag{7.129}$$

As in equation (7.121) there is a factor $(1/\Delta t)$ which multiplies the velocity in the momentum equation, this implies that with (7.129), the local temporal error is of second order. The splitting method thus has first-order global behavior in time. In addition, this method presents a non zero decoupling error, even when the problem becomes stationary. This is a problem for industrial applications for which a stationary solution is often required. Therefore, the temporal behavior of the algorithm must be improved and the decoupling error eliminated.

A second order method in time can be obtained by modifying (7.124) to put it in the form

$$\begin{pmatrix} \boldsymbol{H} & -\boldsymbol{HQD}^{\mathrm{T}} \\ -\boldsymbol{D} & \boldsymbol{0} \end{pmatrix} \begin{pmatrix} \boldsymbol{V}^{\mathrm{t}+\Delta\mathrm{t}} \\ \boldsymbol{P}^{\mathrm{t}+\Delta\mathrm{t}} - \boldsymbol{P}^{\mathrm{t}} \end{pmatrix} = \begin{pmatrix} \boldsymbol{f}^{\mathrm{t}+\Delta\mathrm{t}} + \boldsymbol{D}^{\mathrm{T}}\boldsymbol{P}^{\mathrm{t}} \\ \boldsymbol{0} \end{pmatrix} \tag{7.130}$$

By the application of the block *LU* decomposition to (7.130) with $[\boldsymbol{Q}]$ given by (7.128), we are led to the ***pressure correction method*** (van Kan, 1986) which is a second order scheme. The decoupling error is evaluated by the relation

$$-[\boldsymbol{D}]^{\mathrm{T}}\boldsymbol{P}^{\mathrm{t}+\Delta\mathrm{t}} - (-[\boldsymbol{H}][\boldsymbol{Q}][\boldsymbol{D}]^{\mathrm{T}}(\boldsymbol{P}^{\mathrm{t}+\Delta\mathrm{t}} - \boldsymbol{P}^{\mathrm{t}})) - (-[\boldsymbol{D}]^{\mathrm{T}}\boldsymbol{P}^{\mathrm{t}})$$

$$= \frac{\Delta t}{\rho\beta_2}[\boldsymbol{A}][\boldsymbol{M}]^{-1}[\boldsymbol{D}]^{\mathrm{T}}(\boldsymbol{P}^{\mathrm{t}+\Delta\mathrm{t}} - \boldsymbol{P}^{\mathrm{t}}) = O(\Delta t^2) \tag{7.131}$$

The decoupling error becomes zero for a stationary problem as in this case, $\boldsymbol{P}^{\mathrm{t}+\Delta\mathrm{t}} - \boldsymbol{P}^{\mathrm{t}} = \boldsymbol{0}$.

7.4.4 The non stationary Navier-Stokes problem

The Navier-Stokes equations (7.78) and (7.79) are solved for an incompressible fluid. With respect to the non stationary Stokes problem, the difficulty of the non stationary Navier-Stokes problem comes from the non linear advective part of the inertial terms. This is classically performed by the Galerkin method. For incompressible fluids, in the context of the Galerkin formulation, the integration by parts of the non linear term is not performed. Then the approximation by finite elements defined by equation (7.97) with the summing convention for repeated indices, and using the symbol × to denote the tensor product, gives:

$$\rho\int_{\Omega_{\mathrm{h}}}((\mathbf{Grad}\boldsymbol{v}_{\mathrm{h}})\cdot\boldsymbol{v}_{\mathrm{h}})\cdot\boldsymbol{w}_{\mathrm{h}}\,\mathrm{d}V = \rho\boldsymbol{v}^{\mathrm{j}}\times\boldsymbol{v}^{\mathrm{k}}\int_{\Omega_{\mathrm{h}}}\psi_{\mathrm{i}}^{\mathrm{v}}\psi_{\mathrm{k}}^{\mathrm{v}}\mathbf{grad}\,\psi_{\mathrm{j}}^{\mathrm{v}}\,\mathrm{d}V,\ i, j, k = 1, Nnv \tag{7.132}$$

where $\boldsymbol{w}_{\mathrm{h}} = (\psi_{\mathrm{i}}^{\mathrm{v}}(\boldsymbol{x}), \psi_{\mathrm{i}}^{\mathrm{v}}(\boldsymbol{x}))$, $i = 1, Nnv$. We denote by $N_{\mathrm{L}}(\boldsymbol{v})\boldsymbol{v}$ the non linear operator (7.132) which depends on $\boldsymbol{v}$ such that

$$N_{\mathrm{L}}(\boldsymbol{v})\boldsymbol{v} = \rho(\mathbf{Grad}\,\boldsymbol{v})\cdot\boldsymbol{v} \tag{7.133}$$

Exercise 7.6 permits the verification of the form of the elements of the non linear system (7.132) for a two dimensional problem. With the advection term, a non linear algebraic system involving products of the velocity components appears. To avoid a Newton linearization, the non linear term can be calculated by an explicit extrapolation or by an Adams-Bashforth method (Ortega and Poole, 1981).

Again using the mass $[\boldsymbol{M}]$ and stiffness $[\boldsymbol{A}]$ matrices from the operator (7.105), relations (7.78) and (7.79) discretized in space lead to a semi discrete equation (time is continuous)

$$\rho[\boldsymbol{M}]\frac{\partial \boldsymbol{V}}{\partial t} = -[\boldsymbol{A}]\boldsymbol{V} - [\boldsymbol{N}_{\mathrm{L}}(\boldsymbol{V})]\boldsymbol{V} + [\boldsymbol{D}]^{\mathrm{T}}\boldsymbol{P} + \boldsymbol{f} \tag{7.134}$$

$$-[\boldsymbol{D}]\boldsymbol{V} = \boldsymbol{0} \tag{7.135}$$

Generally, the term $[\boldsymbol{D}]^{\mathrm{T}}\boldsymbol{P}$ is treated at the new time step in the temporal integration. If the first-order implicit Euler scheme is used for the Stokes problem and a second-order Adams-Bashforth scheme (abbreviated, AB2) for the non linear term, one obtains

$$\left.\begin{aligned} \rho[\boldsymbol{M}]\boldsymbol{V}^{\mathrm{t}+\Delta\mathrm{t}} &= \rho[\boldsymbol{M}]\boldsymbol{V}^{\mathrm{t}} - \Delta t[\boldsymbol{A}]\boldsymbol{V}^{\mathrm{t}+\Delta\mathrm{t}} + \Delta t[\boldsymbol{D}]^{\mathrm{T}}\boldsymbol{P}^{\mathrm{t}+\Delta\mathrm{t}} \\ &\quad - \Delta t(\tfrac{3}{2}[\boldsymbol{N}_{\mathrm{L}}(\boldsymbol{V}^{\mathrm{t}})]\boldsymbol{V}^{\mathrm{t}} - \tfrac{1}{2}[\boldsymbol{N}_{\mathrm{L}}(\boldsymbol{V}^{\mathrm{t}-\Delta\mathrm{t}})]\boldsymbol{V}^{\mathrm{t}-\Delta\mathrm{t}}) + \Delta t\boldsymbol{f}^{\mathrm{t}+\Delta\mathrm{t}} \\ -[\boldsymbol{D}]\boldsymbol{V}^{\mathrm{t}+\Delta\mathrm{t}} &= \boldsymbol{0} \end{aligned}\right\} \tag{7.136}$$

This scheme is first-order globally for the temporal error; improving its precision is not trivial. The combination of a second-order implicit Euler scheme with AB2 always yields a scheme that is globally of first-order, despite the fact that each scheme is individually of second-order: this is due to the splitting error.

Another very simple technique consists of applying an extrapolation scheme for the non linear terms. More precisely, the linear and non linear terms of (7.134) are treated by a second-order Euler implicit scheme (7.121):

$$\begin{aligned} \tfrac{3}{2}\rho[\boldsymbol{M}]\boldsymbol{V}^{\mathrm{t}+\Delta\mathrm{t}} &= 2\rho[\boldsymbol{M}]\boldsymbol{V}^{\mathrm{t}} - \tfrac{1}{2}\rho[\boldsymbol{M}]\boldsymbol{V}^{\mathrm{t}-\Delta\mathrm{t}} + \Delta t(-[\boldsymbol{A}]\boldsymbol{V}^{\mathrm{t}+\Delta\mathrm{t}} \\ &\quad + [\boldsymbol{D}]^{\mathrm{T}}\boldsymbol{P}^{\mathrm{t}+\Delta\mathrm{t}} - [\boldsymbol{N}_{\mathrm{L}}(\boldsymbol{V}^{\mathrm{t}+\Delta\mathrm{t}})]\boldsymbol{V}^{\mathrm{t}+\Delta\mathrm{t}} + \boldsymbol{f}^{\mathrm{t}+\Delta\mathrm{t}}) \end{aligned} \tag{7.137}$$

Then, rather than solve an equation that is non linear in $\boldsymbol{V}^{\mathrm{t}+\Delta\mathrm{t}}$ because of the presence of the $\boldsymbol{N}_{\mathrm{L}}$ term, which would require an iterative algorithm, the following extrapolation is used to approximate $\boldsymbol{N}_{\mathrm{L}}$

$$[\boldsymbol{N}_{\mathrm{L}}(\boldsymbol{V}^{\mathrm{t}+\Delta\mathrm{t}})]\boldsymbol{V}^{\mathrm{t}+\Delta\mathrm{t}} = 2[\boldsymbol{N}_{\mathrm{L}}(\boldsymbol{V}^{\mathrm{t}})]\boldsymbol{V}^{\mathrm{t}} - [\boldsymbol{N}_{\mathrm{L}}(\boldsymbol{V}^{\mathrm{t}-\Delta\mathrm{t}})]\boldsymbol{V}^{\mathrm{t}-\Delta\mathrm{t}} + O(\Delta t^2) \tag{7.138}$$

It can be shown that the combination of (7.137) and (7.138) does not produce a splitting error and that the precision of the scheme is globally of second order. These error estimations are obviously valid for algorithms with a constant time step Δt.

It should be underlined that the use of an explicit formula such as (7.138) requires that a numerical stability condition be respected by the value of the time step Δt (cf. sect. 7.5).

Finally, a very good transient algorithm is that of Gresho et al. (1980). It advances in time with an adaptive step to capture all of the dynamic complexity. For ease and conciseness, we will present it for a constant time step. The algorithm is of the predictor-corrector type; it is obtained by combination of two second-order temporal integration schemes. The second order Adams-Bashforth method plays the role of a predictor; its explicit character does not require the solution of the system. The trapezoidal method is the corrector. The matrix systems generated by the implicit trapezoidal method are solved by Newton linearization which requires the construction and solution of a single system per time step.

The trapezoid method applied to an ordinary differential equation

$$\dot{y} = \frac{\mathrm{d}y}{\mathrm{d}t} = f$$

gives

$$y^{t+\Delta t} = y^{t} + \frac{\Delta t}{2}(f(t+\Delta t) + f(t))$$

The unknown $\dot{y}$ can be updated to prepare the next time step by formally inverting the trapezoid rule

$$\dot{y}^{t+\Delta t} = \frac{2}{\Delta t}(y^{t+\Delta t} - y^{t}) - \dot{y}^{t} \tag{7.139}$$

The velocity predictor is calculated by AB2 such that

$$\boldsymbol{V}^{p,\,t+\Delta t} = \boldsymbol{V}^{t} + \frac{\Delta t}{2}(3\dot{\boldsymbol{V}}^{t} - \dot{\boldsymbol{V}}^{t-\Delta t}) \tag{7.140}$$

where the symbol $\dot{\boldsymbol{V}}$ denotes the acceleration which will be evaluated by a formula analogous to (7.139) with $\dot{\boldsymbol{V}}$ playing the role of $\dot{y}$.

The velocity corrector is based on the trapezoid rule (7.139) applied to equations (7.134) and (7.135)

$$\begin{pmatrix} \frac{2}{\Delta t}\rho\boldsymbol{M} + \boldsymbol{A} + \boldsymbol{N}_{L}(\boldsymbol{V}^{t+\Delta t}) & -\boldsymbol{D}^{T} \\ -\boldsymbol{D} & \boldsymbol{0} \end{pmatrix}\begin{pmatrix} \boldsymbol{V}^{t+\Delta t} \\ \boldsymbol{P}^{t+\Delta t} \end{pmatrix} = \begin{pmatrix} \rho\boldsymbol{M}\left(\frac{2}{\Delta t}\boldsymbol{V}^{t} + \dot{\boldsymbol{V}}^{t}\right) + \boldsymbol{f}^{t+\Delta t} \\ \boldsymbol{0} \end{pmatrix} \tag{7.141}$$

The nonlinear system (7.141) is solved by the Newton-Raphson method in one step. We restrict ourselves to a single iteration taking advantage of the fact that, in a temporal integration, the solution at the preceding time step is an excellent approximation of the new solution, which is even further improved by the predictor. Recalling the concepts introduced in section 4.5.2, the Jacobian matrix (or the tangent stiffness) is calculated based on the predictor. Setting $\boldsymbol{x} = (\boldsymbol{V}^{t+\Delta t}, \boldsymbol{P}^{t+\Delta t})^T$ and taking relation (4.101) for solution by Newton's method:

$$\boldsymbol{R}(\boldsymbol{x}) = [\boldsymbol{K}(\boldsymbol{x})]\boldsymbol{x} - \boldsymbol{f}(\boldsymbol{x}) = \boldsymbol{0} \tag{7.142}$$

In the case of (7.141), the matrix $[\boldsymbol{K}(\boldsymbol{x})]$ is that of the left hand side and $\boldsymbol{f}(\boldsymbol{x})$ is the matrix from the right hand side. The correction $\delta\boldsymbol{x}$ is defined by the relation

$$\delta\boldsymbol{x} = \boldsymbol{x} - \boldsymbol{x}^p \tag{7.143}$$

with

$$\boldsymbol{x}^p = (\boldsymbol{V}^{p,\,t+\Delta t}, \boldsymbol{0})^T \tag{7.144}$$

Note that $\boldsymbol{x}^p$ does not contain the predicted pressure field. This is normal, as the pressure variable appears linearly in (7.141). The Jacobian matrix $[\boldsymbol{J}]$ relative to the vector $[\boldsymbol{K}(\boldsymbol{x})]\boldsymbol{x}$ is given by the equation:

$$J_{ij} = \frac{\partial}{\partial x_j}[K_{ik}(\boldsymbol{x})x_k] = K_{ij}(\boldsymbol{x}) + \left(\frac{\partial K_{ik}(\boldsymbol{x})}{\partial x_j}\right)x_k \tag{7.145}$$

Newton's method applied to (7.142) is written

$$\left[\frac{\partial \boldsymbol{R}}{\partial \boldsymbol{x}}\right]\delta\boldsymbol{x} = -\boldsymbol{R}(\boldsymbol{x}^p) \tag{7.146}$$

with the matrix $[\partial\boldsymbol{R}/\partial\boldsymbol{x}] = [\boldsymbol{J}]$.

The linearized matrix $[\boldsymbol{J}]$ corresponding to (7.141) gives, with the index p indicating that the matrix construction is based on (7.140) to evaluate the contribution of the non linear term,

$$[\boldsymbol{J}]_p = \begin{pmatrix} \dfrac{2}{\Delta t}\rho\boldsymbol{M} + \boldsymbol{A} + \boldsymbol{N}_L(\boldsymbol{V}^{p,\,t+\Delta t}) + \boldsymbol{N}_L'(\boldsymbol{V}^{p,\,t+\Delta t}) & -\boldsymbol{D}^T \\ -\boldsymbol{D} & \boldsymbol{0} \end{pmatrix} \tag{7.147}$$

with the matrix $[\boldsymbol{N}_L']$, corresponding to the last term of equation (7.145), defined by the relation

$$\boldsymbol{N}_L'(\boldsymbol{V}^{p,\,t+\Delta t})\delta\boldsymbol{V} = \rho\,\mathbf{Grad}(\boldsymbol{V}^{p,\,t+\Delta t})\cdot\delta\boldsymbol{V} \tag{7.148}$$

The continuous equation analogous to (7.148) is easily obtained from the linearized expression of $(\mathbf{Grad}\, \boldsymbol{v}) \cdot \boldsymbol{v}$. For that, we define the velocity increase by the relation

$$\delta \boldsymbol{v} = \boldsymbol{v}^{t+\Delta t} - \boldsymbol{v}^{p} \tag{7.149}$$

Then, introducing $\boldsymbol{v}^{p} + \delta \boldsymbol{v}$ in the non linear term and only keeping the first-order terms in the increase

$$(\mathbf{Grad}\, \boldsymbol{v}) \cdot \boldsymbol{v} = (\mathbf{Grad}\, \boldsymbol{v}^{p}) \cdot \boldsymbol{v}^{p} + (\mathbf{Grad}\, \delta \boldsymbol{v}) \cdot \boldsymbol{v}^{p} + (\mathbf{Grad}\, \boldsymbol{v}^{p}) \cdot \delta \boldsymbol{v}$$

Taking into account the definition (7.133), the first term on the right hand side is the non linear term of the Navier-Stokes equation evaluated by the predictor and is denoted $\boldsymbol{N}_{L}(\boldsymbol{v}^{p})\boldsymbol{v}^{p}$; the second corresponds to $\boldsymbol{N}_{L}(\boldsymbol{v}_{p})\delta \boldsymbol{v}$; the third is given by $\boldsymbol{N}_{L}'(\boldsymbol{v}^{p})\delta \boldsymbol{v}$.
For a two dimensional problem, the matrix $[\boldsymbol{N}_{L}']$ is of size $2Nnv \times 2Nnv$. It is composed of four blocks according to the following equation:

$$[\boldsymbol{N}_{L}']_{ij}(\boldsymbol{V}) = \begin{bmatrix} \dfrac{\partial [\boldsymbol{N}_{L}]_{ik}(\boldsymbol{V})}{\partial u^{j}} u^{k} & \dfrac{\partial [\boldsymbol{N}_{L}]_{ik}(\boldsymbol{V})}{\partial v^{j}} u^{k} \\ \dfrac{\partial [\boldsymbol{N}_{L}]_{ik}(\boldsymbol{V})}{\partial u^{j}} v^{k} & \dfrac{\partial [\boldsymbol{N}_{L}]_{ik}(\boldsymbol{V})}{\partial v^{j}} v^{k} \end{bmatrix} \tag{7.150}$$

The expression of $[\boldsymbol{N}_{L}'(\boldsymbol{V})]_{ij}$ is found in exercise 7.7, with its components evaluated relative to the element Ω_{h}, for a two dimensional problem.

The complete algorithm for one time step follows the following stages:

1. Assuming $\boldsymbol{V}^{t}$, $\dot{\boldsymbol{V}}^{t}$ and $\dot{\boldsymbol{V}}^{t-\Delta t}$ are known, prediction of the velocities by (7.140).
2. Calculation of the Jacobian matrix (7.147) and the residual $\boldsymbol{R}(\boldsymbol{x}_{p})$ corresponding to (7.141).
3. The velocity and pressure fields at time $t + \Delta t$ are obtained by solving the system

$$[\boldsymbol{J}_{p}](\boldsymbol{x} - \boldsymbol{x}^{p}) = -\boldsymbol{R}(\boldsymbol{x}^{p}) \tag{7.151}$$

4. The acceleration vector is updated via the relation analogous to (7.126)

$$\dot{\boldsymbol{V}}^{t+\Delta t} = \frac{2}{\Delta t}(\boldsymbol{V}^{t+\Delta t} - \boldsymbol{V}^{t}) - \dot{\boldsymbol{V}}^{t} \tag{7.152}$$

5. Return to stage 1 to calculate the next time step.

7.4.5 The stationary equilibrium equations for non Newtonian fluids

The generalized Navier-Stokes, or non Newtonian equations are obtained by inserting the constitutive equation (7.14) with (7.15)-(7.18) and (7.20) into the equilibrium equation (1.42). The system of equations to solve is then written for the stationary case:

$$\rho(\mathbf{Grad}\,\boldsymbol{v})\cdot\boldsymbol{v} = \mathbf{div}\,\boldsymbol{\sigma}+\boldsymbol{F}_{\mathrm{v}} = -\mathbf{grad}\,p+\mathbf{div}\,\boldsymbol{T}_{\mathrm{v}}+\boldsymbol{F}_{\mathrm{v}} \tag{7.153}$$

$$\mathrm{div}\,\boldsymbol{v} = 0 \tag{7.154}$$

$$\boldsymbol{T}_{\mathrm{v}} = \boldsymbol{T}_{\mathrm{v}1}+2\mu_0\boldsymbol{d} \tag{7.155}$$

$$\boldsymbol{T}_{\mathrm{v}1}+\lambda_1\overset{\nabla}{\boldsymbol{T}}_{\mathrm{v}1} = 2\mu_1\boldsymbol{d} \tag{7.156}$$

A weak formulation of the problem (7.153)–(7.156) is obtained using the Galerkin method for the right hand side of (7.153) according to the principle of virtual power (7.96). Crochet, Keunings, Marchal and co-authors have considerably contributed to the numerical simulation of non Newtonian fluid flow. A summary can be found in the review article by Crochet *et al.* (1992). The method consists of retaining the extra-stresses as unknowns of the problem, rather than eliminating them by inserting (7.155) and (7.14) in (7.153) to end up with a more classic problem expressed in $(\boldsymbol{v}, p)$.

In the Galerkin technique, equation (7.153) will be multiplied by a test function $\boldsymbol{w}$, non zero on the boundary of the domain, equation (7.154) by a square integrable scalar function q, and the dynamic viscoelastic extra stress equation (7.156) by the vector test function $\boldsymbol{u}$, also square integrable. Then the mixed variational formulation is written:

Find the unknowns $(\boldsymbol{v}, p, \boldsymbol{T}_{\mathrm{v}1})$ such that

$$\int_{\Omega}(-p\boldsymbol{I}+\boldsymbol{T}_{\mathrm{v}}):\mathbf{Grad}\,\boldsymbol{w}\;\mathrm{d}V+\rho\int_{\Omega}((\mathbf{Grad}\,\boldsymbol{v})\cdot\boldsymbol{v})\cdot\boldsymbol{w}\;\mathrm{d}V$$

$$= \int_{\partial\Omega}\overline{\boldsymbol{T}}\cdot\boldsymbol{w}\;\mathrm{d}S+\int_{\Omega}\boldsymbol{F}_{\mathrm{v}}\cdot\boldsymbol{w}\;\mathrm{d}V \tag{7.157}$$

$$-\int_{\Omega}q\;\mathrm{div}\,\boldsymbol{v}\;\mathrm{d}V = 0 \tag{7.158}$$

$$\int_{\Omega}([\boldsymbol{T}_{\mathrm{v}1}+\lambda_1\overset{\nabla}{\boldsymbol{T}}_{\mathrm{v}1}-\mu_1(\mathbf{Grad}\boldsymbol{v}+(\mathbf{Grad}\boldsymbol{v})^{\mathrm{T}})]\cdot\boldsymbol{u}\;\mathrm{d}V) = 0 \tag{7.159}$$

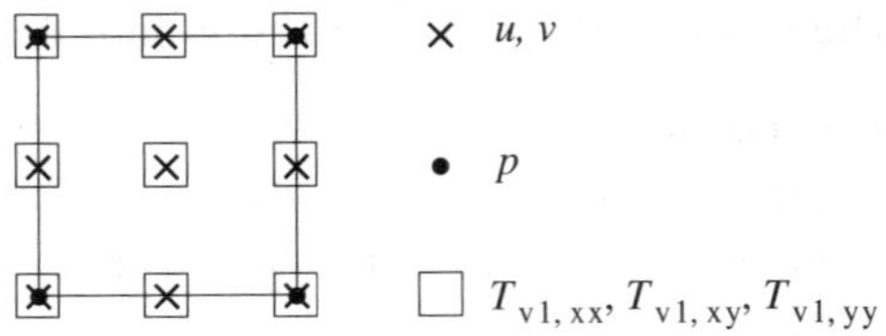

Figure 7.18 MIX1 element.

The discretization of problem (7.157)–(7.159) by the finite element method consists of approximating the velocities and pressures by relations (7.97) already used for the classic Navier-Stokes equations. An approximation of the tensor $\boldsymbol{T}_{v1}$ must be added

$$\boldsymbol{T}_{v1,h}(\boldsymbol{x}) = \sum_{j=1}^{Nnt} \boldsymbol{T}_{v1}^{j}\chi_j(\boldsymbol{x}) \tag{7.160}$$

where the χ_j are the interpolation functions of the extra stresses. One of the first elements proposed for the solution of the discrete version of (7.157)–(7.159) is called MIX1. In two spatial dimensions it is constructed from the quadrangular element *Q2/Q1* already shown in figure 7.17 for the case of a Newtonian fluid. Quadratic interpolants for the extra stresses (7.160) are added to the quadratic velocity and linear pressure interpolants. The result is that $Nnt = Nnv$ and the functions of form χ_j are identical to ψ_j^v. Figure 7.18 shows element MIX1. It can very easily be seen that this element is inconsistent. If $\lambda_1 = 0$, equation (7.159) shows that

$$\boldsymbol{T}_{v1} = \mu_1((\mathbf{Grad}\,\boldsymbol{v}) + (\mathbf{Grad}\,\boldsymbol{v})^{\mathrm{T}})$$

whereas, in the space of finite element approximations defined by MIX1, the approximation *Q2* of $\boldsymbol{T}_{v1,h}$ will be continuous and the linear approximation of $(\mathbf{Grad}\boldsymbol{v})_h$ will only be piecewise continuous. Marchal and Crochet (1987) thus proposed a mixed viscoelastic element which satisfies the inf-sup compatibility condition for discrete approximations of the Stokes problem. This element, shown in figure 7.19, is of the *Q2 / Q1* type for $\boldsymbol{v}, p$. In addition, the element is divided into 4×4 sub elements, on each of which the extra-stress is linearly approximated.

The use of mixed methods is somewhat expensive. For a two dimensional problem, the tensor $\boldsymbol{T}_{v1}$ has three components and the velocity vector has two. The quadrilateral element produces nine nodal variables for a scalar field discretized with biquadratic interpolants, and four for bilinear functions. The sixteen stress sub elements introduce twenty five nodal values per component. Therefore the MIX1 element requires 3(stresses) × 9 + 2(velocities) × 9 + 4(pressures) = 49 nodal values. The Marchal-Crochet element with its stress sub elements has 3 × 25 + 2 × 9 + 4 = 97 nodal values, almost twice as many unknowns.

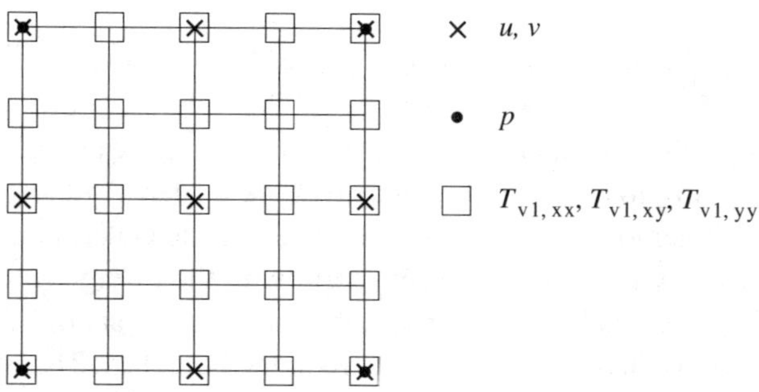

Figure 7.19 Marchal-Crochet viscoelastic element.

The convergence of the simulations poses great difficulty for viscoelastic fluid flows with high values of the Weissenberg number (7.31). In the scientific literature this is called the "high Weissenberg number problem". The difficulty shows up either as the non convergence of Newton's method applied to (7.157)–(7.159), or by the appearance of strong oscillations in the velocity, pressure and extra-stress fields. Tests on finer meshes lead to the same conclusions.

The sensitivity of these numerical simulations can be explained by three main causes. The first is the existence of thin boundary layers where the strong gradients imply very fine meshes. The second comes from stress concentrations in zones near geometric singularities, such as corners (2D) or edges (3D) of the calculation domain. The singularities generate values of stress that go to infinity; refining the mesh reinforces the effect. The exact nature of these singularities is most often unknown, as also their precise influence on the global configuration of the flow. Therefore, it is difficult to employ specific techniques that incorporate the form of the singularity in the numerical method. The third and last cause of imprecision results form the dominant advective character of the viscoelastic constitutive relations. It is well known that the Galerkin method, which is stable and optimal for elliptic problems dominated by viscosity, no longer has these qualities for problems where the fluid transport modeled by the non linear term of the Navier-Stokes equations becomes the preponderant physical phenomena. It is then necessary to use a finite element method of the Petrov-Galerkin type, which we will detail in section 7.5.

7.4.6 The turbulent Navier-Stokes equations

The dynamics of turbulent flows treated with the averaged Navier-Stokes model according to the principles proposed by Reynolds in 1895, are described by the incompressibility relation of the average velocity field (7.34) and the momentum equation obtained by the combination of (7.35), (7.36) and (7.42). We have

$$\rho\frac{\bar{\mathrm{d}}}{\mathrm{d}t}\bar{\boldsymbol{v}} = -\mathbf{grad}(\bar{p}+\tfrac{2}{3}\rho k)+\mathbf{div}(2(\mu+\mu_{\mathrm{T}})\bar{\boldsymbol{d}}) \tag{7.161}$$

to which is added (7.49) and (7.50), as well as the definition of the turbulent viscosity (7.43), to establish the standard $k-\varepsilon$ model.

Equation (7.161) is treated by the classic Galerkin method performing an integration by parts of the viscous term with the divergence theorem. For cases where the velocity is imposed zero on the boundary, equation (7.161) is multiplied by test functions, denoted $\boldsymbol{w}$, that are zero at the boundary. This yields the weak form:

$$\int_{\Omega}\rho\left(\frac{\bar{\mathrm{d}}}{\mathrm{d}t}\bar{\boldsymbol{v}}\right)\cdot\boldsymbol{w}\,\mathrm{d}V = \int_{\Omega}((\bar{p}+\tfrac{2}{3}\rho k)\boldsymbol{I}-2(\mu+\mu_{\mathrm{T}})\bar{\boldsymbol{d}}):\mathbf{Grad}\,\boldsymbol{w}\,\mathrm{d}V \tag{7.162}$$

The spatial discretization of (7.162) can be performed with the Q2 /Q1 element for the velocities and the pressures. Because $\bar{p}$ appears with k in the same term on the right hand side of (7.162), we deduce that an approximation $Q1$ for k is sufficient. This imposes the same approximation for ε. As the turbulent viscosity is dependent on the spatial coordinates, it should be incorporated in the definition of the discrete Laplacian. Its components are:

$$A_{\mathrm{mn}} = \int_{\Omega_{\mathrm{h}}}(\mu+\mu_{\mathrm{T}})\,\mathbf{grad}\,\psi_{\mathrm{m}}^{\mathrm{v}}\cdot\mathbf{grad}\,\psi_{\mathrm{n}}^{\mathrm{v}}\,\mathrm{d}V \quad m,n = 1, Nnv$$

This remark implies that for a transient algorithm the system matrix must be reconstructed for every time step, given that the viscosity is no longer invariant. The reader interested in more detail is referred to the book by Pironneau (1989).

7.5 ADVECTION AND DIFFUSION PHENOMENA

In chapter 2 (sections 2.2 and 2.3), the advection-diffusion equations were introduced for one spatial dimension stationary problems (2.1) and two dimensional non stationary problems (2.55). These equations, discretized by the finite difference method with centered schemes, showed oscillations (often in the neighborhood of downstream boundaries) when the grids were insufficiently fine. Finer grids can be avoided while smoothing the oscillations by "upwinding" the discretization and using the information that comes from the direction of the flow. Unfortunately, this skewing is tainted by numerical diffusion which fixes the effective Reynolds number of the calculation at a value sometimes much less than the physical value of the posed problem.

Remember that a temporal integration can only be performed by an explicit scheme when the stability conditions (2.38) and (2.40) are satisfied, which become in the case of Newtonian viscous fluids

$$\frac{v\Delta t}{h^2} \le \frac{1}{4} \quad \text{and} \quad \frac{v\Delta t}{h} \le 1 \tag{7.163}$$

for a planar problem where h denotes the minimum spatial step in the grid, the first stability condition (7.163) is analogous to that for the diffusion problem treated in chapter 2, whereas the second, known as the CFL condition, states that information can not move a distance greater than a grid spacing during one time step.

Unfortunately, respecting the stability conditions (7.163) does not permit the avoidance of two undesirable numerical phenomena when advection-diffusion problems with dominant advection are treated: numerical dissipation and dispersion.

7.5.1 Numerical dissipation and dispersion

For ease of analysis and comprehension of these phenomena, we will first examine one dimensional model problems. Given the pure advection (simple wave) equation to solve for concentration c

$$\frac{\partial c}{\partial t} + v\frac{\partial c}{\partial x} = 0 \qquad 0 < x < 1 \tag{7.164}$$

with $\quad c(x, t = 0) = g(x)$
and $\quad c(0, t) = f(t)$

The velocity v is the displacement or wave velocity, and is assumed constant. Setting $v = 1$, $g(x) = \exp(-(x/2)^2)$ and $f(t) = \exp(-(t/2)^2)$, the exact solution of problem (7.164) is given by

$$c(x, t) = \exp\left(-\frac{(x-t)^2}{4}\right) \tag{7.165}$$

which corresponds to a Gaussian curve that traverses the interval [0, 1] without being dampened; there is no dissipative term in equation (7.164).

Using the upwind scheme, of first-order in time and space, and assuming the discretization points are numbered from upstream to downstream,

$$\frac{c_i^{t+\Delta t} - c_i^t}{\Delta t} + v\frac{c_i^t - c_{i-1}^t}{h} = 0 \tag{7.166}$$

Expanding each of the terms of (7.166) in Taylor series:

$$\frac{c_i^{t+\Delta t} - c_i^t}{\Delta t} = \frac{\partial c_i^t}{\partial t} + \frac{\Delta t}{2}\frac{\partial^2 c_i^t}{\partial t^2} + O(\Delta t^2)$$

$$\frac{c_i^t - c_{i-1}^t}{h} = \frac{\partial c_i^t}{\partial x} - \frac{h}{2}\frac{\partial^2 c_i^t}{\partial x^2} + O(h^2)$$

Inserting these expressions in (7.166) and using the advection equation (7.164), the modified differential equation is obtained which is solved by the discrete scheme:

$$\frac{\partial c_i^t}{\partial t} + v\frac{\partial c_i^t}{\partial x} = \frac{1}{2}\frac{\partial^2 c_i^t}{\partial x^2}(vh - v^2\Delta t) + O(h^2, \Delta t^2) \tag{7.167}$$

The symbol O in (7.167) signifies that the terms are proportional to h^2 and Δt^2. Note that if $v\Delta t = h$, the scheme (7.167) is of second order. In addition, the upwind scheme introduced a numerical diffusion term which dampens the oscillations which can arise in the calculation. In (7.167), it is seen that the numerical (kinematic) viscosity of the upwind scheme, taking into account (7.163-2), is positive, with the value

$$\nu_{\text{num}} = \frac{v}{2}(h - v\Delta t)$$

The upwind scheme is intrinsically dissipative. If the CFL scheme were not respected, the numerical viscosity would be negative and the scheme would be unstable.

Consider now the implicit Crank-Nicolson scheme which is unconditionally stable and uses the second order centered finite differences:

$$\frac{c_i^{t+\Delta t} - c_i^t}{\Delta t} + \frac{v}{2}\left(\frac{c_{i+1}^{t+\Delta t} - c_{i-1}^{t+\Delta t}}{2h} + \frac{c_{i+1}^t - c_{i-1}^t}{2h}\right) = 0 \tag{7.168}$$

The modified differential equation corresponding to (7.168) is given by:

$$\frac{\partial c_i^t}{\partial t} + v\frac{\partial c_i^t}{\partial x} = \left(-\frac{v^3}{12}\Delta t^2 - \frac{vh^2}{6}\right)\frac{\partial^3 c_i^t}{\partial x^3} + O(\Delta t^3, h^3) \tag{7.169}$$

It is seen that the "viscous" term has disappeared. However, the appearance of the term in $\partial^3 c/\partial x^3$ gives a dispersive equation. This new concept requires some explanation. It is well known that an advection equation (7.164) has a solution in the form of a plane wave

$$c = a\exp(\text{i}(\alpha x - \omega t)) \tag{7.170}$$

where a is a constant, α the spatial frequency – called the wave number – and ω the angular frequency. The phase velocity defined by $\omega(\alpha)/\alpha$ is equal to v in the case of a simple wave and does not depend on the wave number. A wave is said to be dispersive when it is obtained by the superposition of many harmonics of (7.170) with different values of α. If the phase velocity ω/α is not the same for all values of α, that is, if $\omega \neq v\alpha$ with v constant, then the harmonics propagate at different velocities for different values of α; they disperse.

The Fourier stability analysis of a finite difference scheme in chapter 2 ((2.39) and (2.40)) imposed

$$|R_\alpha| \le 1 \tag{7.171}$$

The ratio R_α, also known by the name amplification factor, is defined by the relation

$$R_\alpha = \frac{C_\alpha^{t+\Delta t}}{C_\alpha^t} \text{ for } \alpha \in [0, \pi] \tag{7.172}$$

with the notation $C_\alpha^{t+\Delta t}$ for the amplitude of the harmonic at wave number α.

The scheme (7.168) does not have numerical dissipation, which is eminently desirable for an advection problem. However, our numerical solution will still not be correct, because of the phase error. It can be said in general that the study of the modified differential equation gives information on the type of error introduced by the scheme. If the main truncation error introduces even order derivatives, $\partial^2 c/\partial x^2$, $\partial^4 c/\partial x^4$, ..., it will create amplitude problems, and thus introduces numerical dissipation. If the error is in the odd terms $\partial^3 c/\partial x^3$, $\partial^5 c/\partial x^5$, ..., it affects the phase and provokes numerical dispersion. Applying the Fourier method described in section 2.2.5 to the upwind (7.166) and Crank-Nicolson (CN) (7.168) schemes, we obtain the respective amplification coefficients:

$$R_{\alpha,\,\text{upwind}} = 1 - 2\frac{v\Delta t}{h}\sin^2\frac{\alpha}{2} - \mathrm{i}\frac{v\Delta t}{h}\sin\alpha \tag{7.173}$$

$$R_{\alpha,\,\text{CN}} = \frac{1 - \mathrm{i}(v\Delta t/2h)\sin\alpha}{1 + \mathrm{i}(v\Delta t/2h)\sin\alpha} \tag{7.174}$$

One easily calculates that

$$|R_\alpha|_{\text{upwind}} = \left(1 - 4\left(1 - \frac{v\Delta t}{h}\right)\frac{v\Delta t}{h}\sin^2\frac{\alpha}{2}\right)^{1/2} \tag{7.175}$$

and

$$|R_{\alpha,\,\text{CN}}| = 1 \tag{7.176}$$

It is seen that for $\alpha = \pi$ and with the Courant number defined by $v\Delta t/h$ equal to 1/2, the upstream scheme eliminates all the spatial phenomena of wavelength $2h$ which are the smallest periodic phenomena that can be represented on a grid with spacing h. In this case, the numerical dissipation is maximal and dampens all the small geometric structures of the flow configuration. Conversely, the Crank-Nicolson is not dissipative.

However, both schemes are dispersive, as can be seen by examining the phase angle after a time step Δt. The exact solution shows a phase change of angle $\varphi = -\omega\Delta t = -v\alpha\Delta t$, whereas for the two schemes under consideration

$$\varphi_{\text{upwind}} = \tan^{-1}\frac{(-v\Delta t/h)\sin\alpha}{1-(2v\Delta t/h)\sin^2\alpha/2} \tag{7.177}$$

$$\varphi_{\text{CN}} = \tan^{-1}\frac{(-v\Delta t/h)\sin\alpha}{1-(v^2\Delta t^2/4h^2)\sin^2\alpha} \tag{7.178}$$

Figures 7.20 and 7.21 show the relative errors $\varphi_{\text{scheme}}/\varphi_{\text{exact}}$ for each discretization as a function of α and $v\Delta t/h$. It is seen that for most cases, the ratio $\varphi_{\text{scheme}}/\varphi_{\text{exact}}$ is less than one. This is the sign of a numerical solution with phase delayed with respect to the exact solution.

It was seen in section 2.2.3 that the upwind scheme can be improved by employing a first derivative skewed by a factor θ that is obtained by linear combination of the upstream and downstream first derivatives:

$$c_{\text{i}}' = (1-\theta)c_{\text{i+}}' + \theta c_{\text{i-}}' \tag{7.179}$$

with an optimal θ ((2.26)) given by

$$\theta = \tfrac{1}{2} + \tfrac{1}{2}\coth\frac{Pe_{\text{h}}}{2} - \frac{1}{Pe_{\text{h}}} \tag{7.180}$$

Unfortunately, this scheme can not be generalized in a satisfactory manner to many spatial dimensions. Even if one employs a θ_{x} and θ_{y} in the two spatial direc-

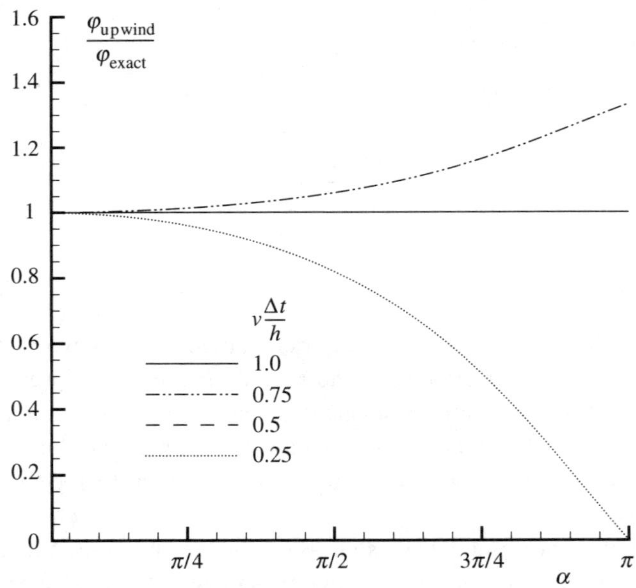

Figure 7.20 Evolution of the ratio $\varphi/\varphi_{\text{exact}}$ as a function of α for the upwind scheme.

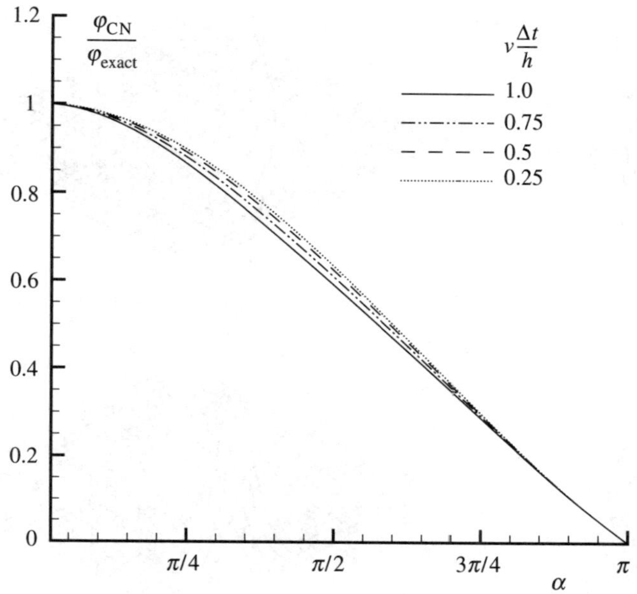

Figure 7.21 Evolution of the ratio φ/φ_{exact} as a function of α for the Crank-Nicolson scheme.

tions for planar problems, with the analogous expressions to (7.180) for each direction, the upstream discretization (and even Crank-Nicolson) is affected by numerical dissipation and dispersion. One can convince oneself by solving a severe test problem on which modelers must check their favorite schemes. It is the case of rigid body rotation, at constant angular velocity, of a pollution spot on a water surface. The flow is planar, described by the equation

$$\frac{\partial c}{\partial t} + u\frac{\partial c}{\partial x} + v\frac{\partial c}{\partial y} = 0 \qquad (7.181)$$

with c the concentration, u and v the given velocity components. The pollution concentration decreases from a maximum value of 1 at the center, to 0 at the edges. After a revolution, the pollution spot should return to the initial condition. Figure 7.22 shows the calculation results for two cases obtained on a domain $[-16,16] \times [-16,16]$ with a 32×32 regular mesh after one complete revolution. The radius of the cone of concentration is 4. The Courant number at the peak of the cone is 1/2 for case (a) and 1 for case (b). For both cases, the isoconcentration lines are shown on the left with continuous lines for positive values 0.1, 0.3, 0.5 and 0.7 of the calculated solution and dotted lines for the negative values –0.05, –0.1 and –0.2. A dashed line at concentration 0.1 of the exact solution is shown for comparison. A three dimensional perspective plot of the concentration is shown on the right. Despite a fourth-order finite difference spatial scheme and a second-order temporal scheme, the dispersed waves of the numerical solution are visible. As the cone

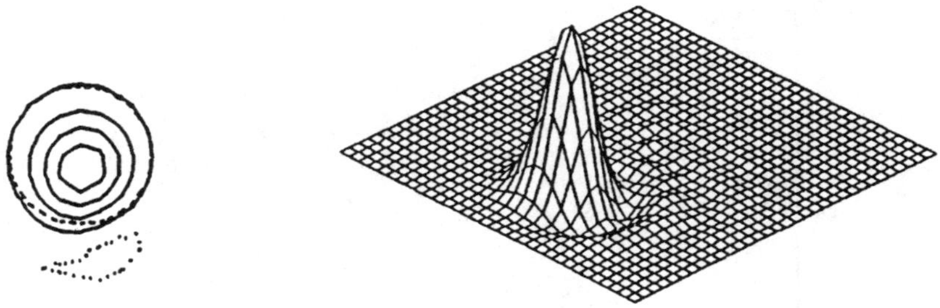

(a) max = 86, min = 9

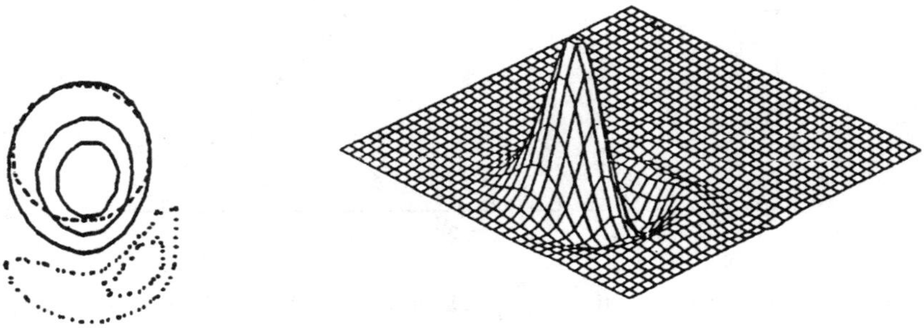

(b) max = 70, min = 14

Figure 7.22 Rotating concentration cone problem.

rotates clockwise, the waves essentially lag in phase. The max and min values shown correspond to the largest positive and negative values in percent. Besides the loss of precision associated with the dispersion, the maximum values are reduced by the numerical dissipation.

To circumvent these defects, we have two options: increase the precision of the schemes or reduce the grid spacing. The latter solution is the most often practiced, notably with adaptive grid methods which are not discussed in this book. The precision of the upwind scheme can be improved by incorporating the main truncation term

$$\frac{\partial c_i^t}{\partial x} = \frac{c_i^t - c_{i-1}^t}{h} + \frac{1}{2h}(c_{i+1}^t - 2c_i^t + c_{i-1}^t)$$

7.5.2 The Petrov-Galerkin method

The Galerkin method leads to centered discretizations. To generate upwind finite element schemes, the Petrov-Galerkin method is implemented, consisting of using test functions φ different from the interpolation functions ψ to project the residual R, (cf. (3.55)–(3.56)). In order to introduce this method in a simple manner, we first treat the one dimensional stationary advection-diffusion problem, for the concentration c of a pollutant or a chemical species defined on the interval [0, 1]:

$$v\frac{\partial c}{\partial x} = D\frac{\partial^2 c}{\partial x^2} \qquad v > 0 \tag{7.182}$$

with the boundary conditions

$$c(0) = 0 \qquad c(1) = 1 \tag{7.183}$$

As shown in chapter 2, the exact solution of the problem (7.182), (7.183) is given by

$$c(x) = \frac{\exp(Pe\ x) - 1}{\exp(Pe) - 1} \tag{7.184}$$

where the Péclet number Pe is defined (with $L = 1$):

$$Pe = \frac{v}{D} \tag{7.185}$$

Using the same notation as in section 3.2.3, the Galerkin linear finite element method applied to relation (7.182) gives the equation (see exercise 7.8)

$$D\left(-\frac{c^{i-1}}{\Delta x_l} + \left(\frac{1}{\Delta x_l} + \frac{1}{\Delta x_r}\right)c^i - \frac{c^{i+1}}{\Delta x_r}\right) + \frac{v}{2}(c^{i+1} - c^{i-1}) = 0 \tag{7.186}$$

On the reference element [–1, +1], the modified weight functions for the Petrov-Galerkin method are defined by the following relations, where a quadratic interpolant equal to zero at $\xi = \pm 1$ is superimposed on the linear basis functions

$$\begin{aligned}
\varphi_1^r(\xi) &= \psi_1^r(\xi) - \frac{3\Theta}{4}(1 - \xi^2) = \tfrac{1}{2}(1 - \xi) - \frac{3\Theta}{4}(1 - \xi^2) \\
\varphi_2^r(\xi) &= \psi_2^r(\xi) + \frac{3\Theta}{4}(1 - \xi^2) = \tfrac{1}{2}(1 + \xi) + \frac{3\Theta}{4}(1 - \xi^2) \\
&-1 \le \xi \le 1
\end{aligned} \tag{7.187}$$

with Θ being the upwinding parameter. When the linear approximations of the solution of (7.182) are assembled, the hat function encountered in chapter 3 is obtained. The function φ is superimposed on this, which gives more importance to the information coming from upstream than from downstream (figure 7.23). Taking the parameter Θ to have the same sign as c^i, the test functions (7.187) are "sucked" upstream. In exercise 7.9, the Petrov-Galerkin discretization of (7.182) is calculated, to obtain:

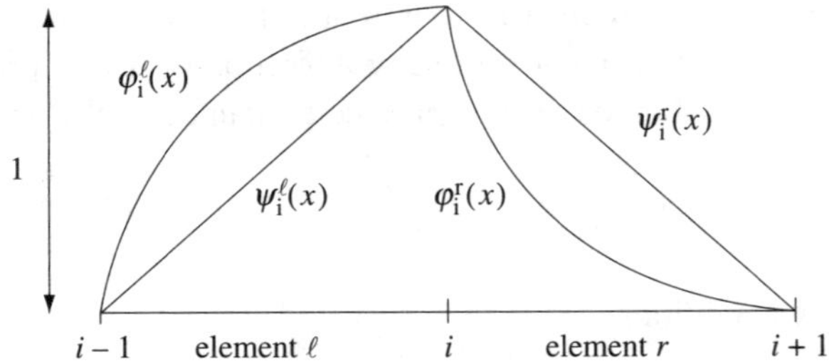

Figure 7.23 Hat function $\psi_i(x)$ and modified function $\varphi_i(x)$ for the Petrov-Galerkin method.

$$D\left(-\frac{c^{i-1}}{\Delta x_\ell}+\left(\frac{1}{\Delta x_\ell}+\frac{1}{\Delta x_r}\right)c^i-\frac{c^{i+1}}{\Delta x_r}\right)$$

$$+v\left(-\frac{1+\Theta}{2}c^{i-1}+\Theta c^i+\frac{1-\Theta}{2}c^{i+1}\right)=0 \tag{7.188}$$

which is identical to the relation found with the FDM.

Then, it can easily be shown that the exact solution of the problem is obtained at the mesh nodes if the parameter Θ equals

$$\Theta = \coth\frac{Pe_h}{2}-\frac{2}{Pe_h} \tag{7.189}$$

This solution is identical to the right hand side of (2.26) obtained for finite differences.

7.5.3 Streamline Upwind Methods (SU and SUPG)

Take again the problem of chemical diffusion given by equation (3.52) and solved by the finite element method in many spatial dimensions. Conforming to the preceding paragraph, the *SU* (streamline upwind) and *SUPG* (streamline upwind Petrov-Galerkin) methods consist of perturbing the test functions to give more importance to the upstream regions of the velocity vector than those downstream and thus to better take into account the hyperbolic character of the operator.

Recall the weak form of (3.52) written ((3.59)):

$$\int_\Omega \varphi \boldsymbol{v}\cdot\mathbf{grad}\,c\;\mathrm{d}V+\int_{\partial\Omega_N}\varphi\alpha(c-\bar{c}_a)\;\mathrm{d}S+\int_\Omega D\,\mathbf{grad}\,\varphi\cdot\mathbf{grad}\,c\;\mathrm{d}V = 0\quad\forall\varphi\in E_{\varphi 0} \tag{7.190}$$

For the advection term corresponding to the first integral of (7.190), the relative test function at node i is taken as

$$\varphi_i = \psi_i+\Theta\,\boldsymbol{v}\cdot\mathbf{grad}\,\psi_i \tag{7.191}$$

With respect to the Galerkin method for which the discrete equation is given by (3.62), an additional advection term is obtained:

$$\int_{\Omega} \Theta(\boldsymbol{v} \cdot \mathbf{grad}\, \psi_{\mathrm{i}})(\boldsymbol{v} \cdot \mathbf{grad}\, c)\, \mathrm{d}V = \int_{\Omega} \Theta\, (\boldsymbol{v} \times \boldsymbol{v})\mathbf{grad}\, c \cdot \mathbf{grad}\, \psi_{\mathrm{i}}\, \mathrm{d}V \quad (7.192)$$

This new term is clearly an anisotropic diffusion term directed with the flow. The associated Péclet number is called the mesh Péclet number, and is given by $Pe_{\mathrm{h}} = h/(\Theta\,|\mathbf{v}|)$ with h, the average parameter characteristic of the size of the elements. As $Pe_{\mathrm{h}} = 2$, is commonly chosen, one finally obtains $\Theta = h/(2|\boldsymbol{v}|)$. The form factor of the element can be taken into account in different ways. For example, it has been proposed for hexahedral elements:

$$\varphi_{\mathrm{i}} = \psi_{\mathrm{i}} + \frac{1}{2|\boldsymbol{v}|^2}\left(\sum_{\mathrm{p}=1}^{3} (h_{\mathrm{p}} v_{\mathrm{p}}^{\mathrm{c}})^2\right)^{1/2} \boldsymbol{v} \cdot \mathbf{grad}\, \psi_{\mathrm{i}} \quad (7.193)$$

where h_{p} is the size of the element in the direction x_{p} and $\boldsymbol{v}^{\mathrm{c}}$ the velocity interpolated at the center (centroid of the element). Figure 7.24 shows the modification of the weight function φ_{i} in one spatial dimension for a linear interpolant ψ_{i} (hat func-

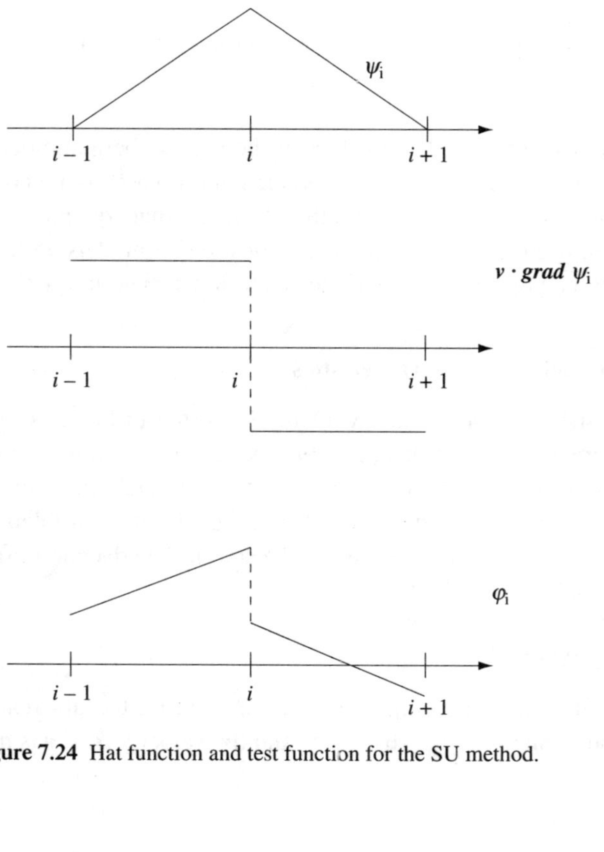

Figure 7.24 Hat function and test function for the SU method.

tion) and a positive velocity. The "streamline upwind" method (SU) is very stable. It should be noted that the modification of the weight function goes to zero as the mesh is refined ($h_p \to 0$). The SU method can be interpreted as a finite element method with a perturbation of $O(h)$. Nonetheless, it has the inconvenience of solving a problem modified with respect to the original. This leads to modification of the weight function on all terms of (3.62), not only the advection term, but also the other terms in the weak form. This consistent formulation is called "streamline-upwind/Petrov-Galerkin" (SUPG). It can be shown that the perturbation of φ_i does not modify the approximation of the diffusive term when the interpolant is (multi) linear.

When the SUPG method is applied to the Navier-Stokes equations the modification of the weight function affects the pressure term. This can cause bad spatial convergence of the method and poor convergence (or even divergence) if an iterative algorithm such as Newton's method is implemented. Therefore, one prefers to stay with the SU method for the Navier-Stokes equations.

The SU method was applied with much success by Marchal and Crochet to viscoelastic flows. In the SU formulation, equation (7.159) becomes

$$\int_{\Omega} \boldsymbol{T}_{v1} + \lambda_1 \overset{\nabla}{\boldsymbol{T}}_{v1} - \mu_1(\mathbf{Grad}\,\boldsymbol{v} + (\mathbf{Grad}\,\boldsymbol{v})^{\mathrm{T}}) \cdot \boldsymbol{u}\, \mathrm{d}V$$

$$+ \Theta \int_{\Omega} (\boldsymbol{v} \cdot \mathbf{Grad}\,\boldsymbol{T}_{v1}) \cdot (\boldsymbol{v} \cdot \mathbf{Grad}\,\boldsymbol{u})\, \mathrm{d}V = 0 \qquad (7.194)$$

With this technique large values of the Weissenberg number could be attained. However, the numerical solutions obtained in this case do not correspond to experimental observations. This qualitative – rather than quantitative – disagreement seems to indicate that the viscoelastic constitutive models are deficient, not representing all the physics that govern the experiments and industrial fluid flows.

7.5.4 The method of characteristics

This method allows the avoidance of some difficulties associated with the methods previously described, notably the stability condition imposed on the time step by the CFL criterion, as well as the artificial diffusion generated by the upwinding techniques so detrimental for pure advection problems.

We restrict ourselves initially to the stationary advection problem for the concentration field

$$\boldsymbol{v} \cdot \mathbf{grad}\, c(\boldsymbol{x}) = f(\boldsymbol{x}) \qquad (7.195)$$

for which the boundary conditions are imposed on the upstream boundary of the calculation domain $\partial\Omega_{\mathrm{entry}}$ (fig. 7.25) and the velocity field $\boldsymbol{v}$ is divergence free.

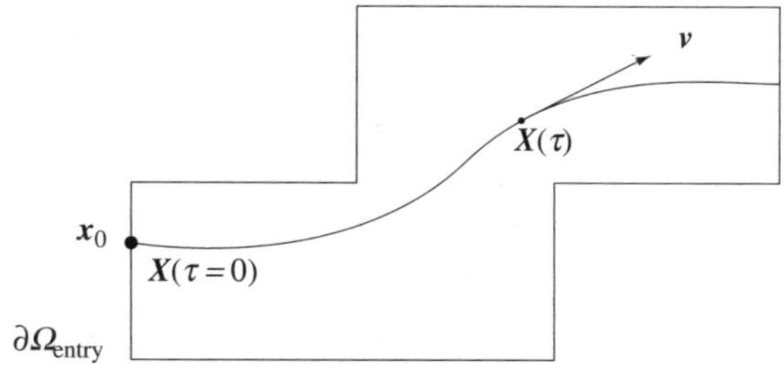

Figure 7.25 Streamline in stationary flow.

For a stationary problem, the streamlines and the material trajectories coincide. Recall that the streamline is the line tangent at all points to the velocity vector. Consider the streamline passing through point $\boldsymbol{x}_0$ of $\partial\Omega_{\text{entry}}$. It is defined by

$$\left\{ \boldsymbol{X} \in \Omega | \exists\ \tau > 0 \text{ such that } \frac{\mathrm{d}\boldsymbol{X}(\tau)}{\mathrm{d}\tau} = \boldsymbol{v}(\boldsymbol{X}(\tau)) \right\} \tag{7.196}$$

and

$$\boldsymbol{x}_0 = \boldsymbol{X}(\tau = 0) \tag{7.197}$$

The parameter τ can be seen as a curvilinear abscissa along the streamline or as the time that a particle takes to go from $\boldsymbol{x}_0$ at $t = 0$ to $\boldsymbol{x} = \boldsymbol{X}(\tau)$ at time τ. At every point on the streamline

$$\boldsymbol{v} \cdot \mathbf{grad} c = v_j \frac{\partial c}{\partial x_j} = \frac{\mathrm{d}X_j(\boldsymbol{x}_0, \tau)}{\mathrm{d}\tau} \frac{\partial c}{\partial x_j} = \frac{\mathrm{d}}{\mathrm{d}\tau}[c(\boldsymbol{X}(\boldsymbol{x}_0, \tau))] = f(\boldsymbol{x}_0, \tau) \tag{7.198}$$

In (7.198), the transformation from the Eulerian flow description to the particular derivative, and the stationarity of the problem have been employed (see chapter 1). It is seen by (7.198) that $\boldsymbol{X}$ represents the material or Lagrangian description of the flow that is obtained when a fluid particle is followed in its trajectory. Equations (7.196) and (7.197) thus describe the position of a particle $\boldsymbol{X}(\tau)$ that was carried by the velocity field $\boldsymbol{v}$ from the initial position $\boldsymbol{x}_0$. Relation (7.198) is easily integrated, giving

$$c(\boldsymbol{x}) = \int_0^{\tau} f(\boldsymbol{X}(\boldsymbol{x}, \tau))\ \mathrm{d}\tau + c(\boldsymbol{X}(\boldsymbol{x}_0)) \tag{7.199}$$

In other words, the concentration field is obtained by integration of the right hand side, f, along the streamline. The practical implementation of the method is shown in figure 7.26. Starting from the current point $\boldsymbol{x}$, which can be a mesh node or

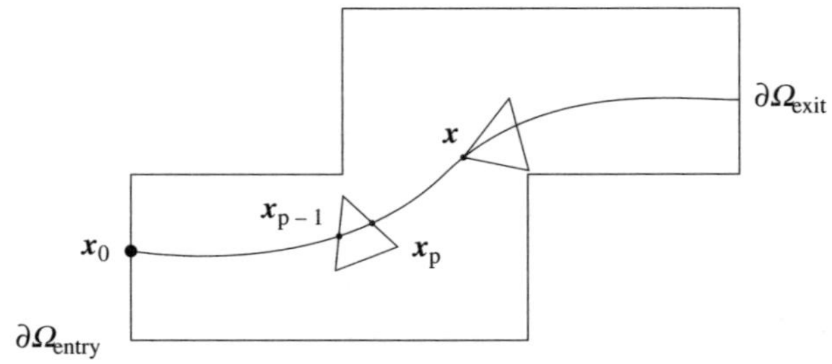

Figure 7.26 practical implementation for integrating along a streamline.

an integration point according to the case, the upstream part of the streamline that passes through $\boldsymbol{x}$ is constructed, in the form of a succession of points $\boldsymbol{x}_p$ corresponding to the intersections with the element edges. A first-order scheme is often used (Euler type) to backtrack the streamline in an element. Then the integral (7.199) is evaluated, accumulating the source term integrals on the intervals $[\boldsymbol{x}_{p-1}, \boldsymbol{x}_p]$ to find the field c.

The preceding considerations can be generalized to the non stationary case

$$\frac{\partial c}{\partial t} + \boldsymbol{v} \cdot \mathbf{grad}\, c = 0 \tag{7.200}$$

The characteristic lines associated with the vector $\boldsymbol{v} = \boldsymbol{v}(\boldsymbol{x}, t)$ are defined by the relations

$$\frac{\mathrm{d}\boldsymbol{X}}{\mathrm{d}t}(t; \boldsymbol{x}, s) = \boldsymbol{v}(t, \boldsymbol{X}(t; \boldsymbol{x}, s)) \tag{7.201}$$

and

$$\boldsymbol{X}(s; \boldsymbol{x}, s) = \boldsymbol{x} \tag{7.202}$$

Equations (7.201) and (7.202) are interpreted by remarking that from a geometric point of view, $\boldsymbol{X}(t; \boldsymbol{x}, s)$ is the position at time t of a particle displaced by the velocity field $\boldsymbol{v}$ and that occupied the position $\boldsymbol{x}$ at time s.

Defining $c(t, \boldsymbol{x}_0)$ by the relation

$$c(t, \boldsymbol{x}_0) = c(t, \boldsymbol{X}(t; \boldsymbol{x}_0, 0)) \tag{7.203}$$

it can be shown (see exercise 7.10) that

$$\frac{\mathrm{d}c}{\mathrm{d}t}(t, \boldsymbol{x}_0) = \left(\frac{\partial c}{\partial t} + \boldsymbol{v} \cdot \mathbf{grad}\, c\right)(t, \boldsymbol{X}(t; \boldsymbol{x}_0, 0)) \tag{7.204}$$

Thus, if a non stationary advection diffusion problem is treated, the temporal total derivative can be approximated by an implicit Euler scheme

$$\frac{\mathrm{d}c}{\mathrm{d}t}(t+\Delta t, \boldsymbol{x}_0) \simeq \frac{c^{\mathrm{t}+\Delta\mathrm{t}}(\boldsymbol{x}_0) - c^{\mathrm{t}}(\boldsymbol{x}_0)}{\Delta t}$$

Setting $\boldsymbol{x}_0 = \boldsymbol{X}(0; \boldsymbol{x}, t+\Delta t)$, one obtains with (7.203),

$$\frac{\mathrm{d}c}{\mathrm{d}t}(t+\Delta t, \boldsymbol{X}(0; \boldsymbol{x}, t+\Delta t)) \simeq \frac{c(t+\Delta t, \boldsymbol{x}) - c(t, \boldsymbol{X}(t; \boldsymbol{x}, t+\Delta t))}{\Delta t} \tag{7.205}$$

One can also choose an implicit Euler scheme to integrate (7.201) rewritten in the form

$$\frac{\mathrm{d}\boldsymbol{X}}{\mathrm{d}t}(t; \boldsymbol{x}, t+\Delta t) = \boldsymbol{v}(t, \boldsymbol{X}(t; \boldsymbol{x}, t+\Delta t)) \tag{7.206}$$

The following approximation of $\boldsymbol{X}(t; \boldsymbol{x}, t+\Delta t)$ is thus obtained

$$\tilde{\boldsymbol{X}}(t, \boldsymbol{x}) = \boldsymbol{x} - \boldsymbol{v}(t+\Delta t, \boldsymbol{x})\Delta t \tag{7.207}$$

The preceding process can be applied to the non stationary Navier-Stokes equations (7.78)–(7.79). The characteristic direction of the flow is defined by $\boldsymbol{X} = \boldsymbol{X}(t\,; \boldsymbol{x}, s)$ given by relations (7.201) and (7.202). Equations (7.78) and (7.79) are then discretized along this characteristic flow direction. Choosing an implicit Euler method to advance from time t to time $t+\Delta t$, one has a Stokes problem to solve

$$\rho \boldsymbol{v}^{\mathrm{t}+\Delta\mathrm{t}} - \mu \Delta t \Delta \boldsymbol{v}^{\mathrm{t}+\Delta\mathrm{t}} + \Delta t\, \mathbf{grad}\, p^{\mathrm{t}+\Delta\mathrm{t}} = \rho \tilde{\boldsymbol{v}} + \Delta t\, \boldsymbol{F}_{\mathrm{v}}^{\mathrm{t}+\Delta\mathrm{t}}$$

$$\mathrm{div}\, \boldsymbol{v}^{\mathrm{t}+\Delta\mathrm{t}} = 0 \tag{7.208}$$

with $\tilde{\boldsymbol{v}} = \boldsymbol{v}(t, \tilde{\boldsymbol{X}}(t, \boldsymbol{x}))$ and $\tilde{\boldsymbol{X}}(t, \boldsymbol{x})$ an approximation sufficiently precise for $\boldsymbol{X}(t\,; \boldsymbol{x}, t+\Delta t)$. With respect to (7.193) where the velocity field was data of the problem, in the case of the Navier-Stokes equations, one resorts to an explicit Euler scheme to calculate

$$\tilde{\boldsymbol{X}}(t, \boldsymbol{x}) = \boldsymbol{x} - \boldsymbol{v}(t, \boldsymbol{x})\Delta t \tag{7.209}$$

The evaluation of the velocity field $\tilde{\boldsymbol{v}}$ in (7.208) is performed by calculating it at the base of the characteristic of time t that passes through $\boldsymbol{x}$ at time $t+\Delta t$. The reader is referred to Pironneau (1989) for further details on the method of characteristics applied to the Navier-Stokes equations.

7.5.5 Treatment of free surfaces

Free surfaces appear often in materials forming processes. Examples include blowing processes (glass, polymers), metal injection in molds, the surfaces in multilayering, films, fibers produced by jets, etc. The main difficulty is in the calculation of the form of the surface which in three dimensions can be curved and distorted. The calculation techniques fall into two large categories. The first consists of capturing the geometry of the surface as well as possible starting from a given initial configuration; an example of this type is the MAC method. The second consists of accepting a certain loss of information: such is the case of the ***volume of fluid method*** (VOF).

MAC method

In the MAC method, massless fluid particles or "markers" are introduced that follow the local velocity. An equation such as (7.201) and (7.202) is integrated for each particle. The velocity of the particle is evaluated by a weighted average of the local velocities defined on the mesh element.

An empty cell does not contain any particles. It is denoted V. A surface cell is occupied by particles, and has at least one empty neighboring cell; it is denoted S. A full cell is completely occupied by particles, and by definition, there can be no adjacent empty cells; it is denoted F (Fig. 7.27).

An arbitrary configuration of the various cells in a free surface flow can be seen in the figure below. For a cell of type S_1, the surface is assumed horizontal $n_x = 0$, $n_y = 1$. Conditions (7.72) and (7.73) are written

$$- p_{\text{viscous}} + 2\mu\frac{\partial v}{\partial y} = -p_{\text{air}} \tag{7.210}$$

$$\frac{\partial u}{\partial y} + \frac{\partial v}{\partial x} = 0 \tag{7.211}$$

In cell S_3, the free surface is assumed vertical $n_x = 1$, $n_y = 0$, while for the corner cell S_2, the surface is taken as inclined at 45° with respect to the horizontal. In the latter case, in addition to the free surface conditions (7.72) and (7.73), the

<table>
<tr><td>V</td><td>V</td><td></td><td></td><td></td><td></td><td></td></tr>
<tr><td>V</td><td>V</td><td>V</td><td></td><td></td><td></td><td></td></tr>
<tr><td>S</td><td>S_1</td><td>S</td><td>V</td><td>V</td><td></td><td></td></tr>
<tr><td>F</td><td>F</td><td>F</td><td>S</td><td>S_2</td><td>V</td><td>V</td></tr>
<tr><td>F</td><td>F</td><td>F</td><td>F</td><td>F</td><td>S</td><td>V</td></tr>
<tr><td>F</td><td>F</td><td>F</td><td>F</td><td>F</td><td>S_3</td><td>V</td></tr>
<tr><td>F</td><td>F</td><td>F</td><td>F</td><td>F</td><td>S</td><td>V</td></tr>
</table>

Figure 7.27 Arbitrary configuration of calculation cells in a free surface flow calculation.

incompressibility equation is taken into account

$$-p_{\text{viscous}} + \mu\left(\frac{\partial u}{\partial y} + \frac{\partial v}{\partial x}\right) = -p_{\text{air}} \tag{7.212}$$

$$\frac{\partial u}{\partial x} = 0 \quad \text{and} \quad \frac{\partial v}{\partial y} = 0 \tag{7.213}$$

It is clear that this technique approximates the free surface by a polygonal line. The surface cells of type S_2 are detrimental for the precision of the method. More sophisticated treatments use surface particles; they are used to establish the form of the surface, approximating it by piecewise cubic spline interpolation, for example. With this implementation surface tension effects can easily be incorporated (7.77).

The volume of fluid method (VOF)

In this method a function F (fraction of fluid volume) is employed, that is equal to one at every point occupied by the fluid, and zero otherwise.

The average value of F in a mesh element then represents the volume fraction of the element occupied by the fluid. In particular, in a full element, $F = 1$ and for an empty element, $F = 0$. The elements for which $0 < F < 1$, contain a free surface. The dynamic equation that governs F is given by

$$\frac{\mathrm{d}F}{\mathrm{d}t} = 0 \tag{7.214}$$

which expresses that the quantity F is transported by the fluid.

In comparison with the MAC method, the VOF method allows the treatment of more complicated interfaces, with ruptures and reconnections. The reader is referred to the specialized literature for implementation details. Note however, that the equation (7.214) being of the advection type, the upwind or SU techniques are the most appropriate for the discretization. The calculation of curvature in this method, necessary to take into account surface tension effects, amounts to the estimation of the normal of the free surface. It is given by $\boldsymbol{n} = \mathbf{grad}\, F/|\mathbf{grad}\, F|$. The curvature is thus expressed by (div $\boldsymbol{n}$).

7.6 EXAMPLES

7.6.1 Filling of a foundry mold

The foundry process for casting metals consists of taking metal in a liquid form and pouring or injecting it into a mold having the desired form for the finished part. Recourse to numerical simulation becomes inevitable as the specifications of the fabrication and mechanical properties become more and more demanding. In addition, the diversification of the products (small parts for automobiles, large ingots for forging) and of materials (cast iron, steel, aluminum, and all the associated alloys) further complicate the task.

The phase during which the mold is filled is crucial for the production of a part in a foundry as it conditions the following phases: the cooling and solidification. In particular, certain defects of the finished product are generated during the filling, such as “cold shuts” (unfilled zones resulting from premature solidification of the metal) or inclusions of oxide skin inside the volume.

The flows encountered in foundries are turbulent. Take the case of the filling of an industrial ingot mold for forging. The molten steel has the following characteristics: $\rho \simeq 7000\ \mathrm{kg\,m^{-3}}$, $\mu \simeq 10^{-3}\ \mathrm{Pa\,s}$. It follows that $\nu \simeq 1.5\ 10^{-7}\ \mathrm{m^2\,s^{-1}}$. For gravity pouring, realistic operating conditions are around 3000 kg $\mathrm{min^{-1}}$ for a feed tube of 7 cm diameter. From that, a characteristic value of the velocity of 1 $\mathrm{m\,s^{-1}}$ is determined. Consequently, the Reynolds number (7.11) is 5.10^5; the flow is turbulent. We are led to employ the $k-\varepsilon$ model to solve the problem.

The geometry of the models can be extremely complex. There are curves, necks, re-entrant corners, etc. Before attacking a large simulation, the programs are tested on simpler cases. An excellent test is the backward facing step shown in figure 7.8, which presents the situation occurring at a sudden enlargement of the mold section. In order to be able to validate the calculation and the model, we take values corresponding to measurements published in the literature; $h = 2H$ the distance $DE = 100H$ and the Reynolds number is taken equal to 124 000. Two models are used: the linear and non linear $k-\varepsilon$ models (7.42), (7.51) (A.-P. Gonze 1994). Figure 7.28 shows the stream function. Recall that the equation "stream function = constant" gives the streamline corresponding to that constant. It is seen that a recircula-

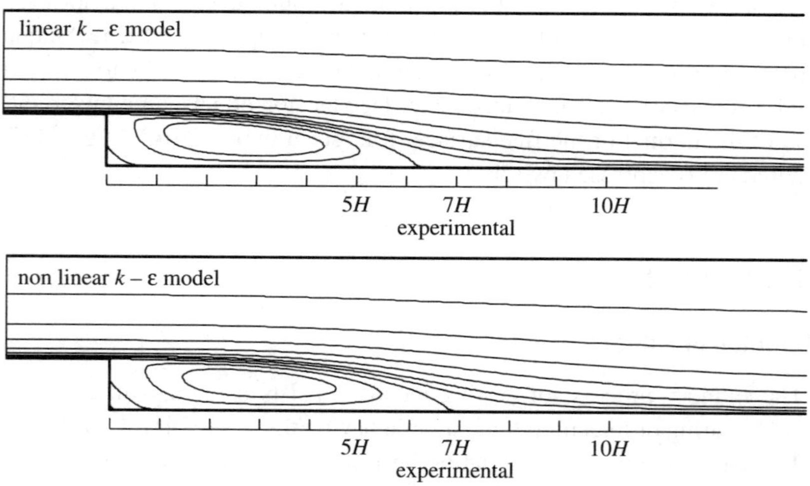

Figure 7.28 Streamlines of the turbulent flow over the backward facing step.

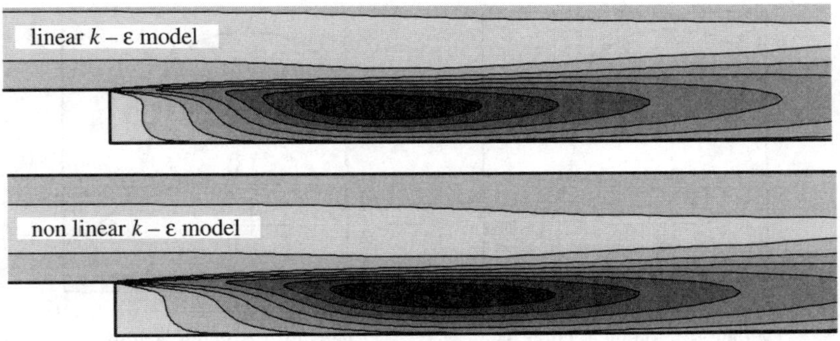

Figure 7.29 Turbulent kinetic energy.

tion zone appears behind the step, whereas the main flow passes over this geometric discontinuity, enlarging its section gradually, and becoming again developed far downstream of the step. The separation line detaches from the corner and reattaches to the wall at a certain distance called the reattachment length. The experimental value of this length is equal to $7H$. The linear model gives $6.2H$ whereas the non linear model predicts $6.9H$. The lower performance of the linear model is attributed to its higher turbulent dissipation than for the non linear model. One also sees two supplementary vortices nested in the lower corner of the step that are captured by the non linear model. Figure 7.29 shows the contour lines of the turbulent kinetic energy. The black areas are those with the highest intensity. They are more extensive in the case of the non linear model, which is the sign of a less dissipative model. In figure 7.30, these calculation results are compared with measurements of the average velocity profile $\bar{v}_1$ as a function of y measured from the upper wall. Note that the agreement is good; however, the calculated velocities are under estimated in the region close to the reattachment point. The behavior of the two models is not fundamentally different.

The presence of the recirculation zone, and more generally, of convection during the solidification of an alloy, can noticeably affect the homogeneity of the final concentration ("macrosegregation" phenomena). In such situations, it is necessary to couple the Navier-Stokes equations with those governing heat exchange and solute transfer (see chapter 5).

The last example is that of the filling of a mold during the casting of a metallic part of complex form. The mathematical model includes the mass conservation equation (1.5), the momentum equations (1.103) modified with a Darcy term to take into account the solidification, and the energy conservation equation (1.62) with the enthalpy as the principle unknown and temperature T in the heat flux. The surface velocities obtained by dividing the velocity of the liquid phase by the liquid fraction at temperature T are used in the momentum equations in order to have the simplest form of the mass conservation equation (1.5). As the liquid fraction goes to zero, the

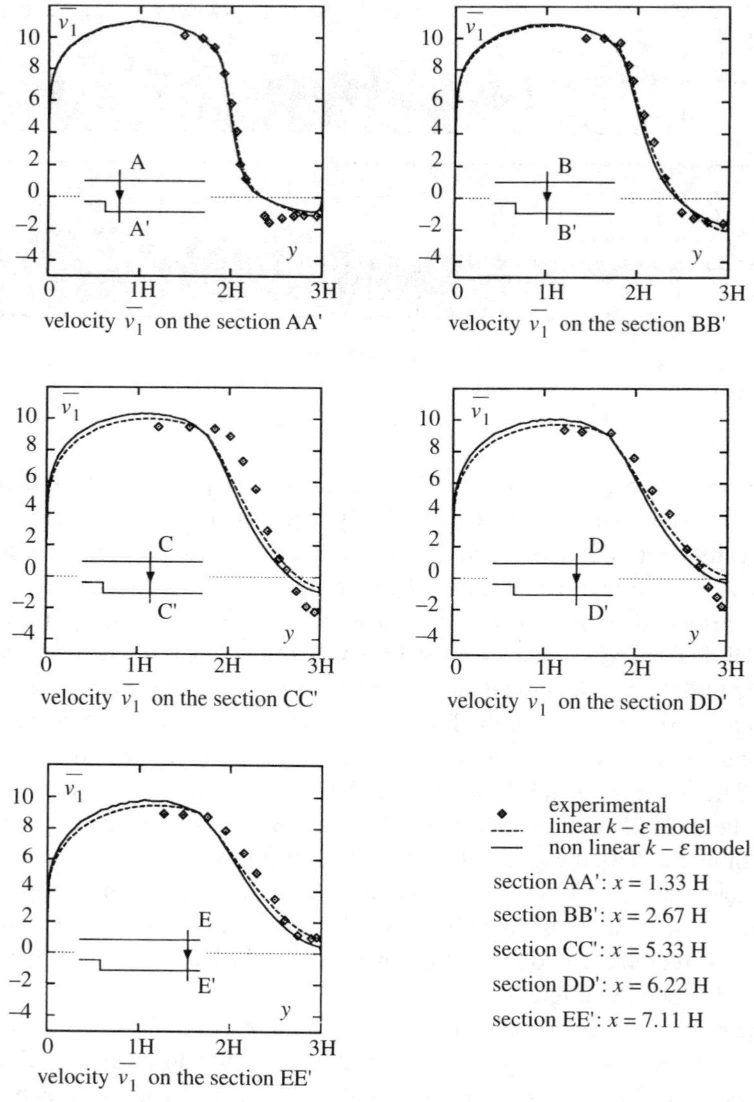

Figure 7.30 Comparison with experiments of the horizontal component of the average velocity field.

Darcy term becomes dominant; each velocity component is null and a static pressure results. The heat transfer incorporates conduction, convection and radiation. The free surface is treated with the VOF method. The numerical discretization uses linear finite elements; a Poisson equation for the pressure is solved by a preconditioned conjugate gradient method. A $k - \varepsilon$ model accounts for the turbulence, if necessary.

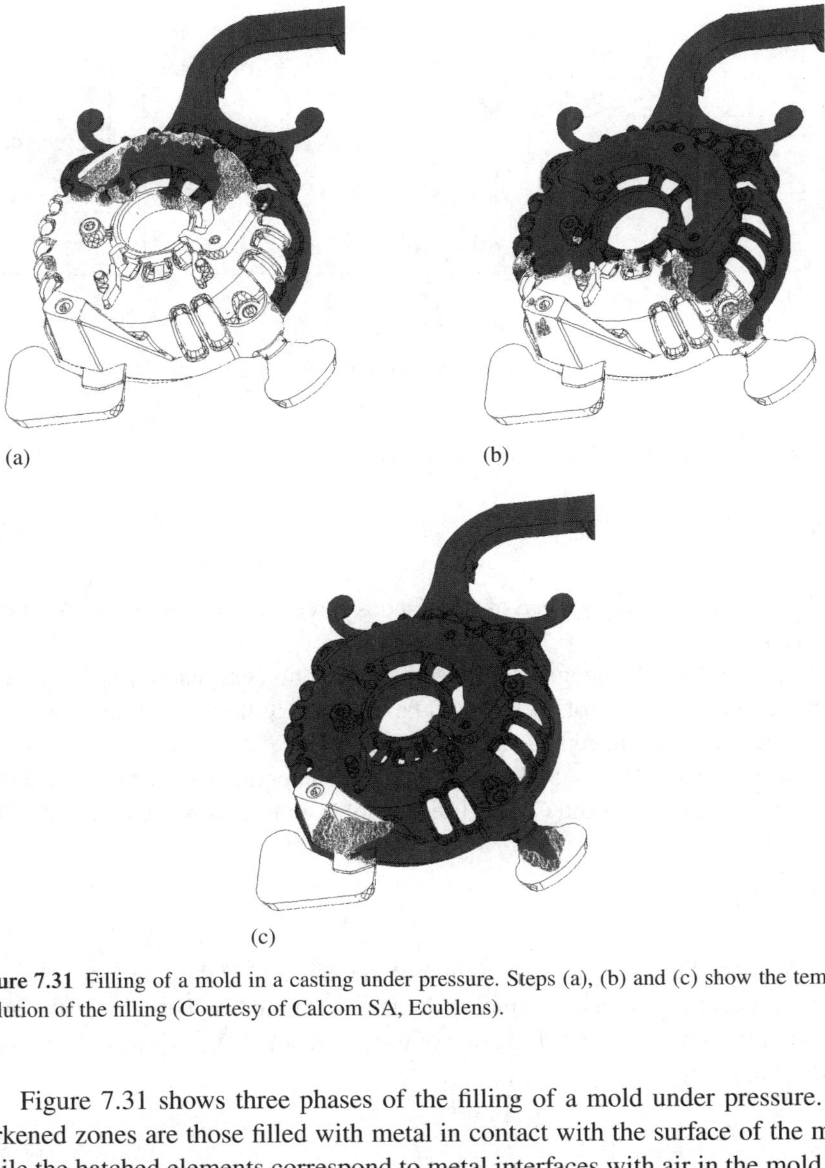

Figure 7.31 Filling of a mold in a casting under pressure. Steps (a), (b) and (c) show the temporal evolution of the filling (Courtesy of Calcom SA, Ecublens).

Figure 7.31 shows three phases of the filling of a mold under pressure. The darkened zones are those filled with metal in contact with the surface of the mold, while the hatched elements correspond to metal interfaces with air in the mold.

7.6.2 Production of thin sheets

Plastic sheets, fibers, threads, etc. are produced by flowing a plastic material through a die. We restrict ourselves in this example to a viscous Newtonian fluid that represents a low molecular weight material.

Nonetheless, we will be interested in the coupled effects of viscosity, inertia and surface tension. In order to take the latter into account we introduce the capil-

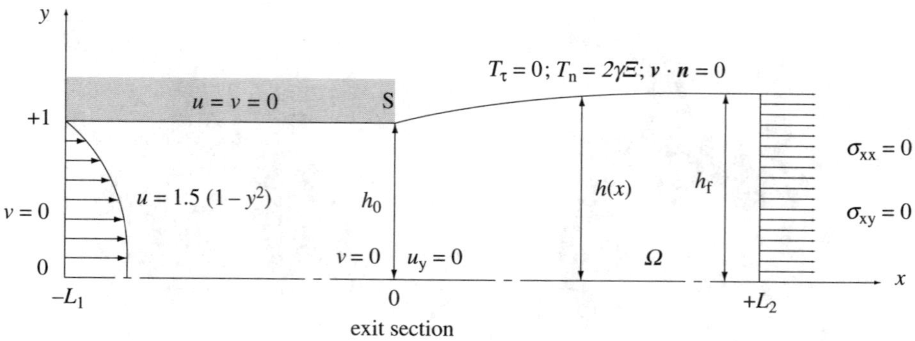

Figure 7.32 Flow in a die.

lary number of the flow defined by the relation

$$Ca = \frac{\mu U}{\gamma} \tag{7.215}$$

which expresses the importance of the viscous forces μ with respect to the surface tension, γ.

In figure 7.32, the geometry of rectangular section extrusion flow is modeled. The reference velocity U of (7.215) will be the average fluid injection velocity, the free surface being assumed attached at the exit point S.

The stationary Navier-Stokes equations are integrated with the free surface conditions (7.73), (7.76) and (7.77). The Reynolds number is based on a flux velocity such as

$$D = 2h_0 U$$

where h_0 is the height of the half section of the tube and D the flow rate. The final height of the jet exiting the calculation domain is denoted h_f. This allows the introduction of the asymptotic swelling rate of the jet, denoted Sw, expressed in percent, and defined by the relation

$$Sw = \frac{(h_f - h_0)}{h_0} \times 100 \tag{7.216}$$

The results presented here were obtained by X. Aubert (1983). Figure 7.33 shows the profiles of the jet for a range of values of the surface tension parameter Ca^{-1} in a slow flow ($Re = 0$). In the absence of inertial forces, the jet dilates less and less as Ca^{-1} grows, and goes to a flat jet for a surface tension going to infinity. Very strong influence of the capillary force is noted for Ca^{-1} between 0 and 5. To have an idea, note that values of Ca^{-1} between 0 and 8 are common in flows implemented in the photographic industry for manufacturing of film substrate.

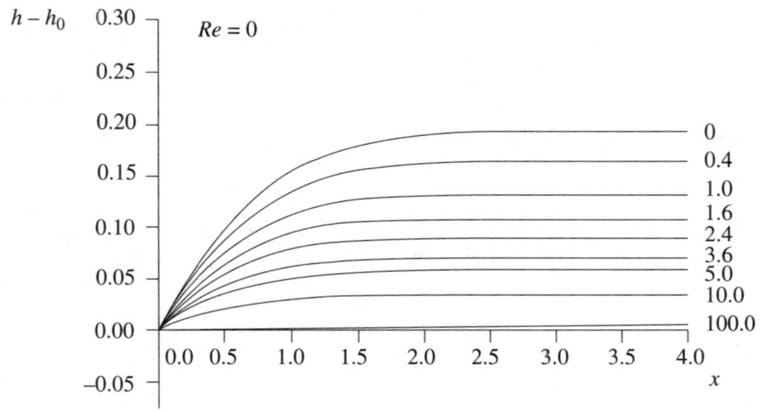

Figure 7.33 Free surface profiles for a planar die problem as a function of the surface tension; $Re = 0$, $Ca^{-1} = 0$ to 100. (8× vertical enlargement).

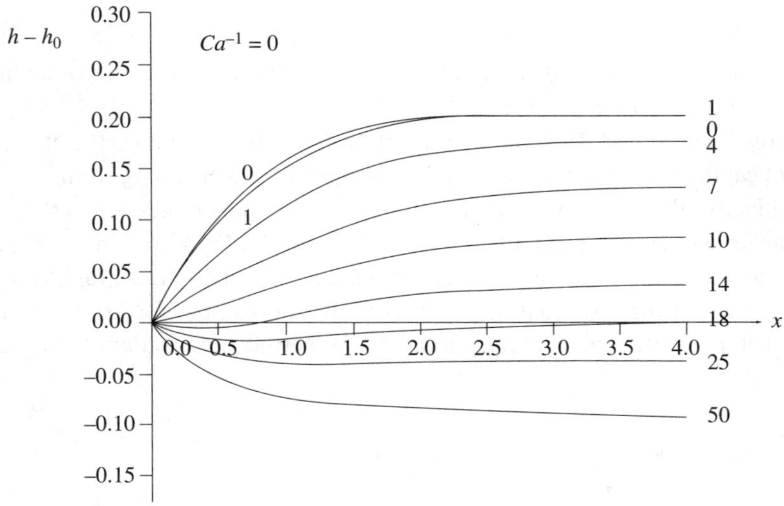

Figure 7.34 Free surface profiles for a planar die problem as a function of the Reynolds number *Re*, $Ca^{-1} = 0$, $Re = 0$ to 50. (8× vertical enlargement).

Figure 7.34 shows the jet profiles as a function of *Re*, $0 \leq Re \leq 50$, in the absence of capillary forces. As *Re* increases, the thickness of the jet diminishes, contracting for values beyond a critical value equal to 18. Note also the inflection of the profile at the separation point when *Re* is greater than 4. The maximum swelling of the jet is obtained for $Re = 1$, and equals $Sw = 19.1\%$.

In figure 7.35, profiles obtained for three values of $\gamma = 0$, 2.22 and 4.44 and for $Re = 18$ are compared. Contrary to what was seen for a slow flow at $Re = 0$, now the

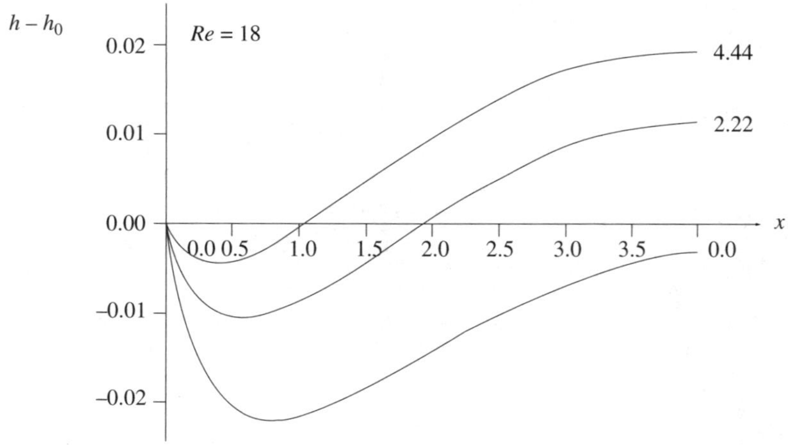

Figure 7.35 Free surface profiles for a planar die problem as a function of the surface tension; $Re = 18$, $\gamma = 0$ to 4.44. (65× vertical enlargement).

effect of the capillary forces is to increase the height of the jet, because of its high curvature at the separation point for null Ca^{-1}.

Figures 7.36 (a) and (b) show a series of profiles obtained under the effect of significant capillary forces ($Ca^{-1} = 4.44$) for Reynolds numbers going from 0 to 25. The combined influence of the inertial and surface tension forces noticeably increases the jet section initially, with a maximum, $Sw = 8\%$, for $Re = 7$ (fig. 7.36(a)). Then the profile declines at the separation point and the jet contracts rapidly with increasing inertial forces, contracting below the horizontal for $Re = 25$ (fig. 7.36(b)). These results show that the interaction of inertial and capillary forces lead

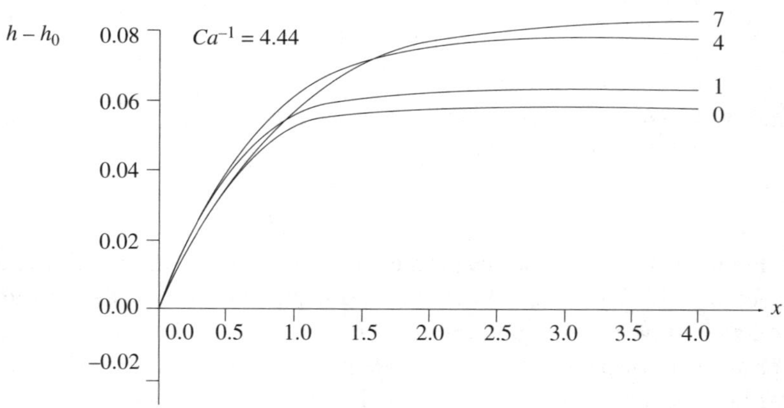

(a)

Figure 7.36(a) Free surface profiles for a planar die problem as a function of the Reynolds number Re; $Ca^{-1} = 4.44$, $Re = 0$ to 7. (25× vertical enlargement).

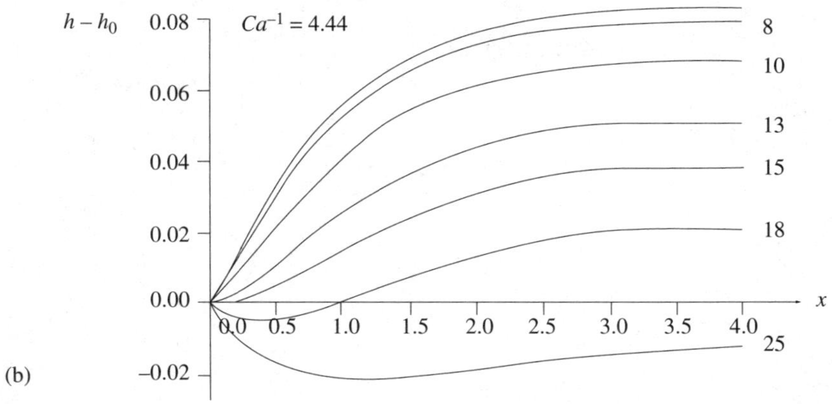

Figure 7.36(b) Free surface profiles for a planar die problem as a function of the Reynolds number *Re*; $Ca^{-1} = 4.44$, $Re = 7$ to 25. (25× vertical enlargement).

to surprising effects when they are of the same order. When the value of *Re* is much larger than that of the surface tension parameter Ca^{-1}, the inertial effects dominate, and one can estimate the height of the jet independently of the capillary forces when *Re* is greater than 100.

7.6.3 Design of a die: an inverse problem

As we have seen in the preceding example, the flow parameters affect the form of the free surface, and thus the extrusion. This is particularly true in the case of polymers where the form of the die is an essential factor to obtain a finished product in the prescribed form. While numerical rheology often ends with the form of the extrusion obtained a posteriori for a given die profile, the design engineer poses the inverse problem which consists of determining the geometry of the die to obtain the given extrusion profile.

Figure 7.37 shows the challenges of the inverse problem.

In this problem, the rheology properties of the material as well as the geometric configurations must be taken into account. As the latter are complex, the use of the finite element method is unavoidable. We will briefly describe the technique based on the work of V. Legat (1992).

We restrict ourselves to a two dimensional case for simplicity, although from the point of view of the method, the reasoning extends to three dimensional problems without difficulty. In figure 7.38, the inverse problem is posed. Given the height corresponding to the desired form of the extrusion, the problem is to find the geometric form of the die, notably its height at the lip. We introduce a vector displacement of the free surface $\delta \boldsymbol{x}_{\mathrm{FS}}$ such that

$$\delta \boldsymbol{x}_{\mathrm{FS}} = h\boldsymbol{d} \tag{7.217}$$

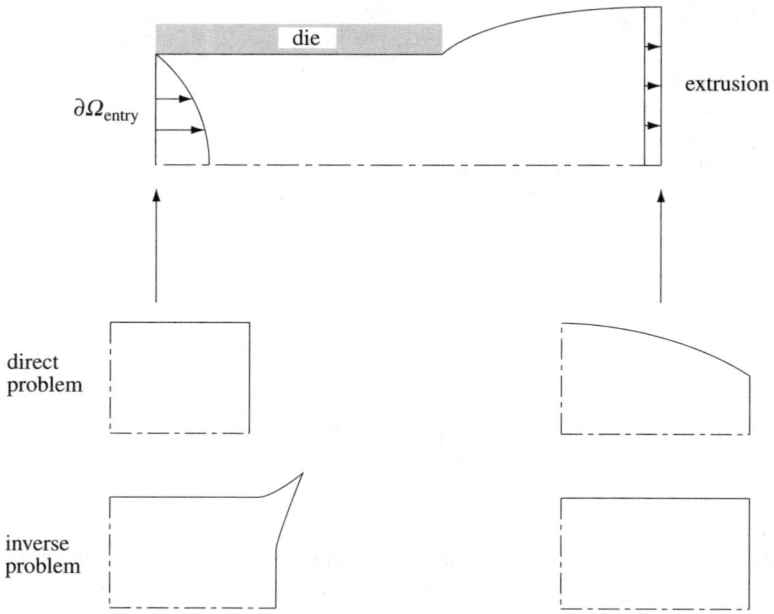

Figure 7.37 Comparison of the direct and inverse problems in the extrusion process.

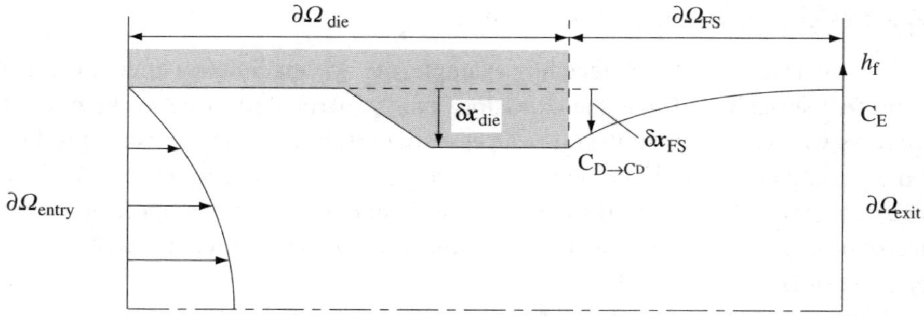

Figure 7.38 The inverse die problem.

with $\boldsymbol{d}$, a direction which can vary from point to point to describe the displacement of the surface. It should be noted that the vector $\boldsymbol{d}$ is given a priori (it is a parameter of the problem) and is not an unknown of the problem. Generally, $\boldsymbol{d}$ is calculated in the direction normal to the initial form (prescribed) of the free surface, for example the horizontal line in the projection of the die upstream. The scalar quantity h, function of $\boldsymbol{x}$, is the variable representing the amplitude of the local displacement of the free surface in the direction $\boldsymbol{d}$. We also have a function for the deformation of the form of the die with respect to the same horizontal

$$\delta \boldsymbol{x}_{\text{die}} = g\boldsymbol{d} \tag{7.218}$$

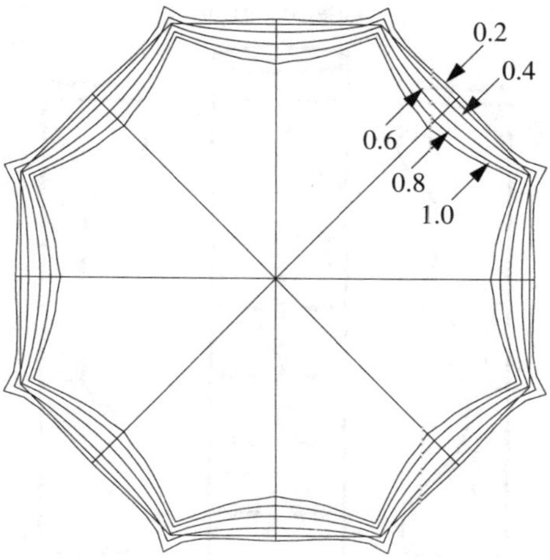

Figure 7.39 Die sections as a function of the power law index for an octagonal extrusion.

At the singular points such as the exit of the die (point C_D), we define two displacements, one for the die and the other for the free surface in order to correctly describe the movement at the corner.

In the direct problem, the form g of the die is given with $h = 0$ imposed at the lip C_D. We obtain h at the exit of the extrusion at point C_E as the result of the free surface problem. For the inverse problem, we fix $h = 0$ at C_E. This relation imposes the form of the extrusion and represents an equation (initial condition) for g. We impose $g = h$ at C_D which constitutes a constraint for h. In the finite element method that is implemented to solve the direct or inverse problem, the functions h and g are the supplementary unknowns of the algorithm. To give an idea, the function h will be obtained from the kinematic constraint of zero flux across the free surface ($\boldsymbol{v} \cdot \boldsymbol{n} = 0$). Figure 7.39 shows the different transverse sections of the die as a function of the index of the power law for a generalized Newtonian fluid of type (7.28), to give a finished extrusion with an octagonal form. The shear-thinning behavior of the fluid reduces the expansion; however, the corner effects are still important, even for small values of the power law index.

7.6.4 Extrusion blow-molding

The extrusion blow-molding process is a good example of the processing of polymer extrusions into semi-finished products. Compared to thermoforming, presented in chapter 6, the polymer is blown in the molten state, after extrusion in the form of a cylindrical tube, called a parison. The process is shown schematically in figure 7.40. It is mainly applied in the fabrication of hollow objects such as bottles and gas tanks.

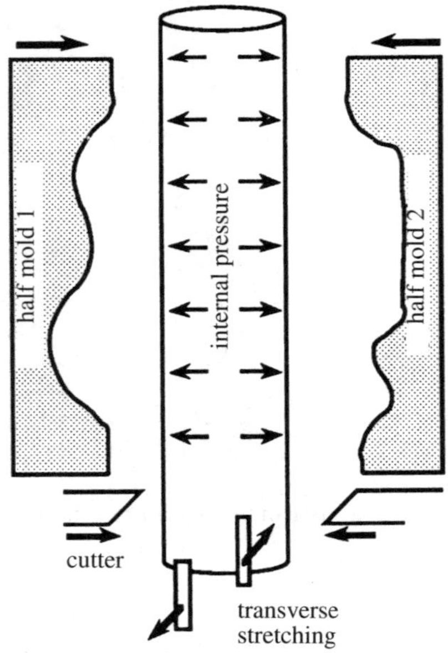

Figure 7.40 Schematic of the extrusion blow-molding process

The question of thickness is the same as for thermoforming. However the constitutive law for the polymers used (mainly high or low density polyethylene) is viscoelastic, because the process is in the molten state.

In the approach presented below, the Maxwell-B law is used. It is given by (7.14), (7.15) with null μ_0, and (7.16). As in the case of thermoforming, curvilinear coordinates are extremely useful, and the mechanical membrane formulation described in paragraph 6.5.2 is employed. This yields a simplified expression of the upper-convective derivative (only the material derivative of the covariant components). In extrusion blow-molding, if z designates the direction of the extrusion, the first curvilinear coordinate θ^1 is taken equal to the initial polar angle of the point in the (x, y) plane, and the second, θ^2, is taken equal to the initial distance z. However, we will not continue the development further because the treatment of the constitutive law in curvilinear coordinates requires an excessively cumbersome tensor analysis.

Similarly as for the simulation of thermoforming, the constitutive law is temporally discretized with an implicit integration scheme and the same spatial discretization is applied as presented before (linear triangular membrane finite elements).

The automatic mesh regeneration technique was integrated in the modeling. With respect to the global representation of remeshing given in paragraph 6.4.5, here it is a matter of remeshing a warped surface discretized in three dimensional space by linear triangular elements. One proceeds by local improvements of the mesh, as shown in figures 7.41 and 7.42. The algorithm consists of scanning all the nodes and edges of the mesh. From a given node or edge, the local sub topology is

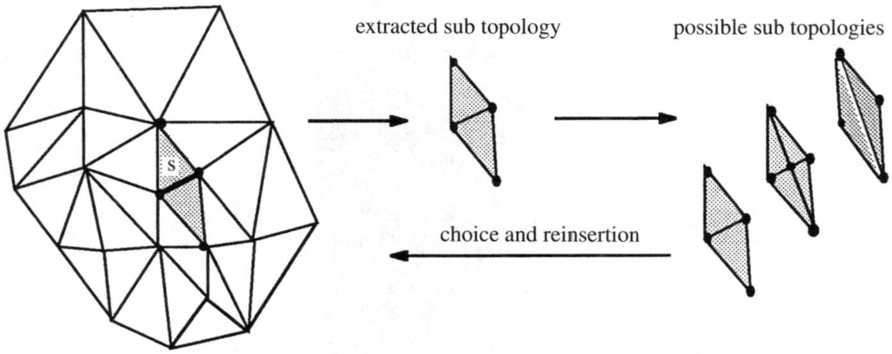

Figure 7.41 Extraction of a local sub topology around an edge. The solutions possible with the addition of a node and change of a connection by inversion of a diagonal are examined.

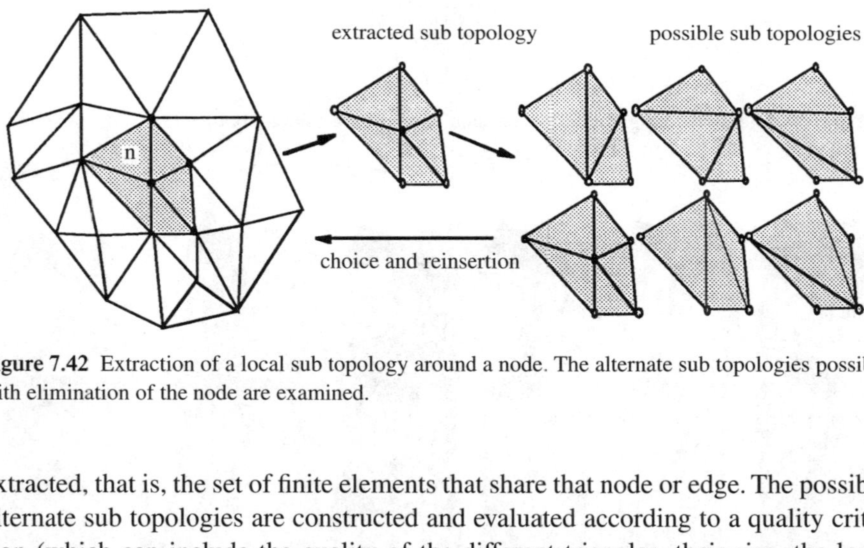

Figure 7.42 Extraction of a local sub topology around a node. The alternate sub topologies possible with elimination of the node are examined.

extracted, that is, the set of finite elements that share that node or edge. The possible alternate sub topologies are constructed and evaluated according to a quality criterion (which can include the quality of the different triangles, their size, the local curvature ...). The best sub topology in the sense of the criterion is retained and inserted in place of the one extracted. The process continues with the extraction of another sub topology and stops when no further changes improve the criterion.

Figures 7.43 and 7.44 illustrate a simulation of blowing of a polyethylene bottle having the particularity of a handle around which the gradients are very severe. In its initial state, the cylinder has a radius of 60 mm, a height of 500 mm and a thickness of 1.5 mm. Because of vertical symmetry, only half of the cylinder is calculated. It is assumed fixed at the upper edge, which is the exit of the die. A null vertical displacement condition is imposed at the lower edge. Note that the form takes into account the swelling at the die exit and the creep observed in real tests.

In figure 7.44, the progressive establishment of contact between the plastic and the mold can be seen after the latter's closing before the blowing actually starts. The

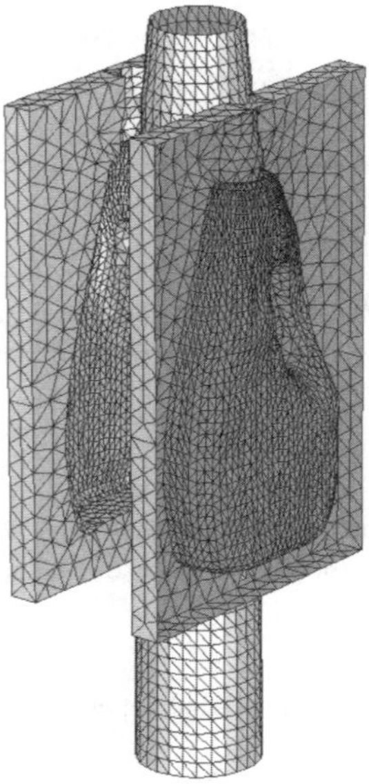

Figure 7.43 Initial configuration for the simulation of blowing of a bottle (Tform3® software).

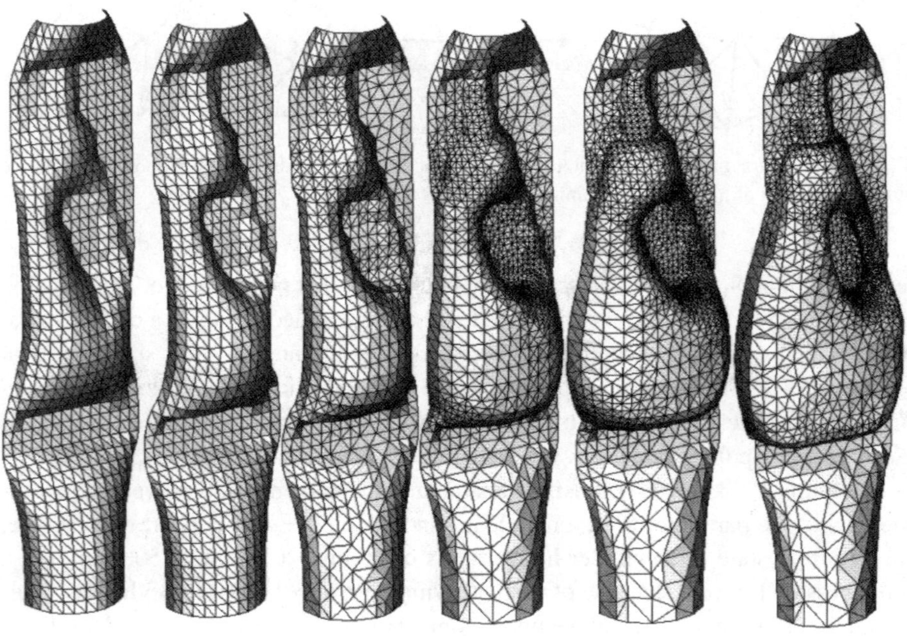

Figure 7.44 Successive deformations with automatic mesh regeneration.

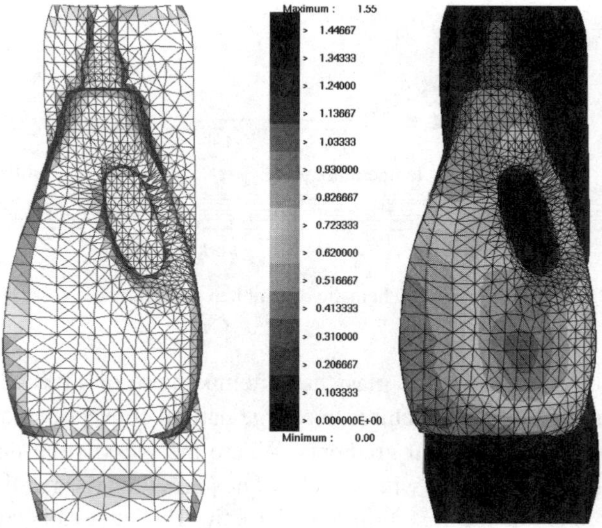

Figure 7.45 Final result of the simulation: deformed mesh and thickness distribution (mm).

good behavior of the mesh regeneration is also apparent, in terms of control of the local size of the elements, being deduced from a geometric error estimator. The calculation is thus performed automatically without operator intervention.

Figure 7.45 shows the thickness profile of the finished calculation. It can be seen in this figure that the form is offset from the side with the handle. This is necessary to avoid a fold in the material that forms when the handle is blown. This phenomenon, revealed by the calculation, is corroborated by experiments. It is then necessary to make a compromise, as the offset has the disadvantage of causing high stretching of the material – and thus reduced thickness – in the zones opposite the handle.

7.6.5 Convection currents in a melting furnace

All the material forming processes are essentially non isothermal (see chapter 5), whether one thinks of plastics manufacturing, metallurgy, or the processing of glass. The last example of this chapter is devoted to the latter technique.

The production of plate glass or hollow glass is performed by melting the raw material in a melting furnace. This raw material, composed of sand, melting agents and recycled glass is called the groisil. It is introduced upstream in the furnace and floats on the molten glass to form a composite layer (fig. 7.46). Laterally positioned oil burners heat the surface; melting the composite layer and entraining the groisil into the mass. The temperature measured at the surface varies from 1100 °C at the insertion up to a maximum of 1550 °C, at about a third of the length of the furnace. The temperature decreases linearly from there to the exit of the furnace where the

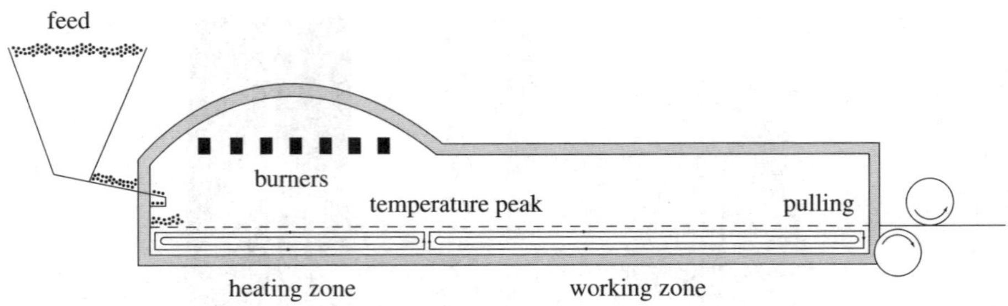

Figure 7.46 Schematic of a molten glass furnace.

glass is pulled. At this point the glass has a temperature of 1100 °C. This peaked surface temperature profile structures two large natural convection cells in the molten glass driven by the thermal gradients. Where the glass is the hottest, it is the least dense; while it is heavier where cooler. The glass thus rises at the temperature peak, and descends at the insertion and the exit. The upstream convection cell, known as the refining zone, is entirely favorable; it transports hot glass under the composite layer. Then descending again under the composite layer, it entrains cold glass that outgasses when reheated, releasing bubbles and purifying. For the glass maker, an important parameter for the quality of the produced glass is the residence time: the time a glass particle passes from the injection until being pulled. The calculated velocity field will allow this evaluation.

The calculation of the temperature field in the mass permits the calculation of a global heat balance for the furnace and thus optimize the energy consumption; energy conservation considerations will always be prescribed in the future.

In addition, the coupled velocity-temperature calculations also give quantitative information on the refractory aluminum blocks that form the walls of the furnace and affects its lifetime (wall erosion, local chemical reactions, glass crystallization, etc.).

The glass, as other molten materials, is a (quasi-) incompressible fluid. At high temperature, it has Newtonian behavior. This is no longer true in the transition zone situated at temperatures between 600 and 800 °C. The thermal effects are accounted for in the calculation of the velocity field with the Boussinesq approximation (see chapter 1). This assumes that the fluid is incompressible for all the terms of the Navier-Stokes and energy equations, except for the body force term linked to gravity.

The dynamic viscosity of glass is a function of the temperature field, in the following form:

$$\mu(T) = \exp\left(A + \frac{B}{T - C}\right) \tag{7.219}$$

where A, B, C are constants depending on the type of glass considered (composition, color).

Little of the heat transfer is by conduction (glass is a poor conductor of heat; windows isolate us from the exterior meteorological conditions), compared to convection and radiation. The latter is a mechanism of the mass that absorbs and reemits radiation. Molten glass behaves as a semi transparent material with respect to infrared radiation. The external radiation from the flames penetrates into the volume a distance further than the mean free path of the radiation (which is of the order of 3 to 7 cm). We use the Rosseland approximation (well known to astrophysicists for radiative heat transfer in stars!) which lumps the radiative transfer with the conduction, but with a characteristic coefficient κ_R that depends on the temperature (eq. (1.78)). This approximation is valid for optically thick media (the depth of glass is around 1.5 m) in comparison with the mean free path ($5 \cdot 10^{-2}$ m). However, this assumption is not at all valid near the walls and the free surface. The practical exploitation of this approximation is by using a real conductivity coefficient combining the two phenomena:

$$\kappa_{\text{real}}(T) = \kappa_c + \kappa_R(T) \tag{7.220}$$

where κ_c is the real conductivity factor of the glass. The convective energy transfer term is modeled according to the equations in chapter 1.

The complete mathematical model includes the Navier-Stokes and energy equations. The incompressibility of the fluid imposes equation (7.1). The constitutive equation is that of a Newtonian viscous fluid (7.6) with the law (7.219) for the dependence of the viscosity on the temperature. The Navier-Stokes equation (1.20) becomes with the Boussinesq approximation:

$$\rho_0 \frac{\mathrm{d}\boldsymbol{v}}{\mathrm{d}t} = -\mathbf{grad}\, p + \mathbf{div}(2\mu \boldsymbol{d}) + \rho_0 \boldsymbol{g}(1 - \beta_T(T - T_0)) \tag{7.221}$$

where the density in the acceleration term is equal to the reference value ρ_0 at the temperature T_0 while in the gravity term, the state equation for density is used

$$\rho = \rho_0(1 - \beta_T(T - T_0)) \tag{7.222}$$

The energy equation is given by (1.80) with $\dot{Q}_T$ null in many cases.

However, when part of the heating is with carbon electrodes (electric "boosting"), the Joule effect can be taken into account with (1.81). One might suspect that because of the very viscous character of the molten glass (around a million times more viscous than water), that the effects of viscous dissipation ($\boldsymbol{\sigma} : \boldsymbol{d}$) in the mass little modify the thermal balance. Calculations including this term in the energy equation show that this term is truly negligible.

Figure 7.47 shows the boundary conditions used. From the kinematic point of view, the lateral walls and the floor of the furnace are fixed and thus the vector velocity is null. We impose on the free surface that it be horizontal at constant height: the vertical component of the velocity is null.

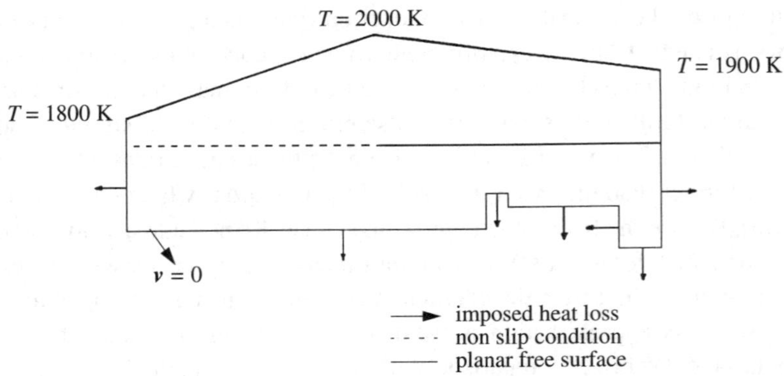

Figure 7.47 Kinematic and thermal boundary conditions.

From the insertion to the temperature peak the composite layer impedes all movement; the horizontal velocity component is null. From the temperature peak to the exit, the surface is in contact with the atmosphere in the furnace and thus can not sustain any stress: $\sigma_{xy} = 0$. From the thermal point of view, at the surface one imposes either the temperature profile, or a more general radiative condition (1.117), with ε_r an average emissivity taking into account the gray nature of the glass surface and the ceiling of the furnace (a net radiation method, explained in chapter 5, could have been implemented for more rigor). A heat flux is imposed through the refractory boundaries on the lateral walls and on the floor.

A finite element method of the Q^2/Q^1 type for the velocities and pressures is implemented. As for the temperature field, experience shows that it is useful to take a biquadratic approximation rather than bilinear. This yields the best results, notably to capture the thermal gradients under the layer of raw material and in the entry zone.

The coupled thermal convection problems introduce two non dimensional numbers. The first is the ***Rayleigh number***, denoted Ra and defined by (see chapter 10):

$$Ra = \frac{g\beta_T \Delta T_c L^3}{\alpha \nu} \tag{7.223}$$

The numerator expresses the "boiling" force: the Archimedes buoyancy that results from the density difference due to the temperature difference $T_1 - T_0$. It is a physical mechanism that tends to destabilize the flow (possibly at rest). In the denominator are two material parameters α and ν, which are associated with the thermal and molecular (viscous) diffusion respectively. These are stabilizing physical mechanisms (heat conduction, for example, tends to equalize the temperatures). The Rayleigh number must exceed a critical value before convection cells appear.

For molten glass, the value is near 10^6, well beyond this critical value. The second non dimensional number is the ***Prandtl number***, denoted *Pr*, and defined by:

$$Pr = \frac{\nu}{\alpha} \tag{7.224}$$

It compares two diffusion processes: that due to viscosity and that due to heat conduction. It is a characteristic of the material. It also gives information on the relative importance of the thickness of the hydrodynamic and thermal boundary layers, denoted δ_{H} and δ_{T} respectively:

$$\frac{\delta_{\mathrm{H}}}{\delta_{\mathrm{T}}} \sim Pr^{1/2} \tag{7.225}$$

For glass, $Pr = 300$. Given that the flow of glass is slow, $Re = 1$, its hydrodynamic boundary layer δ_{H} fills the domain. With (7.225), we conclude that the thermal boundary layer is around 20 times thinner than δ_{H}. Thus the thermal problem is more difficult to resolve spatially, its characteristic scale being so much smaller than that of the hydrodynamic problem. This justifies *a posteriori* approximating the field *T* with *Q*2 elements rather than *Q*1.

In some molten semi-conductors such as silicon and gallium arsenide, $Pr = 10^{-2}$; the hydrodynamic boundary layer is around 10 times thinner than the thermal boundary layer. The latter is large, on the order of the size of the calculation domain as a metal is a good conductor of heat. We thus have a situation opposite that of glass in a furnace. The hydrodynamic problem must be more finely meshed than the thermal problem or, using the same mesh for the coupled problem, *Q*2/*Q*1 is used for $\boldsymbol{v}$ and *p*, and *Q*1 for *T*.

Figure 7.48 shows the structured mesh used for the numerical solution of the molten glass flow problem. Figure 7.49 shows the thermal field with the very strong gradients at the entry, while the velocity field is shown in figure 7.50. The two con-

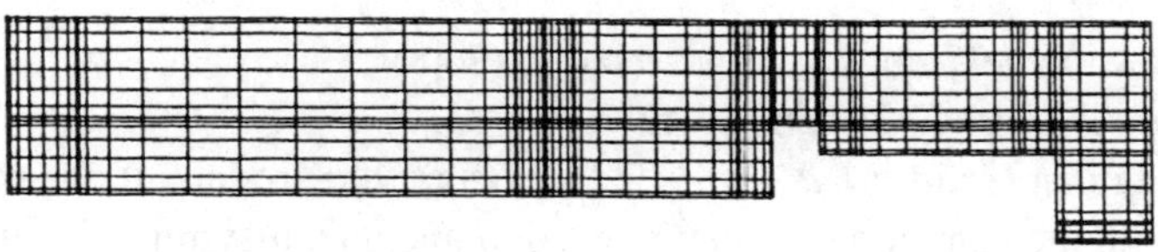

Figure 7.48 Mesh used for the solution in the melting furnace.

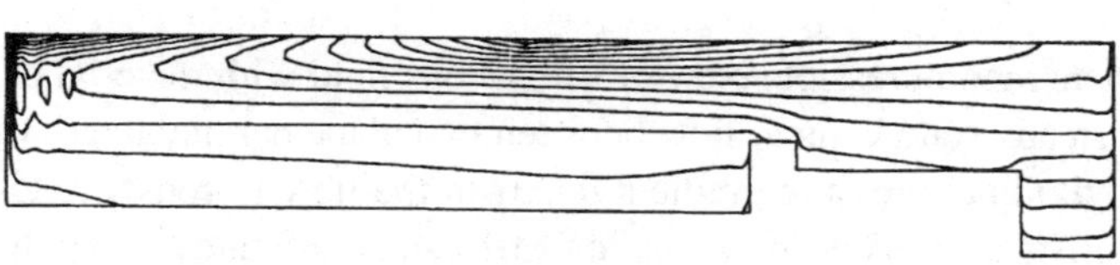

Figure 7.49 Steady state isotherms [1520 K (12.5) 1770 K].

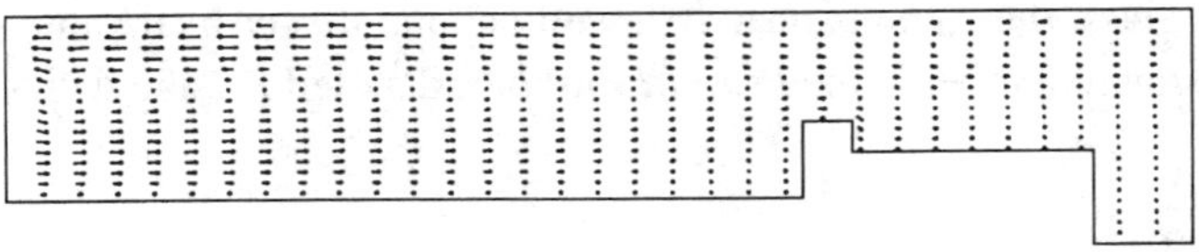

Figure 7.50 Velocity vectors in the glass.

vection cells are clearly seen. In addition, the presence of the dam on the floor of the furnace allows the uncoupling of the two portions of the flow. We note also that in reality it is necessary to take into account the entry effects that are modeled by a constant vertical velocity on the composite layer, as well as the exit, where an average flux velocity is imposed.

7.7 EXERCISES

7.7.1 Given an incompressible Newtonian viscous fluid in stationary laminar flow between two concentric cylinders of infinite length. The inner and outer radii are denoted R_1 and R_2 with angular velocities ω_1, and ω_2 respectively. If ρ and μ are the material characteristics of the viscous fluid, obtain the form of the azimuthal velocity profile (7.13) from the Navier-Stokes equations in cylindrical coordinates (r, θ, z). In the general case, the constants A and B take the values: $A = (\omega_2 R_2^2 - \omega_1 R_1^2)/(R_2^2 - R_1^2)$, $B = (\omega_1 - \omega_2) R_1^2 R_2^2/(R_2^2 - R_1^2)$.

7.7.2 1. Using the conservation of mass (7.1) prove that the non linear term written in index form $v_j\, \partial v_i/\partial x_j$ can be put in the form $(\partial/\partial x_j)(v_i v_j)$, that is $\mathbf{div}(\boldsymbol{v} \times \boldsymbol{v})$

2. Apply the average (7.33) to the tensor product $v \times v$ to prove that

$$\overline{v_i v_j} = \bar{v}_i \bar{v}_j + \overline{v_i' v_j'}$$

7.7.3 Follow the next steps to prove relation (7.48).

1. Replace $\overline{v_i' v_j'}$ with (7.39) and (7.42)
2. Recalling that **Grad** $\boldsymbol{v} = \boldsymbol{d} + \boldsymbol{\omega}$, (1.22 a), obtain the product of a symmetric tensor $\boldsymbol{R}$ with the sum of a symmetric tensor $\bar{\boldsymbol{d}}$ and an antisymmetric tensor $\bar{\boldsymbol{\omega}}$.
3. The result (7.48) is obtained invoking the theorem that proves that the contracted product of a symmetric tensor and an asymmetric tensor is zero.

7.7.4 Given an incompressible Newtonian viscous fluid with material characteristics μ and ρ, in steady state laminar flow between two plane parallel fixed walls. Assume that the longitudinal pressure gradient $\partial p/\partial x$ in the flow is constant. Using the two dimensional Navier-Stokes equations in Cartesian coordinates, show that the velocity profile is parabolic. This solution is known as the planar Poiseuille profile.

7.7.5 Derive the second order implicit Euler formula (7.121). Perform the Taylor series expansion in v^{t} about $v^{t+\Delta t}$ including terms up to second order in Δt^2. Replace the second derivative with a second order centered finite difference approximation. Express the first derivative with respect to time to obtain the desired relation.

7.7.6 By developing expression (7.132) show that for a two dimensional problem ($d = 2$), the right hand side of the x momentum equation takes the form

$$[N_{\mathrm{L}}(V)V]_{\mathrm{ij}} = \rho \int_{\Omega_{\mathrm{h}}} \psi_{\mathrm{i}} \psi_{\mathrm{k}} \left(\frac{\partial \psi_{\mathrm{j}}}{\partial x} u^{\mathrm{k}} + \frac{\partial \psi_{\mathrm{j}}}{\partial y} v^{\mathrm{k}} \right) u^{\mathrm{j}} \, \mathrm{d}V$$

and for the y momentum

$$[N_{\mathrm{L}}(V)V]_{\mathrm{ij}} = \rho \int_{\Omega_{\mathrm{h}}} \psi_{\mathrm{i}} \psi_{\mathrm{k}} \left(\frac{\partial \psi_{\mathrm{j}}}{\partial x} u^{\mathrm{k}} + \frac{\partial \psi_{\mathrm{j}}}{\partial y} v^{\mathrm{k}} \right) v^{\mathrm{j}} \, \mathrm{d}V$$

7.7.7 Using the results of exercise 7.7.6 and equations (7.148) and (7.150), determine the components of $[N_{\mathrm{L}}'(\boldsymbol{v})]_{\mathrm{ij}}$ such that for a two dimensional problem

$$[N_{\mathrm{L}}']_{\mathrm{ij}} = \rho \begin{bmatrix} u^{\mathrm{k}} \int_{\Omega_{\mathrm{h}}} \psi_{\mathrm{i}} \psi_{\mathrm{j}} \frac{\partial \psi_{\mathrm{k}}}{\partial x} \, \mathrm{d}V & u^{\mathrm{k}} \int_{\Omega_{\mathrm{h}}} \psi_{\mathrm{i}} \psi_{\mathrm{j}} \frac{\partial \psi_{\mathrm{k}}}{\partial y} \, \mathrm{d}V \\ v^{\mathrm{k}} \int_{\Omega_{\mathrm{h}}} \psi_{\mathrm{i}} \psi_{\mathrm{j}} \frac{\partial \psi_{\mathrm{k}}}{\partial x} \, \mathrm{d}V & v^{\mathrm{k}} \int_{\Omega_{\mathrm{h}}} \psi_{\mathrm{i}} \psi_{\mathrm{j}} \frac{\partial \psi_{\mathrm{k}}}{\partial y} \, \mathrm{d}V \end{bmatrix}$$

7.7.8 Write the three point discrete scheme for the Galerkin method applied to the stationary advection-diffusion equation (7.182). Use a linear finite element approximation (hat functions from chapter 3). The integrals in the weak formulation are analytically calculated on the reference element. Obtain equation (7.186).

7.7.9 Write the three point discrete scheme for the Galerkin method applied to equation (7.182). Use a linear finite element approximation for the interpolation functions and parabolic finite elements for the test functions (7.187). The integrals in the weak formulation are analytically calculated on the reference element. Obtain equation (7.188).

7.7.10 Prove relation (7.204) taking the partial derivative with respect to time of equation (7.203) and using definitions (7.201) and (7.202).

7.8 BIBLIOGRAPHY

J.F. Agassant, P. Avenas, P. Sergent, P. Carreau, *Polymer Processing, Principles and Modeling*, Hanser Publishers, Munich, 1991.

X. Aubert, *Transformation de coordonnées et schémas compacts aux différences finies appliquées à la résolution des équations de Navier-Stokes*, Thèse de doctorat, Faculté des Sciences Appliquées, Université catholique de Louvain, Louvain-la-Neuve, 1983.

D. Coles, "Transition in Circular Couette Flow", *J. Fluid Mech.*, vol. 21, pp. 385–425, 1965.

M. J. Crochet, A. R. Davies, K. Walters, *Numerical Simulation of a Non-Newtonian Flow*, Elsevier, Amsterdam, 1984.

M. J. Crochet, B. Debbaut, R. Keunings, J. M. Marchal, "Polyflow: a multi-purpose finite element program for continous polymer flows", in *Computer modelling for extrusion and other continous processess*, K. O. Brien Ed., Hanser, Munich, pp. 25–50, 1992.

J. Donea, A. Huerta, *Finite Element Methods for Flow Problems*, Wiley, New York, 2001.

P. Germain, *Cours de mécanique des milieux continus, tome 1, théorie générale*, Masson, Paris, 1973.

V. Girault, P.-A. Raviart, *Finite element methods for Navier-Stokes equations*, Springer, Berlin, 1986.

A.-P. Gonze, *Méthode directe de résolution d'écoulements turbulents discrétisés par les éléments finis*, Thèse de doctorat, Faculté des Sciences Appliquées, Université catholique de Louvain, Louvain-la-Neuve, 1994.

P. Gresho, R.L. Lee, R.L. Sani, "On the time-dependent solution of the incompressible Navier-Stokes equations in two and three dimensions", in *Recent Advances in Numerical Methods in Fluids*, Eds. C. Taylor, K. Morgan, Pineridge Press Ltd, Swansea, pp. 27–79, 1980.

P. Gresho, R. Sani, M.S. Engelman, *Incompressible Flow and the Finite Element Method. Advection-Diffusion and Isothermal Laminar Flow*, J. Wiley, West Sussex, U.K., 1998.

P.M. Gresho, R.L. Sani, *Incompressible Flow and the Finite Elemente Method*, vol. 1 and 2, Wiley, New York, 2001.

F. H. Harlow, J. E. Welch, "Numerical Calculation of Time-Dependent Viscous Incompressible Flow of Fluid with Free Surface", *Phys. Fluids*, vol. 8, pp. 2182–2189, 1965.

D. D. Joseph, *Fluid Dynamics of Viscoelastic Liquids*, Springer, New York, 1990.

V. Legat, *Méthodes d'éléments finis pour le calcul d'écoulements tridimensionnels à surfaces libres*, Thèse de doctorat, Faculté des Sciences Appliquées, Université catholique de Louvain, Louvain-la-Neuve, 1992.

J. M. Marchal, M. J. Crochet, "A new mixed finite element for calculating viscoelastic flow", *J. Non-Newtonian Fluid Mech,*, vol. 26, pp. 77-114, 1987.

J. M. Ortega, W. G. Poole, Jr., *An Introduction to Numerical Methods for Differential Equations*, Pitman, Marksfield, MA, 1981.

J. B. Perot, "An analysis of the fractional step method", *Journal of Computational Physics*, vol.108, p. 51–58, 1993.

O. Pironneau, *Finite Element Methods for Fluids*, J. Wiley, Chichester, 1989.

A. Rodriguez-Villa, *Etude théorique et expérimentale de l'extrusion-soufflage de corps creux en polymère*, Thèse de Doctorat en Sciences et Génie des Matériaux, Ecole des Mines de Paris, 1997.

I.L. Ryhming, *Dynamique des fluides*, Presses Polytechniques et Universitaires Romandes, Lausanne, 1991.

R. Sani, P. Gresho, R. Lee, D. Griffiths et M. Engleman, "The cause and cure (?) of the spurious pressures generated by certain FEM solutions of the incompressible Navier-Stokes equations", *Int. J. of Num. Meth. Fluids*, Vol. 1, p.17 (Part 1), p. 171 (Part 2), 1981.

C.G. SPEZIALE, "Analytical Methods for the Development of Reynolds-Stress Closures in Turbulence", *Annual Review of Fluid Mechanics*, vol. 23, pp. 107–157, 1991.

H. TENNEKES, J. L. LUMLEY, *A First Course in Turbulence*, MIT Press, Cambridge, MA, 1972.

F. THOMASSET, *Implementation of finite element methods for Navier-Stokes equations*, Springer, New-York, 1981.

J. VAN KAN, "A Second-Order Accurate Pressure-Correction Scheme for Viscous Incompressible Flow", *SIAM J. Sci. Stat. Comput.*, vol.7, pp. 870–891, 1986.

O.C. ZIENKIEWICZ, R.L. TAYLOR, Fluid Dynamics, McGraw-Hill, New York, 2001.

CHAPTER 8

INVERSE METHODS

8.1 OBJECTIVES

- Describe the general principles of inverse methods with simple examples.
- Show the equivalence of inverse methods and the common least squares method.
- Extend the principle of the method to more complex situations.

The application of the finite difference and finite element numerical methods to the solution of the conservation equations requires the definition of all the values that appear in the problem. In particular, it is necessary to specify the boundary and interface conditions, the initial conditions – if it is a non stationary problem – and all the material properties. In a ***direct simulation***, the boundary conditions and the properties are given at the start of modeling, using data from the literature for example. The quality of the simulation is judged by its correspondence with experiments performed in parallel. To have "good" agreement, requires that: 1) the important phenomena are correctly taken into account in the physical model via the state and conservation equation formalism; 2) the translation of the model into a computational program is free of programming errors and also of numerical defects (instabilities, oscillations, etc.); 3) the "ingredients" appearing in the model equations (boundary conditions, initial conditions, properties) are as close as possible to reality. If the agreement is judged unsatisfactory, one or more of these aspects must be suspected, and the source of the disagreement identified by work similar to that of a detective.

Assuming that the first two aspects mentioned above are valid, an ***inverse simulation*** consists of determining some of the ingredients appearing in the simulation on the basis of obtaining the *best agreement* possible, in terms of least square errors, between predicted and measured values. For example, the problem may be to find the heat flux at the boundary of a domain or the thermal conductivity of a medium that yields calculated temperatures that are "as close as possible" to those measured by thermocouples at certain points in the domain. One may also seek the mechanical properties of a material based on mechanical tests performed under complex conditions. Inverse methods are a very powerful means to interpret measurements based on conservation and state equations. In this chapter we propose to present the basics of such methods through simple situations, showing the equivalence that can exist with the usual least squares method.

8.2 A SIMPLE LINEAR ONE DIMENSIONAL PROBLEM

8.2.1 Linear one parameter least squares method

Consider first the very simple situation of the one dimensional stationary heat diffusion equation in a medium without advection or phase transformations (fig. 8.1):

$$\frac{d^2T}{dx^2} = 0 \tag{8.1}$$

with the boundary conditions:

$$T(x=0) = \bar{T}_0 \tag{8.2a}$$

$$\kappa\left[\frac{dT}{dx}\right]_{x=L} = q_T \tag{8.2b}$$

κ is the thermal conductivity of the medium. Assuming the temperature $\bar{T}_0$ known at $x = 0$, we propose to determine the flux q_T at $x = L$ based on the measured temperatures, $T_i^m(i = 1, N_m)$, at N_m points in the domain $[0, L]$ ($N_m = 3$ in figure 8.1). Having a single unknown to determine with N_m values, the flux q_T will be adjusted to these measurements according to least squares. This problem is trivial as an analytic solution exists for (8.1). Using the boundary conditions (8.2), we obtain the solution:

$$T(x) = \frac{q_T}{\kappa}x + \bar{T}_0 \tag{8.3}$$

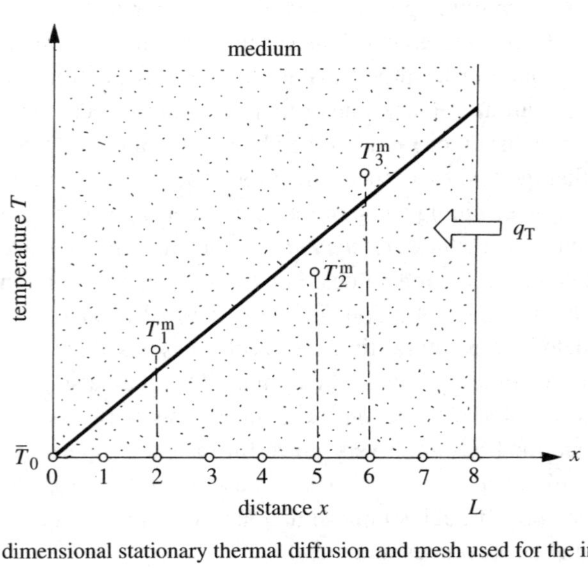

Figure 8.1 One dimensional stationary thermal diffusion and mesh used for the inverse method.

The flux q_T can thus be determined by minimizing the sum S of the squares of the differences between the measured and calculated temperatures, T_i^m and T_i

$$S = \sum_{i=1}^{N_m} [T_i^m - T_i]^2 = \sum_{i=1}^{N_m} \left[T_i^m - \left(\bar{T}_0 + \frac{q_T}{\kappa} x_i\right)\right]^2 \tag{8.4}$$

Contrary to the Einstein convention for repeated indices so often adopted in this book, the sums will most often be explicitly written in this chapter in order to more clearly distinguish the sums over measurement points, time, and parameters. The minimum of S is given by:

$$\frac{dS}{dq_T} = 0 = -2\sum_{i=1}^{N_m} \left[T_i^m - \bar{T}_0 - \frac{q_T}{\kappa} x_i\right] \frac{x_i}{\kappa} \tag{8.5}$$

which yields the value of q_T:

$$q_T = \frac{\sum_{i=1}^{N_m} [T_i^m - \bar{T}_0] x_i / \kappa}{\sum_{i=1}^{N_m} [x_i / \kappa]^2} \tag{8.6}$$

In this case the problem is that of a simple linear regression over N_m measurement points with the imposed value at the origin, $T(x = 0) = \bar{T}_0$ (only the slope is unknown).

8.2.2 Inverse method

Now assume that we do not know the analytic solution (8.3) and that we seek to determine q_T so as to yield a minimum S. For that, we employ a numerical solution, T_k, of the diffusion equation (8.1). For example, choosing the finite difference method (chap. 2) and a regular grid, we obtain the system of equations (2.14):

$$\begin{array}{llllll} 2T_1 & -T_2 & & & = & \bar{T}_0 \\ -T_1 & +2T_2 & -T_3 & & = & 0 \\ & -T_2 & +2T_3 & -T_4 & = & 0 \\ & \ldots & \ldots & \ldots & = & \ldots \\ & & -T_{N-1} & +T_N & = & \dfrac{q_T \Delta x}{\kappa} \end{array} \tag{8.7}$$

where N is the number of grid points ($N = 8$ in figure 8.1). Unlike the system of equations (2.14), in this case the grid includes a point located on the boundary $x = L$, as a Neumann condition is imposed there (sect. 2.2.6). In matrix form, this system of equations can also be written:

$$[\boldsymbol{D}]\boldsymbol{T}(q_T) = \boldsymbol{b}(q_T) \tag{8.8}$$

Note that the term on the right hand side is a function of the flux q_T, as is also the solution vector of temperatures at the grid nodes, $\boldsymbol{T}(q_T)$. Assuming that the solution of the system of equations (8.7) or (8.8) is unknown, how is q_T obtained in these conditions?

In order to avoid interpolation between grid points that will only complicate the expressions, the thermocouples are assumed to be located precisely at certain nodes of the grid, $k = kt(i)$, $i = 1, N_m$. The table kt contains the indices of the nodes of the thermocouples ($kt(1) = 2$, $kt(2) = 5$ and $kt(3) = 6$ in figure 8.1). Consider again the expression for S (8.4); this time the solution, T_i, of the diffusion equation is not explicitly known, but is given by the system of equations (8.7). The minimum of S is still given by the relation:

$$\frac{\mathrm{d}S}{\mathrm{d}q_T} = 0 = -2\sum_{i=1}^{N_m}[T_i^m - T_{kt(i)}]\frac{\mathrm{d}T_{kt(i)}}{\mathrm{d}q_T} = -2\sum_{i=1}^{N_m}[T_i^m - T_{kt(i)}]X_{kt(i)} \tag{8.9}$$

with:

$$X_{kt(i)} = \frac{\mathrm{d}T_{kt(i)}}{\mathrm{d}q_T} \tag{8.10}$$

The coefficient $X_{kt(i)}$ is the ***sensitivity*** or ***susceptibility coefficient*** with respect to the flux q_T at the position of the ith thermocouple (or the $kt(i)$th grid node). These coefficients can easily be obtained for all the grid points by differentiating with respect to q_T the system of equations (8.7). We obtain:

$$\begin{array}{lllll} 2X_1 & -X_2 & & & = 0 \\ -X_1 & +2X_2 & -X_3 & & = 0 \\ & -X_2 & +2X_3 & -X_4 & = 0 \\ & \ldots & \ldots & \ldots & = \ldots \\ & & & -X_{N-1} + X_N & = \dfrac{\Delta x}{\kappa} \end{array} \tag{8.11}$$

or:

$$[\boldsymbol{D}]\boldsymbol{X} = \frac{\mathrm{d}\boldsymbol{b}}{\mathrm{d}q_T} \tag{8.12}$$

It should be noted that by the form of the system of linear equations, the coefficient "the most sensitive" to the flux is X_N, the other coefficients decreasing with increasing distance from the boundary where the flux q_T is applied.

The system of equations (8.7) being linear in q_T, the temperature $T_k(q_T)$ appearing in equation (8.9) can be linearized:

$$T_k(q_T) = T_k(0) + \frac{dT_k}{dq_T} q_T = T_k(0) + X_k q_T \tag{8.13}$$

(the index k has been used rather than $kt(i)$ to simplify the expression). Inserting this relation in equation (8.9), and after rearrangement, we obtain:

$$q_T = \frac{\sum_{i=1}^{N_m} [T_i^m - T_{kt(i)}(0)] X_{kt(i)}}{\sum_{i=1}^{N_m} [X_{kt(i)}]^2} \tag{8.14}$$

Assuming thus that the analytic solution of the problem is unknown, the inverse problem consists in this case of the following steps:

1) Solving the equations (8.7) with $q_T = 0$ in order to find the temperature, $T_k(0) = T_k(q_T = 0)$ $(k = 1, N)$, at all the grid points;
2) Solving the system of equations (8.11) to find the sensitivity coefficients, $X_k (k = 1, N)$, at all the grid points;
3) Inserting into (8.14) the values $T_{k(i)}(0)$ and $X_{kt(i)}$ thus obtained for the thermocouples locations.

In this example it is easy to show that this somewhat complicated procedure is rigorously identical to the analytic solution (8.6). By Gaussian elimination, the solution of the two systems of equations (8.7) with $q_T = 0$ and (8.11) are, respectively:

$$T_k(q_T = 0) = \bar{T}_0 \; \forall k \qquad \text{and} \qquad X_k = \frac{k\Delta x}{\kappa} \tag{8.15}$$

where $k\Delta x = kt(i)\Delta x = x_i$ is the position of the ith thermocouple.

It is seen in this case that the inverse method consists of the solution of two systems of N equations, the first to obtain the temperatures $T_k(0)$ and the second to obtain the sensitivity coefficients X_k. Finally, note that it is not necessary to know how to derive the system of equations (8.7) to obtain the sensitivity coefficients. Step (2) above can be replaced by the calculation of the temperature $T_k(\delta q_T)$, with an arbitrary flux δq_T, using the system of equations (8.7). The coefficients X_k are then given by the differences:

$$X_k \simeq \frac{T_k(\delta q_T) - T_k(0)}{\delta q_T} \tag{8.16}$$

The calculated temperatures $T_k(\delta q_T)$ being linear functions of the flux q_T, the choice of δq_T is arbitrary.

8.3 A NON LINEAR ONE DIMENSIONAL PROBLEM

8.3.1 Non linear least squares method

The equivalence that exists between the inverse method and the least squares method can also be shown for the non stationary chemical diffusion problem treated in section 2.2.4 with the finite difference method (fig. 8.2). The analytic solution of the diffusion equation with an imposed concentration, $\bar{c}_0$, at the extremities of the domain $[0, L]$ is given by (2.30) and is written as follows:

$$c(x, t) = \bar{c}_0 - \frac{4\bar{c}_0}{\pi}\sum_{n=0}^{\infty}\frac{1}{(2n+1)}\sin\left[\frac{2n+1}{L}\pi x\right]\exp\left[-\frac{(2n+1)^2}{L^2}\pi^2 Dt\right] \tag{8.17}$$

where zero initial concentration in the domain $[0, L]$ has been assumed for simplicity. D is the diffusion coefficient of the solute element in the material. We propose to determine the latter on the basis of a certain number of concentration measurements, $c_i^{m,p}$, obtained, for example, by a microprobe at N_m points x_i in the domain and at M_m times t_p (fig. 8.2 where $N_m = 3$ and $M_m = 2$).

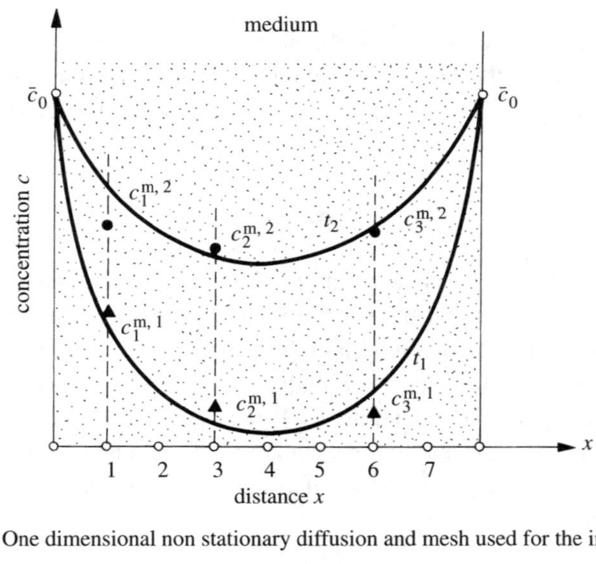

Figure 8.2 One dimensional non stationary diffusion and mesh used for the inverse method.

In a standard least squares method based on the analytic solution (8.17), one seeks the minimum of the sum S of the differences:

$$S = \sum_{i=1}^{N_m} \sum_{p=1}^{M_m} [c_i^{m,p} - c_i^p(D)]^2 \tag{8.18}$$

where the sum is this time over the N_m measurement points and the M_m time steps. The values $c_i^p(D)$ correspond to the values calculated with (8.17) at point x_i, time t_p and a value D of the diffusion coefficient. Proceeding as before, we have:

$$\frac{dS}{dD} = -2\sum_{i=1}^{N_m} \sum_{p=1}^{M_m} [c_i^{m,p} - c_i^p(D)]X_i^p = 0 \tag{8.19}$$

where X_i^p is now the sensitivity coefficient relative to the diffusion coefficient D for the concentration at point x_i and time t_p. It is defined as:

$$X_i^p = \frac{dc_i^p}{dD} = \frac{4\bar{c}_0\pi t_p}{L^2}\sum_{n=0}^{\infty}(2n+1)\sin\left[\frac{2n+1}{L}\pi x_i\right]\exp\left[-\frac{(2n+1)^2}{L^2}\pi^2 D t_p\right] \tag{8.20}$$

The solution of (8.19), non-linear in the coefficient D, can only be performed by an iterative method. Here we use a Newton method, the principles of which are reviewed in figure 8.3. Assuming that the value, $D^{(v)}$, of the diffusion coefficient is

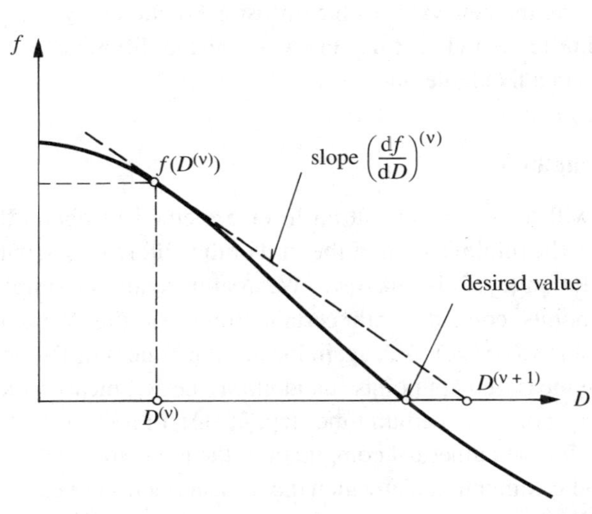

Figure 8.3 Principles of Newton's method for finding a zero of a function of a single parameter, $f(D)$.

known at an iteration, ν, the value $f^{(\nu)} = (dS/dD)^{(\nu)}$ given by (8.19) can be calculated with (8.17) and (8.20). The value $D^{(\nu+1)}$ at the following iteration is thus given by:

$$D^{(\nu+1)} = D^{(\nu)} - \frac{f^{(\nu)}}{[df/dD]^{(\nu)}} \tag{8.21}$$

The derivative $df/dD = d^2S/dD^2$ calculated at iteration ν can also be written as:

$$\left[\frac{df}{dD}\right]^{(\nu)} = \left[\frac{d^2S}{dD^2}\right]^{(\nu)} = -2\sum_{i=1}^{N_m}\sum_{p=1}^{M_m}[c_i^{m,p} - c_i^p(D^{(\nu)})]Z_i^p + 2\sum_{i=1}^{N_m}\sum_{p=1}^{M_m}[X_i^p]^2 \tag{8.22}$$

involving the second derivatives, Z_i^p, of the concentration c_i^p with respect to D:

$$Z_i^p = \frac{d^2c_i^p}{dD^2} = \frac{-4\bar{c}_0\pi^3t_p^2}{L^4}\sum_{n=0}^{\infty}(2n+1)^3\sin\left[\frac{2n+1}{L}\pi x_i\right]\exp\left[-\frac{(2n+1)^2}{L^2}\pi^2Dt_p\right] \tag{8.23}$$

The non linear least squares method consists of the following steps, starting from an initial estimation of the diffusion coefficient, $D^{(0)} = D^{(\nu=0)}$:

1) Calculating the concentrations $c_i^p(D^{(\nu)})$ (eq. (8.17)),
2) Calculating the first derivatives, or sensitivity coefficients $X_i^p(D^{(\nu)})$ (eq. (8.20)),
3) Calculating the second derivatives of the concentration $Z_i^p(D^{(\nu)})$ (eq. (8.23)),
4) Calculating the new value of the diffusion coefficient with equation (8.21),
5) Returning to step (1) setting $\nu := \nu + 1$ if the difference $|D^{(\nu+1)} - D^{(\nu)}|$ is larger than a fixed tolerance.

8.3.2 Inverse method

Now we will try to implement an inverse method to obtain the value of D, always based on the minimization of the sum of the differences S, but assuming that the analytic solution (8.17) is unknown. We assume again for simplify that the N_m measurement points coincide with certain points of the N point regular grid (fig. 8.4); the table $kt(i)$, $i = 1, N_m$, again locates the indices of the grid nodes corresponding to the measurement points. In addition, the M_m measurement times t_p are assumed to correspond to a certain time step, $jt(p)\Delta t$, of a set of M regularly spaced time steps used for the numerical computation (the table $jt(p)$, $p = 1, M_m$ locates the time indices in the computation at which the measurements are performed). In order to have more flexibility in the choice of time steps, Δt, and grid spacing, Δx, we choose an implicit FDM to solve the direct problem.

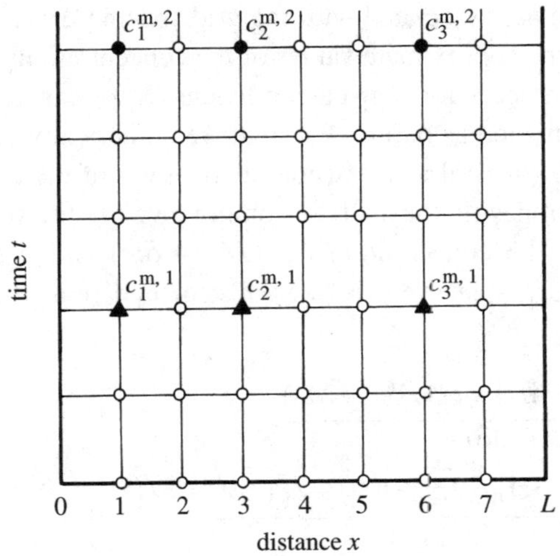

Figure 8.4 Space-time showing the correspondence between mesh points and measurement locations, and between calculation time steps and measurement times.

Starting from an estimation $D^{(0)}$ of the diffusion coefficient, we have seen in chapter 2 that the concentrations c_k^j can be calculated from the concentrations at the preceding time, c_k^{j-1}, by the system of equations (2.35) :

$$\boldsymbol{c}^j = [\boldsymbol{A}]^{-1}[\boldsymbol{b} + \boldsymbol{c}^{j-1}] \tag{8.24}$$

with:

$$[\boldsymbol{A}] = \begin{bmatrix} 1+2Fo & -Fo & & & & \\ -Fo & 1+2Fo & -Fo & & & \\ & -Fo & 1+2Fo & -Fo & & \\ & & \dots & \dots & \dots & \\ & & & -Fo & 1+2Fo & -Fo \\ & & & & -Fo & 1+2Fo \end{bmatrix}$$

$$\boldsymbol{b} = \begin{bmatrix} Fo\bar{c}_0 \\ 0 \\ 0 \\ \dots \\ \dots \\ 0 \\ Fo\bar{c}_0 \end{bmatrix} \quad \text{and} \quad \boldsymbol{c}^{j-1} = \begin{bmatrix} c_1^{j-1} \\ c_2^{j-1} \\ \dots \\ \dots \\ c_{N-1}^{j-1} \\ c_N^{j-1} \end{bmatrix} \tag{8.25}$$

Fo is the Fourier number associated with the grid spacing Δx and the time step Δt ($Fo = D\Delta t/\Delta x^2$). As the approximate values of the concentrations $c_k^j(D)$ can be calculated by finite differences for a given coefficient D, we can apply the same non linear least squares method as before. For that, we numerically calculate the sensitivity coefficients, X_k^j, as well as the second derivatives of the concentrations, Z_k^j, with respect to D. Not having the analytic solution, we use the system of equations (8.24) and calculate the concentrations $c_k^j(D^{(\nu)} + \delta D)$ and $c_k^j(D^{(\nu)} - \delta D)$. The coefficients X_k^j and Z_k^j appearing in Newton's method (eqs. (8.21–22)) are then simply given by:

$$X_k^j = \frac{c_k^j(D^{(\nu)} + \delta D) - c_k^j(D^{(\nu)} - \delta D)}{2\delta D} \tag{8.26a}$$

$$Z_k^j = \frac{c_k^j(D^{(\nu)} + \delta D) - 2c_k^j(D^{(\nu)}) + c_k^j(D^{(\nu)} - \delta D)}{\delta D^2} \tag{8.26b}$$

The inverse method that permits the coefficient D to be found, giving "the best possible agreement" with the experimental measurements, thus consists of:

1) Estimating an initial value of the coefficient $D^{(0)} = D^{(\nu = 0)}$,
2) Numerically calculating the concentrations $c_k^j(D^{(\nu)} + \delta D)$, $c_k^j(D^{(\nu)})$ and $c_k^j(D^{(\nu)} - \delta D)$ with an implicit method (eq. (8.24)),
3) Numerically calculating the first derivatives, X_k^j, and second derivatives, Z_k^j, of the concentrations c_k^j (eq. (8.26)),
4) Deducing the new value of $D^{(\nu + 1)}$ (eq. (8.21)), using the numerical expressions of $c_{kt(i)}^{jt(p)}$, $X_{kt(i)}^{jt(p)}$, $Z_{kt(i)}^{jt(p)}$ at the measurement points $kt(i)$, $i = 1, N_m$, and measurement times $jt(p)$, $p = 1, M_m$.
5) Returning to step (2) if $|D^{(\nu + 1)} - D^{(\nu)}| > \varepsilon_D$ setting $\nu := \nu + 1$.

The value chosen for δD is somewhat arbitrary. It should be sufficiently small to correctly approximate the derivatives with differences. However, the calculation of a derivative by (8.26a) is subject to instability and numerical imprecision. The latter because subtraction causes the loss of significant figures from the mantissa if the values of $c_k^j(D^{(\nu)} + \delta D)$ and $c_k^j(D^{(\nu)} - \delta D)$ are too close. In practice, $\delta D/D \simeq 0.001 - 0.01$ is chosen.

The solution of the direct implicit problem for M time steps (M not necessarily equal to M_m) requires the solution of M systems of equations with N unknowns (8.24). Note that in the case of a constant time step, the matrix $[\mathbf{A}]$ is constant and needs only be factored once by a Gaussian elimination method. In the general case where, for example, non constant mixed boundary conditions are imposed, the matrix is not constant and the linear system must be inverted at each time step. It is seen with the inverse method that at each time step the same system of equations must be solved three times with different values $D^{(\nu)}$, $D^{(\nu)} + \delta D$ and $D^{(\nu)} - \delta D$ (or $Fo^{(\nu)}$, $Fo^{(\nu)} + \delta Fo$ and $Fo^{(\nu)} - \delta Fo$), before being able to calculate X_k^j and Z_k^j, and thus the new value $D^{(\nu + 1)}$.

These calculations can be somewhat simplified by taking a different linearization method than Newton's. Return to the expression (8.19) giving the minimum of the differences, and linearize it taking the sensitivity coefficients at iteration ν and the concentrations $c_k^j(D^{(\nu+1)})$ at the following iteration. The latter can be linearized by taking the first two terms of the Taylor series expansion:

$$\begin{aligned} c_k^j(D^{(\nu+1)}) &\simeq c_k^j(D^{(\nu)}) + \left[\frac{dc_k^j}{dD}\right]^{(\nu)}(D^{(\nu+1)} - D^{(\nu)}) \\ &= c_k^j(D^{(\nu)}) + [X_k^j]^{(\nu)}(D^{(\nu+1)} - D^{(\nu)}) \end{aligned} \tag{8.27}$$

thus introducing again the sensitivity coefficients X_k^j at iteration ν. Inserting this expression in (8.19) for the chosen measurement points and time steps:

$$D^{(\nu+1)} = D^{(\nu)} + \frac{\sum_{i=1}^{N_m}\sum_{p=1}^{M_m}[c_i^{m,p} - c_{kt(i)}^{jt(p)}(D^{(\nu)})]X_{kt(i)}^{jt(p)}(D^{(\nu)})}{\sum_{i=1}^{N_m}\sum_{p=1}^{M_m}[X_{kt(i)}^{jt(p)}(D^{(\nu)})]^2} \tag{8.28}$$

It is easy to show that this linearization method, called the Gauss-Newton method, is equivalent to neglecting the second derivatives, Z_k^j, of the concentrations in Newton's method. It is simpler because only the sensitivity coefficients, $X_k^j(D^{(\nu)})$, appear. They can be numerically calculated by solving the system of equations (8.24) two times, with the values $D^{(\nu)}$ and $D^{(\nu)} + \delta D$, and taking the skewed derivatives:

$$X_k^j(D^{(\nu+1)}) = \frac{c_k^j(D^{(\nu)} + \delta D) - c_k^j(D^{(\nu)})}{\delta D} \tag{8.29}$$

We will generalize this latter method to the case where more than one parameter is to be determined with an inverse method.

8.4 INVERSE METHOD WITH TIME INDEPENDENT PARAMETERS

8.4.1 Inverse method based on least squares

As long as the available measurements are sufficiently numerous and precise, the inverse method described above to determine a single parameter (the heat flux or the diffusion coefficient, for example) can be generalized to the case of many parameters. The method is also not limited to a single conservation equation and can take into account coupled equations. However, for simplicity, we assume in the following that we have a set of equations representing the FDM or FEM formulation of a single conservation equation. If the problem is stationary, this equation can in most cases be written in condensed form:

$$[\boldsymbol{A}]\boldsymbol{T} = \boldsymbol{b} \tag{8.30}$$

while if the problem evolves in time, it becomes:

$$[\boldsymbol{A}]\boldsymbol{T}^{\mathrm{j}} = \boldsymbol{b} + \boldsymbol{T}^{\mathrm{j}-1} \tag{8.31}$$

In the two cases, the equations above are associated with the solution of a direct problem where the boundary conditions, the initial conditions, and the properties are specified. We have also retained "T" for the variable name as the inverse methods are mainly applied to thermal problems. In this case, (8.30) and (8.31) are obtained from the stationary and non stationary heat equations, respectively. But a chemical diffusion or mechanical deformation problem can be treated in the same manner.

Given that certain data for the direct problem are absent, we propose to determine them with experimental measurements based on the best possible agreement between modeling and experimentation. For a first estimate, the missing data are assumed to be time independent. We denote by $\boldsymbol{\beta}$ the vector composed of the N_β missing data, β_s, $s = 1, N_\beta$:

$$\boldsymbol{\beta} = (\beta_1, \beta_2, \ldots, \beta_{N_\beta - 1}, \beta_{N_\beta}) \tag{8.32}$$

For example, if the material has a thermal conductivity with unknown dependence on temperature, the vector $\boldsymbol{\beta} = \boldsymbol{\kappa}$ can be composed of N_β values of thermal conductivity taken at given temperatures (fig. 8.5) or by N_β values of a polynomial expansion of the conductivity. In the first case, the thermal conductivity is given by linear interpolation between these values, while in the second case it is expressed by the polynomial of order $(N_\beta - 1)$:

$$\kappa(T) = \kappa_1 + \kappa_2 T + \kappa_3 T^2 + \ldots + \kappa_{N_\beta} T^{N_\beta - 1} \tag{8.33}$$

This vector of missing data can also be composed of boundary conditions independent of time but functions of the position. For example, in the two dimensional domain in figure 8.6, we have five values, β_s, to determine on three portions of the boundary $\partial\Omega$ of the domain: on $\partial\Omega_2$, the condition is of Cauchy type and the heat transfer coefficient is assumed to depend linearly on y: $h_T = (\beta_1 + \beta_2 y)$; on $\partial\Omega_3$, the heat flux q_T (Neumann condition) depends linearly on x: $q_T = \beta_3 + \beta_4 x$; finally, on $\partial\Omega_4$, the flux $q_T = \beta_5$ is uniform.

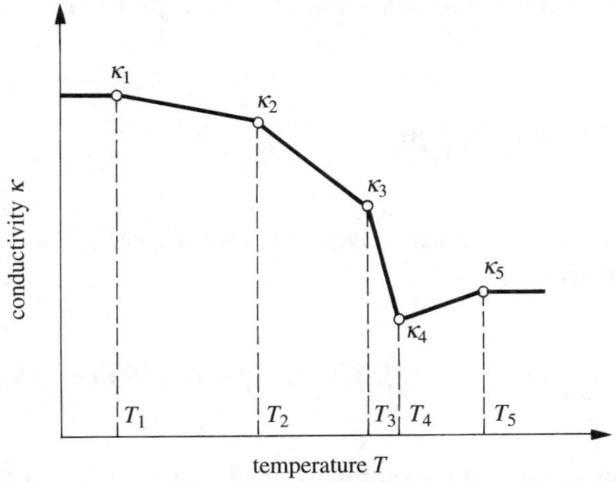

Figure 8.5 Tabulated values of the thermal conductivity $\{\kappa_1, \kappa_2, \ldots, \kappa_{N_\beta}\}$ at temperatures $\{T_1, T_2, \ldots, T_{N_\beta}\}$.

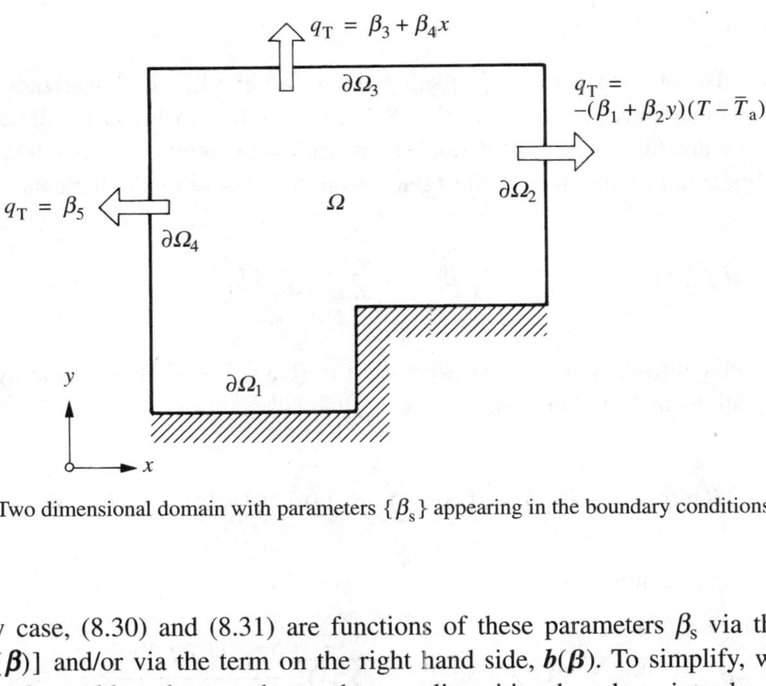

Figure 8.6 Two dimensional domain with parameters $\{\beta_s\}$ appearing in the boundary conditions.

In any case, (8.30) and (8.31) are functions of these parameters β_s via the matrix $[\boldsymbol{A}(\boldsymbol{\beta})]$ and/or via the term on the right hand side, $\boldsymbol{b}(\boldsymbol{\beta})$. To simplify, we assume that the problem does not have other non linearities than those introduced by the parameters to be determined. The solution $T_k^j(\boldsymbol{\beta}) = T_k^j(\beta_1, \beta_2, \ldots, \beta_{N_\beta})$ is then also a function of these parameters. As before, we attempt to determine these parameters based on the minimum difference between measured values, $T_i^{m,p}$, and calculated values, $T_{kt(i)}^{jt(p)}$, at the corresponding nodes of a mesh, $kt(i)$, $i = 1, N_m$, and for certain calculation time steps, $jt(p)$, $p = 1, M_m$. We again minimize the sum S:

$$S = \sum_{i=1}^{N_m} \sum_{p=1}^{M_m} [T_i^{m,p} - T_{kt(i)}^{jt(p)}(\boldsymbol{\beta})]^2 \tag{8.34}$$

Proceeding as before, the derivatives of S with respect to each of the coefficients β_s are set to zero:

$$\frac{\partial S}{\partial \beta_s} = -2\sum_{i=1}^{N_m} \sum_{p=1}^{M_m} [T_i^{m,p} - T_{kt(i)}^{jt(p)}(\boldsymbol{\beta})]\ {}^{s}X_{kt(i)}^{jt(p)} = 0 \qquad \text{for } s = 1, N_\beta \tag{8.35}$$

We see the appearance of the sensitivity coefficient, ${}^{s}X_{kt(i)}^{jt(p)}$, for the calculated temperature at point $kt(i)$ and time $jt(p)$ with respect to the coefficient β_s. It is defined as:

$${}^{s}X_{kt(i)}^{jt(p)} = \frac{\partial T_{kt(i)}^{jt(p)}}{\partial \beta_s} \tag{8.36}$$

To solve the set of N_β equations in β_s, we employ the linearization technique previously encountered in section 8.3.2, taking the sensitivity coefficients at iteration ν and the value of the variable at the following iteration. Expanding the latter in a linear approximation, we introduce again the sensitivity coefficients:

$$T_{kt(i)}^{jt(p)}(\boldsymbol{\beta}^{(\nu+1)}) = T_{kt(i)}^{jt(p)}(\boldsymbol{\beta}^{(\nu)}) + \sum_{r=1}^{N_\beta} {}^{r}X_{kt(i)}^{jt(p)}[\beta_r^{(\nu+1)} - \beta_r^{(\nu)}] + \ldots \tag{8.37}$$

By introducing this last expression in (8.35), we obtain a linear system of N_β equations in N_β unknowns:

$$[\boldsymbol{B}](\boldsymbol{\beta}^{(\nu+1)} - \boldsymbol{\beta}^{(\nu)}) = \boldsymbol{f} \ \text{ or } \ \sum_{r=1}^{N_\beta} B_{sr}(\beta_r^{(\nu+1)} - \beta_r^{(\nu)}) = f_s \quad s = 1, N_\beta \tag{8.38}$$

with the definitions:

$$B_{sr} = \sum_{i=1}^{N_m} \sum_{p=1}^{M_m} {}^{s}X_{kt(i)}^{jt(p)}\ {}^{r}X_{kt(i)}^{jt(p)} = \sum_{i=1}^{N_m} \sum_{p=1}^{M_m} \frac{\partial T_{kt(i)}^{jt(p)}}{\partial \beta_s} \frac{\partial T_{kt(i)}^{jt(p)}}{\partial \beta_r} \tag{8.39a}$$

$$f_s = \sum_{i=1}^{N_m} \sum_{p=1}^{M_m} [T_i^{m,p} - T_{kt(i)}^{jt(p)}(\boldsymbol{\beta}^{(\nu)})]\ {}^{s}X_{kt(i)}^{jt(p)} \tag{8.39b}$$

By inversion of the matrix $[\boldsymbol{B}]$, the system of equations (8.38) allows the increment $\Delta\boldsymbol{\beta} = (\boldsymbol{\beta}^{(\nu+1)} - \boldsymbol{\beta}^{(\nu)})$ to be found. The sensitivity coefficients are numerically calculated using the relations:

$$ {}^{s}X_{k}^{j} = \frac{T_{k}^{j}(\beta_{1}^{(\nu)}, \ldots, \beta_{s}^{(\nu)} + \delta\beta_{s}, \ldots, \beta_{N_\beta}^{(\nu)}) - T_{k}^{j}(\beta_{1}^{(\nu)}, \ldots, \beta_{s}^{(\nu)}, \ldots, \beta_{N_\beta}^{(\nu)})}{\delta\beta_{s}} \tag{8.40} $$

Then, if one can solve a direct problem such as that given by (8.30) or (8.31), the inverse method to determine a set of N_β parameters $(\beta_1, \beta_2, \ldots, \beta_{N_\beta})$ is the following:

1) Starting from an estimation $\boldsymbol{\beta}^{(0)} = \boldsymbol{\beta}^{(\nu=0)} = (\beta_1^{(0)}, \beta_2^{(0)}, \ldots, \beta_{N_\beta}^{(0)})$ at iteration $\nu = 0$,
2) Calculating a set of values of the function $T_k^j(\beta_1^{(\nu)}, \beta_2^{(\nu)}, \ldots, \beta_{N_\beta}^{(\nu)})$ for all the points of the mesh, k, and all the time steps, j, using the values of the parameters of the preceding iteration, ν,
3) Calculating the set of values of the function $T_k^j(\beta_1^{(\nu)}, \beta_2^{(\nu)}, \ldots, \beta_s^{(\nu)} + \delta\beta_s, \ldots, \beta_{N_\beta}^{(\nu)})$ by successively "perturbing" each of the values, $\beta_s^{(\nu)}$, by an increment $\delta\beta_s$ (typically $\delta\beta_s/\beta_s \simeq 0.001 - 0.01$). This step is equivalent to repeating the direct calculation of step (2) above N_β times,
4) Starting from the values calculated in (2) and (3), deduction of the values of the sensitivity coefficients ${}^sX_k^j$ at the measurement points $k = kt(i)$, $i = 1, N_m$, and at the times $j = jt(p)$, $p = 1, M_m$ with eqs. (8.40),
5) Calculating the elements B_{sr} of the matrix $[B]$ (eq. (8.39a)) and the terms f_s (eq. (8.39b)) of the right hand side of (8.38),
6) Solving the system equations (8.38) to obtain the new values of the parameters $\beta_s^{(\nu+1)}$. If these new values are too different from the preceding, recommencing at step (2) with $\nu := \nu + 1$.

It is seen that the method is similar to that presented earlier for a single parameter (Sect. 8.2.2) when the second derivatives appearing in Newton's method were neglected. The problem here is a bit more involved as, for each iteration, it is equivalent to solving a direct problem (steps (2) and (3) above) $(N_\beta + 1)$ times before being able to calculate the sensitivity coefficients and the values of the function appearing in the least squares method. The advantage of such a formalism is that from the moment one has a code for solving a direct problem, it is not difficult to adapt it for steps (2) and (3) of the procedure established above.

The possibility should be mentioned however, of calculating the sensitivity coefficients analytically, which permits the avoidance of the N_β direct calculations of step 3 above which would be necessary to numerically calculate these coefficients. For that, we start again with (8.31) which is written in the form:

$$ \boldsymbol{R}^j = [\boldsymbol{A}]\boldsymbol{T}^j - \boldsymbol{b} - \boldsymbol{T}^{j-1} = \boldsymbol{0} \tag{8.41} $$

Considering the component n of the vector $\boldsymbol{R}^j$, we need to have for every parameter β_s:

$$\frac{\partial R_n^j}{\partial \beta_s} = \sum_{m=1}^{N} \left[\frac{\partial A_{nm}}{\partial \beta_s} T_m^j + A_{nm} \frac{\partial T_m^j}{\partial \beta_s} \right] - \frac{\partial b_n}{\partial \beta_s} - \frac{\partial T_n^{j-1}}{\partial \beta_s} = 0 \tag{8.42}$$

$$\forall s = 1, N_\beta ; \forall n = 1, N$$

Thus, if the derivatives with respect to β_s of the matrix $[\boldsymbol{A}]$ and the right hand side vector $\boldsymbol{b}$, can be analytically calculated, the sensitivity coefficients ${}^sX_m^j$ at the following time step can be directly calculated from (8.42) by solving the linear system of equations:

$$\sum_{m=1}^{N} A_{nm} {}^sX_m^j = -\sum_{m=1}^{N} \left[\frac{\partial A_{nm}}{\partial \beta_s} T_m^j \right] + \frac{\partial b_n}{\partial \beta_s} + {}^sX_n^{j-1} \tag{8.43}$$

This assumes of course that the sensitivity coefficients at the preceding time step, ${}^sX_n^{j-1}$, are known. Note that with a Gaussian elimination method, the matrix $[\boldsymbol{A}]$ has already been factored for the solution of the direct problem. Then, although (8.43) must be solved in addition to the direct problem, it is at marginal cost, since we can now avoid the N_β direct calculations necessary to deduce the sensitivity coefficients by the perturbation method. This method is thus particularly interesting when there are several parameters to be determined or when the direct calculation is very costly.

As in all traditional least squares methods for non linear parameters, it is necessary to pay attention to the convergence of the problem. If the initial estimates are too far from the solution, or if the influence of the parameters on the function values is not sufficiently "pronounced", the solution will not converge! In certain cases when convergence proves to be difficult, it may be necessary to apply Newton's method, and thus to have to calculate also the second derivatives of the function with respect to the parameters. This introduces mixed derivatives which considerably encumber the calculations and multiply the number of direct problems to solve at each iteration. Of course, other linearization techniques can be envisioned (chap. 4), for example, calculating the parameter corrections by relaxation at each iteration.

A method to protect against ridiculous results that is simple enough to implement is presented in the next section; it permits the limitation of the variation range, and thus avoids absurd values of the parameters β_s.

8.4.2 Maximum a posteriori method (MAP)

The *maximum a posteriori* method (MAP) consists of adding another sum to the sum of the squares of the differences of measured and calculated values; this new sum measures the differences of the parameter values, β_r, with respect to *a priori* estimated values, $\bar{\beta}_r$. Normalizing the sums by the corresponding averaged variances, the MAP method amounts to seeking the minimum of:

$$S = \frac{1}{\sigma_T^2} \sum_{i=1}^{N_m} \sum_{p=1}^{M_m} [T_i^{m,p} - T_{kt(i)}^{jt(p)}(\boldsymbol{\beta})]^2 + \sum_{r=1}^{N_\beta} \left[\frac{\beta_r - \bar{\beta}_r}{\sigma_r}\right]^2 \tag{8.44}$$

For simplicity, we have assumed that the error σ_T in each of the measurements of the temperatures $T_i^{m,p}$ was the same. The variation range that is permitted for each of the parameters, β_r, is fixed by σ_r: as σ_r goes to zero, the minimization of S imposes $\beta_r = \bar{\beta}_r$, while for very large σ_r, the optimization with respect to the corresponding parameter will be equivalent to the usual least squares method. The preliminary definition of the estimated values, $\bar{\boldsymbol{\beta}}$, and the standard deviations, $\boldsymbol{\sigma}$, thus allows searching for the values $\boldsymbol{\beta}$, which give the best agreement between the calculated values, $T_{kt(i)}^{jt(p)}$, and measured values, $T_i^{m,p}$, while limiting the variation of the parameters to a certain range.

The derivative of (8.44) with respect to β_s gives:

$$\frac{\partial S}{\partial \beta_s} = -\frac{2}{\sigma_T^2} \sum_{i=1}^{N_m} \sum_{p=1}^{M_m} [T_i^{m,p} - T_{kt(i)}^{jt(p)}(\boldsymbol{\beta})]\, {}^sX_{kt(i)}^{jt(p)} + \frac{2}{\sigma_s^2}(\beta_s - \bar{\beta}_s) = 0 \tag{8.45}$$

Adopting the linearization technique previously presented, it can easily be shown that the value of $\boldsymbol{\beta}^{(\nu+1)}$ at the next iteration is given by:

$$[\boldsymbol{B}]\boldsymbol{\beta}^{(\nu+1)} = \boldsymbol{f} + [\boldsymbol{B}]\boldsymbol{\beta}^{(\nu)} \quad \text{or} \quad \sum_{r=1}^{N_\beta} B_{sr}\beta_r^{(\nu+1)} = f_s + \sum_{r=1}^{N_\beta} B_{sr}\beta_r^{(\nu)} \quad \text{for } s = 1, N_\beta \tag{8.46}$$

with the definitions:

$$B_{sr} = \frac{1}{\sigma_T^2} \sum_{i=1}^{N_m} \sum_{p=1}^{M_m} {}^sX_{kt(i)}^{jt(p)}\, {}^rX_{kt(i)}^{jt(p)} + \frac{1}{\sigma_s^2}\delta_{sr} \tag{8.47a}$$

$$f_s = \frac{1}{\sigma_T^2} \sum_{i=1}^{N_m} \sum_{p=1}^{M_m} [T_i^{m,p} - T_{kt(i)}^{jt(p)}(\boldsymbol{\beta}^{(\nu)})]\, {}^sX_{kt(i)}^{jt(p)} + \frac{1}{\sigma_s^2}(\bar{\beta}_s - \beta_s^{(\nu)}) \tag{8.47b}$$

where δ_{sr} is the Kronecker delta function. Except for the slightly different form of the matrices and vectors (eqs. (8.46) and (8.47)), the MAP search strategy for the parameters $\boldsymbol{\beta}$ is identical to that presented earlier.

8.5 INVERSE METHOD WITH TIME DEPENDENT PARAMETERS

In numerous non stationary problems, it is necessary to determine parameters that are functions of time. For the problems most often studied with inverse methods, those of heat diffusion, it is generally the boundary conditions that vary. It can be the temperature, $\overline{T}(t)$, the flux, $q_T(t)$, or the heat transfer coefficient, $h_T(t)$, at the domain surface the values of which are functions of time. Or, the problem may be to estimate the heat transfer between two media of different temperatures, one of which undergoes a phase transition. The variable thickness of the thin air layer (or vacuum) that exists inevitably between these two media will create non stationary heat transfer conditions. In general, this type of inverse problem is "ill posed". Why?

8.5.1 "Ill posed" nature of the inverse problem

Consider the one dimensional semi infinite domain shown in figure 8.7 for which the surface temperature at $x = 0$, $T_0(t)$, is an unknown function of time to be determined by the inverse method. For that, we have a measurement of the temperature $T_L^m(t) = T^m(x = L, t)$ obtained at a depth L in the material. The flux $q_0(t)$ at the surface could of course have been chosen as the unknown in place of the temperature $T_0(t)$. The domain volume is assumed to be initially at temperature zero. In

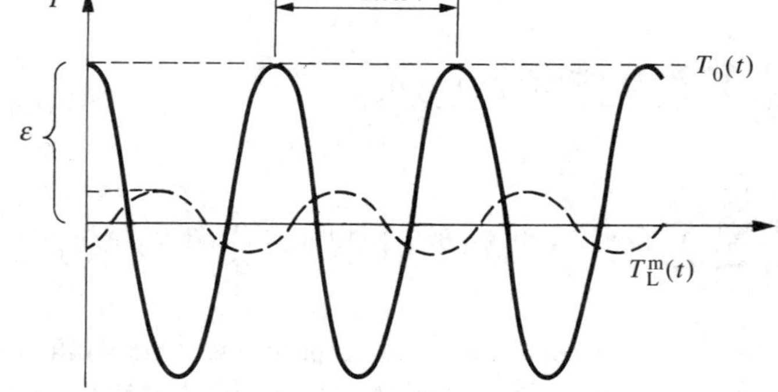

Figure 8.7 Sinusoidal temperature variations at the surface of a semi infinite domain and values measured at depth L.

this idealized situation, we can first solve the direct problem – fix $T_0(t)$ and calculate $T_L^m(t)$ – then invert the solution. Different solutions are proposed for this problem in the excellent book by Carslaw and Jaeger referred to in the bibliography.

To show the ill posed nature of the inverse problem, we choose to impose sinusoidal variations of the temperature on the surface $x = 0$:

$$T_0(t) = \varepsilon \cos(\omega t) \tag{8.48}$$

where ε is the amplitude of this perturbation and $\omega = 2\pi\tau^{-1}$ its frequency (τ is the period). Carslaw and Jaeger show that the steady state oscillating solution measured at distance $x = L$ is given by:

$$T_L^m(t) = \varepsilon \exp\left[-L\left(\frac{\omega}{2\alpha}\right)^{1/2}\right]\cos\left[\omega t - L\left(\frac{\omega}{2\alpha}\right)^{1/2}\right] \tag{8.49}$$

where α is the thermal diffusivity of the medium. It is seen that the imposed perturbation is *damped* exponentially with a damping factor given by the ratio (L/λ), where λ is the thermal diffusion layer associated with the sinusoidally pulsing perturbation $\lambda \simeq \sqrt{2\alpha/\omega}$. It appears clearly that the rapid temperature variations on the surface $x = 0$ will be strongly damped with distance, all the more if the material is a poor conductor of heat. Note also the appearance of a phase shift of the perturbation in the solution (8.49), identical in its form to the damping factor.

Now assume that the small sinusoidal variations of the temperature at depth $x = L$ are measured in an inverse method. Interpreted in terms of temperature variations at the domain surface, these correspond to:

$$T_0(t) = T_L^m(t)\frac{\cos(\omega t)}{\cos\left[\omega t - L\left(\frac{\omega}{2\alpha}\right)^{1/2}\right]}\exp\left[L\left(\frac{\omega}{2\alpha}\right)^{1/2}\right] \tag{8.50}$$

It is seen that in the inverse problem, aside from the phase shift terms, the damping factor is transformed into an amplification factor. Thus, if the temperature measured at $x = L$ has noise of frequency ω and amplitude ε, the noise will be multiplied by this amplification factor for the surface temperature. Even for a metal that is a very good conductor of heat such as pure aluminum ($\alpha \simeq 10^{-4}$ m^2 s^{-1}), noise of 0.1 °C at 50 Hz measured at a depth of 1 cm corresponds to surface temperature variations of 700 °C !

In other words, the characteristic diffusion time, t_c, associated with the distance between the measurement point and the surface being fixed ($t_c \sim L^2/\alpha$), a small temperature variation (noise) measured during a period $\tau \sim \omega^{-1} << t_c$ can only be interpreted by a very large temperature (or flux) variation at the surface. Figure 8.8 shows this phenomenon. Having imposed a flux of 500 kW m^{-2} at the surface, the temperatures in the interior of the one dimensional domain were calculated with

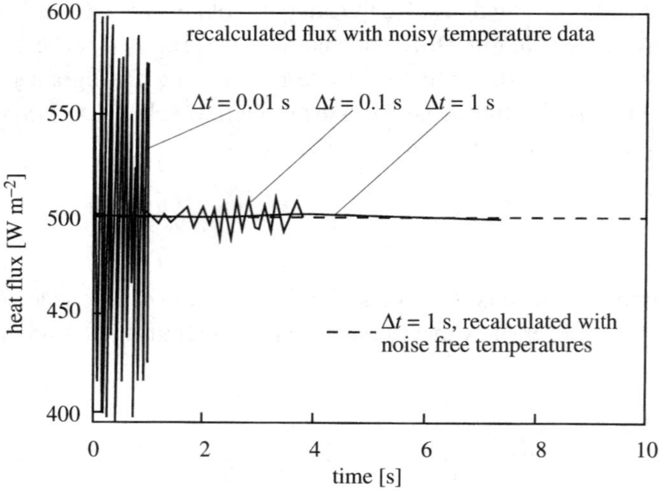

Figure 8.8 Flux value determined by the inverse method ($q_T = 500$ kW m^{-2}) from calculated temperatures with added noise of ± 0.5 °C. The various fluxes were obtained by varying the time step.

the FDM. A random noise of ± 0.5 °C was then added to the calculated temperatures for calculating the flux by the inverse method. While the average flux deduced in this way corresponds to the expected value, it is seen that the noise grows rapidly with a reduction in the time step (that is, an increase in ω).

Many so called *regularization* techniques exist to limit the noise introduced by non stationary inverse methods. The interested reader is referred to the book by Beck, Blackwell and St Clair in the bibliography. In the following section we will only treat the technique at a single time step, assuming that it is sufficiently large with respect to the characteristic diffusion times associated with the measurement points.

8.5.2 Single time step method to determine the flux

In the following we consider the domain shown in figure 8.1. The heat flow is assumed to be uniquely in the x direction (one dimensional problem). Unlike the case illustrated in figure 8.1 where we were interested in the steady state, here we are interested in the temperature evolution over time. We want to determine the values of the flux, q_T^j, at $x = L$ for time steps $j = 1, M_m$ which are assumed regular. For that we have the temperatures, $T_i^{m,j}$, measured at N_m locations in the domain ($i = 1, N_m$) at the same times. For each time step j, the problem is to minimize the sum of the squared differences:

$$S = \sum_{i=1}^{N_m} [T_i^{m,j} - T_{kt(i)}^j(q_T^j)]^2 \tag{8.51}$$

where T_k^j is again the calculated temperature at the mesh points. As in general, the number of mesh nodes is larger than the number of thermocouples, we again employ a table $kt(i)$ to locate the mesh nodes corresponding to the measurement points (fig. 8.1). We proceed as before, differentiating S with respect to q_T^j:

$$\frac{\partial S}{\partial q_T^j} = -2\sum_{i=1}^{N_m}[T_i^{m,j} - T_{kt(i)}^j(q_T^j)]X_{kt(i)}^j$$

$$= -2\sum_{i=1}^{N_m}[T_i^{m,j} - T_{kt(i)}^j(q_T^j)]\frac{\partial T_{kt(i)}^j}{\partial q_T^j} = 0 \tag{8.52}$$

which again introduces the sensitivity coefficient, $X_{kt(i)}^j$, for the temperature, $T_{kt(i)}^j$, with respect to a flux variation, q_T^j. Expanding the temperatures inside the parenthesis in the above equation around the flux value at the previous time step, q_T^{j-1}, it is easy to show that the flux increment Δq_T^j during this time step is given by:

$$\Delta q_T^j = \frac{\sum_{i=1}^{N_m}[T_i^{m,j} - T_{kt(i)}^j(q_T^{j-1})]X_{kt(i)}^j}{\sum_{i=1}^{N_m}[X_{kt(i)}^j]^2} \tag{8.53}$$

The method thus consists of the following process at each time step:

1) Calculating the temperatures, T_k^j at the mesh nodes with the flux, q_T^{j-1}, and the temperatures, T_k^{j-1}, of the previous time step,
2) Calculating the sensitivity coefficients, X_k^j, either by directly differentiating with respect to the flux the system of equations governing the calculation of the temperature field (system of equations analogous to (8.11)), or by perturbing the flux by δq_T and recalculating $T_k^j(q_T^{j-1} + \delta q_T)$,
3) Calculating the flux increment, Δq_T^j (eq. (8.53)), to obtain the new flux, $q_T^j = q_T^{j-1} + \Delta q_T^j$,
4) Updating the calculated temperatures, T_k^j, with the new value of the flux before moving on to the next time step. Note that these temperatures can be updated, either by a new FDM/FEM computation, which is rather expensive, or with the relations:

$$T_k^j(q_T^j) = T_k^j(q_T^{j-1}) + X_k^j \Delta q_T^j \tag{8.54}$$

8.6 EXAMPLES

In this section we give two examples for the application of inverse methods to thermal problems, as well as an extension of this type of method for the determination of rheological parameters.

8.6.1 Determination of boundary conditions

The first example is the determination of the heat transfer coefficient at the surface of a steel cylinder cooled laterally by a water spray (fig. 8.9). Such a configuration is encountered in the surface hardening of a part by quenching. In a first step, the part is heated by induction to transform the initial ferrito-pearlitic structure into austenite up to a certain depth. The surface region of the part is then rapidly cooled with a fine water spray that transforms the austenite into martensite. It is thus possible to obtain increased hardness of the steel at the surface while retaining the ductility of the basic material in the core.

An inverse method has been used to determine the thermal parameters of the spray cooling. The adopted geometry is a cylinder that is thermally insulated on its horizontal faces and receives a fine spray on the lateral surface. The problem is thus assumed to be one dimensional; the temperature is only a function of the radius and time. The cylinder is heated by induction to around 900 °C and then the cooling curves are measured during spray cooling with thermocouples placed at different radii in the part. Their positions are indicated (in millimeters) in the schematic in the upper part of the figure; the measured cooling curves are plotted in figure 8.9(b). The induction heating of the part is highest at the surface, but weak heating of the core ($r = 0$ mm) of the part is noted before the effects of the spray cooling are felt: the corresponding temperature curve displays a maximum after about 10 seconds. A variation of the cooling rate can also be noted around 650 °C due to solid state transformations.

Eight values of the heat transfer coefficient were determined by the inverse method from the measured temperature curves in figure 8.9, and tabulated as a function of the surface temperature of the part. The results are shown in figure 8.10. It is seen that the coefficient is relatively high (around 12 kW m^{-2} K^{-1} at 100 °C). The temperatures at the thermocouple locations recalculated with these boundary conditions are shown by the dashed lines in figure 8.9. The good agreement with the measured curves is not surprising, as the inverse method attempts to do exactly that: minimize the differences!

8.6.2 Determination of the thermal conductivity of an alloy

Another application of inverse methods in thermal studies is shown in figure 8.11. The curves are again for cooling, here measured in an aluminum-4.5wt% copper alloy, initially liquid. A one dimensional thermal configuration has been chosen, the alloy being contained in a highly insulating ceramic cylindrical mold. The lower

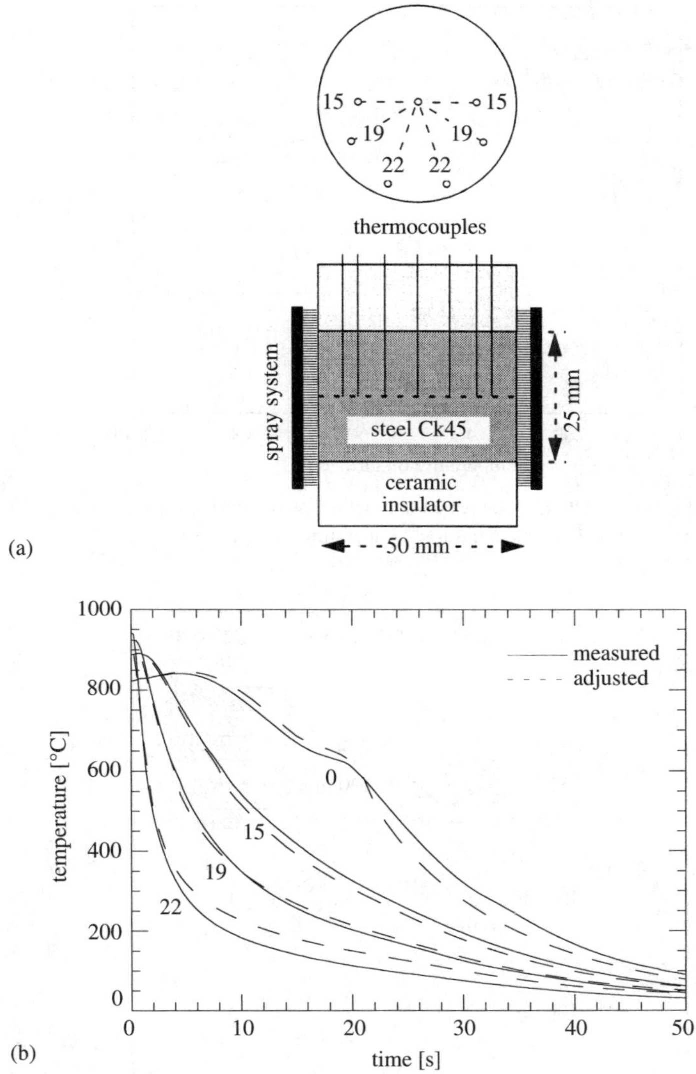

Figure 8.9 (a) Geometry used for the determination of the boundary conditions at the surfaces of a steel cylinder cooled by a fine water spray. The cooling curves (b) shown by the solid lines correspond to the measured temperatures at different locations in the part. Those shown by dashed lines are the temperatures recalculated with the transfer coefficients given in figure 8.10.

part of the mold is composed of a plate of aluminum nitride, a very good heat conductor. After the alloy was melted, the mold was taken out of the oven and placed on a water cooled copper chill. The heat flux extracted by the chill was thus assumed to be one dimensional along the axis of the cylinder. The temperatures recorded by thermocouples at various heights in the ingot are shown in figure 8.11. Three dis-

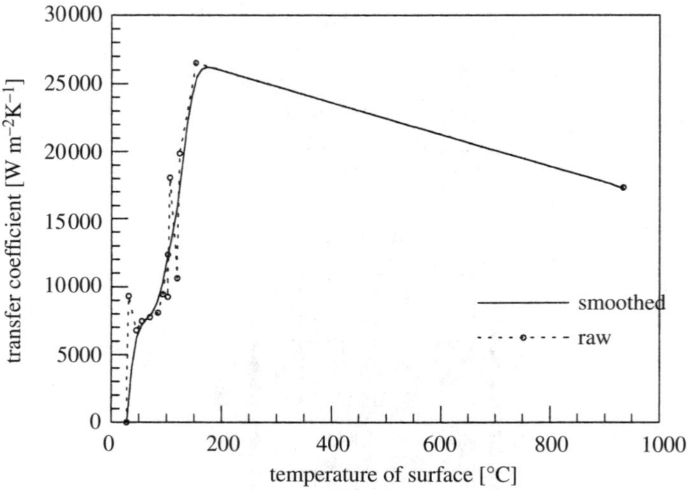

Figure 8.10 Heat transfer coefficient values as a function of the surface temperature of the part determined by inverse method from the measured temperatures in figure 8.9.

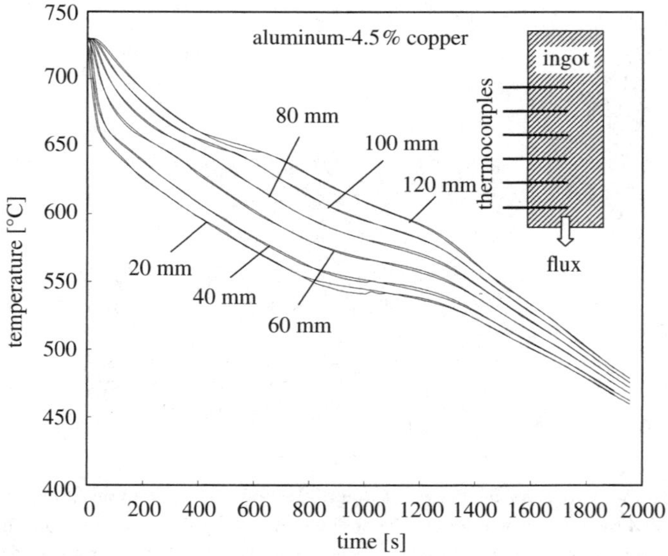

Figure 8.11 Cooling curves measured in a solidifying aluminum-4.5wt% copper ingot at various heights, and curves calculated with the thermal conductivity data in figure 8.12.

tinct regions are visible in these curves. Above 650 °C, the metal is liquid and the cooling rate is relatively high. The latent heat liberated near the liquidus of the alloy appears in the form of a temperature plateau, especially for the cooling curve corresponding to the thermocouple furthest from the chill. After 1500 seconds, the entire ingot has solidified and the cooling rate increases again. Using the temperature of

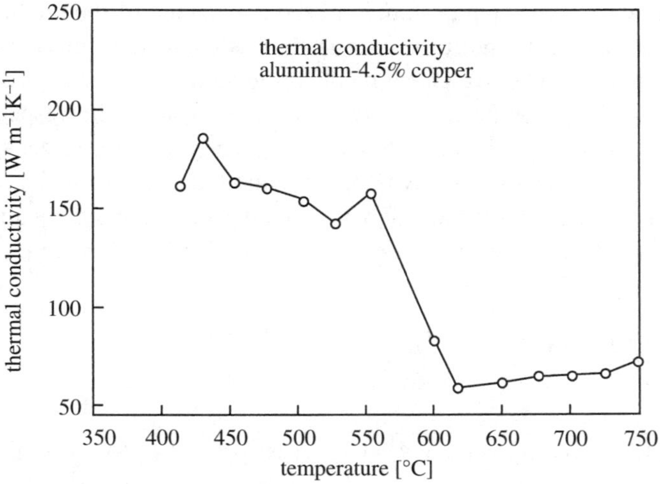

Figure 8.12 Thermal conductivity of an aluminum-4.5wt% copper alloy as determined by the inverse method from the measured temperatures in figure 8.11.

the lowest thermocouple as a boundary condition, and assuming the upper boundary to be adiabatic, the inverse method has been used to determine the tabulated values of the thermal conductivity of the alloy. The results are shown in figure 8.12 and the cooling curves calculated with these values are compared with the measured curves in figure 8.11. It is seen in figure 8.12 that the thermal conductivity of the aluminum liquid is significantly lower than that of the solid. This is reflected in the spacing of the cooling curves in figure 8.11.

8.6.3 Identification of rheological parameters

The identification of rheological parameters of materials constitutes another field for the application of inverse methods. Numerical simulation of forming processes, for which we have given examples in chapter 6, can only be predictive if the material characterization (i.e. the identification of the parameters and the constitutive law) is sufficiently precise. One is thus confronted with one of the two following problems. Either the characterization experiments are elementary and thus do not represent the reality of the process, suffering in addition from assumptions and approximations limiting the validity of their interpretation (which is the case for example of traction, compression or torsion experiments); or they are more complex, taking into account the real forces and deformations that the material experiences during its transformation, and are thus difficult to analyze by classical methods. In both cases, the advantages are apparent for the development of automatic parameter identification methods by inverse analysis that result from the coupling of a finite element simulation code for rheology experiments and a calculation-experiment difference minimization algorithm for the measurable characteristics (forces, torques, etc).

In the following, we give an example of the application of the method to the rheological identification by hot torsion tests. In this context, an inverse analysis program is composed of a finite element model for torsion tests, an analytical calculation of the sensitivity coefficients of the observable values with respect to the parameters to be identified (this calculation comes directly from the direct finite element model), and a Gauss-Newton type minimization algorithm for the error. An example of a viscoplastic constitutive law is given by the following equation for a titanium alloy (Ti-6%Al-4%V), with 8 parameters to identify: K_0, n_0, n_1, c_0, c_1, β, m_0 and m_1.

$$s = 2K(\varepsilon_{eq}, T)(\sqrt{3}\dot{\varepsilon}_{eq})^{m(T)-1}\dot{\boldsymbol{\varepsilon}} \tag{8.55}$$

with

$$K = K_0\sqrt{1-\exp[-(n_0+n_1T)(\varepsilon_{eq}+\varepsilon_{eq}^0)]}\exp[-(c_0+c_1T)\varepsilon_{eq}]\exp\left(\frac{\beta}{T}\right) \tag{8.56}$$

and

$$m = m_0 + m_1 T \tag{8.57}$$

The results of the numerical identification are shown in the following table. The curves in figure 8.13 show the comparison between experimentally measured

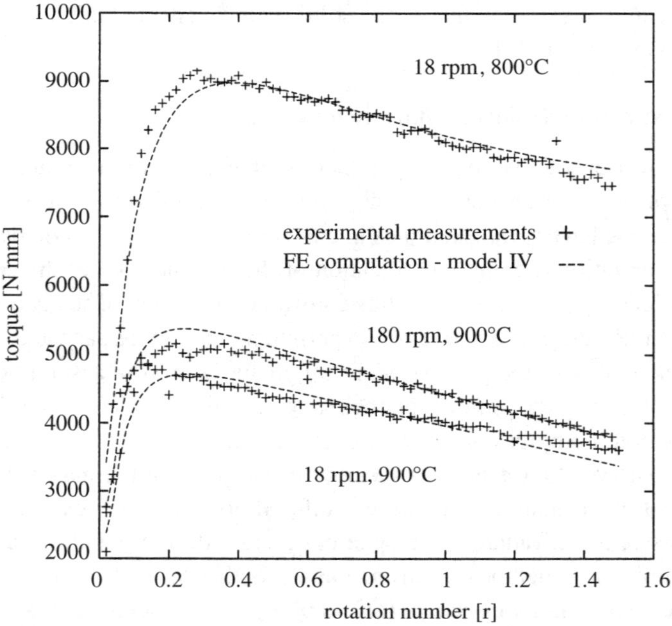

Figure 8.13 Comparison between experimental and calculated torques for three operating conditions: 18 rpm and 800°C, 180 rpm and 900°C, 18 rpm and 900°C.

torques and those calculated by finite elements for the set of parameters obtained by automatic identification.

Table 8.1

	Initial parameters	**Identified parameters**
$K_0 \times 10^2$	0.1	3.222
n_0	10	–314.87
n_1	0	0.323
c_0	0	–11.552
c_1	0	0.0104
β	8000	9443.93
m_0	0.1	0.358
$m_1 \times 10^3$	0	–0.254
number of iterations	–	20
$S[\%]$ = calculation-experiment difference	27.3%	3.3%

In conclusion, the use of inverse methods with measurements gives access to values that are difficult to estimate by other methods (transfer coefficient, thermal conductivity, mechanical properties, etc). In general, such methods offer optimal exploitation of the measurements. For completeness, it should be reiterated that the precision of the values deduced by inverse methods is directly related to that of the measurements, and that the methods are incapable of eliminating measurement imprecision. Rather the inverse is true (!): inverse methods tend to amplify them. Moreover, the simpler the geometry, the more reliable and rapid are the methods. Finally, it should be noted that inverse methods are not limited to thermal and rheology domains; in principle, any direct calculation based on FDM or FEM can be "inverted".

8.7 EXERCISES

8.7.1 Show that the stationary temperature distribution, $T(r)$, of the problem of an infinitely long tube in figure 8.1 is given by:

$$T(r) = \frac{q_T R_o}{\kappa} \ln\left[\frac{r}{R_i}\right] + \bar{T}_0$$

where R_i and R_o are the inner and outer radii of the tube. How is the flux, q_T, adjusted to the experimental points with a least squares method? What effects does this have on the equivalent inverse method?

8.7.2 For a semi infinite plate, show that the non stationary heat diffusion equation has the solution:

$$T(x, t) = T_0 + (\bar{T} - T_0)\mathrm{erfc}\left[\frac{x}{\sqrt{4\alpha t}}\right]$$

where α is the thermal diffusivity, T_0 is the initial temperature of the plate and $\bar{T}$ is the imposed temperature at its surface. The error function erfc(u) is such that:

$$\mathrm{erfc}(u) = \frac{2}{\sqrt{\pi}}\int_{\mathrm{u}}^{\infty} \exp(-v^2)\,\mathrm{d}v$$

How is the thermal diffusivity of the material determined by a least squares method based on a number of experimental points?

8.7.3 An inhomogeneous material initially has a periodic concentration variation in x (one dimensional problem) given by:

$$c(x, t = 0) = \begin{cases} \bar{c}_{\mathrm{a}} & \text{if } 0 \le x \le \lambda f \\ 0 & \text{if } \lambda f < x < \lambda \end{cases}$$

where λ is the spatial period of the inhomogeneity (characteristic of the spacing between precipitates), and f is a parameter between 0 and 1 describing the dimension of the inhomogenities with respect to λ. By separation of variables, $c(x, t) = X(x)\ \tau(t)$, find the concentration distribution in time during a homogenization treatment. How is the material diffusivity deduced by a least squares inverse method based on measurements of concentration?

8.7.4 A cylinder of height, H, much greater than the diameter, $2R$, contains highly radioactive waste giving off thermal power (source) $\dot{Q}$ per unit volume. Taking a Cauchy type boundary condition at the cylinder surface, show that the temperature distribution is given by:

$$T(r) = -\frac{\dot{Q}}{4\kappa}[r^2 - R^2] + \frac{\dot{Q}R}{h^2} + \bar{T}_0$$

where $\bar{T}_0$ is the temperature of the cooling liquid surrounding the cylinder; h is the transfer coefficient and κ is the thermal conductivity of the material. How can the emitted power be deduced by temperature measurements with the least squares method and with an inverse method?

8.7.5 An optical window absorbs infrared light with a coefficient α_r. The intensity is thus equal to $I_0 \exp(-\alpha_r x)$, where I_0 is the intensity of the incident beam. Show that the stationary heat equation in the thickness of the window is given by:

$$\kappa \frac{d^2T}{dx^2} - \frac{dI}{dx} = 0$$

Find the temperature distribution analytically in the window with Cauchy boundary conditions (given the transfer coefficient h). Write the finite difference equation also. How can the absorption coefficient be deduced based on temperature measurements in the optical glass with the least squares method and with an inverse method?

8.8 BIBLIOGRAPHY

J. V. BECK and K. J. ARNOLD, *Parameter Estimation in Engineering and Science*, Wiley, New York, 1977.

J. V. BECK, B. BLACKWELL and C. R. St. CLAIR Jr, *Inverse Heat Conduction – Ill-posed Problems,* Wiley, New York, 1985.

H. S. CARSLAW and J. C. JAEGER, *Conduction of Heat in Solids,* Oxford Univ. Press, 1959.

A. GAVRUS, E. MASSONI and J. L. CHENOT, "An Inverse Analysis using a Finite Element Model for Identification of Rheological Parameters". *J. Materials Processing Technology,* vol. 60, p. 447, 1996.

D. KNUTH, *The Art of Computer Programming, vol. 1, Fundamental Algorithms*, Addison-Wesley, Reading, MA, 1968.

M. RAPPAZ, J.-L. DESBIOLLES, J.-M. DREZE, Ch.-A. GANDIN, A. JACOT and Ph. THÉVOZ, "Application of Inverse Methods to the Estimation of Boundary Conditions and Properties", *in Modeling of Casting, Welding and Advanced Solidification Processes*, Ed. M. Cross and J. Campbell (TMS Publ., Warrendale, 1995) p. 449.

CHAPTER 9

STOCHASTIC METHODS

9.1 OBJECTIVES

- Introduce the concepts of distribution and generation of random numbers.
- Apply the method of random number generation to the integration of a function and the solution of a linear system of equations by Markov chains.
- Present the Monte Carlo method, the Ising and Potts models, and a few applications in materials science.
- Introduce the concept of random walkers for treating diffusion and aggregation problems.
- Give a general idea of deterministic and stochastic cellular automata methods.

Thermodynamics allows the calculation of averages over a large number of atoms or molecules in order to extract macroscopic values. At the atomic scale, each entity is characterized by values that fluctuate (for example, the velocity or magnetic moment), while at the macroscopic scale the corresponding average is a well defined value (the temperature or the magnetization). Developed before the appearance of computers, notably by von Neumann during the second world war, the Monte Carlo method (MC) – whose name is taken from the casino in the principality of Monaco – consists of taking a collection of atoms or molecules with the appropriate interactions, letting them interact among themselves, and extracting averages. First applied to magnetism, the method is now well established in numerous sectors of materials science (phase diagrams, grain growth, phase transformations, polymer chain arrangements, etc). The computer being an ideal tool for the generation of a large number of possible configurations and calculating their averages, other stochastic methods based on random number generation have appeared to complement the MC method for simulations as diverse as the generation of microstructures, crack propagation, and the interaction of dislocations or crystal growth. After a brief introduction to the generation of random numbers, this chapter gives a rapid overview of the most often used stochastic methods (MC, random walkers, cellular automata). These notions should allow the reader to have a sufficient base of understanding to be able to then consult the numerous variants of these methods that appear in the literature.

9.2 GENERATION OF RANDOM NUMBERS

9.2.1 Random numbers

The concept of "random numbers", although generally understood intuitively by the average person through numerous games (dice, roulette, lottery, etc), remains a bit vague. In the rest of this chapter, random numbers mean real numbers uniformly distributed on an interval $[a, b]$ (fig. 9.1). Therefore, a sequence of N real numbers, $S \equiv \{x_1, x_2, x_3, \ldots, x_N\}$, is uniformly distributed in the interval $[a, b]$ if the *distribution*, $p_s(x)$, associated with this sequence is such that:

$$p_s(x)\Delta x = \frac{1}{N}\sum_{k=1}^{N}[H(x+\Delta x - x_k) - H(x - x_k)] = \frac{\Delta x}{b-a} \quad \forall x \in [a, b] \quad (9.1)$$

$p_s(x)\Delta x$ is then the probability that a number in the sequence S falls in the interval $[x, x + \Delta x]$. $H(x - x_k)$ is the ***Heaviside function***, defined as (fig. 9.2):

$$H(x - x_k) = \begin{cases} 0 \text{ if } x < x_k \\ 1 \text{ if } x \geq x_k \end{cases} \quad (9.2)$$

The term in brackets in (9.1) equals 1 if $x < x_k \leq x + \Delta x$, and the sum is thus equivalent to counting all the values x_k falling between x and $(x + \Delta x)$. It should be noted that for real numbers, the notion of a uniformly distributed sequence of numbers can only be strictly satisfied if the sequence is infinite ($N \to \infty$) or if Δx has a finite value. Otherwise it is sufficient to choose an interval $\Delta x < (b - a)/N$ to have some regions $[x, x + \Delta x]$ of the interval $[a, b]$ unoccupied by the numbers in the

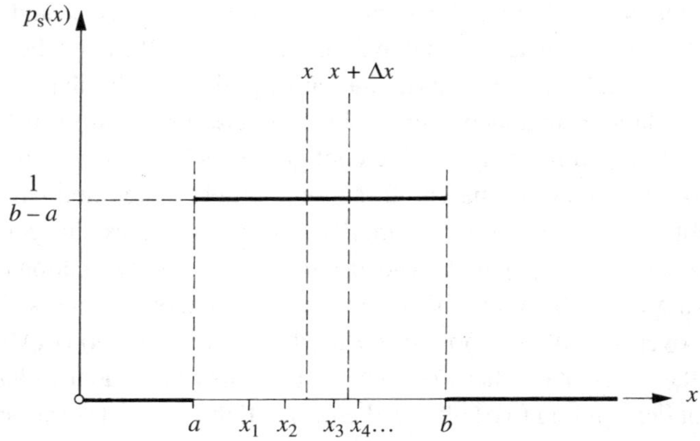

Figure 9.1 Uniform distribution $p_s(x)$ on the interval $[a, b]$.

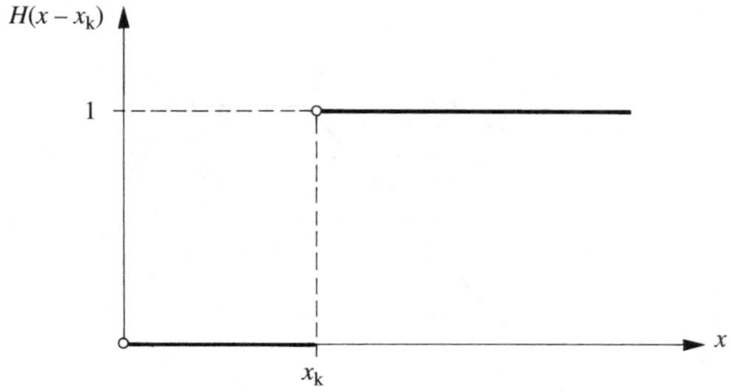

Figure 9.2 Heaviside function $H(x - x_k)$

sequence. In the case of a finite sequence of randomly distributed numbers, it is necessary to choose Δx "sufficiently large" for the statistical fluctuations of the number of values in each interval to be low[1].

If the sequence S satisfying (9.1) is such that $x_1 < x_2 < \ldots < x_N$, it is said to be ***ordered***. If $[a, b]$ is the unit interval $[0, 1]$, the uniformly distributed sequence of real numbers in this interval, denoted S_u, is characterized by $p_u(x) = 1$.

The cumulative distribution, $Q_s(x)$, of a sequence of real numbers is defined as:

$$Q_s(x) = \frac{1}{N}\sum_{k=1}^{N} H(x - x_k) \tag{9.3}$$

$Q_s(x)$ thus counts all the values x_k of the sequence which are smaller than x. With the definitions (9.1) and (9.3), we have of course, $Q_s(x + \Delta x) - Q_s(x) = p_s(x)\Delta x$. In other words, as long as the limit, $\Delta x \to 0$, has a sense:

$$\frac{dQ_s}{dx} = p_s(x) \qquad \text{or} \qquad Q_s(x) = \int_{-\infty}^{x} p_s(x')\,dx' \tag{9.4}$$

If the numbers are uniformly distributed in an interval $[a,b]$ (fig. 9.3), then:

$$Q_s(x) = \begin{cases} 0 \text{ if } x \le a \\ \dfrac{x-a}{b-a} \text{ if } a \le x \le b \\ 1 \text{ if } b \le x \end{cases} \tag{9.5}$$

[1] For fixed Δx, the average number of values, $N_{\Delta x}$, falling in the interval $[x, x + \Delta x]$ is given by the total number of values, N, divided by the number of intervals, $(b - a)/\Delta x$. The fluctuations of $N_{\Delta x}$ are of the order of $\sqrt{(N\Delta x)/(b-a)}$.

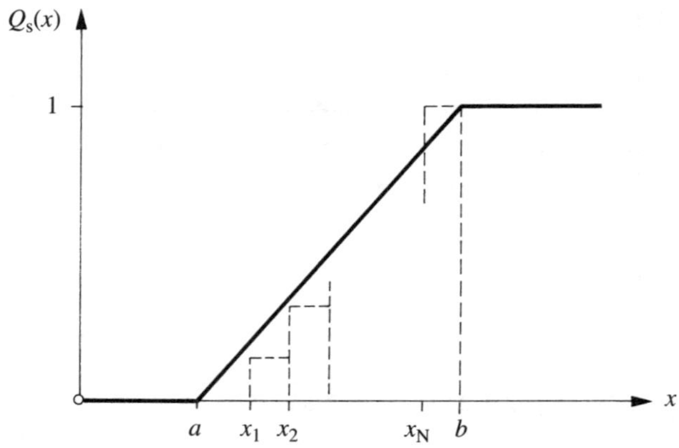

Figure 9.3 Cumulative distribution $Q_s(x)$ corresponding to the uniform distribution on the interval $[a, b]$.

and thus $Q_s(x) = x$ for the uniform distribution on the interval [0, 1]. It is possible to define other distributions besides the uniform distribution of real numbers considered above (sect. 9.2.4).

9.2.2 Generation of random numbers

Contrary to common wisdom, computers do not generate random numbers; they can generate sets of numbers distributed more or less uniformly in a given interval, but in a *given sequence*. For the user, it is as if the numbers of the sequence were not correlated, but they are: the next number calculated by the computer is a function of the previous as they are all part of a ***congruent sequence***, as we will see. Thus, if the random number generating procedure is run a second time, it will give the same sequence. An exterior "randomizing" or stochastic element can be included in the start of the sequence; for example, the first number of the sequence, x_1, on which all the following depend, can be initialized with a value that is dependent on the time at which the user taps a key on the keyboard!

The ***congruent method*** for generating a sequence of uniformly distributed numbers in an interval, $S \equiv \{x_1, x_2, \ldots, x_N\}$, is based on a sequence of integers, $\{i_1, i_2, \ldots, i_N\}$ linked by a recursive relation of the type:

$$i_{k+1} = \mathrm{Mod}[ai_k + c, m] = (ai_k + c) - m\mathrm{Int}\left[\frac{ai_k + c}{m}\right] \tag{9.6}$$

where a, the multiplier, c, the increment, and m, the modulo term, are integers. The value of i modulo m, Mod(i, m), is the integer remainder of the division (i/m) for which the expression in terms of Int(i/m) is given in (9.6). The first term of the sequence, i_1 ($< m$), is called the ***seed*** of the sequence. As the values i_k are less than m, a sequence of real numbers distributed in the interval [0, 1] is obtained by:

$$x_k = \frac{i_k}{m} \tag{9.7}$$

The sequence thus obtained is not only correlated (i_{k+1} is a function of i_k), but also periodic: if one of the terms of the sequence, i_n, becomes equal to one of the preceding numbers, i_m, we fall into a loop: the sequence of numbers $\{i_m, i_{m+1}, \ldots, i_n\}$ is then repeated indefinitely. The frequency, ν, of the sequence is defined as the smallest difference in indices n and m for which $i_n = i_m$. As the integers of the sequence are necessarily less than m, the frequency ν is at best equal to m. One thus wants to choose the value of m as high as possible in order to generate a sequence of numbers that appears uncorrelated. If a combination of integers a, c and m can be found such that $\nu = m$, the sequence is said to have a ***full frequency***. It can be shown that the sequence has full frequency if and only if:

i) c and a are prime numbers,
ii) Mod $[a, d] = 1$ for all prime divisors, d, in m,
iii) Mod $[a, 4] = 1$ if m is a multiple of 4.

Examples:

1) The choice $i_1 = 1$, $a = 5$, $c = 3$ and $m = 10$ for the recursive relation (9.6) generates the sequence: $\{1, 8, 3, 8, 3, 8, \ldots\}$ (frequency $\nu = 2$). The relation (ii) is not satisfied in this case.
2) The choice $i_1 = 1$, $a = 11$, $c = 3$ and $m = 10$ for the recursive relation (9.6) generates the full frequency sequence ($\nu = 10$) $\{1, 4, 7, 0, 3, 6, 9, 2, 5, 8, 1, 4, 7, \ldots\}$. Dividing by $m = 10$, the sequence, S_u, of real number is obtained in the interval [0, 1]: $\{0.1, 0.4, 0.7, 0, 0.3, 0.6, 0.9, 0.2, 0.5, 0.8, 0.1, \ldots\}$.

Good results can be obtained for example with: $m = 2^{35}$, $a = 2^8 + 1$ and $c = 1$. Before starting a program based on random numbers, it may be useful to first perform a statistical analysis of the generated numbers to verify that they are uniformly distributed in the interval [0, 1].

9.2.3 Statistics of a sequence of numbers

To test whether a sequence of real numbers $S_u \equiv \{x_1, x_2, \ldots, x_N\}$ generated by an algorithm such as (9.6) is actually uniformly distributed over the interval [0, 1], there are a number of tests, the ***χ^2 test*** being the most well known.

χ^2 Test

For the case of a finite sequence S_u of real numbers in the interval [0, 1], one would like to verify that the associated distribution p_u is actually "close" to the ideal relation $\bar{p}_u = 1$ (eq. (9.1) with $a = 0$ and $b = 1$). For that, the interval [0, 1] is divided into N_c intervals of size $(1/N_c)$ which do not overlap. The number of values falling in each interval will ideally be $\overline{N}_n = (N/N_c)$, independent of the class n, while the actual number of values of the sequence S_u that fall in each of the classes n is equal to N_n. The χ^2 test consists of summing the normalized differences for each of the classes:

$$\chi^2 = \sum_{n=1}^{N_c} \frac{[N_n - \overline{N}_n]^2}{\overline{N}_n} = \frac{N_c}{N} \sum_{n=1}^{N_c} \left[N_n - \frac{N}{N_c}\right]^2 \tag{9.8}$$

The number of degrees of freedom of this statistics problem, which allows the verification of the assumption that a sequence of numbers is uniformly distributed in [0, 1], is equal to $(N_c - 1)$. The numbers N_n found in each of the intervals are independent of each other except for the last one, for which the value N_{N_c} is given by the complement to N of the sum of the $(N_c - 1)$ other classes. It is sufficient to look at a table of χ^2 tests to judge the confidence that one can have in a uniform distribution (fig. 9.4). For example, a value of χ^2 of around 11 calculated with 21 classes (that is 20 degrees of freedom) allows saying with relative confidence (95 % chance) that the distribution of the numbers in the sequence is effectively uniform. The qualifications "probably significant", "significant" and "highly significant" are generally attributed to probabilities of 95 %, 99 % and 99.9 % resulting from the χ^2 test defined by (9.8).

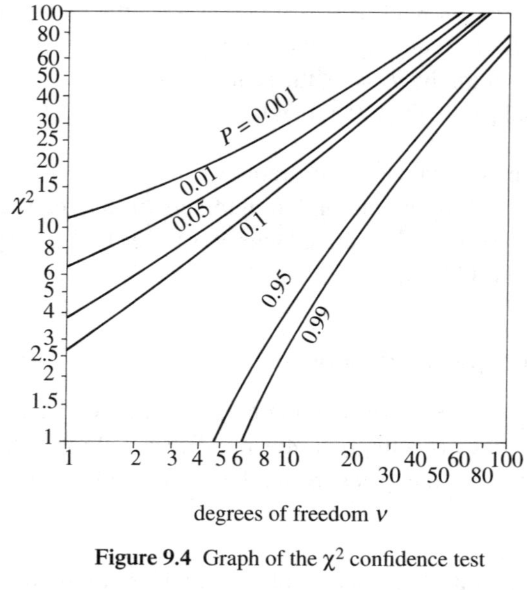

Figure 9.4 Graph of the χ^2 confidence test

9.2.4 Sequence of non uniformly distributed numbers

In some applications, one may desire to use a sequence of random real numbers that are not uniformly distributed on x. This could be the case of spherical particles of a composite material, for example, for which the radii, r_k, have a Gaussian distribution about an average value, $\langle r \rangle$ (fig. 9.5). In such cases, one wants to generate a sequence $S \equiv \{r_1, r_2, \ldots, r_N\}$ with a distribution ideally given by:

$$p(r) = \bar{f}(r) \tag{9.9}$$

where $\bar{f}(r)$ for a Gaussian distribution is given by:

$$\bar{f}(r) = \frac{1}{\sigma\sqrt{2\pi}} \exp\left(-\left[\frac{r - \langle r \rangle}{\sqrt{2}\sigma}\right]^2\right) \tag{9.10}$$

$\langle r \rangle$ is the center of the distribution and σ its standard deviation (fig. 9.6a).

Specific methods have been developed for most standard distributions (see the book by Rubinstein). In general, a sequence S of numbers can be obtained from the base sequence S_u with one of the following methods:

i) inverse transformation
ii) composition method
iii) rejection-acceptation method

Only the first method will be presented below.

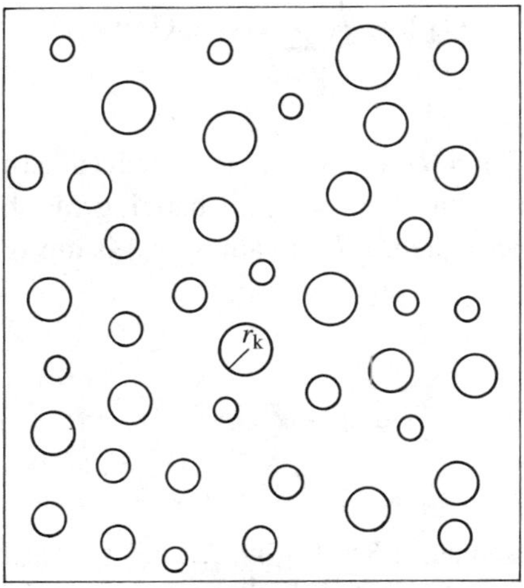

Figure 9.5 Random distribution of fibers in a transverse section generated with a Gaussian radii distribution.

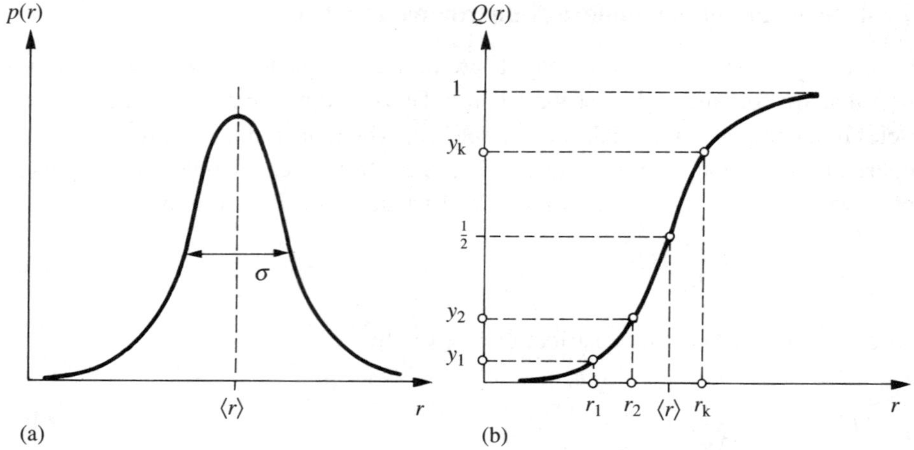

Figure 9.6 a) Gaussian distribution of radii, $p(r)$; b) cumulative distribution $Q(r)$ (error function).

Inverse transformation

By definition, the cumulative distribution, $y = \bar{Q}(r)$, associated with $\bar{f}(r)$ is necessarily a monotonically increasing function, and it is also normalized ($0 \leq y \leq 1$, see figure 9.6). It can be inverted to obtain $\bar{Q}^{-1}(y)$. Now consider for example, a uniform sequence of numbers generated by the congruent method (eqs. (9.6) and (9.7)) in the interval [0, 1]: $S_u \equiv \{y_1, y_2, \ldots, y_N\}$. It is easy to show that the sequence $S \equiv \{r_1, r_2, \ldots, r_N\} \equiv \{\bar{Q}^{-1}(y_1), \bar{Q}^{-1}(y_2), \ldots, \bar{Q}^{-1}(y_N)\}$ follows the distribution $\bar{f}(r)$. According to (9.3), the cumulative distribution of S equals:

$$Q_s(r) = \frac{1}{N}\sum_{k=1}^{N} H(r - r_k) = \frac{1}{N}\sum_{k=1}^{N} H(r - \bar{Q}^{-1}(y_k)) \tag{9.11}$$

But $H(r - \bar{Q}^{-1}(y_k)) = H(\bar{Q}(r) - y_k)$ (the Heaviside function goes from 0 to 1 when $r = \bar{Q}^{-1}(y_k)$ or when $\bar{Q}(r) = y_k$). Changing variables $y = \bar{Q}(r)$ and taking into account that the sequence S_u of values $\{y_k\}$ is uniform (eq. (9.5) with $a = 0$ and $b = 1$):

$$Q_s(r) = \frac{1}{N}\sum_{k=1}^{N} H(y - y_k) = y = \bar{Q}(r) \tag{9.12}$$

The values of the sequence $S \equiv \{r_1, r_2, \ldots, r_N\}$ so constructed are thus distributed according to the cumulative distribution $\bar{Q}(r)$ (or the distribution $\bar{f}(r)$). For a Gaussian distribution (9.10), the cumulative distribution $\bar{Q}(r)$ is none other than the error function, erf(r), shown in figure 9.6, and the values of the desired sequence are thus given by the sequence of values $\mathrm{erf}^{-1}(y_k)$, where the values y_k are uniformly distributed on [0, 1].

9.3 INTEGRATION BY STOCHASTIC METHODS

The generation of random numbers can be used to calculate the integral of a function $g(x)$ on an interval $[a, b]$:

$$I = \int_a^b g(x)\,\mathrm{d}x \tag{9.13}$$

For simplicity, $g(x)$ is assumed positive for every point $x \in [a, b]$. A first integration method, called the ***hit-or-miss method***, consists first of generating points, (x_i, y_i), randomly distributed in the rectangle defined by $[a, b]$ and $[0, c]$, where c is the maximum of the function $g(x)$ in the integration interval (fig. 9.7). For N points so generated, one counts those that fall below the curve (the black circles), i.e., those for which $y_i \le g(x_i)$. If N_t points are located under the curve $g(x)$, the integral I is approximated by:

$$I \simeq c(b-a)\frac{N_t}{N} \tag{9.14}$$

A second method for the estimation of integral I consists of using a ***sampling*** of points $\{x_k;\, k = 1, N\}$ distributed according to a function $f(x)$ chosen *a priori*. A family of points $\{x_k;\, k = 1, N\}$ distributed according to $f(x)$ is such that:

$$f(x)\Delta x \simeq \frac{1}{N}\int_x^{x+\Delta x}\left[\sum_{k=1}^{N}\delta(x-x_k)\right]\mathrm{d}x \tag{9.15}$$

where $\delta(x - x_k)$ is the Dirac delta function centered at x_k. The integral of the Dirac function, $\delta(x - x_k)$, on an interval including the value x_k is equal to 1 by definition. In other words, the integral of the Dirac function is none other than the Heaviside function, $H(x - x_k)$. Then:

$$\int_a^b \delta(x-x_k)\,\mathrm{d}x = H(x-x_k)\Big|_a^b = H(b-x_k) - H(a-x_k) \tag{9.16}$$

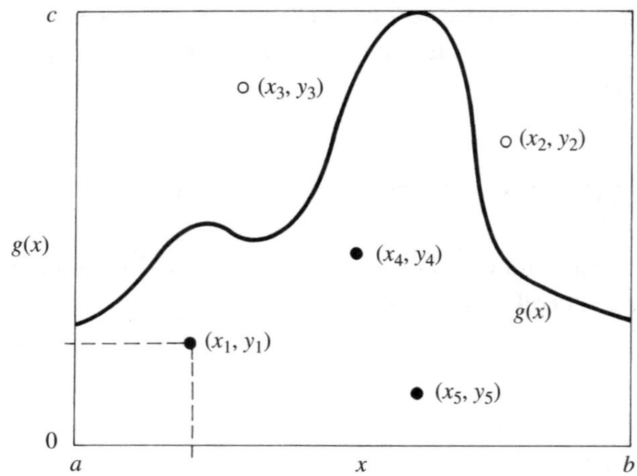

Figure 9.7 Hit-or-miss method for the definite integral of a function on an interval $[a, b]$.

The properties of the Dirac function are such that:

$$\int_a^b g(x)\delta(x - x_k)\, \mathrm{d}x = g(x_k) \qquad \text{if } x_k \in [a, b] \tag{9.17}$$

With (9.15) and (9.17), the integral (9.13) can be written in the form:

$$I = \int_a^b \frac{g(x)}{f(x)} f(x)\, \mathrm{d}x \simeq \frac{1}{N}\sum_{k=1}^{N} \frac{g(x_k)}{f(x_k)} \tag{9.18}$$

If the points x_k used for the sampling are uniformly distributed on the interval $[a, b]$, then $f(x) = 1/(b - a)$ and (9.18) becomes:

$$I = \frac{b-a}{N}\sum_{k=1}^{N} g(x_k) \tag{9.19}$$

It is seen that this method is equivalent to a very simple integration method: if the N points are distributed in an equidistant manner on the interval $[a, b]$, it amounts to associating with each value $g(x_k)$ a rectangle of width $\Delta x = (b - a)/N$ centered on x_k and of height $g(x_k)$ (fig. 9.8). Equivalent to a trapezoidal method, this numerical integration is not very precise, but it nevertheless has an advantage: it is possible to use weighted sample points that are not necessarily uniformly distrib-

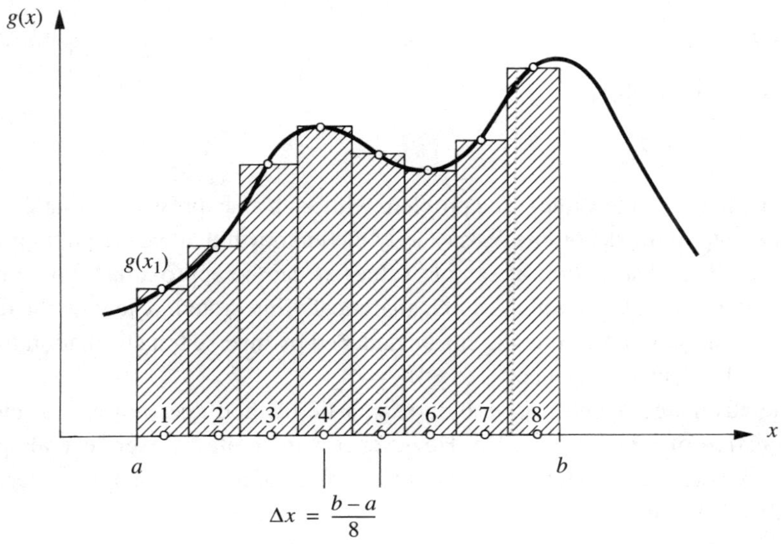

Figure 9.8 Integration of a function by generation of regularly spaced points (trapezoidal method).

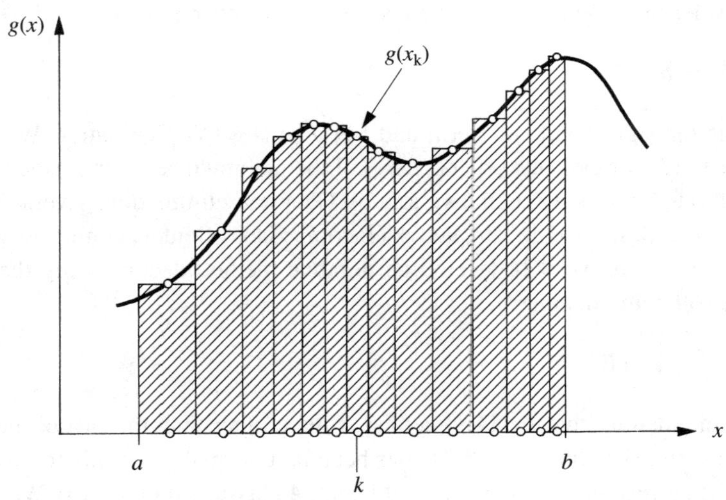

Figure 9.9 Integration of a function with points distributed along x with a weighting function $g(x)$.

uted. For example, more points can be taken in the zones where the function $g(x)$ varies more rapidly (fig. 9.9). It can be shown (see the book by Rubinstein) that the sampling of points x_k that optimizes the calculation of the integral (9.18) (minimizing the variance) is given by the distribution:

$$f(x) = \frac{|g(x)|}{\int_a^b |g(x)|\, dx} \tag{9.20}$$

($|.|$ being the absolute value). The problem with such an expression is that the integral appearing in the denominator is exactly the unknown that we are attempting to determine! This distribution of sample values x_k can be approximated by a curve $h(x)$ "similar" to $g(x)$, but which can be analytically integrated. Knowing the distribution $f(x)$, the sampling values x_k can be deduced by inverse transformation from a uniformly distributed sequence.

The advantage of integration by a stochastic method does not appear clearly for a function of a single variable. However, for integrations over several spatial variables where the integration limits are interdependent ("telescopic" integrals), the method can offer real advantages.

9.4 SOLUTION OF SYSTEMS OF EQUATIONS

Consider the problem of solving a system of N linear equations:

$$[\boldsymbol{A}]\boldsymbol{x} = \boldsymbol{b} \tag{9.21}$$

where $\boldsymbol{b}$ is the right hand side term and $[A]$ a square $(N \times N)$ matrix. We seek to determine the N components $(x_1, x_2, \ldots, x_N)$ of the unknown vector $\boldsymbol{x}$. Aside from the exact methods for solving such a system by Gaussian elimination, several iterative methods were described in chapter 4 (Jacobi, Gauss-Seidel, conjugate gradient methods, etc). Here we will develop an iterative method decomposing the matrix $[A]$ in the following manner:

$$[\boldsymbol{A}] = [\boldsymbol{I}] - [\boldsymbol{B}] \tag{9.22}$$

$[I]$ being the identity matrix. This method is none other than the Jacobi method in which the system of equations (9.21) has been first normalized by dividing each of them by the corresponding diagonal coefficient, A_{ii} (assumed non zero). We are thus led to solve the system:

$$\boldsymbol{x} = [\boldsymbol{B}]\boldsymbol{x} + \boldsymbol{b} \tag{9.23}$$

If the coefficients of $[B]$ are such that: $\sum_j |B_{ij}| < 1$, the following iterative method converges:

$$\boldsymbol{x}^{(\nu+1)} = [\boldsymbol{B}]\boldsymbol{x}^{(\nu)} + \boldsymbol{b} \tag{9.24}$$

Starting from the solution $\boldsymbol{x}^{(0)} = \mathbf{0}$, we have:

$$\boldsymbol{x}^{(1)} = \boldsymbol{b};\ \boldsymbol{x}^{(2)} = [\boldsymbol{B}]\boldsymbol{b} + \boldsymbol{b};\ \boldsymbol{x}^{(3)} = [\boldsymbol{B}]^2\boldsymbol{b} + [\boldsymbol{B}]\boldsymbol{b} + \boldsymbol{b}$$

$$\ldots ;\ \boldsymbol{x}^{(\nu+1)} = \left(\sum_{m=0}^{\nu} [\boldsymbol{B}]^m \boldsymbol{b}\right) \quad (\text{where } [\boldsymbol{B}]^0) = [\boldsymbol{I}]) \quad (9.25)$$

As ν goes to infinity:

$$\sum_{m=0}^{\nu} [\boldsymbol{B}]^m \simeq ([\boldsymbol{I}] - [\boldsymbol{B}])^{-1} = [\boldsymbol{A}]^{-1} \text{ ; which is the desired solution.}$$

For an index i_0 of the vector $\boldsymbol{x}^{(\nu+1)}$, (9.25) is written:

$$x_{i_0}^{(\nu+1)} = b_{i_0} + \sum_{i_1} B_{i_0 i_1} b_{i_1} + \sum_{i_1, i_2} B_{i_0 i_1} B_{i_1 i_2} b_{i_2} + \ldots$$

$$\ldots + \sum_{i_1, \ldots, i_\nu} B_{i_0 i_1} B_{i_1 i_2} \ldots B_{i_{\nu-1} i_\nu} b_{i_\nu} \quad (9.26)$$

This method is rather expensive as, for the calculation of each power m of the matrix $[\boldsymbol{B}]$, N^3 multiplications must be performed (the matrix $[\boldsymbol{B}]^{m-1}$ known, the matrix $[\boldsymbol{B}]^m$ is given by the product $[\boldsymbol{B}]^{m-1}[\boldsymbol{B}]$, which implies the multiplication of the N lines of $[\boldsymbol{B}]^{m-1}$ by the N columns of $[\boldsymbol{B}]$, thus N^3 multiplications).

We will now *estimate* the components of $\boldsymbol{x}^{(\nu+1)}$ by a stochastic method. For that, we introduce the concept of a ***Markov chain***. This consists of first defining a matrix $[\boldsymbol{P}]$ of transition probabilities of an index i to any other index j ($i, j = 1, N$). Thus:

$$[\boldsymbol{P}] = \begin{bmatrix} P_{11} & P_{12} & P_{13} & \ldots & P_{1N} \\ P_{21} & P_{22} & P_{23} & \ldots & P_{2N} \\ P_{31} & P_{32} & P_{33} & \ldots & P_{3N} \\ \ldots & \ldots & \ldots & \ldots & \ldots \\ P_{N1} & P_{N2} & P_{N3} & \ldots & P_{NN} \end{bmatrix} \quad (9.27)$$

These probabilities are necessarily positive or zero, but we impose that they be zero if the corresponding term of the matrix $[\boldsymbol{B}]$ is zero (thus $P_{ij} = 0$ if $B_{ij} = 0$). If all the terms of the matrix $[\boldsymbol{B}]$ are non zero, the *homogeneous* transition probabilities can be defined, on which case $P_{ij} = P$ regardless of the indices i and j. We impose a normalization condition which consists of saying that, from any state i, we are sure to jump to some state j; in other words:

$$\sum_j P_{ij} = 1 \quad \forall i = 1, N \tag{9.28}$$

Now look again at the problem of solving (9.25). Starting from the index i_0, a chain is defined following the transitions probabilities P_{ij} defined in (9.27):

$$i_0 \to i_1 \to i_2 \to i_3 \to \dots \to i_\nu \tag{9.29}$$

For the case of a *homogeneous Markov chain*, the indices i_1, i_2, i_3,..., i_ν are obtained by successive random integers uniformly distributed between 1 and N. In all cases, the probability of having a particular path of indices is given by:

$$p(i_0\, i_1\, i_2\, i_3 \dots\, i_\nu) = P_{i_0 i_1} P_{i_1 i_2} P_{i_2 i_3} \dots P_{i_{\nu-1} i_\nu} \tag{9.30}$$

The particular value, ξ_{i_0}, associated with this index path is of random nature and is thus calculated:

$$\xi_{i_0} = b_{i_0} + \frac{B_{i_0 i_1}}{P_{i_0 i_1}} b_{i_1} + \frac{B_{i_0 i_1} B_{i_1 i_2}}{P_{i_0 i_1} P_{i_1 i_2}} b_{i_2} + \dots + \frac{B_{i_0 i_1} B_{i_1 i_2} \dots B_{i_{\nu-1} i_\nu}}{P_{i_0 i_1} P_{i_1 i_2} \dots P_{i_{\nu-1} i_\nu}} b_{i_\nu} \tag{9.31}$$

Contrary to the Einstein convention for implicit summation for repeated indices often seen in this book, the terms appearing in (9.31) do not imply any sums over variable indices (i_1 or i_2 for example) as they correspond to a particular configuration of indices for the path considered (9.29). Defining the following coefficients:

$$W_0 = 1; W_1 = \frac{B_{i_0 i_1}}{P_{i_0 i_1}}; W_2 = \frac{B_{i_0 i_1} B_{i_1 i_2}}{P_{i_0 i_1} P_{i_1 i_2}} = W_1 \frac{B_{i_1 i_2}}{P_{i_1 i_2}}; \dots; W_\nu = W_{\nu-1} \frac{B_{i_{\nu-1} i_\nu}}{P_{i_{\nu-1} i_\nu}} \tag{9.32}$$

we finally obtain:

$$\xi_{i_0}^{(\nu+1)} = \sum_{m=0}^{\nu} W_m b_{i_m} \tag{9.33}$$

It can be shown that the average value of the ξ_{i_0} obtained when taking a large number of Markov chains is an estimator of ξ_{i_0} defined by (9.26). This average value is given by the sum over all possible configurations of Markov chains weighted by the corresponding probability:

$$\langle \xi_{i_0}^{(\nu+1)} \rangle = \langle b_{i_0} \rangle + \langle \frac{B_{i_0 i_1}}{P_{i_0 i_1}} b_{i_1} \rangle + \langle \frac{B_{i_0 i_1} B_{i_1 i_2}}{P_{i_0 i_1} P_{i_1 i_2}} b_{i_2} \rangle + \dots$$

$$= b_{i_0} + \sum_{i_1} \frac{B_{i_0 i_1}}{P_{i_0 i_1}} P_{i_0 i_1} b_{i_1} + \sum_{i_1, i_2} \frac{B_{i_0 i_1} B_{i_1 i_2}}{P_{i_0 i_1} P_{i_1 i_2}} P_{i_0 i_1} P_{i_1 i_2} b_{i_2} + \dots \qquad (9.34)$$

as the probability of obtaining the first index i_1 is given by $P_{i_0 i_1}$, that to obtain the indices i_1 and i_2 consecutively for the third term being equal to $P_{i_0 i_1} \ P_{i_1 i_2}$, and so on. We end up with the desired value x_{i_0} given by (9.26).

The stochastic method for solving the system of linear equations (9.21) is thus the following:

- Generate M Markov chains (s): $i_0 \to i_1^s \to i_2^s \to i_3^s \to \dots \to i_\nu^s$ ($s = 1, M$) following the table of probabilities P_{ij}. For the case of a homogeneous Markov chain with coefficients $B_{ij} \neq 0$, these probabilities are equal to $1/N$, N being the number of equations to solve.
- Calculate the values W_m^s corresponding to the various chains.
- Algebraically average the values of $\xi_{i_0}^{(\nu+1),\, s}$ thus obtaining for each of the chains:

$$\langle \xi_{i_0} \rangle = b_{i_0} + \frac{1}{M} \sum_{s=1}^{M} W_1^s b_{i_1^s} + \dots + \frac{1}{M} \sum_{s=1}^{M} W_k^s b_{i_k^s} \simeq x_{i_0}^{(\nu+1)}$$

$$\text{with } W_m^s = W_{m-1}^s \frac{B_{i_{m-1}^s i_m^s}}{P_{i_{m-1}^s i_m^s}}$$

REMARKS

- With the same algorithm, the inverse matrix $[\boldsymbol{A}]^{-1}$ can be calculated by considering particular right hand side vectors. Setting $\boldsymbol{b} = (0, 0, \dots, 1, 0, \dots, 0)$ (*i.e.*, $b_j = \delta_{jj_0}$ with j_0 successively taking the values 1, 2, …, N), the corresponding value obtained for ξ according to (9.33) is none other than the coefficient $A_{i_0 j_0}^{-1}$. The calculation of the coefficients W_m^s being independent of the particular vectors $\boldsymbol{b}$, all the values $A_{i_0 j_0}^{-1}$ can be calculated for a fixed value of i_0 by successively choosing different values of b_j. Contrary to other inversion methods presented in chapter 4, the stochastic method allows the calculation of the coefficients of the inverse matrix $[\boldsymbol{A}]^{-1}$ *separately*. It thus offers a real advantage if one is interested in knowing only one of the components, $A_{i_0 j_0}^{-1}$, of the inverse matrix or x_{i_0} of the solution vector. On advanced computer architectures with many processors, such a method is intrinsically parallelizable.
- The Markov chain method of course lends itself to the solution of systems of equations resulting from the formulation of the conservation equations in finite

differences or finite elements. As the associated matrices with such problems are sparse (i.e., the non zero coefficients on a line are very few), the corresponding Markov chains will generally be very short as the probabilities P_{ij} are set to zero every time $B_{ij} = 0$. It will be seen in section 9.6 that this method is thus equivalent to that of random walkers used for the calculation of the concentration or the temperature *at a single point* of a mesh in a diffusion problem, without having to solve a system of equations.

9.5 MONTE CARLO METHOD

9.5.1 Ising model

The generation of random numbers, a stochastic process by nature, allows the calculation of *averages* and thus the determination of probabilities. For example, the generation of random numbers between 0 and 1 (sequence S_u) gives an average value of 0.5. Therefore it is not surprising that the Monte Carlo method (MC) was applied to problems in *thermodynamics*. This field of science is in its essence a result of averages over large numbers of fluctuating microscopic entities (atoms or molecules). Using the MC method to generate numerous random configurations of atoms, magnetic moments or electric dipoles, one can attempt to extract averages and thus information at the macroscopic scale (respectively, the free energy, the magnetization or the polarization).

The best known example is the ***Ising model*** for a population of magnetic moments (fig. 9.10). It is simple in the sense that each spin can take one of two possible values: $+^1/_2$ and $-^1/_2$, according to whether the spin is aligned with or opposed to the direction (z) of an external applied magnetic field, $\boldsymbol{H}$ (quantification axis). The total magnetic energy associated with a given configuration of this family of spins is the sum of two contributions: the interaction with the external field and the interaction between the spins. The first term is minimized when all the spins are aligned with $\boldsymbol{H}$, while the second has an energy minimum if the spins alternate (successively pointing up, and then down). If S_i denotes the z component of the spin occupying the i position of the lattice, the configuration energy is written as:

$$E = -\sum_{j \neq k}\sum_{k} J_{jk} S_j S_k - g\mu_B\mu_0 H \sum_{k} S_k \tag{9.35}$$

J_{jk} is an ***exchange coefficient*** between the spins which tends to make them opposed ($J_{jk} < 0$), μ_B is the ***Bohr magneton***, the characteristic magnetic moment of an electron, and g is the ***Landé factor*** (approximately equal to 2 and μ_0 the magnetic permeability of vacuum). As the exchange integral between two spins of the lattice is a rapidly decreasing function of the distance between them, the double summation over j and k ($N(N-1)$ terms) can be reduced to a simple sum over the four nearest neighbors of each of the N lattice sites ($4N$ terms). Assuming the exchange component J to be constant between the nearest neighbors:

$$E = -\sum_{j=1}^{4}\sum_{k} J S_k S_{v(k,j)} - g\mu_B\mu_0 H \sum_{k} S_k \tag{9.36}$$

where $v(k, j)$ is a table giving the neighbors to the East, North, West and South of a node k ($k = 1, N$; $j = 1, 4$). As the calculation domain on which the spins are placed (fig. 9.10) only represents a very small portion of a crystal lattice, boundary effects are eliminated by introducing *periodic boundary conditions*. For that, each site on the boundary of the domain has in place of its missing neighbor, the corresponding site on the opposite boundary. Then the rectangular domain of figure (9.10) reconnects to itself as a torus. In the example shown in figure 9.10, the eastern neighbor of site 64 is number 57, while that to the north is number 8.

Random numbers can be used to distribute the spins over the nodes of the lattice: with a coin toss, heads or tails would correspond to spin up or down. If real numbers, x_i, uniformly distributed in the interval [0, 1] are generated by a congruent sequence, it is sufficient to define the rule: spin down for $x_i < 0.5$, spin up for $x_i \geq 0.5$. The energy E of the configuration can be calculated with (9.36). Thermodynamics tells us that the magnetization $\langle M \rangle$ of such a system of spins at tempera-

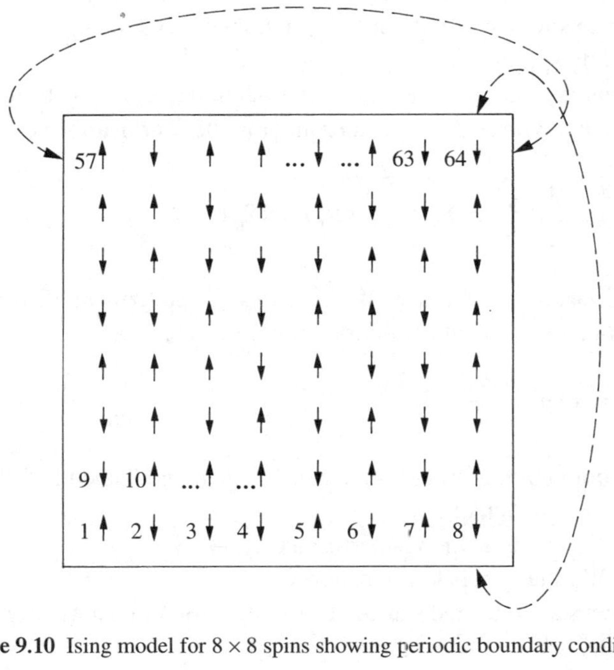

Figure 9.10 Ising model for 8 × 8 spins showing periodic boundary conditions.

ture T is an average over all the possible configurations ν (Boltzmann distribution):

$$\langle M\rangle = \frac{\sum_{\nu} M_{\nu} \exp\left[\frac{-E_{\nu}}{k_B T}\right]}{\sum_{\nu} \exp\left[\frac{-E_{\nu}}{k_B T}\right]} \tag{9.37}$$

where k_B is the Boltzmann constant, E_ν the energy of the configuration ν thus generated (eq. (9.36)) and M_ν the corresponding magnetization, given by:

$$M_{\nu} = g\mu_B \sum_i S_i^{\nu} \tag{9.38}$$

For N spins distributed over a lattice ($N = 64$ in figure 9.10), the number of possible configurations is given by 2^N! A calculation of the resulting magnetization from an average over such a number of configurations becomes rapidly impossible and one generally resorts to a much simpler algorithm known by the name of its inventor, ***the Metropolis algorithm***.

9.5.2 Metropolis algorithm

The Metropolis method generates states that are already distributed according to a Boltzmann distribution. The method consists of the following steps:

(1) The spins are first randomly distributed over the lattice nodes.

(2) A lattice site is selected randomly, number k, and the inversion of the corresponding spin is considered $S_k \rightarrow -S_k$.

(3) The energy variation, ΔE_k, of the system associated with the inversion is calculated with (9.36), considering only the four nearest neighbors:

$$\Delta E_k = 2\left[J \sum_{j=1}^{4} S_k S_{\nu(k,j)} + g\mu_B \mu_0 H S_k\right] \tag{9.39}$$

(4) The transition probability, W_k, given by the Boltzmann distribution for the variation of the local configuration is calculated:

$$W_k = \exp\left[\frac{-\Delta E_k}{k_B T}\right] \tag{9.40}$$

(5) A random number x between 0 and 1 is generated, and the following transition rule is adopted:
if $x < W_k$, the transition is performed $S_k \rightarrow -S_k$,
if $x \geq W_k$, the spin is left unchanged.

(6) The steps are repeated starting from step 2 for another randomly chosen lattice site.

The succession of N executions of steps (2) to (6) for N randomly selected sites is called a ***Monte Carlo step***. Not all the lattice sites are visited during a MC step, and it is difficult to link a MC step to the physical time controlling the evolution of the spin population. Note that the Boltzmann distribution having already been taken into account for the local transitions of the spins, the average magnetization of the system is directly given by the configuration when equilibrium is reached.

Note: The transition probability can also be defined for a spin inversion $S_k \to -S_k$ as being:

$$W_k^* = \frac{\exp\left[\frac{-\Delta E_k}{k_B T}\right]}{1 + \exp\left[\frac{-\Delta E_k}{k_B T}\right]} \tag{9.41}$$

This definition offers two advantages: first, this probability is strictly limited to be between 0 and 1, which was not the case for W_k when ΔE_k becomes negative; second, it seems that equilibrium states are more rapidly attained.

9.5.3 Extension of the method

The same type of algorithm can be applied to calculate the phase diagrams for simple binary systems if the system energy is written as:

$$E = \sum_{j \neq k} \sum_{k} \phi_{jk}^{AA} c_j^A c_k^A + \phi_{jk}^{BB} c_j^B c_k^B + 2\phi_{jk}^{AB} c_j^A c_k^B \tag{9.42}$$

where c_k^A represents the concentration of element A at site k of the MC lattice (c_k^A = 1 or 0 according to whether the site is occupied or not by an atom A), c_k^B is the concentration of element B at the same site ($c_k^B = 1 - c_k^A$), and the functions ϕ describe the interactions of pairs of atoms A and B (fig. 9.11). It is seen that this problem is made completely analogous to that of the Ising model by setting:

$$c_k^A = \tfrac{1}{2} + S_k \quad \text{and} \quad c_k^B = \tfrac{1}{2} - S_k \quad (\text{with } S_k = \pm\tfrac{1}{2})$$

Such an approach allows studying the order parameters of solid solutions in cubic crystalline structure phases. However, if the global domain concentrations of elements A and B are to be conserved, the local configuration variations can not be performed independently. For example, an atom at site k can be exchanged with one of its neighbors $v(k, j)$: if they are of the same type, there is no variation of the local energy. If they are of different types (fig. 9.12), the energy variation can be calcu-

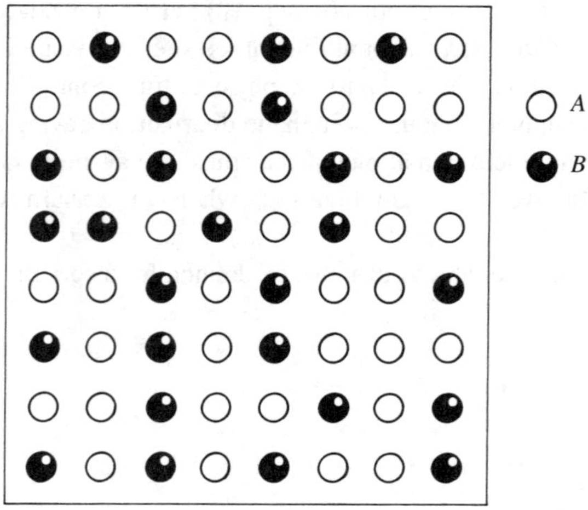

Figure 9.11 Solid solution of atoms A and B.

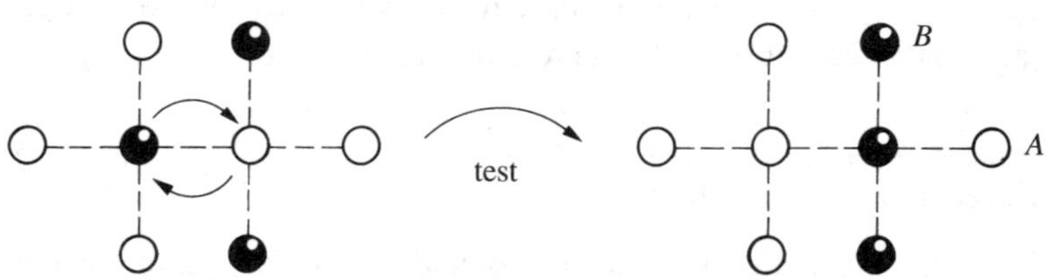

$$E_{av} = 6\phi^{AB} + \phi^{AA} \qquad E_{ap} = 3\phi^{AA} + 2\phi^{AB} + 2\phi^{BB}$$

$$\Delta E = 2\phi^{AA} + 2\phi^{BB} - 4\phi^{AB}$$

Figure 9.12 Initial and final configurations with associated energies for the permutation of two atoms A and B.

lated by considering the interactions between the nearest neighbors affected by the permutation (in this case, there are 7 coupling energies).

The two examples presented above (magnetism, solid solution) are clearly linked to the atomic arrangement of the crystal lattice. The spatial discretization for the MC method is thus natural. However, even with a grid of $100 \times 100 \times 100$ sites in 3 dimensions (10^6 sites), we only describe a portion of material measuring a dozen nanometers! To describe aspects at the scale of the microstructure ($> \mu$m), the MC method must be adapted such that each site no longer represents an atomic position but rather a portion of material containing billions of atoms. The spatial discretization in this case is no longer naturally introduced, and can bias the results.

A typical application of the MC method in microstructure problems is the study of grain growth and recrystallization. It is known that due to the effect of the interface tensions existing between the grains of a sample, their average size nor-

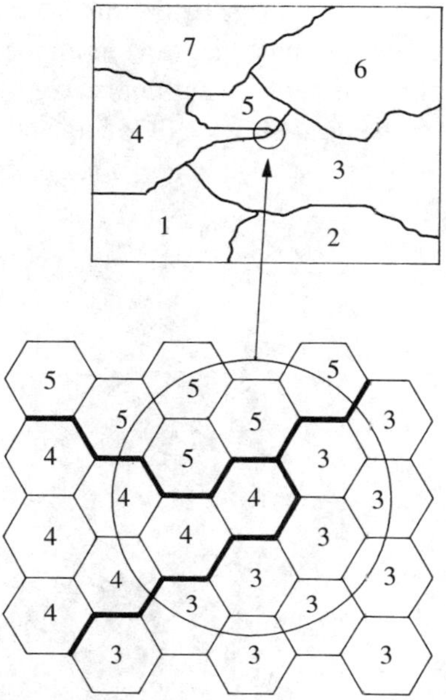

Figure 9.13 Extension of the Monte Carlo method to the growth of grains with a hexagonal cell lattice.

mally varies as the square root of time. Such an evolution can be studied with the MC method as described in the following.

The space is first discretized into a cell lattice; the center of each cell is a MC site. To avoid biasing the results too much with this artificial partitioning of the material, a hexagonal grid is chosen in two dimensions rather than squares (fig. 9.13): there are thus six nearest neighbors for each cell instead of four. Each cell k of the lattice is characterized by an integer index, I_k. A grain is defined as a connected set of lattice sites having the same index I. An initial distribution of grains in the domain is thus characterized by lattice sites k which are surrounded by six neighbors with the same index ($I_{v(k,\,j)} = I_k$ for every site $v(k, j)$ neighboring k ($j = 1, 6$)), and sites that have at least one neighbor with a different index. In the first situation, the lattice site represents an internal portion of a grain, while the second type of site is clearly situated at a grain boundary. The growth of grains being determined by their interfacial energy, γ, which is assumed isotropic for simplicity, the energy of the set of grains is defined by:

$$E = -\gamma \sum_{j=1}^{6} \sum_{k} [\delta(I_k, I_{v(k,\,j)}) - 1] \quad \text{with } \delta(I_k, I_{v(k,\,j)}) = \begin{cases} 1 \text{ if } I_{v(k,\,j)} = I_k \\ 0 \text{ if } I_{v(k,\,j)} \neq I_k \end{cases} \tag{9.43}$$

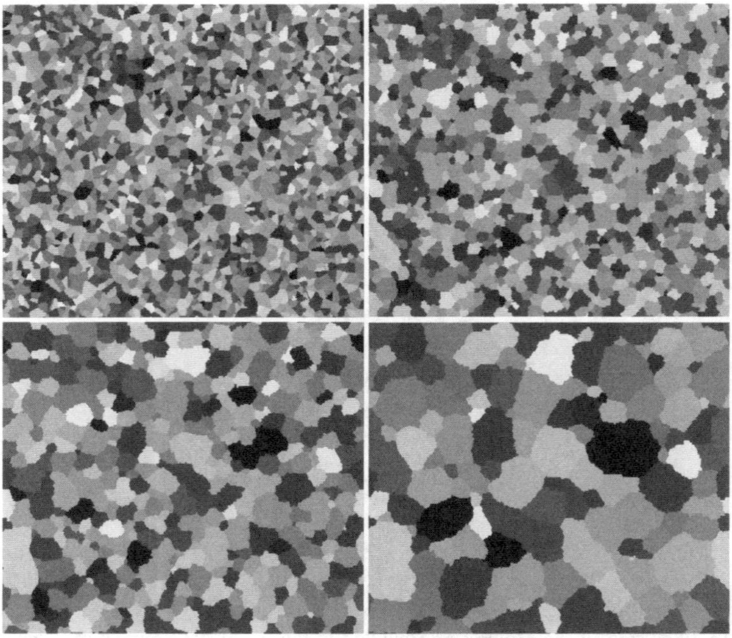

Figure 9.14 Evolution of a family of grains after 50, 500 and 1500 Monte Carlo steps.

The k summation is over the N sites of the MC lattice while the j sum is over the six nearest neighbors of each cell. Such a model (called the Potts model) coupled with a Metropolis algorithm permits the calculation of the structure of the grains during the MC time steps (fig. 9.14). For each cell selected at random, a transition of the index is tested, from I_k to an index $I_{v(k,j)}$ of one of its neighbors, chosen at random ($j = 1, 6$). The local interface energy difference between the two configurations obtained is used to calculate the transition probability according to (9.40). Two problems are encountered in such an extension of the original method:

- As already mentioned, it is not easy to align the MC step (the random selection of N sites for the Metropolis algorithm) to physical time.
- The artificial discretization of the domain in a regular lattice of cells introduces an anisotropy (of order 6 for the hexagonal grid) and influences the orientation of the calculated grain boundaries.

9.6 RANDOM WALKERS METHOD

9.6.1 Non stationary one dimensional case

Diffusion is the result of random jumps by atoms. Why not use random jumps to calculate diffusion profiles rather than solve Fick's equation? This is the principle of ***random walkers***.

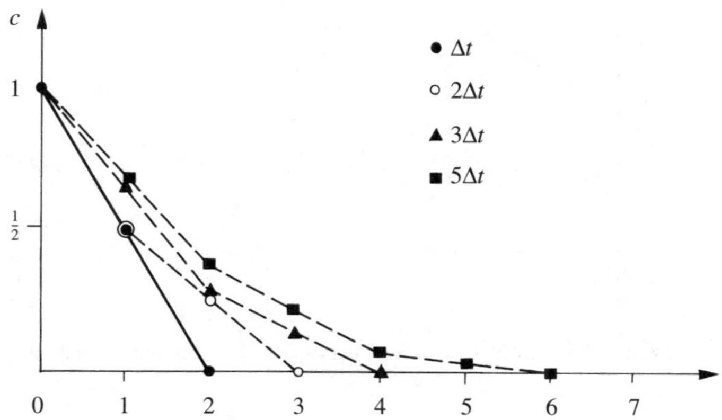

Figure 9.15 Calculated concentration profiles with an explicit scheme and $Fo = 0.5$ for different time steps and an imposed concentration condition. The concentration, expressed in number of atoms per m^3, can be made non dimensional by multiplying it by the characteristic volume of an atomic site, thus giving an occupation probability.

We first present the method for the simple case of one dimensional interstitial diffusion of atoms B between regularly spaced atoms A (fig. 9.15). We assume that on the left boundary there is an inexhaustible reservoir of atoms B (a Dirichlet condition with the imposed concentration normalized to 1). We choose a time step, Δt, such that $Fo = D\Delta t/a^2 = 0.5$, where a is the interatomic distance. This situation corresponds, in microscopic diffusion theory, to a jump probability for atoms B equal to 1 at each time step Δt. The random walk can now start, taking the following rules into account:

- At each time step, the atoms B must jump to one of the two neighboring sites, with a probability equal to 0.5 for each.
- The boundary where the concentration is imposed (node 0) is an inexhaustible source of atoms: at each time step, an atom from this boundary is "released" into the domain (at this site) and can thus jump with a probability of 0.5 to the first site of the lattice where the concentration is imposed. However, every atom or walker that returns to this boundary of the domain is absorbed.

At the first time step, only one walker can jump from the boundary to the first site, with a probability of 0.5. The concentration profile thus has the form $\{1, {}^1/_2, 0, 0, 0, \ldots\}$ for nodes 0, 1, 2, 3, 4, … At $t = 2\Delta t$, the half-walker at 1 can return to the boundary or jump to node 2. As a new B atom is emitted at the boundary at each time step, the concentration profile now has the form $\{1, {}^1/_2, {}^1/_4, 0, 0, \ldots\}$. At $t = 3\Delta t$, an atom at the boundary can jump to site 1 (probability 0.5), the half-atom at position 1 can jump to the boundary or to site 2, the quarter-atom at position 2 can jump to sites 1 or 3. The result is $\{1, ({}^1/_2 + {}^1/_8), {}^1/_4, {}^1/_8, 0, \ldots\} = \{1, {}^5/_8, {}^2/_8, {}^1/_8, 0, \ldots\}$

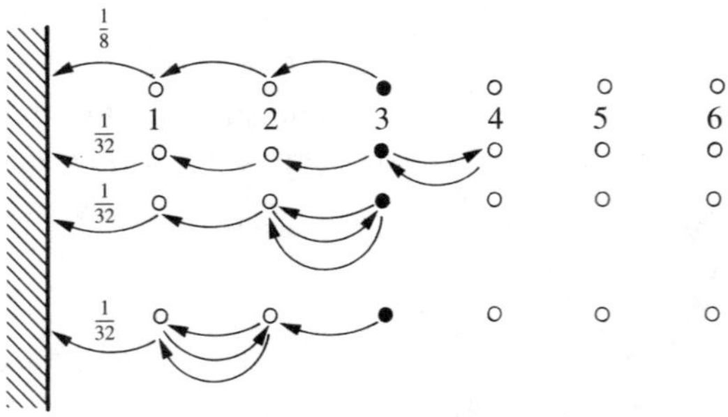

Figure 9.16 Random walkers emitted at node 3 that can reach the boundary in 5 jumps or less.

for the concentrations at 0, 1, 2, 3, 4, ... Thus, from neighbor to neighbor, the atoms are diffused by random jumps inside the domain. For $t = 5\Delta t$, the profile $\{1, {}^{22}/_{32}, {}^{12}/_{32}, {}^{7}/_{32}, {}^{2}/_{32}, {}^{1}/_{32},...\}$ is obtained.

The procedure adopted here is none other than the calculation by explicit differences (chap. 2) of the diffusion profile with $Fo = 0.5$, approached from the angle of stochastic jumps. This method would offer little interest if it were not possible to *invert* the method. It was seen in chapters 2 and 3 that the finite difference or finite element methods required the solution of a system of equations for each time step, that is, the knowledge of the concentrations at *all* the nodes for each time step. With the random walker method, the concentration at a point can be found for any time step without having to calculate the concentrations at the other nodes. The method is as follows.

Consider node 3 at time step $5\Delta t$ (concentration ${}^{7}/_{32}$). Random walkers are emitted from this node and they make five random jumps corresponding to the five time steps. After these jumps the *initial concentration* at the arrival point is cumulated: 1 for the boundary, 0 otherwise. With the particular initial conditions that we have chosen, it is sufficient to count only the paths that lead to the boundary from node 3 (fig. 9.16). There are four:

- $3 \rightarrow 2 \rightarrow 1 \rightarrow 0$ (probability $({}^{1}/_{2})^3 = {}^{1}/_{8}$);
- $3 \rightarrow 4 \rightarrow 3 \rightarrow 2 \rightarrow 1 \rightarrow 0$ (probability $({}^{1}/_{2})^5 = {}^{1}/_{32}$);
- $3 \rightarrow 2 \rightarrow 3 \rightarrow 2 \rightarrow 1 \rightarrow 0$ (probability ${}^{1}/_{32}$);
- $3 \rightarrow 2 \rightarrow 1 \rightarrow 2 \rightarrow 1 \rightarrow 0$ (probability ${}^{1}/_{32}$).

The total, ${}^{1}/_{8} + 3\ {}^{1}/_{32} = {}^{7}/_{32}$, is the value that we were seeking!

9.6.2 Two dimensional case

The generalization of the algorithm to a stationary two dimensional problem is easy. Consider first the case shown in figure 9.17 where the concentration is imposed on the whole boundary, but is not necessarily uniform. The stationary solution of the ***Laplace equation***:

$$\Delta c(x, y) = \frac{\partial^2 c}{\partial x^2} + \frac{\partial^2 c}{\partial y^2} = 0 \quad \text{in } \Omega \text{ with } c(x, y) = \bar{c}(x, y) \text{ on } \partial\Omega \tag{9.44}$$

based on the finite difference method and a regular grid $\Delta x = \Delta y$ amounts to solving a set of equations of the type (chap. 2):

$$c_{i,j} = \tfrac{1}{4}c_{i+1,j} + \tfrac{1}{4}c_{i-1,j} + \tfrac{1}{4}c_{i,j+1} + \tfrac{1}{4}c_{i,j-1} \text{ for } (i, j) \text{ inside } \Omega \tag{9.45}$$

The indices i and j are used to locate the x and y positions of the nodes. For a node next to at least one boundary node, the corresponding concentration on the right hand side of (9.45) is of course replaced by the imposed concentration $\bar{c}_{i\pm 1, j\pm 1}$.

By analogy with what was done above, the concentration at point (i, j) can be calculated by emitting N random walkers (ν) from this point and letting them move on the grid nodes with a jump probability of $^1/_4$ for each direction (East, North, West, South). These walkers stop when they touch a boundary node (i_ν, j_ν), and the corresponding imposed concentration, $\bar{c}_{i_\nu, j_\nu}$, is then cumulated. If the number of random walkers is sufficiently high, then the concentration defined by:

$$c_{i,j} = \frac{1}{N}\sum_{\nu=1}^{N} \bar{c}_{i_\nu, j_\nu} \tag{9.46}$$

is a good approximation of the solution of the stationary diffusion equation. In figure 9.17, only two random walkers emitted from site (9, 7) are shown: the first arrives at the boundary with imposed concentration $\bar{c}_1$ after two jumps (probability

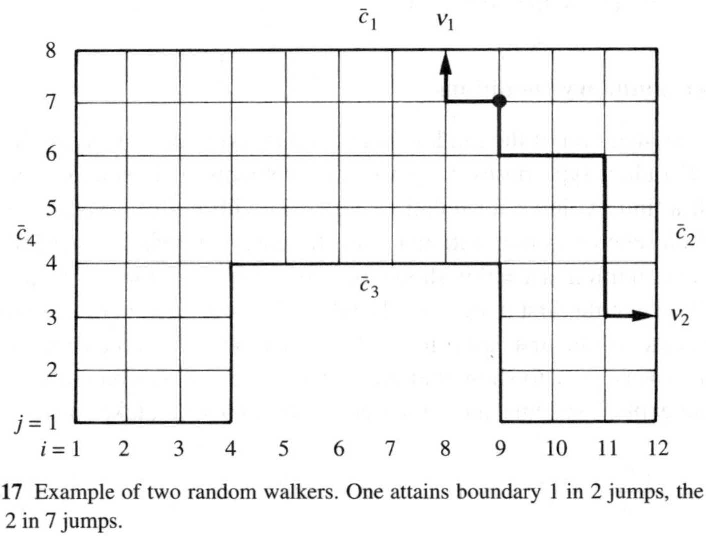

Figure 9.17 Example of two random walkers. One attains boundary 1 in 2 jumps, the other attains boundary 2 in 7 jumps.

of this path: $(^1/_4)^2$) while the second arrives after 7 jumps (associated probability: $(^1/_4)^7$) at the boundary with concentration $\bar{c}_2$. The closer a given boundary is to the point where the walkers are emitted, the higher the probability that the latter will arrive there. The weighting over numerous walkers then gives a good approximation of the concentration at a given point of the domain, without having to solve a system of equations as for the FDM or the FEM (avoiding a matrix factorization or an iterative method). We see the analogy between the solution of a system of equations yielding the solution at all the mesh points and obtaining the solution for a particular point by a Markov chain.

For the non stationary problem, it is sufficient to rewrite the corresponding diffusion equation formulated in an explicit scheme with $Fo = {^1/_4}$ (chap. 2 without an advection term): it is identical to (9.45) except that the concentration on the left, $c_{i,j}$, is taken at time $t + \Delta t$ while those of the neighbors on the right are taken at time t. The random walker method to calculate the concentration at a particular point (i, j) after M time steps ($t = M\,\Delta t$) is then as follows:

- Emit N walkers (ν) from the point (i, j) which will each make M random jumps, with equal probability in the directions $\pm x$, $\pm y$ $(^1/_4)$.
- If a walker reaches the boundary before the end of the M jumps, the imposed concentration at this node $\bar{c}_{i_\nu, j_\nu}$ is cumulated. If it is at an interior node (i_ν, j_ν) of the domain after the M jumps, the initial concentration of this node $c^0_{i_\nu, j_\nu}$ is cumulated. The concentration $c^M_{i,j}$ is then given by:

$$c^M_{i,j} = \frac{1}{N}\left[\sum_{\nu=1}^{N_1} \bar{c}_{i_\nu, j_\nu} + \sum_{\nu=1}^{N_2} c^0_{i_\nu, j_\nu}\right] \tag{9.47}$$

where N_1 and N_2 are the numbers of walkers arriving at the boundary and remaining inside the domain, respectively, ($N_1 + N_2 = N$).

9.6.3 Other boundary conditions

The generalization of the random walkers method to boundary conditions that are not of Dirichlet type poses no particular problems. It is sufficient to recommence with a finite volume formulation (chap. 2) with uniform volume elements, and an explicit scheme. A non stationary one dimensional diffusive problem with an imposed flux condition at $x = 0$ is shown in figure 9.18. Contrary to what was done in chapter 2, where the first node was placed on the boundary, here it is considered to be at the center of the first cell (finite volume method). The dimension of the cells being Δx, it is very easy to show that the non stationary diffusion equation formulated with an explicit scheme and a Fourier number, $Fo = 0.5$, takes the form:

$$c_1^{t+\Delta t} = \frac{c_1^t + c_2^t}{2} + Q \qquad \text{for node 1} \tag{9.48a}$$

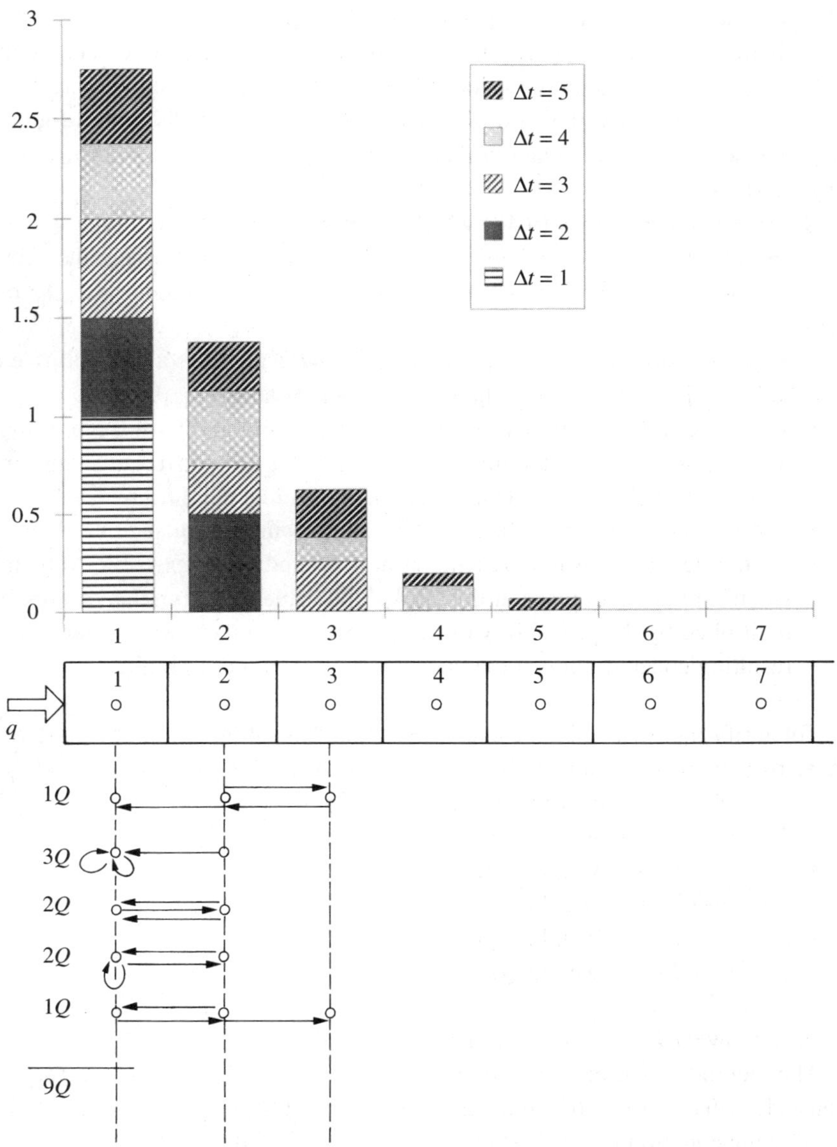

Figure 9.18 Concentration profiles calculated with an explicit scheme and *Fo* = 0.5 for different time steps and an imposed flux condition.

$$c_i^{t+\Delta t} = \frac{c_{i-1}^t + c_{i+1}^t}{2} \qquad \text{for the other nodes} \tag{9.48b}$$

where $Q = q\Delta t/\Delta x$. Starting form zero concentration everywhere in the domain, this means that the first cell receives a solute quantity Q at each time step, while all the arithmetic means appearing in equations (9.48) represent the diffusion between cells as before. Thus, at the fourth time step, the concentration profile in the domain normalized by Q is given by the values $\{^{19}/_8, ^9/_8, ^3/_8, ^1/_8, 0, \ldots\}$ at nodes 1, 2, 3, ... (fig. 9.18). It is seen that the sum of the concentrations at all the nodes ($4Q$) correctly represents the solute quantity that penetrated into the domain during the first four time steps.

Starting from equations (9.48) which are valid for a Neumann type boundary condition, rules for random walkers can now be established that permit the calculation of the concentration at any point and any time, without reference to the other values:

- At the interior of the volume (node $i \geq 2$), the walker has a probability equal to 0.5 of jumping left or right to a neighboring node (eq. (9.48b)).
- When the walker arrives at position 1, it has a probability of 0.5 of jumping in place and of 0.5 of returning to node 2 (eq. (9.48a)), but it no longer remains "stuck" at the boundary as for a Dirichlet condition.
- For time step $N\Delta t$, the walker emitted at any point must make $(N - 1)$ jumps.
- Each time that the walker reaches node 1 (including the possible "rebounds" when it remains in place at this node), the value Q is cumulated. This sum, multiplied by the probability of the corresponding path, will be added to the resulting concentration at the node emitting the random walker.

To clarify the rules, take the example of the concentration at node 2 after four time steps (*9/8Q*). The paths with three jumps that touch node 1 (the boundary) at least once each have probability 1/8 (fig. 9.18). These are:

- $2 \rightarrow 3 \rightarrow 2 \rightarrow 1$ (weight Q)
- $2 \rightarrow 1 \rightarrow 1 \rightarrow 1$ (weight $3Q$)
- $2 \rightarrow 1 \rightarrow 2 \rightarrow 1$ (weight $2Q$)
- $2 \rightarrow 1 \rightarrow 1 \rightarrow 2$ (weight $2Q$)
- $2 \rightarrow 1 \rightarrow 2 \rightarrow 3$ (weight Q)

In sum, we have $9Q$ with each path having a probability of *1/8*.

The method generalizes to two or three dimensions with mixed boundary conditions. Note that a closed (or adiabatic) boundary condition is equivalent to writing the following equation for the first node of the grid in figure 9.18 for $Fo = 0.5$ (finite volume method):

$$c_1^{t+\Delta t} = \frac{c_1^t + c_2^t}{2} \tag{9.49}$$

Thus a random walker that arrives at such a boundary (node 1) has the same probability of $^1/_2$ of jumping to the neighboring node 2 or of jumping in place. There is of course no accumulation of flux for such a situation. In general, it is sufficient to write the explicit diffusion equation with $Fo = 0.5$ (cf eqs. (9.48) or (9.49)) to find the laws governing the jump probabilities of the random walkers and the effect of the right hand side.

To conclude this brief introduction to the random walkers technique, recall that it is none other than the transcription of the Markov chain technique encountered in section 9.4 to the case of an explicit finite difference matrix with $Fo = 0.5$. Finally, we note that this method was employed with success to calculate ***the phenomenon of diffusion limited aggregation***. In this case, the walkers – or atoms, more precisely – are randomly placed in a rectangular domain, finely divided into cells, with a certain average concentration. Each atom can jump at each time step to the neighboring cell if it is empty. Periodic or closed boundary conditions are applied at the limits of the domain. Finally, a "dead" atom is placed at the center of the domain: this one does not move in time. Every time a mobile atom touches a dead atom, it also becomes immobile. The result of such a procedure is shown in figure 9.19: it is an arborescent structure with many branches. The center of this object (called a ***fractal***) evolves little; the action occurs at the periphery of the branches. This procedure can also be visualized as a *cellular automata*, the subject of the next section.

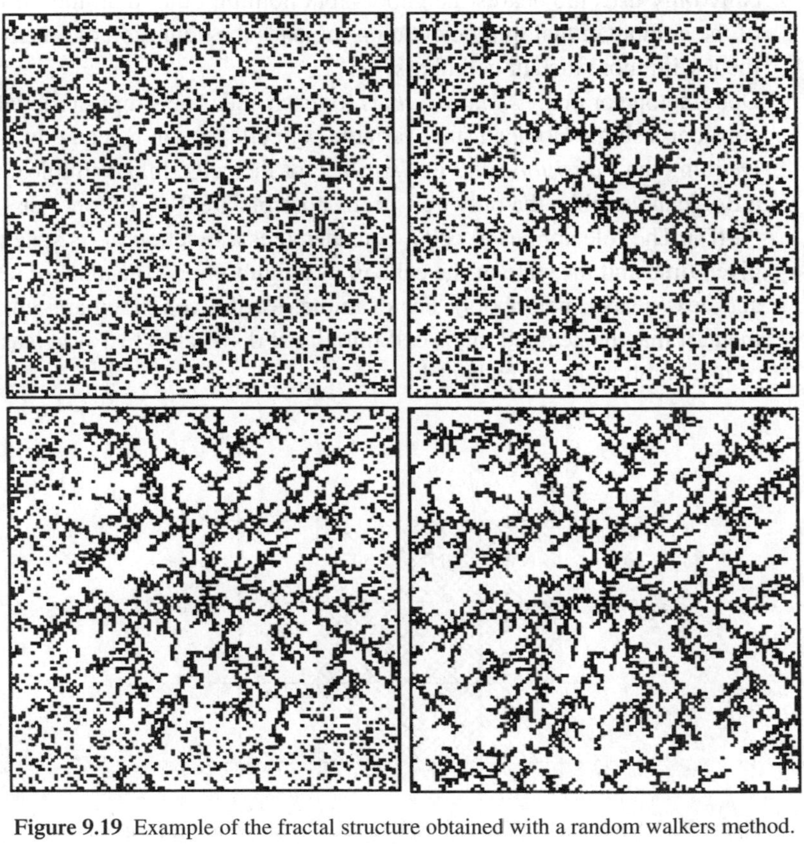

Figure 9.19 Example of the fractal structure obtained with a random walkers method.

9.7 CELLULAR AUTOMATA METHOD

9.7.1 Generalities

Introduced by von Neumann towards the end of the 1940s, cellular automata (CA) aim to reproduce complex physical phenomena with simple rules defined at a microscopic level. The idea is already present in the Monte Carlo method described in section 9.5, and the random walkers of section 9.6. The similarities do not end there, and it is sometimes difficult to distinguish between these methods. In general, cellular automata consist of the following steps:

1. Dividing space into a lattice of cells, as for the MC method. In two dimensions, a square or hexagonal lattice can be taken, whereas in three dimensions a cubic lattice is generally adopted (fig. 9.20).
2. Defining a *neighborhood* for the cells. For the square lattice, the four nearest neighbors can be taken (cells to the North (N), West (W), South (S), East (E)) or the eight nearest and next nearest neighbors (the corner cells NW, NE, SW and SE, are added). Respectively, these are the ***von Neumann*** and ***Moore environments***. For the hexagonal cell lattice, one typically only uses the nearest neighbors, but nothing other than the difficulty of the computational implementation impedes the definition of an environment with more distant neighbors.
3. Defining one or many *state variables* for each cell. The simplest model is to consider a single Boolean state variable per cell that can be equal to 0 or 1.
4. Defining the *transition rules* or *life rules* for the cells. In the simple case of a Boolean variable, the only transitions possible are $0 \to 1$ ("on") or $1 \to 0$ ("off"). The rule applies to cell i, considered as a function of its state at the given time and the state of the cells in its neighborhood (nearest, or nearest and next nearest neighbors). For example, for the von Neumann environment, it may be decided that a cell initially at 0 switches to 1 if at least two of its neighbors are already at 1.

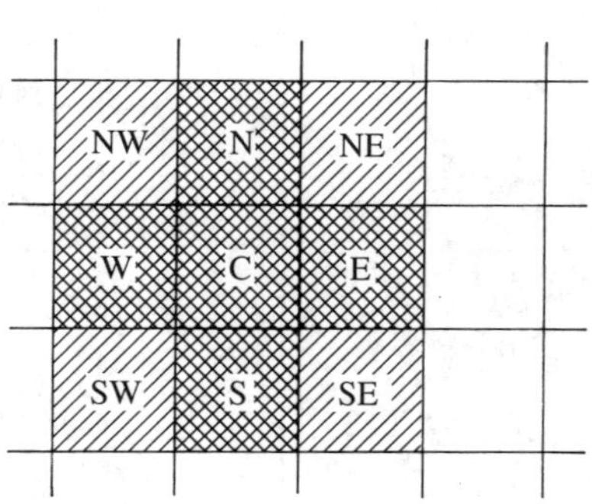

Figure 9.20 Square cell lattice and the Moore and von Neumann environments.

> **Attention:** In general, the new state of the cells is decided on the basis of the current state of the ensemble of cells and updating occurs only after all the cells have been recalculated. However, in other applications it may be decided to update the state of the cell as soon as their new state is known.

It is apparent that a CA is simple, and yet it permits the definition of extremely rich transition rules. Consider for example the von Neumann environment of a square lattice of cells (4 nearest neighbors). At a given time, there are $2^5 = 32$ possible configurations of the environment of a cell (table 9.1).

Table 9.1 Table of the possible configurations for a von Neumann environment.

N	W	S	E	C	C_{new}	N	W	S	E	C	C_{new}
0	0	0	0	0		1	0	0	0	0	
0	0	0	0	1		1	0	0	0	1	
0	0	0	1	0		1	0	0	1	0	
0	0	0	1	1		1	0	0	1	1	
0	0	1	0	0		1	0	1	0	0	
0	0	1	0	1		1	0	1	0	1	
0	0	1	1	0		1	0	1	1	0	
0	0	1	1	1		1	0	1	1	1	
0	1	0	0	0		1	1	0	0	0	
0	1	0	0	1		1	1	0	0	1	
0	1	0	1	0		1	1	0	1	0	
0	1	0	1	1		1	1	0	1	1	
0	1	1	0	0		1	1	1	0	0	
0	1	1	0	1		1	1	1	0	1	
0	1	1	1	0		1	1	1	1	0	
0	1	1	1	1		1	1	1	1	1	

As there are 32 possible cases for the new value of the cell C, there are 2^{32} ($2^{(2^5)}$) possibilities for putting 0s and 1s in these cases, and thus more than 4 billion possible rules for the CA. With the Moore environment, the possibilities become $2^{(2^9)} = 2^{512} \simeq 10^{154}$, as Margolin and Toffoli (bibliography) noted, more than the square of the estimated number of particles in the universe! Many of these rules are without any physical significance – in addition there are symmetry relations – but some play a very particular role in materials science, notably in phase transformations. Note also that some CA rules are not based on the individual states of the cells in the environment, but on their sum: in this case, the number of possible rules is 2^5 and 2^9 for the von Neumann and Moore environments, respectively. One then speaks of ***totalizer CA***. The transition rules can be *deterministic or stochastic*. In the first case the transition (or non-transition) of the central cell is performed according to the defined rules. In the second case, a random number will again be generated to decide the change. We give below a few illustrative examples for CA methods.

9.7.2 CA examples

Growth of a fractal structure

We start with a Moore configuration where all the cells are liquid (index = 0) except one cell at the center, which plays the role of a nucleation point (index = 1). The transition rule is then the following: an initially liquid cell becomes solid if one and only one neighboring cell is solid, otherwise it remains unchanged. The result of this simple transition rule, which has no real physical significance, is shown in figure 9.21 after 221 growth steps.

Reheating or annealing

A clever algorithm was discovered by Vichniac to simulate the annealing of a structure under the influence of curvature with CA. For this, the 0s and 1s are redistributed in the neighborhood of an interface without changing the proportions of the

Figure 9.21 The fractal calculated with a Moore environment CA (with permission of R. D'Souza).

"on" and "off" cells globally. A totalizer CA is then implemented with the following rule:

Let $I_{tot} = I_C + I_N + I_W + I_S + I_E + I_{NE} + I_{NW} + I_{SW} + I_{SE}$, the total number of "on" cells in the Moore environment, including of course, the center cell ($0 \le I_{tot} \le 9$). Then:

if $I_{tot} = 6, 7, 8, 9$ $\quad I_{C_{new}} = 1$
if $I_{tot} = 5$ $\quad I_{C_{new}} = 0$
if $I_{tot} = 4$ $\quad I_{C_{new}} = 1$
if $I_{tot} = 0, 1, 2, 3$ $\quad I_{C_{new}} = 0$

The result employing such a transition rule is shown in figure 9.22: starting from an initial structure of "on" cells (light gray zones), the regions with positive curvature tend to be eroded (bumps), those with negative curvature (hollow) to be filled (exercise). The result of this ***annealing*** of the structure is shown in figure 9.22 for two successive time steps. Certain small islands tend however to persist; this disadvantage can be eliminated by using a *stochastic* CA rather than a *deterministic* one. In this case the stochastic transition rules for the center cell are given by:

if $I_{tot} = 6, 7, 8, 9$ $\quad I_{C_{new}} = 1$
if $I_{tot} = 5$ $\quad I_{C_{new}} = IT(x < 0.5)$
if $I_{tot} = 4$ $\quad I_{C_{new}} = IT(x \ge 0.5)$
if $I_{tot} = 0, 1, 2, 3$ $\quad I_{C_{new}} = 0$

where *IT*(test) is the Boolean result of the test in parenthesis (*IT* = 1 if the test is verified, *IT* = 0 otherwise), and *x* is a random number between 0 and 1.

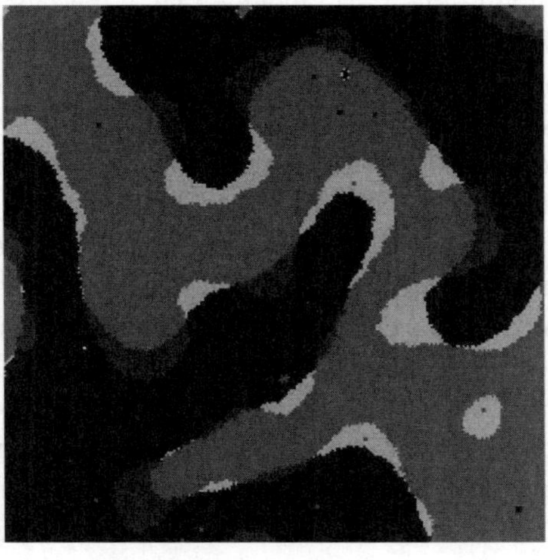

Figure 9.22 Annealing of a structure with the Vichniac CA algorithm (with permission of R. D'Souza).

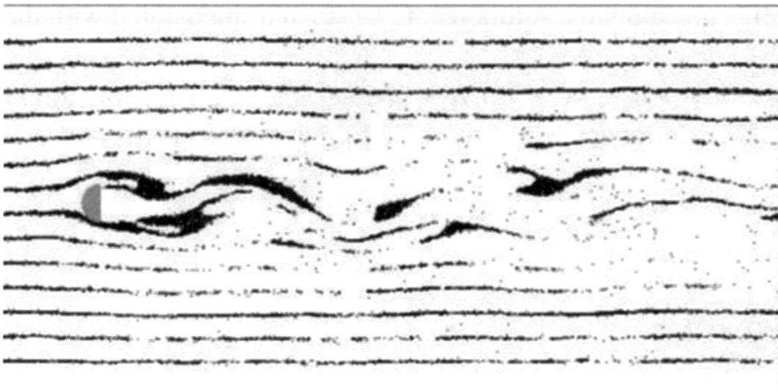

Figure 9.23 Fluid flow calculated around a blunt body with a CA (with permission of R. D'Souza).

Other rules

It is easy enough to imagine that stochastic CA can also be used to simulate diffusive phenomena (which is then close to random walkers), of growth or even of annealing of structures (fractals, grains, etc). Defining the cell environments that vary in time, it is even possible to model the trajectory of particles in space, including their collisions. Such methods can be used to model fluid flow. Figure 9.23 shows the flow calculated with a CA around a blunt body. For more details on this type of method, the reader is referred to the book by Margolin and Toffoli.

9.8 EXAMPLES

We have seen in this chapter the principle stochastic methods using generation of random numbers: the Monte Carlo method (MC), the random walkers method, and the Cellular Automata method (CA). Of course they are not the only methods, and numerous variants exist. In general, the application of these methods to materials science problems can pose some problems, the most important often being to know how to link the methods to the physics of the problem. Mention has already been made of the difficulty of linking a MC step to time, for example. As for the finite difference or finite element methods, the spatial discretization for all these methods introduces a *numerical anisotropy*. Thus, a diffusion equation for an isotropic medium formulated with a cubic grid becomes orthotropic (diffusion along a diagonal does not occur at the same speed as along an axis of the grid).

Despite some handicaps, stochastic methods offer understanding of phenomena at the microstructure or atomic scale, which makes them attractive notably for phase transformations. They allow a visualization of the microstructure and also provide the material science engineer a real *numerical metallographic* tool relating three dimensional microstructures to their observation in two dimensional sections (*numerical stereology*).

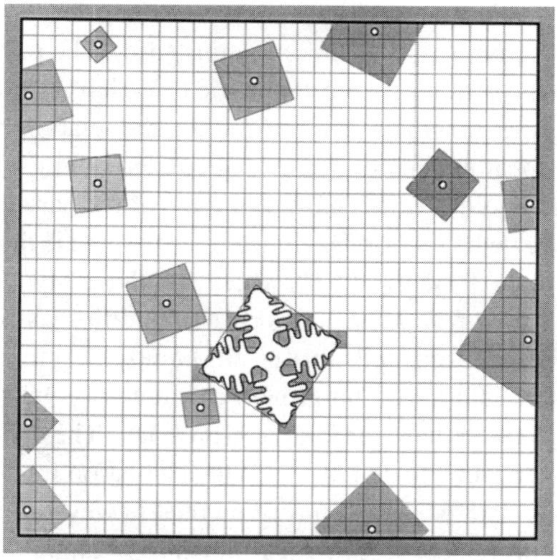

Figure 9.24 Schematic of two dimensional dendrite growth in a uniform temperature plate, and subdivision into cells.

The following example touches upon the stochastic modeling of grain formation in dendritic alloys during solidification. It is based on a physical rather than Boolean CA, i.e., each spatial cell has several real variables. We first describe this technique for a small two dimensional volume element assumed to be at uniform, decreasing temperature (fig. 9.24). This could be a portion of a metal plate cast between two ceramic walls. The square cells used for this CA have a state index, $I_{\mathrm{k}}^{\ell s}$, which is initially set to zero, signifying that they are all liquid. In a procedure where the temperature, $T(t)$, is known for the time steps, the cells can be transformed into "solid" cells (actually into "mushy" cells, as the alloy is dendritic) by two different mechanisms: by spontaneous heterogeneous nucleation of a new grain, or by "capture" by a neighboring cell in a growth phase. A certain number of cells can nucleate following a prescribed law. For the first mechanism, the number of cells δN nucleating at each time step δt is calculated in a deterministic way according to a nucleation law. These nucleation cells are chosen randomly among the set of cells in the domain. Note that different laws can be prescribed for the volume of the metal (cells with neighbors on all sides) and for the surface in contact with the ceramic mold (boundary cells). A boundary index, $I_{\mathrm{k}}^{\mathrm{b}}$, is thus attributed to each cell in addition to the liquid/solid index.

If a new solid seed forms in a given cell, k, its index $I_{\mathrm{k}}^{\ell s}$ changes from the liquid state (0) to a positive integer ($I_{\mathrm{k}}^{\ell s} \geq 1$) the value of which characterizes the orientation of the seed (i.e., in two dimensions, the orientation of one of its crystallographic directions <100> with respect to the horizontal axis). This value is of course chosen at random among a maximum number of possible orientation classes. The cell then enters a growth phase as long as it possesses at least one neighboring liquid cell (a Moore environment was chosen in this case). The portion of the dendrite tips with respect to the center of the cell k in the directions <100>, $R_k(t)$, is calculated by integrating the growth kinetic, $v(\Delta T)$, i.e., the relation between the growth velocity, v, and the local undercooling, ΔT. The undercooling is defined as the difference between the liquidus of the alloy and the real temperature. Such a dendrite growth kinetic can be deduced from analytical theories and is thus known; it is often expressed as a law in the form $v = A\Delta T^n$, where A and n are two constants related to the nature of the alloy. When the network of dendrites associated with $R_k(t)$ reaches one of the neighboring cells that is still liquid, it captures it, transmitting to it its index $I_k^{\ell s}$ (its orientation). The growth calculation is thus repeated from the center of this new cell with the same kinetic growth law. The dendritic growth of the grain is propagated from cell to cell, the growth calculation for a cell stopping when its eight neighbors are all "solid". It must be noted however, that the growth algorithm as presented introduces a numerical anisotropy along the axes of the CA, which does not respect the anisotropy of the dendritic growth along the privileged <100> directions. Variants of the algorithm have been proposed to eliminate this problem (see bibliography).

The results of such a calculation are shown in figure 9.25 for a small square solidifying from the boundary under three different conditions ((a), (b) and (c)) and for four times during solidification. The black cells in the center correspond to the liquid portion. Starting from the numerous grains formed at the surface of the square, growth toward the center is columnar or basaltic. A selection mechanism, naturally reproduced by the CA, eliminates the grains whose orientation <100> is not aligned with the perpendicular to the mold surface. The alloy solidifies in an entirely columnar morphology in case (a). Cases (b) and (c) show that for identical thermal conditions, the columnar growth can be stopped by grains nucleating in the bulk of the liquid volume when the latter is inoculated. Note that each gray level corresponds to a given crystallographic orientation. Such a CA has been extended to non uniform thermal situations (coupled to finite element thermal calculations), and more recently, to three dimensional situations. Figure 9.26 compares the real grains structure in a longitudinal cut of an aluminum-7wt% silicon ingot cast in a directional manner on a copper chill with the corresponding cut calculated with a three dimensional CA for the same thermal conditions. It is seen that the model is perfectly capable of reproducing the different columnar and equiaxied zones observed in the ingot, with only three adjustable parameters related to nucleation. The CA also yields the cooling curves in this case, including the out-of-equilibrium effects such as recalescence (increase in temperature for a negative enthalpy variation), as well as the evolution of the *texture* of the grains in the columnar zone.

(a) (b) (c)

Figure 9.25 Grain growth calculated with a CA for a metallic alloy solidified in a square with three different inoculation conditions (a,b and c).

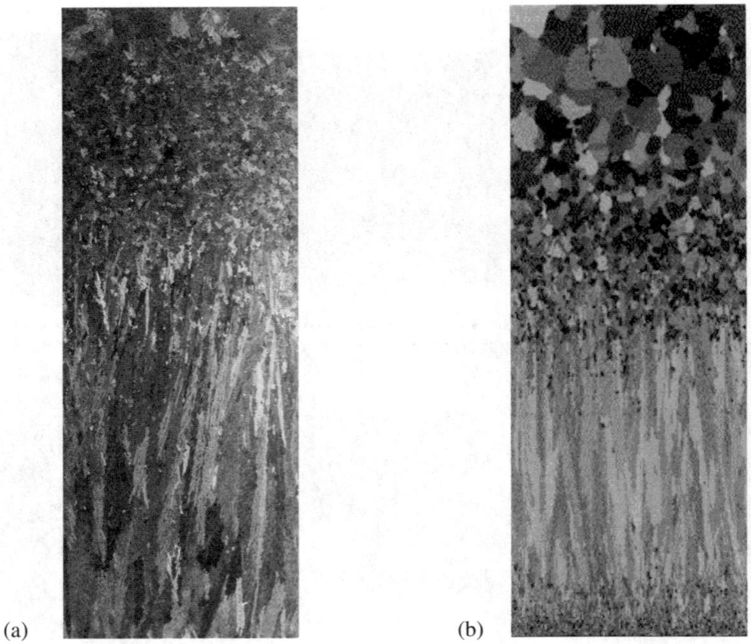

Figure 9.26 Grain structure observed in a longitudinal section of a directionally solidified aluminum-7wt% silicon alloy (a), and calculated with a CA coupled with finite element thermal calculations (b).

9.9 EXERCISES

9.9.1 Does there exist a combination of random numbers (a, c, m) for relation (9.6) that can generate a sequence of random integers reproducing a six faced die toss? If yes, is this sequence truly random? How can a die toss be simulated otherwise?

9.9.2 In the analysis of microstructures it is often necessary to relate volumetric, superficial and linear entities (stereology). Taking a spherical particle of radius R and considering a random planar section, what is the probability that the circle intersected by the plane has a radius r? How can this distribution be generated from random numbers uniformly distributed on [0, 1]? What is the average radius of all the randomly intersected circles?

9.9.3 Given a cubic domain for which the temperatures at the external faces are imposed to be T_1, T_2, T_3, T_4, T_5 and T_6. Find the steady state temperature at the center with a method inspired by random walkers.

9.9.4 What is the number of life rules possible in a hexagonal structure cellular automata considering only the nearest neighbors?

9.9.5 Consider a 45° diagonal of cells arranged as stair steps in a square lattice cellular automata. Scanning the cells from left to right and from top to bottom, how does the Vichniac algorithm erode the steps? What major disadvantage does this algorithm have with respect to the physical problem of annealing?

9.10 BIBLIOGRAPHY

Ch.-A. GANDIN and M. RAPPAZ, "A Coupled Finite Element-Cellular Automaton Model for the Prediction of Dendritic Grain Structures in Solidification Processes", *Acta Met. Mater.*, vol. 42, 2233, 1994.

Handbook of Numerical Heat Transfer, Ed. W. J. Minkowycz, E. M. Sparrow, G. E. Schneider and R. H. Pletcher, John Wiley & Sons, New York, 1988.

Monte Carlo Methods in Statistical Physics, Ed. K. Binder, Topics in Current Physics, Springer Verlag, Berlin, 1979.

R. Y. RUBINSTEIN, *Simulation and the Monte Carlo Method,* J. Wiley & Sons, New York, 1981.

T. TOFFOLI et N. MARGOLIN, *Cellular Automata Machines,* MIT Press, Boston, 1987.

CHAPTER 10

APPENDICES

10.1 TABLE OF SYMBOLS

Symbol	Description	Unit
$\boldsymbol{a}$	acceleration vector	$m\ s^{-2}$
a_r	chemical reaction rate	s^{-1}
$\boldsymbol{b}$	boundary condition vector, "right hand side" vector	–
Bi	Biot number	–
bw	bandwidth of a matrix	–
c	volumetric concentration	mole m^{-3}
$\boldsymbol{C}$	right Cauchy-Green deformation tensor	–
C_μ, $C_{\varepsilon 1}$, $C_{\varepsilon 2}$	constants of the $k - \varepsilon$ turbulence model	–
c_p	mass specific heat at constant pressure	$J\ kg^{-1}\ K^{-1}$
d	distance	m
d	dimension of the problem (1, 2 or 3)	–
$\boldsymbol{d}$	strain rate or velocity gradient tensor (symmetric part)	s^{-1}
D	diffusion coefficient	$m^2\ s^{-1}$
$\boldsymbol{D}^{e\ell}$	elasticity tensor	Pa ($N\ m^{-2}$)
E	elastic modulus	Pa
$\boldsymbol{E}$	electric field	$V\ m^{-1}$
f	volumetric fraction	–
f	plasticity criteria function	–
$\boldsymbol{f}$	right hand side vector	–
$\boldsymbol{F}_v$	volumetric force	$N\ m^{-3}$
$\boldsymbol{F}$	force	N
Fo	Fourier number	–
$\boldsymbol{F}$	deformation gradient tensor	–
g	contact indicator function	–
$\boldsymbol{g}$	gravity vector	$m\ s^{-2}$
$\boldsymbol{g}_i$	ith basis vector in curvilinear coordinates	–
Gr	Grasshof number	–
h	heat transfer coefficient	$W\ m^{-2}\ K^{-1}$
H	mass specific enthalpy	$J\ kg^{-1}$
H	Heaviside function	–
$\boldsymbol{H}$	magnetic field vector	$A\ m^{-1}$
$\boldsymbol{I}$	identity tensor	–
$\boldsymbol{j}_E$	electric current density	$A\ m^{-2}$
$\boldsymbol{j}_i$	solute flux associated with the concentration c_i ($-D\,\mathbf{grad}\,c_i$)	mole $m^{-2}\ s^{-1}$
$\boldsymbol{j}_T$	heat flux ($-\kappa\,\mathbf{grad}\,T$)	$W\ m^{-2}$
J_2	second invariant of the deviatoric stress tensor	Pa^2
k	turbulent kinetic energy per unit mass	$J\ kg^{-1}$
k_i	partition coefficient associated with the concentration c_i	–
K	viscoplastic consistency	$Pa\ s^m$
k_B	Boltzmann constant (1.381×10^{-23})	$J\ K^{-1}$
$[\boldsymbol{K}]$	stiffness matrix	
L, l	length	m
L_f	latent heat of fusion	$J\ kg^{-1}$

L_v	latent heat of vaporization	$J\,kg^{-1}$
m	strain rate sensitivity coefficient	–
$\bar{m}$	Tresca friction coefficient	–
m_i	liquidus slope associated with the concentration c_i	K or K $\%^{-1}$
M	time step number	–
n	index of refraction	–
n	work hardening coefficient	–
n	grain density	m^{-3}
nne	number of nodes per finite element	–
$nipe$	number of integration points per finite element	–
$\boldsymbol{n}$	outgoing unit normal of a domain	–
N	number of unknowns per node	–
N	number of grains	–
N	number of nodes in finite differences	–
N_1, N_2	differences of the normal stresses of a viscoelastic fluid	Pa
Ne	number of elements in a finite element mesh	–
N^{free}	number of unconstrained degrees of freedom in finite elements	–
N^{imp}	number of imposed degrees of freedom in finite elements	–
Nn	number of nodes in finite elements	–
N_A	Avogadro number (6.022×10^{23})	$mole^{-1}$
$\boldsymbol{p}$	momentum density ($= \rho \boldsymbol{v}$)	$kg\,m^{-2}\,s^{-1}$
p	pressure	Pa
p_f	viscoplastic friction coefficient	–
Pe	Péclet number	–
Pr	Prandtl number	–
q_T	heat flux	$W\,m^{-2}$
$\dot{Q}_i$	volumetric solute production term	$mole\,m^{-3}\,s^{-1}$
$\dot{Q}_T$	volumetric heat production term	$W\,m^{-3}$
R	radius	m
$\boldsymbol{R}$	residual vector	–
$\boldsymbol{R}$	Reynolds stress tensor	Pa
R	perfect gas constant (8.316)	$J\,mole^{-1}\,K^{-1}$
Re	Reynolds number	–
S	surface	m^2
$\boldsymbol{s}$	deviatoric stress tensor	Pa
t	time	s
$\delta t, \Delta t$	time increment	s
T	temperature	K, °C
T_E	eutectic temperature	K, °C
T_m	melting temperature	K, °C
T_S	solidus	K, °C
T_L	liquidus	K, °C
ΔT	undercooling or variation temperature	K, °C
$\boldsymbol{T}_v$	viscoelastic extra-stress tensor	Pa
$\boldsymbol{T}_{v1}$	viscoelastic part of $\boldsymbol{T}_v$	Pa
$\boldsymbol{T}$	stress vector ($= \boldsymbol{\sigma} \cdot \boldsymbol{n}$)	Pa
$\boldsymbol{u}$	displacement vector, components (u_x, u_y, u_z) (u_1, u_2, u_3)	m
$\boldsymbol{u}_*$	friction velocity	$m\,s^{-1}$
U	reference velocity	$m\,s^{-1}$
$\boldsymbol{v}$	velocity vector, components (v_x, v_y, v_z) or (v_1, v_2, v_3)	$m\,s^{-1}$
V	volume	m^3
ΔV	volume element	m^3
∂V	volume element surface	m^2
w	mass fraction	wt%
w	integration point weighting in finite elements	–
$\dot{w}$	power density	$W\,m^{-3}$
W	work or energy	J
We	Weissenberg number	–

$\dot{W}$	power	W
$\boldsymbol{x}$	position vector, components (x_1, x_2, x_3) or (x, y, z)	m
α	chemical transfer coefficient	m s^{-1}
α	thermal diffusivity $(= \kappa/(\rho c_p))$	m^2 s^{-1}
α_f	viscoplastic friction coefficient	–
α_r	absorption coefficient	–
α_T	one dimensional thermal dilatation coefficient	K^{-1}
β_T	volumetric thermal dilatation coefficient	K^{-1}
β_i	solutal dilatation coefficient associated with chemical species (i)	m^3 $mole^{-1}$
γ	surface tension	N m^{-1}
Γ	Gibbs Thomson coefficient	K m
$\partial\Omega$	domain boundary	–
δ	boundary layer thickness	m
δ	Dirac function	–
φ	test function	–
φ	viscoplastic potential	–
$\boldsymbol{\Phi}$	functional	–
ε	turbulent kinetic energy dissipation rate	m^2 s^{-3}
ε_r	gray body radiative emissivity	–
$\boldsymbol{\varepsilon}$	strain tensor	–
$\dot{\boldsymbol{\varepsilon}}$	strain rate tensor	s^{-1}
$\dot{\varepsilon}_{eq}$	generalized strain rate or equivalent strain rate	s^{-1}
$\dot{\gamma}$	shear rate	s^{-1}
$\dot{\gamma}_{eq}$	generalized or equivalent shear rate	s^{-1}
λ	matrix profile	–
$\lambda^{p\ell}$	plastic multiplier	–
λ_1	viscoelastic fluid relaxation time	s
μ	chemical potential	J $mole^{-1}$
μ	dynamic viscosity	Pa s (kg m^{-1} s^{-1})
μ_B	Bohr magneton $(\mu_B = 9.27\ 10^{-24})$	A m^2
μ_0	magnetic permeability of vacuum $(\mu_0 = 1.256\ 10^{-6})$	Vs A^{-1} m^{-1}
μ_0	Newtonian dynamic viscosity of a non Newtonian fluid	Pa s
μ_1	dynamic viscosity of the viscoelastic component of the extra-stress of a non Newtonian fluid	Pa s
μ_T	turbulent dynamic viscosity	Pa s
μ_f	Coulomb friction coefficient	–
ν_P	Poisson coefficient	–
ν	kinematic viscosity $(= \mu/\rho)$	m^2 s^{-1}
ν_T	turbulent kinematic viscosity	m^2 s^{-1}
θ	angle	rad
θ^i	material or curvilinear coordinate i	–
θ	integration scheme or decentering parameter	–
κ	thermal conductivity	W m^{-1} K^{-1}
κ	von Karman constant (0.41)	–
K	permeability	m^2
ψ_n	interpolation function at node n	–
Ψ	stream function	m^2 s^{-1}
ω	successive relaxation method parameter	–
ω	angular velocity	rad s^{-1}
$\boldsymbol{\omega}$	rotation rate tensor	s^{-1}
$\boldsymbol{\omega}$	vorticity vector	s^{-1}
Ω	domain	–
ρ	specific mass	kg m^{-3}
ρ	spectral radius of a matrix	–
ρ	descent (gradient) methods parameter	–

ρ_E	electrical resitivity	W m
ρ_p	penalty coefficient	–
σ_B	Stefan Boltzmann constant (5.67×10^{-8})	W m^{-2} K^{-4}
$\boldsymbol{\sigma}$	stress tensor	Pa
σ_0	flow stress	Pa
σ_{eq}	von Mises equivalent stress	Pa
$\sigma_k, \sigma_\varepsilon$	form factors of the $k-\varepsilon$ model equations	–
$\boldsymbol{\tau}$	unit tangent vector	–
$\boldsymbol{\tau}$	shear stress vector	Pa
Ξ	curvature	m^{-1}
χ'_p, χ''_p	penalty coefficients	–
$\boldsymbol{\chi}$	internal stress tensor	Pa
$\boldsymbol{\xi}$	local coordinates vector inside a finite element	–

Indices

a	ambient
c	chemical, critical, kinematic or contact
E	eutectic
i	internal
k	kinetic
ℓ	liquid
L	liquidus
n	normal component
0	nominal or initial
r	curvature
s	solid
S	solidus
T	thermal
x, y, z	components
τ	tangential component
∞	at infinity

Exponents

e	in element e
eℓ	elastic
n	at node n (finite elements)
pℓ	plastic
r	relative to the reference element, in finite elements
s	slip
t	at time t
t + Δt	at time $t + \Delta t$
th	thermal
vp	viscoplastic
α, β, γ	phases
(ν)	at iteration ν
*	interface between two phases
$\underline{*}$	virtual
$\bar{.}$	imposed value

Notations

[.]	matrix
$[\,.\,]^T$	transposed matrix
$\lvert\,.\,\rvert$	norm of a vector or a tensor
$\boldsymbol{v}, \boldsymbol{\tau}$	vector or tensor
$\bar{.}$	temporal average
$\langle\,.\,\rangle$	spatial average
$\mathbb{R}$	real space

Operators

$\cdot$	scalar product
$\otimes$	vector product
$\times$	tensor product
:	contracted product of two tensors of second order ($\boldsymbol{A}:\boldsymbol{B} = \sum_{i,j} A_{ij}B_{ij}$)
grad	gradient of a scalar field (vector field)
Grad	gradient of a vector field (tensor field)
det	determinant of a matrix
div	divergence of a vector field (scalar field)
div	divergence of a tensor field (vector field)
curl	curl of a vector field
Tr	trace of a second order tensor ($\mathrm{Tr}\ \boldsymbol{\sigma} = \sum_i \sigma_{ii}$)
$[I]$	identity matrix
δ_{ij}	Kronecker delta function (1 if $i = j$, 0 otherwise)
Δ	Laplacian
$\frac{d}{dt}$	particular, material, or total derivative
$\frac{\partial}{\partial t}$	partial derivative with respect to time

Coordinate systems

x, y, z	or	x_1, x_2, x_3	Cartesian coordinates
r, θ, z			polar (cylindrical) coordinates
r, θ, ϕ			spherical coordinates
ξ, η, ζ	or	ξ_1, ξ_2, ξ_3	local coordinates in finite elements

10.2 VECTOR CALCULUS

10.2.1 Cartesian coordinates

Let $(\boldsymbol{e}_x, \boldsymbol{e}_y, \boldsymbol{e}_z)$ be an orthonormal Cartesian coordinate system. The differential operators will be expressed as follows:

- ***Divergence*** of a vector field, $\boldsymbol{v}(x, y, z)$:

$$\mathrm{div}\ \boldsymbol{v} = \frac{\partial v_x}{\partial x} + \frac{\partial v_y}{\partial y} + \frac{\partial v_z}{\partial z} \tag{10.1}$$

- ***Divergence*** of a tensor field, $\boldsymbol{\sigma}(x, y, z)$:

$$\mathbf{div}\ \boldsymbol{\sigma} = \left[\frac{\partial \sigma_{xx}}{\partial x} + \frac{\partial \sigma_{xy}}{\partial y} + \frac{\partial \sigma_{xz}}{\partial z}\right]\boldsymbol{e}_x + \left[\frac{\partial \sigma_{yx}}{\partial x} + \frac{\partial \sigma_{yy}}{\partial y} + \frac{\partial \sigma_{yz}}{\partial z}\right]\boldsymbol{e}_y + \left[\frac{\partial \sigma_{zx}}{\partial x} + \frac{\partial \sigma_{zy}}{\partial y} + \frac{\partial \sigma_{zz}}{\partial z}\right]\boldsymbol{e}_z \tag{10.2}$$

- ***Gradient*** of a scalar field, $f(x, y, z)$:

$$\mathbf{grad}\, f = \frac{\partial f}{\partial x}\boldsymbol{e}_\mathrm{x} + \frac{\partial f}{\partial y}\boldsymbol{e}_\mathrm{y} + \frac{\partial f}{\partial z}\boldsymbol{e}_\mathrm{z} \tag{10.3}$$

- ***Gradient*** of a vector field, $\boldsymbol{v}(x, y, z)$, and its symmetric part:

$$\mathbf{Grad}\,\boldsymbol{v} = \begin{bmatrix} \frac{\partial v_\mathrm{x}}{\partial x} & \frac{\partial v_\mathrm{x}}{\partial y} & \frac{\partial v_\mathrm{x}}{\partial z} \\ \frac{\partial v_\mathrm{y}}{\partial x} & \frac{\partial v_\mathrm{y}}{\partial y} & \frac{\partial v_\mathrm{y}}{\partial z} \\ \frac{\partial v_\mathrm{z}}{\partial x} & \frac{\partial v_\mathrm{z}}{\partial y} & \frac{\partial v_\mathrm{z}}{\partial z} \end{bmatrix} \tag{10.4}$$

$$\tfrac{1}{2}[\mathbf{Grad}\,\boldsymbol{v} + (\mathbf{Grad}\,\boldsymbol{v})^\mathrm{T}] = \begin{bmatrix} \frac{\partial v_\mathrm{x}}{\partial x} & \frac{1}{2}\left(\frac{\partial v_\mathrm{x}}{\partial y} + \frac{\partial v_\mathrm{y}}{\partial x}\right) & \frac{1}{2}\left(\frac{\partial v_\mathrm{x}}{\partial z} + \frac{\partial v_\mathrm{z}}{\partial x}\right) \\ & \frac{\partial v_\mathrm{y}}{\partial y} & \frac{1}{2}\left(\frac{\partial v_\mathrm{z}}{\partial y} + \frac{\partial v_\mathrm{y}}{\partial z}\right) \\ \text{symmetric} & & \frac{\partial v_\mathrm{z}}{\partial z} \end{bmatrix} \tag{10.5}$$

- ***Curl*** of a vector field, $\boldsymbol{v}(x, y, z)$:

$$\boldsymbol{curl}\,\boldsymbol{v} = \left[\frac{\partial v_\mathrm{z}}{\partial y} - \frac{\partial v_\mathrm{y}}{\partial z}\right]\boldsymbol{e}_\mathrm{x} + \left[\frac{\partial v_\mathrm{x}}{\partial z} - \frac{\partial v_\mathrm{z}}{\partial x}\right]\boldsymbol{e}_\mathrm{y} + \left[\frac{\partial v_\mathrm{y}}{\partial x} - \frac{\partial v_\mathrm{x}}{\partial y}\right]\boldsymbol{e}_\mathrm{z} \tag{10.6}$$

- ***Laplacian*** of a scalar field, $f(x, y, z)$:

$$\Delta f = \mathrm{div}\,(\mathbf{grad}\, f) = \frac{\partial^2 f}{\partial x^2} + \frac{\partial^2 f}{\partial y^2} + \frac{\partial^2 f}{\partial z^2} \tag{10.7}$$

10.2.2 Cylindrical coordinates

Let $(\boldsymbol{e}_\mathrm{r}, \boldsymbol{e}_\theta, \boldsymbol{e}_\mathrm{z})$ be an orthonormal cylindrical coordinate system. The differential operators will be expressed as follows:

- ***Divergence*** of a vector field, $\boldsymbol{v}(r, \theta, z)$:

$$\mathrm{div}\,\boldsymbol{v} = \frac{1}{r}\frac{\partial}{\partial r}(r v_\mathrm{r}) + \frac{1}{r}\frac{\partial v_\theta}{\partial \theta} + \frac{\partial v_\mathrm{z}}{\partial z} \tag{10.8}$$

or

$$\mathrm{div}\,\boldsymbol{v} = \frac{\partial v_\mathrm{r}}{\partial r} + \frac{v_r}{r} + \frac{1}{r}\frac{\partial v_\theta}{\partial \theta} + \frac{\partial v_\mathrm{z}}{\partial z} \tag{10.9}$$

- ***Divergence*** of a tensor field, $\boldsymbol{\sigma}(r, \theta, z)$:

$$\begin{aligned}\mathbf{div}\ \boldsymbol{\sigma} = & \left[\frac{\partial \sigma_{\mathrm{rr}}}{\partial r} + \frac{1}{r}\frac{\partial \sigma_{\mathrm{r\theta}}}{\partial \theta} + \frac{\partial \sigma_{\mathrm{rz}}}{\partial z} + \frac{\sigma_{\mathrm{rr}} - \sigma_{\theta\theta}}{r}\right]\boldsymbol{e}_{\mathrm{r}} \\ & + \left[\frac{\partial \sigma_{\mathrm{r\theta}}}{\partial r} + \frac{1}{r}\frac{\partial \sigma_{\theta\theta}}{\partial \theta} + \frac{\partial \sigma_{\theta\mathrm{z}}}{\partial z} + \frac{2\sigma_{\mathrm{r\theta}}}{r}\right]\boldsymbol{e}_{\theta} \\ & + \left[\frac{\partial \sigma_{\mathrm{rz}}}{\partial r} + \frac{1}{r}\frac{\partial \sigma_{\theta\mathrm{z}}}{\partial \theta} + \frac{\partial \sigma_{\mathrm{zz}}}{\partial z} + \frac{\sigma_{\mathrm{rz}}}{r}\right]\boldsymbol{e}_{\mathrm{z}} \end{aligned} \tag{10.10}$$

- ***Gradient*** of a scalar field, $f(r, \theta, z)$:

$$\mathbf{grad}\ f = \frac{\partial f}{\partial r}\boldsymbol{e}_{\mathrm{r}} + \frac{1}{r}\frac{\partial f}{\partial \theta}\boldsymbol{e}_{\theta} + \frac{\partial f}{\partial z}\boldsymbol{e}_{\mathrm{z}} \tag{10.11}$$

- ***Gradient*** of a vector field, $\boldsymbol{v}(r, \theta, z)$, and its symmetric part:

$$\mathbf{Grad}\ \boldsymbol{v} = \begin{bmatrix} \dfrac{\partial v_{\mathrm{r}}}{\partial r} & \dfrac{1}{r}\dfrac{\partial v_{\mathrm{r}}}{\partial \theta} - \dfrac{v_{\theta}}{r} & \dfrac{\partial v_{\mathrm{r}}}{\partial z} \\ \dfrac{\partial v_{\theta}}{\partial r} & \dfrac{1}{r}\dfrac{\partial v_{\theta}}{\partial \theta} + \dfrac{v_{\mathrm{r}}}{r} & \dfrac{\partial v_{\theta}}{\partial z} \\ \dfrac{\partial v_{\mathrm{z}}}{\partial r} & \dfrac{1}{r}\dfrac{\partial v_{\mathrm{z}}}{\partial \theta} & \dfrac{\partial v_{\mathrm{z}}}{\partial z} \end{bmatrix} \tag{10.12}$$

$$\tfrac{1}{2}[\mathbf{Grad}\ \boldsymbol{v} + (\mathbf{Grad}\ \boldsymbol{v})^{\mathrm{T}}] = \begin{bmatrix} \dfrac{\partial v_{\mathrm{r}}}{\partial r} & \tfrac{1}{2}\left(\dfrac{\partial v_{\theta}}{\partial r} - \dfrac{v_{\theta}}{r} + \dfrac{1}{r}\dfrac{\partial v_{\mathrm{r}}}{\partial \theta}\right) & \tfrac{1}{2}\left(\dfrac{\partial v_{\mathrm{r}}}{\partial z} + \dfrac{\partial v_{\mathrm{z}}}{\partial r}\right) \\ & \dfrac{1}{r}\dfrac{\partial v_{\theta}}{\partial \theta} + \dfrac{v_{\mathrm{r}}}{r} & \tfrac{1}{2}\left(\dfrac{1}{r}\dfrac{\partial v_{\mathrm{z}}}{\partial \theta} + \dfrac{\partial v_{\theta}}{\partial z}\right) \\ \text{symmetric} & & \dfrac{\partial v_{\mathrm{z}}}{\partial z} \end{bmatrix} \tag{10.13}$$

- ***Curl*** of a vector field, $\boldsymbol{v}(r, \theta, z)$:

$$\boldsymbol{curl}\ \boldsymbol{v} = \left[\frac{1}{r}\frac{\partial v_{\mathrm{z}}}{\partial \theta} - \frac{\partial v_{\theta}}{\partial z}\right]\boldsymbol{e}_{\mathrm{r}} + \left[\frac{\partial v_{\mathrm{r}}}{\partial z} - \frac{\partial v_{\mathrm{z}}}{\partial r}\right]\boldsymbol{e}_{\theta} + \left[\frac{1}{r}\frac{\partial}{\partial r}[r v_{\theta}] - \frac{1}{r}\frac{\partial v_{\mathrm{r}}}{\partial \theta}\right]\boldsymbol{e}_{\mathrm{z}} \tag{10.14}$$

- ***Laplacian*** of a scalar field, $f(r, \theta, z)$:

$$\Delta f = \mathrm{div}\,(\mathbf{grad}\ f) = \frac{1}{r}\frac{\partial}{\partial r}\left[r\frac{\partial f}{\partial r}\right] + \frac{1}{r^2}\frac{\partial^2 f}{\partial \theta^2} + \frac{\partial^2 f}{\partial z^2} \tag{10.15}$$

10.2.3 Spherical coordinates

Let ($\boldsymbol{e}_{\mathrm{r}}$, $\boldsymbol{e}_{\theta}$, $\boldsymbol{e}_{\phi}$) be an orthonormal spherical coordinate system. The differential operators will be expressed as follows:

- *Divergence* of a vector field, $\boldsymbol{v}(r, \theta, \phi)$:

$$\operatorname{div}\boldsymbol{v} = \frac{1}{r^2}\frac{\partial}{\partial r}(r^2 v_{\mathrm{r}}) + \frac{1}{r\sin\theta}\frac{\partial}{\partial\theta}(v_{\theta}\sin\theta) + \frac{1}{r\sin\theta}\frac{\partial v_{\phi}}{\partial\phi} \tag{10.16}$$

- *Divergence* of a tensor field, $\boldsymbol{\sigma}(r, \theta, \phi)$:

$$\begin{aligned}\mathbf{div}\,\boldsymbol{\sigma} = {} & \left[\frac{\partial\sigma_{\mathrm{rr}}}{\partial r} + \frac{1}{r}\frac{\partial\sigma_{\mathrm{r}\theta}}{\partial\theta} + \frac{1}{r\sin\theta}\frac{\partial\sigma_{\mathrm{r}\phi}}{\partial\phi} + \frac{1}{r}(2\sigma_{\mathrm{rr}} - \sigma_{\theta\theta} - \sigma_{\phi\phi} + \sigma_{\mathrm{r}\theta}\operatorname{cotg}\theta)\right]\boldsymbol{e}_{\mathrm{r}} \\ & + \left[\frac{\partial\sigma_{\mathrm{r}\theta}}{\partial r} + \frac{1}{r}\frac{\partial\sigma_{\theta\theta}}{\partial\theta} + \frac{1}{r\sin\theta}\frac{\partial\sigma_{\theta\phi}}{\partial\phi} + \frac{1}{r}(3\sigma_{\mathrm{r}\theta} + (\sigma_{\theta\theta} - \sigma_{\phi\phi})\operatorname{cotg}\theta)\right]\boldsymbol{e}_{\theta} \\ & + \left[\frac{\partial\sigma_{\mathrm{r}\phi}}{\partial r} + \frac{1}{r}\frac{\partial\sigma_{\theta\phi}}{\partial\phi} + \frac{1}{r\sin\theta}\frac{\partial\sigma_{\phi\phi}}{\partial\phi} + \frac{1}{r}(3\sigma_{\mathrm{r}\phi} + 2\sigma_{\theta\phi}\operatorname{cotg}\theta)\right]\boldsymbol{e}_{\mathrm{z}}\end{aligned} \tag{10.17}$$

- *Gradient* of a scalar field, $f(r, \theta, \phi)$:

$$\mathbf{grad}\,f = \frac{\partial f}{\partial r}\boldsymbol{e}_{\mathrm{r}} + \frac{1}{r}\frac{\partial f}{\partial\theta}\boldsymbol{e}_{\theta} + \frac{1}{r\sin\theta}\frac{\partial f}{\partial\phi}\boldsymbol{e}_{\phi} \tag{10.18}$$

- *Gradient* of a vector field, $\boldsymbol{v}(r, \theta, z)$, and its symmetric part:

$$\mathbf{Grad}\,\boldsymbol{v} = \begin{bmatrix} \dfrac{\partial v_{\mathrm{r}}}{\partial r} & \dfrac{1}{r}\dfrac{\partial v_{\mathrm{r}}}{\partial\theta} - \dfrac{v_{\theta}}{r} & \dfrac{1}{r\sin\theta}\dfrac{\partial v_{\mathrm{r}}}{\partial\phi} - \dfrac{v_{\phi}}{r} \\ \dfrac{\partial v_{\theta}}{\partial r} & \dfrac{1}{r}\dfrac{\partial v_{\theta}}{\partial\theta} + \dfrac{v_{\mathrm{r}}}{r} & \dfrac{1}{r\sin\theta}\dfrac{\partial v_{\theta}}{\partial\phi} - \dfrac{\operatorname{cotg}\theta}{r}v_{\phi} \\ \dfrac{\partial v_{\phi}}{\partial r} & \dfrac{1}{r}\dfrac{\partial v_{\phi}}{\partial\theta} & \dfrac{1}{r\sin\theta}\dfrac{\partial v_{\phi}}{\partial\phi} + \dfrac{v_{\theta}}{r}\operatorname{cotg}\theta + \dfrac{v_{\mathrm{r}}}{r} \end{bmatrix} \tag{10.19}$$

$$\tfrac{1}{2}[\mathbf{Grad}\,\boldsymbol{v} + (\mathbf{Grad}\,\boldsymbol{v})^{\mathrm{T}}]$$

$$= \begin{bmatrix} \dfrac{\partial v_{\mathrm{r}}}{\partial r} & \tfrac{1}{2}\left(\dfrac{1}{r}\dfrac{\partial v_{\mathrm{r}}}{\partial\theta} + \dfrac{\partial v_{\theta}}{\partial r} - \dfrac{v_{\theta}}{r}\right) & \tfrac{1}{2}\left(\dfrac{1}{r\sin\theta}\dfrac{\partial v_{\mathrm{r}}}{\partial\phi} + \dfrac{\partial v_{\phi}}{\partial r} - \dfrac{v_{\theta}}{r}\right) \\ & \dfrac{1}{r}\dfrac{\partial v_{\theta}}{\partial\theta} + \dfrac{v_{\mathrm{r}}}{r} & \tfrac{1}{2}\left(\dfrac{1}{r}\dfrac{\partial v_{\phi}}{\partial\theta} + \dfrac{1}{r\sin\theta}\dfrac{\partial v_{\theta}}{\partial\phi} - \dfrac{\operatorname{cotg}\theta}{r}v_{\phi}\right) \\ \text{symmetric} & & \dfrac{1}{r\sin\theta}\dfrac{\partial v_{\phi}}{\partial\phi} + \dfrac{v_{\mathrm{r}}}{r} + \dfrac{v_{\theta}}{r}\operatorname{cotg}\theta \end{bmatrix} \tag{10.20}$$

- ***Curl*** of a vector field, $v(r, \theta, \phi)$:

$$\mathbf{curl}\ \boldsymbol{v} = \frac{1}{r\sin\theta}\left[\frac{\partial}{\partial\theta}[v_\phi \sin\theta] - \frac{\partial v_\theta}{\partial\phi}\right]\boldsymbol{e}_r + \left[\frac{1}{r\sin\theta}\frac{\partial v_r}{\partial\phi} - \frac{1}{r}\frac{\partial}{\partial r}[rv_\phi]\right]\boldsymbol{e}_\theta$$

$$+\left[\frac{1}{r}\frac{\partial}{\partial r}[rv_\theta] - \frac{1}{r}\frac{\partial v_r}{\partial\theta}\right]\boldsymbol{e}_\phi \qquad (10.21)$$

- ***Laplacian*** of a scalar field, $f(r, \theta, \phi)$:

$$\Delta f = \mathrm{div}(\mathbf{grad}\ f) = \frac{1}{r^2}\frac{\partial}{\partial r}\left[r^2\frac{\partial f}{\partial r}\right] + \frac{1}{r^2\sin^2\theta}\frac{\partial}{\partial\theta}\left[r^2\sin^2\theta\frac{\partial f}{\partial\theta}\right]$$

$$+\frac{1}{r^2\sin^2\theta}\frac{\partial^2 f}{\partial\phi^2} \qquad (10.22)$$

10.2.4 Divergence (or Gauss) theorem

The integral on the surface $\partial\Omega$ of a domain Ω, of the normal component of a vector field $\boldsymbol{v}(x, y, z)$ can be obtained through a volume integral

$$\int_{\partial\Omega} \boldsymbol{v}\cdot\boldsymbol{n}\ \mathrm{d}S = \int_{\Omega} \mathrm{div}\ \boldsymbol{v}\ \mathrm{d}V \qquad (10.23)$$

where $\boldsymbol{n}$ is the unit normal vector pointing outside of the domain.

10.3 GAUSS INTEGRATION METHOD

This integration method is based upon Hermite interpolation. Considering the values f_i and f_i' of a function $f(x)$ and its derivative $f'(x)$ at some sampling points x_i, this interpolation takes the following form:

$$f(x) = \sum_{i=1}^{n} f_i h_i(x) + \sum_{i=1}^{n} f_i' g_i(x) \qquad (10.24)$$

with the conditions:

$$f(x_i) = f_i \quad \text{and} \quad f'(x_i) = f_i' \qquad (10.25)$$

From (10.24) and (10.25) we obtain:

$$h_i(x_k) = \delta_{ik} = \begin{cases} 1 \text{ if } i = k \\ 0 \text{ if } i \neq k \end{cases} \qquad (10.26)$$

$$g_i(x_k) = 0 \qquad h_i'(x_k) = 0 \quad \forall i, k \tag{10.27}$$

$$g_i'(x_k) = \delta_{ik} \tag{10.28}$$

To obtain the solution of this interpolation, it is first necessary to introduce the following expressions:

$$Q_n(x) = \prod_{i=1}^{n}(x - x_i) = (x - x_1)(x - x_2)\dots(x - x_{n-1})(x - x_n) \tag{10.29}$$

and

$$p_i(x) = \frac{Q_n(x)}{(x - x_i)Q_n'(x_i)} = \frac{(x - x_1)(x - x_2)\dots(x - x_{i-1})(x - x_{i+1})\dots(x - x_n)}{(x_i - x_1)(x_i - x_2)\dots(x_i - x_{i-1})(x_i - x_{i+1})\dots(x_i - x_n)} \tag{10.30}$$

Note that the polynomials $p_i(x)$ are the same as those used for Lagrange interpolation. With these notations, the polynomials $g_i(x)$ and $h_i(x)$ are then given by

$$g_i(x) = (x - x_i)p_i^2(x) \tag{10.31}$$

$$h_i(x) = [1 - 2(x - x_i)p_i'(x_i)]p_i^2(x) \tag{10.32}$$

Noting that $p_i(x_k) = \delta_{ik}$, one can easily check that the polynomials defined by (10.31) satisfy the relations (10.26–28). The integration of Hermite interpolation yields:

$$\int_a^b f(x)\,\mathrm{d}x = \sum_{i=1}^{n} f_i \int_a^b h_i(x)\,\mathrm{d}x + \sum_{i=1}^{n} f_i' \int_a^b g_i(x)\,\mathrm{d}x \tag{10.33}$$

The Gauss integration method consists of *choosing the tabulation points* x_i in such a way that:

$$G_i = \int_a^b g_i(x)\,\mathrm{d}x = \int_a^b (x - x_i)p_i^2\,\mathrm{d}x = 0 \tag{10.34}$$

These points being defined by (10.34), we have:

$$\int_a^b f(x)\,\mathrm{d}x = \sum_{i=1}^{n} f_i H_i \tag{10.35}$$

where

$$H_i = \int_a^b h_i(x)\,\mathrm{d}x = \int_a^b p_i^2(x)\,\mathrm{d}x - 2p_i'(x_i)\int_a^b (x-x_i)p_i^2(x)\,\mathrm{d}x = \int_a^b p_i^2(x)\,\mathrm{d}x \tag{10.36}$$

(the second integral in the above expression vanishes, due to the prescribed condition (10.34)).

Without loss of generality, one can replace the integral on the interval $[a, b]$ by an integral on the elementary interval $[-1, +1]$. This can be done with the change of variable:

$$u = \frac{2x-(a+b)}{b-a} \tag{10.37}$$

In the interval $[-1, +1]$, the values x_i that are roots of G_i (10.34) are none other than the zeros of the Legendre polynomial of degree n, $p_n(x)$. These values being fixed, the coefficients H_i are independent of $f(x)$. In particular, one can choose $f(x) = p_k(x)$, which yields :

$$\int_a^b p_k(x)\,\mathrm{d}x = \sum_{i=1}^{n} p_k(x_i)H_i = \sum_{i=1}^{n} \delta_{ki}H_i = H_k \tag{10.38}$$

and by (10.36) :

$$H_k = \int_{-1}^{1} p_k^2(x)\,\mathrm{d}x = \int_{-1}^{1} p_k(x)\,\mathrm{d}x \tag{10.39}$$

Example

Consider $n = 2$. The two values x_1 and x_2 that are roots of G_i are necessarily symmetrical with respect to zero in the interval $[-1, +1]$; we denote them $\pm x_1$. From (10.34) we have:

$$\int_{-1}^{1} (x-x_1)\left[\frac{x+x_1}{2x_1}\right]^2 \mathrm{d}x = \frac{1}{4x_1^2}\left[\tfrac{2}{3}x_1 - 2x_1^3\right] = 0 \tag{10.40}$$

which leads to: $x_1 = +\frac{1}{\sqrt{3}}$ and $x_2 = -\frac{1}{\sqrt{3}}$.

The values of coefficients H_1 and H_2 are then:

$$H_1 = \int_{-1}^{1} \frac{x - 1/\sqrt{3}}{-2(1/\sqrt{3})}\,\mathrm{d}x = 1 = H_2 \tag{10.41}$$

The generalization is straightforward and results in the following table:

n	**Zeros**	**Coefficients** H_i
2	$\pm\frac{1}{\sqrt{3}}$	1
3	0 $\pm\frac{\sqrt{3}}{\sqrt{5}}$	$\frac{8}{9}$ $\frac{5}{9}$
4	± 0.339981 ± 0.861136	0.652145 0.347855
5	0 ± 0.538469 ± 0.906180	0.568889 0.478629 0.236927

As an example, we use these coefficients to evaluate the following integral numerically:

$$J = \int_{1}^{3} \frac{\mathrm{d}x}{x} = |\ln x|_1^3 = \ln 3 \quad (= 1.098612)$$

By means of a Gauss integration with three points ($n = 3$), we obtain:

$$J = \int_{1}^{3} \frac{\mathrm{d}x}{x} = \int_{-1}^{+1} \frac{\mathrm{d}u}{u+2} = \int_{-1}^{+1} f(u)\,\mathrm{d}u \simeq \tfrac{5}{9}f\left(\frac{\sqrt{3}}{\sqrt{5}}\right) + \tfrac{8}{9}f(0) + \tfrac{5}{9}f\left(-\frac{\sqrt{3}}{\sqrt{5}}\right)$$
$$= 1.098039$$

A Simpson method with $n = 3$ (using four equidistant points) gives:

$$J = \frac{3h}{8}[f(-1) + 3f(\tfrac{-1}{3}) + 3f(\tfrac{1}{3}) + f(1)] = 1.104762 \qquad (h = \tfrac{2}{3})$$

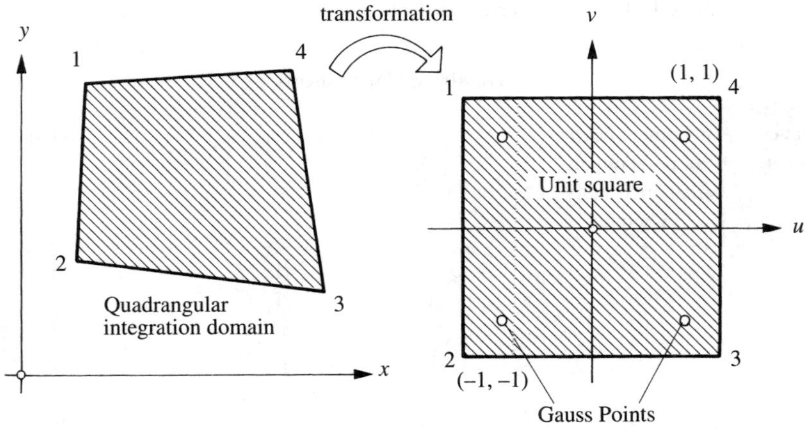

Figure 10.1 Spatial transformation for a two-dimensional domain and location of the Gauss points in the reference element.

One can see that the Gauss integration is much more precise than Simpson integration. This method is classically used in numerical simulation, especially when it comes to integrate functions over the finite elements, in either two or three dimensions. The ***Gauss points*** are then fixed with respect to the reference element, as illustrated in figure 10.1 in the case of a quadrangle.

The most frequent integration formulas in two dimensions are given in table 10.1.

Table 10.1 Most frequently used integration formulas in two dimension.

Quadrangular elements				
number of points	maximum degree of polynomials that are exactly integrated	coordinates ξ	η	weight
4	3	$-1/\sqrt{3}$ $1/\sqrt{3}$ $1/\sqrt{3}$ $-1/\sqrt{3}$	$-1/\sqrt{3}$ $-1/\sqrt{3}$ $1/\sqrt{3}$ $1/\sqrt{3}$	1 1 1 1
9	5	0 $-\sqrt{3}/\sqrt{5}$ 0 $\sqrt{3}/\sqrt{5}$ $\sqrt{3}/\sqrt{5}$ $\sqrt{3}/\sqrt{5}$ 0 $-\sqrt{3}/\sqrt{5}$ $-\sqrt{3}/\sqrt{5}$	0 $-\sqrt{3}/\sqrt{5}$ $-\sqrt{3}/\sqrt{5}$ $-\sqrt{3}/\sqrt{5}$ 0 $\sqrt{3}/\sqrt{5}$ $\sqrt{3}/\sqrt{5}$ $\sqrt{3}/\sqrt{5}$ 0	64/81 25/81 40/81 25/81 40/81 25/81 40/81 25/81 40/81
Triangular elements				
number of points	maximum degree of polynomials that are exactly integrated	coordinates ξ	η	weight
1	1	1/3	1/3	1/2
3	2	1/6 2/3 1/6	1/6 1/6 2/3	1/6 1/6 1/6
4	3	1/3 1/5 3/5 1/5	1/3 1/5 1/5 3/5	–27/96 25/96 25/96 25/96
7	5	1/3 $a = (6+\sqrt{15})/21$ $1-2a$ a $b = 4/7 - a$ $1-2b$ b	1/3 a a $1-2a$ b b $1-2b$	9/80 $\alpha = (155+\sqrt{15})/2400$ α α $\beta = 31/240 - \alpha$ β β

10.4 NON DIMENSIONAL NUMBERS

Name	Definition	Signification
Biot (thermal)	$Bi = \frac{hL}{\kappa}$	Ratio between the temperature jump at an interface and the temperature variations inside a material
Biot (solutal)	$Bi = \frac{\alpha_i L}{D_i}$	Ratio between the concentration jump at an interface and the variations of concentration of the chemical species i inside a material
Nüsselt	$Nu = \frac{hL}{\alpha}$	Normalized thermal gradient in a flow (α is the thermal diffusivity, and L the characteristic length in flow direction)
Sherwood	$Sh = \frac{\alpha_i L}{D_i}$	Normalized solutal gradient in a flow (D_i is the diffusion coefficient, L the characteristic length in flow direction, and α_i the chemical transfer coefficient of species i)
Péclet (thermal/solutal)	$Pe = \frac{Lv}{\alpha} / \frac{Lv}{D_i}$	Ratio between the domain dimension and the size of the thermal boundary layer α/v ($\alpha = \kappa/\rho c_p$), or the size of the solutal boundary layer, D_i/v
Reynolds	$Re = \frac{Lv}{\nu}$	Ratio between the inertia and viscous forces in a Newtonian flow
Fourier (thermal/solutal)	$Fo = \frac{\alpha t}{L^2} / \frac{D_i t}{L^2}$	Ratio between time and the diffusion time associated with the length L for the thermal or solutal diffusion (L^2/α or L^2/D)
Prandtl	$Pr = \frac{\nu}{\alpha}$	Ratio between the kinematic viscosity and the thermal diffusivity
Schmidt	$Sc = \frac{\nu}{D_i}$	Ratio between the kinematic viscosity and the chemical diffusion coefficient
Lewis	$Le = \frac{\alpha}{D_i}$	Ratio between the thermal diffusivity and the chemical diffusion coefficient
Grashof (thermal)	$Gr = \frac{g\beta_T \Delta T L^3}{\nu^2}$	Ratio between the buoyancy forces associated with natural thermal convection and the viscous forces
Grashof (solutal)	$Gr = \frac{g\beta_i \Delta c_i L^3}{\nu^2}$	Ratio between the buoyancy forces associated with natural solutal convection and the viscous forces
Rayleigh	$Ra = \frac{g\beta_T \Delta T L^3}{\nu\alpha}$	Product of Grashof and Prandtl numbers

10.5 INTERPRETATION OF THE TERMS OF THE ELEMENTARY STIFFNESS MATRIX FOR A DIFFUSION PROBLEM ON A TRIANGULAR LINEAR FINITE ELEMENT

Consider a triangular element of a finite element mesh with linear interpolation functions (fig. 3.8). It has been seen in chapter 3 that the elementary non-consistent mass matrix is only composed of diagonal terms, the values of which correspond to one third of the triangle surface (zones that are delimited by the three medians of the triangle, as indicated in fig. 3.23). In this section, we will give a "physical" interpretation of the terms of the elementary stiffness matrix for a diffusion problem.

With this intention, we again write the expressions from chapter 3 (for example 3.7.5) giving the values of K^e_{ij} as the functions of the coordinates (x_i, y_i) of the triangle nodes:

$$K^e_{ij} = \frac{D}{2\det[\boldsymbol{J}^e]}[\beta_i\beta_j + \gamma_i\gamma_j] \tag{10.42}$$

where D is the diffusion coefficient,

$$\beta_i = (y_j - y_k) \tag{10.43}$$

$$\gamma_i = (x_k - y_j) \tag{10.44}$$

((i, j, k): circular permutation of $(1, 2, 3)$)

Referring to fig. 10.2, K^e_{ij} can also be written:

$$K^e_{ij} = \frac{D}{2\det[\boldsymbol{J}^e]}(\boldsymbol{r}_i \cdot \boldsymbol{r}_j) \tag{10.45}$$

where $\boldsymbol{r}_i$ is the vector joining two apexes j and k of the triangle in the counter clockwise direction. Consider now node number 1, also denoted A in fig. 10.2. The fluxes of solute – suppose for example that it is a chemical diffusion problem – that are exchanged between this node and its two neighbors can be expressed by

$$q_1 = K^e_{11}c_1 + K^e_{12}c_2 + K^e_{13}c_3 = \frac{D}{2\det[\boldsymbol{J}^e]}[\boldsymbol{r}_1 \cdot \boldsymbol{r}_1 c_1 + \boldsymbol{r}_1 \cdot \boldsymbol{r}_2 c_2 + \boldsymbol{r}_1 \cdot \boldsymbol{r}_3 c_3] \tag{10.46}$$

It is now necessary to take into account some geometrical data of the triangle, specifically:

$\det[\boldsymbol{J}^e] = r_1 h_1$ where h_1 is the height of the triangle to node 1 (A)

$\boldsymbol{r}_1 \cdot \boldsymbol{r}_i = r_1\, r_{i,1}$ where $r_{i,1}$ is the projection of $\boldsymbol{r}_i$ upon $\boldsymbol{r}_1$

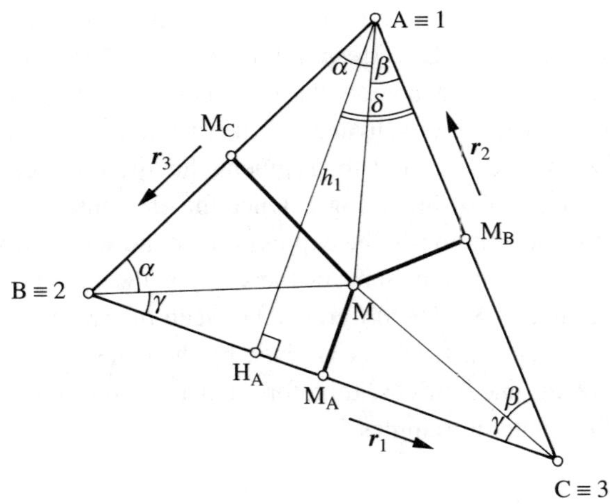

Figure 10.2 Triangle and mediatrices.

Given that $r_{1,\,1} = -\,(r_{2,\,1} + r_{3,\,1}) = \overline{\mathrm{H_AC}} + \overline{\mathrm{H_AB}}$, where $\mathrm{H_A}$ is the projection of the apex A on the base BC, we obtain:

$$q_1 = D\frac{\overline{\mathrm{H_AC}}}{2h_1}(c_1 - c_2) + D\frac{\overline{\mathrm{H_AB}}}{2h_1}(c_1 - c_3) \tag{10.47}$$

We are going to show now that the triangles $\mathrm{AH_AC}$ and $\mathrm{AM_CM}$ are similar, as are $\mathrm{AH_AB}$ and $\mathrm{AM_BM}$, where M is the center of the mediatrices and $\mathrm{M_B}$ and $\mathrm{M_C}$ the bases of the mediatrices on the edges AC and AB, respectively. By construction of the mediatrices, the angles of the triangle are such that:

$$(\alpha + \beta) + (\beta + \gamma) + (\gamma + \alpha) = 180 \qquad \text{or} \qquad \alpha + \beta + \gamma = 90 \tag{10.48}$$

In addition, the triangle $\mathrm{AH_AC}$ having a right angle at $\mathrm{H_A}$, we also have:

$$\delta + \beta + \gamma = 90 \tag{10.49}$$

From the relations (10.48) and (10.49), we conclude that $\alpha = \delta$. Since $\mathrm{AH_AC}$ and $\mathrm{AM_CM}$ are both right triangles, they are similar. We can then write:

$$\frac{\overline{\mathrm{H_AC}}}{h_1} = \frac{\overline{\mathrm{M_CM}}}{\overline{\mathrm{M_CA}}} = 2\frac{\overline{\mathrm{M_CM}}}{\overline{\mathrm{AB}}} = 2\frac{\overline{\mathrm{M_CM}}}{r_3} \tag{10.50}$$

The exchanged fluxes with node 1 can then be written:

$$q_1 = -D\frac{c_2 - c_1}{r_3}\overline{\mathrm{M_CM}} - D\frac{c_3 - c_1}{r_2}\overline{\mathrm{M_BM}} \tag{10.51}$$

We thus see the appearance of the solute fluxes, which are given by the product of the diffusion coefficient and the corresponding concentration gradients. As in the finite volume or finite difference method, these gradients are expressed as the difference between nodal concentrations divided by the distance separating them. These fluxes are multiplied by the length of the segments having for ends the center and the bases of the mediatrices of the triangle. Once the elementary contributions are assembled for a finite element mesh, the expressions of the solute fluxes will be rigorously identical to those already seen in the case of the direct finite difference method (Sect. 2.4.3, fig. 2.18). The difference between the two methods lies in the fact that the finite element mass matrix is given by the center of the medians, and not by the center of the mediatrices: therefore a strict equivalence will be found only in the case of equilateral triangles.

INDEX

Printing: Mercedes-Druck, Berlin
Binding: Stein+Lehmann, Berlin